T0329305

Visualizing More Quaternions

Visualizing More Quaternions

Andrew J. Hanson

MORGAN KAUFMANN PUBLISHERS

AN IMPRINT OF ELSEVIER

For information on all Morgan Kaufmann publications
visit our website at https://www.elsevier.com/books-and-journals

Publisher: Mara Conner
Acquisitions Editor: Chris Katsaropoulos
Editorial Project Manager: Andrea Gallego Ortiz
Production Project Manager: Surya Narayanan Jayachandran
Cover Designer: Matthew Limbert

Typeset by VTeX

Working together
to grow libraries in
developing countries

www.elsevier.com • www.bookaid.org

To My Mother
Elisabeth M. Hanson
1917–2016

Contents

Preface

My first attempt to explore and expose to others the vast potential of quaternions happened largely because the late Steve Cunningham, as chair of the SIGGRAPH computer graphics association, strongly encouraged me to give some tutorial lectures on 4D graphics entitled "Computer Graphics beyond the Third Dimension" at the SIGGRAPH 1998 conference. Over a number of years, sections of those lectures evolved into a series of papers and further lectures on 4D topics, and quaternions eventually became a principal focus of my work. Ultimately my research in the quaternion domain acquired critical mass, leading to my getting to know Morgan-Kaufmann editor Tim Cox, who then persuaded me to write my first book, *Visualizing Quaternions* (Elsevier, 2006). When I completed that book, deep into my first chaotic years serving as a Department Chair, I was very glad to be done with it and to return to the real world with one less very complicated task to worry about. But I was satisfied with the thought I had covered just about everything of current interest that there was to be said about quaternions, and thereafter I often indulged in a somewhat sacrilegious joke that went something like this:

> "What Hamilton didn't know about quaternions would fill a book
> – my book!"

Little did I suspect that I would become the subject of my own joke: it turns out that what I have learned about quaternions after writing that first book would itself fill a book – and that is the book I have now written. But now I am many years older and hopefully somewhat wiser, because it is clear that there is much, much, more to know about quaternions and their applications than would ever fit in these two books. All I can wish for is that some of these additional insights that I have learned and included here will be of use and inspiration to others; I hope that those of you who read these pages will enjoy the kinds of mathematical tricks and antics that quaternions facilitate as much as I have enjoyed them myself.

Notes to the reader: The first book of this pair, *Visualizing Quaternions*, also contains material that is intended to be elementary and intuitive for those without previous exposure to quaternions. This book, like its predecessor, collects a number of published results along with previously unpublished material, but in general this book is quite a bit more technical. Thus the reader who has not had previous contact with the world of quaternions should read carefully Chapter 2 in order to be prepared for much of the subsequent contents. In a handful of locations, I have added brief "advanced" side comments, marked with the symbol ‡, for readers with certain technical backgrounds. When it seemed that an algorithm needed to be presented in a computer program in order better to expose its structure, I have chosen *Mathematica*, which has an algorithm-like readability, and which is also the language in which nearly every computation and figure in this book was implemented. The reader should also be aware that the author is an unapologetic pedagogue, and a number of chapters are composed more in the manner of "how do we figure out what is going on here," than in the manner of a catalog of textbook facts: I have found that this habit has frequently served to reveal the evolution my own understanding in a way that I enjoyed, often leading unexpectedly to novel insights despite a sometimes indirect path.

Acknowledgments: I am indebted to many people for their help, influence, and encouragement to undertake this second volume dedicated to the exploration of quaternions. Chris Katsaropoulos, Senior Acquisitions Editor at Elsevier, played a critical part in encouraging me to pursue another book on this subject, and guiding me through the process of converting the idea into reality 18 years after I started writing my first book on this subject. The patience and support of my wife Patricia L. Foster also played a great part in helping me through many long days of struggling with the challenge of writing this book. I was also very fortunate to have had my daughter Sonya M. Hanson as an inspiration and collaborator; she was completely responsible for reviving my acquaintance with Berthold K.P. Horn's quaternion approach to solving the point-cloud matching problem, which led me to the discovery of the broadly dispersed literature and novel techniques introduced in the coordinate alignment chapters; she provided much useful information and advice that helped lead that challenging investigation to its conclusion, and we joined forces collaborating on the understanding of many new features of quaternions in pose estimation tasks, as well as the adjugate-based rotation-to-quaternion process. On the related subject of protein geometry, I was also fortunate to have my former student Sidarth Thakur as an essential collaborator in the work presented in the chapter on quaternion maps of protein structure. Others who helped with making our treatment of that subject matter more accurate include Tuli Mukhopadhyay, Predrag Radivojac, Fuxiao Xin, and Yuzhen Ye. I am also grateful to Mathieu Desbrun for essential discussions on spherical geometry, and to Robert Hanson (no relation), who has been an enthusiastic supporter of the movement to popularize the exploitation of quaternions in biochemistry, and was especially helpful through his efforts to provide quaternion frame-mapping support in systems like Jmol. Simon Billinge at Columbia University and a number of anonymous referees were also important in many ways in helping me improve the depth and quality of the quaternion-based coordinate alignment work.

Nigel Hitchin, Martin Roček, and Tamas Hausel all played essential parts in introducing me to the world of the ADE discrete groups and Felix Klein's classic work on the subject. Valuable input was provided also by insightful discussions with Chen Lin, Randall Bramley, Roger Germundsson, Michael Trott, and Conor Cosnett. Daniel Weiskopf played an essential role in refining my dubious understanding of relativistic image-based rendering, and encouraged me through his collaboration on our SIGGRAPH 2001 tutorial, "Visualizing Relativity," to work out the details of quaternion-inspired special relativity. Special appreciation is due to Gerardo Ortiz, Amr Sabry, and Yu-Tsung Tai for patiently introducing me to the field of quantum computing. I am grateful in particular to Chris Doran, who for many years has exchanged a wide range of scientific knowledge with me and has especially provided many insights into Clifford algebra/geometric algebra, as well as to Charles Livingston for his guidance in understanding complex geometry and certain related aspects of group theory. Further thanks are due to Ji-Ping Sha, with whose collaboration we implemented the exploitation of a quaternion expansion that led to the chapter on an explicit isometric embedding of the A_1 gravitational instanton metric. Finally, I would like to offer my heartfelt gratitude to "fellow quaternion traveler" Berthold K.P. Horn, who generously offered all manner of ideas, suggestions, and feedback throughout a long correspondence that began only in the gap between the first and second book, but which has been a continuing source of inspiration.

A. J. H.
Bloomington, Indiana
September, 2023

Biography

Andrew J. Hanson

The author is an Emeritus Professor of Computer Science at Indiana University. He earned a bachelor's degree in Chemistry and Physics from Harvard University in 1966 and a PhD in Theoretical Physics from MIT in 1971. His interests range from general relativity to computer graphics, artificial intelligence, and bioinformatics; he is particularly concerned with applications of quaternions and with exploitation of higher-dimensional graphics to the visualization of complex scientific contexts such as Calabi–Yau spaces and rolling 4D dice. He is the co-discoverer of the Eguchi–Hanson "gravitational instanton" Einstein metric (1978), author of *Visualizing Quaternions* (Elsevier, 2006), and designer of a variety of graphical user interfaces for interacting with 4D virtual worlds, including the domain of quaternions themselves.

"I make pictures of things no one has ever seen before."

$$z_1{}^5 + z_2{}^5 = 1$$

2D cross-section of the Calabi–Yau quintic 6-manifold[a]

$$z_0{}^5 + z_1{}^5 + z_2{}^5 + z_3{}^5 + z_4{}^5 = 0$$

This surface is projected from its natural 4D embedding to 3D. The 25 distinct colored patches are labeled by pairs of integers (k_1, k_2), each running from $(0, \ldots, 4)$, denoting the pair of fifth roots of unity from which they are constructed by multiplying the fundamental domain.[b]

---

[a] https://commons.wikimedia.org/w/index.php?title=File:CalabiYau5.jpg&oldid=471153027.

[b] A.J. Hanson, "A Construction for Computer Visualization of Certain Complex Curves," in "Computers and Mathematics" column, ed. Keith Devlin, of *Notices of the American Mathematical Society*, **41**, No. 9, pp. 1156–1163 (American Math. Soc., Providence, November/December, 1994).

Review

While it would be useful for readers of this book to be familiar with our first book on this subject, *Visualizing Quaternions* (Hanson, 2006), that should not be necessary, as our intention here in Part 1 is to provide a self-contained summary of the basic background, ideas, and algorithms necessary to follow the narrative. That being said, the first book contains some excellent introductory material that is less technical and less mathematical, and that material, for some readers, might be attractive as an introduction to the subject. Quaternions are a beautiful and elegant mathematical construction, with far-reaching consequences, contexts, and applications. The basics reviewed in these first chapters will be fundamental to understanding the rest of the book. The reader should therefore be aware that, hereafter, we shall not be shy about assuming familiarity with the material here in Part 1.

> *Think of the quaternion mathematics as a language in which epic poetry is written: there are fascinating things to come, and this introduction is your dictionary for the language in which the forthcoming ideas are experienced and enjoyed, the means by which these timeless thoughts are expressed in symbols and revealed as images!*

Introduction

1

1.1 Introductory remarks

In this book, we expand on the methods for exploiting quaternions introduced in our previous book on the subject, *Visualizing Quaternions* (Hanson, 2006). In order to make this work more self-contained, we will first revisit some of the fundamental historical developments and properties of quaternions. Then we will provide a detailed review of the basic properties of quaternions that are needed from the earlier book; familiarity with these fundamentals will be essential and will be assumed in later chapters. As the reader will discover, this book covers a great deal of material that is of a more sophisticated nature than many of the dominant themes in the first book; the selected topics that are covered again here typically appear because there are new insights to offer. Readers should be aware that there are materials in the first book, and not covered here, that might also be of interest, and which may also be more accessible for someone encountering quaternions for the first time. What we will see again here is that the ways in which quaternion methods provide tools for solving certain problems are very powerful, and we have chosen topics that, for the author at least, embody the excitement that follows from that power.

1.2 Hamilton's walk

The story of quaternions began in the mid-19th century as William Rowan Hamilton (see Fig. 1.1) sought relentlessly to find a way to generalize complex numbers, well-known by then to have very deep relationships to 2D space, to a system that would realize analogous relations in 3D space.

Hamilton struggled for years attempting to make sense of an unsuccessful algebraic system containing one real and two "imaginary" parts. But finally, in 1843, at the age of 38, Hamilton had a brilliant stroke of imagination, and invented in a single instant the idea of a *three-part* "imaginary" system that became the quaternion algebra. According to Hamilton, he was walking with his wife along the Royal Canal in Dublin, on his way to chair a meeting of the Royal Irish Academy, when the thought struck him. Concerning that moment, he later wrote to his son Archibald:

"On the 16th day of [October] – which happened to be a Monday, and a Council day of the Royal Irish Academy – I was walking in to attend and preside, and your mother was walking with me, along the Royal Canal, to which she had perhaps driven; and although she talked with me now and then, yet an under-current of thought was going on in my mind, which gave at last a result, whereof it is not too much to say that I felt at once the importance. An electric circuit seemed to close; and a spark flashed forth, the herald (as I foresaw, immediately) of many long years to come of definitely directed

Visualizing More Quaternions. https://doi.org/10.1016/B978-0-32-399202-2.00009-5

FIGURE 1.1

Sir William Rowan Hamilton, 4 August 1805 — 2 September 1865. (History of Mathematics web pages of the University of St. Andrews, Scotland.)

thought and work, by myself if spared, and at all events on the part of others, if I should even be allowed to live long enough distinctly to communicate the discovery. Nor could I resist the impulse – unphilosophical as it may have been – to cut with a knife on a stone of Brougham Bridge, as we passed it, the fundamental formula with the symbols, i, j, k; namely,

$$i^2 = j^2 = k^2 = ijk = -1$$

which contains the Solution of the Problem..."

Hamilton was apparently so overwhelmed by his discovery that he feared he might die before he had a chance to tell anyone, and thus, for safety, carved the equations into the nearest wall, the side of a bridge arching over the canal along which he and his wife were walking! While Hamilton's own carving soon disappeared with the weather, a commemorative plaque was later placed on the bridge, shown in Fig. 1.2 in the company of the author.

A feature of quaternions that we will use throughout this book is that they are closely related to 3D rotations, a fact apparent to Hamilton almost immediately, but first published by Hamilton's contemporary Arthur Cayley (Cayley, 1845). Hamilton's quaternion multiplication rule, which contains three subrules identical to ordinary complex multiplication, expresses a deep connection between unit-length 4-vectors and rotations in three Euclidean dimensions. Curiously, the rule could in principle have been discovered directly by seeking such a connection, since quaternion-like relations appear buried in rotation-related formulas that predate the discovery of quaternions by Hamilton; Rodrigues appears to actually have been the first, in 1840 (Rodrigues, 1840; Gray, 1980), to write down an equivalent version of the equations that Hamilton scratched on Broome Bridge in 1843 (Altmann, 1989; Biedenharn and

FIGURE 1.2

The author in 2012, on his first visit to the Hamilton plaque on Broome Bridge in Dublin, Ireland, many years after publishing *Visualizing Quaternions* (Hanson, 2006). The plaque commemorates the legendary location where Hamilton conceived of the idea of quaternions, and reportedly carved the fundamental equations.

Louck, 1984; Klein, 1884). (We remark that Hamilton himself in his correspondence referred to the bridge in question as "Brougham Bridge," but that most sources agree that the bridge was named after a local official named William Broome.)

The appearance of three copies of the complex multiplication rule and the involvement of three Euclidean dimensions is not accidental; complex multiplication faithfully represents the 2D rotation transformation, and each complex multiplication subrule in the quaternion rule will be seen to correspond to a rotation mixing two of the three possible orthogonal 3D axes.

Hamilton proceeded to develop the features of quaternions in depth over the next decade, and published his classic book *Lectures on Quaternions* (Hamilton, 1853) exactly 10 years later in 1853; the still more extensive *Elements of Quaternions* (Hamilton, 1866) was not published until 1866, shortly after Hamilton's death in 1865. The banner of quaternions was then picked up by Peter Tait, who had diplomatically awaited Hamilton's passing before publishing his own, perhaps more readable, opus, *Elementary Treatise on Quaternions* (Tait, 1867), in 1867; Tait followed the *Treatise* with *Introduction to Quaternions* in 1873 (Tait, 1873), and produced a major revision of the *Treatise* in 1890 (Tait, 1890). Tait is usually credited with formulating the equations of a gyroscope in terms of quaternions. From there, the story continues in many directions that we shall not pursue here (see, for example, Altmann (1986, 1989), Crowe (1994), or van der Waerden (1976)).

1.3 Philosophical thoughts

There have been many ideas and insights developed both in Hamilton's time and in modern times about the underlying significance of quaternions and the ways in which they may, and may not, be generalized or extended.

The question "Can a quaternion be a vector?" Among the historical thoughts that have been put forward is the observation that the quaternions themselves are not technically vectors, though there is an often-exploited isomorphism between the action of quaternion conjugation on a pure three-element imaginary quaternion, with vanishing real part, and the action of 3D rotations on a 3D vector. As Altmann (1986, 1989) emphasized, a quaternion is an element of the Lie group $\mathbf{SU(2)}$, and would therefore seem well-suited only to represent the action of the rotation group; hence a pure imaginary quaternion would not obviously be interpretable as a 3D vector. That is, one should be wary of the fact that the pure quaternion typically used to perform vector-related operations is in fact a rotation by π radians about an axis through a point on the celestial sphere; as a mathematical object, it is, technically speaking, a spin $1/2$ representation of a rotation, a viewpoint that Hamilton could not appreciate without modern mathematical insights. In Appendix E, we include for completeness a derivation of 3D rotations from quaternions that does not require any such correspondence.

Some insights into quaternions as vectors. However, there are interesting paths to including 3D vectors directly in the quaternion framework that shed some unusual lights on the arguments of Altmann and others, and we will examine these in depth in later chapters. One way to include 3-vectors and Euclidean translations in a quaternion framework is to employ *dual quaternions*, which are now commonly exploited in modern robotics. Dual quaternions are a representation of $\mathbf{SE}(3)$, the six-degree-of-freedom group of translations and rotations in 3D Euclidean space; these objects consist of a pair of three-degree-of-freedom parameters, one corresponding to the usual rotational quaternion and the other, dimensionally incompatible, corresponding to a spatial translation with units of length. So that is one way to incorporate Hamilton's goal to merge 3-vectors into the quaternion framework, and in fact the formalism was essentially already being studied by Clifford in 1873, Hamilton having passed away in 1865. A second insight comes from another unexpected direction, which is the classic treatment of the representations of the *Lorentz group* of Einstein's theory of special relativity. The standard method of defining quantum fields of various masses and spins follows from considering an *imaginary* quaternion as the spatial part of a four-parameter spacetime coordinate element defined by the 2×2 unitary matrices of the group $\mathbf{SU(2)}$. In this formalism the Hamilton quaternion just acts on the *spatial part* of a full spacetime point $[t, x, y, z]$; complexified quaternions in the closely related group $\mathbf{SL(2, C)}$ contain the full six parameters of the Lorentz group, three rotational and three for the boost. When these $SL(2, C)$ group elements act by conjugation on the $[t, x, y, z]$ spacetime, they produce exactly the transformations of special relativity. In this widely used relativistic field theory context, Hamilton's quaternions appear almost as an accidental side-effect of leaving time out of a compelling framework that includes the entire spacetime of special relativity: even the minus sign of the Minkowski metric appears naturally here, leading us to the conjecture that relativistic spacetime itself might have been buried, awaiting revelation, in Hamilton's discovery.

What comes after a quaternion? Another interesting question is the issue of why quaternions only seem relevant to 3D and 4D rotations, where we explore the lesser-known 4D case later in its own chapter. But what happened to higher dimensions? There seems to be no particular distinction between the features of rotations in higher dimensions: each lower-dimensional rotation is realized as a subspace of the next higher-dimensional rotation! Here the answer appears in what is known in mathematics as *Clifford algebra*, also referred to as *geometric algebra* in certain communities in theoretical physics. Working backwards from the relatively straightforward construction of N-dimensional Euclidean rotations, embodied in the special orthogonal rotation groups $SO(N)$, one can determine that there are *double coverings* of each orthogonal group called the *spin groups*, $Spin(N)$; these groups are generated by basis elements that do not commute, very distinct from the bases of a standard Euclidean vector space. Quaternions have properties that are identical to those of the $Spin(3)$ group generated by a 3D Clifford algebra, and also identical to the equivalent special unitary group $SU(2)$, whose basis is the set of Pauli matrices from fundamental physics: 4D doubled-parity versions of these matrices are the Dirac matrices used to construct the wave equation of the relativistic spin $1/2$ electron. One does not have to understand all the details of this context to appreciate that there are powerful concepts involving Clifford algebras that are valid in all dimensions, and which collapse to quaternions in the simplest nontrivial case, our ordinary 3D spatial world.

Basic quaternion formulas

Our purpose in this chapter is to provide a compact summary of the basic quaternion formulas that we will utilize in this book (see also our first book (Hanson, 2006)). More specialized applications and extensions will be treated in other chapters, but those treatments will assume that the formulas here are already known to the reader.

We define a *quaternion* to be a point

$$q = (q_0, q_1, q_2, q_3) = (q_0, \mathbf{q}) \tag{2.1}$$

in 4D Euclidean space with unit norm, $q \cdot q = q_0{}^2 + \mathbf{q} \cdot \mathbf{q} = 1$, and so geometrically it is a point on the unit 3-sphere $\mathbf{S}^3$. Unless otherwise noted, all quaternions in this book will be assumed to have unit norm, which, for the most part, has the effect of restricting our attention to applications involving 3D rotations. The first term, q_0, plays the role of a real number, and the last three terms, written often as the 3-vector $\mathbf{q}$, play the role of a generalized imaginary number, and so are treated differently from the first: in particular the *complex conjugation operation* is taken to be

$$\bar{q} = (q_0, -\mathbf{q}) . \tag{2.2}$$

(Some of the literature may employ the alternative notation q^* for complex conjugation instead of the bar notation.) Quaternions possess a multiplication operation denoted by $\star$ and defined as follows:

$$q \star p = (q_0 p_0 - q_1 p_1 - q_2 p_2 - q_3 p_3, \ q_1 p_0 + q_0 p_1 - q_3 p_2 + q_2 p_3,$$
$$q_2 p_0 + q_3 p_1 + q_0 p_2 - q_1 p_3, \ q_3 p_0 - q_2 p_1 + q_1 p_2 + q_0 p_3) \tag{2.3}$$
$$= (q_0 p_0 - \mathbf{q} \cdot \mathbf{p}, \ q_0 \mathbf{p} + p_0 \mathbf{q} + \mathbf{q} \times \mathbf{p}) . \tag{2.4}$$

This is equivalent to the algebra resulting from writing out the quaternion product using Hamilton's imaginaries,

$$\left. \begin{array}{c} q = q_0 + q_1 \mathbf{I} + q_2 \mathbf{J} + q_3 \mathbf{K} \\ \text{with } \mathbf{I}^2 = \mathbf{J}^2 = \mathbf{K}^2 = \mathbf{IJK} = -1 \\ \mathbf{I} = \mathbf{JK} = -\mathbf{KJ} \\ \mathbf{J} = \mathbf{KI} = -\mathbf{IK} \\ \mathbf{K} = \mathbf{IJ} = -\mathbf{JI} \end{array} \right\} . \tag{2.5}$$

Visualizing More Quaternions. https://doi.org/10.1016/B978-0-32-399202-2.00010-1

The quaternion algebra can also be conveniently expressed as a matrix multiplication operation,

$$q \star p = \mathbf{Q}(q) \cdot p = \begin{bmatrix} q_0 & -q_1 & -q_2 & -q_3 \\ q_1 & q_0 & -q_3 & +q_2 \\ q_2 & +q_3 & q_0 & -q_1 \\ q_3 & -q_2 & +q_1 & q_0 \end{bmatrix} \cdot \begin{bmatrix} p_0 \\ p_1 \\ p_2 \\ p_3 \end{bmatrix} = (q_0 p_0 - \mathbf{q} \cdot \mathbf{p}, \ q_0 \mathbf{p} + p_0 \mathbf{q} + \mathbf{q} \times \mathbf{p}), \quad (2.6)$$

where $\mathbf{Q}(q)$ is an orthonormal matrix implementing left quaternion multiplication that can be useful. Some applications involve right quaternion multiplication, by which we mean *acting-to-the-left*, and the rightward-acting matrix $\widetilde{\mathbf{Q}}(q)$ that realizes that matrix multiplication operation is

$$p \star q = \widetilde{\mathbf{Q}}(q) \cdot p = \begin{bmatrix} q_0 & -q_1 & -q_2 & -q_3 \\ q_1 & q_0 & +q_3 & -q_2 \\ q_2 & -q_3 & q_0 & +q_1 \\ q_3 & +q_2 & -q_1 & q_0 \end{bmatrix} \cdot \begin{bmatrix} p_0 \\ p_1 \\ p_2 \\ p_3 \end{bmatrix} = (q_0 p_0 - \mathbf{q} \cdot \mathbf{p}, \ q_0 \mathbf{p} + p_0 \mathbf{q} - \mathbf{q} \times \mathbf{p}), \quad (2.7)$$

where we see that only the lower-right 3×3 rectangular block's elements are different, with the three off-diagonal pairs of imaginary components changing their sign (which is equivalent to changing the sign of the 3D cross product in the algebraic form notation).

In addition, we can actually represent the entire quaternion and its multiplication in a consistent way by writing quaternions themselves as 4×4 orthogonal matrices instead of as unit 4-vectors. We see that if we define a matrix like $\mathbf{Q}$ above for each quaternion, then we can write

$$p \star q = \mathbf{P} \cdot \mathbf{Q} \qquad (2.8)$$

$$= \begin{bmatrix} p_0 & -p_1 & -p_2 & -p_3 \\ p_1 & p_0 & -p_3 & p_2 \\ p_2 & p_3 & p_0 & -p_1 \\ p_3 & -p_2 & p_1 & p_0 \end{bmatrix} \cdot \begin{bmatrix} q_0 & -q_1 & -q_2 & -q_3 \\ q_1 & q_0 & -q_3 & q_2 \\ q_2 & q_3 & q_0 & -q_1 \\ q_3 & -q_2 & q_1 & q_0 \end{bmatrix} \qquad (2.9)$$

$$= \begin{bmatrix} r_0 & -r_1 & -r_2 & -r_3 \\ r_1 & r_0 & -r_3 & r_2 \\ r_2 & r_3 & r_0 & -r_1 \\ r_3 & -r_2 & r_1 & r_0 \end{bmatrix}, \qquad (2.10)$$

where each component of the quaternion $r = (r_0, r_1, r_2, r_3)$ is just the corresponding component of the quaternion product $p \star q$ given in Eq. (2.4). This exactly parallels the behavior of the quaternion represented as a 2×2 *complex* matrix $[q]$ in the group $\mathbf{SU}(2)$, built from the Pauli matrices, that we will encounter in later chapters.

Connection with the fourth dimension. The 4×4 matrices $\mathbf{Q}(q)$ and $\widetilde{\mathbf{Q}}(q)$ are each orthonormal, which means that the quaternion multiplication $q \star p$ is *literally* a rotation matrix acting on p in the 4D Euclidean space in which quaternions are embedded: the unit quaternion p is effectively a *direction* in 4D space that is rotated to a *new direction* in 4D space (a new position on the sphere $\mathbf{S}^3$) when it is acted on by another quaternion q.

The matrices $\mathbf{Q}(q)$ and $\widetilde{\mathbf{Q}}(q)$ each have only three free parameters, while 4D orthogonal matrices (which are *4D rotations*) have six free parameters, so while neither $\mathbf{Q}(q)$ nor $\widetilde{\mathbf{Q}}(q)$ includes all 4D rotations, *together* they cover all six 4D rotation parameters. What $\mathbf{Q}$ by itself does include is precisely the expansion of the quaternion multiplication in 2×2 complex Pauli matrix form into its smallest equivalent real matrix parameterization, and that of course has only three real parameters. Another way to look at this is to note that we only need two parameters, latitude and longitude, to define any point on the Earth, even though the full rotation group has three parameters. The three free quaternion parameters in $Q(q)$ are basically 4D analogs of latitude and longitude, able to rotate a given 4D point on $\mathbf{S}^3$ to any other point on $\mathbf{S}^3$ without needing all six parameters of the full 4D rotation group.

Some important identities. Choosing exactly one of the three imaginary components in both q and p to be nonzero gives back the classic complex algebra $(q_0 + iq_1)(p_0 + ip_1) = (q_0 p_0 - q_1 p_1) + i(q_0 p_1 + p_0 q_1)$, so there are three copies of the complex numbers embedded in the quaternion algebra; the difference is that in general the final term $\mathbf{q} \times \mathbf{p}$ in Eq. (2.4) changes sign if one reverses the order, making the quaternion product *order-dependent*, unlike the complex product. Nevertheless, like complex numbers, the quaternion algebra satisfies the nontrivial "multiplicative norm" relation

$$\|q\| \, \|p\| = \|q \star p\| \,, \tag{2.11}$$

where $\|q\|^2 = q \cdot q = \mathrm{Re}\,(q \star \bar{q})$, i.e., quaternions are one of the four possible Hurwitz algebras (real, complex, quaternion, and octonion) (Hurwitz, 1923; Wikipedia, 2023j).

Quaternions also obey a number of interesting scalar triple-product identities that are generalizations of the 3D vector identities $A \cdot (B \times C) = B \cdot (C \times A) = C \cdot (A \times B)$, along with $A \times B = -B \times A$. The corresponding quaternion identities, which we will exploit when examining the properties of quaternion frames, are

$$\begin{aligned} r \cdot (q \star p) &= q \cdot (r \star \bar{p}) = p \cdot (\bar{q} \star r) \,, \\ \bar{r} \cdot (\bar{p} \star \bar{q}) &= \bar{q} \cdot (p \star \bar{r}) = \bar{p} \cdot (\bar{r} \star q) \,, \end{aligned} \tag{2.12}$$

where the complex conjugate entries are the natural consequences of the sign changes occurring only in the (imaginary) 3D part.

It can be shown that conjugating a vector $\mathbf{x} = (x, y, z)$, written as a purely "imaginary" quaternion $(0, \mathbf{x})$, by a quaternion q and its inverse $\bar{q}$ is isomorphic to the construction of a 3D Euclidean rotation $R(q)$ generating all possible elements of the special orthogonal group $\mathbf{SO}(3)$. If we compute

$$q \star (c, x, y, z) \star \bar{q} = (c, R(q) \cdot \mathbf{x}) \,, \tag{2.13}$$

we see that only the purely imaginary part is affected, whether or not the arbitrary real constant c vanishes. Technically, this invariance of the q_0 term under conjugation by a quaternion and its inverse implements an operation under which *the identity element* $q_{ID} = [1, 0, 0, 0]$ *is invariant*; the rest of the quaternion elements wander in the space around this identity element. We will see in Chapter 24 on relativity that this fact has profound implications.

The fundamental quaternion rotation formula. The result of collecting the coefficients of the vector term is a *proper* orthonormal 3D rotation matrix quadratic in the quaternion elements that can be written

$$R_{ij}(q) = \delta_{ij}\left(q_0^2 - \mathbf{q}^2\right) + 2q_i q_j - 2\epsilon_{ijk} q_0 q_k$$

$$R(q) = \begin{bmatrix} q_0^2 + q_1^2 - q_2^2 - q_3^2 & 2q_1 q_2 - 2q_0 q_3 & 2q_1 q_3 + 2q_0 q_2 \\ 2q_1 q_2 + 2q_0 q_3 & q_0^2 - q_1^2 + q_2^2 - q_3^2 & 2q_2 q_3 - 2q_0 q_1 \\ 2q_1 q_3 - 2q_0 q_2 & 2q_2 q_3 + 2q_0 q_1 & q_0^2 - q_1^2 - q_2^2 + q_3^2 \end{bmatrix} , \qquad (2.14)$$

with determinant $\det R(q) = (q \cdot q)^3 = +1$, and ϵ_{ijk} is the totally antisymmetric Levi-Civita symbol (see Appendix L for more details).

> *Eq. (2.14) is probably the most important formula in this book, defining the fundamental relationship between quaternions, 3D rotations, and 3D coordinate frame triads.*

The formula for $R(q)$ is technically a two-to-one mapping from quaternion space to the 3D rotation group because $R(q) = R(-q)$; changing the sign of the quaternion preserves the rotation matrix. Note also that the identity quaternion $q_{\mathrm{ID}} = (1, 0, 0, 0) \equiv q \star \bar{q}$ corresponds to the identity rotation matrix, as does $-q_{\mathrm{ID}} = (-1, 0, 0, 0)$. The 3×3 matrix $R(q)$ is not only fundamental to the quaternion study of 3D rotations in space, but is also essential to the study of 3D orientation frames representing how an object sits in 3D space: the *columns* of $R(q)$ can be thought of precisely as a quaternion representation of the *frame triad* describing the orientation of a body in 3D space, i.e., the columns are the vectors of the frame's local x-, y-, and z-axes relative to an initial identity frame.

Multiplying a quaternion p by the quaternion q to get a new quaternion $p' = q \star p$ simply *rotates* the frame corresponding to p by the matrix Eq. (2.14) written in terms of q. This has nontrivial implications for 3D rotations, and tells us that quaternion multiplication corresponds exactly to multiplication of two *independent* 3×3 orthogonal rotation matrices, and in fact

$$R(q \star p) = R(q) \cdot R(p) . \qquad (2.15)$$

This collapse of repeated rotation matrices to a single rotation matrix with multiplied quaternion arguments can be continued indefinitely. The property Eq. (2.15) means precisely that the 3×3 rotation matrix $R(q)$ is a *representation* of the group action of quaternion multiplication.

If we choose the following specific three-parameter form of the quaternion q preserving $q \cdot q = 1$,

$$q(\theta, \hat{\mathbf{n}}) = \left(\cos(\theta/2), \hat{n}_1 \sin(\theta/2), \hat{n}_2 \sin(\theta/2), \hat{n}_3 \sin(\theta/2)\right) = \left(\cos(\theta/2), \hat{\mathbf{n}} \sin(\theta/2)\right) , \qquad (2.16)$$

where we require $\hat{\mathbf{n}} \cdot \hat{\mathbf{n}} = 1$, we can then easily show from Eq. (2.14) that $R(q) = R(\theta, \hat{\mathbf{n}})$ is precisely the "axis-angle" form of the 3D spatial rotation matrix

$$R(q) = R(\theta, \hat{\mathbf{n}}) =$$
$$\begin{bmatrix} \cos\theta + (1 - \cos\theta)\hat{n}_1^2 & (1 - \cos\theta)\hat{n}_1\hat{n}_2 - \sin\theta\,\hat{n}_3 & (1 - \cos\theta)\hat{n}_1\hat{n}_3 + \sin\theta\,\hat{n}_2 \\ (1 - \cos\theta)\hat{n}_1\hat{n}_2 + \sin\theta\,\hat{n}_3 & \cos\theta + (1 - \cos\theta)\hat{n}_2^2 & (1 - \cos\theta)\hat{n}_2\hat{n}_3 - \sin\theta\,\hat{n}_1 \\ (1 - \cos\theta)\hat{n}_1\hat{n}_3 - \sin\theta\,\hat{n}_2 & (1 - \cos\theta)\hat{n}_2\hat{n}_3 + \sin\theta\,\hat{n}_1 & \cos\theta + (1 - \cos\theta)\hat{n}_3^2 \end{bmatrix} . \qquad (2.17)$$

The matrix $R(\theta, \hat{\mathbf{n}})$ rotates a Euclidean 3-vector by an angle θ about the direction $\hat{\mathbf{n}}$, where we can parameterize the unit vector $\hat{\mathbf{n}}$ in terms of its longitude–latitude angles (α, β) as

$$\hat{\mathbf{n}} = (\cos\alpha \sin\beta, \ \sin\alpha \sin\beta, \ \cos\beta) .$$

(Technically, the latitude is $\lambda = \pi/2 - \beta$, but these coordinates facilitate the construction below.) Note that in Eq. (2.16), we will always use the *full* rotation angle parameter θ as the argument of $q(\theta, \hat{\mathbf{n}})$ in order to match Eq. (2.17), not the quaternion half-angle internal parameter.

The form Eq. (2.17) for the 3D rotation $R(\theta, \hat{\mathbf{n}})$ exposes the fact that, since the direction $\hat{\mathbf{n}}$ is fixed, $\hat{\mathbf{n}}$ is a real eigenvector (the only one) of R, and that its eigenvalue is equal to unity. Eq. (2.16) can further be understood in terms of its construction. We exploit the knowledge that we can tilt the North pole into the direction $\hat{\mathbf{n}}$ using a rotation by the angle β, with $\hat{n}_z = \cos \beta$, about the direction

$$(\hat{m}_x, \hat{m}_y) = \frac{(-\hat{n}_y, \hat{n}_x)}{\sqrt{\hat{n}_x^2 + \hat{n}_y^2}} = (-\sin \alpha, \cos \alpha)$$

perpendicular to $(\hat{n}_x, \hat{n}_y)$ by defining

$$\begin{aligned} q_{xy}(\beta) &= \left(\cos(\beta/2),\ \hat{m}_x \sin(\beta/2),\ \hat{m}_y \sin(\beta/2),\ 0\right) \\ &= (\cos(\beta/2),\ -\sin \alpha \sin(\beta/2),\ \cos \alpha \sin(\beta/2),\ 0) \ . \end{aligned}$$

If we start with the knowledge of $\hat{\mathbf{n}}$, we first *reverse* our tilt using $\bar{q}_{xy}(\beta)$ to move our axis to the North pole, followed by a spin about the z-axis by the angle θ,

$$q_{z\text{-axis}}(\theta) = (\cos(\theta/2), 0, 0, \sin(\theta/2)) \ ,$$

and then tilt the z-axis back to the $\hat{\mathbf{n}}$ direction to produce

$$q(\theta, \hat{\mathbf{n}}) = q_{xy}(\beta) \star q_{z\text{-axis}}(\theta) \star \bar{q}_{xy}(\beta) \ . \tag{2.18}$$

This quaternion product rotates the direction $\hat{\mathbf{n}}$ back to align with $\hat{\mathbf{z}}$, spins about $\hat{\mathbf{z}}$ by θ, and then returns the $\hat{\mathbf{z}}$ direction back to $\hat{\mathbf{n}}$, resulting in Eq. (2.16), understood as the composite of simple rotation components. The concept behind the conjugation construction of Eq. (2.18) will reappear as a useful technique in later applications.

The slerp. Relationships among quaternions can be studied using the `slerp`, the *spherical linear interpolation* (Shoemake, 1985), which smoothly parameterizes the points on the shortest ("geodesic") quaternion path between two constant quaternions, q_0 and q_1, as

$$\text{slerp}(q_0, q_1; s) \equiv q(s)[q_0, q_1] = q_0 \frac{\sin((1-s)\phi)}{\sin \phi} + q_1 \frac{\sin(s\,\phi)}{\sin \phi} \ . \tag{2.19}$$

Here $\cos \phi = q_0 \cdot q_1$ defines the angle ϕ between the two given quaternions, while $q(s = 0) = q_0$ and $q(s = 1) = q_1$. The "long" geodesic can be obtained for $1 \leqslant s \leqslant 2\pi/\phi$. For small ϕ, Eq. (2.19) reduces to the standard linear interpolation $(1 - s) q_0 + s q_1$. The unit norm is preserved, $q(s) \cdot q(s) = 1$ for all s, so $q(s)$ is always a valid quaternion and $R(q(s))$ defined by Eq. (2.14) is always a valid 3D rotation matrix. We note that one can formally write Eq. (2.19) as an exponential of the form

$$q_0 \star (\bar{q}_0 \star q_1)^s \ , \tag{2.20}$$

but since this requires computing a logarithm and a matrix exponential whose most efficient reduction to a practical computer program is Eq. (2.19), this is mostly of pedagogical interest.

What are quaternions?

For readers who have already learned a bit about quaternions, as well as those who have little or no previous familiarity, it is fair to ask a top-level question,

> **What are quaternions?**

Here we will talk about various ways that people have formulated and looked at quaternions to fill in some background about the multiple, and possibly confusing, ways that quaternions have been described and put into context. Many more detailed mathematical properties of quaternions will be presented in later chapters, so not every idea will be fully explained here, and that is intentional. Those who are already prepared for the more complete mathematical framework that we will use for detailed calculations may want to jump ahead. However, the enumeration of equivalent descriptions that appear in this chapter may provide useful context, particularly because later we will often choose one preferred notation, that of the topological space $\mathbf{S}^3$, which is best suited to our dominant theme of *visualization*, and we will rarely repeat a given calculation in the other possible alternative quaternion notations.

The executive summary of the many faces of quaternions is the following:

1. **Quaternions are an extension of the complex numbers.** This is the way that Hamilton himself was thinking when he developed the concept of a quaternion. We will almost always restrict ourselves to unit-length quaternions, which Hamilton called *versors*: this class of quaternions corresponds directly to rotations in 3D. Expanding the exponential of $i = \sqrt{-1}$ times a real number θ produces the expansion $e^{i\theta} = \cos\theta + i\sin\theta$, which automatically has unit absolute value. With Hamilton's extended imaginaries, $\mathbf{I}^2 = \mathbf{J}^2 = \mathbf{K}^2 = \mathbf{IJK} = -1$, exponentials of entirely imaginary (or "pure") quaternions using the Hamilton imaginaries $(x\mathbf{I} + y\mathbf{J} + z\mathbf{K})$, with no real or scalar component, produce arbitrary unit length quaternions, and are strongly parallel to the case of exponentials of imaginary numbers.

2. **Quaternions are certain 2×2 complex matrices.** Modern physics makes extensive use of the *Pauli matrices*,

$$\sigma_x = \begin{bmatrix} 0 & 1 \\ 1 & 0 \end{bmatrix}, \qquad \sigma_y = \begin{bmatrix} 0 & -i \\ i & 0 \end{bmatrix}, \qquad \sigma_z = \begin{bmatrix} 1 & 0 \\ 0 & -1 \end{bmatrix},$$

which are the basis of a mathematical framework of matrices equivalent to quaternions. Such noncommuting complex matrices are an essential part of quantum physics, and the so-called "spin one-half" of the electron and other similar elementary particles is very closely related to quaternions. The matrix exponential of a linear combination of Pauli matrices, with the identification

$$\mathbf{I} = -i\sigma_x, \quad \mathbf{J} = -i\sigma_y, \quad \mathbf{K} = -i\sigma_z,$$

produces a set of four matrices that can be identified with the elements of a quaternion.

Visualizing More Quaternions. https://doi.org/10.1016/B978-0-32-399202-2.00011-3

3. **Quaternions are a complexified pair of complex numbers.** Many mathematical treatments frame a quaternion as a unimodular pair of complex numbers, $z_1 = x_1 + iy_1$ and $z_2 = x_2 + iy_2$, with $\|z_1\|^2 + \|z_2\|^2 = 1$, assembled into a quaternion $q = z_1 + z_2 \mathbf{J}$ with the Hamilton $\mathbf{J}$, which has $\mathbf{J}^2 = -1$. But this notation can be misleading, since if the usual $i = \sqrt{-1}$ is used in writing each complex number z, simply requiring $\mathbf{J}^* = -\mathbf{J}$ and $\mathbf{J}^2 = -1$ is insufficient to define a quaternion point on the 3-sphere $\mathbf{S}^3$, or to produce the noncommutativity of the quaternion algebra. This representation only works if the "ordinary" imaginary number "i" is replaced by the Hamilton $\mathbf{I}$ that does not commute with $\mathbf{J}$, and performing the transformation $i\mathbf{J} \rightarrow \mathbf{IJ} = \mathbf{K}$, obeying the conjugation relations of $\mathbf{K}$. Reducing the mathematical abstraction of this basis to an actual Pauli matrix representation, or the equivalent Hamilton $\{\mathbf{I}, \mathbf{J}, \mathbf{K}\}$ representation, in fact reproduces the correct quaternion algebra only if we replace $z = x + iy$ by $z = x + \mathbf{I}y$, or, equivalently in matrix form, $z = xI_2 - iy\sigma_1$, so we have

$$q = z_1 + z_2\mathbf{J} = q_0 + \mathbf{I}q_1 + (q_2 + \mathbf{I}q_3)\mathbf{J}$$
$$= q_0 + \mathbf{I}q_1 + \mathbf{J}q_2 + \mathbf{K}q_3$$
$$= q_0 I_2 - iq_1\sigma_1 - iq_2\sigma_2 - iq_3\sigma_3 \, .$$

4. **Quaternions form the defining representation of the Lie group SU(2).** SU(2) is the group of 2×2 unit-determinant complex matrices whose complex conjugate transpose (Hermitian conjugate) is their inverse. One can represent quaternions using the Pauli matrices without directly using concepts of group theory, but a complete mathematical context must necessarily mention the essential role that the *group properties* of quaternions play in modern mathematics and theoretical physics. Through the relationship of the group **SU(2)** to the construction of the group **SO(3)** of proper rotations in 3D space, we have a rigorous mathematical path from quaternions to 3D rotations. The exponentiation properties of *pure* quaternions correspond to the properties of the *Lie algebra* conventionally denoted as **su(2)**.

5. **Quaternions are a basis for rotations realized in the context of projective transformations on the complex plane.** Later in this book we will examine the Kleinian groups, exhaustively studied in Klein (1884). His treatment first works forwards, examining the actions of 3D rotations on the surface of the two-sphere $\mathbf{S}^2$ and how those actions project points on $\mathbf{S}^2$ to the complex plane, omitting the North pole $(0, 0, +1)$ that projects to infinity. Then, working backwards, he shows how linear fractional transformations on the complex plane containing half-angles typical of quaternion elements correspond to rotations when lifted to the corresponding action on the 3D points of $\mathbf{S}^2$. The key element is that the topological $\mathbf{S}^2$ embedded in 3D space corresponds to the complex projective space $\mathbf{CP}^1$, also known as the Riemann sphere. The finite complex plane parameter $z = x + iy$ parameterizing $\mathbf{CP}^1$ transforms the plane to itself via *linear fractional transformations* taking the form

$$z' = \frac{az + b}{cz + d} \qquad \text{where the matrix} \qquad L(a, b, c, d) = \begin{bmatrix} a & b \\ c & d \end{bmatrix} \qquad (3.1)$$

is a non-singular complex matrix. If we analyze the requirements of the actions of rotations in the projections back and forth from the Riemann sphere represented as $\mathbf{S}^2$ embedded in 3D and the complex plane $z = x + iy$, we find that a particular subspace of quaternion parameters in $L(a, b, c, d)$ perfectly implements 3D rotations. These parameters are the complex conjugates of the elements of the Pauli matrix representation of the quaternion algebra. With this remarkable tool connecting quaternion rotations to classical complex analysis, Klein opened a whole new panorama of opportunities for studying mathematical properties related to quaternions.

6. **Quaternions are one of the four possible normed division algebras.** A normed division algebra obeys a rule of the form $\|x \star y\| = \|x\|\|y\|$. From a very famous result attributed to Hurwitz, there are only four possible normed division algebras: real numbers, complex numbers, quaternions, and octonions. Given a generalized number x that is a sum of N real numbers, $N - 1$ of which are coefficients of basis elements that, multiplied by one another, produce either another basis element or a real number, there are three requirements for a normed division algebra: excepting zero, each such number must have an inverse, each such number must have a norm $\|x\|^2$ that is the sum of the squares of the N real numbers, and the norm of the product of two numbers (denoted by "$\star$") must be the product of the norms, $\|x \star y\| = \|x\|\|y\|$. The reader is invited to write this out for complex numbers: even for that simple case, it is an amazing identity and a very special constraint. Hurwitz's theorem declares that the only possible values of N are 1, 2, 4, and 8. $N = 4$ is the quaternions.

7. **Quaternions correspond to the four even-dimensional (scalar and quadratic) products of 3D *Clifford algebra* elements.** The Clifford algebra representation of a 3-vector V is a linear combination of single Clifford basis elements that have an algebra distinct from usual vectors. Rotations in 3D correspond to the double conjugation of a Clifford 3-vector by real unit-length 2-vectors, say A and B; one such action $A \star V \star A$ is a reflection in the plane perpendicular to A, while acting again by B produces a rotation in 3D space about the axis $A \times B$, and the angle between A and B is *one half* of the total 3D rotation angle experienced by the vector. Remarkably, this action can be reorganized to reveal that the quaternion for this rotation corresponds precisely to the Clifford-valued pair (A, B) and can be written in the Clifford basis as

$$q = q(A \cdot B, A \times B) = A \cdot B + (A \times B) \cdot [e_2 e_3, e_3 e_1, e_1 e_2].$$

Roughly speaking, one might consider the Clifford algebra elements (A, B) to correspond to *square roots* of the quaternions, since it takes two Clifford algebra elements to parameterize a quaternion.

8. **Quaternions of fixed length are a topological space that is a hypersphere, written as $\mathbf{S}^3$.** We deal almost exclusively with quaternions that correspond to 3D rotations, and are thus unit-length 4-vectors q in real Euclidean space $\mathbb{R}^4$, with $q \cdot q = 1$. We will assume this constraint throughout. Any quaternion whatsoever is thus a *point* on the $\mathbf{S}^3$ of unit radius embedded in $\mathbb{R}^4$; we may also think of such a point as a vector from the origin (the origin is not in the sphere itself) to the point in the sphere. This geometry permits us to define similarities and distances in a way unavailable to other methods of encoding rotations because of the well-defined Riemannian metric on this sphere. Connecting to some of the other alternative ways of writing a quaternion, we can see that each point on the sphere also corresponds to one instance of the group $\mathbf{SU}(2)$, while one such point and its negative correspond to *two* equivalent instances of the group $\mathbf{SO}(3)$. Thus the group $\mathbf{SU}(2)$ itself is the topological manifold $\mathbf{S}^3$, and one can deduce that the group $\mathbf{SO}(3)$ is topologically $\mathbf{RP}^3$, the real projective space of dimension three, constructed by identifying opposite points on $\mathbf{S}^3$.

> *We will use this correspondence with $\mathbf{S}^3$ and the way quaternion multiplication manifests itself in this space for most of our approaches in this book to the task of* **visualizing quaternions.**

3.1 Quaternions are extended complex numbers

By the time Hamilton produced the concept of quaternions, complex numbers were well known, and were written in the form

$$z = x + iy, \quad \text{where } i^2 = -1 \text{ implied that } z^2 = x^2 - y^2 + 2ixy,$$

so that the general multiplication rule became

$$z_1 z_2 = x_1 x_2 - y_1 y_2 + i(x_1 y_2 + x_2 y_1).$$

The notation that Hamilton invented successfully generalized the concept of complex numbers extrapolated from a 1D imaginary to a 3D system of imaginaries. (For deep reasons following from Hurwitz's theorem, only 1D, 3D, and 7D imaginaries actually have useful mathematical properties; Hamilton hit a dead end studying 2D imaginaries for many years.) It was thus natural to choose a notation that generalized $i^2 = -1$ as closely as possible, and in particular to implement multiplication as the product of *sums* of terms analogous to $z = x + iy$. Hamilton thus introduced the notation for a number with one real part but *three* imaginary parts as a four-term sum written as

$$q = q_0 + Iq_1 + Jq_2 + Kq_3.$$

The first constraint was that this expression should obey the expectation to act like a normal imaginary number if the other two imaginary coefficients vanished, and in addition there had to be rules about how they multiplied each other: Hamilton's proposal, carved into the side of Broome Bridge, was

$$
\begin{array}{lll}
I^2 = -1, & J^2 = -1, & K^2 = -1, \\
JK = I, & KI = J, & IJ = K, \\
KJ = -I, & IK = -J, & JI = -K, \\
& IJK = -1. &
\end{array}
\tag{3.2}
$$

Thus all that was needed to do quaternion arithmetic was to use ordinary multiplication plus the above rules for the three imaginaries. As a result, the full generality of quaternion multiplication for Hamilton looked like this:

$$
\begin{aligned}
p \star q = \{ & p_0 q_0 - p_1 q_1 - p_2 q_2 - p_3 q_3 + p_0(Iq_1 + Jq_2 + Kq_3) + q_0(Ip_1 + Jp_2 + Kp_3) \\
& + I(p_2 q_3 - p_3 q_2) + J(p_3 q_1 - p_1 q_3) + K(p_1 q_2 - p_2 q_1).
\end{aligned}
$$

Here we notice that quaternion multiplication, unlike complex multiplication, is *order-dependent* (non-commutative), and so in general $p \star q \neq q \star p$; we introduce the symbol $\star$ to remind us that the order matters, and we will use this notation throughout for all manifestations of quaternion multiplication to distinguish it from commutative ordinary multiplication.

3.2 Quaternions are two-by-two matrices

The Hamilton "imaginaries" have properties that are identical to certain complex 2×2 matrices, and unit-length quaternions correspond in their algebraic properties to specific 2×2 matrices, the *Pauli matrices* used to represent quantum mechanical features such as spin one-half elementary particles like electrons. In particular, these are equivalent ways to write a unit quaternion:

$$q = q_0 + \mathbf{I}q_1 + \mathbf{J}q_2 + \mathbf{K}q_3$$
$$= q_0 I_2 - iq_1\sigma_x - iq_2\sigma_y - iq_3\sigma_z \,, \tag{3.3}$$

where I_2 is the 2×2 identity matrix and the σ's are the Pauli matrices

$$I_2 = \begin{bmatrix} 1 & 0 \\ 0 & 1 \end{bmatrix}, \qquad \sigma_x = \begin{bmatrix} 0 & 1 \\ 1 & 0 \end{bmatrix}, \qquad \sigma_y = \begin{bmatrix} 0 & -i \\ i & 0 \end{bmatrix}, \qquad \sigma_z = \begin{bmatrix} 1 & 0 \\ 0 & -1 \end{bmatrix}.$$

The following are essential properties of these matrices:

$$\sigma_x{}^2 = \sigma_y{}^2 = \sigma_z{}^2 = -i\,\sigma_x\sigma_y\sigma_z = I_2 = \begin{bmatrix} 1 & 0 \\ 0 & 1 \end{bmatrix},$$
$$\sigma_i\sigma_j = I_2\,\delta_{ij} + i\epsilon_{ijk}\,\sigma_k \,,$$

where i, j, and k can each take the values (x, y, z) and ϵ_{ijk} is the totally antisymmetric Levi-Civita symbol (see Appendix L). We can trivially see that since

$$\sigma_x\sigma_y\sigma_z = i$$
$$i\sigma_x i\sigma_y i\sigma_z = +1$$
$$(-i\sigma_x)(-i\sigma_y)(-i\sigma_z) = -1 \,,$$

these matrices have the same properties as Hamilton's imaginaries only if we choose the $-i\sigma$ correspondence. Choosing the positive sign violates the requirement for the matrices to match the $\mathbf{IJK} = -1$ identity. Thus the factor of $-i$ multiplying the individual Pauli matrices is very important; some treatments are careless with this sign, resulting in multiplication rules for the inverse quaternions. We remark also that the gamma matrices of the relativistic Dirac equation for the electron are 4×4, and are thus distinct from the 2×2 Pauli matrices and quaternions, though they are all of course closely related.

3.3 Quaternions are a complexified pair of complex numbers

The basic structure of quaternions as Hamilton treated them looks like three separate complex numbers, one for each of the three dimensions, sharing a single real component, and mixing among themselves to obtain the noncommutative properties of 3D rotations. However, the unit-length condition $q \cdot q = 1$ appearing in the Hamilton approach can be phrased in a different way, using a pair of complex numbers obeying $\|z_1\|^2 + \|z_2\|^2 = 1$ that is preferred by some for various application-related reasons. The implementation reduces exactly to Hamilton's imaginaries, but it has a different look to it. It is tempting just to say that we can let $z_1 = x_1 + iy_1$ and $z_2 = x_2 + iy_2$, with $i = \sqrt{-1}$ being the usual basis for imaginary numbers, and then pick a different imaginary, let us label it $\mathbf{J}$, and write a quaternion as

$$q = z_1 + z_2\mathbf{J} \,. \tag{3.4}$$

Then, with $\mathbf{J}^2 = -1$ and $\bar{\mathbf{J}} = -\mathbf{J}$, we get something like this that looks quite reasonable:

$$\|q\|^2 = (z_1 + z_2\mathbf{J})(\bar{z}_1 - \mathbf{J}\bar{z}_2) \stackrel{?}{=} \|z_1\|^2 + \|z_2\|^2 .$$

Unfortunately, if the usual $i = \sqrt{-1}$ that commutes with $\mathbf{J}$ is used, this calculation fails.

The actual solution to describing quaternions as a pair of complex numbers is thus far less simple than one might hope, and much closer to simply rearranging the Hamilton imaginaries in a somewhat artificial way. One must replace the "i" in $z = x + iy$ by Hamilton's $\mathbf{I}$, where the usual single-term complex arithmetic of $z = x + \mathbf{I}y$, with $\mathbf{I}^2 = -1$, is still true, but Eq. (3.4) is now much more complicated: $\mathbf{J}$, for one thing, can *only* appear on the right of z_2, so

$$z_2\mathbf{J} = x_2\mathbf{J} + \mathbf{I}\mathbf{J}y_2 = x_2\mathbf{J} + y_2\mathbf{K} ,$$

and the imaginary $\mathbf{I}$ now appearing in z_1 and z_2 *must not* commute with $\mathbf{J}$, and obeys $\mathbf{I}\mathbf{J} = -\mathbf{J}\mathbf{I} = \mathbf{K}$. Imposing all those conditions, we finally have a working version of the two-complex-variable syntax,

$$q = z_1 + z_2\mathbf{J} = x_1 + \mathbf{I}y_1 + (x_2 + \mathbf{I}y_2)\mathbf{J} ,$$

to represent a quaternion, and now

$$\|q\|^2 = |z_1|^2 + |z_2|^2 = 1$$

can be imposed within this consistent representation.

3.4 Quaternions are the group SU(2)

The 2×2 Pauli matrix form of the full four-element quaternion and its algebra are exactly the defining matrices of the *Lie group* **SU(2)**, the complex matrices of determinant one whose Hermitian conjugates are their inverses. The commutation relations of the Pauli matrices,

$$[\sigma_i, \sigma_j] \equiv \sigma_i\sigma_j - \sigma_j\sigma_i = 2i\epsilon_{ijk}\sigma_k ,$$

embody the algebra of the group. We have already mentioned the relationship between quaternions and the 3D rotation group **SO(3)**: this arises from the rigorous properties of the group **SU(2)**. In brief, while all representations of the group **SU(2)** have three free parameters corresponding to the three degrees of freedom specifying 3D rotations, there are an infinite number of matrices, labeled by integers and half-integers "j," that refer to elements of this group that are matrices of dimension $(2j + 1)$. Quaternions are effectively representations corresponding to the 2D matrices of **SU(2)** with $j = 1/2$. The familiar 3×3 matrices of **SO(3)**, acting on 3-vectors in Euclidean 3D space, have $j = 1$. In Appendix E, we derive the explicit functional form of these matrices, which can be written in terms of either unit quaternions or Euler angles for any j; all can be written as polynomials of order $2j$ in the quaternions. Obviously, with $j = 1/2$, we get the quaternions themselves as the 2×2 Pauli matrix formula linear in q; with $j = 1$, we get a formula for **SO(3)** that is quadratic in q, and so on for any integer multiple of $1/2$.

The fact that two quaternion points correspond to the same **SO**(3) rotation follows directly from the fact that every element in **SO**(3) is quadratic in q, and thus plugging in $q \to -q$ gives the same numbers. Similar relations hold for any j under the substitutions $q \to \exp(2\pi i k/(2j))\, q$ for $0 \leqslant k \leqslant 2j - 1$; we mention this to emphasize that *there is nothing unusual* about q and $-q$ giving the same rotation in **SO**(3). That multiplicity is simply a fact of the group theory of building higher-level representations from polynomials of the defining representation. One final fact is that while the dimension of the rotation group parameters is *three* (corresponding to the Euler angles or the three independent components of the unit quaternions), the matrix dimension of the smallest representation, **SU**(**2**), is two; since **SO**(3) has matrix dimension three, it has a special role as the *adjoint representation*, whose matrix has the same dimension as the space of group parameters itself.

3.5 Quaternion actions can be realized in complex projective space

While exploring all possible ways that rotations could be applied to map regular polygons on the surface of the sphere S^2 into themselves,[1] mathematician Felix Klein (1884), citing existing classic work by B. Riemann and C. Newmann, thoroughly studied the relation of complex analysis to quaternion-parameterized rotations.

Klein's treatment related the quaternion representation of rotations to complex analysis starting from the idea that an ordinary sphere S^2 is topologically a one-dimensional complex projective space CP^1 that can be projected, minus the neighborhood of the topmost point, to the infinite complex plane parameterized by $z = x + i y$. As illustrated in Fig. 3.1, this projection from the point $\hat{\mathbf{n}} = [n_x, n_y, n_z]^t$ (with $\hat{\mathbf{n}} \cdot \hat{\mathbf{n}} = 1$) on the unit sphere S^2 to the complex plane takes the form

$$z_{\text{fp}} = x + i y = \frac{n_x + i n_y}{1 - n_z} = \frac{n_x + i n_y}{1 - \sqrt{1 - n_x^2 - n_y^2}} \,. \tag{3.5}$$

We will also need to consider the diametrically opposite *conjugate* point $\tilde{\hat{\mathbf{n}}} = -\hat{\mathbf{n}}$ on S^2, which projects to the distinct point

$$\tilde{z}_{\text{fp}} = \tilde{x} + i \tilde{y} = \frac{-n_x - i n_y}{1 + n_z} \,. \tag{3.6}$$

The reverse transformation from the complex point $z_{\text{fp}} = x + i y$ back to its source unit vector $\hat{\mathbf{n}}$ on the Riemann sphere is

$$n_x = \frac{2x}{1 + x^2 + y^2}, \qquad n_y = \frac{2y}{1 + x^2 + y^2}, \qquad n_z = \frac{-1 + x^2 + y^2}{1 + x^2 + y^2}, \tag{3.7}$$

where we verify that $z = 0$ corresponds to the South pole at $(0, 0, -1)$, and $z \to \infty$ is the missing North pole $(0, 0, +1)$ for this coordinate chart of CP^1.

Rotations as Linear Fractional Transformations in the Complex Plane. Rotations about a fixed direction $\hat{\mathbf{n}}$ on the unit sphere transform any 3D point on S^2 to a circular path in S^2. That path is mapped

---

[1] We will treat these topics extensively in Chapters 25, 26, 27, and 28.

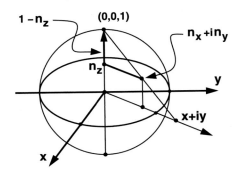

FIGURE 3.1

The projection from the point $\hat{\mathbf{n}}$ on the unit sphere, $n_x^2 + n_y^2 + n_z^2 = 1$, to the complex plane. The similar triangles allow us to define the projection to the (x, y) plane based on the 3D coordinates $\hat{\mathbf{n}}$, through the perpendicular axis $(0, 0, 1)$; the similar-triangle ratio $(x + iy)/1 = (n_x + in_y)/(1 - n_z)$ defines the projection.

by stereographic projection to a unique corresponding path on the complex plane; this is accomplished via **PGL(2, C)** transformations on the homogeneous coordinates $\{z_1, z_2\}$ of the complex projective space $\mathbf{CP}^1$ of the general form

$$\begin{bmatrix} z_1' \\ z_2' \end{bmatrix} = \begin{bmatrix} a & b \\ c & d \end{bmatrix} \cdot \begin{bmatrix} z_1 \\ z_2 \end{bmatrix} = \begin{bmatrix} az_1 + bz_2 \\ cz_1 + dz_2 \end{bmatrix}. \tag{3.8}$$

The projective space $\mathbf{CP}^1$ is typically represented by two charts, each a complex plane; this construction is basically identical to covering the Earth's non-trivial $\mathbf{S}^2$ manifold with a pair of hemispheres. The charts of $\mathbf{CP}^1$ are each described by a corresponding local inhomogeneous complex variable z built from ratios of the homogeneous variables $[z_1, z_2]$ employed in Eq. (3.8), that is

$$z = \frac{z_1}{z_2} \quad \text{for } z_2 \neq 0 \quad \text{or} \quad z = \frac{z_2}{z_1} \quad \text{for } z_1 \neq 0. \tag{3.9}$$

By default we choose the first chart for z (the North Pole chart), so that every point $\hat{\mathbf{n}}$ on the unit sphere maps to the complex plane at $x + iy$ except $(0, 0, 1)$, which corresponds to the point at infinity; the latter can be covered by choosing the South pole chart if necessary. For any rotation about the fixed axis $\hat{\mathbf{n}}$, there is a conjugate pair of fixed points Eq. (3.5) and Eq. (3.6) at z_{fp} and $\tilde{z}_{\mathrm{fp}}$, respectively.

Rotations in the complex plane are implemented as a subset of the linear fractional form defined using the matrix from Eq. (3.8) defined as

$$L(a, b, c, d) = \begin{bmatrix} a & b \\ c & d \end{bmatrix}. \tag{3.10}$$

This produces the linear fractional transformation on the homogeneous variables $[z_1, z_2]$ of $\mathbf{CP}^1$ as

$$\frac{z_1'}{z_2'} = \frac{az_1 + bz_2}{cz_1 + dz_2}. \tag{3.11}$$

Choosing the standard chart for the inhomogeneous complex plane variable $z = z_1/z_2$, the corresponding linear fractional transformation that we use as the framework for z-plane rotations becomes

$$z' = \frac{az+b}{cz+d} .$$
(3.12)

One would guess that to implement rotations on the complex plane using 2×2 matrices, we would simply take the axis-angle form of the quaternion,

$$q(\theta, \hat{\mathbf{n}}) = \left(\cos(\theta/2), n_x \sin(\theta/2), n_y \sin(\theta/2), n_z \sin(\theta/2)\right),$$

and define the matrix L to be the corresponding $\mathbf{SU(2)}$ matrix representing 3D rotations,

$$L(a,b,c,d) \overset{??}{=} [Q(q)] = q_0 I_2 - i\mathbf{q} \cdot \boldsymbol{\sigma} = \begin{bmatrix} q_0 - iq_z & -iq_x - i(-iq_y) \\ -iq_x - i(+iq_y) & q_0 + iq_z \end{bmatrix}$$
(3.13)

$$= \begin{bmatrix} \cos(\theta/2) - i\sin(\theta/2)n_z & -i\sin(\theta/2)\left(n_x - in_y\right) \\ -i\sin(\theta/2)\left(n_x + in_y\right) & \cos(\theta/2) + i\sin(\theta/2)n_z \end{bmatrix}.$$
(3.14)

How can we check to see if this works? We know that the complex plane transform must have rotational fixed points in the complex plane corresponding to the fixed axes $\pm\hat{\mathbf{n}}$ on $\mathbf{S}^2$. These fixed points do not change under the transformation Eq. (3.12), so the two solutions of the quadratic equation

$$z = \frac{az+b}{cz+d}$$
(3.15)

with matrix elements Eq. (3.14) should be the fixed points Eq. (3.5) and Eq. (3.6). $\boxed{\textit{This is false.}}$ As noted by Klein (1884), the correct linear fractional transformation corresponding to a standard right-handed Hamilton quaternion rotation is in fact the *complex conjugate* of the postulated $\mathbf{SU(2)}$ rotation Eq. (3.14), that is

$$L(a,b,c,d) = \left[Q(q)^*\right] = \begin{bmatrix} q_0 + iq_z & +iq_x - q_y) \\ +iq_x + q_y) & q_0 - iq_z \end{bmatrix}.$$
(3.16)

We see that the matrix Eq. (3.16) simply replaces $(q_x, q_z) \to (-q_x, -q_z)$ leaving q_y unchanged, equivalent to a parity-preserving change of coordinates rotated by $180°$ about the $\hat{\mathbf{y}}$ axis.

One can easily verify (possibly needing the identity $(1 + n_z)/(n_x - in_y) = (n_x + in_y)/(1 - n_z)$) that the two solutions of the complex plane quadratic fixed point constraint Eq. (3.15) with $L(a,b,c,d) = \left[Q(q)^*\right]$ are precisely the fixed points Eq. (3.5) and Eq. (3.6).

In Fig. 3.2 we illustrate the action of rotations about a fixed axis $\hat{\mathbf{n}}$ (a) upon lattice points on the complex plane and (b) upon the circular paths of points on $\mathbf{S}^2$ projected to the complex plane. The two corresponding fixed points on $\mathbf{S}^2$ under that family of rotations are marked by red dots in the complex plane. Note that righthanded paths on $\mathbf{S}^2$ correspond to *both* right-handed paths, near the $-\hat{\mathbf{n}}$ fixed point, and left-handed paths near the $+\hat{\mathbf{n}}$ fixed point, with a straight line path through infinity dividing the two regions.

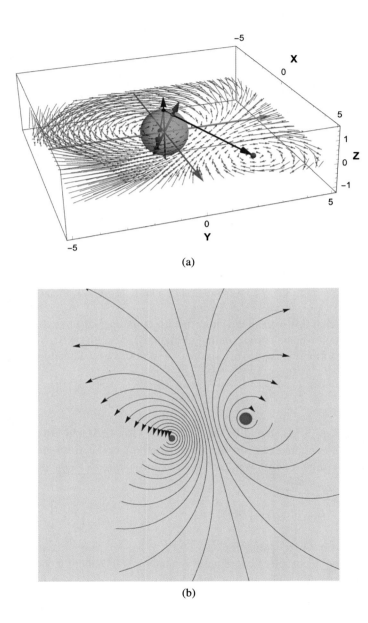

(a)

(b)

FIGURE 3.2

Rotation paths on the sphere S^2, projected through the top point $(0, 0, 1)$ to the corresponding paths on the complex plane. One sees clearly both the usual fixed rotation axis $\hat{n}$ and the conjugate fixed point $-\hat{n}$ as the centers of clockwise and counterclockwise swirls in the complex plane. (a) Shows the paths of points in a regular lattice in the complex plane acted upon by 3D rotations around a fixed axis $\hat{n}$ on the sphere S^2. (b) Shows the paths in the complex plane corresponding to the projections of circular arcs on S^2 to the plane.

3.6 Quaternions are one of the four normed division algebras

Completely independent of Hamilton, it is likely that quaternions would eventually have appeared spontaneously when pure mathematicians started examining in detail the kinds of arithmetic that was allowed in such domains as the matrix elements of groups. When one starts to ask what kinds of numbers are useful, a logical starting point is to ask questions about generalizing complex numbers, which is pretty much exactly what Hamilton did in 1843. Hamilton had tried and failed to find a consistent system with two imaginaries instead of just the one $i = \sqrt{-1}$ of ordinary complex numbers. When he tried *three* imaginaries, he immediately came up with the consistent system of quaternion algebra. Later the same year James Graves discovered *octonions*, a system with *seven* imaginaries, and they were rediscovered independently and published in 1845 by Arthur Cayley. For a long time thereafter, it was unclear what further systems of generalized imaginary numbers might lie unknown.

Eventually, it was understood what essential properties these four systems had in common, with one element (reals), two elements (complex), four elements (quaternions), and eight elements (octonions). We start with a generalized number x that is a sum of N real numbers, $N - 1$ of which are multiplied by special basis elements. These elements have the property that, when multiplied by one another using some rule "$\star$," they produce either another basis element or a real number. The properties that are required of such a candidate system to form a normed division algebra are as follows:

- **A norm.** This is essentially the absolute value, a nonnegative real number giving the length of any number in the system as the square root of the sum of the squares of the N real numbers in x, written as $\|x\|$.
- **A multiplicative inverse.** Excluding the number with all coefficients vanishing, the "zero" element with vanishing norm, every number x in these systems possesses another number $\bar{x}/\|x\|^2$ that is its inverse,

$$x \star \frac{\bar{x}}{\|x\|^2} = 1 \,,$$

where $\bar{x}$ is the element with all the generalized imaginaries reversing sign.
- **The norm of a product is the product of the norms.** This is a remarkable feature of the four working number systems, which we can write, for any x and y, as the requirement that

$$\|x \star y\| = \|x\| \, \|y\| \,.$$

From the very beginning, it was observed that for the four special number systems, one could use this property of the norms to prove that, e.g., for octonions, the product of sums of eight perfect squares was equal to the sum of eight other perfect squares. Similarly, quaternions relate sums of four squares, and complex numbers relate sums of two perfect squares. If you write this out even for ordinary complex numbers, you will see that it is a bit of surprise, as well it should be, because these are very special properties.

Number systems having these characteristics are known as *normed division algebras*. It is now known, thanks to the rigorous result of a theorem attributed to Hurwitz, that if you try to form generalized complex numbers using N real numbers with $N - 1$ of these multiplying generalized imaginaries, there are no normed division algebras except for $N = 1, 2, 4$, and 8, corresponding to the reals, complex numbers, quaternions, and octonions. In fact, things get progressively more complicated as N

increases, because, although the reals and complexes have order-independent (commutative) multiplication rules, quaternions are noncommutative but associative, and octonions are neither commutative nor associative: for octonions, $a \star (b \star c) \neq (a \star b) \star c$. This is why we use the "$\star$" symbol for symbolic quaternion multiplication, to remind ourselves that this is not ordinary multiplication but noncommutative multiplication, so we must carefully specify the *order* in which operations are carried out.

3.7 Quaternions are a special case of Clifford algebras

On the one hand, quaternions are very special, being one of the four possible normed division algebras allowed by the Hurwitz theorem: real, complex, quaternion, and octonion number systems. However, all of the properties of quaternions also arise as a particular special case of a completely general system that works for all dimensions. This system, which realizes the special properties of complex numbers and quaternions more or less as an accident of lower dimensionality, is that of 2D and 3D *Clifford algebras*. Setting aside a few of the technical details (see Appendix F), we define a Clifford 3-vector as

$$V = v_1 e_1 + v_2 e_2 + v_3 e_3 \,,$$

where the basis elements e_i obey a very strange relationship,

$$e_i e_j + e_j e_i = -2\delta_{ij} \,.$$

(Note that some authors prefer a plus sign here.) What this means in practice is that enclosing a given Clifford 3-vector V by another unit-norm Clifford 3-vector, A, with $A \cdot A = 1$, produces a *reflection* about the plane perpendicular to A:

$$A \star V \star A = V - 2A(A \cdot V) \,,$$

as shown in Fig. 3.3(a). If we repeat this operation, and reflect a second time about a new plane perpendicular to the unit vector B, the result is not a reflection, but a *rotation* applied to the Clifford algebra components of the vector V,

$$B \star (A \star V \star A) \star B = R(\theta, \hat{\mathbf{n}}) \cdot V \,,$$

as shown in Fig. 3.3(b). But the remarkable thing is that the angle of the rotation is

$$\theta = 2 \arccos(A \cdot B) \quad \rightarrow \quad A \cdot B = \cos\frac{\theta}{2} \,,$$

and the fixed axis of this rotation is $\hat{\mathbf{n}} = A \times B / \|A \times B\| = A \times B / \sin(\theta/2)$, so

$$A \times B = \hat{\mathbf{n}} \sin(\arccos A \cdot B) = \hat{\mathbf{n}} \sin\frac{\theta}{2} \,.$$

That means in fact that the rotation $R(\theta, \hat{\mathbf{n}})$ is generated by the quaternion

$$q(A, B) = \left(\cos\frac{\theta}{2}, \hat{\mathbf{n}} \sin\frac{\theta}{2} \right) = (A \cdot B, A \times B) \,, \tag{3.17}$$

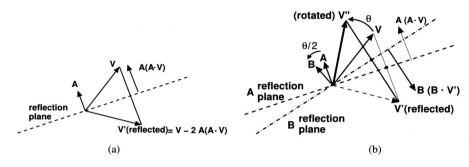

FIGURE 3.3

(a) A single Clifford reflection of a vector in the direction of **A**. (b) A double Clifford reflection has the feature that the angle between the reflection directions **A** and **B** is *half* the angle of the 3D vector rotation, implying a deep relation to quaternions.

or, in terms of the Clifford 3-vector basis $\{e_1, e_2, e_3\}$, the Clifford quaternion is written as

$$q = A \cdot B \,[\text{identity}] + (A \times B) \cdot [e_2e_3, \, e_3e_1, \, e_1e_2] \,.$$

Thus the quaternion is composed precisely of the scalar and vector parts, the dot product and the cross product, of a double Clifford reflection.

One may observe that, since 3D rotations are quadratic polynomials in the quaternion, quaternions are a kind of generalized *square root* of 3D rotations. But quaternions themselves are quadratic polynomials in the Clifford reflection parameters, so one might say that the Clifford reflections form the *square root* of quaternions, and therefore are a kind of *fourth root* of 3D rotations.

3.8 Quaternions are points on the 3-sphere $\mathbf{S}^3$

This book and its predecessor are dedicated to the fundamental fact that we can represent quaternions themselves as points in space, which makes it possible to draw pictures of them that facilitate our understanding of their meaning and applications. Every alternative way that we have of representing quaternions can be transformed to writing down points in 4D Euclidean space that have unit distance from the origin, and thus form a topological manifold of dimension three, the hypersphere or 3-sphere, written as $\mathbf{S}^3$. That is, we can write any quaternion in the form

$$q = (q_0, q_1, q_2, q_3) = (q_0, \mathbf{q})$$

with

$$q \cdot q = (q_0)^2 + (q_1)^2 + (q_2)^2 + (q_3)^2 = (q_0)^2 + \mathbf{q} \cdot \mathbf{q} = 1 \,,$$

which is the algebraic equation of a point on $\mathbf{S}^3$ embedded in $\mathbb{R}^4$, 4D Euclidean space.

There are many ways to explore and take advantage of this intrinsic geometry of the unit quaternion. Our approach is to combine the *geometry*, which is where the points of a structure lie in space, with the *topology*, which is how the continuous global structure containing those points is constructed from local pieces of that structure connected by a family of local relationships. The geometry we handle by taking the actual mathematical points of q and projecting them along a 4D viewing ray into a 3D volume, each point of which is ambiguous as it includes an entire ray along the view direction. Then we can separate neighboring groups of points into topologically separate sets, which for $\mathbf{S}^3$ are the two solid balls of quaternions: one is the solid ball whose center is at a logical "North pole," say $q_0 = +1$, extending downward just beyond the logical "Equator," which is the 2-sphere $\mathbf{S}^2$ at $q_0 = 0$ or $\mathbf{q} \cdot \mathbf{q} = 1$, and the other is the solid ball at the logical "South pole," say $q_0 = -1$, extending upward to just beyond the logical "Equator." If we want to continuously connect all possible quaternions together, we need only choose paths that let us know when we are close to the equator, and change coordinate systems as we pass through the equator – this is the classical topological method for describing a nontrivial manifold (of which the sphere is one) with an "atlas" of overlapping distinct coordinate patches or "charts."

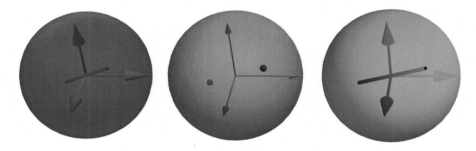

FIGURE 3.4

Basic entire $\mathbf{S}^3$ visual. The entire quaternion 3-sphere $\mathbf{S}^3$, a 3-manifold embedded in $\mathbb{R}^4$, can be visualized in a 3D image consisting of an open solid ball around the North pole, that is, $q_{\text{ID}} = (1, 0, 0, 0)$, shown on the right in cyan, the equatorial $\mathbf{S}^2$ in orange as the closure at $q = (0, x, y, z)$ with $x^2 + y^2 + z^2 = 1$, and the open solid ball around the South pole, $-q_{\text{ID}} = (-1, 0, 0, 0)$, shown on the left in magenta. The lines in the North and South solid balls are the path of the quaternion $q = (\cos\theta/2, \hat{\mathbf{n}}\sin\theta/2)$ for a diagonal rotation axis $\hat{\mathbf{n}} = (1/\sqrt{3}, 1/\sqrt{3}, 1/\sqrt{3})$; the path intersects the central equatorial spherical surface at two points, $\pm\hat{\mathbf{n}}$.

Our goal is to understand the global nature of possibly complicated sets of quaternions. We will advocate an approach to covering the entire quaternion space in a viewable way by first labeling the volume with $q_0 > 0$ as the Northern hemisphere, containing the North pole, q_{ID}, that is, $q_0 = 1$, and the $q_0 < 0$ volume as the Southern hemisphere, with $-q_{\text{ID}}$, that is, $q_0 = -1$, designated as the South pole; note that since the quaternion $\mathbf{S}^3$ is a 3-manifold, these "polar hemispheres" are solid volumetric balls, sharing a balloon-skin 2-sphere Equator $\mathbf{S}^2$. As we often have special phenomena happening exactly at the Equator, $q_0 = 0$ or $\mathbf{q} \cdot \mathbf{q} = 1$, we often adopt a complete graphical display that has three parts: the Northern hemisphere an open solid ball with 3D coordinates $\mathbf{q}$ for the quaternion with $0 < q_0 \leqslant 1$, the $\mathbf{S}^2$ Equator with $q_0 \equiv 0$ sitting in the middle, and the Southern hemisphere a separate open solid ball for 3D points $\mathbf{q}$ with $0 > q_0 \geqslant -1$. This representation is illustrated in Fig. 3.4, showing a quaternion circle $(\cos\theta/2, \hat{\mathbf{n}}\sin\theta/2)$ as a continuous line across the entire quaternion $\mathbf{S}^3$ volume through both q_{ID} and $-q_{\text{ID}}$, hitting the equator $\|\mathbf{q}\| = 1$ at just the two points marked on the Equatorial sphere in the center. This circle is a straight line because every q_0 component is implicit, understood in the $\mathbf{q}$

spaces as the invisible dependent value defined by $\pm\sqrt{1 - \mathbf{q} \cdot \mathbf{q}}$. Every possible point of any quaternion lying in $\mathbf{S}^3$ thus has a unique visible location, and any path or motion can also be tracked through these three viewable sectors. We have the choice also, if we want to view periodic paths in q_0 more clearly, to replace one of the $\mathbf{q}$ axes by q_0 and consider the replaced variable instead as the invisible implicit value. More details will be worked out in Chapter 4.

3.9 Synopsis

We now have seen how to imagine a quaternion in all these ways:

1. **Hamilton imaginaries:** $q = q_0 + \mathbf{I}q_1 + \mathbf{J}q_2 + \mathbf{K}q_3$.
2. **Pauli matrices:** $q = q_0 I_2 - iq_1\sigma_x - iq_2\sigma_y - iq_3\sigma_z$, where the $\sigma_{\{x,y,z\}}$ are the 2×2 complex Pauli matrices used in physics.
3. **A Pair of Complex Numbers:** A quaternion can be considered as a pair of complex numbers $q = z_1 + z_2\mathbf{J}$ satisfying the $\mathbf{S}^3$ constraint $\|z_1\|^2 + \|z_2\|^2 = 1$.
4. **Group theory:** $q \in \mathbf{SU}(2)$, where the adjoint representation of the ordinary 3D rotation group $\mathbf{SO}(3)$ is double-covered by a 3×3 quadratic form in q.
5. **A Linear Fractional Transformation:** The action of a quaternion producing a 3D rotation can be written as a linear fractional transformation in the complex plane with a particular choice of 2×2 matrix parameters.
6. **Normed division algebra:** Quaternions are the dimension $N = 4$ members of the set of four possible normed division algebras with $N = 1, 2, 4$, or 8 that satisfy $\|x \star y\| = \|x\| \|y\|$.
7. **Double reflection of a 3D Clifford algebra:** Two unit-length 3-vectors written as Clifford algebra vectors A and B generate a 3D rotation via a double reflection, and the quaternion corresponding to that rotation is a quadratic form in the 3D Clifford algebra basis with coefficients $A \cdot B$ and $A \times B$.
8. **Hypersphere geometry:** $q = (q_0, q_1, q_2, q_3)$, with $q \cdot q = 1$, defines a point on the (round) 3-sphere $\mathbf{S}^3$ in 4D Euclidean space with the standard constant-curvature metric; this context facilitates making informative computer graphics representations of quaternion properties and applications.

In each case, the multiplication rules for two quaternions (or any number of quaternions, remembering that the multiplication is order-dependent) all obey exactly the same algebra that Hamilton originally wrote down. With the exception of selected applications in which alternative notations are useful, in this book we will always use the $\mathbf{S}^3$ context for our notation. That is, we write the quaternion as either a 4-vector $q = (q_0, q_1, q_2, q_3)$ or as a 4-vector split for notational convenience into a "real" part q_0 and an "imaginary" or "pure quaternion" 3-vector part $\mathbf{q}$, so $q = (q_0, \mathbf{q})$. Thus we preferentially think of the quaternion multiplication algebra as

$$p \star q = (p_0 q_0 - \mathbf{p} \cdot \mathbf{q}, \; p_0\mathbf{q} + q_0\mathbf{p} + \mathbf{p} \times \mathbf{q}) \; .$$

The $\mathbf{I}, \mathbf{J}, \mathbf{K}$ and $\mathbf{SU}(2)$ (Pauli) matrix multiplications give *exactly* that form as well, provided one is careful to use the sign $\mathbf{I} = -i\sigma_x$, etc.

Frequently asked questions: The answers to the following common questions about quaternions are "yes" on all counts:

- Yes, quaternions can be thought of as generalizing complex numbers to have three imaginary units $\{\mathbf{I}, \mathbf{J}, \mathbf{K}\}$ whose square is (-1) instead of just one $i = \sqrt{-1}$, but they have a new, and order-dependent, behavior when they multiply each other.
- Yes, (unit) quaternions are essentially elements of the matrix group $\mathbf{SU(2)}$, the special unitary group of 2D unitary complex matrices with unit determinant.
- Yes, (unit) quaternions, like the group $\mathbf{SU(2)}$, have a quadratic form that corresponds exactly to 3D special orthogonal rotation matrices $R(q)$, the elements of the group $\mathbf{SO}(3)$. In addition, quaternions are a double covering of $\mathbf{SO}(3)$, that is, each point in $\mathbf{SO}(3)$ corresponds to two distinct points in quaternion space, so $R(q)$ is the same rotation as $R(-q)$.
- Yes, (unit) quaternions span the topological space of the 3-sphere, $\mathbf{S}^3$, which also corresponds exactly to the topological space of the group $\mathbf{SU(2)}$.

What is quaternion visualization?

This is a book that claims to be about "visualizing quaternions." Now that we have a bit of context about the question "What *is* a quaternion?", the reader might well ask another equally relevant question:

> ## What is quaternion visualization?

We now know that quaternions manifest themselves in many different ways. A quaternion is always a quaternion, but different ways of describing it may each be better adapted to different objectives. Writing down an ephemeris for the moon, a complicated assembly of equations giving the location of the moon in a particular coordinate system for a particular range of times on the calendar, is important if you want to send a rocket to the moon. However, it does not easily translate into an intuitive image that you can use to understand a given detail of the complicated path of the moon during a given month; the information is essentially there in the ephemeris, but it is not as well suited to the "explain to yourself what the moon is doing" task as a visual representation might be. Though we will certainly *do* complicated calculations in this book whenever it is appropriate, we will focus on *enjoying an informative and intuitive visual context* for those calculations whenever we can. That is our underlying goal, and that is what we mean by visualization.

$\mathbf{S}^3$ **is the key.** The first important step in visualizing a quaternion is thus to choose an appropriate representation that adapts itself to graphical illustration. There is only one quaternion representation that is directly convertible to graphics: the geometric expression of a quaternion as a unit-length 4-vector, a point on the surface of the hypersphere $\mathbf{S}^3$ embedded in 4D Euclidean space, and that will be the context of most of our "quaternion visualization" applications. We may exploit alternative methods, e.g., to facilitate comparisons, from time to time, but basically most will trace back to $\mathbf{S}^3$.

But exactly how do we do this? Graphics images on a page in a book are only 2D, and even allowing for the use of successful methods from many disciplines that simulate pictures of objects and scenes in a 3D world using the 2D page as a "photograph," pictures of a 4D world seem quite challenging. Nevertheless, that is exactly what we shall do: we will make pictures on the 2D page that exploit the methods of 3D graphics, and yet represent images of a 4D scene in ways that retain both geometric and topological features. There are special properties of spheres that allow us to do that, with characteristics that we can learn to recognize in 2D images; certain interactive 3D graphics techniques can also be exploited when available. Our method will thus be to find ways to represent 4D scenes as familiar 3D camera-like images on the 2D page or display screen.

Only one thing is asked of the reader: a small amount of practice is required to interpret these graphical representations in a way that permits the mental extraction of 4D coordinates from our graphics in

Visualizing More Quaternions. https://doi.org/10.1016/B978-0-32-399202-2.00012-5

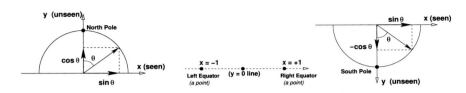

The Northern Hemicircle of S^1 The 2-Point Equator of S^1 The Southern Hemicircle of S^1

FIGURE 4.1

Schematic of the 2-hemicircle approach to visualizing S^1. Any point (x, y) in the Northern hemicircle, with $y > 0$, is uniquely defined by a single x-coordinate, since we know that $y = +\sqrt{1 - x^2}$. If y is exactly $= 0$, the value of $x = \pm 1$ uniquely determines that point in S^1's very simple equator. Points in x with $y < 0$ then obviously describe unique points on the circle with $y = -\sqrt{1 - x^2}$ in the Southern hemicircle.

a deterministic and natural way. To do this, we invoke a way of thinking about circles and spheres that goes back to the earliest days of round-Earth cartography.

Spheres as a pair of hemispheres. Now it is time to get specific about how we look at a quaternion in space as a graphical entity. There are a couple of techniques that we can use, and some variants that involve interaction instead of static images. Since this is a book, we focus on what we think are the most informative approaches using static quaternion imagery, and will return later to the question of interaction.

The fundamental idea is that a sphere is technically not describable as a *function*. A sphere is a nontrivial *manifold*. The basic example of a trivial manifold is a graph in Euclidean space, for which there is one global coordinate system that we might write as $z = f(x, y)$. Any sphere, S^1 (the circle), S^2 (the spherical surface of a globe or a balloon), the quaternion hypersphere S^3, etc., must be described by no fewer than *two local coordinate systems*. Conceptually, a standard choice of these two coordinate systems corresponds to a North pole view and a South pole view. Luckily, we can actually work our way up to how to draw quaternion objects in S^3 by starting with simpler lower-dimensional spheres, learning how to look at them as pairs of hemispheres, and extending that concept to quaternions.

4.1 Fundamental idea: the circle S^1

Here is how it works for a circle S^1. Consider Fig. 4.1, and look first on the left at the *Northern open hemicircle or half-circle*, the part of the circle $x^2 + y^2 = 1$ embedded in the 2D Euclidean space $\mathbb{R}^2$ that satisfies $y > 0$. Every point on this Northern part except $y = 0$, $x = \pm 1$ has a corresponding point with the opposite sign of y, that is, the *Southern open hemicircle or half-circle*. This whole circle is not a function, as it cannot be written as a function $y = f(x)$ that is single-valued for all x: it is a *manifold* that has more than one part of its topological space in y for each choice of x, and, furthermore, there is a particular rule for sewing these together if we allow $y = 0$, or in practice, $y > -\epsilon$ for the North, and $y < \epsilon$ for the South, to give a small band of overlap, as illustrated in Fig. 4.2. We remark that in some cases we may want even smaller ranges, such as a single quadrant, for each coordinate patch or "chart."

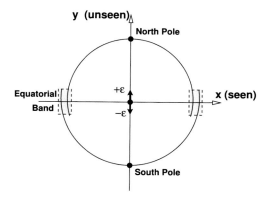

FIGURE 4.2

While no single coordinate system covers an entire nontrivial manifold like a circle or a sphere, it is standard practice to use an *atlas* of multiple single-valued coordinate patches or "charts" related where they overlap. For the circle $\mathbf{S}^1$, we label that region in this figure as the *equatorial band*, where $|y| \leqslant \epsilon$ and both coordinate systems are defined.

The viewer's task. We promised that the reader would need only a small amount of practice to gain insight into quaternion sphere geometry from the pictures we will draw as visualizations. Almost every necessary idea appears already when we try to draw the properties of a *circle* using only the projections of its top and bottom half-circles (which we may call "hemicircles" in analogy to "hemispheres") onto a horizontal line. We study this in two parts:

- **Breaking a circle into single-valued pieces.** In Fig. 4.1, we showed the decomposition of an ordinary circle, i.e., the graph of $x^2 + y^2 = 1$ in the xy-plane, into single-valued pieces that fully
  describe the *entire* circle without ambiguity. But this cannot be done without separating, for example, the North pole and South pole regions into separate single-valued maps, with an overlap in a
  band around the equator, that is, $|y| < \epsilon$, centered at the exact equator defined by $y = 0$, as illustrated
  in Fig. 4.2. In that region there are two alternative coordinate systems describing the same point, and
  that is what allows a hypothetical bug to cross from the Northern hemicircle to the Southern hemicircle while maintaining a grip on reality. This is precisely the process mathematicians use to maintain
  their sanity in complicated geometric contexts, and it is known as an "atlas" of single-valued overlapping patches ("charts") on the manifold, and the "transition functions" that tell you how to glide
  without confusion from one to the other. The reader can look up these terms to find additional technical details (see, e.g., Berger, 1987), but the description just given is hopefully sufficient for this
  narrative.
- **Training the eye to compensate for predictable distortion.** Next, we have to learn how to extract
  distances from a picture of the circle, going another level beyond just setting up valid topological
  charts describing the multivalued nature of the circle. To accomplish this, we need to be precise
  about the nature of our *projection* of the semicircular polar spaces onto the line at $y = 0$. The local
  space seen by a bug living on the circle is completely uniform, no matter where the bug is sitting.
  But from a vertical point of view, looking down through the space where the bug lives and dropping

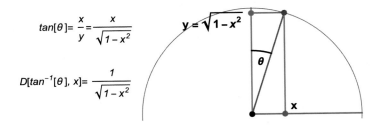

FIGURE 4.3

We illustrate the geometry of part of a hemicircle or a slice of a hemisphere as it relates to a picture taken from the North pole, and how the angle θ of the tilt from vertical relates to a point at a distance x from the vertical axis. For incremental changes in x, the adjacent side of the angle has length $y = \sqrt{1 - x^2}$. The tangent of θ is the opposite side of the right triangle divided by the adjacent side, so we calculate the change in the angle as x goes from the origin to the edge of the circle at $x = \pm 1$. The size of the subtended arc length for the same increment in x increases dramatically as x increases from $0 \to 1$, with minimum distortion, $d\theta/dx = 1$, at $x = 0$.

each point onto the flat linear space that we can *see* on a given chart, we realize the *only* spots where our view is as uniform as the bug's are right at the North pole in one part of our depiction, and right at the South pole in the other part. All other points are progressively more compressed, with normal distances for the bug corresponding to smaller and smaller distances for us as we move towards the equator. We must learn to take the visible point-to-point distances projected onto the $y = 0$ horizontal line (the x-axis image) and *translate* them to the intrinsic local *geometric distances* perceived by a crawling bug.

Now we attempt to show both the quantitative and qualitative character of the distortion we see looking down on a circular space, projecting, as though taking a photograph, onto the line at $y = 0$ from a point $[x, y]$ on the circle $x^2 + y^2 = 1$ in a way that drops the point's y-coordinate from view, leaving only the corresponding x-coordinate. This distortion can be very precisely computed from the equation of the circle itself.

Any point on the Northern hemicircle will have coordinates $(x, y = \sqrt{1 - x^2})$. Shifting our viewpoint sideways to show the whole plane containing a hemicircle, we can see the geometry illustrated in Fig. 4.3. The x dependence of the angle θ, measured downward from the pole, is then

$$\theta = \arctan\left(\textbf{opposite} = x\ , \textbf{adjacent} = \sqrt{1 - x^2}\right) = \arctan\left(\frac{x}{\sqrt{1 - x^2}}\right). \tag{4.1}$$

Therefore the incremental change in θ for an incremental displacement in the x-coordinate is just the derivative

$$\frac{d\theta}{dx} = \frac{d \arctan(x/\sqrt{1 - x^2})}{dx} = \frac{1}{\sqrt{1 - x^2}}. \tag{4.2}$$

Fig. 4.4 illustrates this dramatic change in density as $x \to \pm 1$. The key is that, once one knows the radial distance $x = \sin\theta$ and the direction from the pole on the photograph, one knows precisely not only the x-position of the point, the horizontal distance from one of the poles, but also *the exact*

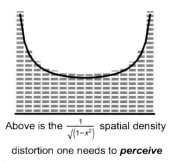

Above is the $\frac{1}{\sqrt{(1-x^2)}}$ spatial density

distortion one needs to **perceive**

This line of projections
is what you **actually see**

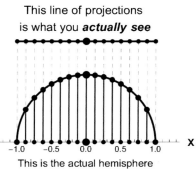

This is the actual hemisphere
with its unseen height coordinates:
$hidden = \sqrt{(1-x^2)}$

FIGURE 4.4

The density plot of the S^1 display written in terms of one of the pole-view semicircles. The dots projected to the straight line on the x-axis are taken at equal intervals in x, but when projected to the semicircle, the angular distance intervals increase dramatically towards the edges, where the x-axis coordinates approach the equatorial points $x = \pm 1$.

value of the unseen height $y = \sqrt{1 - x^2} = \cos\theta$. In summary, given any measured point (x, y) in the interior of the circular hoop viewed from the top, we can summarize our knowledge as follows:

$$x = \sin\theta, \quad \text{Euclidean location on the projection line}$$

where   (4.3)

$$y = \cos\theta = \sqrt{1 - x^2}, \quad \text{hidden Euclidean height of point at } x .$$

4.2 A richer example: the 2-sphere S^2

We can learn more now by going up to the next dimension. The balloon-like S^2 is defined as the solution to $x^2 + y^2 + z^2 = 1$ embedded in the 3D Euclidean space $\mathbb{R}^3$, but now the Northern hemisphere obeys

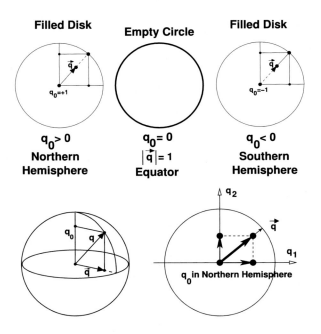

FIGURE 4.5

Schematic of the 2-hemisphere approach to visualizing $\mathbf{S}^2$, with a North pole disk spliced across the circular equator at $z = 0$, $x^2 + y^2 = 1$, to combine with the independent South pole disk.

$z \geqslant 0$ and the Southern hemisphere obeys $z \leqslant 0$. Looking down from a distant point on the positive z-axis, we see the *Northern hemisphere* as a filled disk with its farthest visible edges being the equator at $z = 0$; this is half of our two-valued full manifold. Looking up from a distant point on the negative z-axis, we see the *Southern hemisphere*, another filled disk bounded by the equator. If we tell our crawling bug about the mystical land in the ribbon $|z| \leqslant \epsilon$, the bug can match up the Northern coordinate system and the Southern coordinate system and understand how to get between the two hemispheres. The geometric distortion for a particular plane through the z-axis is basically the same as for $\mathbf{S}^1$, and can be visualized again using Fig. 4.3. As we show in Fig. 4.5, we can draw the two hemispheres of $\mathbf{S}^2$ side-by-side, showing in each hemisphere only what is visible on each filled disk as seen by a distant polar camera. That is, although the full spherical surface cannot be drawn with any single set of coordinates that match the geometry, we can *double up* and draw them side-by-side, with annotation that tells you where you should appear in the Southern hemisphere when you cross the equator *from any given point* in the Northern hemisphere.

Perceiving distance viewing Earth as the 2-sphere. We can practice using our familiarity with the Earth as a mental model for the $\mathbf{S}^2$ case. If we are in a space ship and take a picture looking straight down at the North or South pole of the Earth, the distances between points on the photograph near the poles are very nearly the distances perceived if we lived at the pole. However, as we travel away from the pole, moving towards the equator, the distances in the photograph compared to the distances we

FIGURE 4.6

The density plot of the $\mathbf{S}^2$ display written in terms of one of the polar-view hemispheres. The circles on the disk image projected upward from the hemisphere are latitude lines taken at equal intervals of latitude. The circles near the pole have similar radial distances from one another, as the projection is nearly flat and undistorted around the pole. As the radius $r = \sqrt{1 - x^2 - y^2}$ increases from $r = 0 \rightarrow r = 1$, the circles of constant latitude appear successively closer together, approaching infinite density as r approaches the equator.

experience at that latitude of the Earth become progressively smaller, until, very near the equator, a tiny distance on the photograph is a huge distance on the Earth. This distortion is systematic and predictable. Following a mathematical model analogous to the $\mathbf{S}^1$ case, Eqs. (4.1), (4.2), and (4.3), we can compute exactly the actual spherical arc length along a longitude line relative to a radial displacement $r = \sqrt{x^2 + y^2}$ from the North pole in the photograph, which we take to be centered at $(0, 0)$ in our photograph's coordinate system. Any point on the 3D surface of the Earth with radius R thus has coordinates $(x, y, \sqrt{R^2 - x^2 - y^2})$. So, choosing units that set $R = 1$ and choosing our viewpoint so the 2D projected point (x, y) is in the plane of the figure, with $r = \sqrt{x^2 + y^2}$, we can rewrite the coordinates as $(r, 0, \sqrt{1 - r^2})$ and arrive at the same simplified geometry illustrated in Fig. 4.3. The r dependence of the angle $\theta(r)$, measured downward from the pole, is then

$$\theta = \arctan\left(\mathbf{opposite} = r, \mathbf{adjacent} = \sqrt{1 - r^2}\right) = \arctan\left(\frac{r}{\sqrt{1 - r^2}}\right). \tag{4.4}$$

From Eq. (4.4), we can compute the change in subtended polar angle θ for an incremental change in r as a function of the $r = \sqrt{x^2 + y^2}$ to be

$$\frac{d\theta}{dr} = \frac{d\arctan(r/\sqrt{1 - r^2})}{dr} = \frac{1}{\sqrt{1 - r^2}}. \tag{4.5}$$

Incremental distances in the Earth coordinates approach become very large at the tangential edges $r \rightarrow 1$ at the boundary of our visible area. We illustrate the appearance of this distortion from the top-down view in Fig. 4.6 and from oblique views in Fig. 4.7. Then we can summarize our knowledge as

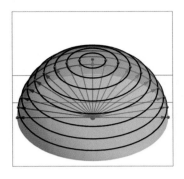 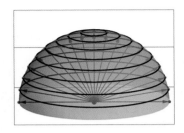

FIGURE 4.7

Oblique and sideways perspectives of the S^2 hemisphere density plot. The circles on the disk image projected upward from the hemisphere are latitude lines taken at equal intervals of latitude. As the radial coordinate $r = \sqrt{1 - x^2 - y^2}$ approaches the equator at $r = 1$, the circles of constant latitude appear successively closer together, even though the actual spherical distance between neighboring circles is the same.

follows:

$$x, y = r\cos\phi, r\sin\phi \quad \text{Euclidean location on the 2D photograph disk}$$
where
$$r = \sqrt{x^2 + y^2}\,,$$
$$\phi = \arctan(y/x)\,.$$

(4.6)

The (scaled by R) height of a point on the Earth above the Equator that we should perceive at distance r is just

$$z = \sqrt{1 - x^2 - y^2} = \sqrt{1 - r^2}\,.$$

4.3 Seeing quaternions as points on the 3-sphere S^3

We now turn to the realistic quaternion case, for which our quaternion geometry is always expressible as a point in $\mathbb{R}^4$ that has unit distance from the origin, corresponding to the hypersphere or 3-sphere S^3. Thus we can write an arbitrary quaternion as

$$q = (q_0, q_1, q_2, q_3) = (q_0, \mathbf{q})$$

and our equation analogous to $x^2 + y^2 = 1$ for S^1 and $x^2 + y^2 + z^2 = 1$ for S^2 is

$$q \cdot q = (q_0)^2 + (q_x)^2 + (q_y)^2 + (q_z)^2 = (q_0)^2 + \mathbf{q} \cdot \mathbf{q} = 1\,.$$

Following our investigation of the simpler spheres, we obviously need to split S^3 into its two hemispheres, each of which is describable by a single-valued function, and also to identify the object that plays the role of the *equator*, whose neighborhood contains the complete set of adjacent points sharable by the two hemispheres, or as the closure of each hemisphere if we take the hemispheres as open sets each approaching but not including $q_0 = 0$. Let us start with a table summarizing what happened with $\mathbf{S}^1$ and $\mathbf{S}^2$ and perform a dimensional induction to identify what we expect for $\mathbf{S}^3$:

Dimension	Sphere	Northern half	In-between Equator	Southern half
1	$\mathbf{S}^1$	Half-circle $y \geqslant 0$	$\mathbf{S}^0$: Two points $x = \pm 1$ on the x-axis line at $y = 0$	Half circle $y \leqslant 0$
2	$\mathbf{S}^2$	Filled disk $z \geqslant 0$	$\mathbf{S}^1$: Circle $x^2 + y^2 = 1$ in xy-plane at $z = 0$	Filled disk $z \leqslant 0$
3	$\mathbf{S}^3$	Solid ball $q_0 \geqslant 0$	$\mathbf{S}^2$: Sphere $(q_x)^2 + (q_y)^2 + (q_z)^2 = 1$ on surface at $q_0 = 0$	Solid ball $q_0 \leqslant 0$

The quaternion hemispheres. We deduce that, just as the Earth has two hemispheres that can be flattened into the images of two disks, side-by-side, to show, with some distortion, every point on the Earth's $\mathbf{S}^2$, the quaternion $\mathbf{S}^3$ has two hemispheres that can be flattened into *two solid balls*, drawn side-by-side in 3D, to show, with some distortion, every point of the quaternion $\mathbf{S}^3$. This geometric fact is what allows us to visualize quaternions:

> If we know the point $\mathbf{q} = [q_x, q_y, q_z]$ in the North (South) solid ball, we know the fourth quaternion component is $q_0 = +\sqrt{1 - \mathbf{q} \cdot \mathbf{q}}$ $\left(q_0 = -\sqrt{1 - \mathbf{q} \cdot \mathbf{q}} \right)$.

The viewer's task for seeing quaternions. We promised that the viewer would need particular learned skills to interpret both the quaternion topology and the geometry, including the perception of relative distances between points in a typical quaternion graphic display. We now understand that one such skill is the ability to recognize and mentally compensate for the $1/\sqrt{1 - \mathbf{q} \cdot \mathbf{q}}$ density distortion that increases with proximity to the quaternion equator, and another is the ability to deduce the value of the knowable but hidden fourth quaternion variable that we must compute for ourselves at each projected point $\mathbf{q}$ to be

$$q_0 = \pm \sqrt{1 - \mathbf{q} \cdot \mathbf{q}}.$$

The 3-sphere quaternion case. For the $\mathbf{S}^3$ case, we now generalize to $r \to \sqrt{x^2 + y^2 + z^2} \equiv \|\mathbf{q}\|$, and $q_0 = \pm\sqrt{1 - \|\mathbf{q}\|^2}$ depending on the hemisphere. The North pole view's filled disk for $\mathbf{S}^2$ visualizing becomes a filled solid ball, with increasing density corresponding to spheres with radii increasingly close to $\|\mathbf{q}\| = 1$ or $q_0 = 0$, as illustrated in the solid-ball nested-sphere volume rendering in Fig. 4.8.

4.4 Interactively following a quaternion path of incremental rotations

The relationship between a sequence of incremental rotations and the quaternion corresponding to the rotation reveals yet another way of intuitively exposing how quaternions work. Consider an interactive 3D design program that allows one, by some means, to interactively specify the orientation of an object in a 3D scene, or the orientation of an entire 3D scene in the view of a camera image. Assume also that this placement needs to be done starting from a world identity frame, and incrementally changing that frame to reach the desired frame. Note that there are many methods for carrying out such incremental

FIGURE 4.8

The density plot of the $\mathbf{S}^3$ displaying one of the polar-view hemispheres. The concentric spheres are at equal angular tilts from the pole in local space, becoming closer in the projection as we approach the outer edge. As the radius $r = \sqrt{1 - x^2 - y^2 - z^2}$ increases from $r = 0 \to r = 1$, the spheres appear successively closer together, approaching infinite density as r approaches the equator, i.e., as $q_0 \to 0$.

rotations (see, e.g., Chen et al., 1988; Shoemake, 1994; Hanson, 1992, 1995), and in Chapter 5, we go into detail about a specific context-independent method known as the "rolling ball" algorithm (Hanson, 1992).

The first thing to observe is that we start with a 3D world identity frame corresponding to the positive identity quaternion,

$$q_{\text{ID}} = (1, 0, 0, 0) \quad \to \quad I_3 = \begin{bmatrix} 1 & 0 & 0 \\ 0 & 1 & 0 \\ 0 & 0 & 1 \end{bmatrix}. \tag{4.7}$$

We then encode each individual motion as changing from the current quaternion to a new quaternion, e.g., by identifying the axis left fixed by the rotation control algorithm and multiplying the current quaternion (starting with q_{ID}) by the corresponding form of Eq. (2.16) to get a new, nearby, quaternion. Eq. (2.14) then gives us immediately the rotation matrix to apply to our scene to get the new orientation, in any of several ways:

$$q_{n+1} = q(\theta, \hat{\mathbf{n}}) \star q_n , \tag{4.8}$$
$$R_{n+1} = R(q_{n+1}) = R(q(\theta, \hat{\mathbf{n}})) \cdot R_n . \tag{4.9}$$

The basic idea is then to track the vector (imaginary) part $\mathbf{q}$ of the evolving quaternion $q = (q_0, \mathbf{q})$ in a unit sphere overlaid at the center of the sphere as it changes, illustrated in Fig. 4.9(a) for the rotation of a simple coordinate frame triad. We typically display both the current state, a 3D vector from the origin to the point $\mathbf{q} = \hat{\mathbf{n}} \sin(\theta/2)$ inside the unit sphere, and the *history*, the set of points or a connected curve showing where $\mathbf{q}(t)$ has been for a chosen segment of recent incremental rotations, shown in

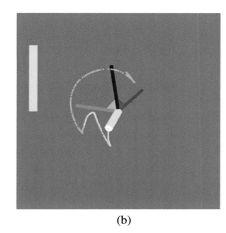

(a)   (b)

FIGURE 4.9

Appearance of an interactive quaternion visualization application. (a) Using an interface such as the rolling ball (Chapter 5), the moving orientation of the coordinate frame triad is displayed with the quaternion vector part **q** represented as a thicker tube with color coded by the sign of q_0. The tube color matches the q_0-valued bar on the left, which turns yellow for $q_0 > 0$ and blue for $q_0 < 0$. (b) On the right, the interactive record of the quaternion path is sampled and preserved, again color coded with the sign of q_0, and changing color whenever $q_0 = 0$.

Fig. 4.9(b). One use for this path is to pick individual anchor points to use for a multipoint `slerp`, which can then form the basis for an interactive design of a smooth rotation sequence.

However, one of the interesting intuitive features of this process is that, as one rotates the scene through a full 360-degree rotation, the scene returns to its original appearance, but the quaternion *does not*: the incrementally modified quaternion will take the value $q = -q_{\mathrm{ID}} = (-1, 0, 0, 0)$, and only returns to q_{ID} after two full turns, or 720 degrees. Thus we want to track carefully all components of the quaternion, and keeping track of **q** is not enough. Since we only have 3D space imaging on a graphics screen, we need to display some supplementary information.

Our preferred strategy is to provide *redundant cues*, the first of which is to color the 3D path of q by a *color coding* with distinct colors showing when $q_0 > 0$ (say yellow) for a specific plotted point on the curve of **q** evolution, and a different color (say blue) when $q_0 < 0$. We supplement that by a sidebar that is a like a thermometer taking values $-1 \leqslant q_0 \leqslant +1$, having zero length when $q_0 = 0$, showing a yellow bar from 0 up to q_0 when $q_0 > 0$, and a blue bar downward from 0 down to $-|q_0|$ when $q_0 < 0$. Fig. 4.9(b) shows an example with a long history trail changing color as it crosses back and forth between $q_0 > 0$ and $q_0 < 0$ during an investigation of possible orientations, showing the current explicit q_0 value and its color bar on the left. Another example with an alternative 3D projection to (q_0, q_y, q_z) and tracking a series of repetitive controller motions is shown in Fig. 4.10.

This interactive visualization method is very easy to program (see Chapter 5 for more implementation details), and is a very intuitive tool to actually watch the double covering of the quaternion in action with your own physical experience.

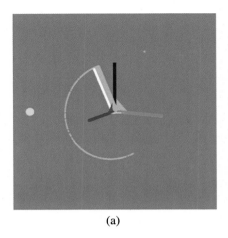
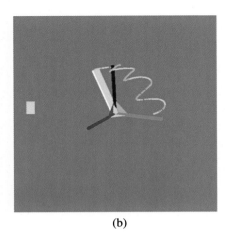

(a)   (b)

FIGURE 4.10

Interactive quaternion visualization application, showing paths of rotation actions. (a) The 3D (q_0, q_y, q_z) path of a coordinate frame rotation about the $\hat{z}$-axis. (b) The path of a rolling ball action in the xy-plane (see Chapter 5) that incrementally produces the same global $\hat{z}$-axis rotation. Note the points where the path changes from blue to yellow, indicating that incremental quaternion values have crossed over $q_0 = 0$, with $q_0 > 0$ in blue and $q_0 > 0$ in yellow.

4.5 Displaying every point of the quaternion 3-sphere S^3

We can exploit our understanding of spherical geometry from the exercises in earlier sections in many ways. For example, S^3 is a very smooth manifold, with a Riemannian metric and constant curvature, making it easy to find geodesic or shortest-distance paths in the manifold from one quaternion to another; that in turn allows us to rigorously determine consistent distance measures from one *orientation* to another via the geodesics between the corresponding quaternions. Another feature is that quaternion multiplication, which effectively composes rotations in 3D space, which we will normally write as

$$p \star q = \{p_0 q_0 - \mathbf{p} \cdot \mathbf{q}, \; p_0 \mathbf{q} + q_0 \mathbf{p} + \mathbf{p} \times \mathbf{q}\},$$

can be rephrased as a 4×4 matrix multiplication $P \cdot q$, where

$$P = \begin{bmatrix} p_0 & -p_1 & -p_2 & -p_3 \\ p_1 & p_0 & -p_3 & p_2 \\ p_2 & p_3 & p_0 & -p_1 \\ p_3 & -p_2 & p_1 & p_0 \end{bmatrix}.$$

Quaternion multiplication is equivalent to applying the rotation matrix P to a quaternion in tandem with the corresponding 3D rotation in ordinary 3D space, and permits us to follow a 4D rotation of the quaternion around the hypersurface S^3 in any chosen projection from ordinary Euclidean 4D space. We already know that such quaternion graphics can be implemented with variations on the hemisphere technique: as shown schematically in Fig. 4.11, any quaternion can be accurately depicted using only

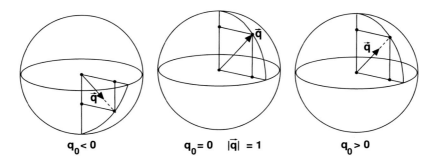

$q_0 < 0$ $q_0 = 0$ $|\vec{q}| = 1$ $q_0 > 0$

FIGURE 4.11

(Left) The South pole solid ball "hemisphere" of $\mathbf{S}^3$ with $q_0 < 0$. The observed value of $\mathbf{q}$, with any sign or direction, determines the unseen value of q_0 to be $q_0 = -\sqrt{1 - \mathbf{q} \cdot \mathbf{q}}$. (Center) The pure imaginary ($q_0 = 0$ and $\|\mathbf{q}\| = 1$) components of the quaternion are represented by the vector to the spherical point $\mathbf{q} = (q_x, q_y, q_z)$. These points with $q_0 = 0$ lie exactly between the North and South solid balls, and they appear only here, not on the surfaces of the North or South solid balls. (Right) The North pole solid ball "hemisphere" of the 3-sphere $\mathbf{S}^3$ with $q_0 > 0$ has the hidden q_0 component completely determined by the equation $q_0 = +\sqrt{1 - \mathbf{q} \cdot \mathbf{q}}$. Note that in fact one can choose to omit any *other* single element of q or a rotated q, keeping in mind that the omitted fourth element is always computable from $q \cdot q = 1$. Every quaternion point can be displayed on this triple canvas without ambiguity.

the $\mathbf{q}$ component drawn in one of two solid balls, which share a single ordinary spherical shell as their effective equator, through which we can pass smoothly from one ball to the other to understand every single possible point in quaternion geometry visually. We typically separate out the "equatorial 2-sphere" at exactly $q_0 = 0$, excluding it from both the Northern open solid ball and the open Southern solid ball, and displaying it separately as its own $\mathbf{S}^2$.

Summary of the full triple-ball $\mathbf{S}^3$ display method. In this method, illustrated in Fig. 4.12, we display the entire quaternion space, including *both* of the two equivalent, opposite-sign quaternions corresponding to each 3D rotation group matrix, by dividing up the space of quaternions into three parts:

- **Red: Southern hemisphere:** $q_0 < 0$. Just as an ordinary 2-sphere $\mathbf{S}^2$ has a Northern and a Southern hemisphere that each project to a round disk, bounded by the equator at zero latitude, when viewed from a distance, the quaternion 3-sphere $\mathbf{S}^3$ has exactly analogous properties. The difference is that the 2D case with a hemisphere that is a disk bounded by the equatorial circle is replaced by a *solid ball*, with the "South pole" in the center at $q_0 = -1$, and a boundary that is a closed 2-sphere $\mathbf{S}^2$ instead of a ring. We will freely refer to this solid-ball hyperhemisphere as a "hemisphere" since it has virtually the same properties as the familiar hemisphere, just moved up by one dimension.
- **Orange: Equatorial 2-sphere dividing North from South:** $q_0 = 0$. Since the choice of hemisphere in which to draw a quaternion with $q_0 = 0$ is ambiguous, we completely resolve this ambiguity by creating a separate plot only for quaternions that lie exactly on the $q_0 = 0$ equator. Because all of the discrete quaternions corresponding to the Kleinian groups that we will study later have significant elements with $q_0 = 0$, this is a very important subset warranting this special treatment.

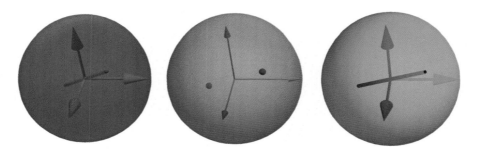

FIGURE 4.12

Using the coordinate system of Fig. 4.11, with color coding reddish $\rightarrow q_0 < 0$, orange $\rightarrow q_0 = 0$, and bluish $\rightarrow q_0 > 0$, we draw the *circle* $q(t) = (\cos(t/2), \hat{\mathbf{n}} \sin(t/2))$ in the 4D plane defined by the q_0-axis and the $\hat{\mathbf{n}} = (0, 1/\sqrt{3}, 1/\sqrt{3}, 1/\sqrt{3})$-axis. This is a pure circle in Euclidean 4D space, but we see it running through the $q_0 = +1$ North pole at the center of the right-hand blue solid ball, touching two points on the orange spherical (equatorial) surface where $q_0 \rightarrow 0$, and then passing back through the red solid ball in both directions to meet at the $q_0 = -1$ South pole at the center of that solid ball, $q = (-1, 0, 0, 0)$. Keep in mind that $q \cdot q = 1$ throughout the curve, as it must for a circular arc, and that the curve moves in *both* directions along $\hat{\mathbf{n}}$ in, say, the right-hand solid ball, as it would for the t variable in a simple hemicircle $(t, \sqrt{1 - t^2})$. Every single quaternion point is displayed uniquely; there is no overlapping.

FIGURE 4.13

If we prefer to actually see the circular intuitive shape of the circle $q(t) = (\cos(t/2), \hat{\mathbf{n}} \sin(t/2))$, we can reformat the display of Fig. 4.12 to move the q_0-axis out from the center, making the q_z-axis implicit instead. The q_z-axis will now have circles in its plane fall on a straight line as in Fig. 4.12 instead of circles in the plane containing q_0: there is a tradeoff. This display singles out the (q_0, q_x, q_y) 3D space, assigning $q_z < 0$ points to the left sphere, and $q_z > 0$ points to the right sphere, with $q_z = 0$ (which occurs at the poles of the circle where $|q_0| = 1$) in the center spherical shell.

- **Blue: Northern hemisphere:** $q_0 > 0$. The Northern hemisphere of $\mathbf{S}^3$ is a second solid ball, bounded by the same $\mathbf{S}^2$ equator as the Southern hemisphere. The actual North pole is at the center of this solid ball, at $q_0 = 1$, the exact analog of having the Earth's North Pole at $z = 1$ or latitude $90°$ in idealized Earth coordinates.

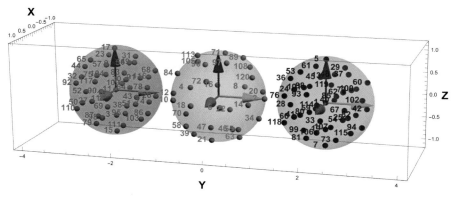

FIGURE 4.14

This is the most complicated display we might choose to use to show the sign-doubled elements of the discrete quaternion icosahedral group E_8 studied in Chapter 26. The group has 60 unique elements; here we include the quaternion sign doubling to display all 120 possible group elements. The 15 edge flips, 30 with sign doubling, all have $q_0 = 0$, and lie on the orange equator, while the rest of the 45 vertex and face group elements with $q_0 \neq 0$ collect in reflected pairs on the left and right.

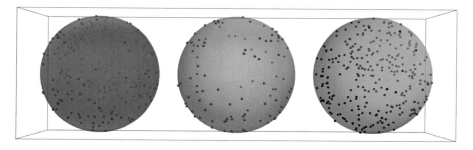

FIGURE 4.15

This is a sample of how a randomly distributed cloud of orientations would appear, with $q_0 < 0$ split out on the left, a hypothetical set of $q_0 = 0$ samples on the equatorial $\mathbf{S}^2$ in the middle, and $q_0 > 0$ samples on the right. The point is that absolutely every unique quaternion, including sign-reversed pairs, has its own point in this display. All quaternions that exist can be singled out, e.g., for interactive selection, in this type of cloud-data display.

4.6 Examples of quaternion visualization strategies

We conclude with a collection of examples that are employed throughout our various applications of quaternions to visualize what is going on.

- Triple-sphere display in (x, y, z) projection. In Fig. 4.12, we show the path of a quaternion curve $q(\theta, \hat{\mathbf{n}})$ for a constant fixed rotation axis $\hat{\mathbf{n}}$ in (x, y, z)-coordinates; the circular path cuts a straight

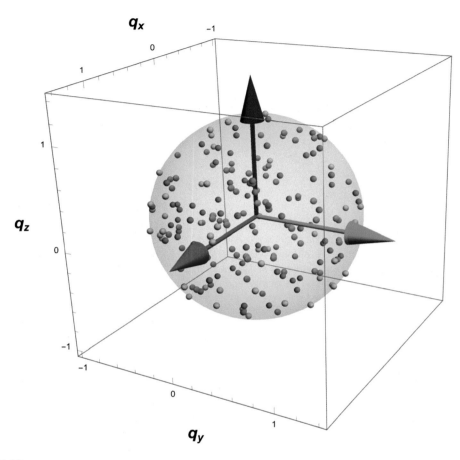

FIGURE 4.16

This method displays **q** in the (x, y, z)-axis coordinates, where as usual we can determine the magnitude of the missing q_0 component from $\sqrt{1 - \mathbf{q} \cdot \mathbf{q}}$, but now we condense the display into a single 3D volume by marking the sign of the hidden q_0 component with a color coding, with red for $q_0 < 0$, blue for $q_0 > 0$, and orange for $q_0 = 0$ points lying exactly on the equatorial $\mathbf{S}^2$. Thus in principle the correct value of the entire quaternion at any point can be determined by inspection. This method is most useful for sparse data: two very different quaternions can have very similar **q** values that may hide one another in this display method.

line through the spheres, with the position of the curve locations on the equatorial $\mathbf{S}^2$ being simply $\pm\hat{\mathbf{n}}$.

- Triple-sphere display in (q_0, x, y) projection. In Fig. 4.13 we show the path of a quaternion curve $q(\theta, \hat{\mathbf{n}})$ for a constant fixed rotation axis $\hat{\mathbf{n}}$ in (q_0, x, y)-coordinates; the circular path now traces exactly a circle in this hybrid projection.

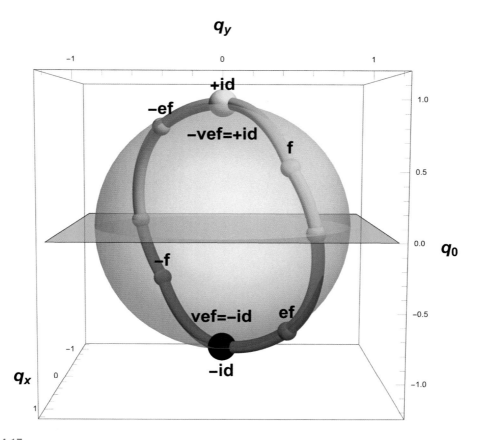

FIGURE 4.17

This method pulls the q_0-coordinate out from its hiding place for full display, substituting for another coordinate or a rotated combination of the (x, y, z)-coordinates. Since many types of quaternion transformations or interpolations naturally are cyclic in q_0 or pass smoothly through one or both of the $q_0 = \pm 1$ identities, using this coordinate system, with clear annotation such as the orange plane here for the $q_0 = 0$ locus, can more clearly expose the features of a circular behavior.

- Triple-sphere display of a set of discrete quaternion rotations. In Fig. 4.14, we show the discrete quaternion group locations from Chapter 26 for the icosahedral group E_8. Each group element is distinct, and the proximity of any two elements can be easily extracted.
- Triple-sphere display of a statistical distribution of quaternions. In Fig. 4.15, we show a random distribution of quaternions centered at a particular quaternion point. Note that we do not need to double the quaternions, though that is possible to ensure that all possible paths to the same 3D rotation include both a positive quaternion and its negative.
- Single-sphere display in (x, y, z) projection. In Fig. 4.16, we exhibit the merged-hemisphere strategy, using different colors to display points or parts of a curve that have different signs of the omitted fourth coordinate (here nominally q_0, but that choice is arbitrary).

- Single-sphere display in (q_0, q_x, q_y) projection. In Fig. 4.17, we show an important alternative displaying how a circular path with q_0 having cosine-like behavior can be paired with a plane in (x, y, z) chosen to expose a sine-like behavior. This coordinate choice can be adjusted specifically to expose a quaternion circle in the display subsystem as it passes through both the positive $q_0 = 1$ identity and the negative $q_0 = -1$ identity. The example here in Fig. 4.17 is used later in Chapter 27 on discrete quaternion groups to show a cycling three-part quaternion multiplication with the special property that its path passes through both identities, and also passes twice through the orange plane marking the 2D manifold $q_0 = 0$.

Interacting with quaternions as hyperspheres

5

One of the things we may want to do when studying 3D rotations and their 4D quaternion counterparts is to interact with them. Ideally, an interactive computer system for this purpose should be able to smoothly change the viewpoint of the simulated world to build intuition about the geometry of the scene we are looking at. By rotating the object at will, we can see all sides and aspects of the 3D objects and scenes being explored.

There are several traditional ways to accomplish this. One is to simply take the Euler angles or the $(\theta, \hat{\mathbf{n}}(\alpha, \beta))$ axis-angle variables. The latter are especially attractive in principle as the quaternion parameterized using Eq. (2.16) generates the corresponding rotation Eq. (2.17). However, varying the axis-angle variables in individual control panels is awkward because one has to keep switching among controllers to explore the whole space; furthermore, it is difficult to keep track of where in the orientation space you are, as the effect of any controller parameter adjustment depends on the current orientation.

Context-dependent controllers. Another popular family of orientation control strategies employs a more powerful multidimensional controller. Chen et al. (1988) introduced a widely used method, the "Virtual Sphere," that incrementally rotates an object about axes that change from lying in the xy-plane to an axis pointing out of the screen plane as one moves away from the image center. The "arcball rotation" method of Shoemake (1994) selects points on the sphere and rotates smoothly between them. Both of these require the user to pay attention to the context of the controller, and divert attention from the *object being rotated* to the *controller position itself*. This family of methods can be classified as *context-dependent* orientation controllers.

Context-free rotation control. However, there is another option, which is the *context-free* orientation controller system that we call the *rolling ball* method (Hanson, 1992), which can in principle be applied not only to 3D orientation control, but also 4D orientation control if one wishes to explore the quaternion space itself (Cross and Hanson, 1994; Yan et al., 2012). This is the method we will discuss in detail in this chapter: its notable feature is that the *action* of the controller depends only on its incremental motion, not on the location of the controller, e.g., a pointer on the 2D screen or a 3D wand. You can take your attention off the controller completely and focus only on how the scene's appearance changes as you apply smooth incremental rotations to it.

5.1 The 3D rolling ball
In this section we first review the details of the context-free 3D rolling ball controller. The basic mental model is that the parameters change in a way that simulates a natural motion like holding a baseball in

Visualizing More Quaternions. https://doi.org/10.1016/B978-0-32-399202-2.00013-7

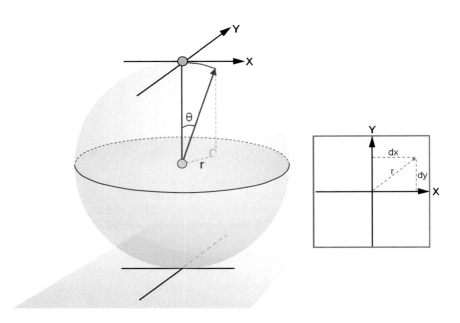

FIGURE 5.1

The 3D rolling ball approach to 3D orientation control. A 2D controller moving in the xy-plane rotates the world about an axis in that world xy-plane that is perpendicular to the direction of the incremental motion, independent of the current *location* of the controller.

your hand and rotating it to examine the pattern of the seams. Alternatively, one can imagine rotating the object of interest as though it were encased in a rotatable glass ball with the object fixed at its center. As illustrated in Fig. 5.1, if you look at the point at the top (North pole) of this virtual glass ball, the *rolling ball* control simply tilts that point in any direction you choose, incrementally hiding those parts of the object in the direction you tilt the center point, and incrementally exposing unseen parts of the object on the opposite side. One can thus instantly understand what is being made visible and what is becoming hidden as a consequence of the rotation. After each rotation, a new point on the virtual glass ball is at the top, and one starts afresh, with no need to maintain *any* knowledge of context. This contrasts with other clever, but controller-position-dependent, methods that rotate about a particular axis under some tracked conditions instead of always tilting the current top point.

The equations of the 3D rolling ball algorithm are quite simple, and, in a moment, we will show how they are extendable to a 4D rolling ball orientation control method for inspecting quaternion $\mathbf{S}^3$ geometry. We begin by looking straight down on the virtual glass ball containing our object at the center, locating the nearest point (typically at the middle of the screen), and inputting an arbitrary small displacement of the center in screen coordinates, namely a 2D vector (dx, dy). Now let $r = (dx^2 + dy^2)^{1/2}$ be the Euclidean length of this (very short) vector, and choose a scaling size R to define the virtual radius of the glass sphere. Then if the standard rotation by an angle θ about a unit-length direction $\hat{\mathbf{n}} = (n_x, n_y, n_z)$ is written as $R(\theta, \hat{\mathbf{n}})$, then $R_{\mathbf{3Droll}}(dx, dy)$ is defined by computing θ and $\hat{\mathbf{n}}$ as follows:

$$\tan\theta = \frac{r}{R} \qquad\Longleftrightarrow\qquad \cos\theta = \frac{R}{\sqrt{r^2+R^2}} \qquad \sin\theta = \frac{r}{\sqrt{r^2+R^2}}\,, \qquad (5.1)$$

$$\hat{\mathbf{n}} = \left[\frac{-dy}{r},\frac{dx}{r},0\right]^{\mathrm{t}} \qquad\Longleftrightarrow\qquad \hat{\mathbf{n}}\cdot[dx,dy,0]^{\mathrm{t}} = 0\,. \qquad (5.2)$$

The corresponding rotation can also be constructed incrementally by initially assigning $(dx/r,\,dy/r) = (\cos a,\sin a)$, so $\hat{\mathbf{n}} = (-\sin a,\cos a,0)$, performing a rotation of $(dx,dy,0)$ back to the (x,z)-plane with $R(-a,\hat{\mathbf{z}})$, next tilting by the angle θ using $R(\theta,\hat{\mathbf{y}})$, and then undoing the first rotation, leading to

$$R_{\mathbf{3Droll}}(dx,dy) = R(+a,\hat{\mathbf{z}})\cdot R(\theta,\hat{\mathbf{y}})\cdot R(-a,\hat{\mathbf{z}})\,. \qquad (5.3)$$

(Recall Eq. (2.18) from Chapter 2.) This rotation will always tilt the invisible z-axis "pointing straight at you" in the direction $(dx,\,dy)$ that you choose, allowing you to explore any part of a 3D object. Its explicit form, with $c = \cos\theta$ and $s = \sin\theta$, is

$$R_{\mathbf{3Droll}} = \begin{bmatrix} c\,(dx/r)^2 + (dy/r)^2 & (c-1)\,(dx/r)(dy/r) & (dx/r)\,s \\ (c-1)\,(dx/r)(dy/r) & (dx/r)^2 + c\,(dy/r)^2 & (dy/r)\,s \\ -(dx/r)\,s & -(dy/r)\,s & c \end{bmatrix}\,. \qquad (5.4)$$

Using the fact that $n_x^2 + n_y^2 = (dx/r)^2 + (dy/r)^2 = 1$ gives the computationally less expensive form

$$R_{\mathbf{3Droll}} = \begin{bmatrix} 1 + (c-1)\,(dx/r)^2 & (c-1)\,(dx/r)(dy/r) & (dx/r)\,s \\ (c-1)\,(dx/r)(dy/r) & 1 + (c-1)\,(dy/r)^2 & (dy/r)\,s \\ -(dx/r)\,s & -(dy/r)\,s & c \end{bmatrix}\,,$$

where we note that since $n_i\,s = (dx_i/r)(r/(r^2+R^2)^{1/2})$, we can eliminate the r from these terms for further efficiency.

Group theory magically adds a third degree of freedom. There is, however, one big potential problem: this method has only *two* parameters, (dx,dy). We know 3D rotations have *three* parameters. Have we lost something? Thanks to the subtle properties of order-dependent, or non-Abelian, groups, to which 3D rotations belong, we can *always* find a sequence of these two-parameter rotations that will take us to *any* given orientation in 3D. This is *not* true of two-parameter rotation methods that use, say, the global latitude and longitude of a point on a sphere to implement rotations. The basic fact is that if we make a *box* motion with our controller, rotating first about the positive $\hat{\mathbf{x}}$-axis in a downward stroke, then the positive $\hat{\mathbf{y}}$-axis with a rightward stroke, then closing the box with reversed strokes,

Step 1 + Step 2 :	Positive $\hat{\mathbf{x}}$: $\Downarrow$	Positive $\hat{\mathbf{y}}$: $\Longrightarrow$
Step 3 + Step 4 :	Negative $\hat{\mathbf{x}}$: $\Uparrow$	Negative $\hat{\mathbf{y}}$: $\Longleftarrow$,

the world will rotate *in the opposite direction*. The fundamental properties of the order dependence, or noncommutativity, of small cycles of rotations guarantee that a clockwise circuit of small strokes will rotate the world *counterclockwise*, and a counterclockwise circuit of small strokes will rotate the world *clockwise*, as illustrated in Fig. 5.2.

This fundamental property requires us to learn one more thing: we can easily move to any desired orientation of our scene provided we achieve rotations about the direction perpendicular to the screen

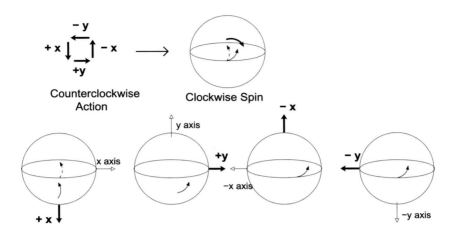

FIGURE 5.2

Illustrating the effect of the action of a cyclic motion of a controller moving a ball counterclockwise in the xy-plane. Due to the order dependence of 3D rotation sequences, and the inalterable *sign* of the group-theoretical coefficients of that action, the orientation of the top surface of the ball (here sticking out of the plane towards the viewer) will *counter-rotate*, spinning in a clockwise direction, leaving the z-axis fixed. Therefore, the xy-plane 2D controller can not only control rotations leaving the x-axis or y-axis fixed, but also use cyclic motions to access rotations acting on the missing z-fixed direction sticking out of the page towards the viewer. All three degrees of freedom of 3D rotations are thus accessible from a two-degree-of-freedom controller. Similar properties in 4D allow all six degrees of freedom to be accessed by a much simpler three-degree-of-freedom controller.

by *counter-rotating*, using a cycle of small strokes in the *opposite sense* of the direction we wish to rotate the world. In return for accepting this odd fact, we can get *three context-free degrees of freedom* from a *2D controller*.

More remarkably, we will find in 4D that the group properties of 4D rotations permit us to use a *3D controller* to obtain any arbitrary orientation in the fourth dimension's *six*-degree-of-freedom parameter space! In the next section, we show how the 4D version of the rolling ball allows us to do this.

5.2 The 4D rolling ball

Four dimensions is peculiarly available to interactive exploration because the entire space of possible 4D orientations can be explored with an incremental 3D vector. That is, the precise analog of the 2D vector that controls the 3D rolling ball orientation-exploration algorithm is a 3D vector that controls the 4D rolling ball. The best way to think of the 4D rolling ball conceptually is to realize that the analog of looking at a 3D ball on a 2D screen, with the x- and y-axes visible and the z-axis *pointing at you* and therefore *invisible*, is to imagine a 4D object projected to 3D space along the w-axis, which is "pointing at you" and therefore invisible. The x-, y-, and z-axes are *visible* in the projection from 4D

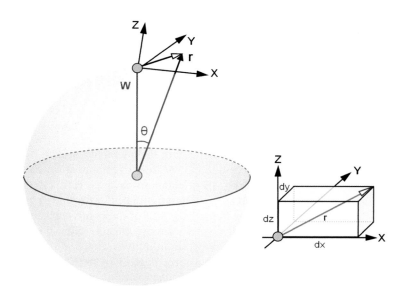

FIGURE 5.3

The 4D rolling ball approach to 4D orientation control. A 3D controller moving in a cycle inside an (x, y, z) box rotates the world about the plane of the controller cycle in that world, independent of the current *location* of the controller.

to a 3D world along the w-axis. Thus the perfect orientation-exploration procedure is to specify a 3D vector (dx, dy, dz) that *tilts* the w-axis slightly in the plane defined by the direction (dx, dy, dz) and the w-axis itself. That makes a new piece of the geometry appear "on the opposite side" so to speak, and hides an old piece of the geometry on the side the vector is moving towards, exactly the same as for the 3D rolling ball, except that we must think of a 3D projection screen as our canvas instead of a 2D projection screen.

The 4D rolling ball matrix can literally be guessed by induction from Eq. (5.4) for the 3D rolling ball matrix, and it takes the following form:

$$R_{4\text{Droll}} = \begin{bmatrix} 1 + (c-1)(dx/r)^2 & (c-1)(dx/r)(dy/r) & (c-1)(dx/r)(dz/r) & (dx/r)s \\ (c-1)(dx/r)(dy/r) & 1 + (c-1)(dy/r)^2 & (c-1)(dy/r)(dz/r) & (dy/r)s \\ (c-1)(dx/r)(dz/r) & (c-1)(dy/r)(dz/r) & 1 + (c-1)(dz/r)^2 & (dz/r)s \\ -(dx/r)s & -(dy/r)s & -(dz/r)s & c \end{bmatrix}. \quad (5.5)$$

Here dx/r with $r^2 = dx \cdot dx$ is a 3D vector derived from the 3D controller input $dx = (dx, dy, dz)$, and $c = \cos\theta$, $s = \sin\theta$ with $\theta = \arctan(r/R)$ exactly as in the 3D case; this geometry is illustrated in Fig. 5.3. An exercise for the reader is to verify this formula by taking a 3D unit vector $\hat{n}$ embedded in 4D, and rotate it first in the yz-plane so the z component is eliminated and then rotate that result in the xy-plane to give a pure x vector; applying a rotation in the xw-plane to perform our "tilt" of the w-axis, and then undoing the previous xy and yz rotations gives the formula. Symbolically,

$$R_{4\text{Droll}} = R^{-1}(yz) \cdot R^{-1}(xy) \cdot R(\theta, wx) \cdot R(xy) \cdot R(yz) .$$

More explicitly, we can take $\mathbf{dx}/r = (\cos a, \sin a \cos b, \sin a \sin b)$, and compute the form of the 4D rolling ball from

$$R_{4\text{Droll}} = R(b, yz) \cdot R(a, xy) \cdot R(\theta, wx) \cdot R^{-1}(a, xy) \cdot R^{-1}(b, yz) .$$

Finding all the 4D degrees of freedom. The main task of the interactive system is of course to create natural ways with existing controllers to access the 4D degrees of rotational freedom. However, while 3D orientations have three degrees of freedom, 4D orientations have *six*. How can our three-parameter 4D rolling ball formula of Eq. (5.5), which apparently only rotates in the three planes (xw, yw, zw), generate the other three degrees of freedom? In fact, any *pair* of directions such as (xw, yw) will produce a rotation in the *plane of the pair*, e.g., (x, y), if we move the controller in *circular cycles* in that plane. Note that, just as in 3D, planar controller cycles in a given direction produce planar *counter-rotation* of the world's orientation. Thus, since our three degrees of freedom (xw, yw, zw) embody three *pairs* of possible cycles generating (yz, zx, xy)-plane rotations, we actually can produce, incrementally, rotations in all six planes (xw, yw, zw, yz, zx, xy), and therefore any possible 4D orientation can be reached. In particular, the entire quaternion space can be interactively explored using either a 3D virtual-reality wand that produces incremental 3D motion-direction data, or a 2D controller with a modifier mode to simulate the third degree of freedom. On a touchscreen, the motion of a two-finger spin around a fixed center can be exploited to access the third controller parameter (Yan et al., 2012).

5.3 The ND rolling ball

While we will not need controllers for dimensions greater than four in our treatment here, it is straight-forward to extend the idea to rotation controllers for arbitrary Euclidean dimensions. Using either a direct derivation method as noted above or deducing the obvious ND form once we know the 3D and 4D forms, we find the ND rolling ball formula (Hanson, 1995) with an $(N-1)$-vector $\hat{\mathbf{n}} = (n_1, \ldots, n_{N-1})$ providing the control direction to be

$$R_{\text{NDroll}}(c, s; \hat{\mathbf{n}}) = \begin{vmatrix} 1 + (c-1)n_1^2 & (c-1)n_1n_2 & \cdots & (c-1)n_1n_{N-1} & s\,n_1 \\ (c-1)n_1n_2 & 1 + (c-1)n_2^2 & \cdots & (c-1)n_2n_{N-1} & s\,n_2 \\ \vdots & \vdots & \vdots & \vdots & \vdots \\ (c-1)n_1n_{N-1} & (c-1)n_2n_{N-1} & \cdots & 1 + (c-1)n_{N-1}^2 & s\,n_{N-1} \\ -s\,n_1 & -s\,n_2 & \cdots & -s\,n_{N-1} & c \end{vmatrix} . \quad (5.6)$$

The counting for obtaining the additional degrees of freedom works out as well. If we write the basic rotation matrix in a single plane (x_i, x_N) defined by n_i in the formula as the R_{iN}, then it can be shown that the group algebra of $\mathbf{SO}(N)$ gives a commutator $[R_{iN}, R_{jN}] = -R_{ij}$. This means that, by applying the $(N-1)$ ND rolling ball rotation matrices R_{iN} repeatedly to one another, we can generate the $(N-1)(N-2)/2$ degrees of freedom in the R_{ij} by incremental motions. We thus can *always* use an $(N-1)$-degree-of-freedom controller to achieve all of the

$$(N-1) + \frac{1}{2}(N-1)(N-2) = \frac{1}{2}N(N-1)$$

degrees of freedom required to explore the *entire* ND orientation space.

Quaternions and uniform rotation distributions

Because of their intimate relationship with the constant curvature embedding of the topological space $\mathbf{S}^3$, quaternions provide an ideal method to obtain statistically uniform distributions of 3D rotations. However, in practice, this has not always been obvious, and many different *ad hoc* methods have been used to produce distributions of rotations. Therefore we will study in this chapter the methods by which uniform rotation distributions can be defined, along with rigorous tools for measuring and evaluating sampling methods.

6.1 Points on spheres

The general concept of a sphere can be extended to any dimension N, and it is useful to examine point distributions on spheres in general before we specialize to the quaternion case of $\mathbf{S}^3$. First, we confirm our notation: when we refer to an N-sphere, denoted as $\mathbf{S}^N$, the dimension N refers to the intrinsic local, embedding-independent, dimension you would perceive if you were a small observer living at a point on a very large N-sphere: on Earth, the sphere we live on is effectively a 2-manifold; we are able to move North and South, or East and West, so the Earth's surface is a 2-sphere, $\mathbf{S}^2$. A 1-sphere $\mathbf{S}^1$ is a circle, with only one direction (left or right) in which we could move. On the quaternion 3-sphere $\mathbf{S}^3$, however, we can move in *three* independent directions.

Suppose we now look at the *whole* sphere instead of our small hypothetical neighborhood of a point – if we keep on going in the same direction, we get back to where we started. To visualize this, we describe the whole N-sphere as embedded in a Euclidean space of dimension $(N + 1)$, written as $\mathbb{R}^{N+1}$, with coordinates $[x_0, x_1, \ldots, x_N]$. A precise description of an N-sphere is then given by the equation

$$x_0{}^2 + x_1{}^2 + \cdots + x_N{}^2 = 1 \,, \tag{6.1}$$

which is a single constraint on $N + 1$ variables, and thus has only N degrees of freedom, consistent with our introductory remarks. (Note: We will generally assume a default unit radius for simplicity. For a sphere of arbitrary radius R, we change the "1" on the right-hand side to R^2.) In Fig. 6.1, we show the three lowest-dimensional examples, starting with $\mathbf{S}^0$, which satisfies $x^2 = 1$, and thus has only the two (0D) points $x = \pm 1$ as its solutions.

Each point on the sphere $\mathbf{S}^3$ can be thought of as a quaternion, and, up to an overall sign, each quaternion q has three free parameters representing a unique 3D rotation $R(q)$. Because geodesic arc lengths between quaternion points on $\mathbf{S}^3$ are equal for corresponding rotations differing by twice that angle of rotation, distributions of *quaternions* that are statistically equidistant from their neighbors will

Visualizing More Quaternions. https://doi.org/10.1016/B978-0-32-399202-2.00014-9

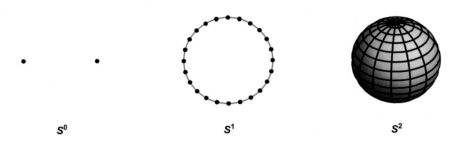

FIGURE 6.1

The spheres of dimension zero, where $\mathbf{S}^0$ is just the two points $x = \pm 1$, dimension one, where the circle $\mathbf{S}^1$ is described by $x^2 + y^2 = 1$, and dimension two, where the surface $\mathbf{S}^2$ is embedded as $x^2 + y^2 + z^2 = 1$ in 3D Euclidean space.

generate *rotations* that are also statistically equidistant from their neighbors. So our task is to produce uniform distributions of quaternions.

6.2 Theoretical basis for uniformity of spherical distributions

Before we can begin examining statistical distribution algorithms for their appropriateness to use for uniform quaternion distributions, and hence uniform rotation samples, we require an evaluation tool. There are a variety of methods one could use for this purpose, but we have found that one particular method is especially easy to use and easy to apply to recognizing features of uniformity, and this is the method proposed by Shoemake (1992) and Diaconis and Shahshahani (1987). We will focus on this approach because of its clarity and ease of application. The basic algorithm is applicable to proposed spherical distributions of any dimension, and has the following structure:

- Generate a candidate spherical data sample of K points on $\mathbf{S}^N$ in $\mathbb{R}^{N+1}$.
- Choose one of the $N + 1$ Cartesian axes, say the axis u selected from the set $(1, \ldots, N + 1)$, and for each sample point, find the value of its coordinate with index u and collect those K values into one list.
- Compute the histogram of this collection of u-axis values.
- **Compare the shape of the histogram** to that predicted by the single-axis slice integrals for the particular $\mathbf{S}^N$.
- Repeat for all $N + 1$ possible values of the axis u if the results might not be symmetric.

We now proceed to study the details, and we will see that, for the quaternion sphere $\mathbf{S}^3$ embedded in four dimensions, all four Cartesian axes of a uniform quaternion distribution will have a histogram that is exactly a half disk with a circular arc boundary.

Distribution of the coordinate slice density for each $\mathbf{S}^N$. We note that in each dimension, if we take any point $\mathbf{p}$ on an $\mathbf{S}^N$ (with $\mathbf{p} \cdot \mathbf{p} = 1$), then we can project every point $\mathbf{v}_i$ in the distribution onto the arbitrarily oriented axis $[-\mathbf{p}, \mathbf{p}]$ as a scalar value $x_i = \mathbf{v}_i \cdot \mathbf{p}$. This generates a density distribution on that axis that varies with N, and we can *algebraically* compute the theoretical N-volume measure of each projected sphere slice. So in order to extract the expected profile of a uniform density, we must evaluate the integral of the abstract N-volume slice measure projected incrementally onto an axis element dx at a distance x from the center of the sphere. For the simplest case of a circle $x^2 + y^2 = 1$ projected to the x-axis, we compute the arc-slice component at x, with $\tan\theta = y/x$, to be the arc length $rd\theta = r(d\theta/dx)dx$, where

$$\frac{d\theta}{dx} = \frac{d\arctan\left(\frac{y}{x}\right)}{dx} \tag{6.2}$$

$$= \frac{d\arctan\left(\frac{\sqrt{1-x^2}}{x}\right)}{dx} = -\frac{1}{\sqrt{1-x^2}}. \tag{6.3}$$

(We defined θ to be 90 degrees on the y-axis, at $x = 0$, so as x increases towards $x = 1$, the changes in θ are negative, and progressively larger.) This is the functional form of the change in the projected arc length with x for the circle, $N = 1$, and for each higher N, we can just multiply by the $(N-1)$-th power of the radius $y(x) = \sqrt{1-x^2}$ to get the theoretical density distribution on a single axis. Thus the result for the N-volume of the sphere contained in a slice cut by one projection axis is

$$\mathbf{S}^N \text{ [axis-projected density]:} = \frac{\left(\sqrt{1-x^2}\right)^{N-1}}{\sqrt{1-x^2}} = \left(\sqrt{1-x^2}\right)^{N-2}. \tag{6.4}$$

This measure is proportional to the total N-dimensional incremental volume in a slice of $\mathbf{S}^N$ at Cartesian coordinate x. In Fig. 6.2, we plot these curves for a selection of N values. Remarkably, for $N = 2$, the 2-sphere $\mathbf{S}^2$ has a constant point density for unbiased samples, while $\mathbf{S}^3$ has a unique decreasing density curve as $x \to \pm 1$,

$$\mathbf{S}^3 \text{ slice density} = \sqrt{1-x^2}. \tag{6.5}$$

Written as a Cartesian curve, this is $[x, \sqrt{1-x^2}]$, which is just the equation for the top half of a circle. We can thus use comparison of axis densities with Eq. (6.5) to check whether a proposed statistical quaternion sample satisfies the uniform quaternion density requirements.

6.3 Elements of the uniform random sphere

Uniformity by culling to a uniform neighborhood. One might be tempted simply to run a uniform random number generator across each of the $N + 1$ embedding dimensions and normalize the resulting Euclidean points to unit length, projecting the points onto the surface of $\mathbf{S}^N$. We start in Fig. 6.3(a) with a plot of a 2D uniform distribution on the two dimensions of a square, and superimpose an outline of the $\mathbf{S}^1$ circular line onto which we want to project each point by normalizing its radial distance to unity. The diagonal densities will obviously be higher than the left-right or top-bottom projected densities. So if we normalize those square-allocated points along the radial direction to lie on the unit circle, we will have unacceptable bumps at the corners. If we use the truncated-to-the-disk-interior

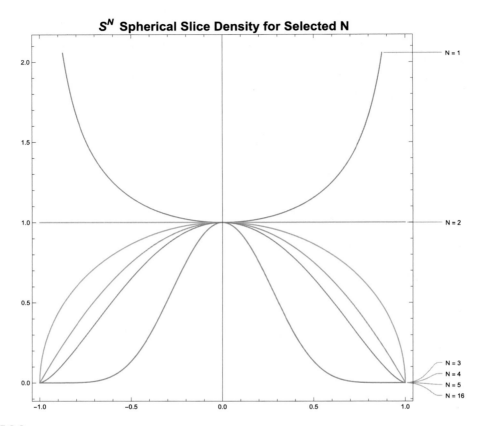

FIGURE 6.2

Plot of the spherical $(N-1)$-volume of a slice of $\mathbf{S}^N$ as a function of the distance interval, from the South pole at the left to the North pole at the right. The volume density at the poles is a maximum for the circle $N=1$, but the relative polar density rapidly decreases with increasing N.

distribution in Fig. 6.3(b), these corner anomalies disappear, and we can deduce that normalizing to the unit circle will be statistically uniform. However, we can see another problem, which is that points too near the origin could end up being "normalized" by dividing by zero; so we should impose the gray disk of forbidden points as well to prevent division by small numbers. We note that in Fig. 6.3(a), the densities projected along the square's *vertical* and *horizontal* axes match the projections of the corresponding circularly symmetric density; that means that, although in those regions the densities match, there will obviously be a nonmatching bump in the rectangular single-axis distribution projected along diagonal axes *between* the vertical and horizontal. This pattern is obviously repeated for any dimension of rectangular sampling on Cartesian axes.

Following the intuition of Fig. 6.3(b) for a uniform $\mathbf{S}^1$ distribution derived from normalizing a filled disk sample region embedded in $\mathbb{R}^2$, we propose an exact analog as our first, and most intuitive, construction method for uniform quaternions. We build a uniform 4D solid hypercube distribution of

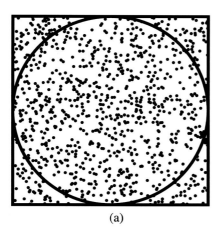 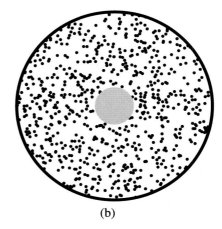

(a)   (b)

FIGURE 6.3

(a) Uniform random numbers on a square projected to the unit circle will be biased. (b) Accepting proposed uniform random numbers from a square distribution only if they lie within a circular (preferably punctured) disk will project uniformly to the unit circle.

Table 6.1 **Spherical subspace culling algorithm extracting a uniform distribution of points on any sphere $\mathbf{S}^N$ from a random uniform distribution of points in a hypercube with axes ranges $= -1 \leqslant x \leqslant +1$ in $\mathbb{R}^{N+1}$.**

```
genUniformNSphereSample[n, dimN] :=
            Variables:   (out = {}, num = 0, rmin=0.2, test),
                  While[num < n,
                        test = RandomReal[{-1, 1}, dimN + 1]
                        If[rmin <= Norm[test] <= 1.0,
                              num++
                              AppendTo[out, Normalize[test]]]]
            Return(out)
```

the form

$$\text{RandomReal[\{-1,1\} ,\{(size), 4\}] ,}$$

discard all 4D points v with $v \cdot v > 1$, then, as in Fig. 6.3(b), chop out a small ball of radius $\sqrt{v \cdot v} < r_{min}$ to avoid divisions by small numbers, and finally normalize, pushing all points with $1 \geqslant \|v\| \geqslant r_{min}$ to $v \cdot v = 1$ after normalization. The normalization process produces a uniform distribution on the $\mathbf{S}^3$ hypersurface. Thus a standard algorithm for accomplishing a uniformly random distribution of points on $\mathbf{S}^3$ for quaternions, or in fact for any $\mathbf{S}^N$, is presented in Table 6.1.

It is instructive to use the uniform-sampling algorithm of Table 6.1 to verify experimentally the shape of the theoretical curves for $\mathbf{S}^N$ examples with different values of N in Fig. 6.2. We create data samples for $N = 1, 2, 3$ and histogram the point density results for one axis. The resulting histograms

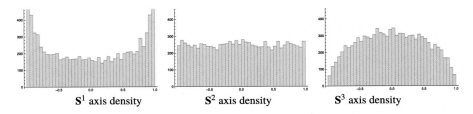

FIGURE 6.4

Histograms of the single-axis sample densities for uniform point distributions on $\mathbf{S}^1$, $\mathbf{S}^2$, and $\mathbf{S}^3$, with 10,000 sample points and bin size 50. The results clearly distinguish the profiles for different dimensions and are in quantitative agreement with the analytic plots in Fig. 6.2.

in Fig. 6.4 produce distributions that are consistent with the analytic predictions. Finally, we produce a complete plot and uniformity evaluation for our canonical spherical culling method of Table 6.1 for a large quaternion data distribution, shown in Fig. 6.5. Again, this seems to be an ideal uniform sample of quaternion points.

Remark: There are circumstances in which for some reason one wants to use a fixed-interval sampling, analogous to a checkerboard on a plane. For the circle $\mathbf{S}^1$, it is obvious from Fig. 6.1 that simply choosing evenly spaced angles θ produces uniform fixed-interval samples by choosing the points $(\cos\theta, \sin\theta)$. For $\mathbf{S}^2$, astronomers have developed approaches such as HEALPIX (see https://healpix.sourceforge.io/ and https://github.com/michitaro/healpix) that provide good approximations to regular point locations sampled on the sky, and support some applications of spherical harmonics; we will not deal with this case in detail, referring the interested reader to (e.g., Snyder, 1993; Yershova et al., 2010).

6.4 Attractive incorrect quaternion sampling

Now that we have one algorithm that clearly matches the features required for uniform distributions on arbitrary spheres, we turn our attention first to alternative tempting but defective ways to try to sample rotations uniformly.

Signature of nonuniformity of 4D rectangular uniform distribution. First, let us go back to Fig. 6.3(a) and extend it to four dimensions, creating a uniform 4D distribution in a hypercube using the uniform rectangular random number generator

```
RandomReal[ {-1,1} ,{( size ), 4}] .
```

Now we verify that if we try to make a quaternion distribution by simply normalizing all these points to have unit distance from the origin, we fail to get a uniform distribution in $\mathbf{S}^3$. Applying our single-axis density plot, we get the results in Fig. 6.6. The figure clearly shows that each axis has the same

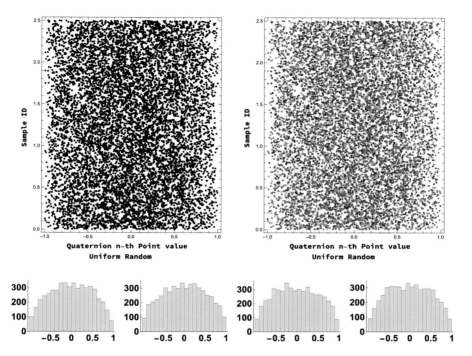

FIGURE 6.5

The uniform distribution of Table 6.1 for 5000 uniform 4D points on $\mathbf{S}^3$. Top: The locations by axis, merged in BW, and color coded by (q_0, q_1, q_2, q_3). Bottom: Histogramming the coordinates of each of the four axes gives the same distribution on each axis, and these profiles are in agreement with the predicted axis histogram shape predicted for a uniform $\mathbf{S}^3$ distribution shown in Fig. 6.2.

double bump distribution, the signature of the "corner excess" of out-of-$\mathbf{S}^3$ samples in each dimension that we illustrated intuitively in Fig. 6.3(a). This method is clearly an incorrect approach to quaternion sampling, illustrating the defect of failing to cull to the interior of a unit sphere.

Signature of nonuniformity of uniform Euler angle sampling. For the quaternion 3-sphere $\mathbf{S}^3$, another first thought might be to attempt a uniform random sampling of each of the three Euler angles to get a uniform sampling of 3D rotations. If that succeeded, it could also easily be applied to the quaternion domain because any given XYZ permutation of Euler angle definitions has an exact quaternion counterpart. Consider, for example, the **ZYZ** Euler angle definition, which applies a spin by ψ about the **Z**-axis, followed on the left by a latitude-like tilt from the North pole about the **Y**-axis to the angle θ, ending by a third rotation acting from the left on the existing rotations with a longitude-like rotation about the *original* **Z**-axis at the North pole by an angle ϕ. This gives us a rotation matrix that looks like

$$R(\phi, \theta, \psi) = R(\phi, \mathbf{Z}) \cdot R(\theta, \mathbf{Y}) \cdot R(\psi, \mathbf{Z}) =$$

$$\begin{bmatrix} \cos(\phi)\cos(\theta)\cos(\psi) - \sin(\phi)\sin(\psi) & -\sin(\phi)\cos(\psi) - \cos(\phi)\cos(\theta)\sin(\psi) & \cos(\phi)\sin(\theta) \\ \sin(\phi)\cos(\theta)\cos(\psi) + \cos(\phi)\sin(\psi) & \cos(\phi)\cos(\psi) - \sin(\phi)\cos(\theta)\sin(\psi) & \sin(\phi)\sin(\theta) \\ -\sin(\theta)\cos(\psi) & \sin(\theta)\sin(\psi) & \cos(\theta) \end{bmatrix}. \quad (6.6)$$

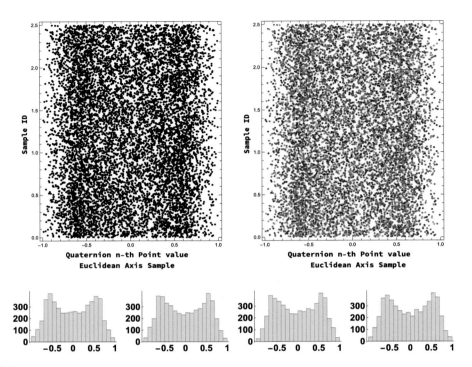

FIGURE 6.6

The biased $\mathbf{S}^3$ distribution produced from a normalized uniform Euclidean distribution on four axes. Top: The locations by axis, merged in BW, and color coded by (q_0, q_1, q_2, q_3). Bottom: Histogramming the coordinates of each of the four axes (q_0, q_1, q_2, q_3) gives the a nonuniform double peaked distribution on each axis.

The corresponding Euler angle quaternion form is

$$q(\phi, \theta, \psi) =$$

$$\left[\; \cos\left(\tfrac{\theta}{2}\right)\cos\left(\tfrac{\phi+\psi}{2}\right), \quad -\sin\left(\tfrac{\theta}{2}\right)\sin\left(\tfrac{\phi-\psi}{2}\right), \quad \sin\left(\tfrac{\theta}{2}\right)\cos\left(\tfrac{\phi-\psi}{2}\right), \quad \cos\left(\tfrac{\theta}{2}\right)\sin\left(\tfrac{\phi+\psi}{2}\right) \; \right]. \quad (6.7)$$

Unfortunately, with uniformly distributed angles, this turns out to be extremely biased, with a high density near $\theta = 0$, for example. The histograms of the Cartesian axis densities whose shape needs to be a semicircle for uniform $\mathbf{S}^3$ samples take instead the form shown in Fig. 6.7, with concave instead of convex profiles falling off from the origin towards the limits of the coordinate values.

6.5 More uniform quaternions: Gaussian and Shoemake/Diaconis

We have a fondness for the spherical-volume culling approach of Table 6.1 for uniform quaternion distributions, but there are two more perfectly respectable algorithms that we can use that are more

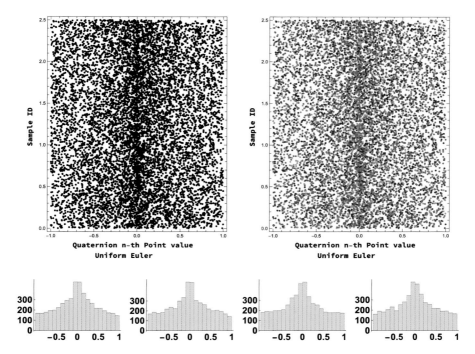

FIGURE 6.7

Uniform sampling of 5000 Euler angles fails to give a uniform distribution on the axis density histogram. Top: The locations by axis, merged in BW, and color coded by (q_0, q_1, q_2, q_3). Bottom: Histogramming the coordinates of each of the four axes (q_0, q_1, q_2, q_3) gives a nonuniform centrally peaked distribution on each axis.

economical for *really big* data sets because they do not have to throw away discarded out-of-range data, and keep all of their generated data.

Spherical Gaussian normal distribution. Now we show that, with a different method of choosing a conventional distribution of random numbers, we can immediately get a set of uniformly distributed quaternions. The basic fact is that Gaussian distributions, or "normal distributions," have the remarkable feature that the product of two or more Gaussian distributions, unlike Euclidean uniform distributions, is *spherically symmetric*, as illustrated in Fig. 6.8. Since the product of Gaussian distributions is radially distributed without a preferred direction, we can simply take the distribution with zero mean and unit standard deviation,

```
RandomVariate[NormalDistribution[0, 1], {( size ), 4} ,
```

punch out a spherical neighborhood of the origin if desired, and normalize. The results are shown in Fig. 6.9. The new distribution is now perfectly spherically symmetric according to the Shoemake criterion, and hence can be used as a source of uniformly sampled quaternions in the 3-sphere $\mathbf{S}^3$.

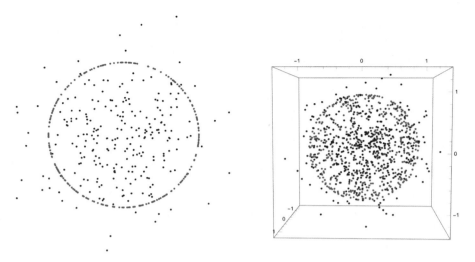

FIGURE 6.8

Gaussian distributions in 2D and 3D (and any dimension) are rotation invariant about the origin, and have statistically uniform distributions when normalized to unit radius (shown in red), unlike Euclidean uniform distributions.

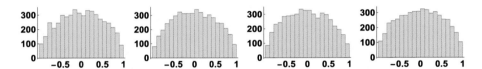

FIGURE 6.9

The Gaussian or normal distribution applied to yield a 5000-point normalized uniform radial distribution on four axes. Histogramming the coordinates (q_0, q_1, q_2, q_3) gives a uniform smooth distribution on each of the four axes.

Shoemake/Diaconis et al. method. A seminal paper (Diaconis and Shahshahani, 1987) that has not received a lot of attention was revisited by Shoemake in studying the concept of uniform distributions on nonflat manifolds, of which our quaternion problem turns out to be a classic example. Diaconis and Shahshahani (1987) specifically derived methods for obtaining a uniform distribution from almost any Lie group, of which the **SU(2)** group underlying quaternions happens to be one. Shoemake (1992) cites Diaconis et al. and then derives his very elegant version for uniform quaternions based on their **SU(2)** method. The Shoemake algorithm is implemented by taking a uniform random distribution of the same size on the interval $[0, 1]$ for every variable (a, b, c) in the equation

$$q(a, b, c) = (\sqrt{a} \cos(2\pi b), \sqrt{a} \sin(2\pi b), \sqrt{1-a} \cos(2\pi c), \sqrt{1-a} \sin(2\pi c)) . \tag{6.8}$$

Table 6.2   The Shoemake algorithm for a random distribution of points on S^3.

shoemakeRandomQuat[n] :=
       Variables: (a,b,c, out)
              out = Table[
                     a = RandomReal[{0,1}]
                     b = RandomReal[{0,1}]
                     c = RandomReal[{0,1}]
                     { Sqrt[a]* Cos[2*b*pi] , Sqrt[a]* Sin[2*b*pi],
                                   Sqrt[1 - a] *Cos[2 *c* pi], Sqrt[1 - a] *Sin[2* c *pi] },
                                          n]
              Return(out)

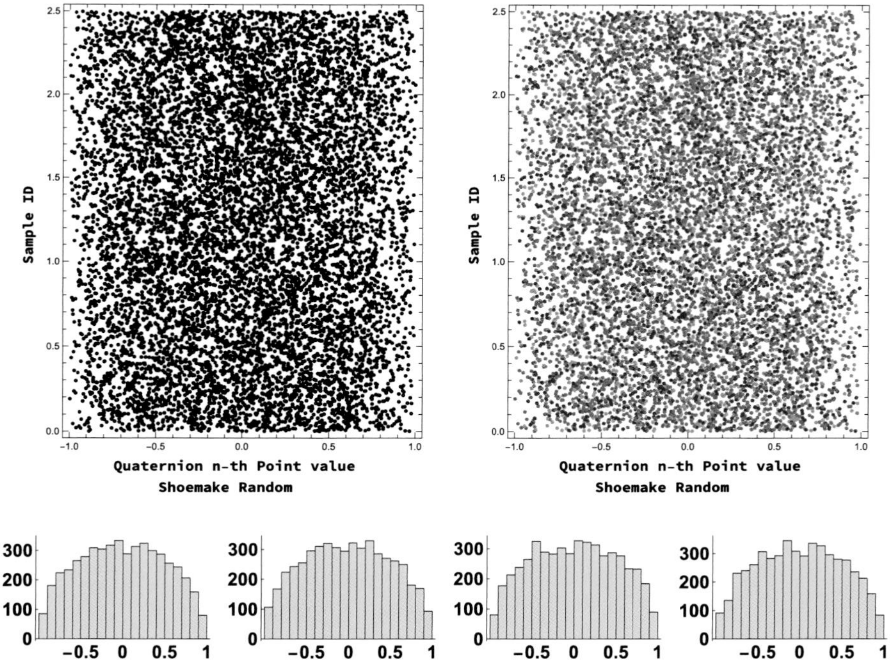

FIGURE 6.10

The uniform Shoemake distribution of Table 6.2 for 5000 4D points on S^3. Top: The locations by axis, merged in BW, and color coded by (q_0, q_1, q_2, q_3). Bottom: Histogramming the coordinates of each of the four q_i-axes gives the same distribution on each axis, in agreement with the uniform shape prediction of Fig. 6.2.

Table 6.2 gives the fundamental Shoemake algorithm that can be used in place of our standard algorithm in Table 6.1, avoiding the extra cost of discarded samples. This is illustrated in Fig. 6.10 to be uniform on S^3.

Orientation matching[☆]

The application of certain quaternion methods to solving orientation optimization problems goes back at least to the 1960s when they were first applied to aerospace problems. Here in Part 2 we present an historical review and details of quaternion methods as they apply to the orientation alignment task for point-cloud matching and coordinate frame cloud matching. Much of the material in these chapters is adapted from a treatment published by the author in (Hanson, 2020).

Here we will focus first on the simplified 2D point-cloud alignment problem to build some intuition with exact solutions, and then present an extensive treatment of the real 3D world and the quaternion approach to the 3D point-cloud alignment problem. Chapter 9 deals with the very useful Bar-Itzhack method (Bar-Itzhack, 2000) that extends the same technology used in 3D cloud alignment to the extraction of the best quaternion that corresponds to a given possibly imperfect rotation matrix; this is significant because the most common methods for extracting a rotation matrix's quaternion assume perfect rotation matrices. Turning from point clouds to the domain of orientation frames attached to points in a cloud, we address the task of aligning collections of orientation frames. Later in Part 6, Chapter 20, we generalize the entire 3D quaternion framework in these chapters to 4D, expanding our

---

[☆] The chapters in this part are based largely on Hanson (2020).

scope to include 4D Euclidean rotations written as quaternion pairs, extending our 3D methods to 4D for point-cloud matching and frame alignment, and finally adapting the Bar-Itzhack method to find the best quaternions corresponding to 4D rotation matrices.

Introduction: 2D cloud alignment problem☆

Rotations abound in applications that describe and discover relations among objects in scene data. Since rotations can be expressed with quaternions, we might naturally look for quaternion approaches to these problems. In this first chapter of Part 2, we will introduce the problem in the simple terms afforded by a 2D environment; our purpose here is to build intuition that will be useful when we investigate 3D and 4D in later chapters.

7.1 Context of the alignment task

We begin with a context in which we are given a cloud of reference points, and a rotation that acts on those points to give a target cloud that will be assumed to be described by either perfect or noisy measurements. The goal is to use only the reference points and the measured target points to determine the *optimal candidate* for the rotation that generated the target points.

All of our useful algebraic approaches to finding the acting rotation will depend upon knowing the correspondences between reference points and target points: the related unordered problem is of course also important, but we will consider that to be beyond our scope here.

The least-squares loss matching problem in 2D. We denote our reference data by a set of 2D points $\mathbf{X}$ with elements $\{\mathbf{x}_i = [x_i, y_i]\}$, and the target data set $\mathbf{U}$ with elements $\{\mathbf{u}_i = [u_i, v_i]\}$. $\mathbf{U}$ is assumed to be produced by acting on $\mathbf{X}$ with a rotation matrix parameterized by a 2D subspace of the quaternions that we parameterize as (a, b), with

$$R(\theta) = \begin{bmatrix} \cos\theta & -\sin\theta \\ \sin\theta & \cos\theta \end{bmatrix}, \tag{7.1}$$

$$R(a, b) = \begin{bmatrix} a^2 - b^2 & -2ab \\ 2ab & a^2 - b^2 \end{bmatrix}. \tag{7.2}$$

Here the reduced-dimension 2D quaternions in Eq. (7.2) have the expected form $a = \cos(\theta/2)$, $b = \sin(\theta/2)$, reducing via the half-angle formula to Eq. (7.1). In Fig. 7.1, we show a sample 2D (origin-centered) reference set in red, and the action of a random rotation matrix taking the reference set to the target set, with the possibility of adding noise after the rotation acts. The least-squares loss function

---

☆ This chapter derives much of its content from Hanson (2020).

Visualizing More Quaternions. https://doi.org/10.1016/B978-0-32-399202-2.00016-2

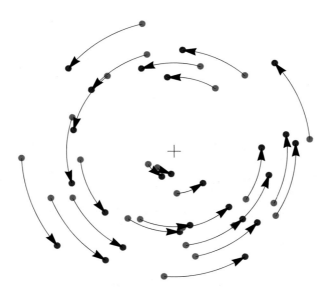

FIGURE 7.1

The cloud-to-cloud matching problem in 2D. Given a reference cloud (in red), we rotated it to generate a target cloud (in blue), optionally adding noise. The task is to find the best candidate for the rotation given the reference cloud and the evidence provided by the measured locations of the corresponding points in the target cloud.

problem then takes the form

$$\mathbf{S}_{(2D\ Match)} = \sum_{k=1}^{K} \| R(a, b) \cdot \mathbf{x}_k - \mathbf{u}_k \|^2 \tag{7.3}$$

$$= constants - 2 \sum_{k=1}^{K} \mathbf{u}_k \cdot R \cdot \mathbf{x}_k \ . \tag{7.4}$$

We thus have the choice of *minimizing* the least-squares loss Eq. (7.3) or *maximizing* the positive form of the least-squares cross-term appearing in Eq. (7.4). We adopt the latter for our demonstration because it allows us to reduce the *entire impact* of the numerical data to two numbers. We proceed by defining the *cross-covariance* matrix

$$E_{ij} = \sum_{k=1}^{K} [x_k]_i \, [u_k]_j \tag{7.5}$$

for $i, j = \{1, 2\}$, allowing us to phrase the matching task as the maximization of

$$\Delta = \sum_{k=1}^{K} (R(\theta) \cdot x_k) \cdot u_k = \sum_{i=1, j=1}^{2} R^{ji} E_{ij} = (E_{xx} + E_{yy}) \cos\theta + (E_{xy} - E_{yx}) \sin\theta \tag{7.6}$$

over the values of $R(\theta)$ in Eq. (7.1).

We see already that only two data-containing quantities enter into the problem, and we can split these out as a symmetric and an antisymmetric term,

$$C = E_{xx} + E_{yy} \, ,$$
$$S = E_{xy} - E_{yx} \, .$$

We can either differentiate with respect to θ and set $\Delta'(\theta) = 0$, or simply observe directly that $\Delta(\theta)$ is largest when the vector $(\cos\theta, \sin\theta)$ is parallel to its coefficients; both arguments lead to the explicit *least-squares* solution

$$\tan\theta = \frac{E_{xy} - E_{yx}}{E_{xx} + E_{yy}} = \frac{S}{C} \, , \tag{7.7}$$

$$(\cos\theta, \sin\theta) = \left(\frac{C}{\sqrt{C^2 + S^2}}, \frac{S}{\sqrt{C^2 + S^2}} \right) . \tag{7.8}$$

We notice the parallels between the elements of the 2D cross-covariance matrix and unnormalized trigonometric functions. We will take explicit advantage of this correspondence in later chapters.

The 2D quaternion eigensystem solution to the matching problem. But of course what we want to do is to learn more about how quaternions can illuminate the cloud matching problem, so now let us rewrite Eq. (7.6) in terms of quaternions instead of the rotation angle. We find

$$\Delta(a, b) = \sum_{k=1}^{K} (R(a, b) \cdot x_k) \cdot u_k = \text{tr}(R(a, b) \cdot E = C \, (a^2 - b^2) + S \, (2ab)$$

$$= [a \, b] \cdot M \cdot \begin{bmatrix} a \\ b \end{bmatrix} , \tag{7.9}$$

where we now define the *profile matrix* as

$$\mathbf{M} = \begin{bmatrix} C & S \\ S & -C \end{bmatrix} . \tag{7.10}$$

From Eq. (7.9), we see that Δ is maximized by the maximal eigenvalue of M, and its corresponding maximal quaternion eigenvector values of $[a_{\text{opt}}, b_{\text{opt}}]$. The eigenvalues are trivial, with the values $\lambda = \pm\sqrt{C^2 + S^2}$. The maximal eigenvalue takes the plus sign, so the maximal quaternion eigenvector is any normalized column of the characteristic equation's adjugate $A(C, S)$, defined with an inconsequential sign change from the actual mathematical definition to improve readability, as

$$A(C, S) = -\text{Adjugate} \begin{bmatrix} C - \lambda & S \\ S & -C - \lambda \end{bmatrix} = \begin{bmatrix} \lambda + C & S \\ S & \lambda - C \end{bmatrix} . \tag{7.11}$$

Here we notice an anomaly that we will investigate in more detail in Chapter 12: we cannot normalize both columns of the adjugate matrix quaternion eigenvectors if $C = +\lambda$ or $C = -\lambda$, since in each case we also have $S = 0$, one column vanishes, and we cannot normalize a 0-vector. Assuming that we have taken this singularity into account, in the well-behaved cases, we can normalize the first column to find the quaternion solution, up to a sign,

$$a_{opt} = \cos(\theta/2) = \frac{\lambda + C}{\sqrt{2\lambda(\lambda + C)}} = \sqrt{\frac{\lambda + C}{2\lambda}},$$

$$b_{opt} = \sin(\theta/2) = \frac{S}{\sqrt{2\lambda(\lambda + C)}} = \text{sign}\, S \sqrt{\frac{\lambda - C}{2\lambda}}.$$

(7.12)

Alternately, normalizing the eigenvector in the second column, we find, up to a sign,

$$a_{opt} = \cos(\theta/2) = \frac{S}{\sqrt{2\lambda(\lambda - C)}} = \text{sign}\, S \sqrt{\frac{\lambda + C}{2\lambda}},$$

$$b_{opt} = \sin(\theta/2) = \frac{\lambda - C}{\sqrt{2\lambda(\lambda - C)}} = \sqrt{\frac{\lambda - C}{2\lambda}}.$$

(7.13)

Note the crucial (sign S) factor; it tells us that (mostly) the two columns of the adjugate are the same up to a sign, but that care is needed to create a valid quaternion solution. Going back to our original 2D rotation matrix in Eq. (7.2) and substituting either Eq. (7.12) or Eq. (7.13), we recover the same rotation as that found in Eq. (7.8) by directly solving the least-squares problem, namely

$$R_2(\theta) = \begin{bmatrix} \dfrac{C}{\sqrt{C^2 + S^2}} & -\dfrac{S}{\sqrt{C^2 + S^2}} \\ \dfrac{S}{\sqrt{C^2 + S^2}} & \dfrac{C}{\sqrt{C^2 + S^2}} \end{bmatrix}$$

(7.14)

$$= \begin{bmatrix} \cos\theta & -\sin\theta \\ \sin\theta & \cos\theta \end{bmatrix}.$$

(7.15)

These results are interesting to study because, despite the complexity of the general solution, the intrinsic algebraic structure of any RMSD problem is entirely characterized by a planar rotation such as that described by Eq. (7.14). Among the other important properties of this algebraic least-squares (actually maximized cross-term) solution is that there are no restrictions on the data: these same formulas are valid for any pairs of reference:target data sets, with or without noise. In later chapters, we will find that solutions that are robust to noise can be more difficult to obtain exactly in higher dimensions.

7.2 Remark: some alignment problems are very simple

All rotations of the type we will be trying to optimize reduce to a rotation in a 2D plane, which in 3D is defined by the plane perpendicular to the eigenvector $\hat{\mathbf{n}}$ of the rotation matrix Eq. (2.14). Data sets that are highly linear, determining a robust straight line from least-squares, can even circumvent the full alignment problem entirely: a very good rotation matrix can be calculated from the direction $\hat{\mathbf{x}}$ determined by the line fitted to the data set $\{\mathbf{x}_i\}$, and the similar direction $\hat{\mathbf{u}}$ corresponding to the reference data set $\{\mathbf{u}_i\}$. An optimal rotation matrix in 3D then just requires a rotation by the angle determined by $\hat{\mathbf{x}} \cdot \hat{\mathbf{u}} = \cos\theta$ in the plane determined by $\hat{\mathbf{x}} \times \hat{\mathbf{u}} \approx \hat{\mathbf{n}}$, or simply

$$R(\theta, \hat{\mathbf{n}}) = R(\arccos(\hat{\mathbf{x}} \cdot \hat{\mathbf{u}}), \widehat{\hat{\mathbf{x}} \times \hat{\mathbf{u}}}).$$

(7.16)

Note the parallels between Eq. (7.16) and the Clifford algebra quaternion form in Eq. (3.17), and the contrast between $A \cdot B = \cos(\theta/2)$ there and $\hat{\mathbf{x}} \cdot \hat{\mathbf{u}} = \cos\theta$ here. Eq. (7.16) is easily generalized to any dimension by isolating just the projections of vectors to the plane determined by $\hat{\mathbf{x}}$ and $\hat{\mathbf{u}}$, and rotating in that 2D basis. Of course *any* 3D rotation reduces to a rotation by θ in a plane determined by the eigenvector $\hat{\mathbf{n}}$ of the rotation matrix because of Eq. (2.17), but we conclude from this argument that if we had access to initial data of the proper form, we could find $\hat{\mathbf{n}}$ and the rotation angle θ by inspection, and then immediately use Eq. (7.16) to solve the entire cloud matching problem.

The 3D quaternion-based alignment problem☆

8

In this chapter, we study quaternion methods for obtaining solutions to the problem of finding global rotations that optimally align pairs of corresponding lists of 3D cloud-like spatial data. Aligning matched sets of spatial point data is a universal problem that occurs in a wide variety of applications involving generic objects such as protein residues, parts of composite object models, satellite alignment, and camera calibration. There are multiple methods for solving this problem, and we will point them out in our introduction to the rich historical context of the alignment task. The quaternion methods have some advantages in conceptual clarity and the ability, at least in principle, to support writing down exact differentiable algebraic solutions.

8.1 Introduction

We have already reviewed the simplified 2D matching problem in the previous chapter. We now continue by examining the possible quaternion-based approaches to the optimal alignment problem for matched sets of rotated objects in 3D space, which may be referred to in its most generic sense as the "generalized orthogonal Procrustes problem" (Golub and van Loan, 1983), and is also known as the "root mean square deviation" (RMSD) problem.

We devote some attention to identifying the surprising breadth of domains and literature where the various approaches, including particularly quaternion-based methods, have appeared; in fact the number of times in which quaternion-related methods have been described independently without cross-disciplinary references is rather interesting, and exposes some challenging issues related to the wide dispersion of both historical and modern scientific literature relevant to these subjects.

In the following, we first present a technical overview and then review the diverse bodies of literature regarding the extraction of 3D rotations that optimally align matched pairs of Euclidean point data sets. It is important for us to remark that we have repeatedly become aware of additional literature in the course of investigating this topic, and it is entirely possible that other worthy references have been overlooked.

8.2 Spatial alignment of matched 3D point sets

We now turn to the details of the full 3D RMSD problem, which seeks a global rotation R that rotates a reference set $\mathbf{X}$ in such a way as to minimize the squared Euclidean differences relative to a matched set of previously rotated point test data $\mathbf{U}$.

☆ This chapter derives much of its content from Hanson (2020).

Visualizing More Quaternions. https://doi.org/10.1016/B978-0-32-399202-2.00017-4

77

We will see that for the matching problem in any dimension D, all of the required information from the two point clouds is reduced to the content of the $D \times D$ cross-covariance matrix of the pair $(\mathbf{X}, \mathbf{U})$ of N columns of D-dimensional vectors, namely $E = \mathbf{X} \cdot \mathbf{U}^t$, and a maximization problem over possible rotation matrices for the elementary form $\operatorname{tr}(R \cdot E)$. We note that we will consider cases where E could have almost any origin.

One solution to this problem valid for arbitrary dimensions uses the decomposition of the general matrix E into an orthogonal matrix O and a symmetric matrix S that takes the form $E = O \cdot S = O \cdot (E^t \cdot E)^{1/2}$, giving $R_{\mathrm{opt}} = O^{-1} = (E^t \cdot E)^{1/2} \cdot E^{-1}$; note that there exist several equivalent forms (see, e.g., Green (1952); Horn et al. (1988)). General solutions may also be found using singular value decomposition (SVD) methods, starting with the decomposition $E = U \cdot S \cdot V^t$, where S is now diagonal and U and V are orthogonal matrices, to give the result $R_{\mathrm{opt}} = V \cdot D \cdot U^t$, where D is the identity matrix up to a possible sign in one element (see, e.g., Kabsch (1976, 1978); Golub and van Loan (1983); Markley (1988)).

In addition to these general methods based on traditional linear algebra approaches, a significant body of literature exists for three dimensions that exploits the relationship between 3D rotation matrices and quaternions, and rephrases the task of finding R_{opt} as a *quaternion eigensystem* problem, which we have already demonstrated in its simplest form for 2D cloud matching. This approach notes that, using the quadratic quaternion form $R(q)$ in Eq. (2.14) for the rotation matrix, one can rewrite the measure as $\operatorname{tr}(R(q) \cdot E) \to q \cdot M(E) \cdot q$, where the *profile matrix* $M(E)$ is a traceless, symmetric 4×4 matrix consisting of linear combinations of the elements of the 3×3 matrix E. Finding the largest eigenvalue λ_{opt} of $M(E)$ determines the optimal quaternion eigenvector q_{opt} and thus the solution $R_{\mathrm{opt}} = R(q_{\mathrm{opt}})$.

The 3D quaternion framework will be our main topic here.

Historical overview. Although our focus is the quaternion eigensystem context, we first note that one of the original approaches to the RMSD task exploited the singular-value decomposition directly to obtain an optimal rotation matrix. This solution appears to date at least from the work of Thompson (1958) and Schut (1960), as cited in Haralick et al.'s (1989) SVD treatment, though Schönemann's thesis (1966) is frequently cited due to its being credited for the SVD method, e.g., in the book of Golub and van Loan (1983). (Similar work by Cliff (1966) appears later in the same journal issue as Schönemann (1966).) Applications of the SVD method to alignment in the aerospace literature appear, for example, in the context of Wahba's problem (Wikipedia, 2023l; Wahba, 1965), and are used explicitly, e.g., in Markley (1988), while the introduction of the SVD for the alignment problem in molecular chemistry generally is attributed to Kabsch (Wikipedia, 2023k; Kabsch, 1976).

We believe that the quaternion eigenvalue approach itself was first noticed around 1968 by Davenport (1968) in the context of Wahba's problem, and then studied in 1973 with extensive details of its abstract properties by Sansò (1973), as noted in Haralick et al. (1989). The quaternion method was rediscovered in 1983 by Hebert (1983); Faugeras and Hebert (1983, 1986) in the context of machine vision, and then found independently a third time in 1986 by Horn (1987).

An alternative quaternion-free approach by Horn et al. (1988) with the optimal rotation of the form $R_{\mathrm{opt}} = (E^t \cdot E)^{1/2} \cdot E^{-1}$ appeared in 1988, but this basic form was apparently known elsewhere as early as 1952 (Green, 1952; Gibson, 1960) .

Much of the recent activity has occurred in the context of the molecular alignment problem, starting from a basic framework put forth by Kabsch (1976, 1978). So far as we can determine, the matrix eigenvalue approach to molecular alignment was introduced in 1988 without actually mentioning quaternions

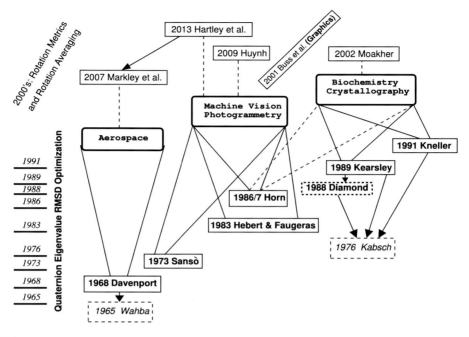

FIGURE 8.1

The quaternion eigensystem method for computing the optimal rotation matching two 3D spatial data sets was discovered independently and published without cross-references at least six times in at least three distinct literatures. Downward arrows point to the introduction of the abstract problem, and upward rays indicate domains of publications specifically citing the quaternion method. Horn eventually appeared routinely in the crystallography citations, and reviews such as Flower (1999) and Förstner (1999) introduced multiple cross-field citations. Several fields have included activity on quaternion-related rotation metrics and rotation averaging with varying degrees of cross-field awareness.

by name in Diamond (1988), and was subsequently refined to specifically incorporate quaternion methods in 1989 by Kearsley (1989). In 1991 Kneller (1991) independently described a version of the quaternion-eigenvalue-based approach that is widely cited as well. A concise and useful review can be found in Flower (1999), in which the contributions of Schönemann, Faugeras and Hebert, Horn, Diamond, and Kearsley are acknowledged and all cited in the same place, while the contemporary review of Förstner (1999) includes Sansò (1973).

A graphical summary of the discovery chronology in various domains is given in Fig. 8.1. Most of these treatments mention using numerical methods to find the optimal eigenvalue, though several references, starting with Horn (1987), point out that 16th-century algebraic methods for solving the quartic polynomial characteristic equation, discussed in the next paragraph and in Appendix J, could also be used to determine the explicit symbolic and differentiable versions of the eigenvalues. In our treatment we will study the explicit form of these algebraic solutions for the 3D problem, taking advantage of several threads of the literature. Chapter 20 extends these methods to the 4D problem.

The story of the quartic. The actual solution to the quartic equation, and thus the solution of the characteristic polynomial of the 4D eigensystem of interest to us, was first published in 1545 by Gerolamo Cardano (Wikipedia, 2023a) in his book *Ars Magna*. The intellectual history of this fact is controversial and narrated with varying emphasis in diverse sources. It seems generally agreed upon that Cardano's student Lodovico Ferrari was the first to discover the basic method for solving the quartic in 1540, but his technique was incomplete as it only reduced the problem to the cubic equation, for which no solution was publicly known at that time, and that apparently prevented him from publishing it. The complication appears to be that Cardano had actually learned of a method for solving the cubic already in 1539 from Niccolò Fontana Tartaglia (legendarily in the form of a poem), but had been sworn to secrecy, and so could not reveal the final explicit step needed to complete Ferrari's implicit solution. Where it gets controversial is that at some point between 1539 and 1545, Cardano learned that Scipione del Ferro had found the same cubic solution as the one of Tartaglia that he had sworn not to reveal, and furthermore that del Ferro had discovered his solution before Tartaglia did. (Some sources speculate that Tartaglia may have actually learned of the solution in some way traceable to del Ferro.) Cardano interpreted learning of del Ferro's solution as releasing him from his oath of secrecy to Tartaglia (which Tartaglia did not appreciate), allowing him to publish the complete solution to the quartic, incorporating the cubic solution into Ferrari's result. Sources claiming that Cardano "stole" Ferrari's solution may perhaps be exaggerated, since Ferrari did not have access to the cubic solutions, and Cardano did not conceal his sources; exactly who "solved" the quartic is thus philosophically complicated, but Cardano does seem to be the one who combined the multiple threads needed to express the equations as a single complete formula.

Other interesting observations were made later, for example, by Descartes in 1637 (Descartes, 1637 (1954)), and in 1733 by Euler (Euler, 1733; Bell, 2008 (1733)). For further descriptions, one may consult, e.g., Abramowitz and Stegun (1970) and Boyer and Merzbach (1991), as well as the narratives in Weisstein (2019a,b). Additional amusing pedagogical investigations of the historical solutions may be found in several expositions by Nickalls (1993, 2009).

Supplementary background. A very informative treatment of the features of the quaternion eigenvalue solutions has been given by Coutsias et al. (2004), and expanded in 2019 by Coutsias and Wester (2019). Coutsias et al. not only take on a thorough review of the quaternion RMSD method, but also explore the relationship between the linear algebra of the SVD method and the quaternion eigenvalue system; we review some of these connections with SVD in Appendix N, and the subject comes up again when we turn to the 4D problem in Chapter 20. Coutsias et al. also exhaustively enumerate the special cases involving mirror geometries and degenerate eigenvalues that may appear rarely, but must be dealt with on occasion. Efficiency is another area of potential interest, and Theobald et al. in Theobald (2005); Liu et al. (2010) argue that among the many variants of numerical methods that have been used to compute the optimal quaternion eigenvalues, Horn's original proposal to use Newton's method directly on the characteristic equations of the relevant eigenvalue systems may well be the best approach.

There is also a rich body of literature dealing with distance measures among representations of rotation frames themselves, some dealing directly with the properties of distances computed with rotation matrices or quaternions, e.g., Huynh (2009), and others combining discussion of the distance measures with associated applications such as rotation averaging or finding "rotational centers of mass," e.g., Brown and Worsey (1992); Park and Ravani (1997); Buss and Fillmore (2001); Moakher (2002); Markley et al. (2007); Hartley et al. (2013). The specific computations explored in Chapter 10 on the optimal alignment of matched pairs of orientation frames make extensive use of the quaternion-based and rotation-based measures discussed in these treatments. Further details are given in their own chapters, Chapter 18 and Chapter 19.

Application context. Finding global rotations that optimally align pairs of corresponding lists of spatial data is significant in diverse application domains. Among these are aligning spacecraft (see, e.g., Wahba (1965); Davenport (1968); Markley (1988); Markley and Mortari (2000)), obtaining correspondence of registration points in 3D model matching (see, e.g., Faugeras and Hebert (1983, 1986)), matching structures in aerial imagery (see, e.g., Horn (1987); Horn et al. (1988); Huang et al. (1986); Arun et al. (1987); Umeyama (1991); Zhang (2000)), and alignment of matched molecular and biochemical structures (see, e.g., Kabsch (1976, 1978); MacLachlan (1982); Lesk (1986); Diamond (1988); Kearsley (1989, 1990); Kneller (1991); Coutsias et al. (2004); Theobald (2005); Liu et al. (2010); Coutsias and Wester (2019)). There are several alternative approaches that in principle produce the same optimal global rotation to solve a given alignment problem, and the SVD and $(E^{\mathrm{t}} \cdot E)^{1/2} \cdot E^{-1}$ methods apply to any dimension. Here we critically examine the quaternion eigensystem decomposition approach to studying the rotation matrices appearing in the 3D RMSD optimization problem. Starting from the exact quartic algebraic solutions to the eigensystems arising in these optimization problems, we direct attention to the elegant algebraic forms of the eigenvalue solutions appropriate for these applications. Chapter 10 treats the related problem of matching *orientation frame data*, since 3D orientation frames can *themselves* be expressed as quaternions, and the more complicated extension to 4D data is presented in Chapter 20.

8.3 Reviewing the 3D spatial alignment RMSD problem

We now review the basic ideas of spatial data alignment, and then specialize to 3D (see, e.g., Wahba (1965); Davenport (1968); Markley (1988); Markley and Mortari (2000); Kabsch (1976, 1978); MacLachlan (1982); Lesk (1986); Faugeras and Hebert (1983); Horn (1987); Huang et al. (1986); Arun et al. (1987); Diamond (1988); Kearsley (1989, 1990); Umeyama (1991); Kneller (1991); Coutsias et al. (2004); Theobald (2005)). We will then employ quaternion methods to reduce the 3D spatial alignment problem to the task of finding the optimal quaternion eigenvalue of a certain 4×4 matrix. This is the approach we have discussed in the introduction, and it can be solved using numerical or algebraic eigensystem methods. In a separate section, we will explore in particular the classical quartic equation solutions for the exact algebraic form of the entire four-part eigensystem, with additional detail provided in Appendix J.

Aligning matched data sets in Euclidean space. We begin with the general least-squares form of the RMSD problem, which is solved by minimizing the optimization measure over the space of rotations, which we will convert to an optimization over the space of unit quaternions. We take as input one data array with N columns of D-dimensional points $\mathbf{X} = \{x_k\}$ as the *reference* structure, and a second array of N columns of *matched* points $\mathbf{U} = \{u_k\}$ as the *test* or *target* structure. Our task is to find the global **SO**(D) rotation matrix R_D that has to be applied to the reference data to achieve the minimum value of the cumulative quadratic distance measure

$$\mathbf{S}_D = \sum_{k=1}^{N} \|R_D \cdot x_k - u_k\|^2 . \tag{8.1}$$

We assume, as is customary, that any overall translational components have been eliminated by displacing both data sets to their centers of mass (see, e.g., Faugeras and Hebert (1983); Coutsias et al. (2004)). When this measure is minimized with respect to the rotation R_D, the optimal R_D will rotate the reference set $\{x_k\}$ to be as close as possible to the measured test data $\{u_k\}$. Here we will focus on 3D data sets, targeting the quaternion approach. The entire problem is solved in 4D later in Chapter 20.

Quaternion minimization of the full loss expression. In 3D, one can convert the full unsimplified least-squares measure Eq. (8.1) directly into a quaternion optimization problem using the method of Hebert (1983); Faugeras and Hebert (1983, 1986). Although we focus below on the cross-term maximal eigenvalue approach introduced by (Davenport, 1968; Horn, 1987), we summarize here for completeness the original equivalent *minimal* eigenvalue approach.

Starting with the 3D Euclidean minimizing distance measure Eq. (8.1), we can exploit Eq. (2.14) for $R(q)$, along with the normed division algebra property Eq. (2.11), to produce an alternative quaternion eigenvalue problem whose *minimal* eigenvalue determines the eigenvector q_{opt} specifying the matrix that rotates the reference data $\{x_k\}$ into closest correspondence with the target data $\{u_k\}$.

Adopting the convenient notation $\mathbf{x} = (0, x_1, x_2, x_3)$ and $\mathbf{u} = (0, u_1, u_2, u_3)$ for pure imaginary quaternions, we employ the following steps:

$$
\begin{aligned}
\mathbf{S}_3 &= \sum_{k=1}^{N} \| R_3(q) \cdot x_k - u_k \|^2 \\
&= \sum_{k=1}^{N} \| q \star \mathbf{x}_k \star \bar{q} - \mathbf{u}_k \|^2 = \sum_{k=1}^{N} \| q \star \mathbf{x}_k \star \bar{q} - \mathbf{u}_k \|^2 \|q\|^2 \\
&= \sum_{k=1}^{N} \| q \star \mathbf{x}_k - \mathbf{u}_k \star q \|^2 \quad \text{by Eq. (2.11): } \|p \star q\| = \|p\| \, \|q\| \\
&= \sum_{k=1}^{N} \| A(\mathbf{x}_k, \mathbf{u}_k) \cdot q \|^2 = \sum_{k=1}^{N} q \cdot A_k^{\,t} \cdot A_k \cdot q \\
&= \sum_{k=1}^{N} q \cdot B_k \cdot q = q \cdot B \cdot q \ .
\end{aligned}
\tag{8.2}
$$

Here we may write, for each k, the matrix $A(\mathbf{x}_k, \mathbf{u}_k)$ as

$$
A_k = \begin{bmatrix} 0 & -a_1 & -a_2 & -a_3 \\ a_1 & 0 & s_3 & -s_2 \\ a_2 & -s_3 & 0 & s_1 \\ a_3 & s_2 & -s_1 & 0 \end{bmatrix}_k ,
$$

(8.3)

where, with "a" for "antisymmetric" and "s" for "symmetric,"

$$
a_{\{1,2,3\}} = \{x_1 - u_1, \, x_2 - u_2, \, x_3 - u_3\} ,
$$

$$
s_{\{1,2,3\}} = \{x_1 + u_1, \, x_2 + u_2, \, x_3 + u_3\} ,
$$

and, again for each k,

$$B_k = A_k{}^{\mathrm{t}} \cdot A_k$$

$$= \begin{bmatrix} a_1{}^2 + a_2{}^2 + a_3{}^2 & a_3s_2 - a_2s_3 & a_1s_3 - a_3s_1 & a_2s_1 - a_1s_2 \\ a_3s_2 - a_2s_3 & a_1{}^2 + s_2{}^2 + s_3{}^2 & a_1a_2 - s_1s_2 & a_1a_3 - s_1s_3 \\ a_1s_3 - a_3s_1 & a_1a_2 - s_1s_2 & a_2{}^2 + s_1{}^2 + s_3{}^2 & a_2a_3 - s_2s_3 \\ a_2s_1 - a_1s_2 & a_1a_3 - s_1s_3 & a_2a_3 - s_2s_3 & a_3{}^2 + s_1{}^2 + s_2{}^2 \end{bmatrix}_k , \qquad (8.4)$$

and thus

$$B = \sum_{k=1}^{N} B_k . \qquad (8.5)$$

Using the full squared-difference minimization measure Eq. (8.1) requires the global minimal value, so the solution for the optimal quaternion in Eq. (8.2) is the eigenvector of the *minimal* eigenvalue of B in Eq. (8.5). This is the approach used by Faugeras and Hebert in the earliest application of the quaternion method to scene alignment of which we are aware.

While it is important to be aware of this method, from here onward we will focus on the alternative form exploiting only the nonconstant cross-term appearing in Eq. (8.1), as does most of the recent molecular structure literature. The cross-term requires the determination of the *maximal* eigenvalue rather than the *minimal* eigenvalue of the corresponding data matrix. Direct numerical calculation verifies that, though the minimal eigenvalue of Eq. (8.5) differs from the maximal eigenvalue of the cross-term approach, the exact same optimal eigenvector is obtained, a result that can presumably be proven algebraically but that we will not need to pursue here.

Converting from least-squares minimization to cross-term maximization. We choose from here onward to focus on an equivalent method based on expanding the measure given in Eq. (8.1), removing the constant terms, and recasting the RMSD least-squares minimization problem as the task of maximizing the surviving cross-term expression (Davenport, 1968; Horn, 1987; Diamond, 1988; Kearsley, 1989; Kneller, 1991). For D spatial dimensions, this takes the general form

$$\Delta_D = \sum_{k=1}^{N} (R_D \cdot x_k) \cdot u_k = \sum_{a=1,b=1}^{D} [R_D]_{ba} E_{ab} = \mathrm{tr}\, R_D \cdot E , \qquad (8.6)$$

where

$$E_{ab} = \sum_{k=1}^{N} [x_k]_a [u_k]_b = [\mathbf{X} \cdot \mathbf{U}^{\mathrm{t}}]_{ab} \qquad (8.7)$$

is the *cross-covariance matrix* of the data, $[x_k]$ denotes the D-dimensional vector in the kth column of $\mathbf{X}$, and the range of the indices (a, b) is the spatial dimension D. The SVD and matrix-square-root methods (Schönemann, 1966; Golub and van Loan, 1983; Horn et al., 1988) can be used in any dimension to obtain the optimal aligning rotation directly from E_{ab}, but from here on we will focus on the quaternion eigensystem method for 3D data.

Quaternion transform from the 3D cross-term to the profile matrix. We now restrict our attention to the *3D cross-term form* of Eq. (8.6) with pairs of 3D point data related by a proper rotation. The key step is to substitute Eq. (2.14) for $R(q)$ into Eq. (8.6), and pull out the terms corresponding to pairs of components of the quaternions q. In this way the 3D expression is transformed into the 4×4 matrix $M(E)$ sandwiched between two identical quaternions (*not* a conjugate pair), of the form

$$\Delta(q) = \operatorname{tr} R(q) \cdot E = [q_0, q_1, q_2, q_3] \cdot M(E) \cdot [q_0, q_1, q_2, q_3]^t \equiv q \cdot M(E) \cdot q . \tag{8.8}$$

Here $M(E)$ is the traceless, symmetric 4×4 matrix

$$M(E) = \begin{bmatrix} E_{xx} + E_{yy} + E_{zz} & E_{yz} - E_{zy} & E_{zx} - E_{xz} & E_{xy} - E_{yx} \\ E_{yz} - E_{zy} & E_{xx} - E_{yy} - E_{zz} & E_{xy} + E_{yx} & E_{zx} + E_{xz} \\ E_{zx} - E_{xz} & E_{xy} + E_{yx} & -E_{xx} + E_{yy} - E_{zz} & E_{yz} + E_{zy} \\ E_{xy} - E_{yx} & E_{zx} + E_{xz} & E_{yz} + E_{zy} & -E_{xx} - E_{yy} + E_{zz} \end{bmatrix} \tag{8.9}$$

built from our original 3×3 cross-covariance matrix E_{ab} defined by Eq. (8.7) and using index labels $a, b = \{x, y, z\}$. We will refer to $M(E)$ from here on as the *profile matrix*, as it essentially reveals key features of the optimization function and its relationship to the matrix E. Note that in some literature, matrices related to the cross-covariance matrix E may be referred to as "attitude profile matrices," and one also may see the term "key matrix" referring to $M(E)$.

The bottom line is that if one decomposes Eq. (8.9) into its eigensystem, the measure Eq. (8.8) is maximized when the unit-length quaternion vector q is the eigenvector of $M(E)$'s largest eigenvalue. The RMSD optimal-rotation problem thus reduces to finding the maximal eigenvalue λ_{opt} of $M(E)$ (which we emphasize depends only on the numerical data). Plugging the corresponding eigenvector q_{opt} into Eq. (2.14), we obtain the rotation matrix $R(q_{\mathrm{opt}})$ that solves the problem. The resulting proximity measure relating $\{x_k\}$ and $\{u_k\}$ is simply

$$\left. \begin{aligned} \Delta_{\mathrm{opt}} &= q_{\mathrm{opt}} \cdot M(E) \cdot q_{\mathrm{opt}} \\ &= q_{\mathrm{opt}} \cdot (\lambda_{\mathrm{opt}} \, q_{\mathrm{opt}}) \\ &= \lambda_{\mathrm{opt}} \end{aligned} \right\} , \tag{8.10}$$

and does not require us to actually compute q_{opt} or $R(q_{\mathrm{opt}})$ explicitly if all we want to do is compare various test data sets to a reference structure.

---

‡ *Advanced: Notes on improper rotations and enantiomers.* In the interests of conceptual and notational simplicity, we have made a number of assumptions. For one thing, in declaring that Eq. (2.14) describes our sought-for rotation matrix, we have presumed that the optimal rotation matrix will always be a proper rotation, with det $R = +1$. Also, as mentioned, we have omitted any general translation problems, assuming that there is a way to translate each data set to an appropriate center, e.g., by subtracting the center of mass. The global translation optimization process is treated in Faugeras and Hebert (1986); Coutsias et al. (2004), and discussions of center-of-mass alignment, scaling, and point weighting are given in much of the original literature; see, e.g., Horn (1987); Coutsias et al. (2004); Theobald (2005). Finally, in real problems, structures such as molecules may appear in mirror-image or enantiomer form, and such issues were introduced early on by Kabsch (1976, 1978). There can also be particular symmetries, or very close approximations to symmetries, that can make some of our natural assumptions about the good behavior of the profile matrix invalid, and many of these issues, including ways to treat degenerate cases, have been carefully studied; see, e.g., Coutsias et al. (2004);

Coutsias and Wester (2019). The latter authors also point out that if a particular data set $M(E)$ produces a negative smallest eigenvalue λ_4 such that $|\lambda_4| > \lambda_{\text{opt}}$, this can be a sign of a reflected match, and the *negative* rotation matrix $R_{\text{opt}} = -R(q(\lambda_4))$ may actually produce the best alignment. These considerations may be essential in some applications, and concerned readers are referred to the original literature for details.

Illustrative example. We can visualize the transition from the initial data $\Delta(q_{\text{ID}}) = \text{tr } E$ to the optimal alignment $\Delta(q_{\text{opt}}) = \lambda_{\text{opt}}$ by exploiting the `slerp` geodesic interpolation Eq. (2.19) from the identity quaternion q_{ID} to q_{opt} given by

$$q(s) = \text{slerp}(q_{\text{ID}}, q_{\text{opt}}, s) \,, \tag{8.11}$$

and applying the resulting rotation matrix $R(q(s))$ to the test data, ending with $R(q_{\text{opt}})$ showing the best alignment of the two data sets. In Fig. 8.2, we show a sample reference data set in red, a sample test data set in blue connected to the reference data set by blue lines, an intermediate partial alignment, and finally the optimally aligned pair. The yellow arrow is the *spatial part of the quaternion solution*, proportional to the eigenvector $\hat{\mathbf{n}}$ (fixed axis) of the optimal 3D rotation matrix $R(q) = R(\theta, \hat{\mathbf{n}})$, and whose length is $\sin(\theta/2)$, sine of half the rotation angle needed to perform the optimal alignment of the test data with the reference data. In Fig. 8.3, we visualize the optimization process in an alternative way, showing random samples of $q = (q_0, \mathbf{q})$ in $\mathbf{S}^3$, separated into the "Northern hemisphere" 3D unit-radius ball in (a) with $q_0 \geqslant 0$, and the "Southern hemisphere" 3D unit-radius ball in (b) with $q_0 < 0$. Statistically, each ball contains a similar distribution covering just one set of rotations. The dot sizes of the randomly distributed quaternions are *scaled* by $\Delta(q) = \text{tr } R(q) \cdot E$ and located at their corresponding spatial ("real") quaternion points $\mathbf{q}$ in the solid balls. The yellow arrows, equivalent negatives of each other, show the spatial part $\mathbf{q}_{\text{opt}}$ of the optimal quaternion q_{opt}, and the tips of the arrows clearly fall in the middle of the mirror pair of clusters of the largest values of $\Delta(q)$. Note that the lower left dots in (a) continue smoothly into the larger lower left dots in (b), which is the center of the optimal quaternion in (b), and vice versa for the upper right dots in (b). Further details of such methods of displaying quaternions are discussed in Chapter 4.

8.4 Algebraic solution of the eigensystem for 3D spatial alignment

At this point, one can simply use the traditional *numerical* methods to solve Eq. (8.8) for the maximal eigenvalue λ_{opt} of $M(E)$ and its eigenvector q_{opt}, thus solving the 3D spatial alignment problem of Eq. (8.6). Alternatively, we can also exploit *symbolic* methods to study the properties of the eigensystems of 4×4 matrices M algebraically to provide deeper insights into the structure of the problem, and that is the subject of this section. Additional detail and features for matrices more general than the profile matrix M that is our focus here are supplied in Appendix J.

Theoretically, the algebraic form of our eigensystem is a textbook problem following from the 16th-century-era solution of the quartic algebraic equation given, e.g., in Abramowitz and Stegun (1970). Our objective here is to explore this textbook solution in the specific context of its application to eigensystems of 4×4 matrices and its behavior relative to the properties of such matrices. The real, symmetric,

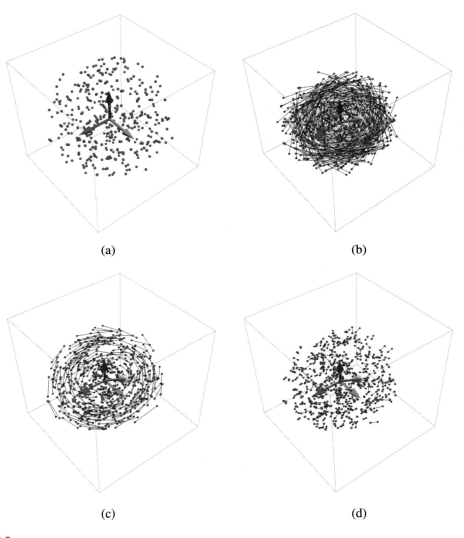

(a)   (b)

(c)   (d)

FIGURE 8.2

(a) A typical 3D spatial reference data set. (b) The reference data in red alongside the test data in blue, with blue lines representing the Euclidean distances connecting each test data point with its corresponding reference point. (c) Using Eq. (8.11) to obtain partial alignment at $s = 0.75$. (d) The optimal alignment for this data set at $s = 1.0$. The yellow arrow is the axis of rotation specified by the optimal quaternion's spatial components.

traceless profile matrix $M(E)$ in Eq. (8.9) appearing in the 3D spatial RMSD optimization problem must necessarily possess only real eigenvalues, and the properties of $M(E)$ permit some particular simplifications in the algebraic solutions that we will discuss. The quaternion RMSD literature varies

q ≥ 0   q < 0

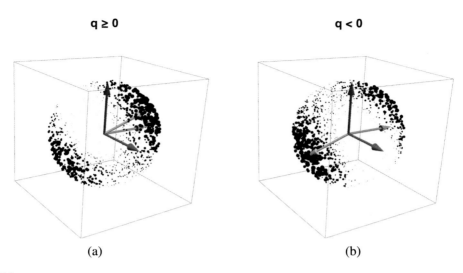

(a)   (b)

FIGURE 8.3

The values of $\Delta(q) = \operatorname{tr} R(q) \cdot E = q \cdot M(E) \cdot q$ represented by the sizes of the dots placed randomly in the "Northern" and "Southern" 3D solid balls spanning the entire hypersphere $\mathbf{S}^3$ with (a) containing the $q_0 \geqslant 0$ sector and (b) containing the $q_0 < 0$ sector. We display the data dots at the locations of their spatial quaternion components $\mathbf{q} = (q_1, q_2, q_3)$, and we know that $q_0 = \pm\sqrt{1 - \mathbf{q} \cdot \mathbf{q}}$ so the $\mathbf{q}$ data uniquely specify the full quaternion. Since $R(q) = R(-q)$, the points in *each* ball actually represent all possible unique rotation matrices. The spatial component of the maximal eigenvector is shown by the yellow arrows, which clearly end in the middle of the maximum values of $\Delta(q)$. Note that, in the quaternion context, diametrically opposite points on the spherical surface are identical rotations, so the cluster of larger dots at the upper right of (a) is, in the entire sphere, representing the same data as the "diametrically opposite" lower left cluster in (b), both surrounding the tips of their own yellow arrows. The smaller dots at the upper right of (b) are contiguous with the upper right region of (a), forming a single cloud centered on $\mathbf{q}_{\mathrm{opt}}$, and similarly for the lower left of (a) and the lower left of (b). The whole figure contains two distinct clusters of dots (related roughly by $q \rightarrow -q$) centered around $\pm\mathbf{q}_{\mathrm{opt}}$.

widely in the details of its treatment of the algebraic solutions, ranging from no discussion at all, to Horn, who mentions the possibility but does not explore it, to the work of Coutsias et al. (2004); Coutsias and Wester (2019), who present an exhaustive treatment, in addition to working out the exact details of the correspondence between the SVD eigensystem and the quaternion eigensystem, both of which in principle embody the algebraic solution to the RMSD optimization problem. (See Appendix N for more details.) In addition to the treatment of Coutsias et al., other approaches similar to the one we will study are due to Euler (Euler, 1733; Bell, 2008 (1733)), as well as a series of papers on the quartic by Nickalls (1993, 2009).

Eigenvalue expressions. We begin by writing down the determinant of the characteristic equation of the profile matrix, which by definition produces a polynomial whose solutions are the eigenvalues of the problem. The eigenvalue equation is

$$\det[M - eI_4] = e^4 + e^3 p_1 + e^2 p_2 + e p_3 + p_4 = 0, \tag{8.12}$$

where e denotes a generic eigenvalue, I_4 is the 4D identity matrix, and the p_k are homogeneous polynomials of degree k in the elements of M. For the special case of a traceless, symmetric profile matrix $M(E)$ defined by Eq. (8.9), the $p_k(E)$ coefficients simplify and can be expressed as the following functions either of M or of E:

$$
\left.
\begin{aligned}
p_1(E) &= -\operatorname{tr}[M] = 0 \\
p_2(E) &= -\frac{1}{2}\operatorname{tr}[M \cdot M] = -2\operatorname{tr}[E \cdot E^{\mathrm{t}}] \\
&= -2\left(E_{xx}^2 + E_{xy}^2 + E_{xz}^2 + E_{yx}^2 + E_{yy}^2 + E_{yz}^2 + E_{zx}^2 + E_{zy}^2 + E_{zz}^2\right) \\
p_3(E) &= -\frac{1}{3}\operatorname{tr}[M \cdot M \cdot M] = -8\det[E] \\
&= 8\left(E_{xx} E_{yz} E_{zy} + E_{yy} E_{xz} E_{zx} + E_{zz} E_{xy} E_{yx}\right) \\
&\quad - 8\left(E_{xx} E_{yy} E_{zz} + E_{xy} E_{yz} E_{zx} + E_{xz} E_{zy} E_{yx}\right) \\
p_4(E) &= \det[M] = 2\operatorname{tr}[E \cdot E^{\mathrm{t}} \cdot E \cdot E^{\mathrm{t}}] - \left(\operatorname{tr}[E \cdot E^{\mathrm{t}}]\right)^2
\end{aligned}
\right\} . \tag{8.13}
$$

Interestingly, the polynomial $M(E)$ is arranged so that $-p_2(E)/2$ is the (squared) Frobenius norm of E, and $-p_3(E)/8$ is its determinant. Our task now is to express the four eigenvalues $e = \lambda_k(p_1, p_2, p_3, p_4)$, $k = 1, \ldots, 4$, usefully in terms of the matrix elements, and also to find their eigenvectors; we are of course particularly interested in the maximal eigenvalue λ_{opt}.

Rotation invariance. The forms in Eq. (8.13) have some important properties in certain cases. In most realistic cases, there is a list of distinct sample data sets, with each instance having target data related to its reference data by a rotation, *and* the target data are somewhat noisy, so that they do not correspond exactly to a pure rotation applied to the reference data. However, a particularly interesting feature appears if there is no noise, or if the noise is essentially negligible. In this case, the self-covariance of the reference data completely determines the optimal eigenvalue of the profile matrix for each sample, and, since the solution's loss function for each sample is exactly the eigenvalue, data sets with the same reference data cannot be distinguished. The *eigenvectors* corresponding to one single reference set's eigenvalue are of course distinct, and, treated as quaternions, they determine the optimal rotation that must be applied to align the reference data with each distinct target sample.

This feature of the "ideal" noise-free situation can be proven by looking at Eq. (8.13) and noting first that $-p_2(E)/2$ is the (squared) Frobenius norm of E, and $-p_3(E)/8$ is its determinant, while p_4 is quadratic in the matrix E. If we now write the sample data array to explicitly show its derivation from the reference data, that is, $u_k{}^a = R(q)_{ab}x_k{}^b$ for all $k \in \{1, \ldots K\}$, then the cross-covariance matrix becomes

$$E_{ab} = \sum_{k=1}^{N} [x_k]_a R(q)_{bc} [x_k]_c = \left[R(q)_{bc}\mathbf{X}_c \cdot \mathbf{X}_a^{\mathrm{t}}\right]. \tag{8.14}$$

We now take the three distinct components of Eq. (8.13) and write them out as

$$
\left.\begin{aligned}
\det[E] &= \det \sum_k R_{bc} x_k{}^c x_k{}^a \\
&= \det R_{bc} \sum_k x_k{}^c x_k{}^a \\
&= \det[R] \det E[X, X] \equiv \det E_0 \\
\operatorname{tr}[E \cdot E^{\mathrm{t}}] &= \sum_{k,k'} R_{bc} x_k{}^c x_k{}^a R_{bd} x_{k'}{}^d x_{k'}{}^a \\
&= \sum_k x_k{}^c x_k{}^a \sum_{k'} x_{k'}{}^c x_{k'}{}^{ab} \\
&= \operatorname{tr} E_0 \cdot E_0{}^{\mathrm{t}} \\
\operatorname{tr}[E \cdot E^{\mathrm{t}} \cdot E \cdot E^{\mathrm{t}}] &= \operatorname{tr}[E_0 \cdot E_0{}^{\mathrm{t}} \cdot E_0 \cdot E_0{}^{\mathrm{t}}]
\end{aligned}\right\} . \tag{8.15}
$$

Thus we know that, modulo assuming irrelevant errors in the data, the entire characteristic equation Eq. (8.12) is independent of the rotation matrix applied to obtain the sample data, and thus the eigenvalues of the profile matrix $M(E)$ are independent of the applied rotations embodied in the sample data; only the self-covariance of the reference data

$$
E_0 = \mathbf{X} \cdot \mathbf{X}^{\mathrm{t}} \tag{8.16}
$$

enters into the determination of the eigenvalues of $M(E)$, and since E_0 is symmetric, the last three elements of the top row and left-hand column of $M(E_0)$ *vanish*, making the maximum eigenvalue just

$$
[\text{max eigenvalue}](M(E_0)) = \operatorname{tr} E_0 .
$$

Approaches to algebraic solutions. Eq. (8.12) can be solved directly using the quartic equations published by Cardano in 1545 (see, e.g., Abramowitz and Stegun (1970); Weisstein (2019b); Wikipedia (2023a)), which are incorporated into the Mathematica function

$$
\texttt{Solve[myQuarticEqn[e] == 0, e, Quartics} \to \texttt{True]} \tag{8.17}
$$

that immediately returns a suitable algebraic formula. At this point we defer detailed discussion of the general solution to Appendix J, and instead focus on a particularly symmetric version of the solution and the form it takes for the eigenvalue problem for traceless, symmetric 4×4 matrices such as our profile matrices $M(E)$. For this purpose, we look for an alternative solution by considering the following traceless ($p_1 = 0$) Ansatz:

$$
\left.\begin{aligned}
\lambda_1(p) &\overset{?}{=} \sqrt{X(p)} + \sqrt{Y(p)} + \sqrt{Z(p)} & \lambda_2(p) &\overset{?}{=} \sqrt{X(p)} - \sqrt{Y(p)} - \sqrt{Z(p)}) \\[2mm]
\lambda_3(p) &\overset{?}{=} -\sqrt{X(p)} + \sqrt{Y(p)} - \sqrt{Z(p)} & \lambda_4(p) &\overset{?}{=} -\sqrt{X(p)} - \sqrt{Y(p)} + \sqrt{Z(p)}
\end{aligned}\right\} . \tag{8.18}
$$

This form emphasizes some additional explicit symmetries that are connected to the SVD method and the role of cube roots in the quartic algebraic solutions (see, e.g., Coutsias and Wester (2019) and Appendix N) . We can turn it into an equation for $\lambda_k(p)$ to be solved in terms of the matrix parameters

$p_k(E)$ as follows: First we eliminate e by expanding the eigenvalue polynomial

$$(e - \lambda_1)(e - \lambda_2)(e - \lambda_3)(e - \lambda_4) = 0$$

in a power series to express the matrix data expressions p_k directly in terms of totally symmetric polynomials of the eigenvalues in the form (Abramowitz and Stegun, 1970)

$$\left.\begin{aligned}
p_1 &= -\lambda_1 - \lambda_2 - \lambda_3 - \lambda_4 \\
p_2 &= \lambda_1\lambda_2 + \lambda_1\lambda_3 + \lambda_2\lambda_3 + \lambda_1\lambda_4 + \lambda_2\lambda_4 + \lambda_3\lambda_4 \\
p_3 &= -\lambda_1\lambda_2\lambda_3 - \lambda_1\lambda_2\lambda_4 - \lambda_1\lambda_3\lambda_4 - \lambda_2\lambda_3\lambda_4 \\
p_4 &= \lambda_1\lambda_2\lambda_3\lambda_4
\end{aligned}\right\} . \tag{8.19}$$

Next we substitute our expression Eq. (8.18) for the λ_k in terms of the $\{X, Y, Z\}$ functions into Eq. (8.19), yielding a completely different alternative to Eq. (8.12) that *also* will solve the 3D RMSD eigenvalue problem if we can invert it to express $\{X(p), Y(p), Z)p)\}$ in terms of the data $p_k(E)$ as presented in Eq. (8.13):

$$\left.\begin{aligned}
p_1 &= 0 \\
p_2 &= -2(X + Y + Z) \\
p_3 &= -8\sqrt{XYZ} \\
p_4 &= X^2 + Y^2 + Z^2 - 2(YZ + ZX + XY)
\end{aligned}\right\} . \tag{8.20}$$

We already see the critical property in p_3 that, while p_3 itself has a deterministic sign from the matrix data, the possibly variable signs of the square roots in Eq. (8.18) have to be constrained so their product $\sqrt{XYZ}$ agrees with the sign of p_3. Manipulating the quartic equation solutions that we can obtain by applying the library function Eq. (8.17) to Eq. (8.20), and restricting our domain to real traceless, symmetric matrices (and hence real eigenvalues), we find solutions for $X(p)$, $Y(p)$, and $Z(p)$ of the following form:

$$F_f(0, p_2, p_3, p_4) = +\frac{1}{6}\left(r(p)\cos_f(p) - p_2\right) . \tag{8.21}$$

Here $f = (x, y, z)$ corresponds to F_f being $X(p)$, $Y(p)$, or $Z(p)$, and the $\cos_f(p)$ terms share the fundamental angle $\Phi = \arg(a + ib)$, differing only by a cube-root phase:

$$\cos_x(p) = \cos\left(\frac{\Phi}{3}\right) , \quad \cos_y(p) = \cos\left(\frac{\Phi}{3} - \frac{2\pi}{3}\right) , \quad \cos_z(p) = \cos\left(\frac{\Phi}{3} + \frac{2\pi}{3}\right) . \tag{8.22}$$

For the sake of implementers, we recall that $\arg(a + ib) = \text{atan2}(b, a)$ in the C mathematics library, or ArcTan$[a, b]$ in Mathematica. The specific utility functions appearing in the equations for our traceless

$p_1 = 0$ case are

$$
\left.
\begin{aligned}
r^2(0, p_2, p_3, p_4) &= p_2{}^2 + 12p_4 = \sqrt[3]{a^2 + b^2} = (a + ib)^{1/3}(a - ib)^{1/3} \\
a(0, p_2, p_3, p_4) &= p_2{}^3 + \tfrac{1}{2}\left(27p_3{}^2 - 72p_2 p_4\right) \\
b^2(0, p_2, p_3, p_4) &= r^6(p) - a^2(p) \\
&= \frac{27}{4}\left(16p_4 p_2{}^4 - 4p_3{}^2 p_2{}^3 - 128p_4{}^2 p_2{}^2 + 144p_3{}^2 p_4 p_2 - 27p_3{}^4 + 256p_4{}^3\right)
\end{aligned}
\right\}.
\tag{8.23}
$$

The function $b^2(p)$ has the essential property that, for real solutions to the cubic, which imply the required real solutions to our eigenvalue equations (Abramowitz and Stegun, 1970), we must have $b^2(p) \geqslant 0$. That important property allowed us to convert the bare solution into terms involving $\{(a + ib)^{1/3}, (a - ib)^{1/3}\}$ whose sums form the manifestly real cube-root-related cosine terms in Eq. (8.22).

Final eigenvalue algorithm. While Eqs. (8.21) and (8.22) are well defined, square roots must be taken to finish the computation of the eigenvalues postulated in Eq. (8.18). In our special case of symmetric, traceless matrices such as $M(E)$, we can always choose the signs of the first two square roots to be positive, but the sign of the $\sqrt{Z}$ term is nontrivial, and in fact is the sign of $\det[E]$. The form of the solution in Eqs. (8.18) and (8.21) that works specifically for all traceless symmetric matrices such as $M(E)$ is given by our equations for $p_k(E)$ in Eq. (8.13), along with Eqs. (8.21)–(8.23), *provided* we modify Eq. (8.18) using $\sigma(p) = \mathrm{sign}\,(\det[E]) = \mathrm{sign}\,(-p_3)$ as follows:

$$
\left.
\begin{aligned}
\lambda_1(p) &= \sqrt{X(p)} + \sqrt{Y(p)} + \sigma(p)\sqrt{Z(p)} & \lambda_2(p) &= \sqrt{X(p)} - \sqrt{Y(p)} - \sigma(p)\sqrt{Z(p)} \\
\lambda_3(p) &= -\sqrt{X(p)} + \sqrt{Y(p)} - \sigma(p)\sqrt{Z(p)} & \lambda_4(p) &= -\sqrt{X(p)} - \sqrt{Y(p)} + \sigma(p)\sqrt{Z(p)}
\end{aligned}
\right\}.
\tag{8.24}
$$

The particular order of the numerical eigenvalues in our chosen form of the solution Eq. (8.24) is found in regular cases to be uniformly nonincreasing in numerical order for our $M(E)$ matrices, so $\lambda_1(p)$ is always the leading eigenvalue. This is our preferred symbolic version of the solution to the 3D RMSD problem defined by $M(E)$.

Remark: We have experimentally confirmed the numerical behavior of Eq. (8.21) in Eq. (8.24) with 1,000,000 randomly generated sets of 3D cross-covariance matrices E, along with the corresponding profile matrices $M(E)$, producing numerical values of p_k inserted into the equations for $X(p)$, $Y(p)$, and $Z(p)$. We confirmed that the sign of $\sigma(p)$ varied randomly, and found that the algebraically computed values of $\lambda_k(p)$ corresponded to the standard numerical eigenvalues of the matrices $M(E)$ in all cases, to within expected variations due to numerical evaluation behavior and expected occasional instabilities. In particular, we found a maximum per-eigenvalue discrepancy of about 10^{-13} for the *algebraic* methods relative to the standard *numerical* eigenvalue methods, and a median difference of 10^{-15}, in the context of machine precision of about 10^{-16}.

Eigenvectors for 3D data via the adjugate. The eigenvectors corresponding to any eigenvalue $\lambda(E)$ can be computed in a variety of ways. The brute force method is to write down the characteristic matrix

$\chi(E)$ whose vanishing determinant solved the eigenvalue equation,

$$\chi(E) = [M(E) - \lambda(E)I_4] \,, \tag{8.25}$$

take a generic unnormalized eigenvector, and choose one element to be unity, e.g., $\mathbf{v} = [1, \ v_2, \ v_3, \ v_4]$. All that remains is to solve the corresponding eigenvector equation,

$$\chi(E) \cdot \mathbf{v} = [M(E) \cdot \mathbf{v} - \lambda \, \mathbf{v}] = 0 \,, \tag{8.26}$$

by Cramer's rule directly for the three remaining elements $[v_2, \ v_3, \ v_4]$ of $\mathbf{v}$ as a function of a selected eigenvalue λ. This fails if the chosen element vanishes, and thus cannot be set to unity; the algorithm then chooses another element as unity, and repeats as necessary to find a valid normalizable eigenvector.

However, we prefer an appealing alternative solution to this standard problem of finding a valid eigenvector, namely the *adjugate matrix method*, which we will encounter in several contexts in subsequent chapters. Previewing that method here specifically for cloud-matching, we simply take the characteristic equation Eq. (8.25) of the eigenvalue problem, and compute its 4×4 *adjugate matrix* directly for any valid eigenvalue λ, so we have

$$\text{Adj}(E) = (\text{adjugate})(\chi(E)) = (\text{adjugate})[M(E) - \lambda(E)I_4] \,. \tag{8.27}$$

Then $\chi(E) \cdot \text{Adj}(E) = \det(\chi(E)) = 0$, where $\det(\chi(E)) = 0$ is true by definition for any valid eigenvalue. Expanding in the columns of $\text{Adj}(E)$, we see that the columns of $\text{Adj}(E)$ embody four unnormalized copies of the same eigenvector of λ. Any failure of the usual algorithm to allow setting a given component of $\mathbf{v}$ to unity is reflected in an unnormalizable column of $\text{Adj}(E)$, so another column must be chosen. We therefore incorporate our four copies of the eigenvector of our optimal eigenvalue into the 4×4 matrix $A(\chi) = -\text{Adj}(E)$, where (since the sign of an unnormalized eigenvector is inconsequential) we choose the minus sign to align with later appearances of this eigenvector system. $A(\chi)$ thus defines *four identical but independently scaled* (and possibly vanishing) multiples of the eigenvector $q_{\text{opt}} = [q_0, q_1, q_2.q_3]$, namely the columns

$$A(\chi) = \begin{bmatrix} A_1 & A_2 & A_3 & A_4 \end{bmatrix}$$

obeying the eigensystems

$$\chi \cdot A_i = M \cdot A_i - \lambda_{\text{opt}} A_i = \det \chi = 0 \,,$$

for $i = (1, 2, 3, 4)$. However, since the profile matrix M is *symmetric*, then the adjugate matrix $A(\chi)$ is *also symmetric*, and each of the four *rows* of $A(\chi)$ is also proportional to the *same eigenvector* as the columns. Therefore $A(\chi)$ is a symmetric rank 1 matrix, and every row and column is proportional to the eigenvector q_{opt}. Hence there is no choice: the adjugate matrix of $\chi(\lambda_{\text{opt}})$ must be a symmetric matrix with each row being the eigenvector q_{opt} multiplied by each column's eigenvector components, or, dropping an arbitrary overall constant, we must have

$$A(\chi) = \left\{ q_0 \begin{bmatrix} q_0 \\ q_1 \\ q_2 \\ q_3 \end{bmatrix}, \ q_1 \begin{bmatrix} q_0 \\ q_1 \\ q_2 \\ q_3 \end{bmatrix}, \ q_2 \begin{bmatrix} q_0 \\ q_1 \\ q_2 \\ q_3 \end{bmatrix}, \ q_3 \begin{bmatrix} q_0 \\ q_1 \\ q_2 \\ q_3 \end{bmatrix} \right\} \,. \tag{8.28}$$

It is obvious now that we can define the quadratic "adjugate quaternion variables" $q_{ij} = q_i q_j$, drop any overall constant, and define the solution to the eigensystem of the profile matrix $M(E)$ from the adjugate matrix of Eq. (8.25), computed directly from the problem's input data from the cross-covariance matrix E,

$$\text{adjugate variable system: } A(\chi(E)) = \begin{bmatrix} q_{00} & q_{01} & q_{02} & q_{03} \\ q_{01} & q_{11} & q_{12} & q_{13} \\ q_{02} & q_{12} & q_{22} & q_{23} \\ q_{03} & q_{13} & q_{23} & q_{33} \end{bmatrix}. \tag{8.29}$$

We select our optimal quaternion q_{opt} by normalizing the column with the largest diagonal in Eq. (8.29), and construct the optimal rotation matrix by computing $R_{\text{opt}} = R(q_{\text{opt}})$. While we used the context of 3D cloud matching here to produce the adjugate matrix Eq. (8.29), we will elaborate on this construction in several alternative contexts, including Chapter 9 and Chapter 12.

Note that this solution can often have $q_0 < 0$, and that whenever the problem in question depends on the sign of q_0, such as calculating a `slerp` starting at q_{ID}, one should choose the sign of q appropriately to give the desired sign of q_0; some applications may also require an element of statistical randomness, in which case one might randomly pick a sign for q.

Remark: Yet another approach to computing eigenvectors that, surprisingly, *almost* entirely avoids any reference to the original matrix, but needs only its eigenvalues and minor eigenvalues, has recently been rescued from relative obscurity by Denton et al. (2019). (The authors uncovered a long list of noncross-citing literature mentioning the result dating back at least to 1934.) If for a real, symmetric 4×4 matrix M we label the set of four eigenvectors v_i by the index i and the four components of any single such 4-vector by a, the squares of each of the 16 corresponding components take the form

$$\left([v_i]_a\right)^2 = \frac{\prod_{j=1}^{3} \left(\lambda_i(M) - \lambda_j(\mu_a)\right)}{\prod_{k=1; k \neq i}^{4} \left(\lambda_i(M) - \lambda_k(M)\right)}. \tag{8.30}$$

Here the μ_a are the 3×3 minors obtained by removing the ath row and column of M, and the $\lambda_j(\mu_a)$ comprise the list of three eigenvalues of each of these minors. Attempting to obtain the eigenvectors by taking square roots is of course hampered by the nondeterministic sign; however, since the eigenvalues $\lambda_i(M)$ are known, and the overall sign of each eigenvector v_i is arbitrary, one needs to check at most eight sign combinations to find the one for which $M \cdot v_i = \lambda_i(M) v_i$, solving the problem. Note that the general formula extends to Hermitian matrices of any dimension.

8.5 Conclusion

Our objective has been to explore quaternion-based treatments of the RMSD data-comparison problem as developed in the work of Davenport (1968), Sansò (1973), Faugeras and Hebert (1983), Horn (1987),

Diamond (1988), Kearsley (1989), and Kneller (1991), among others, and to bring attention to the exact algebraic solutions. We studied the intrinsic properties of the RMSD problem for comparing spatial data in quaternion-accessible domains, and we examined the nature of the solutions for the eigensystems of the 3D spatial RMSD problem. In the following chapters, we show how to effectively discover a quaternion from a rotation matrix, find methods to solve the corresponding 3D quaternion orientation-frame alignment problem, and examine solutions for the combined 3D spatial and orientation-frame RMSD problem. This entire set of problems is then extended to 4D in Chapter 20.

Quaternion of a 3D rotation from the Bar-Itzhack method☆

9.1 Classic methods for finding the quaternion of a 3D rotation

A standard problem that has been around ever since the relation between quaternions and 3D rotations was discovered is the task of finding the quaternion q, given a particular numerical rotation matrix $R(m)$, such that $R(q) = R(m)$. If the rotation matrix is well defined, orthonormal, and with negligible measurement error, the corresponding quaternion is unique up to the usual sign ambiguity, and must exist. However, it has been known for a very long time that this extraction required special care because of possible divide-by-zero circumstances that could arise. The classic algorithm used to solve this problem reduces to the task of recovering the best axis-angle parameters describing $R(m)$. This is a subtle multi-step process (see, e.g., Shepperd (1978); Shuster and Natanson (1993); Sarabandi and Thomas (2019); Section 16.1 of Hanson (2006)). In order to account for all possible anomalies, however rare, the classical procedure must check for assorted zeroes, conducting several separate checks for small numbers. A description of the explicit classical algorithm and sample implementations are given in Appendix G.

This algorithm reliably generates the axis-angle parameters $\cos\theta$, $\sin\theta$, and $\hat{\mathbf{n}}$ needed to extract the values of $q(\theta, \hat{\mathbf{n}})$ from the rotation matrix $R(m)$ provided that it has *accurate* numerical data, i.e., it is perfectly described by the form of $R(\theta, \hat{\mathbf{n}})$ in Eq. (2.17).

Our purpose in this section is to present an algorithm that determines the *optimal* quaternion q_{opt} corresponding to a measured matrix $R(m)$ that may be only an *approximation* to a rotation. Then the corresponding exactly orthonormal 3D rotation matrix $R(q_{\text{opt}})$ is the best possible approximation for an imperfectly defined $R(m)$. This objective can be achieved using the variational method of Bar-Itzhack (2000). We note that the Bar-Itzhack method can be viewed either as a variant of the RMSD cloud matching algorithm of Chapter 8, with a transposed rotation matrix substituted for the cross-covariance matrix $E(\mathbf{X}, \mathbf{U})$, or as the single-rotation version of the rotation averaging methods to be explored in Chapter 18.

The adjugate matrix and eigenvectors in cloud matching. The standard method for finding the optimal quaternion corresponding to a rotation aligning two 3D point clouds uses the quaternion eigenvector corresponding to the maximal eigenvalue of the 4×4 *profile matrix M* as detailed in Chapter 8 (see also, e.g., Horn, 1987; Hanson, 2020). There we chose the adjugate matrix approach as a reliable method of extracting the optimal eigenvector from the profile matrix M. Given the characteristic matrix

---

☆ This chapter derives some of its content from Hanson (2020).

χ of M,

$$\chi = [M - \lambda\, I_4]\,, \tag{9.1}$$

and the maximal eigenvalue λ_{opt} solving $\det \chi = 0$, we determined that an appropriate quaternion eigenvector q_{opt} can always be selected from columns of the adjugate matrix $A(\chi(M, \lambda_{\mathrm{opt}}))$. The subtlety in this procedure is that it is the *normalization* step that introduces the possible singularities that are avoided by the Shepperd algorithm supplied in Appendix G. In the next section, we will see an alternative path to the adjugate variables in the process of using rotation matrices themselves as optimization targets.

9.2 The Bar-Itzhack approach

While the algorithm in Appendix G is reliable and widely used for perfect input rotation data, we will now describe a more versatile and elegant method, applicable also to noisy rotation input data, that we first became aware of from the work of Itzhack Bar-Itzhack (2000). The basic insight of this approach is that the same RMSD profile matrix method that we used in Chapter 8 to find the optimal quaternion rotation for a 3D cloud matching problem, typically using the adjugate procedure, can be adapted to find the best quaternion approximating an isolated noisy rotation.

Bar-Itzhack observed that if we simply replace the cross-covariance data matrix E_{ab} in the 3D quaternion-based matching problem by the transpose $Q^{\mathrm{t}}(m)$ of a noisy, roughly orthogonal numerical rotation matrix, the perfect quaternion q_{opt} whose $R_{\mathrm{opt}} = R(q_{\mathrm{opt}})$ *most closely* corresponds to this numerical matrix $Q(m)$ can be found by solving our now-familiar maximal quaternion eigenvalue problem. The initially unknown optimal rotation matrix (technically its quaternion) computed by maximizing the similarity measure turns out to be equivalent to the solution of a single-element quaternion barycenter problem. In summary, our goal is to find a perfect quaternion q_{opt} that produces a pure rotation $R_{\mathrm{opt}} = R(q_{\mathrm{opt}})$ that is as close to $Q(m)$ as possible, even if $Q(m)$ is badly damaged by noise.

We start with the Frobenius measure describing the proximity of two rotation matrices corresponding to the quaternion q for the unknown quaternion and the numeric matrix $Q(m)$ containing the known 3×3 rotation matrix data:

$$\mathbf{S}_{\mathrm{BI}} = \|R(q) - Q(m)\|_{\mathrm{Frob}}^2 = \mathrm{tr}\left([R(q) - Q(m)] \cdot [R^{\mathrm{t}}(q) - Q^{\mathrm{t}}(m)]\right) \tag{9.2}$$

$$= \mathrm{tr}\left(I_3 + I_3 - 2\left(R(q) \cdot Q^{\mathrm{t}}(m)\right)\right) \tag{9.3}$$

$$= \mathrm{constant} - 2\,\mathrm{tr}\, R \cdot Q^{\mathrm{t}}\,. \tag{9.4}$$

Obviously the cross-covariance matrix E from Chapter 8 has automatically been replaced by the transpose of our target rotation matrix data. We convert Eq. (9.4) from a least-squares-style minimization task to a maximization task by pulling out the cross-term as usual, to arrive at

$$\Delta_{\mathrm{BI}} = \mathrm{tr}\, R(q) \cdot Q^{\mathrm{t}}(m) = q \cdot K(m) \cdot q\,, \tag{9.5}$$

where $Q(m)$ is (perhaps approximately) an orthogonal matrix of numerical data, and $K(m)$ is analogous to the profile matrix $M(E)$ from the 3D:3D cloud matching problem of Chapter 8. Now $R(q)$ is an abstract rotation matrix quadratic in q, which allowed the construction of the matrix form with $K(m)$ in

Eq. (9.5). The components m of the matrix $Q(m)$ are supposed to be a reasonable numerical approximation to a rotation matrix, and thus the product $T = R \cdot Q^t$ should also be a reasonable approximation to an **SO**(3) rotation matrix. Hence that product itself corresponds closely to some axis $\hat{\mathbf{n}}$ and angular difference θ, where (supposing we knew R's exact quaternion q_{opt})

$$\mathrm{tr}\, R(q_{opt}) \cdot Q(m)^t = \mathrm{tr}\, T(\theta, \hat{\mathbf{n}}) \approx 1 + 2\cos\theta\,.$$

The maximum should obviously occur when T is near the identity matrix, with the ideal value at $\theta = 0$, corresponding to $R \approx Q$. Thus if we find the maximal quaternion eigenvalue λ_{opt} of the profile matrix $K(m)$ in Eq. (9.5), our closest solution will be represented by the corresponding normalized eigenvector q_{opt},

$$q_{opt} = \text{eigenvector of maximal eigenvalue of } K(m)\,. \tag{9.6}$$

This numerical solution for q_{opt} will correspond to the pure rotation matrix $R_{opt} = R(q_{opt})$ closest to the targeted numerical rotation matrix, solving the problem. To complete the details of the computation, we replace the elements E_{ab} in Eq. (8.9) by encoding the elements m of $Q(m)$ in a more convenient form as matrix with columns $\mathbf{X} = (x_1, x_2, x_3)$, $\mathbf{Y}$, and $\mathbf{Z}$. Scaling by $1/3$, we obtain a 4×4 profile matrix K whose elements in terms of a known numerical matrix $Q = [\mathbf{X}|\mathbf{Y}|\mathbf{Z}]$ (transposed in the algebraic expression for K due to the Q^t) are

$$K(Q) = \frac{1}{3} \begin{bmatrix} x_1 + y_2 + z_3 & y_3 - z_2 & z_1 - x_3 & x_2 - y_1 \\ y_3 - z_2 & x_1 - y_2 - z_3 & x_2 + y_1 & x_3 + z_1 \\ z_1 - x_3 & x_2 + y_1 & -x_1 + y_2 - z_3 & y_3 + z_2 \\ x_2 - y_1 & x_3 + z_1 & y_3 + z_2 & -x_1 - y_2 + z_3 \end{bmatrix}. \tag{9.7}$$

Noise-free idealization. Determining the explicit algebraic eigensystem of Eq. (9.7) is a possible but nontrivial task. However, we learn a great deal by working through the ideal noise-free case, for which any orthogonal 3D rotation matrix can also be expressed in terms of quaternions via Eq. (2.14). To avoid confusion with q appearing in Eq. (9.5), we use a new quaternion r and replace the form $Q = [\mathbf{X}|\mathbf{Y}|\mathbf{Z}]$ that produced Eq. (9.7) by the pure rotation matrix $Q(r)$. This yields an alternate useful algebraic form

$$K(r) =$$
$$\frac{1}{3} \begin{bmatrix} 3r_0^2 - r_1^2 - r_2^2 - r_3^2 & 4r_0r_1 & 4r_0r_2 & 4r_0r_3 \\ 4r_0r_1 & -r_0^2 + 3r_1^2 - r_2^2 - r_3^2 & 4r_1r_2 & 4r_1r_3 \\ 4r_0r_2 & 4r_1r_2 & -r_0^2 - r_1^2 + 3r_2^2 - r_3^2 & 4r_2r_3 \\ 4r_0r_3 & 4r_1r_3 & 4r_2r_3 & -r_0^2 - r_1^2 - r_2^2 + 3r_3^2 \end{bmatrix}. \tag{9.8}$$

This equation then allows us to quickly prove for noise-free data that K has the correct properties to solve for the appropriate quaternion corresponding to $Q(r)$. First we note that, writing the eigensystem of $K(r)$ in terms of its characteristic equation $\chi(K)$ as

$$\det \chi(K(r)) = \det[K(r) - \lambda \times I_4] = 0 \tag{9.9}$$

$$\rightarrow \quad \lambda^4 + p_1\lambda^3 + p_2\lambda^2 + p_3\lambda + p_4 = 0\,, \tag{9.10}$$

we find that the coefficients p_n of the eigensystem in this error-free quaternion form are simply constants,

$$p_1 = 0, \qquad p_2 = -\tfrac{2}{3}, \qquad p_3 = -\tfrac{8}{27}, \qquad p_4 = -\tfrac{1}{27} \cdot$$

Computing the eigenvalues and eigenvectors using the symbolic quaternion form, we see that the eigenvalues are constant, with maximal eigenvalue exactly one, and the eigenvectors are almost trivial, with the maximal eigenvector being the quaternion r_{opt} that corresponds to the (numerical) rotation matrix:

$$\lambda = \{1, \ -\frac{1}{3}, \ -\frac{1}{3}, \ -\frac{1}{3}\}, \tag{9.11}$$

$$r_{opt} = \left\{ \begin{bmatrix} r_0 \\ r_1 \\ r_2 \\ r_3 \end{bmatrix}, \begin{bmatrix} -r_1 \\ r_0 \\ 0 \\ 0 \end{bmatrix}, \begin{bmatrix} -r_2 \\ 0 \\ r_0 \\ 0 \end{bmatrix}, \begin{bmatrix} -r_3 \\ 0 \\ 0 \\ r_0 \end{bmatrix} \right\}. \tag{9.12}$$

The first column is the quaternion r_{opt}, with $\Delta_{BI}(r_{opt}) = 1$. (This would be 3 if we had not divided by 3 in the definition of K.)

Alternate version. From the quaternion barycenter work of Markley et al. (2007), we know that Eq. (9.8) actually has a much simpler form with the same unit eigenvalue and natural quaternion eigenvector. If we simply take Eq. (9.8) multiplied by 3, add the constant term $I_4 = (r_0^2 + r_1^2 + r_2^2 + r_3^2)I_4$, and divide by 4, we get a more compact quaternion form of the matrix, namely

$$K'(r) = \begin{bmatrix} r_0^2 & r_0 r_1 & r_0 r_2 & r_0 r_3 \\ r_0 r_1 & r_1^2 & r_1 r_2 & r_1 r_3 \\ r_0 r_2 & r_1 r_2 & r_2^2 & r_2 r_3 \\ r_0 r_3 & r_1 r_3 & r_2 r_3 & r_3^2 \end{bmatrix}. \tag{9.13}$$

While this profile matrix could certainly be confused with the quaternion adjugate variable matrix in Chapter 8, Eq. (8.29), this is the *profile* matrix, *not* the eigenvector-encoding adjugate matrix. Due to the properties of eigensystem transformations, while Eq. (9.13) has different eigenvalues from the profile matrix of Eq. (9.8), its eigenvectors are *identical* to those of Eq. (9.12), and the full profile eigensystem has been transformed from $K(r)$ to $K'(r)$ with the result

$$\lambda = \{1, \ 0, \ 0, \ 0\}, \tag{9.14}$$

$$r_{opt} = \left\{ \begin{bmatrix} r_0 \\ r_1 \\ r_2 \\ r_3 \end{bmatrix}, \begin{bmatrix} -r_1 \\ r_0 \\ 0 \\ 0 \end{bmatrix}, \begin{bmatrix} -r_2 \\ 0 \\ r_0 \\ 0 \end{bmatrix}, \begin{bmatrix} -r_3 \\ 0 \\ 0 \\ r_0 \end{bmatrix} \right\}. \tag{9.15}$$

As elegant as this is, in practice, our numerical input data are from the 3×3 input rotation matrix itself, and not the ideal quaternions, so we will almost always just use the numbers in Eq. (9.7) to solve the numerical problem if any error is present.

The adjugate completion of the noise-free quaternion eigenvector solution. In typical noise-free applications, *the solution is immediate, requiring only trivial algebra.* The maximal eigenvalue is always known in advance to be unity for any valid rotation matrix, so to compute the quaternion eigenvector solution, we need only the (perfect) numerical matrix Eq. (9.7) with unit eigenvalue. We simply identify and normalize any nonsingular column of the adjugate matrix of the characteristic equation with (maximal) unit eigenvalue,

$$\chi(r) = [K'(r) - I_4] \,.$$

The adjugate matrix of this characteristic matrix $\chi(r)$, which we know contains four scaled copies of the maximal eigenvector q_{opt}, is made easily computable from the trick of replacing $I_4 \to$ $\left(r_0{}^2 + r_1{}^2 + r_2{}^2 + r_3{}^2\right) I_4$, yielding the adjugate matrix

$$\mathrm{Adj}(\chi(r)) = - \begin{bmatrix} r_0{}^2 & r_0 r_1 & r_0 r_2 & r_0 r_3 \\ r_0 r_1 & r_1{}^2 & r_1 r_2 & r_1 r_3 \\ r_0 r_2 & r_1 r_2 & r_2{}^2 & r_2 r_3 \\ r_0 r_3 & r_1 r_3 & r_2 r_3 & r_3{}^2 \end{bmatrix}. \tag{9.16}$$

That is, the adjugate matrix containing the scaled eigenvectors for the maximal eigenvalue $\lambda_{\mathrm{opt}} = 1$ in the error-free case is *the negative of the profile matrix*, Eq. (9.13)! Since the sign is an inconsequential overall scale, we see that any normalizable column of Eq. (9.16) solves the problem of determining the best quaternion q_{opt} matching the rotation matrix, so we know $R_{\mathrm{opt}} = R(q_{\mathrm{opt}})$ is the best match for Q.

Noise-accommodating solution. In the case of noisy data, the *closest rotation matrix* is determined in exactly the same way as the solution for q_{opt} in the noisy 3D cloud matching process of Chapter 8. That is, we simply compute $\lambda_{\mathrm{experimental}}$ determined from the numerical eigensystem of $K(Q(x, y, z))$ given by Eq. (9.7), write down the characteristic equation

$$\chi(Q) = \left[K\left(Q(x, y, z)\right) - \lambda_{\mathrm{experimental}} I_4\right], \tag{9.17}$$

compute the adjugate matrix of $\chi(Q)$, and select its best column to normalize for the eigenvector q_{opt} determining $R_{\mathrm{opt}} = R(q_{\mathrm{opt}}) \approx Q(m)$.

9.3 Optimal quaternions from projection matrix data

We conclude with a somewhat surprising observation regarding extraction of information from *partial* rotation matrix data. Consider the top two rows of a rotation matrix, e.g., the 2×3 submatrix $R(q)_{\{1,2\},\{1,2,3\}}$,

$$P(q) = \begin{bmatrix} q_0{}^2 + q_1{}^2 - q_2{}^2 - q_3{}^2 & 2q_1 q_2 - 2q_0 q_3 & 2q_1 q_3 + 2q_0 q_2 \\ 2q_1 q_2 + 2q_0 q_3 & q_0{}^2 - q_1{}^2 + q_2{}^2 - q_3{}^2 & 2q_2 q_3 - 2q_0 q_1 \end{bmatrix}, \tag{9.18}$$

with the equivalent numerical data 2×3 matrix corresponding to the top two data rows $T \to$ $S_{\{1,2\},\{1,2,3\}}([\mathbf{X}|\mathbf{Y}|\mathbf{Z}])$, or

$$T(m) = \begin{bmatrix} x_1 & y_1 & z_1 \\ x_2 & y_2 & z_2 \end{bmatrix} = \begin{bmatrix} m_{11} & m_{12} & m_{13} \\ m_{21} & m_{22} & m_{23} \end{bmatrix}. \tag{9.19}$$

These matrices are effectively the *orthographic projection matrices* that arise naturally in attempting to analyze a least-squares description of the task of finding a quaternion q_{opt} corresponding to the rotation matrix $R(q_{opt})$ that is most likely to have produced a certain parallel projected image. Here we will consider only the restricted problem of finding the best match to a projection matrix P.

Subject to the condition that the original 3×3 matrix, of which we now have access to only two rows, is either *error-free*, or *close enough to error-free* for our purposes, we can assume that if we can find the closest quaternion for the first two rows, we can calculate the missing third row of the rotation matrix simply by *computing the cross product*, which, for quaternions obeying $q \cdot q = 1$, has exactly enough degrees of freedom to exactly reproduce the bottom row, $\mathbf{D}(q)$, of Eq. (2.14) from the top two rows, $\mathbf{P}_1(q)$ and $\mathbf{P}_2(q)$:

$$\mathbf{D}(q) = \text{third row of } R(q) = \mathbf{P}_1(q) \times \mathbf{P}_2(q) \tag{9.20}$$

$$= \begin{bmatrix} -2(q_0 q_2 - q_1 q_3) & 2(q_0 q_1 + q_2 q_3) & q_0^2 - q_1^2 - q_2^2 + q_3^2 \end{bmatrix}. \tag{9.21}$$

Note that if the first two rows $\mathbf{P}_1$ and $\mathbf{P}_2$ are orthonormal, $\mathbf{D}$ will be automatically normalized.

The reduced profile matrix allows significant simplification. We begin by examining the projection-matrix variant of Eq. (9.5), which is a maximization measure,

$$\Delta_{2 \times 3} = \text{tr } P(q) \cdot T^t(\mathbf{x}_1, \mathbf{x}_2) = q \cdot K_2(x) \cdot q. \tag{9.22}$$

Here $T^t(\mathbf{x}_1, \mathbf{x}_2)$ is a pair of (perhaps approximately) orthonormal rows of a projection matrix, and $K_2(x)$ is analogous to the profile matrix $M(E)$. What is remarkable is that, if we write out the terms of Δ,

$$\begin{aligned}
\Delta(\mathbf{x}_1, \mathbf{x}_2) = & q_0^2 (x_1 + y_2) + q_1^2 (x_1 - y_2) + q_2^2 (y_2 - x_1) + q_3^2 (-x_1 - y_2) \\
& + 2 q_3 q_0 (x_2 - y_1) + 2 q_1 q_2 (x_2 + y_1) + 2 q_2 q_0 z_1 - 2 q_1 q_0 z_2 + 2 q_1 q_3 z_1 + 2 q_2 q_3 z_2,
\end{aligned} \tag{9.23}$$

we see that *all* pairs of $q_i q_j$ terms appear. Therefore $K_2(x)$ is a fully populated 4×4 profile matrix, which, to the best of our knowledge, first appeared in Eq. (1) of Bar-Itzhack (2000),

$$K_2(\mathbf{x}_1, \mathbf{x}_2) = \begin{bmatrix} x_1 + y_2 & -z_2 & z_1 & x_2 - y_1 \\ -z_2 & x_1 - y_2 & x_2 + y_1 & z_1 \\ z_1 & x_2 + y_1 & y_2 - x_1 & z_2 \\ x_2 - y_1 & z_1 & z_2 & -x_1 - y_2 \end{bmatrix}. \tag{9.24}$$

The eigenvalues are immediately computable, and take the form[1]

$$\lambda = \begin{bmatrix} \sqrt{\mathbf{x}_1 \cdot \mathbf{x}_1 + \mathbf{x}_2 \cdot \mathbf{x}_2 + 2\sqrt{(\mathbf{x}_1 \cdot \mathbf{x}_1)(\mathbf{x}_2 \cdot \mathbf{x}_2) - (\mathbf{x}_1 \cdot \mathbf{x}_2)^2}} \\ \sqrt{\mathbf{x}_1 \cdot \mathbf{x}_1 + \mathbf{x}_2 \cdot \mathbf{x}_2 - 2\sqrt{(\mathbf{x}_1 \cdot \mathbf{x}_1)(\mathbf{x}_2 \cdot \mathbf{x}_2) - (\mathbf{x}_1 \cdot \mathbf{x}_2)^2}} \\ -\sqrt{\mathbf{x}_1 \cdot \mathbf{x}_1 + \mathbf{x}_2 \cdot \mathbf{x}_2 - 2\sqrt{(\mathbf{x}_1 \cdot \mathbf{x}_1)(\mathbf{x}_2 \cdot \mathbf{x}_2) - (\mathbf{x}_1 \cdot \mathbf{x}_2)^2}} \\ -\sqrt{\mathbf{x}_1 \cdot \mathbf{x}_1 + \mathbf{x}_2 \cdot \mathbf{x}_2 + 2\sqrt{(\mathbf{x}_1 \cdot \mathbf{x}_1)(\mathbf{x}_2 \cdot \mathbf{x}_2) - (\mathbf{x}_1 \cdot \mathbf{x}_2)^2}} \end{bmatrix}. \tag{9.25}$$

---

[1] We thank B.K.P. Horn for correspondence on this point.

The top row is the maximum eigenvalue λ_{opt}, and from that we can compute the corresponding eigenvector, which is straightforward to compute by whatever means one chooses, but we omit the algebraic forms, which are long and not very informative.

To summarize the importance of this section, we have shown that we do not need the full 3D data set to obtain an adequate alignment solution: for the Bar-Itzhack rotation matrix problem, we only need a 2×3 portion of the target rotation matrix. This can be exploited for the task of aligning a 3D reference cloud with its 2D orthographic image: we will see in Chapter 15 that, for low noise, we have sufficient information to compute a quaternion that determines not only the 2×3 projection matrix $P(q)$, but the whole 3×3 rotation matrix $R(q)$. If we are able to obtain or compute only a valid $P(q)$, the cross product of those two rows reliably generates a candidate third row of the full rotation matrix, and this is adequate for the orthographic projection problem with relatively small errors. One needs less information to deduce a rotation from the data than one would think.

9.4 Summary of finding quaternions from rotation data

It is important to note, as emphasized by Bar-Itzhack (2000), that if there are *significant errors* in the numerical matrix $Q(m)$, then the actual nonunit maximal eigenvalue of $K(m)$ can be computed numerically or algebraically as usual, and then that eigenvalue's eigenvector determines the *closest* normalized quaternion to the errorful rotation matrix $Q(m)$, which can be very useful since such a quaternion *always* produces a valid rotation matrix.

In any case, *up to an overall sign*, the optimal quaternion eigenvector q_{opt} corresponds to the target numerical rotation matrix

$$[closest\ rotation](Q(m) = R(q_{opt})\ .$$

In some circumstances, one is looking for a uniform statistical distribution of quaternions, in which case the overall sign of q could be chosen randomly.

The Bar-Itzhack approach solves the problem of extracting the quaternion of an arbitrary numerical 3D rotation matrix, even with noise, in a fashion that involves no singularities and only trivial testing for special cases, thus essentially making the traditional methods obsolete. We note that the extension of Bar-Itzhack's method to the case of 4D rotations is given in Chapter 20, Section 20.4, and that further quaternion adjugate applications are studied starting in Chapter 12.

The 3D quaternion-based frame alignment problem ☆

10

We now turn to the orientation frame problem, exploiting the fact that 3D orientation frames can *themselves* be expressed as quaternions. We consider data such as lists of orientations of roller coaster cars, or lists of residue orientations in a protein, ordered pairwise in some way, but *without* specifically considering any spatial location or nearest-neighbor ordering information. In N-dimensional space, the *columns* of any $\mathbf{SO}(N)$ orthonormal $N \times N$ rotation matrix R_N are what we mean by an orientation frame, since these columns are the directions pointed to by the axes of the identity matrix after rotating something from its defining identity frame to a new attitude; note that no spatial location information whatever is contained in R_N, though one may wish to choose a local center for each frame if the construction involves coordinates such as amino acid atom locations. See also Chapter 16 for an extensive application of quaternion frames to protein structure.

In 2D, 3D, and 4D, there exist two-to-one quadratic maps from the topological spaces $\mathbf{S}^1$, $\mathbf{S}^3$, and $\mathbf{S}^3 \times \mathbf{S}^3$ to the rotation matrices R_2, R_3, and R_4. These are the quaternion-related objects that we will use to obtain elegant representations of the frame data alignment problem. In 2D, a frame data element can be expressed as a complex phase, while in 3D the frame corresponds to a unit quaternion (see Hanson (2006); Hanson and Thakur (2012)). In 4D the frame is described by a *pair* of unit quaternions, and is treated in Chapter 20.

Various proximity measures for such orientation data have been explored in the literature (see, e.g., Park and Ravani (1997); Moakher (2002); Huynh (2009); Huggins (2014)), and the general consensus is that the most rigorous measure minimizes the sums of squares of geodesic arc lengths between pairs of quaternions. This ideal proximity measure is highly nonlinear compared to the analogous spatial alignment RMSD measure, but fortunately there is an often-justifiable linearization, the chord angular distance measure; we will present several alternative solutions exploiting this approximation that closely parallel our spatial RMSD formulation.

In the following, we start with a basic introduction to the way that coordinate frames manifest themselves in the quaternion framework, and then proceed into the details of the frame alignment optimization problem, working out exact algebraic solutions for the eigenvalues that can be used as an alternative to traditional numerical methods.

☆ This chapter derives much of its content from Hanson (2020).

Visualizing More Quaternions. https://doi.org/10.1016/B978-0-32-399202-2.00019-8

10.1 Introduction to quaternion orientation frames

Remark: Readers unfamiliar with the use of complex numbers and quaternions to obtain elegant representations of 2D and 3D orientation frames are particularly encouraged to study this section, while readers conversant with these concepts may wish to move on to the next section.

As preparation for our investigation into the frame matching problem, we begin with a review of what we mean by a coordinate frame, and how coordinate frames are related to quaternions. As is our custom, we start with a simplified 2D problem expressed in terms of complex variables, and then generalize to the full quaternion-based 3D problem to answer the question:

> **What is a quaternion frame?**

2D complex frames. We will first present a bit of intuition about coordinate frames in two dimensions that may help some readers envision what is going on. If we take the special case of a quaternion representing a rotation in the 2D (x, y)-plane, the 3D rotation matrix Eq. (2.14) reduces to what is essentially a quaternion form of a 2D rotation that looks like

$$R_2(a, b) = \begin{bmatrix} a^2 - b^2 & -2ab \\ 2ab & a^2 - b^2 \end{bmatrix}, \tag{10.1}$$

where $a = \cos(\theta/2)$ and $b = \sin(\theta/2)$. From Eq. (2.17) with $\hat{\mathbf{n}} = \hat{\mathbf{z}}$, the z-axis fixed, or from the half-angle formulas, we see that this reduces to the standard right-handed 2D rotation

$$R_2(\theta) = \begin{bmatrix} \cos\theta & -\sin\theta \\ \sin\theta & \cos\theta \end{bmatrix}. \tag{10.2}$$

As shown in Fig. 10.1(a), we can use θ to define a unit direction in the complex plane defined by $z = \exp i\theta$, and then the *columns* of the matrix $R_2(\theta)$ naturally correspond to a unique associated 2D coordinate frame diad; an entire collection of points z and their corresponding frame diads are depicted in Fig. 10.1(b).

Next, we look at the reduced quaternion (a, b) frames and associate them with the complex variable $z = \exp(i\theta)$ as follows:

$$(a, b) = (\cos(\theta/2), \sin(\theta/2)), \tag{10.3}$$

$$u = (a + ib) = \sqrt{z} = e^{i\theta/2}. \tag{10.4}$$

Our simplified 2D quaternion thus describes the *square root* of the usual Euclidean frame given by the columns of $R_2(\theta)$. Thus the pair (a, b) (the reduced quaternion) itself corresponds to a frame. In Fig. 10.2(a), we show how a given "quaternion frame," i.e., the columns of $R_2(a, b)$, corresponds to a point $u = a + ib$ in the complex plane. Diametrically *opposite* points (a, b) and $(-a, -b)$ now correspond to the *same* frame! Fig. 10.2(b) shows the corresponding frames for a large collection of points (a, b) in the complex plane, and we see the new and unfamiliar feature that the frames make

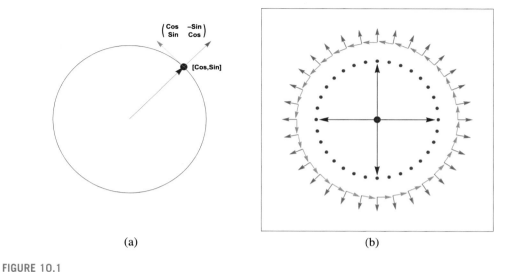

(a)   (b)

FIGURE 10.1

(a) Any standard 2D coordinate frame corresponds to the columns of an ordinary rotation matrix, and is associated to the point $(\cos\theta, \sin\theta)$ on a unit circle. (b) The standard 2D coordinate frames with x-axis $[\cos\theta, \sin\theta]^t$ and y-axis $[-\sin\theta, \cos\theta]^t$ associated with each sampled unit vector $[\cos\theta, \sin\theta]^t$ on a circle.

two full rotations on the complex circle instead of just one as in Fig. 10.1(b). The last step is to notice that in Fig. 10.2(b) we can represent the set of frames in one half of the complex circle, $a \geqslant 0$ shown in magenta, as distinct from those in the other half, $a < 0$ shown in dark blue; for any value of b, the vertical axis, there is a *pair* of a's with opposite signs and colors. For example, we could omit the a-coordinates and display only the color-tagged b-coordinates in Fig. 10.2(b) projected to the vertical axis, realizing that if we know the sign-correlated coloring, we can determine both the magnitude and the sign of the dependent variable $a = \pm\sqrt{1 - b^2}$.

> This is our first indication that *a point in space can describe the axes of a coordinate frame*, and that in fact, we can omit one dimension of the display and still uniquely determine the full coordinates of the point, and thus the corresponding frame.

3D quaternion frames. Starting from the 2D context, we can now get a clear intuitive picture of what we mean by a "quaternion frame" before diving into the complexities of the quaternion frame matching problem. We progress to using a full quaternion to represent an arbitrary 3D frame triad via Eq. (2.14), which we repeat here for convenience:

$$\begin{bmatrix} q_0^2 + q_1^2 - q_2^2 - q_3^2 & 2q_1q_2 - 2q_0q_3 & 2q_1q_3 + 2q_0q_2 \\ 2q_1q_2 + 2q_0q_3 & q_0^2 - q_1^2 + q_2^2 - q_3^2 & 2q_2q_3 - 2q_0q_1 \\ 2q_1q_3 - 2q_0q_2 & 2q_2q_3 + 2q_0q_1 & q_0^2 - q_1^2 - q_2^2 + q_3^2 \end{bmatrix}. \tag{10.5}$$

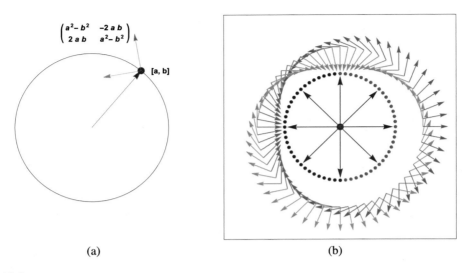

FIGURE 10.2

(a) The quaternion point (a, b), in contrast, corresponds via the double-angle formula to coordinate frames that rotate twice as rapidly as (a, b) progresses around the unit circle that is a simplified version of quaternion space. (b) The set of 2D frames associated with the entire circle of quaternion points (a, b); each diametrically opposite point corresponds to an identical frame. For later use in displaying full quaternions, we show how color coding can be used to encode the sign of one of the coordinates on the circle.

To make it clear that this is a 3D coordinate frame triad, think first of the identity matrix

$$\begin{bmatrix} 1 & 0 & 0 \\ 0 & 1 & 0 \\ 0 & 0 & 1 \end{bmatrix}.$$

The first column is simply the unit vector pointing along the direction of the x-axis, the first element of a frame triad defining the orientation of the coordinate system itself; the second and third columns are the y- and z-axis unit vectors. When we define a quaternion, the basic rotation construction of Eq. (10.5) transforms the identity frame to a new frame by matrix multiplication acting to the right ("left-multiplication"), so the *new* x-axis is the first column of the rotation matrix, that is, the first column of Eq. (10.5) if we write it in terms of the quaternion mapping. Thus the second and third columns are simply the new y- and z-axes: a single quaternion point, a unit-length vector in 4D Euclidean space, corresponds *exactly* to a triad of orthonormal 3D vectors, the columns of Eq. (10.5), describing an arbitrary orientation of an object in 3D space.

Example. We can in fact display the points defining a quaternion frame uniquely inside one single sphere, in parallel to what we observed in the 2D case. If we display only a quaternion's 3-vector part $\mathbf{q} = (q_x, q_y, q_z)$ along with a color specifying the sign of q_0, we implicitly know both the magnitude and sign of

$$q_0 = \pm\sqrt{1 - q_x{}^2 - q_y{}^2 - q_z{}^2}\,,$$

and a point on such a 3D plot therefore uniquely depicts *any* quaternion, and *any* orientation frame. Another alternative we have employed in Chapter 4 is to use *two* solid balls, one a "Northern hemisphere" for the $q_0 > 0$ components and the other a "Southern hemisphere" for the $q_0 < 0$ components, where the equator at $q_0 = 0$ can be displayed separately, or merged with either solid ball, depending on the requirements of one's data. Any of these methods may be useful in different contexts.

We illustrate all this in Fig. 10.3(a), which shows a typical collection of quaternion reference frame data displaying only the **q** components of $(q_0, \mathbf{q})$; the $q_0 \geqslant 0$ data are mixed with the $q_0 < 0$ data, but are distinguished by their color coding. In Fig. 10.3(b), we show the frame triads resulting from applying Eq. (10.5) to each quaternion point and plotting the result at the associated point **q** in the display.

10.2 Overview of 3D quaternion frame alignment

We turn now to the problem of aligning corresponding sets of 3D orientation frames, just as we already studied the alignment of sets of 3D spatial coordinates by performing an optimal rotation. We will make essential use of the concept of a single point, a single quaternion 4-vector q, as a perfect description of a 3D orientation frame via Eq. (10.5). There will be more than one feasible method. We might assume we could just define the "quaternion frame alignment" problem by converting any list of frame orientation matrices to quaternions and writing down the quaternion equivalents of the RMSD treatment in Eq. (8.1) and Eq. (8.6). However, unlike the linear Euclidean problem, the preferred quaternion optimization function technically requires a *nonlinear* minimization of the squared sums of geodesic arc lengths connecting the points on the quaternion hypersphere $\mathbf{S}^3$. The task of formulating this ideal problem as well as studying alternative approximations is the subject of its own branch of the literature, often known as the *quaternionic barycenter* problem or the *quaternion-averaging* problem (see, e.g., Brown and Worsey (1992); Buss and Fillmore (2001); Moakher (2002); Markley et al. (2007); Huynh (2009); Hartley et al. (2011, 2013) and also Chapter 18). We will focus on L_2 norms (the aforementioned sums of squares of arc lengths), although alternative approaches to the rotation averaging problem, such as employing L_1 norms and using the Weiszfeld algorithm to find the optimal rotation numerically, have been advocated, e.g., by Hartley et al. (2011). The computation of optimally aligning rotations, based on plausible exact or approximate measures relating collections of corresponding pairs of (quaternionic) orientation frames, is now our task.

Choices for the forms of the measures encoding the distance between orientation frames have been widely discussed; see, e.g., Park and Ravani (1997); Moakher (2002); Markley et al. (2007); Huynh (2009); Hartley et al. (2011, 2013); Huggins (2014). Since we are dealing primarily with quaternions, we will start with two measures dealing directly with the quaternion geometry, the geodesic arc length, and the chord length, and later on examine some advantages of starting with quaternion-sign-independent rotation matrix forms. The arc length has the advantage of being the rigorous analog of the Euclidean distance measure, and the disadvantage of lacking a closed algebraic form that works in a least-squares optimization problem. The chord length is only an approximation to the actual sphere-based geodesic distance between quaternions, but it has the advantage of supporting closed form algebraic treatments of the least-squares optimization problem.

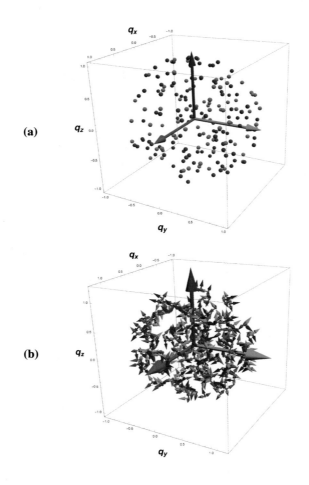

FIGURE 10.3

(a) The 3D portions of the quaternion reference frame data $q = (q_0, q_x, q_y, q_z)$, using different colors for $q_0 \geqslant 0$ and $q_0 < 0$ in the unseen direction. Since $|q_0| = \sqrt{1 - (q_x^2 + q_y^2 + q_z^2)}$, the complete quaternion can in principle be determined from the 3D display. (b) 3D orientation frame triads for a sparser set of quaternion points (q_0, q_x, q_y, q_z) displayed at their associated vector locations $\mathbf{q} = (q_x, q_y, q_z)$.

3D geodesic arc length distance. First, we recall that the matrix Eq. (10.5) has three orthonormal columns that define a quadratic map from the quaternion 3-sphere $\mathbf{S}^3$, a smooth connected Riemannian manifold, to a 3D orientation frame. The squared geodesic arc length distance between two quaternions lying on the 3-sphere $\mathbf{S}^3$ is generally agreed upon as the measure of orientation frame proximity whose properties are the closest in principle to the ordinary squared Euclidean distance measure Eq. (8.1) between points (Huynh, 2009), and we will adopt this measure as our starting point. We begin by writing down a frame-frame distance measure between two unit quaternions q_1 and q_2, corresponding precisely to two orientation frames defined by the columns of $R(q_1)$ and $R(q_2)$. We define the geodesic arc length as an angle α on the hypersphere $\mathbf{S}^3$ computed geometrically from $q_1 \cdot q_2 = \cos \alpha$. As pointed out by Huynh (2009) and Hartley et al. (2013), the geodesic arc length between a reference quaternion q_1 and a test quaternion q_2 has a sign ambiguity and can take two values, since $R(+q_2) = R(-q_2)$. We want the minimum angle (or the maximal $|q_1 \cdot q_2|$). Furthermore, to work on a spherical manifold instead of a plane, we need basically to cluster the ambiguous points in a deterministic way. Starting with the bare angle between two quaternions on $\mathbf{S}^3$, $\alpha = \arccos(q_1 \cdot q_2)$, where we recall that either quaternion can technically have either sign and still describe the same orientation matrix, we define a *pseudometric* (Huynh, 2009) for the geodesic arc length distance as

$$d_{\text{geodesic}}(q_1, q_2) = \min(\alpha, \pi - \alpha): \ 0 \leqslant d_{\text{geodesic}}(q_1, q_2) \leqslant \frac{\pi}{2} , \tag{10.6}$$

illustrated in Fig. 10.4. An efficient implementation of this is to take

$$d_{\text{geodesic}}(q_1, q_2) = \arccos(|q_1 \cdot q_2|) , \tag{10.7}$$

which we see is effectively an L_1 measure.

We now seek to define an ideal minimizing L_2 orientation frame measure, comparable to our minimizing Euclidean RMSD measure, but constructed from geodesic arc lengths on the quaternion hypersphere instead of Euclidean distances in space. Thus to compare a test quaternion frame reference set $\{p_k\}$ to a test data set $\{r_k\}$, we propose the geodesic-based least-squares measure

$$\mathbf{S}_{\text{geodesic}} = \sum_{k=1}^{N} (\arccos |(q \star p_k) \cdot r_k|)^2 = \sum_{k=1}^{N} (\arccos |q \cdot (r_k \star \bar{p}_k)|)^2 , \tag{10.8}$$

where we have used the identities of Eq. (2.12). When $q = q_{\text{ID}}$, the individual measures correspond to the square of Eq. (10.7); otherwise "$q \star p_k$" is the exact analog of "$R(q) \cdot x_k$" in Eq. (8.1), and denotes the quaternion rotation q acting on the entire set $\{p_k\}$ to rotate it to a new orientation that we want to align optimally with the test frames $\{r_k\}$. Analogously, for points on a sphere, the arccosine of an inner product is equivalent to a distance between points in Euclidean space.

Note on numerics. For improved numerical behavior in the computation of the quaternion inner product angle between two quaternions, one may prefer to convert the arccosine to an arctangent form, $\alpha = \arctan(dx, dy) = \arctan(\cos \alpha, |\sin \alpha|)$ (remember the C math library uses the opposite argument order atan2(dy, dx)), with the parameters

$$\cos(\alpha) = |\text{Re} (q_1 \star q_2^{-1})| = |q_1 \cdot q_2|, \quad |\sin(\alpha)| = \left\| \text{Im} (q_1 \star q_2^{-1}) \right\| = \left\| -[q_1]_0 \mathbf{q}_2 + [q_2]_0 \mathbf{q}_1 - \mathbf{q}_1 \times \mathbf{q}_2 \right\| ,$$

which is somewhat more stable.

Adopting the solvable chord measure. Unfortunately, the geodesic arc length measure does not fit into the linear algebra approach that we were able to use to obtain exact solutions for the Euclidean data alignment problem treated so far. Thus we are led to investigate instead a very close approximation to $d_{\text{geodesic}}(q_1, q_2)$ that *does* correspond closely to the Euclidean data case and does, with some

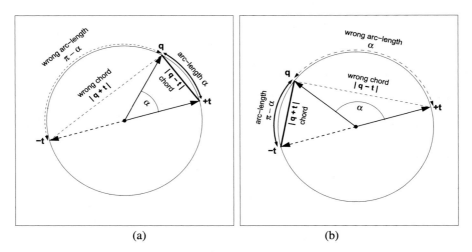

(a)   (b)

FIGURE 10.4

Geometric context involved in choosing a *quaternion distance* that will result in the correct *average rotation matrix* when the quaternion measures are optimized. Because the quaternion vectors represented by t and $-t$ give the same rotation matrix, one must choose $|\cos\alpha|$ or the *minima*, that is, $\min(\alpha, \pi - \alpha)$ or $\min(\|q - t\|, \|q + t\|)$, of the alternative distance measures to get the *correct* items in the arc length or chord measure summations. (a) and (b) represent the cases when the first or second choice should be made, respectively.

contingencies, admit exact solutions. This approximate measure is the *chord distance*, whose individual distance terms analogous to Eq. (10.7) take the form of a closely related pseudometric (Huynh, 2009; Hartley et al., 2013),

$$d_{\text{chord}}(q_1, q_2) = \min(\|q_1 - q_2\|, \|q_1 + q_2\|) : \ 0 \leqslant d_{\text{chord}}(q_1, q_2) \leqslant \sqrt{2} . \tag{10.9}$$

We compare the geometric origins for Eq. (10.7) and Eq. (10.9) in Fig. 10.4. Note that the crossover point between the two expressions in Eq. (10.9) is at $\pi/2$, and the hypotenuse of the right isosceles triangle at that point has length $\sqrt{2}$.

The solvable approximate optimization function analogous to $\|R \cdot x - u\|^2$ that we will now explore for the quaternion frame alignment problem will thus take the form that must be minimized as

$$S_{\text{chord}} = \sum_{k=1}^{N} (\min(\|(q \star p_k) - r_k\|, \|(q \star p_k) + r_k\|))^2 . \tag{10.10}$$

We can convert the sign ambiguity in Eq. (10.10) to a deterministic form like Eq. (10.7) by observing, with the help of Fig. 10.4, that

$$\|q_1 - q_2\|^2 = 2 - 2q_1 \cdot q_2 , \qquad \|q_1 + q_2\|^2 = 2 + 2q_1 \cdot q_2 . \tag{10.11}$$

Clearly $(2 - 2|q_1 \cdot q_2|)$ is always the smallest of the two values. Thus minimizing Eq. (10.10) amounts to maximizing the now-familiar cross-term form, which we can write as

$$\left.\begin{aligned}
\Delta_{\text{chord}}(q) &= \sum_{k=1}^{N} |(q \star p_k) \cdot r_k| \\
&= \sum_{k=1}^{N} |q \cdot (r_k \star \bar{p}_k)| \\
&= \sum_{k=1}^{N} |q \cdot t_k|
\end{aligned}\right\}. \tag{10.12}$$

Here we have used the identity $(q \star p) \cdot r = q \cdot (r \star \bar{p})$ from Eq. (2.12) and defined the quaternion displacement or "attitude error" (Markley et al., 2007)

$$t_k = r_k \star \bar{p}_k, \tag{10.13}$$

that frames our optimization process in a form typical of a rotation averaging problem. Note that we could have derived the same result using Eq. (2.11) to show that $\|q \star p - r\| = \|q \star p - r\| \|p\| = \|q - r \star \bar{p}\|$.

There are several ways to proceed to our final result at this point. The simplest is to pick a neighborhood in which we will choose the samples of q that include our expected optimal quaternion, and adjust the sign of each data value t_k to $\widetilde{t}_k$ by the transformation

$$\widetilde{t}_k = t_k \, \text{sign}(q \cdot t_k) \;\; \rightarrow |q \cdot t_k| = q \cdot \widetilde{t}_k . \tag{10.14}$$

The neighborhood of q matters because, as argued by Hartley et al. (2013), even though the allowed range of 3D rotation angles is $\theta \in (-\pi, \pi)$ (or quaternion sphere angles $\alpha \in (-\pi/2, \pi/2)$), convexity of the optimization problem cannot be guaranteed for collections outside local regions centered on some θ_0 of size $\theta_0 \in (-\pi/2, \pi/2)$ (or $\alpha_0 \in (-\pi/4, \pi/4)$): beyond this range, local basins may exist that allow the mapping Eq. (10.14) to produce distinct local variations in the assignments of the $\{\widetilde{t}_k\}$ and in the solutions for q_{opt}. Within considerations of such constraints, Eq. (10.14) now allows us to take the summation outside the absolute value, and write the quaternion frame optimization problem in terms of maximizing the cross-term expression

$$\left.\begin{aligned}
\Delta_{\text{chord}}(q) &= \sum_{k=1}^{N} q \cdot \widetilde{t}_k \\
&= q \cdot V(t)
\end{aligned}\right\}, \tag{10.15}$$

where $V = \sum_{k=1}^{N} \widetilde{t}_k$ is the analog of the Euclidean RMSD profile matrix M. However, since this is *linear* in q, we have the remarkable result that, as noted in the treatment of Hartley et al. (2013) regarding the quaternion L_2 chordal distance norm, the solution is immediate, being simply

$$q_{\text{opt}} = \frac{V}{\|V\|}, \tag{10.16}$$

since that immediately maximizes the value of $\Delta_{\text{chord}}(q)$ in Eq. (10.15). This gives the maximal value of the measure as

$$\Delta_{\text{chord}}(q_{\text{opt}}) = \|V\|, \tag{10.17}$$

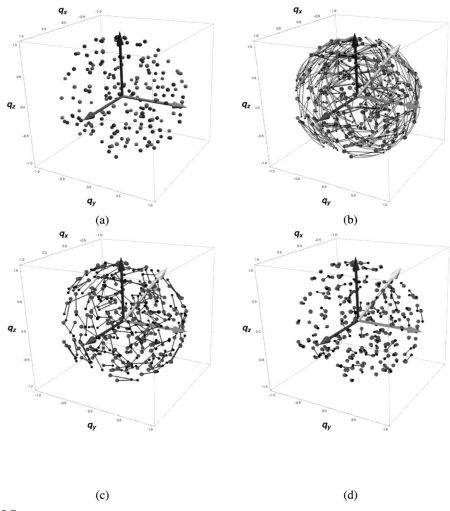

FIGURE 10.5

3D components of a quaternion orientation data set. (a) A quaternion reference set, color coded by the sign of q_0. (b) *Exact* quaternion arc length distances (green) vs. chord distances (black) between the test and reference points. (c) Part-way from starting state to the aligned state, at $s = 0.5$. (d) The final best alignment at $s = 1.0$. The yellow arrow is the direction of the quaternion eigenvector; when scaled, the length is the sine of half the optimal rotation angle.

and thus $\|V\|$ is the exact orientation frame analog of the spatial RMSD maximal eigenvalue λ_{opt}, except it is far easier to compute.

Illustrative example. Using the quaternion display method described in Section 10.1 and illustrated in Fig. 10.3, we present in Fig. 10.5(a) a representative quaternion frame reference data set and in

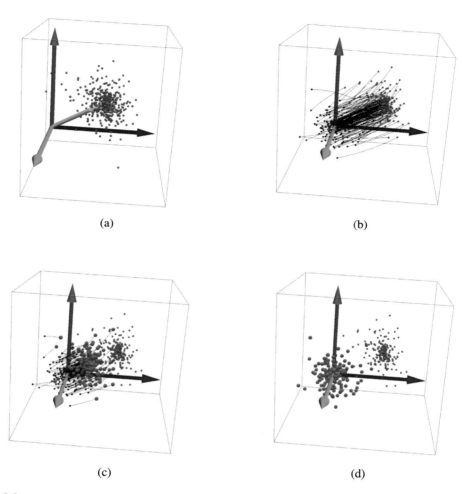

(a)

(b)

(c)

(d)

FIGURE 10.6

3D components of the rotation-average transformation of the quaternion orientation data set, with each point denoting the *displacement* between each pair of frames as a single quaternion, corresponding to the rotation taking the test frame to the reference frame. (a) The cluster of points $t_k = r_k \star \bar{p}_k \to \tilde{t}_k$ derived from the frame matching problem using just the curved arcs in Fig. 10.5(b). If there were no alignment errors introduced in the simulation, these would all be a single point. The yellow arrow is the quaternion solution to the chord distance centroid of this cluster, and is identical to the optimal quaternion rotation transforming the test data to have the minimal chord measure relative to the reference data. (b) Choosing a less cluttered subset of the data in (a), we display the geodesic paths from the initial quaternion displacements $\tilde{t}_k$ to the origin-centered set with minimal chord measure distance relative to the origin. This is the result of applying the inverse of the quaternion q_{opt} to each $\tilde{t}_k$. Note that the paths are curved geodesics lying properly within the quaternion sphere. (c,d) Rotating the cluster using a `slerp` between the quaternion barycenter of the initial misaligned data and the optimally aligned position, which is centered at the origin.

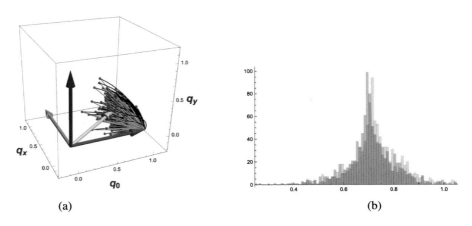

(a)   (b)

FIGURE 10.7

(a) Projecting the geodesic vs. chord distances from the origin to sampled points in a set of frame-displacement data $t_k = r_k \star \bar{p}_k \to \tilde{t}_k$. Since the **q** spatial quaternion paths project to a straight line from the origin, we use the (q_0, q_1, q_2)-coordinates instead of our standard **q**-coordinates to expose the curvature in the arc length distances to the origin. (b) Comparing the distribution of arccosine values of the rigorous geodesic arc length cost function and the chord length method, sampled with a uniform distribution of random quaternions over $\mathbf{S}^3$. The arc length method has a different distribution, as expected, and produces a very slightly better barycenter. However, the optimal quaternions for the arc length vs. chord length measure for this simulated data set differ by less than a hundredth of a degree, so drawing the positions of the two distinct optimal quaternions would not reveal any difference in image (a).

Fig. 10.5(b) the relationship of the arc and chord distances for each point in a set of arc and chord distances (see Fig. 10.4) for each point pair in the quaternion space. In Fig. 10.5(c,d), we show the results of the quaternion frame alignment process using conceptually the same `slerp` of Eq. (8.11) to transition from the raw state at $q(s=0) = q_{\text{ID}}$ to $q(s=0.5)$ for (c) and $q(s=1.0) = q_{\text{opt}}$ for (d). The yellow arrow is the axis of rotation specified by the vector part of the optimal quaternion.

As illustrated in Fig. 10.6, with compatible sign choices, the $\tilde{t}_k$'s cluster around the optimal quaternion, which is clearly consistent with being the barycenter of the quaternion differences; this is intuitively the place to which all the quaternion frames need to be rotated to form an optimally centered cluster. As before, the yellow arrow is the axis of rotation specified by the spatial part of the optimal quaternion. Next, Fig. 10.7 addresses the question of how the rigorous arc length measure is related to the chord length measure that can be treated using the same methods as the spatial RMSD optimization. In parallel to Fig. 10.5(b), Fig. 10.7(a) shows essentially the same comparison for the $\tilde{t}_k$ quaternion-displacement version of the same data. In Fig. 10.7(b), we show the histograms of the chord distances to a sample point, the origin in this case, versus the arc length or geodesic distances. They obviously differ, but in fact for plausible simulations, the arc length numerical optimal quaternion barycenter does not differ significantly from the chord length counterpart. More detailed evaluation of the performance is given in Section 10.4.

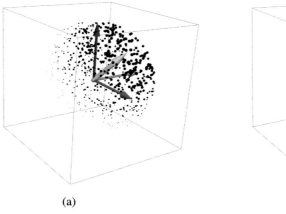
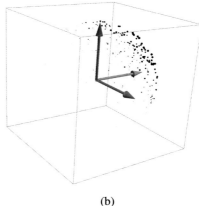

(a)   (b)

FIGURE 10.8

The values of $\Delta = q \cdot V$ represented by the sizes of the dots placed at a random distribution of quaternion points. We display the data dots at the locations of their spatial quaternion components $\mathbf{q} = (q_1, q_2, q_3)$. (a) The Northern hemisphere of $\mathbf{S}^3$, with $q_0 \geqslant 0$. (b) The Southern hemisphere, with $q_0 < 0$. We implicitly know that the value of q_0 is $\pm\sqrt{1 - q_0^2}$. While each ball alone actually represents all possible unique rotation matrices, our cost function covers the *entire* space of quaternions, so q and $-q$ are distinct. The spatial component of the maximal eigenvector is shown by the yellow arrow, which clearly ends in the middle of the maximum values of Δ. The small cloud at the edge of (b) is simply the rest of the complete cloud around the tip of the yellow arrow as q_0 passes through the "equator" at $q_0 = 0$.

Basins of attraction exercise. Next, in Fig. 10.8, we display the values of $\Delta_{\text{chord}} = q \cdot V$ that parallel the RMSD version in Fig. 8.3. The dots show the size of the cost $\Delta(q)$ at randomly sampled points across the entire $\mathbf{S}^3$, with $q_0 \geqslant 0$ in (a) and $q_0 < 0$ in (b). We have all the signs of the $\widetilde{t}_k$ chosen to be centered in an appropriate local neighborhood, so, unlike the quadratic Euclidean RMSD case, there is only one value for q_{opt} which is in the direction of V. Finally, in Fig. 10.9 we present an intuitive sketch of the convexity constraints for the frame optimization task related to Hartley et al. (2013). We start with a set of data in (a) (with both $(q, -q)$ partners), that consists of three local clouds that can be smoothly deformed from dispersed to coinciding locations. Panels (b) and (c) both contain a uniform sample of quaternion sample points q spread over all of quaternion space, shown as magenta dots, with positive and negative q_0 plotted on top of each other. Then each sample q is used to compute *one* set of mappings $t_k \to \widetilde{t}_k$, and the *one* value of $q_{\text{opt}} = V(\widetilde{t})/\|V\|$ that results. The black arrows show the relation of q_{opt} to each original sample q, effectively showing us their *votes* for the best quaternion average. Panel (b) has the clusters positioned far enough apart that we can clearly see that there are several basins of attraction, with no unique solution for q_{opt}, while in (c), we have interpolated the three clusters to lie in the same local neighborhood, roughly in a ball of quaternion radius $\alpha < \pi/4$, and we see that almost all of the black arrows vote for one unique q_{opt} or its equivalent negative. This seems to be a useful exercise to gain intuition about the nature of the basins of attraction for the quaternion averaging problem that is essential for quaternion frame alignment.

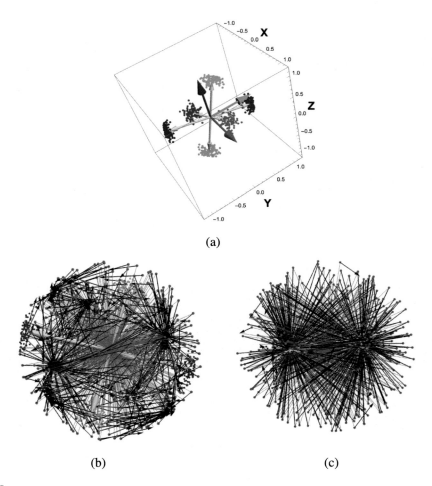

(a)

(b)   (c)

FIGURE 10.9

The behavior of the basins of attraction for the $t_k \to \tilde{t}_k$ map is shown here, starting in (a) with the $(q, -q)$ pairs for three movable clusters of quaternion frame data, each having a well-defined *local* quaternion average $q_{opt} = V(\tilde{t})/\|V\|$ shown as the yellow arrows with their $q \to -q$ equivalents. Next we merge all three samples into one data set that can be smoothly interpolated between the data being outside the $\alpha = \pi/4$ safe zone to all being together within that geometric boundary in quaternion space. Panel (b) shows the results of taking 500 uniform samples of q and computing the set $\{\tilde{t}_k\}$ for *each* sample q, placed at the magenta dots, and *then* computing the resulting q_{opt}; the black arrows follow the line from the sample point to the resultant q_{opt}. Clearly in (b), where the clusters are in their initial widely dispersed configuration, the black arrows (the "votes" for the best q_{opt}) collect in several different basins of attraction, signifying the absence of a global solution. We then interpolate all the clusters close to each other, and show the new results of the voting in (c). Now almost all of the samplings of the full quaternion space converge to point their arrows densely to the two opposite values of q_{opt}, and there is just one effective basin of attraction.

Alternative chord measure approach parallel to the Euclidean case. Having understood the chordal distance approach for the orientation alignment problem in terms of the pseudometric Eq. (10.9) and the measure Eq. (10.12) transformed into the form Eq. (10.15) involving the corrected quaternion displacements $\{\widetilde{r}_k\}$, we now observe that we can also express the problem in a form much closer to our Euclidean RMSD optimization problem. Returning to the form

$$S_{\text{chord}} = \sum_{k=1}^{N} \| q \star p_k - r_k \|^2 ,$$
(10.18)

we see that we can effectively transform the sign of *only* $p_k \to \widetilde{p}_k$ using the same test as Eq. (10.14) to make Eq. (10.18) valid as it stands; we then proceed, in the same fashion as the spatial alignment problem but with the modification required by Eq. (10.12), to convert to a cross-term form as follows:

$$\Delta_{\text{chord}}(q) = \sum_{k=1}^{N} |(q \star p_k) \cdot r_k| = \sum_{k=1}^{N} (q \star \widetilde{p}_k) \cdot r_k$$

$$= \sum_{a=0,b=0}^{3} Q(q)_{ba} \sum_{k=1}^{N} [\widetilde{p}_k]_a \, [r_k]_b$$

$$= \operatorname{tr} Q(q) \cdot W .$$
(10.19)

Here $W = \widetilde{\mathbf{P}} \cdot \mathbf{R}^t$ is essentially a cross-covariance matrix in the quaternion data elements and $Q(q)$ is the quaternion matrix of Eq. (2.4). Since $Q(q)$ is linear in q, we can simply pull out the coefficients of q, yielding

$$\Delta_{\text{chord}}(q) = q \cdot V(W) ,$$
(10.20)

where V is a 4-vector corresponding to the profile matrix in the spatial problem:

$$V(W) = \begin{bmatrix} +W_{00} + W_{11} + W_{22} + W_{33} \\ +W_{01} - W_{10} + W_{23} - W_{32} \\ +W_{02} - W_{20} + W_{31} - W_{13} \\ +W_{03} - W_{30} + W_{12} - W_{21} \end{bmatrix} .$$
(10.21)

This is of course exactly the same as the quaternion difference transformation Eq. (10.13), expressed as a profile matrix transformation, and Eq. (10.20) leads, assuming consistent data localization, to the same optimal unit quaternion

$$q_{\text{opt}} = \frac{V}{\|V\|} ,$$
(10.22)

that maximizes the value of Δ_{chord} in Eq. (10.15), and the maximal value of the measure is again $\Delta_{\text{chord}}(q_{\text{opt}}) = \|V\|$.

10.3 Alternative matrix forms of the linear vector chord distance

If the signs of the quaternions representing orientation frames are well behaved, and the frame problem is our only concern, Eqs. (10.15) and (10.16) provide a simple solution to finding the optimal global rotation. If we are anticipating wanting to combine a spatial profile matrix $M(E)$ with an orientation problem in a single 4×4 matrix, or we have problems defining a consistent quaternion sign, there are two further choices of orientation frame measure we may consider.

Matrix form of the linear vector chord distance. While Eq. (10.15) (or Eq. (10.20)) does not immediately fit into the eigensystem-based RMSD matrix method used in the spatial problem, it can in fact be easily transformed from a system linear in q to an equivalent matrix system *quadratic* in q. Since any power of the optimization measure will yield the same extremal solution, we can simply *square* the right-hand side of Eq. (10.15) and write the result in the form

$$\Delta_{\text{chord-sq}} = (q \cdot V)(q \cdot V)$$

$$= \sum_{a=0,b=0}^{3} q_a V_a V_b q_b$$

$$= q \cdot \Omega \cdot q \,, \tag{10.23}$$

where $\Omega_{ab} = V_a V_b$ is a 4×4 symmetric matrix with $\det \Omega = 0$, and $\text{tr}\,\Omega = \sum_a V_a^2 \neq 0$. The eigensystem of Ω is just defined by the eigenvalue $\|V\|^2$, and combination with the spatial eigensystem can be achieved either numerically or algebraically using the trace $\neq 0$ case of our quartic solution. The process differs dramatically from what we did with Δ_{chord}, but the forms of the eigenvectors are necessarily *identical*. Thus it is in fact possible to merge the quaternion frame alignment system for Δ_{chord} into the matrix method of the spatial RMSD using Eq. (10.23).

Fixing sign problem with quadratic rotation matrix chord distance. However, there is another approach that has a very natural way to incorporate manifestly *sign-independent* quaternion chord distances into our general context, and which has a very interesting close relationship to Δ_{chord}. The method begins with the observation that full 3D rotation matrices like Eq. (2.14) can be arranged to rotate the set of reference frames $\{p_k\}$ to be as close as possible to their test frames $\{r_k\}$ by employing a measure that is a particular product of rotation matrices. The essence is to notice that the trace of any 3D rotation matrix expressed in axis-angle form (rotation about a fixed axis $\hat{\mathbf{n}}$ by θ) can be expressed in two equivalent forms:

$$\text{tr}\,R(\theta, \hat{\mathbf{n}}) = 1 + 2\cos\theta \,, \tag{10.24}$$

$$\text{tr}\,R(q) = 3q_0^2 - q_1^2 - q_2^2 - q_3^2 = 4q_0^2 - 1 \,, \tag{10.25}$$

and therefore traces of rotation matrices can be turned into maximizable functions of the angles appearing in the trace. Noting that the squared Frobenius norm of a matrix M is the trace $(\text{tr}\,M \cdot M^t)$, we begin with the goal of minimizing a Frobenius norm of the form

$$\|R(q) \cdot R(p_k) - R(r_k)\|_{\text{Frob.}}^2 \,,$$

and then convert from a minimization problem in this norm to a maximization of the cross-term as usual. The result is, remarkably, an explicitly symmetric and traceless profile matrix in the quaternions. We thus begin with this form of the orientation frame measure (see, e.g., Huynh (2009); Moakher (2002); Hartley et al. (2013)),

$$\Delta_{\text{RRR}} = \sum_{k=1}^{N} \text{tr}\left[R(q) \cdot R(p_k) \cdot R^{-1}(r_k) \right] = \sum_{k=1}^{N} \text{tr}[R(q \star p_k \star \bar{r}_k)] \tag{10.26}$$

$$= \sum_{k=1}^{N} \text{tr}[R(q) \cdot R(p_k \star \bar{r}_k)] = \sum_{k=1}^{N} \text{tr}\left[R(q) \cdot R^{-1}(r_k \star \bar{p}_k) \right] , \tag{10.27}$$

where $\bar{r}$ denotes the complex conjugate or inverse quaternion. We note that due to the correspondence of Δ_{RRR} with a cosine measure (via Eq. (10.24)), this must be *maximized* to find the optimal q, so both Δ_{chord} and Δ_{RRR} correspond naturally to the cross-term measure we used for Euclidean point data, which we can refer to as Δ_x when necessary to distinguish it.

We next observe that the formulas for Δ_{RRR} and the pre-summation arguments of Δ_{chord} are related as follows:

$$\sum_{k=1}^{N} \text{tr}[R(q) \cdot R(p_k) \cdot R(\bar{r}_k)] = \sum_{k=1}^{N} \left(4 ((q \star p_k) \cdot r_k)^2 - (q \cdot q)(p_k \cdot p_k)(r_k \cdot r_k) \right) , \tag{10.28}$$

where of course the last term reduces to a constant since we apply the unit-length constraint to all the quaternions, but it is algebraically essential to the construction. The odd form of Eq. (10.28) is not a typographical error: the conjugate $\bar{r}$ of the reference data must be used in the $R \cdot R \cdot R$ expression, and the ordinary r must be used in both terms on the right-hand side. We conclude that using the $R \cdot R \cdot R$ measure and replacing the argument of Δ_{chord} by its square *before* summing over k are equivalent maximizing measures that eliminate the quaternion sign dependence. Now using the quaternion triple-term identity $(q \star p) \cdot r = q \cdot (r \star \bar{p})$ of Eq. (2.12), we see that each term of Δ_{RRR} reduces to a quaternion product that is a quaternion difference, or a "quaternion displacement" $t_k = r_k \star \bar{p}_k$, i.e., the rotation mapping each individual test frame to its corresponding reference frame,

$$\left.\begin{aligned}
\Delta_{\text{RRR}} = \sum_{k=1}^{N} \text{tr}[R(q) \cdot R(p_k) \cdot R(\bar{r}_k)] &= \sum_{k=1}^{N} \left(4 ((q \star p_k) \cdot r_k)^2 - (q \cdot q)(p_k \cdot p_k)(r_k \cdot r_k) \right) \\
&= \sum_{k=1}^{N} \left(4 (q \cdot (r_k \star \bar{p}_k))^2 - 1 \right) \\
&= 4 \sum_{a,b} q_a \left(\sum_{k=1}^{N} [t_k]_a \, [t_k]_b \right) q_b - N \\
&= 4 q \cdot A(t) \cdot q - N
\end{aligned}\right\} . \tag{10.29}$$

Here the 4×4 matrix $A(t)_{ab} = \sum_{k=1}^{N} [t_k]_a \, [t_k]_b$ is the alternative (equivalent) profile matrix that was introduced by Markley et al. (2007) and Hartley et al. (2013) for the chord-based *quaternion averaging* problem. We can therefore use either the measure Δ_{RRR} or

$$\Delta_A = q \cdot A(t) \cdot q \tag{10.30}$$

as our rotation-matrix-based sign-insensitive chord distance optimization measure. Exactly like our usual spatial measure, these measures must be *maximized* to find the optimal q. It is, however, important to emphasize that the optimal quaternion will *differ* for the $\Delta_{chord} \sim \Delta_{chord\text{-}sq}$ measures vs the $\Delta_{RRR} \sim \Delta_A$ measures, though they will normally be very similar. More details are explored in Section 10.4.

Details of rotation matrix form. We now recognize that the sign-insensitive measures are all very closely related to our original spatial RMSD problem, and all can be solved by finding the optimal quaternion eigenvector q_{opt} of a 4×4 matrix. The procedure for $\Delta_{chord\text{-}sq}$ and Δ_A follows immediately, but it is useful to work out the options for Δ_{RRR} in a little more detail.

Choosing Eq. (10.26) has the remarkable feature of producing, via Eq. (2.14) for $R(q)$, an expression *quadratic* in q, with a symmetric, traceless profile matrix $U(p, r)$ that is *quartic* in the quaternion elements p_k and r_k. This variant of the chord-based quaternion frame alignment problem thus falls into the same category as the standard RMSD problem, and permits the application of the same exact solution (or, indeed, the traditional numerical solution method if that is more efficient). The profile matrix equation is unwieldy to write down explicitly in terms of the quaternion elements quartic in $\{p, r\}$, but we actually have several options for expressing the content in a simpler form. One is to write the matrices in abstract canonical 3×3 form, e.g.,

$$R(p) = [P] = \begin{bmatrix} p_{xx} & p_{xy} & p_{xz} \\ p_{yx} & p_{yy} & p_{yz} \\ p_{zx} & p_{zy} & p_{zz} \end{bmatrix}. \tag{10.31}$$

The *columns* of this matrix are just the three axes of each data element's frame triad. This is often exactly what our original data look like, for example, if the residue orientation frames of a protein are computed from cross products of atom-atom vectors (see, e.g., Chapter 16). Then we can define for each data element the 3×3 matrix

$$[T_k] = R(p_k) \cdot R(\bar{r}_k)] = R(p_k \star \bar{r}_k) = R^{-1}(t_k),$$

so we can write T either in terms of a 3×3 matrix like Eq. (10.31) derived from the actual frame-column data, or in terms of Eq. (2.14) and the quaternion frame data $t_k = r_k \star \bar{p}_k$. We then may write the frame measure in general as

$$\Delta_{RRR} = \sum_{k=1}^{N} \text{tr}(R(q) \cdot T_k) = \sum_{a=1, b=1}^{3} R_{ba}(q) T_{ab}, \tag{10.32}$$

where the frame-based cross-covariance matrix is simply $T_{ab} = \sum_{k=1}^{N} [T_k]_{ab}$. As before, we can easily expand $R(q)$ using Eq. (2.14) to convert the measure to a 4D linear algebra problem of the form

$$\Delta_{RRR} = \sum_{a=0, b=0}^{3} q_a \cdot U_{ab}(p, r) \cdot q_b = q \cdot U(p, r) \cdot q. \tag{10.33}$$

Here $U(p, r) = U(T)$ has the same relation to T as $M(E)$ does to E in Eq. (8.9). We may choose to write the profile matrix $U = \sum_k U_k$ appearing in Δ_{RRR} either in terms of the individual k-th components of the numerical 3D rotation matrix $T = R^{-1}(t)$ or using the composite quaternion $t = r \star \bar{p}$:

$$U_k(T) \equiv U(t_k)$$

$$= \begin{bmatrix} T_{xx} + T_{yy} + T_{zz} & T_{yz} - T_{zy} & T_{zx} - T_{xz} & T_{xy} - T_{yx} \\ T_{yz} - T_{zy} & T_{xx} - T_{yy} - T_{zz} & T_{xy} + T_{yx} & T_{xz} + T_{zx} \\ T_{zx} - T_{xz} & T_{xy} + T_{yx} & -T_{xx} + T_{yy} - T_{zz} & T_{yz} + T_{zy} \\ T_{xy} - T_{yx} & T_{xz} + T_{zx} & T_{yz} + T_{zy} & -T_{xx} - T_{yy} + T_{zz} \end{bmatrix}_k \quad (10.34)$$

$$= \begin{bmatrix} 3t_0^2 - t_1^2 - t_2^2 - t_3^2 & 4t_0t_1 & 4t_0t_2 & 4t_0t_3 \\ 4t_0t_1 & -t_0^2 + 3t_1^2 - t_2^2 - t_3^2 & 4t_1t_2 & 4t_1t_3 \\ 4t_0t_2 & 4t_1t_2 & -t_0^2 - t_1^2 + 3t_2^2 - t_3^2 & 4t_2t_3 \\ 4t_0t_3 & 4t_1t_3 & 4t_2t_3 & -t_0^2 - t_1^2 - t_2^2 + 3t_3^2 \end{bmatrix}_k .$$
$$(10.35)$$

Both Eq. (10.34) and Eq. (10.35) are *quartic* (and identical) when expanded in terms of the quaternion data $\{p_k, r_k\}$. To compute the necessary 4×4 numerical profile matrix U, one need only substitute the appropriate 3D frame triads or their corresponding quaternions for the k-th frame pair and sum over k. Since the orientation frame profile matrix U is symmetric and traceless just like the Euclidean profile matrix M, the same solution methods for the optimal quaternion rotation q_{opt} will work without alteration in this case, which is probably the preferable method for the general problem.

Remark: Reduction to adjugate form. If we are willing to give up the traceless property of the form of the profile matrix Eq. (10.35), we see that by adding $t_0^2 + t_1^2 + t_2^2 + t_3^2 = 1$ to the diagonal and dividing by 4, we recover exactly the *adjugate form* of the profile matrix encountered in Chapter 9, Eq. (9.13):

$$U(t_k) = \begin{bmatrix} t_0^2 & t_0t_1 & t_0t_2 & t_0t_3 \\ t_0t_1 & t_1^2 & t_1t_2 & t_1t_3 \\ t_0t_2 & t_1t_2 & t_2^2 & t_2t_3 \\ t_0t_3 & t_1t_3 & t_2t_3 & t_3^2 \end{bmatrix}_k . \quad (10.36)$$

As before, this is just the profile matrix, not technically the adjugate matrix extracting the eigenvectors of the characteristic equation, but they are closely related. This form appears in the quaternion averaging literature, and of course in the A matrix of Eq. (10.29), and is deeply related to the other applications of the quaternion adjugate variables in later chapters.

Evaluation. The details of evaluating the properties of our quaternion frame alignment algorithms, and especially comparing the chord approximation to the arc length measure, are tedious and are studied separately next in Section 10.4. The top-level result is that, even for quite large rotational differences, the mean differences between the arc length measure's numerical optimal angle and the various chord approximations are very small.

10.4 Evaluating the 3D orientation frame solution

The validity of our approximate chord measures for determining the optimal global frame rotation can be evaluated by comparing their outcomes to the precise geodesic arc length measure of Eq. (10.8). The latter is tricky to optimize, but choosing appropriate techniques, e.g., using the `scipy.optimize.minimize()` Python utility or the Mathematica `FindMinimum[ ]` utility, it is possible to determine good numerical solutions without writing custom code; in our experiments, fluctuations due to numerical precision limitations were noticeable, but presumably conventional conditioning techniques, which we have not attempted to explore, could improve that significantly. We employed a collection of 1000 simulated quaternion data sets of length 100 for the reference cases, then imposed a normal distribution of random noise on the reference data, followed by a global rotation of all those noisy data points distributed around 45 degrees to produce a corresponding collection of corresponding quaternion test data sets to be aligned. (Observe that, due to noise, we do *not* expect the optimal rotation angles to match the exact global rotations, though they will be nearby.)

We then collected the optimal quaternions for the following cases:

(a) **Arc length (numerical).** This is the "gold standard," modulo the occasional data pair that seems to challenge the numerical stability of the computation (which was to be expected). We obtained data set (a) of quaternions that numerically minimized the nonlinear geodesic arc-length-squared measure of Eq. (10.8); this is in principle the best estimate one can possibly get for the optimal quaternion rotations to align a set of 3D test frame triads with a corresponding set of reference frame triads. There is no known way to find this set of optimal quaternions using our linear algebra methods.

(b) **Chord length (numerical and algebraic).** This approach, designated as data set (b), is based on the approximation to Eq. (8.1) illustrated in Fig. 10.4, replacing the arc length by the chord length, which amounts to removing the arccosine and using the effective maximal cosines $(t \to \tilde{t})$ to define the measure. The form given in Eq. (10.18) is a minimization problem that is exactly the quaternion analog of the RMSD problem definition in Eq. (8.1) for spatial data, with the additional constraint that all the spatial data must be unit-length 4-vectors (which have only three degrees of freedom) instead of arbitrary 3-vectors. In addition, the convergence condition for clustering of the data within a ball should in principle be satisfied for the optimal solution of Eq. (10.18) to be global; our data simulation pushes these limits, but in practice the convergence is typically satisfied. Just as Eq. (8.1) and its cross-term form Eq. (8.6) give exactly the same results for spatial data when the measures are minimized and maximized, respectively, the orientation-problem equations Eq. (10.18) and Eq. (10.19) do the same for the quaternion measure. Finally, the two cross-term forms Eq. (10.20) and Eq. (10.23) give the same optimal quaternions, with the interesting fact that Eq. (10.20) yields the optimal quaternion from a linear equation, and Eq. (10.23) gives an identical result from a quadratic matrix equation that works the same way as the RMSD matrix optimization, except that the symmetric profile matrix is no longer traceless.

Thus there are in fact four ways of looking at the chord length measure and obtaining exactly the same optimal quaternions, and we have checked these using two numerical optimizations and two algebraic optimizations. These options are:

- **Minimizing Euclidean chord-length-squared.** Here we write the chord approximation to the quaternion frame alignment problem using Eq. (10.18), which is exactly parallel to the RMSD problem employing Eq. (8.1), modulo the sign ambiguity issue. We test this by performing a numerical minimization.

- **Maximizing the chord length cross-term.** Just as the RMSD cross-term maximization problem Eq. (8.8) is equivalent to the RMSD least-squares minimization problem of Eq. (8.2), we can use maximization of the quaternion cross-term Eq. (10.19) equivalently with the minimization of the chord length Eq. (10.18). We test this by performing a numerical maximization.

- **Linear reduction of the chord length cross-term.** Pulling out the linear coefficients of each quaternion component in Eq. (10.19) generates Eq. (10.20), where the 4-vector $V_a(W)$ of Eq. (10.21) plays the role of the RMSD profile matrix $M_{ab}(E)$ in Eq. (8.8). Here we test the optimization by algebraically solving the linear expression Eq. (10.20).

- **Quadratic equivalent matrix form of the chord length cross-term.** Finally, there is in fact a maximal matrix eigenvalue problem Eq. (10.23) that works like Eq. (8.8) by squaring Eq. (10.20) to get a matrix problem $q \cdot \Omega \cdot q$ with $\Omega_{ab} = V_a V_b$. Despite the presence of a nonvanishing trace, the maximal quaternion eigenvectors are the same as the other three cases above. This produces the same optimal quaternion solutions as solving the (much, much simpler) linear problem of Eq. (10.20). This can also be checked algebraically.

(c) $(\text{tr}\, R(q) \cdot R(p) \cdot R(\bar{r}))$ **Chord length (algebraic).** Finally, the most rigorous method if consistency of quaternion signs cannot be guaranteed is to use a measure in which algebraic squares occur throughout and enforce rigorous sign independence. This is our data set (c). Such measures must of necessity be *quartic* in the quaternion test and reference data, and thus are distinct from the measures of (b) that are *quadratic* in the data elements. This $(\text{tr}\, R(q) \cdot R(p) \cdot R(\bar{r}))$ measure is the form that is most easily integrated into the combined rotational-translational problem treated in the next section, because the combined matrices are both symmetric and traceless like the original RMSD profile matrices. Furthermore, it is obvious from Eq. (10.28) that this measure is exactly the same as the one obtained from Eq. (10.19) if we squared *each term in k* before summing the cross-term data elements in option (b). Thus, whichever actual formula we choose, we appear to have exhausted the options for quaternion-sign-independent quartic measures for the orientation data problem.

The task now is simply to evaluate how close the optimal quaternion solutions for the arc length measure (a) are to the quadratic chord length measures (b) and the quartic chord length measures (c). In addition, we would like to know how close the fragile but very elegant quadratic measures (b) are to the rigorously sign-insensitive quartic measures (c); we expect them to be similar, but we do not expect them to be identical.

To quantify the closeness of the measures, we took the magnitude of the inner products between competing optimal quaternions for the same data set, which is essentially a cosine measure, took the arccosines, and converted to degrees. The results were histogrammed for 1000 random samples consisting of $N = 100$ data points, and are presented in Fig. 10.10. The means and standard deviations of the optimal total rotations relative to the identity frame for the three cases are given below.

Measure type	Mean (deg)	Std dev (deg)
(a) Arc length	44.8062	11.2307
(b) Chord quadratic	44.8063	11.2308
(c) Chord quartic	44.8065	11.2310

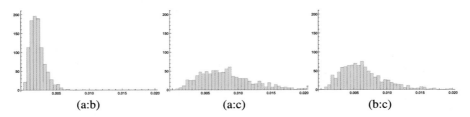

(a:b)   (a:c)   (b:c)

FIGURE 10.10

Spectrum in degrees of angular differences between optimal quaternion alignment rotations for quaternion frames. (a:b): (a) vs. (b), true arc length vs. approximate quadratic chord length measure. (a:c): (a) vs. (c), true arc length vs. approximate quartic chord length measure. (b:c): (b) vs. (c), approximate quadratic vs. approximate quartic chord length measure.

One can see that our simulated data set involved a large range of global rotations, and that all three methods produced a set of rotations back to the optimal alignment that are not significantly different statistically. We thus expect very little difference in the histograms of the case-by-case optimal quaternions produced by the three methods. The mean differences illustrated in the figures are summarized below.

Figure (pair)	Mean (deg)	Std dev (deg)
Fig. 10.10 (a:b)	0.0021268	0.0011284
Fig. 10.10 (a:c)	0.0084807	0.0044809
Fig. 10.10 (b:c)	0.0063539	0.0033526

We emphasize that these numbers are in degrees for 1000 simulated samples with a distribution of global angles having a standard deviation of 11 degrees. Thus we should have no issues using the chord approximation, though it does seem that the $q \cdot V$ measure is significantly better both in accuracy and simplicity of computation.

The combined point-frame alignment problem☆

Since we now have precise alignment procedures for both 3D spatial cloud pairs and 3D frame triad pairs (using the exact measure for the former and the approximate chord measure for the latter), we devote this chapter to considering the full six-degree-of-freedom alignment problem for combined point location and frame orientation data from a single structure. As always, this problem can be solved either by numerical eigenvalue methods or in closed algebraic form using the eigensystem formulation of both alignment problems presented in the previous chapters. There are clearly appropriate domains of this type, e.g., any protein structure in the PDB database can be converted to a list of residue center locations combined with their local frame triads (see Chapter 16 and Hanson and Thakur, 2012). Nevertheless, the combined alignment problem merging 3D spatial data and 3D orientation frame data involves a number of issues and subtleties. Here we explore various options and evaluate their performance to assist anyone who might think of attempting a combined alignment problem. The overall result of the investigation in this chapter is that our best available solutions to the combined alignment problem for simultaneously rotated point cloud data and rotated frame triad data are not obviously very illuminating.

11.1 Combined optimization measures

The RMSD point-to-point matching technology is the starting point of our combined optimization measure, and we begin by discussing how best to work with the more subtle orientation frame alignment technology, and how to merge it most clearly into the context of the RMSD problem.

Thus we start by examining the Δ_{RRR} measure since its profile matrix is traceless and manifestly independent of the quaternion signs, although there is no serious obstacle to using the $\Delta_{\mathrm{frame\text{-}sq}}$ measure if the data are properly prepared and one prefers the simpler measure. We will choose the notation Δ_x for the RMSD cross-term alignment measure, and we will let Δ_f stand for whatever orientation frame measure we have chosen. Any combined measure that merges the spatial and frame measures in some way will be denoted by Δ_{xf}.

Dimensional incompatibility. A significant aspect of establishing a combined measure including the point measure Δ_x and the frame orientation measure Δ_f is the fact that the measures are *dimensionally incompatible*. We *cannot* directly combine the corresponding data minimization measures $\Delta_x(q_x) = \epsilon_x$ and $\Delta_f(q_f) = \epsilon_f$ because the spatial measure has dimensions of $(\text{length})^2$ and the frame measure is

---

☆ This chapter derives much of its content from Hanson (2020).

essentially a dimensionless trigonometric function (the arc distance measure produces (radians)^2, which is still incompatible).

While it should be obvious that combining a rotational measure with a measure having dimensions of spatial length requires an arbitrary, problem-specific, interpolating constant with dimensions of length to produce a compatible measure, there has been some confusion on this point. Applications in the molecular entropy literature, for example, may combine Euclidean distance (as opposed to RMSD's dimensional cross-covariance) with frame orientation measures. The need for dimension-compensating constants was recognized and incorporated, for example, in the work of Fogolari et al. (2016).

Constructing possible measures. Our approach to defining a valid heuristic combined measure has three components:

- **Normalize the profiles.** The numerical sizes of the maximal eigenvalues of the Δ_x and the Δ_f systems can easily differ by orders of magnitude. Since scaling the profile matrices changes the eigenvalues *but not the eigenvectors*, it is perfectly legitimate to start by dividing the profiles by their maximal eigenvalues before beginning the combined optimization, since this accomplishes the sensible effect of assigning maximal eigenvalues of exactly unity to both of our scaled profile matrices.

- **Interpolate between the profiles.** To allow an arbitrary sensible weighting distinguishing between a location-dominated measure and an orientation-dominated measure, we simply incorporate a linear interpolation parameter $t \in [0, 1]$, with $t = 0$ singling out Δ_x and the pure (unit-eigenvalue) location-based RMSD, and $t = 1$ singling out Δ_f and the pure orientation (unit-eigenvalue) quaternion frame solution.

- **Scale the frame profile.** Finally, we incorporate the mandatory dimensional scaling adjustment by incorporating one additional (nominally dimensional) parameter σ that scales the orientation parameter space described by Δ_f to be more or less important than the "canonical" spatial dimension component Δ_x, which we leave unscaled. That is, with $\sigma = 0$ only the spatial measure survives, with $\sigma = 1$ the normalized measures have equal contributions, and with $\sigma > 1$ the orientation measure dominates (this effectively undoes the original frame profile eigenvalue scaling).

We thus start with a combined spatial-rotational measure of the form

$$
\begin{aligned}
\Delta_{\text{initial}} &= (1-t) \sum_{a=1,b=1}^{3} R^{ba}(q) E_{ab} + t\sigma \sum_{a=1,b=1}^{3} R^{ba}(q) S_{ab} \\
&= (1-t)\,\text{tr}\,(R(q) \cdot E) + t\sigma\,\text{tr}\,(R(q) \cdot S) \\
&= \sum_{a=0,b=0}^{3} q_a\,[(1-t)M_{ab}(E) + t\sigma\,U_{ab}(S)]\,q_b \\
&= q \cdot [(1-t)M(E) + t\sigma\,U(S)] \cdot q \,,
\end{aligned}
\tag{11.1}
$$

where E_{ab} and S_{ab}, with $M_{ab}(E)$ and $S_{ab}(E)$, are chosen to have their reference and target data in compatible order. We then impose the unit-eigenvalue normalization on $M(E)$ and $U(S)$, giving our final measure as

$$
\Delta_{xf}(t, \sigma) = q \cdot \left[(1-t)\frac{M(E)}{\epsilon_x} + t\sigma\frac{U(S)}{\epsilon_f} \right] \cdot q \,.
\tag{11.2}
$$

Because of the dimensional incompatibility of Δ_x and Δ_f, we have to treat the ratio

$$\lambda^2 = \frac{t\sigma}{1-t}$$

as a dimensional constant such as that adopted by Fogolari et al. (2016) in their entropy calculations, so if t is dimensionless, then σ carries the dimensional scale information.

From the profile matrix of Eq. (11.2), we now extract our optimal rotation solution using the same equations as always, solving for the maximal eigenvalue and its eigenvector either numerically or algebraically, leading to the equivalent of Eq. (8.10), as we have solved the standard RMSD maximal eigenvalue problem. The result is a parameterized eigensystem

$$\left.\begin{array}{c} \epsilon_{\mathrm{opt}}(t,\sigma) \\ q_{\mathrm{opt}}(t,\sigma) \end{array}\right\} \tag{11.3}$$

yielding the optimal values $R(q_{\mathrm{opt}}(t,\sigma))$, $\Delta_{xf} = \epsilon_{\mathrm{opt}}(t,\sigma)$ based on the data $\{E, S\}$ no matter what we take as the values of the two variables (t,σ).

Properties of the combined optimization. Substantially different features arise in the solutions depending on how close the optimal rotations were for the initial, separate, systems Δ_x and Δ_f. We now choose a selection of simulated data sets with the following choices of approximate initial global rotations of the test data sets relative to the reference data.

Table 11.1 Values of the mean rotational displacements applied to the sample data for the spatial points vs. the sample data for frame orientations. Large displacements were chosen for some cases to stress the system.

Data ID	(space, orientation)
Data Set 1	$(22°, -22°)$
Data Set 2	$(22°, -11°)$
Data Set 3	$(22°, 0°)$
Data Set 4	$(22°, 11°)$
Data Set 5	$(22°, 21°)$

In Fig. 11.1, we plot the trajectory of the maximal combined similarity measure for Data Set 1 as a function of t, showing the behavior for $\sigma = 1.0, 0.80$, and 1.15. Fig. 11.2 shows a more comprehensive representation of the continuous behavior with σ, and in both figures, we see that the true optima are *at the end points*, $t = 0, 1$, the locations associated with the pure profile eigenvector solutions $q_x(\mathrm{opt})$ and $q_f(\mathrm{opt})$. There is no *better* optimal eigenvector (i.e., global rotation) for any intermediate value of t. In some circumstances, however, it might be argued that it is appropriate to choose the *distinguished value* of t at the minimum of the curve $\Delta_{xf}(t, \sigma = 1)$. As we shall see in a moment, just as in Fig. 11.1 for $\sigma = 1$, this point is generally within a few percent of $t = 0.5$. As the spatial and orientation optima get closer and closer, the curves in t become much flatter and less distinguished, while the variation in σ is qualitatively the same as in Fig. 11.2.

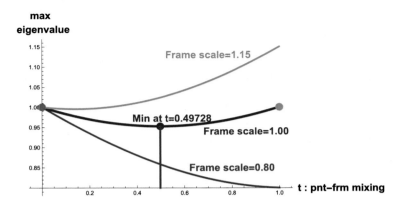

FIGURE 11.1

The blue curve is the path of the composite eigenvalue for Data Set 1 (the value of the similarity measure $\Delta_{xf}(t, 1)$) in the interpolation variable t with equally weighted space and orientation data, i.e., $\sigma = 1$. It has maxima only at the "pure" extremes at $t = 0, 1$, but there is a minimum that occurs, for these data, not at $t = 1/2$, but very nearby at $t = 0.49728$. Increasing the influence of the spatial data by taking $\sigma = 0.8$ gives the red curve, and increasing the influence of the orientation data by taking $\sigma = 1.15$ gives the green curve.

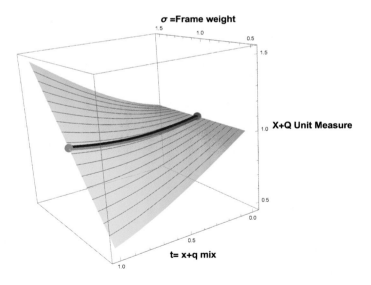

FIGURE 11.2

The $\Delta(t, \sigma)$ similarity measure surface for Data Set 1 as a function of the interpolation parameter t and the relative scaling of the orientation term with σ, with the slightly concave curve at $\sigma = 1$ in the middle. The other data sets look very much like this one.

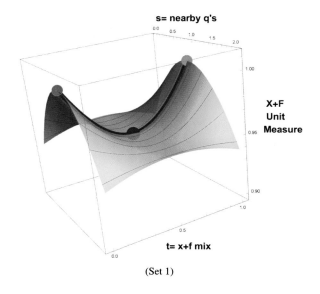

(Set 1)

FIGURE 11.3

The $\Delta_{xf}(t, 1)$ similarity measure surface for Data Set 1, with x-angle = 22 degrees, f-angle = −22 degrees, and fixed $\sigma = 1$, showing the deviation with the quaternion varying perpendicularly around the solution $q(t)$, starting at the identity quaternion at $s = 0$, as a function of the interpolation parameter t. Since $q(t)$ is the maximal eigenvector, all variations in q peak there. Both have distinguished central points at $t \approx 0.5$.

Finally, we examine one more amusing visualization of the properties of the composite solutions, restricting ourselves to $\sigma = 1$ for simplicity, and examining the "sideways warp" in the quaternion eigenvector $q_{opt}(t, \sigma = 1)$ in Eq. (11.3). We examine what happens to the combined similarity measure Eq. (11.2) if we smoothly interpolate from the identity matrix (that is, the quaternion $q_{ID} = (1, 0, 0, 0)$) through the optimal solution for each t and beyond the optimum by the same amount, using the slerp interpolation defined in Eq. (2.19), i.e., $q(s) = \text{slerp}(q_{ID}, q_{opt}(t, \sigma = 1); s)$. Fig. 11.3 shows Data Set 1, with the largest relative spatial vs. orientation angular differences, and Fig. 11.4 corresponds to the intervening Data Sets 2, 3, 4, and 5, with the data set parameters given above in Table 11.1. Data Set 5 in particular is perhaps the most realistic example, having nearly identical spatial and angular rotations, and we see negligible differences between the spatial and angular structures. These graphics also show how the local, nonoptimal, neighboring quaternion values peak in s at the optimal ridge going from $t = 0$ to $t = 1$. The red dot is the maximum of Δ_x at $t = 0$, the green dot is the maximum of Δ_f at $t = 1$, and the blue dot, specific to each data set, is the distinguished point at the *minimum* of $\Delta_{xf}(t, \sigma = 1)$ in t, which for our data sets are always within 1% of $t = 0.5$. We observe that for equal and opposite rotations, the midpoint coincides almost exactly with the identity quaternion that occurs at the left and right boundaries of the plot. In other respects, the data in these figures show that we do not have *maxima* in the middle of the interpolation in t, but we do have a distinguished value, always very near $t = 0.5$, that could be used as a baseline for a hybrid translational-rotational rotation choice.

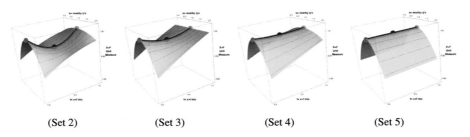

| (Set 2) | (Set 3) | (Set 4) | (Set 5) |

FIGURE 11.4

The $\Delta_{xf}(t, 1)$ similarity measures with $q(s)$ interpolated from the identity through the optimum for Δ_{xf} and past to the identity mirror point, for Data Sets 2, 3, 4, and 5, where Data Set 5 has the x-angle and the f-angle only 1 degree apart, as we might have for real experimental data.

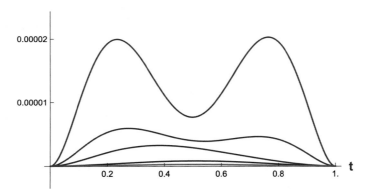

FIGURE 11.5

Here we see how close a simple slerp(t) between the extremal optimal eigenvectors $q_{\mathrm{opt}}(t = 0, \sigma = 1) = q_x(\mathrm{opt})$ and $q_{\mathrm{opt}}(t = 1, \sigma = 1) = q_f(\mathrm{opt})$ is to the rigorous result where we optimized $q_{\mathrm{opt}}(t, \sigma = 1)$ *for all t*. The differences are *relative to the unit eigenvalue*, and thus are of order thousandths of a percent, decreasing significantly as the global rotations applied to the space and orientation data approach one another. The largest deviation is for Data Set 1, which interestingly has a third minimum near the center in t; for the highly similar data in Data Set 5, the difference shown in red had to be magnified by 100 even to show up on the graph.

11.2 Practical simplification of the composite measure

Upon inspection of Eq. (11.2), one wonders what happens if we simply use the slerp defined in Eq. (2.19) to interpolate between the *separate* spatial and orientation frame optimal quaternions. While the eigenvalues that correspond to the two scaled terms M/ϵ_x and U/ϵ_f in Eq. (11.2) are both unity, and thus differ from the eigenvalues of M and U, the individual normalized eigenvectors $q_{x:\mathrm{opt}}$ and $q_{f:\mathrm{opt}}$ are the same. Thus, if we are happy with simply using a hand-tuned fraction of the combination of the two corresponding rotations, we can just choose a composite rotation $R(q(t))$ specified by

$$q(t) = \text{slerp}(q_{x:\text{opt}}, q_{f:\text{opt}}; t) \tag{11.4}$$

to study the composite six-degree-of-freedom alignment problem. In fact, if we simply plug this $q(t)$ into Eq. (11.2) for any t (and $\sigma = 1$), we find *negligible differences* between the quaternions $q(t)$ and $q_{\text{opt}}(t, 1)$ as a function of t. In Fig. 11.5, we plot the continuous differences of the similarity functions, which we recall are scaled to have a maximal eigenvalue equal to unity. We suggest in addition that any particular effect of $\sigma \neq 1$ could be achieved at some value of t in the interpolation. We thus conclude that, for all practical purposes, we might as well use Eq. (11.4) with the parameter t adjusted to achieve the objective of Eq. (11.2) to study composite translational and rotational alignment similarities.

Quaternion adjugate methods for pose estimation ☆

3

This part is closely related to the preceding Part 2, and has again as its basic task the discovery of rotations relating matched clouds of observed data. However, here we look very carefully at the possible pitfalls that are hidden in the process of extracting a quaternion from measured orientation data in a cloud matching problem, and explore some related fundamental features of machine learning. We are then led naturally to investigate problems in pose estimation, where for high-quality data, we find closed form rotation solutions as ratios of determinants for cloud matching and parallel projection tasks, as well as a potentially interesting similar solution for perspective projections.

☆ These chapters are largely inspired by collaborative work with Sonya M. Hanson related to arXiv Hanson and Hanson (2022).

Exploring the quaternion adjugate variables☆

We now study the nuances of computing quaternion data from measured data items that in some way determine an orientation. The foundational problem starts with a measured, assumed errorful, 3D rotation matrix, and the task is to determine the *optimal true rotation* that best approximates that potentially noisy rotation data in terms of a quaternion. We point out that any process that computes a *quaternion function* has inescapable flaws because quaternions lie on the $\mathbf{S}^3$ *manifold*, and cannot be described by a single function. We will argue that quaternions are most effectively discovered in a way consistent with the nontrivial $\mathbf{S}^3$ geometry by exploiting the multivalued construct of the *quaternion adjugate variables* to parameterize a full set of charts forming an atlas for $\mathbf{S}^3$.

In this chapter, we will concentrate on computing the optimal quaternion corresponding to a given exact or inexact measurement of a rotation matrix. This will incorporate the important related problem, already introduced in Chapter 9, of computing the closest possible orthonormal 3D rotation matrix corresponding to a matrix that is not a valid rotation, but is believed to be a useful approximation. Implications of the concepts studied in this chapter for machine learning are briefly explored in Chapter 13. Then in Chapters 14 and 15, we will exploit the unique algebraically simplifying properties of the 2D and 3D quaternion adjugate variables to obtain exact symbolic solutions to the least-squares problems for error-free data produced by 3D:3D cloud matching, 3D:orthographic-2D projection matching, and 3D:perspective-2D projection matching.

12.1 Overview of the quaternion extraction problem

The optimal-quaternion extraction problem is universal, and occurs in many application domains, reflecting the appealing fact that unit quaternions form a smooth manifold that parameterizes rotations free of Euler angle issues such as gimbal lock (see, e.g., Hanson (2006, 2020)). Our investigation is motivated particularly by the so-called "quaternion discontinuity problem" that has been noted in a variety of machine learning contexts involving rotations by, e.g., Saxena et al. (2009); Zhou et al. (2019); Peretroukhin et al. (2020); Zhao et al. (2020); Xiang and Li (2020). Understanding such potential issues is important, as the determination of orientation and pose is widespread in machine learning applications, including self-driving vehicles and drone navigation, as well as general problems of understanding 3D space, evaluating 3D models, and the extraction of 3D information from 2D data. Our first goal is to show explicitly how understanding the topological properties of the quaternion, con-

---

☆ Much of this chapter is derived from Hanson and Hanson (2022).

Visualizing More Quaternions. https://doi.org/10.1016/B978-0-32-399202-2.00022-8

sidered as a multi-sector manifold, can resolve the questions regarding quaternion discontinuities and related anomalies that have been observed.

Traditional computational algorithms (see, e.g., Shepperd (1978); Hanson (2006); Sarabandi et al. (2020) and Appendix G) for extracting a corresponding quaternion from a 3D rotation matrix have always included specific methods to account for possible singularities and discontinuities in the mapping, but have not always been clearly incorporated into applications. One of our specific contributions is to revisit an underappreciated linear algebra construction known formally as the *adjugate matrix*, essentially the numerator of the cofactor-based inverse of a matrix; the adjugate suggests an alternative set of quaternion-related variables that has surprising use cases, greatly clarifying how the traditional quaternion extraction algorithms avoid singularities, suggesting a new appreciation of a variational method for quaternion extraction due to Bar-Itzhack (2000) that we introduced in Chapter 9, and enabling unique ways of applying least-squares methods to solving 3D pose extraction, both analytically and in machine learning contexts. We are thus able to explain the origin of the discontinuity problem and its resolution using multivalued charts based on the quaternion adjugate variables.

12.2 Fundamental background

The arguments we present in this chapter rely on a short list of key background concepts, most already having made an appearance, that we review here:

- **Quaternions parameterize a rotation in terms of a point on a topological 3-sphere.** Any 4D vector q with unit length, $q \cdot q = 1$, is a quaternion point on the unit 3-sphere $\mathbf{S}^3$, and corresponds exactly to a *proper* 3D rotation matrix $R = R(q)$ through the basic formula of Eq. (2.14). We remind ourselves that this fundamental equation is quadratic in q, with $R(q) = R(-q)$, so every possible rotation is represented *twice* in the manifold $\mathbf{S}^3$. Therefore every possible rotation appears *once* in a hyperhemisphere of $\mathbf{S}^3$, the solid 3-ball $\mathbf{B}^3$, which, upon identifying diametrically opposite points on $\mathbf{B}^3$ to produce the projective space $\mathbf{RP}^3$, becomes the topological group $\mathbf{SO}(3)$.
- **Quaternions encompass the axis-angle rotation parameterization.** The axis-angle representation $R = R(\theta, \hat{\mathbf{n}})$ given in Eq. (2.17) parameterizes any 3D rotation in terms of the eigenvector $\hat{\mathbf{n}}$ of R, the direction left fixed by the rotation, and the angular size θ of that rotation, and is described by a doubled quaternion parameterization $q(\theta, \hat{\mathbf{n}}) = \big(\cos(\theta/2), \, \hat{\mathbf{n}} \sin(\theta/2)\big)$, where $0 \leqslant \theta < 4\pi$.
- **Classical error-robust quaternion extraction is hard.** Extracting a corresponding quaternion from a numerical rotation matrix $R(m)$, with measured numerical matrix elements m_{ij}, can be defined precisely as the task of recovering the axis-angle parameters from $R(m)$. This is well known to be a subtle multi-step process (see, e.g., Shepperd, 1978; Hanson, 2006; Sarabandi et al., 2020). In order to account for all possible anomalies, however rare, the classical procedure follows an elaborate algorithm that we include in a variety of implementations in Appendix G. Note that this procedure does not gracefully handle finding the optimal quaternion that *best represents* an error-containing measurement $R(m)$, but this can be achieved using the variational method of Bar-Itzhack (2000), which we introduced in Chapter 9.
- **The adjugate matrix approach to eigenvectors is important.** Standard methods for finding the rotation aligning a rotated 3D point cloud with its reference cloud (see, e.g., Horn, 1987; Hanson, 2020, and Chapter 8) determine the optimal quaternion by finding the maximal eigenvalue of a

certain 4×4 matrix. The normalized eigenvector of that maximal eigenvalue is q_{opt}, the quaternion determining the optimal aligning rotation matrix $R_{\mathrm{opt}} = R(q_{\mathrm{opt}})$. Buried in the last step of this routine linear algebra calculation, we have in fact a mandatory process that closely parallels the Shepperd algorithm (see Appendix G). The seldom-mentioned ambiguity in the process of going from a symmetric real matrix M and one of its eigenvalues λ to a well-behaved corresponding eigenvector can be resolved in a clear form by computing the *adjugate matrix* of the characteristic eigenvalue matrix $\chi(M, \lambda)$, and extracting a well-defined normalized eigenvector by selecting from the four adjugate columns the one that is best behaved under normalization: the existence of at least one adjugate column with a normalizable solution is guaranteed for unit-length quaternions.

Overview of the chapter. We will use the relationships between quaternions and rotation matrices to show there are no singularity-free single functions relating a quaternion to a measured rotation matrix, but that an adjugate matrix, listing four alternatives (or eight, taking into account the quaternion sign ambiguity) describing the entire quaternion manifold $\mathbf{S}^3$, always contains at least one valid quaternion describing a particular domain of rotation matrix parameters. We will illustrate our arguments beginning with a simplified and intuitive 2D rotation framework that exhibits essentially all the relevant properties. We explore several aspects of the 2D problem in preparation for a parallel treatment of the application-relevant 3D quaternion rotation case, including a direct solution for a quaternion in terms of rotation matrix elements and a variational method (Bar-Itzhack, 2000) for both precise and noisy data. This whole procedure is then repeated for the 3D case.

12.3 2D rotations and the quaternion map

All of the basic features we wish to expose occur in 2D rotations, so we begin with two dimensions as a case study and then treat the more relevant problem of quaternions and 3D rotations in Section 12.4. The context in the back of our minds is the exploration of data sets of (reference, sample) pairs of matched ND point clouds with elements $(\mathbf{x}, \mathbf{u})$ related by a single rotation matrix R,

$$\mathbf{u} = R \cdot \mathbf{x} + \langle \mathit{noise} \rangle \, .$$

The data set $\mathbf{X} = \{\mathbf{x}_k\}$ takes the role of a standard K-element template or reference, and $\mathbf{U} = \{\mathbf{u}_k\}$ the role of a matched sample cloud that was created by applying an unknown rotation R to the reference cloud. Our goal is to take the pairwise matched data sets $(\mathbf{x}, \mathbf{u})$ as input, and then identify the best possible rotation R that acts to transform the reference set $\mathbf{X}$ into alignment with $\mathbf{U}$. Although this problem has well-known singular value decomposition (SVD) and matrix-square-root solutions in all dimensions (Schönemann, 1966; Golub and van Loan, 1983; Horn et al., 1988), and quaternion closed form algebraic solutions in 2D, 3D, and 4D (Hanson, 2020), it has been observed in the machine learning literature that there can be issues such as discontinuities that appear when one needs to find a quaternion q that represents a given rotation R, and attempts to learn $q(R)$. We will resolve these discrepancies in 2D, a first step towards confirming our trust in quaternions as the representation of choice for generic 3D optimal rotation discovery problems.

12.3.1 Direct solution of the 2D problem

We begin with the basic structure of 2D rotations, already given a fairly complete treatment in Chapter 7, by writing down the two equivalent forms, suggested by Eqs. (2.14) and (2.16), of an arbitrary 2D rotation matrix R. The transformation rotating the x-axis counterclockwise (right-handed) in the direction of the y-axis by an angle θ is parameterized by $c = \cos\theta$, $s = \sin\theta$ or the dimension-reduced quaternion $a = \cos(\theta/2)$, and $b = \sin(\theta/2)$, where $c^2 + s^2 = a^2 + b^2 = 1$, and we note that $0° \leqslant \theta < 720°$. The quaternion variables (a, b) cover the (a, b) circle only once, while the (c, s) pair covers that circle twice. Our candidate 2D rotations are then

$$R(c, s) = \begin{bmatrix} c & -s \\ s & c \end{bmatrix}, \tag{12.1}$$

$$R(a, b) = \begin{bmatrix} a^2 - b^2 & -2ab \\ 2ab & a^2 - b^2 \end{bmatrix}. \tag{12.2}$$

We easily verify that $\det R(c, s) = \det R(a, b) = 1$, and also that $R \cdot R^t = I_2$, where I_2 is the 2D identity matrix. The most important property of $R(c, s)$ is that its matrix elements correspond to a *numerically measurable* rotation matrix, as do noisy versions of $R(c, s)$ that we will look at shortly, while we will see that $R(a, b)$ has some intriguing ambiguities.

Now suppose we erase the formulas for R in terms of θ, and think only of the algebraic expressions in Eqs. (12.1) and (12.2), assuming that we have some sound way of measuring this 2D rotation to determine the numerical values of (c, s). Then we can find expressions for the now-abstract variables (a, b) in several ways. We begin by noting that the constraints $R(a, b) = R(c, s)$ and $R(a, b) \cdot R(c, s)^t = I_2$ are equivalent and produce the same equations,

$$\begin{aligned} a^2 - b^2 &= c, \\ 2ab &= s. \end{aligned} \tag{12.3}$$

We now take an important step: assuming the constraint $a^2 + b^2 = 1$, we can eliminate either a or b, and complete our solution in terms of the measured rotation transformation parameters (c, s) in two very distinct ways: we can retain the quadratic forms of (a, b) and solve for those as follows:

$$\left.\begin{array}{c} \text{Eliminate } b: \\ a^2 - b^2 = 2a^2 - 1 = c \\ 2ab = s \end{array}\right\} \quad \text{Solve for } (a^2, ab) \quad \rightarrow \quad a^2 = \frac{1+c}{2}, \quad ab = \frac{s}{2}, \tag{12.4}$$

$$\left.\begin{array}{c} \text{Eliminate } a: \\ a^2 - b^2 = 1 - 2b^2 = c \\ 2ab = s \end{array}\right\} \quad \text{Solve for } (ab, b^2) \quad \rightarrow \quad ab = \frac{s}{2}, \quad b^2 = \frac{1-c}{2}. \tag{12.5}$$

These are *unnormalized* forms. Or we can include the normalization and solve for (a, b) individually:

$$\left.\begin{array}{c} \text{Eliminate } b: \\ a^2 - b^2 = 2a^2 - 1 = c \\ 2ab = s \end{array}\right\} \quad \text{Solve for } (a, b) \quad \rightarrow \quad a = \pm\frac{\sqrt{1+c}}{\sqrt{2}}, \quad b = \pm\frac{s}{\sqrt{2}\sqrt{1+c}}, \tag{12.6}$$

$$\left.\begin{array}{c}\text{Eliminate } a\text{:}\\a^2 - b^2 = 1 - 2b^2 = c\\2ab = s\end{array}\right\} \text{ Solve for } (a,b) \quad \to \quad a = \pm\frac{s}{\sqrt{2}\sqrt{1-c}}, \quad b = \pm\frac{\sqrt{1-c}}{\sqrt{2}}.\qquad(12.7)$$

The second set of solutions, Eqs. (12.6) and (12.7), is seen to be the same as the result of normalizing Eqs. (12.4) and (12.5), and these normalized forms are algebraically identical if we multiply them by the ratios $\sqrt{1-c}/\sqrt{1-c}$ or $\sqrt{1+c}/\sqrt{1+c}$, respectively. But, at certain points, this clearly requires multiplying by 0/0! So we see that the first normalized solution is impossible for $a \sim 0$, or $c \sim -1$, a perfectly legal rotation, and the second solution is impossible for $b \sim 0$, or $c \sim +1$, also perfectly legal. In addition, *both signs* in Eqs. (12.6) and (12.7) are valid, as we have the same rotation $R(a,b)$ if $(a,b) \to (-a,-b)$. The problem, actually an important *feature*, is that one normalized solution fails in one experimentally measurable domain, and the other fails in a *different* experimentally measurable domain. *Both must be considered together*, along with their opposite signs, in order to completely cover the full multivalued 720-degree range of θ parameterizing (a,b). Those familiar with the long-standing quaternion extraction method of Shepperd (1978) may recognize some basic features appearing in a novel context here, and in the full quaternion treatment later on: there are in effect conditions on which rotation matrix elements can be trusted to produce a valid quaternion.

12.3.2 Graphical illustration of the 2D rotation case

Fig. 12.1 shows how Eqs. (12.4) and (12.5) describe unit circles passing through the origin, with *distinct centers* at $(1,0)$ and $(0,1)$, while their normalizations, Eqs. (12.6) and (12.7) for a and b, are unit *half-circles* centered at $(0,0)$, covering the positive x-axis and the positive y-axis, respectively. The unnormalized solutions, in blue and red, pass straight through the origin (black dot) but their corresponding normalized quaternions, in green and magenta, respectively, blow up half-way through their angular domain. The two domains of the (a,b) solutions overlapping in the first quadrant work *together* to cover each other's singular normalization locations. This shows us unequivocally how the variables (a,b) and their multiple solutions describe a *manifold*, a topological space that cannot be described by a single function, but requires overlapping descriptions or *charts*. Incorporating both signs of the circles of Eqs. (12.4) and (12.5), passing through the origin but with *distinct centers* at $(1,0)$, $(0,1)$, $(-1,0)$, and $(0,-1)$, gives the small-angle initial picture illustrated in Fig. 12.2(a). The nonsingular almost-half-circles (one sees a "half-pie" shape) resulting from normalization cover the entire range of four progressively overlapping domains that *together* describe the possible values of (a,b) over the whole range $0° \leqslant \theta < 720°$ with four local nonsingular options. The four quarter circles in Fig. 12.2(b) (extendable *almost* to the half circles in Fig. 12.2(c)) form the set of *charts* of the (2D) quaternion manifold's *atlas*. We shall see that the four separate quadrants starting with $(a > 0, b > 0)$, or equivalently $(a > b, a > -b)$, etc., are piecewise well behaved.

Note that the flat bottoms of the half-quaternion circles correspond to the singularity at the origin of the full pre-normalization circles centered at $(\pm 1, 0)$ and $(0, \pm 1)$. Given the variables (a,b), the singularities in their solutions in terms of (c,s) occur when one variable or the other vanishes, $(a,b) \to (0, \pm 1)$ and $(a,b) \to (\pm 1, 0)$. We shall see in the 3D quaternion case that similar domain restrictions occur for 14 combinations of the same types of zero loci where a zero-scaled legal combination of quaternion elements vanishes, mandating multiple overlapping representations to escape the zeroes.

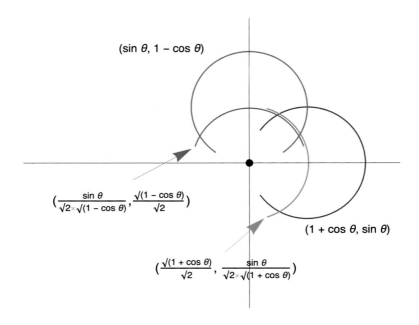

FIGURE 12.1

The overlapping (a, b) regions in the positive quadrants. The normalized x-axis region is in green, derived from the unnormalized region in blue, centered at $(1, 0)$); this sector is regular at $(1, 0)$, and singular at ± 180 degrees (remember that for (a, b), the range of θ is 720 degrees). The normalized y-axis region is in magenta, derived from the unnormalized region in red, centered at $(0, 1)$), which is regular at 180 degrees but singular at 0 degrees and 360 degrees. They overlap in the neighborhood of 90 degrees, so, together with their reversed-sign counterparts, they are regular over the entire 720 degrees parameter range of (a, b).

12.3.3 Variational approach: the Bar-Itzhack method in 2D

Next, we examine the question of solving the equations $R(c, s) = R(a, b)$ for (a, b) using variational methods, and that will lead us again to the approach of Bar-Itzhack (2000) introduced in Chapter 9 for full 3D quaternion rotations. We begin with what is essentially a least-squares formulation, expressing the difference between the two matrices using the Frobenius norm,

$$
\begin{aligned}
L_{\textbf{Frobenius}} &= \text{tr}\left((R(a, b) - R(c, s)) \cdot (R(a, b) - R(c, s))^{\text{t}}\right) \\
&= \text{tr}\left(I_2 + I_2 - 2R(a, b) \cdot R(c, s)^{\text{t}}\right) .
\end{aligned}
\tag{12.8}
$$

At this point we can discard the constants and rephrase the problem of minimizing the least-squares version of the Frobenius norm by a maximization of the cross-term, which we choose to write as

$$
\Delta_{\textbf{F}} = \text{tr}\, \frac{1}{2} R(a, b) \cdot R(c, s)^{\text{t}} .
\tag{12.9}
$$

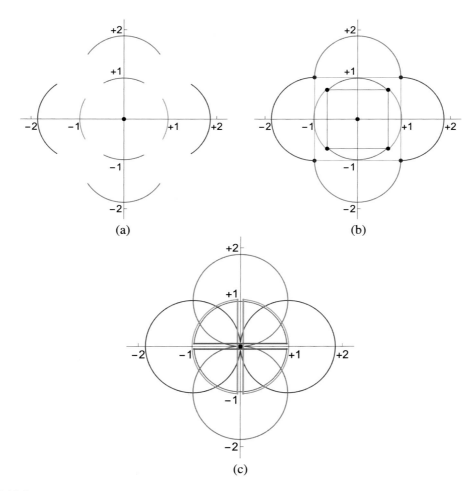

FIGURE 12.2

Each of the four unnormalized maps that cover the full quaternion space has a singularity in the normalization.
How quaternion space in 2D must be multiply covered by alternative solutions depending on the actual 2D
cosine and sine rotation parameters. The blue circles are the paths of $\pm(1+c, s)$, mapping to the green half-
circles in (a, b), failing at $c = -1$, $a = 0$. The red circles are the paths of $\pm(s, 1-c)$, mapping to the magenta
half-circles in (a, b), which fail at $c = +1$, $b = 0$. The curves along the positive axes, extending from -45 degrees to 135 degrees actually cover the whole rotation space, while all four together, passing from (a) to the
full cover in (b), span the entire quaternion space, but no single solution suffices. In (b) note particularly the
circumscribed and inscribed squares that mark the optimal ranges of the normalized projections that cover
exactly the entire quaternion circle. Panel (c) shows the singular limits as the outer circles close in on the
origin, and the divide-by-zero at that limit stops the normalized circle half-way through, indicated by the four
"half-pie" shapes, two in each color.

The task is now to maximize $\Delta_{\mathbf{F}}$ by finding $(a, b)_{\text{opt}}$ such that

$$(a, b)_{\text{opt}} = \underset{(a, b)}{\text{argmax}} \left(\text{tr} \frac{1}{2} R(a, b) \cdot R(c, s)^{\text{t}} \right) \tag{12.10}$$

Regroup quadratic terms in (a, b) into left and right vectors:

$$= \underset{(a, b)}{\text{argmax}} \left([a, b] \cdot K(c, s) \cdot [a, b]^{\text{t}} \right) = \underset{(a, b)}{\text{argmax}} \left([a \ \ b] \cdot \begin{bmatrix} c & s \\ s & -c \end{bmatrix} \cdot \begin{bmatrix} a \\ b \end{bmatrix} \right). \tag{12.11}$$

We refer to the matrix $K(c, s) = \{\{c, s\}, \{s, -c\}\}$, which has eigenvalues $\lambda = \pm 1$, as the *profile matrix* of the Bar-Itzhack optimization problem. The task of maximizing $\Delta_{\mathbf{F}}$ is equivalent to identifying the maximal eigenvalue of the matrix $K(c, s)$, which is just $\lambda = +1$, and determining the corresponding eigenvector, which is $(a, b)_{\text{opt}} = (\cos(\theta/2), \sin(\theta/2))$; this choice for (a, b) minimizes the Frobenius norm Eq. (12.8), and in fact sets it to zero in this simplified example. But this misses the crucial information noted in Eqs. (12.6) and (12.7). A more formal and generalizable way to find the eigenvector, which can have any nonzero scale without changing the eigenvalue, is to form the characteristic equation $\chi(c, s)$ by subtracting the maximal eigenvalue $\lambda = +1$ from the diagonal,

$$\chi(c, s) = K(c, s) - (+1)I_2 = \begin{bmatrix} c - 1 & s \\ s & -c - 1 \end{bmatrix}, \tag{12.12}$$

and extract the eigenvector from $\chi(c, s)$. We now encounter the main point of this approach: the multiple forms of the solutions for (a, b) that we found by direct calculation in Section 12.3.1 now appear *automatically* in the variational version. The crucial fact is that, given that the determinant of χ vanishes, $\det \chi \equiv 0$, *both adjugate columns* of the matrix χ are *unnormalized eigenvectors* of the given maximal eigenvalue. Each individual column of the adjugate, dotted into the corresponding column of its source matrix, yields the determinant. We recall from the introduction that for an $N \times N$ square matrix,

$$M \cdot \text{Adj}(M) = (\det M)I_N .$$

Since, in this case, the matrix χ has vanishing determinant, the columns of the adjugate of χ are eigenvectors and embody $N = 2$ copies of the *same* eigenvector with the chosen eigenvalue $\lambda = 1$:

$$\chi \cdot \text{Adj}(\chi) = \det(\chi)I_2 = 0 \tag{12.13}$$
$$= (K - \lambda I_2) \cdot \text{Adj}(\chi) \tag{12.14}$$

so

$$K \cdot \text{any column of Adj}(\chi) = \lambda \times \text{any column of Adj}(\chi) . \tag{12.15}$$

For our case, we see that the two copies of the (unnormalized) maximal eigenvector are the two columns of the adjugate of χ:

$$\text{Adjugate}(\chi) = \left\{ \begin{bmatrix} -1 - c \\ -s \end{bmatrix} , \begin{bmatrix} -s \\ -1 + c \end{bmatrix} \right\}. \tag{12.16}$$

Since the eigenvectors are insensitive to overall sign and scale, we are free to multiply by $(-1/2)$ to get a more convenient form of the adjugate eigenvectors, which is

$$\text{AdjEigVectors}(\chi) = \left\{ \frac{1}{2} \begin{bmatrix} 1+c \\ s \end{bmatrix} , \frac{1}{2} \begin{bmatrix} s \\ 1-c \end{bmatrix} \right\} \quad (12.17)$$

Exposing the quaternion relation using θ:

$$= \left\{ \begin{bmatrix} a^2 \\ ab \end{bmatrix} , \begin{bmatrix} ab \\ b^2 \end{bmatrix} \right\}. \quad (12.18)$$

The problems here are by now familiar: the adjugate matrix of unnormalized eigenvectors Eq. (12.17) is unnormalizable when $c = \pm 1$ since then $s = 0$ and whole columns vanish; from Eq. (12.18) we see that is equivalent to having either quaternion component vanishing, $a = 0$ or $b = 0$. (Unit norm means they cannot both vanish, so there is always one normalizable solution.) Outside this singular domain, we can proceed to compute a general normalized solution, and we find the rotation matrix variational-method versions of Eq. (7.12) and Eq. (7.13),

$$\text{Normalized adjugate eigenvectors} (\chi) = \left\{ \begin{bmatrix} \frac{\sqrt{1+c}}{\sqrt{2}} \\ \frac{s}{\sqrt{2}\sqrt{1+c}} \end{bmatrix} , \begin{bmatrix} \frac{s}{\sqrt{2}\sqrt{1-c}} \\ \frac{\sqrt{1-c}}{\sqrt{2}} \end{bmatrix} \right\} \quad (12.19)$$

$$= \left\{ \begin{bmatrix} \frac{\sqrt{1+c}}{\sqrt{2}} \\ \text{sign}\,s \frac{\sqrt{1-c}}{\sqrt{2}} \end{bmatrix} , \begin{bmatrix} \text{sign}\,s \frac{\sqrt{1+c}}{\sqrt{2}} \\ \frac{\sqrt{1-c}}{\sqrt{2}} \end{bmatrix} \right\}. \quad (12.20)$$

The first column of Eq. (12.19) is singular at $c = -1$, the second column at $c = +1$, both completely legal points, but neither normalized adjugate column is a valid quaternion-like 2-vector for the *entire range* of the data (c, s). Eq. (12.20) does not escape the singularity because one would have to multiply by 0/0 to obtain those algebraic forms for all parameter values. Finally, we recover exactly the numerical version of the symbolic 2D rotation matrix we started with,

$$R_{\text{opt}}(m) = \begin{bmatrix} a^2 - b^2 & -2ab \\ 2ab & a^2 - b^2 \end{bmatrix} = \begin{bmatrix} \frac{c}{\sqrt{c^2+s^2}} & -\frac{s}{\sqrt{c^2+s^2}} \\ \frac{s}{\sqrt{c^2+s^2}} & \frac{c}{\sqrt{c^2+s^2}} \end{bmatrix} = \begin{bmatrix} c & -s \\ s & c \end{bmatrix}. \quad (12.21)$$

Note incidentally how the "sign s" factors are necessary to get the correct R_{opt} if we use the divisor-rationalized form Eq. (12.20).

From Eq. (12.18), we can see clearly that both columns normalize to the eigenvector (a, b) since $a^2 + b^2 = 1$. But that eigenvector is multiplied by a in the first case, so no normalization is possible as $a \to 0$, and in the second column no normalization is possible as $b \to 0$. *Both pre-normalization columns* of the adjugate matrix must be included to cover the entire space of rotations. Due to the $(a, b) \to (-a, -b)$ equivalence, the full topological space of (a, b) of course actually has four natural components, and we in fact will need to work in each quadrant separately to meet the requirements of machine learning protocols.

12.3.4 Bar-Itzhack errorful measurement strategy

We have thus far assumed that measurements of a rotation matrix resulted in a perfect orthonormal matrix. That strategy allowed us to clearly expose the requirement to use a multivalued formula to find the 2D reduced quaternion (a, b) from the parameters (c, s) of an ideal orthonormal measured rotation matrix. The same basic approach is valid also for inaccurate measurements that report rotation matrix elements that are not orthonormal. The basic ideas appear in the original work of Shepperd (1978) and more explicitly in Bar-Itzhack (2000), reviewed in Hanson (2020). Some alternatives are suggested, e.g., by Sarabandi et al. (2018, 2020). While many of the algebraic manipulations in this subsection have in fact already appeared in Chapter 7, and the basic Bar-Itzhack methods are presented in Chapter 9, we review the 2D framework here to initialize the broader context that follows.

To see how to work with inaccurate rotation matrix measurements, we introduce the "measured matrix data" $R(m)$,

$$R(m) = \begin{bmatrix} m_{11} & m_{12} \\ m_{21} & m_{22} \end{bmatrix}. \tag{12.22}$$

Comparing this to our ideal quadratic quaternion-like target for the same matrix, Eq. (12.2), we might wonder if having two variables in $R(a, b)$ and four variables in $R(m)$ is a problem. While solving directly for (a, b) as we did in the noise-free case seems difficult, in fact, using the eigensystem method, we find that only two combinations of those four variables matter.

We now argue that for noisy data, the Bar-Itzhack optimization approach is easier to understand and justify. We begin with Eq. (12.22) and insert it into the Frobenius norm for the distance between $R(m)$ and Eq. (12.2) for $R(a, b)$, yielding

$$\begin{aligned} S_{\mathbf{F}} &= \text{tr}\,(R(a, b) - R(m))) \cdot (R(a, b) - R(m)))^{\text{t}} \\ &= (a^2 - b^2 - m_{11})^2 + (2ab + m_{12})^2 + (2ab - m_{21})^2 + (a^2 - b^2 - m_{22})^2 \\ &= 2 + \sum_{i,j} \left(m_{ij}\right)^2 - 2(a^2 - b^2)\,(m_{11} + m_{22}) + 4ab\,(m_{12} - m_{21})\ . \end{aligned} \tag{12.23}$$

We strip the constants and change the sign to turn the problem of minimizing $S_{\mathbf{F}}$ to the equivalent problem of maximizing the cross-term, which we convert as before to the matrix product

$$\begin{aligned} \Delta_{\mathbf{F}} &= (a^2 - b^2)\,(m_{11} + m_{22}) + 2ab\,(m_{21} - m_{12}) \\ &= [a\ b] \cdot \begin{bmatrix} (m_{11} + m_{22}) & (m_{21} - m_{12}) \\ (m_{21} - m_{12}) & -(m_{11} + m_{22}) \end{bmatrix} \cdot \begin{bmatrix} a \\ b \end{bmatrix} = [a\ b] \cdot K(m) \cdot \begin{bmatrix} a \\ b \end{bmatrix}, \end{aligned}$$

which, as promised, contains only two independent linear combinations of the elements of $R(m)$. Regardless of the presence of error, the profile matrix $K(m)$ retains this form. The task of solving our optimization problem

$$(a, b)_{\text{opt}} = \underset{(a, b)}{\text{argmax}}\ [a\ b] \cdot K(m) \cdot \begin{bmatrix} a \\ b \end{bmatrix} \tag{12.24}$$

is then reduced to finding the eigenvector $[a, b]_{opt}$ corresponding to the maximal eigenvalue λ_{max} of $K(m)$. Since $\text{tr}[K(m)] = 0$, the eigenvalues are an opposite-sign pair, with the maximal eigenvalue being the positive choice

$$\lambda_{max} = \sqrt{(m_{11} + m_{22})^2 + (m_{12} - m_{21})^2} \,. \tag{12.25}$$

Note that the appearance of only the antisymmetric part of the off-diagonal terms in $R(m)$ is an inevitable consequence of the optimization of the Frobenius norm Eq. (12.23).

The final step is to cover the entire manifold of solutions for the multivalued (a, b) using the adjugate matrix $A(m)$ of the maximal eigensystem as a function solely of the measured elements of $R(m)$:

$$A(m) = -\text{Adjugate}([K(m) - \lambda_{max} I_2]) =$$
$$= \left\{ \begin{bmatrix} \lambda_{max} + (m_{11} + m_{22}) \\ m_{21} - m_{12} \end{bmatrix}, \begin{bmatrix} m_{21} - m_{12} \\ \lambda_{max} - (m_{11} + m_{22}) \end{bmatrix} \right\}. \tag{12.26}$$

We see immediately the automatic appearance of a pair of solutions for $(a, b)_{opt}$ that have singularities in different places when normalized, thus covering, with their negative counterparts, the entire manifold of (a, b). Since the eigenvectors represented by the columns of the adjugate matrix are insensitive to rescaling, we can transform Eq. (12.26) to a more readable form by multiplying by 1/2, and denoting the terms whose error-free forms are basically the cosine and sine of the rotation angle by

$$\text{cc} = \frac{1}{2}(m_{11} + m_{22}) \,, \qquad \text{ss} = \frac{1}{2}(m_{21} - m_{12}) \,, \qquad \lambda = \sqrt{\text{cc}^2 + \text{ss}^2} = \frac{1}{2}\lambda_{max} \,. \tag{12.27}$$

We find the unnormalized eigenvector pairs

$$\text{Adjugate eigenvectors} = \left\{ \pm \begin{bmatrix} \sqrt{\text{ss}^2 + \text{cc}^2} + \text{cc} \\ \text{ss} \end{bmatrix}, \pm \begin{bmatrix} \text{ss} \\ \sqrt{\text{ss}^2 + \text{cc}^2} - \text{cc} \end{bmatrix} \right\} \tag{12.28}$$

$$= \left\{ \pm \begin{bmatrix} \lambda + \text{cc} \\ \text{ss} \end{bmatrix}, \pm \begin{bmatrix} \text{ss} \\ \lambda - \text{cc} \end{bmatrix} \right\}. \tag{12.29}$$

As usual, the normalized versions are equivalent except at two singular points, where they complement each other, one always being a computable normalized eigenvector:

$$A(m) = \boxed{\begin{array}{l} \text{normalized} \\ \text{adjugate} \\ \text{eigenvectors} \end{array}}$$

$$= \left\{ \begin{bmatrix} a_1 \\ b_1 \end{bmatrix}, \begin{bmatrix} a_2 \\ b_2 \end{bmatrix} \right\} = \left\{ \pm \begin{bmatrix} \dfrac{\sqrt{\lambda + \text{cc}}}{\sqrt{2}\sqrt{\lambda}} \\ \dfrac{\text{ss}}{\sqrt{2}\sqrt{\lambda(\lambda + \text{cc})}} \end{bmatrix}, \pm \begin{bmatrix} \dfrac{\text{ss}}{\sqrt{2}\sqrt{\lambda(\lambda - \text{cc})}} \\ \dfrac{\sqrt{\lambda - \text{cc}}}{\sqrt{2}\sqrt{\lambda}} \end{bmatrix} \right\}. \tag{12.30}$$

As before, we can build the rotation matrix directly from the optimal (a, b) solutions in Eq. (12.30), with the result

$$R_{\text{opt}}(m) = \begin{bmatrix} \dfrac{\text{cc}}{\sqrt{\text{cc}^2 + \text{ss}^2}} & -\dfrac{\text{ss}}{\sqrt{\text{cc}^2 + \text{ss}^2}} \\ \dfrac{\text{ss}}{\sqrt{\text{cc}^2 + \text{ss}^2}} & \dfrac{\text{cc}}{\sqrt{\text{cc}^2 + \text{ss}^2}} \end{bmatrix}, \tag{12.31}$$

which of course coincides again with the noise-insensitive form Eq. (7.14) from Chapter 7. Thus we have seen three distinct contexts for the use of 2D quaternions that are each a prototype of a perfect optimal solution for a matching problem in these domains:

- **Chapter 7: 2D cloud matching.** The data appearing in the profile matrix and adjugate matrix are $C = E_{xx} + E_{yy}$ and $S = E_{xy} - E_{yx}$ from the cross-covariance matrix of the cloud and target point data.
- **Chapter 12: Matching a perfect 2D rotation.** The data appearing in the profile matrix and adjugate matrix are the numerical values of $c = \cos\theta$ and $s = \sin\theta$ from the provided perfect rotation matrix.
- **Chapter 12: Matching a flawed 2D rotation.** The data appearing in the profile matrix and adjugate matrix are the numerical values of $\text{cc} = (m_{11} + m_{22})/2$ and $\text{ss} = (m_{21} - m_{12})/2$ based on the matrix elements of the provided imperfect rotation matrix.

This consistent formula provides us with optimal quaternions and their corresponding rotation matrix solutions for this wide variety of applications, and it remains the optimal solution for both noise-free and noisy data.

Note that in essence $\text{cc} \approx c = \cos\theta$, so the singularities are as expected near $c = 1$ and $c = -1$, or $\text{ss} \approx 0$ and $\text{cc} \approx \pm 1$. Furthermore, we see that the numerical eigenvalue λ appears throughout in just such a way that if $\lambda = 1$, we have precisely the previous solution Eq. (12.19) for the normalized set of maximal eigenvectors. It is important to recognize the enormous difference between Eq. (12.19) and Eq. (12.30); the eigenvalue λ, generally $\neq 1$, appears in a completely unexpected nontrivial way in each component of the eigenvectors, where we simply had ones before, and only appears because the Bar-Itzhack method was rigorously employed in the computation of the optimal solution.

> Our point is that we find a pair of 2D eigenvectors (a, b), with complementary normalization singularities, and which, together, produce a perfectly orthonormal 2D rotation matrix via Eq. (2.14) that is the best possible approximation to the noisy numerical data of Eq. (12.22).

12.3.5 Summary

So far in this section, we have worked through the simple case of 2D rotations by an angle θ, in parallel with how that corresponds to the 2D simplification of the quaternion-parameterized rotation matrix in Eq. (2.14) to a 2D version written in terms of the reduced quaternions (a, b). This has led us to the understanding that in order to solve for (a, b) in terms of the elements of a 2D measured rotation matrix, we must have two separate sectors, one regular at $(1, 0)$ and singular at $(0, 1)$, and the other reversed. To account for the full quaternion space where both (a, b) and $(-a, -b)$ correspond to the same 2D rotation $R(a, b)$, we in fact need four sectors covering the full circular manifold of the quaternion space.

We have also seen that the variational techniques that reveal the manifold properties of the quaternion space allow us to detect the defects of relying on single functions; they also give us a way to find the *optimal* exact rotation corresponding to a noisy set of rotation matrix data. In the next section, we perform a parallel analysis for 3D rotations and full quaternions. We will again find that the clearest understanding of this problem is based on the adjugate matrix arising in the Bar-Itzhack optimization algorithm, and this method is again superior for choosing the closest rotation to a matrix that is a noisy approximation to an actual rotation.

12.4 3D rotations and the quaternion map

We now turn to the realistic case of interest, how to correctly determine a quaternion corresponding to a measured 3D rotation matrix. As in Section 12.3, we will begin with a direct derivation using only the symbolic forms of the quaternion rotation problem, followed by the Bar-Itzhack variational version of the same problem, which will exhibit some new features. Again, the multivalued target required for, e.g., a rigorous formulation of a machine learning process, will be expressed in terms of an adjugate matrix. We conclude with the corresponding variational treatment of a noisy inexact rotation measurement, producing the nearest true rotation matrix that can be computed from the noisy measured approximate rotation matrix data.

12.4.1 Direct solution of the 3D problem

A *proper* orthonormal 3D rotation matrix can be written as a quadratic form $R(q)$ in the quaternion elements $q = (q_0, q_1, q_2, q_3)$, with $q \cdot q = 1$, and identified with the axis-angle form $R(\theta, \hat{\mathbf{n}})$ as follows:

$$R(q) = R(\theta, \hat{\mathbf{n}}), \tag{12.32}$$

where $R(q)$ and $R(\theta, \hat{\mathbf{n}})$ were given in Eqs. (2.14) and (2.17), and the corresponding axis-angle quaternion parameterization $q(\theta, \hat{\mathbf{n}})$ is in Eq. (2.16).

We can easily rearrange Eq. (2.17) (or just apply $q(\theta, \hat{\mathbf{n}})$) to produce an explicit solution for the 10 quadratic forms in q. With $\hat{\mathbf{n}} = [n_1, n_2, n_3]$, we can compose a 4×4 symmetric matrix that we identify as the *adjugate variable* matrix:

$$
\begin{bmatrix}
q_0{}^2 & q_0 q_1 & q_0 q_2 & q_0 q_3 \\
q_0 q_1 & q_1{}^2 & q_1 q_2 & q_1 q_3 \\
q_0 q_2 & q_1 q_2 & q_2{}^2 & q_2 q_3 \\
q_0 q_3 & q_1 q_3 & q_2 q_3 & q_3{}^2
\end{bmatrix}
= \frac{1}{2}
\begin{bmatrix}
1+c & s\,n_1 & s\,n_2 & s\,n_3 \\
s\,n_1 & (1-c)\,n_1{}^2 & (1-c)\,n_1 n_2 & (1-c)\,n_1 n_3 \\
s\,n_2 & (1-c)\,n_1 n_2 & (1-c)\,n_2{}^2 & (1-c)\,n_2 n_3 \\
s\,n_3 & (1-c)\,n_1 n_3 & (1-c)\,n_2 n_3 & (1-c)\,n_3{}^2
\end{bmatrix}. \tag{12.33}
$$

Eq. (12.33) is the full quaternion analog of the unnormalized 2D Eqs. (12.4) and (12.5), where we note that $(c, s) = (\cos\theta, \sin\theta)$ are in terms of the full angle, not the half-angle quaternion form. In Eqs. (12.6) and (12.7), we wrote out the explicit normalized quaternions and noted the two distinct singularities at $c = +1$, $c = -1$. Here we choose to analyze the normalization factors separately to clarify the analysis. Obviously each column of Eq. (12.33) is normalized to a *symbolically correct*

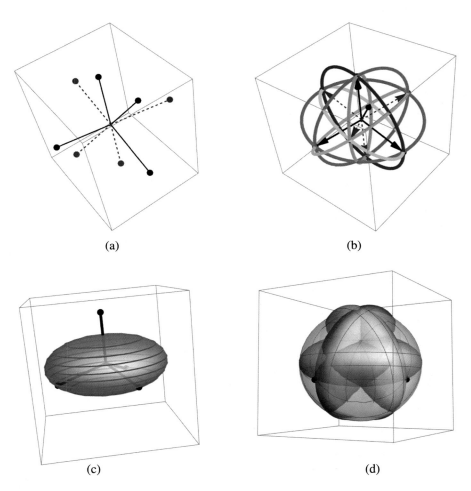

(a)   (b)

(c)   (d)

FIGURE 12.3

Visualizing the 14 quaternion singularity-restricted domains. (a) Regions with three zeroes are just the four pairs of points at each end of the projected 4D axes, $q_0 = \pm1$, $q_1 = \pm1$, $q_2 = \pm1$, and $q_3 = \pm1$. The positive points correspond with the vertices of ends of the 4D axes, while the pairs are intuitively both required because the remaining quaternion manifold in this degenerate case is actually the 0-sphere $\mathbf{S}^0$, the two-point solution of $x^2 = 1$. (b) Regions with two zeroes are six topological circles $\mathbf{S}^1$. A single circle projected to 3D from the unit quaternion with $q_0 = q_1 = 0$ is just the curve $(0, 0, x, y,)$ with $x^2 + y^2 = 1$. The six circles are shown in a tetrahedral projection of the unit axes from 4D to 3D, with circles in the planes $q_0 = q_1 = 0$, $q_0 = q_2 = 0$, $q_0 = q_3 = 0$, $q_2 = q_3 = 0$, $q_3 = q_1 = 0$, and $q_1 = q_2 = 0$. (c) Finally, quaternion regions restricted by a single zero are topological spheres $\mathbf{S}^2$, so here we show a single sphere projected to 3D (and a bit squashed) from the unit quaternion with $q_0 = 0$; the remaining set of quaternion values with this normalization singularity is the manifold $(0, x, y, z)$ with $x^2 + y^2 + z^2 = 1$. (d) All four single-zero spheres, with $q_0 = 0$, $q_1 = 0$, $q_2 = 0$, and $q_3 = 0$. Each sphere lies within a 3-space perpendicular to one of the 4D axes, and appears in the 3D projection as a flattened sphere aligned with one of the faces of the tetrahedron formed by the projected 4D axes.

quaternion $q = (q_0, q_1, q_2, q_3)$ by normalizing, in columnwise sequence, by

$$
\left.
\begin{array}{ll}
\text{column 0:} & \dfrac{1}{q_0} = \dfrac{1}{\cos(\theta/2)} = \sqrt{\dfrac{2}{1+c}} \\[2ex]
\text{column 1:} & \dfrac{1}{q_1} = \dfrac{1}{n_1 \sin(\theta/2)} = \dfrac{1}{n_1}\sqrt{\dfrac{2}{1-c}} \\[2ex]
\text{column 2:} & \dfrac{1}{q_2} = \dfrac{1}{n_2 \sin(\theta/2)} = \dfrac{1}{n_2}\sqrt{\dfrac{2}{1-c}} \\[2ex]
\text{column 3:} & \dfrac{1}{q_3} = \dfrac{1}{n_3 \sin(\theta/2)} = \dfrac{1}{n_3}\sqrt{\dfrac{2}{1-c}}
\end{array}
\right\}.
\tag{12.34}
$$

We observe that there are singularities in the normalization factors at new locations *in addition* to the $c = \pm 1$ singularities appearing in the 2D case. This is easy to understand: our expression for $q(\theta, \hat{\mathbf{n}})$ is arbitrary, and any permutation of the parameter elements is equally valid, so the singularities *should* be spread among the components without singling out any given one. We can conclude that the 4D analog of the 2D set of $(a, b) \to \{(\pm 1, 0), (0, \pm 1)\}$ singularities in the normalization is in fact this set of 14 distinct restrictions on the legal remaining quaternion degrees of freedom due to the 14 ways in which one, two, or three elements of q can vanish in the optimal eigenvector:

$$
\left.
\begin{array}{l}
\text{one zero} \to \{(0, x, y, z), (x, 0, y, z), (x, y, 0, z), (x, y, z, 0)\} \\[1ex]
\text{two zeroes} \to \{(0, 0, x, y), (0, x, 0, y), (0, x, y, 0), (x, 0, 0, y), (x, 0, y, 0), (x, y, 0, 0)\} \\[1ex]
\text{three zeroes} \to \{(\pm 1, 0, 0, 0), (0, \pm 1, 0, 0), (0, 0, \pm 1, 0), (0, 0, 0, \pm 1)\}
\end{array}
\right\},
\tag{12.35}
$$

where $x^2 + y^2 + z^2 = 1$ in the first line and $x^2 + y^2 = 1$ in the second line to preserve $q \cdot q = 1$. Since the first two cases are spheres $\mathbf{S}^2$ and $\mathbf{S}^1$, the sign option $\pm$ in the third case is included; in fact, a more general way to think of the points ± 1 is as just another sphere, the 0-sphere $\mathbf{S}^0$ solving $x^2 = 1$. The key of course is that, since $q \cdot q = 1$, there always has to be at least one element whose magnitude is $\geqslant 1/2$, and thus we can always find a column (or, equivalently, a row) that is normalizable. These 14 restricted regions of the quaternion solution due to vanishing 4-vector elements are illustrated in Fig. 12.3.

Perhaps another useful way to look at the algebraic implications of these normalization anomalies is simply to write down the 14 possible 4×4 quaternion adjugate matrices $[q_\mu q_\nu]$ that result for any combination of single, double, and triple choices of indices for the zeroes of q_k from the set $k \in \{0, 1, 2, 3\}$:

$$
123: \begin{bmatrix} \pm 1 & 0 & 0 & 0 \\ 0 & 0 & 0 & 0 \\ 0 & 0 & 0 & 0 \\ 0 & 0 & 0 & 0 \end{bmatrix},\quad
023: \begin{bmatrix} 0 & 0 & 0 & 0 \\ 0 & \pm 1 & 0 & 0 \\ 0 & 0 & 0 & 0 \\ 0 & 0 & 0 & 0 \end{bmatrix},\quad
013: \begin{bmatrix} 0 & 0 & 0 & 0 \\ 0 & 0 & 0 & 0 \\ 0 & 0 & \pm 1 & 0 \\ 0 & 0 & 0 & 0 \end{bmatrix},\quad
012: \begin{bmatrix} 0 & 0 & 0 & 0 \\ 0 & 0 & 0 & 0 \\ 0 & 0 & 0 & 0 \\ 0 & 0 & 0 & \pm 1 \end{bmatrix},
$$

$$
01: \begin{bmatrix} 0 & 0 & 0 & 0 \\ 0 & 0 & 0 & 0 \\ 0 & 0 & x^2 & xy \\ 0 & 0 & xy & y^2 \end{bmatrix},\quad
02: \begin{bmatrix} 0 & 0 & 0 & 0 \\ 0 & x^2 & 0 & xy \\ 0 & 0 & 0 & 0 \\ 0 & xy & 0 & y^2 \end{bmatrix},\quad
03: \begin{bmatrix} 0 & 0 & 0 & 0 \\ 0 & x^2 & xy & 0 \\ 0 & xy & y^2 & 0 \\ 0 & 0 & 0 & 0 \end{bmatrix},
$$

$$
23:\begin{bmatrix} x^2 & xy & 0 & 0 \\ xy & y^2 & 0 & 0 \\ 0 & 0 & 0 & 0 \\ 0 & 0 & 0 & 0 \end{bmatrix},\quad
31:\begin{bmatrix} x^2 & 0 & xy & 0 \\ 0 & 0 & 0 & 0 \\ xy & 0 & y^2 & 0 \\ 0 & 0 & 0 & 0 \end{bmatrix},\quad
12:\begin{bmatrix} x^2 & 0 & 0 & xy \\ 0 & 0 & 0 & 0 \\ 0 & 0 & 0 & 0 \\ xy & 0 & 0 & y^2 \end{bmatrix},
$$

$$
0:\begin{bmatrix} 0 & 0 & 0 & 0 \\ 0 & x^2 & xy & xz \\ 0 & xy & y^2 & yz \\ 0 & xz & yz & z^2 \end{bmatrix},\quad
1:\begin{bmatrix} x^2 & 0 & xy & xz \\ 0 & 0 & 0 & 0 \\ xy & 0 & y^2 & yz \\ xz & 0 & yz & z^2 \end{bmatrix},\quad
2:\begin{bmatrix} x^2 & xy & 0 & xz \\ xy & y^2 & 0 & yz \\ 0 & 0 & 0 & 0 \\ xz & yz & 0 & z^2 \end{bmatrix},\quad
3:\begin{bmatrix} x^2 & xy & xz & 0 \\ xy & y^2 & yz & 0 \\ xz & yz & z^2 & 0 \\ 0 & 0 & 0 & 0 \end{bmatrix}.
$$

$$(12.36)$$

These matrices are essentially the algebraic version of the graphics in Fig. 12.3. We finish up the treatment of the singularities with some more graphics.

Graphical illustration of 3D subspace singularities. We can get a more complete intuitive feeling for what is going on with the multiple valid regions for the quaternion solution representations by drawing representative spaces corresponding to the unnormalized (and potentially singular) solutions, then making pointwise normalization maps from those spaces to the actual quaternion subspaces, and finally noting where validity of the normalization map fails.

In 2D, these maps were fairly simple to see in Fig. 12.1 and Fig. 12.2, where a simple unnormalized circle mapped to a half-circle in the 2D reduced quaternion plane. Going one step higher in complexity, we can produce images in 2D that correspond to a quaternion map, but one dimension lower. Fig. 12.4 illustrates pairs of 2D spheres embedded in $\mathbb{R}^3$ centered at $(\pm 1, 0, 0)$, $(0, \pm 1, 0)$, and $(0, 0, \pm 1)$, shown in red. Taking each point on the red spheres and applying the normalization operation, we obtain the six green hemispheres creating an embedded cube that partitions the full green sphere centered at the origin $(0, 0, 0)$. As the red spheres' points approach the origin, the normalization approaches a divide-by-zero at the equator of each green hemisphere, and the map can go no farther.

Strategy for depicting the full quaternion map. Despite its 4D intrinsic nature, quaternion geometry can be depicted in a fairly accurate way if we are willing to follow some analogies between lower-dimensional and higher-dimensional spheres. First, we show in Fig. 12.5(a) an ordinary sphere $\mathbf{S}^2$ embedded in 3D Euclidean space $\mathbb{R}^3$, with the three orthogonal axes $\hat{\mathbf{x}}$, $\hat{\mathbf{y}}$, and $\hat{\mathbf{z}}$, projected in the familiar way to a 2D image. Even though the image is a dimension lower than the actual 3D object being depicted, we are accustomed to interpreting this image as a 3D object. Now rotate the sphere as in Fig. 12.5(b) so that the three axes are projected equally onto the 2D image, with the ends of the axes forming the vertices of an equilateral triangle. Now we see that this projection corresponds to one hemisphere of $\mathbf{S}^2$ flattened into a disk containing all three positive axes, and the back hemisphere as a second disk containing all three negative axes. It is clear that if we create two separate images as in Fig. 12.5(c,d), *every single point* on the manifold $\mathbf{S}^2$ can be seen in the two separate hemispherical images. We can do the same thing with a full quaternion map using a *solid ball* containing a 3D quadruple of positive axes (with the four axis ends being the vertices of a tetrahedron), paired with a matching solid ball containing the symmetric projections of the four negative axes. Every point of the quaternion sphere is visible in the two solid balls, exactly analogous to the two filled disks for the hemispheres of $\mathbf{S}^2$ in Fig. 12.5(c).

The full quaternion map from the unnormalized representation to the normalized true quaternion sector is divided into eight distinct regions, in opposite-signed pairs that represent equivalent rotations

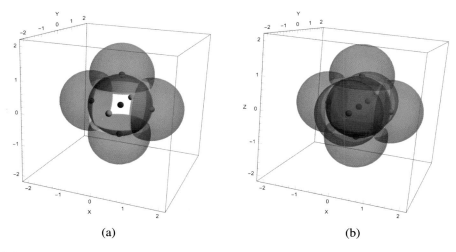

(a)   (b)

FIGURE 12.4

3D subspace showing three axes of singularities. In this 3D subspace of quaternion space, there are partial spheres instead of partial circles, but the singularity occurs in the same way, as the sphere closes in on the origin, normalization is impossible. (a) The x- and y-axes coincide with the four red spheres that produce partial coverings for the inner green quaternion-subspace spheres. (b) Adding in the z-axis to show the full story of how this subspace of quaternions is covered in a nonsingular fashion by three pairs of partial spheres, arranged on the surface of a cube. This geometry is exactly analogous to the semicircular arcs and quarter-circular arcs around the outer and inner embedded squares in Fig. 12.2(b).

due to the identification $R(q) = R(-q)$. Instead of portions of ordinary spheres as in Fig. 12.4, we have portions of hyperspheres centered at $(\pm 1, 0, 0, 0)$, $(0, \pm 1, 0, 0)$, $(0, 0, \pm 1, 0)$, and $(0, 0, 0, \pm 1)$. Instead of being partial hemispherical surfaces, these are now solid balls, each corresponding to a portion of a set of overlapping hemispheres of the quaternion manifold $\mathbf{S}^3$. These are difficult to draw, but an attempt can be made by projecting the axes of the 4D space into 3D in the symmetric directions of the vertices of a tetrahedron. In Fig. 12.6(a), we show first a collection of slices of the solid ball at various radii in 4D, aligned with one axis, for a single choice of the eight unnormalized and normalized maps. Then in Fig. 12.6(b), we reduce the number of samples of the solid balls to one, but show a representative pair of unnormalized and normalized slices for *the four positive unit 4D axes*; there is another opposite sign counterpart for each of these four that is omitted for clarity.

Selecting a solution on a nonsingular manifold sector. A standard choice for selecting a well-defined solution from the 4×4 matrix of alternate quaternion solutions, in parallel to the algorithm of Shepperd (1978), is to note that the *diagonal* of the left-hand side of Eq. (12.33) is simply $\{q_0{}^2, q_1{}^2, q_2{}^2, q_3{}^2\}$, so if we identify the ordinal location k of the maximal diagonal $q_k{}^2$, that is the column we normalize:

$$\text{Normalizable solution:} \quad q_{\text{opt}} = \frac{\pm 1}{|q_k|} (q_0 q_k, q_1 q_k, q_2 q_k, q_3 q_k) \,. \tag{12.37}$$

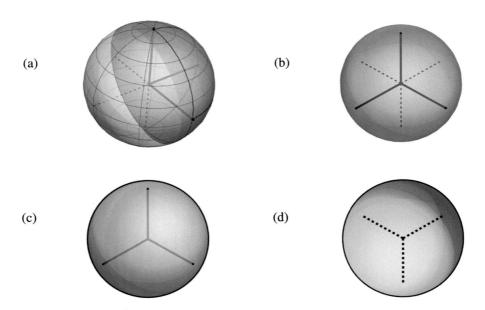

FIGURE 12.5

Analog of S^3 projection with the S^2 full and hemispherical projection. (a) The two-sphere S^2 contains three orthogonal axes in its 3D space projection, shown obliquely here to expose the positive (solid) and negative (dashed) ends of the coordinate axes using a general viewpoint. (b) If we look straight down the diagonal, we see the three positive-valued axes pointing towards us, with ends the vertices of an equilateral triangle; the three negative-valued axes, as dashed lines, point away from us. (c,d) If we simply display the 2D disk with the positive axes in a planar image separately from the 2D disk with the negative axes, we can see a (flattened) depiction in which *every single point* of S^2 is visible and distinct. In order to make every single point of the quaternion hypersphere S^3 visible in our images, we will simply put the *four* axes of 4D space at the symmetrical vertices of a tetrahedron, and use a pair of solid balls (simulating 3D space using 3D graphics images) instead of a pair of filled disks.

The sign is of course arbitrary, though a standard choice is to choose $q_0 > 0$ when possible. The significance of this for our problem is that, because the quaternion sphere S^3 is a topological manifold that cannot be described by a single function, any algorithm that needs to find *a universally applicable quaternion* must produce a *list of four candidates*, remembering that there are 14 ways to fail normalizing any single one, and choose a normalizable candidate from that list to produce a usable quaternion via Eq. (12.37).

12.4.2 Variational approach: Bar-Itzhack in 3D

The variational method we presented for finding $(a, b)_{\text{opt}}$ from experimental 2D rotation matrix data extends straightforwardly to the 3D case to determine q_{opt}. Here we elaborate on selected aspects of our introduction to Bar-Itzhack (2000) methods in Chapter 9.

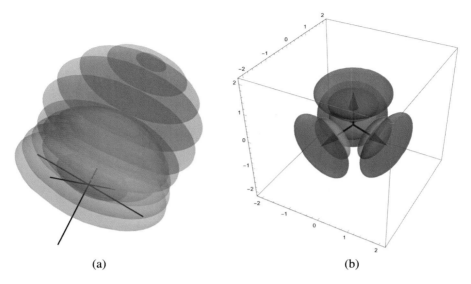

(a)   (b)

FIGURE 12.6

3D subspace showing three axes of singularities. In this 3D subspace of quaternion space, there are partial spheres instead of partial circles, but the singularity occurs in the same way, as the sphere closes in on the origin, normalization is impossible. The four axes correspond to the q_0, q_1, q_2, and q_3 quaternion directions projected down symmetrically to the directions of the four vertices of a 3D regular tetrahedron. While there are eight actual axes in four pairs, corresponding to the two pairs of axes $x = \pm 1$ and $y = \pm 1$ in Fig. 12.2, here for readability we can only show the four positive axes directions whose manifold patches cover the solid ball that is the "Northern hemisphere" of $\mathbf{S}^3$; the omitted "Southern hemisphere" is the second solid ball that, sewn onto the Northern hemisphere along the $\mathbf{S}^2$ equator, completes the full $\mathbf{S}^3$ manifold. (a) A single pair of solid-ball patches, the larger corresponding to the $+q_0$ direction of the solution space q_0 (q_0, q_1, q_2, q_3); the smaller ball is the normalized version collapsing to a patch on the actual $\mathbf{S}^3$ patch that normalizes without singularity in the neighborhood of $q_0 \approx +1$. (b) The four pairs that cover the nonsingular patches around $q_0 \approx +1$, $q_1 \approx +1$, $q_2 \approx +1$, and $q_3 \approx +1$. The actual mathematical balls correspond to a volume rendered solid, which is difficult to portray in a figure, so multiple level sets are shown for each ball to depict the continuous volume with a finite sampling of spherical surfaces.

The full 3D problem consists of finding the 4D quaternion $q_{\mathrm{opt}} = (q_0, q_1, q_2, q_3)$ such that $R(q_{\mathrm{opt}})$ best describes a measured 3D numerical rotation matrix R. We begin as before with the ideal symbolic case, which is now $R = R(\theta, \hat{\mathbf{n}})$. We then exploit quaternion features to minimize the Frobenius norm of the difference between the two matrices, where our initial optimization measure is

$$
\begin{aligned}
\mathrm{L_{Frobenius}} &= \mathrm{tr}\left((R(q) - R(\theta, \hat{\mathbf{n}})) \cdot (R(q) - R(\theta, \hat{\mathbf{n}}))^{\mathrm{t}} \right) \\
&= \mathrm{tr}\left(I_3 + I_3 - 2R(q) \cdot R(\theta, \hat{\mathbf{n}})^{\mathrm{t}} \right) .
\end{aligned}
\tag{12.38}
$$

At this point we can discard the constants and rephrase the problem of minimizing the least-squares version of the Frobenius norm in terms of maximizing the cross-term, which we choose to write as

$$\Delta_{\mathbf{F}} = \operatorname{tr} R(q) \cdot R(\theta, \hat{\mathbf{n}})^t = q \cdot K(\theta, \hat{\mathbf{n}}) \cdot q . \tag{12.39}$$

Expanding $R(q)$ using the form of Eq. (2.17) allows us to rewrite the trace in Eq. (12.39) as a matrix product using the *profile matrix*

$$K_0(\theta, \hat{\mathbf{n}}) = \begin{bmatrix} 2c+1 & 2s\,n_1 & 2s\,n_2 & 2s\,n_3 \\ 2s\,n_1 & 2(1-c)\,n_1^2 - 1 & 2(1-c)\,n_1 n_2 & 2(1-c)\,n_1 n_3 \\ 2s\,n_2 & 2(1-c)\,n_1 n_2 & 2(1-c)\,n_2^2 - 1 & 2(1-c)\,n_2 n_3 \\ 2s\,n_3 & 2(1-c)\,n_1 n_3 & 2(1-c)\,n_2 n_3 & 2(1-c)\,n_3^2 - 1 \end{bmatrix} . \tag{12.40}$$

The task is now to maximize $\Delta_{\mathbf{F}}$ by finding q_{opt} such that

$$q_{\text{opt}} = \underset{q}{\operatorname{argmax}} \left(\operatorname{tr} R(q) \cdot R(\theta, \hat{\mathbf{n}})^t \right) = \underset{q}{\operatorname{argmax}} \left(q \cdot K(\theta, \hat{\mathbf{n}}) \cdot q \right) . \tag{12.41}$$

Since $K_0(\theta, \hat{\mathbf{n}})$ is a symmetric real matrix, the maximum of $q \cdot K_0 \cdot q$ is achieved by picking out the maximal eigenvalue, so the corresponding q_{opt} is the maximal eigenvector. The eigenvalues of $K_0(\theta, \hat{\mathbf{n}})$ are $\lambda = (3, -1, -1, -1)$, so all we need to do is find the eigenvector of $\lambda = 3$. However, we will first use some linear algebra manipulations to simplify the appearance of our equations. We note that:

- Since the characteristic equation we used to solve for the eigenvalues is $\det(K_0 - \lambda I_4) = 0$, *adding a constant* to K_0 simply adds the same constant to the eigenvalues, while *scaling K_0* simply scales the eigenvalues by a constant.
- The eigenvectors of any well-behaved eigensystem can be computed from the *adjugate matrix* of the vanishing-determinant characteristic equation into which a valid eigenvalue has been substituted. This follows from the fact that for any matrix, $M \cdot \operatorname{Adj}(M) = \det M\, I_4$; the latter equation also indicates that for a characteristic matrix $\chi(\lambda_0)$, the four columns of $\operatorname{Adj}(\chi)$ are all formally eigenvectors of the *same eigenvalue* λ_0.
- The eigenvectors themselves can be multiplied by any nonvanishing scale factor without changing the eigenvalue equation, since the eigenvector equation is homogeneous in the eigenvector itself.

Therefore, we can replace our original matrix $K_0(\theta, \hat{\mathbf{n}})$ in Eq. (12.40) by adding one copy of the identity matrix and dividing by 4 to yield a new matrix,

$$K(\theta, \hat{\mathbf{n}}) = \frac{1}{4} \left(K_0(\theta, \hat{\mathbf{n}}) + 1 \times I_4 \right) = \frac{1}{2} \begin{bmatrix} 1+c & s\,n_1 & s\,n_2 & s\,n_3 \\ s\,n_1 & (1-c)\,n_1^2 & (1-c)\,n_1 n_2 & (1-c)\,n_1 n_3 \\ s\,n_2 & (1-c)\,n_1 n_2 & (1-c)\,n_2^2 & (1-c)\,n_2 n_3 \\ s\,n_3 & (1-c)\,n_1 n_3 & (1-c)\,n_2 n_3 & (1-c)\,n_3^2 \end{bmatrix} , \tag{12.42}$$

whose eigenvalues are now $(1, 0, 0, 0)$, so the maximal eigenvalue is now $\lambda_{\text{opt}} = 1$, but whose normalized eigenvectors are preserved. $K(\theta, \hat{\mathbf{n}})$ has some interesting properties. First, we see that we have simply rediscovered Eq. (12.33), except that now we perceive it in a new light, as the core matrix of a variational eigensystem whose maximal eigenvector determines q_{opt}. Furthermore, whichever of the two forms of $K(\theta, \hat{\mathbf{n}}) = K(q)$ we use, the eigenvectors corresponding to the eigenvalues are just

$$\left\{ \begin{bmatrix} q_0 \\ q_1 \\ q_2 \\ q_3 \end{bmatrix} \begin{bmatrix} -q_1 \\ q_0 \\ 0 \\ 0 \end{bmatrix} \begin{bmatrix} -q_2 \\ 0 \\ q_0 \\ 0 \end{bmatrix} \begin{bmatrix} -q_3 \\ 0 \\ 0 \\ q_0 \end{bmatrix} \right\}, \tag{12.43}$$

and the quaternion itself, $q = (q_0, q_1, q_2, q_3)$, is trivially the maximal eigenvector with $\lambda_{opt} = 1$.

However, there is a deeper meaning in the eigensystem generated by the Bar-Itzhack variational method that tells us everything that is important about the nontrivial manifold in which quaternions live. *Given the profile matrix K*, we can compute the maximal eigenvector corresponding to $\lambda_{opt} = 1$ simultaneously in four different ways by writing down the characteristic equation of K with $\lambda = 1$ and computing the 4×4 *adjugate matrix*. We can use any equivalent form we like, but $K(q)$ is particularly simple: first we examine

$$\text{characteristic equation:} \; = \; \chi \; = \; K(q) - 1 \times I_4 \tag{12.44}$$
$$= K(q) - (q \cdot q) \times I_4, \tag{12.45}$$

and then remarkably, when we compute the formal adjugate of the maximal eigenvalue's characteristic equation for $K(q)$, we find that the matrix we call $A(\chi)$ is simply the negative of adjugate(χ), or *the value of $+K(q)$ itself*:

$$A(\chi) = -\text{Adjugate}(\chi) \; = \; +K(q) \; = \; + \begin{bmatrix} q_0^2 & q_0 q_1 & q_0 q_2 & q_0 q_3 \\ q_0 q_1 & q_1^2 & q_1 q_2 & q_1 q_3 \\ q_0 q_2 & q_1 q_2 & q_2^2 & q_2 q_3 \\ q_0 q_3 & q_1 q_3 & q_2 q_3 & q_3^2 \end{bmatrix}. \tag{12.46}$$

Recall that the minus sign can be removed and the positive quadratic matrix employed to represent the family of alternative unnormalized maximal eigenvectors, since the eigenvector equation is insensitive to the scale of the eigenvectors. We already know these solutions for q_{opt} are correct, since each column (or row, as it is symmetric) is proportional to the maximal eigenvector $q = (q_0, q_1, q_2, q_3)$. However, in addition we observe a repetition of our observation in Section 12.4.1 that the four columns of superficially equivalent solutions are *not* equivalent, but indicate that any of the 14 combinations of appearances of zeroes in one, two, or three of the quaternion components (q_0, q_1, q_2, q_3) renders the entire column useless for computing the correct quaternion corresponding to the measured rotation matrix, and another quadratic column with nonsingular normalization must be used for the calculation.

12.4.3 Bar-Itzhack variational approach to 3D noisy data

We have found the solution for the quaternion manifold's four solutions (eight including sign ambiguity) in terms of a perfect orthogonal measurement of the rotation data. As noted by Bar-Itzhack (2000), the same basic procedure can be used for measured rotation matrices with errors, and the resulting quaternion produces a perfect orthonormal rotation matrix that is the *optimal approximation* to the provided errorful data. Methods for dealing with the anomalous behavior produced by extreme errors have been studied, e.g., by Sarabandi et al. (2018), but this is not our target subject, so we will assume we are dealing with data that do not demand extreme treatments. Our starting point is a measured 3×3 matrix $R(m)$ that is assumed to originate from a 3D rotation matrix, but cannot be guaranteed to be orthonormal due to measurement error. We write the unconstrained components of this input data matrix as

$$R(m) = \begin{bmatrix} m_{11} & m_{12} & m_{13} \\ m_{21} & m_{22} & m_{23} \\ m_{31} & m_{32} & m_{33} \end{bmatrix} . \tag{12.47}$$

We set up the Bar-Itzhack variational problem starting with a symbolic rotation in the quadratic quaternion form $R(q)$ given in Eq. (2.14), and write down the cross-term of the Frobenius norm to define a maximization problem that will be our optimization target:

$$\Delta_{\mathbf{F}}(q, m) = \operatorname{tr} R(q) \cdot R(m)^{\mathrm{t}} = q \cdot K_0(m) \cdot q . \tag{12.48}$$

The initial profile matrix $K_0(m)$ resulting from rearranging the quadratic quaternion terms into the form of a scalar-valued symmetric matrix multiplication takes the form

$$K_0(m) = \begin{bmatrix} m_{11} + m_{22} + m_{33} & m_{32} - m_{23} & m_{13} - m_{31} & m_{21} - m_{12} \\ m_{32} - m_{23} & m_{11} - m_{22} - m_{33} & m_{12} + m_{21} & m_{13} + m_{31} \\ m_{13} - m_{31} & m_{12} + m_{21} & -m_{11} + m_{22} - m_{33} & m_{23} + m_{32} \\ m_{21} - m_{12} & m_{13} + m_{31} & m_{23} + m_{32} & -m_{11} - m_{22} + m_{33} \end{bmatrix} . \tag{12.49}$$

Note that $K_0(m)$ is traceless, and since $K_0(m)$ is a real symmetric matrix, it will have real eigenvalues. The eigenvector q_{opt} of $K_0(m)$'s maximal eigenvalue $\lambda_{\mathrm{opt}}(m)$ will maximize $\Delta_{\mathbf{F}}(q, m)$. This maximal eigensystem will solve the optimization problem

$$q_{\mathrm{opt}} = \underset{q}{\operatorname{argmax}} \left(\operatorname{tr} R(q) \cdot R(m)^{\mathrm{t}} \right) = \underset{q}{\operatorname{argmax}} (q \cdot K_0(m) \cdot q) \tag{12.50}$$

with

$$\lambda_{\mathrm{opt}} = \Delta_{\mathbf{F}}(q_{\mathrm{opt}}, m) = \left(q_{\mathrm{opt}} \cdot K_0(m) \cdot q_{\mathrm{opt}} \right) . \tag{12.51}$$

However, to be a proper quaternion, the optimizing value of the eigenvector q of λ_{opt} will have to be normalized to become q_{opt}, and we have argued throughout that this is not always possible, and must be dealt with using the quaternionic manifold $\mathbf{S}^3$ with eight covering coordinate patches instead of relying on a single function.

Fortunately, we know from the exact-rotation-data case in the preceding section how to deal correctly with this issue in the Bar-Itzhack context. We note that while it was useful in the exact case to get a very clean set of formulas by performing an eigenvector-preserving transformation of the form

$$K(m) = \operatorname{scale} \times (K_0(m) + \operatorname{constant} \times I_4) \tag{12.52}$$

to adjust our maximal eigenvalue to the identity, in the general case, we cannot find a single pair of constants that will be all that useful, though one might like to normalize to obtain a unit maximal eigenvalue by dividing by the value of K_0's maximal eigenvalue $\lambda_{\mathrm{opt}}(m)$. We will assume that if there is some reason to readjust $K_0(m)$ to a form $K(m)$ with the same eigenvectors, up to scaling, preserving the corresponding maximal eigenvalue $\lambda_{opt}(m)$ up to an additive constant, we may do so, and thus we will continue with that abstract $K(m)$ to complete our argument.

Obviously what we have to do first is find $\lambda_{opt}(m)$. Standard numerical eigensystem software packages can easily accomplish this, and, for symmetric real matrices up to 4×4 in size, one can even calculate the maximal eigenvalue analytically using Cardano's solution of fourth-degree polynomials (see, e.g., Appendix J or Hanson (2020) for a review). The last step is then to form the characteristic equation's matrix as before, using now the *numerical* eigenvalue, giving

$$\text{characteristic matrix: } = \chi(m) = K(m) - \lambda_{opt}(m) \times I_4 \tag{12.53}$$
$$= K(m) - (q_{opt} \cdot K(m) \cdot q_{opt}) \times I_4 ,$$

where $\det \chi(m) \equiv 0$.

Finally, our full quaternion space covering manifold solution is determined from the adjugate of Eq. (12.53), which is an entirely numerical matrix having the following (sign-adjusted) relation to the sign-doubled set of eight possible quaternion formulas:

$$A(\chi(m)) = -\text{Adjugate}(\chi(m)) = \begin{bmatrix} q_0^2 & q_0 q_1 & q_0 q_2 & q_0 q_3 \\ q_0 q_1 & q_1^2 & q_1 q_2 & q_1 q_3 \\ q_0 q_2 & q_1 q_2 & q_2^2 & q_2 q_3 \\ q_0 q_3 & q_1 q_3 & q_2 q_3 & q_3^2 \end{bmatrix} . \tag{12.54}$$

Recall that the adjugate is determined only up to a nonvanishing scale of either sign, and that one picks the ordinal index k with the largest diagonal value q_k^2 in the adjugate matrix, as indicated in Eq. (12.37), and normalizes that column to obtain q_{opt}. Finally, one calculates the *optimal pure rotation approximation* to the numerically measured $R(m)$ using this quaternion selected from the adjugate matrix,

$$R_{opt}(m) = R(q_{opt}) , \tag{12.55}$$

and our treatment of how to compute a quaternion from any ideal or noisy rotation matrix data is done.

Quaternion-related machine learning*

Machine learning methods are important throughout many applications that involve understanding, processing, and analyzing spatial orientation. Among these are applications that require the determination of unknown rotation matrices from the input data, including automatic navigation, robotics, image understanding, and machine vision. Rotations, in turn, benefit from being represented in terms of quaternions, but, as we have seen in the previous chapter, this procedure is complicated by the fact that quaternions live on the nontrivial manifold $\mathbf{S}^3$. It is generally accepted (see, e.g., Hornik et al., 1989) that neural networks can in principle provide arbitrarily accurate approximations to any *function*; therefore obtaining a quaternion, which is not describable as a pure function, as an answer from a neural network to determine a rotation matrix is likely to fail without mitigating procedures. In this chapter, we complete this picture by working out some elementary examples and exploring this relationship between the nontrivial topological manifold of quaternions and machine learning. There are of course an enormous number of complex related applications, but here we limit ourselves to the modest goal of providing a few simple network demonstrations that help to expose how it all works.

13.1 Overview

One of the typical application areas for quaternions in machine learning has been the class that includes 3D point-cloud matching, a problem that we have dealt with extensively starting in Chapter 8. A number of authors have investigated the basic example with one cloud of 3D points $\mathbf{X}$ as the *reference* data, and a pairwise matched set $\mathbf{U}$ of *test* data presumed to have been created noisily by a quaternion q_{init} incorporated into a rotation matrix $R_{\text{init}} = R(q_{\text{init}})$ that was applied to the reference set. Including the possibility of incorporating noise, we may write this as

$$R_{\text{init}} \cdot \mathbf{X} + \langle noise \rangle = \mathbf{U} . \tag{13.1}$$

Assuming we are given this matched pair of collections of K 3D points, the brute force input to the network would be

$$\text{data points } \mathbf{X}, \mathbf{U} \;\Rightarrow\; [\textbf{Input Layer: dim} = 2 \times 3 \times K] . \tag{13.2}$$

Then, with some network defined to make a trainable transition, the output of the network would be

$$[\textbf{Output Layer: dim} = 4] \;\Rightarrow\; \textbf{quaternion } q_{\text{out}}.$$

---

* Some of this chapter is patterned after Hanson and Hanson (2022) and Lin et al. (2023).

Visualizing More Quaternions. https://doi.org/10.1016/B978-0-32-399202-2.00023-X

The training process would then simply provide as training feedback the expected q_{init} that rotated the original data, applying one of several available loss functions comparing the training value q_{init} to the output layer's q_{out}.

Cloud points versus the cross-covariance matrix. However, one can argue that an input layer as large and complicated as Eq. (13.2), requiring a very large number of cases to properly train, is an unnecessary expense: in fact we know from Chapter 8 that all the data theoretically sufficient to solve the matching problem are contained in the 3×3 *cross-covariance matrix*

$$E_{ab}(\mathbf{X}, \mathbf{U}) = \sum_{k=1}^{K} x_a{}^k u_b{}^k = \mathbf{X} \cdot \mathbf{U}^{\mathrm{t}}. \tag{13.3}$$

All the solutions to the 3D point-cloud matching problem studied in Chapter 8 depend only on polynomials of the nine components of E, and therefore, in our exercises below, we will choose this more parsimonious approach to supplying input to a network designed to apply machine learning to the RMSD point-cloud matching task.

$\mathbf{R_{init}}$ **versus** $\mathbf{R_{opt}}$. There is one more issue concerning the integrity of training a network on a problem like finding the quaternion rotation for a point-cloud matching task. While it is tempting to identify the quaternion q_{init} and its rotation $R_{init} = R(q_{init})$ provided to the test-data simulator as the "ground truth" standard against which the network output should be evaluated for success, this is valid only for *noise-free* problems. Dealing with noisy data has been one of our main themes, and once noise is introduced, the initial rotation parameters are *incorrect training targets* from the standpoint of computability: there is no deterministic way to determine q_{init} or R_{init} from the measured data that we encode in a noisy cross-covariance matrix E. However, we know perfectly well from Chapter 8 that there is an exact *optimal* solution, q_{opt} or $R_{opt} = R(q_{opt})$, that is provably computable in closed form, and which solves the least-squares problem for 3D point-cloud matching even in the presence of noise. The loss function computed with R_{init} will always be greater than or equal to the loss computed with R_{opt}. Therefore, at least for this particular problem, it is demonstrably incorrect to train to the data-defining rotation parameters instead of $R(q_{opt})$ if the task is to find the best aligning rotation.

13.2 Reports of quaternion deficiencies for machine learning

Several implementations of the optimal rotation discovery problem have reported that quaternions may be a deficient representation for this machine learning task. Zhou et al. (2019), for example, describe excessive errors, singularities at certain angles, and discontinuous behavior. They propose to avoid the quaternion defects using alternative methods of encoding the rotation directly, e.g., using 5D or 6D subsets of the nine-dimensional 3D rotation matrix itself, bypassing quaternions altogether. Other authors, including Saxena et al. (2009); Peretroukhin et al. (2020); Zhao et al. (2020), invoke other approaches that avoid using the apparently problematic explicit quaternion representation. The method of Peretroukhin et al. (2020), for example, suggests an improvement based on a heuristic that learns a 4×4 matrix related to quaternions, and post-processes that multiple-dimensional representation by finding its minimal eigenvalue; the corresponding eigenvector turns out to be trainable to match the needed quaternion. Another particularly appropriate method that directly uses multiple alternative quaternion representations instead of avoiding them has been pointed out in Xiang and Li (2020), with an extensive elaboration published by Xiang (2021).

Analysis of previous results. Starting with concepts in Chapter 12, introduced in Hanson and Hanson (2022) and Lin et al. (2023), we can understand the basic features that have been observed. The simplest to understand is that, as we will show in our experiments below, training to a quaternion with data that cover the complete rotation spectrum simply must be inadequate: the nontrivial quaternion topology cannot be accommodated. But it is easy to demonstrate that *restricting* the training domain to source quaternions that are actually single functions, such as rotations restricted to the positive quaternion sector with quaternion signs $(+, +, +, +)$, resolves the problem, as this method trains to *one single-valued function*, and all works perfectly. Another observation is that the adoption of a symmetric 4×4 matrix A as a training target, with a post-process that extracts a quaternion as the eigenvector of the *minimal eigenvalue* of the A matrix, is equivalent to the classic Hebert–Faugeras method (Hebert, 1983; Faugeras and Hebert, 1983, 1986) that produces a closed form solution of the RMSD problem. However, the problem of the quaternion singularities is actually still present in this method, but hidden from sight in the eigensystem library implementation that uses either the optimal adjugate method or something like the Shepperd method to account in an unseen way for the multiple-valued manifold properties that we have been exposing. Any other method that selects different quaternion sectors based on conditions imposed on individual quaternion components must be equivalent to our optimal quaternion adjugate method incorporating the full topological manifold properties of $\mathbf{S}^3$ that we have derived deterministically and analytically in several different ways in previous chapters.

Our takeaway lesson is that the results of machine learning can easily be misinterpreted; careful analysis of the geometry and analytic properties of the problem being addressed is required to obtain reliable behavior.

13.3 Exercises in machine learning with quaternions

We now present a set of elementary but informative neural network exercises that illustrate many of the features we have discussed. We employ a very clean network patterned after the "sanity check" configuration described in the paper by Zhou et al. (2019) to assure ourselves that we are dealing with comparable mathematics when drawing our conclusions. We make no pretense to the development of more sophisticated versions of networks and machine learning methods: our scope is confined to the simplest nontrivial demonstrations that could exhibit the behavior that is of interest here.

Design. Our basic network design has four layers with dimension 128 and intervening leaky ReLU's. Our first experiment will be to produce batches of uniform random quaternions $q = q_{\text{init}}$ and compute the corresponding noise-free 3×3 rotation matrices $R_{\text{init}}(q)$ to employ both as the input to the first layer, and as the training data provided to the loss function. The input layer will therefore always be $\{3, 3\} \times 128$. The output layer will vary across a range of dimensions corresponding to representations that can be mapped ("decoded") to a 3×3 rotation matrix for comparison with the provided training target matrix. For example, a typical four-element quaternion system with output dimension $D = 4$ would need to use Eq. (2.14) as its decoding equation to provide a rotation matrix to the loss function. Our network in general thus has the structure in this table, represented diagrammatically in Fig. 13.1.

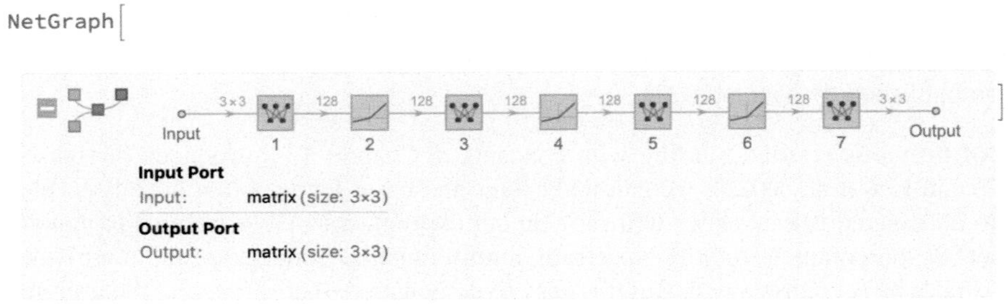

```
NetGraph[
```

FIGURE 13.1

The basic network structure used for our experiments, with four layers, internal size 128, and intervening leaky ReLUs. The output dimension is variable depending on the rotation representation being tested and converted to a valid rotation matrix for the `loss` function. In this example, the network has the fundamental structure of an autoencoder taking as input a distribution of rotations, and producing as output a set of variables that a decoder transforms back to a possible rotation that is combined with the input to compute a loss.

Sanity test network			
Layer	**Type**	**"Input"**	**"Output"**
Input: 1	Linear layer	3×3	128
2	Leaky ReLU	—	—
3	Linear layer	128	128
4	Leaky ReLU	—	—
5	Linear layer	128	128
6	Leaky ReLU	—	—
Output: 7	Linear layer	128	D

13.3.1 Loss layer data encoding

Again following the sanity test procedure documented by Zhou et al., we choose an informative list of parameterizations for which we know a way to generate a standard 3D rotation matrix, create a network that has a compatible output vector $\mathbf{v}$ for each of these parameter sets, and build a family of loss layers that compute a rotation matrix $R(\mathbf{v})$ and combine it with the provided training target matrix R_{init} to compute a Frobenius loss,

$$\text{“loss”} = \text{tr}(R_{\text{init}} - R(\mathbf{v})) \cdot (R_{\text{init}} - R(\mathbf{v}))^{\text{t}} . \tag{13.4}$$

For the self-training of the rotation matrix to itself, no encoding is required, and one can just use the default loss layer in Fig. 13.2, whose core is simply the `MeanSquaredLossLayer[]`. For custom parameter encoding, one uses the appropriate variants of Fig. 13.3 with a hand-coded `FunctionLayer`.

The rotation representations and their encoding maps that we send to the loss function for our studies are the following:

- **Self-train R (3×3).** The output layer is identical in form to the input 3×3 rotation matrix:

$$\mathbf{v}_{\text{out}} = R(3 \times 3) , \tag{13.5}$$

$$R(\mathbf{v}) = \mathbf{v} . \tag{13.6}$$

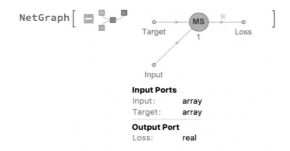

FIGURE 13.2

The generic loss layer structure, with the "loss" element automatically implemented as a mean squared error using `MeanSquaredLossLayer[]`. For our other experiments, we replace that function by other functions that convert the final layer's output to a rotation matrix that is compared to the target matrix using the Frobenius norm for the least-squares loss algorithm.

- **Quaternion (4).** Compute the input trained to a distribution of rotations corresponding to certain restrictions on the supplied quaternion q_{init} specified in the following item, and compute the training rotation $R_{\text{init}} = R(q_{\text{init}})$ as usual from Eq. (2.14):

$$\mathbf{v}_{\text{out}} \rightarrow (q_0, q_1, q_2, q_3) \,, \tag{13.7}$$

$$R(\mathbf{v}) = R(q) = \begin{bmatrix} q_0{}^2 + q_1{}^2 - q_2{}^2 - q_3{}^2 & 2q_1q_2 - 2q_0q_3 & 2q_1q_3 + 2q_0q_2 \\ 2q_1q_2 + 2q_0q_3 & q_0{}^2 - q_1{}^2 + q_2{}^2 - q_3{}^2 & 2q_2q_3 - 2q_0q_1 \\ 2q_1q_3 - 2q_0q_2 & 2q_2q_3 + 2q_0q_1 & q_0{}^2 - q_1{}^2 - q_2{}^2 + q_3{}^2 \end{bmatrix} . \tag{13.8}$$

- **Quaternion (4) output with constrained input.** In order to confirm our hypothesis that simple training to quaternions is successful only when the domain of the target domain is computable, we employ three conditions on the training cases:
  - *Supply training rotations from the entire* $\mathbf{S}^3$ *of quaternions.* For these data, all quaternion signs are distributed equally, over all 16 hyperquadrants of $\mathbf{S}^3$. This algorithm simply computes a quaternion from a random distribution (see Chapter 6) and supplies the corresponding rotation computed from Eq. (2.14).
  - *Supply training rotations only from the* $q_0 > 0$ *hemisphere of the quaternion* $\mathbf{S}^3$. Only rotations produced by quaternions q that have $q_0 > 0$ out of the full random distribution are used to produce the training target $R(q)$. This checks to see whether the quaternion *sign ambiguity* is the issue preventing successful training: in this case *all rotations* occur, but only half the available quaternion spectrum.
  - *Supply training rotations from only a guaranteed single-function sector of the quaternion manifold.* There are 16 such sectors, with quaternion signs $(\pm q_0, \pm q_1, \pm q_2, \pm q_3)$, assuming all the $q_i \geqslant 0$. It suffices to check any one, and we select from the random distribution the $(+, +, +, +)$

sector, namely those quaternions with $(q_0 \geqslant 0, q_1 \geqslant 0, q_2 \geqslant 0, q_3 \geqslant 0)$, and we provide rotations $R(q)$ *only* from that subset of the random distribution for training.

- **Euler angles (3).** The Euler angles are another common method of representing rotations, starting with three angles as the output, and applying the usual Euler rotation form:

$$\mathbf{v}_{\text{out}} \to (\theta, \phi, \psi), \tag{13.9}$$

$$R(\mathbf{v}) = R(\theta, \phi, \psi) =$$

$$\begin{bmatrix} \cos(\theta)\cos(\psi)\cos(\phi) - \sin(\psi)\sin(\phi) & -\cos(\theta)\sin(\psi)\cos(\phi) - \cos(\psi)\sin(\phi) & \sin(\theta)\cos(\phi) \\ \cos(\theta)\cos(\psi)\sin(\phi) + \sin(\psi)\cos(\phi) & \cos(\psi)\cos(\phi) - \cos(\theta)\sin(\psi)\sin(\phi) & \sin(\theta)\sin(\phi) \\ \sin(\theta)(-\cos(\psi)) & \sin(\theta)\sin(\psi) & \cos(\theta) \end{bmatrix}. \tag{13.10}$$

- **R6 (6).** This is the 6D partial rotation matrix advocated by Zhou et al. (2019). The purpose of this data set is to verify that training to just the top two rows of a rotation matrix, and using the cross product of those two rows to compute the bottom row, is sufficient to obtain a good training output. Since this is in fact a function of the actual rotation, as well as a linear combination of a limited set of adjugate quaternion variables, this should conform to the constraints for a trainable target. The encoding function is simply the cross product:

$$\mathbf{v}_{\text{out}} \to \mathbf{R6}(\mathbf{v}) = \begin{bmatrix} \mathbf{R}_1 \\ \mathbf{R}_2 \end{bmatrix} = \begin{bmatrix} R_{11} & R_{12} & R_{13} \\ R_{21} & R_{22} & R_{23} \end{bmatrix}, \tag{13.11}$$

$$R(\mathbf{v}) = \begin{bmatrix} \mathbf{R}_1 \\ \mathbf{R}_2 \\ \mathbf{R}_1 \times \mathbf{R}_2 \end{bmatrix}. \tag{13.12}$$

- **Quaternion adjugate matrix A (10).** This is the standard we adhere to, since we claim that this is essentially a matrix of identical quaternion solutions scaled, in turn, by (q_0, q_1, q_2, q_3) to make them expressible in terms of a function of the rotation input. If we explicitly require a quaternion, we can select the column that is most normalizable before creating a quaternion by normalization. However, since the rotation matrix is actually already quadratic in the quaternions, we can use the adjugate directly to compute the encoded rotation sent to the loss function. Technically, this may also require scaling, e.g., by the trace of the adjugate, to generate a proper rotation matrix, but this can also lead to instabilities. With sufficient training, the loss function itself should be able to enforce normalization. Thus

$$\mathbf{v}_{\text{out}} \to (\mathbf{v}_1, \mathbf{v}_2, \dots, \mathbf{v}_9, \mathbf{v}_{10}), \tag{13.13}$$

$$A(\mathbf{v}) = \begin{bmatrix} v_1 & v_2 & v_3 & v_4 \\ v_2 & v_5 & v_6 & v_7 \\ v_3 & v_6 & v_8 & v_9 \\ v_4 & v_7 & v_9 & v_{10} \end{bmatrix}, \tag{13.14}$$

$$R(\mathbf{v}) = \begin{bmatrix} v_1 + v_5 - v_8 - v_{10} & 2v_6 - 2v_4 & 2v_7 + 2v_3 \\ 2v_6 + 2v_4 & v_1 - v_5 + v_8 - v_{10} & 2v_9 - 2v_2 \\ 2v_7 - 2v_3 & 2v_9 + 2v_2 & v_1 - v_5 - v_8 + v_{10} \end{bmatrix}. \tag{13.15}$$

Batch and loss layer training details. The network training step is supplied with two major utilities, a batch generator, and a loss function, which is implemented as a network layer. The batch genera-

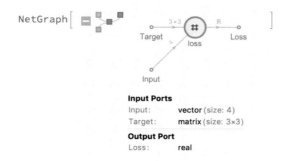

Input Ports
Input: **vector** (size: 4)
Target: **matrix** (size: 3×3)

Output Port
Loss: **real**

FIGURE 13.3

The loss layer structure for the quaternion-to-rotation decoder. This layer takes the size 4 quaternion output from the final layer, computes the transformation from Eq. (2.14) to a rotation matrix $R(q_{out})$, and computes the loss by combining that with the provided target matrix using the Frobenius norm least-squares loss algorithm. Each example rotation-determining output will have its own tailored loss in this graph.

tor produces the requested number of batches as described above, generating random rotations whose quaternion sources may be restricted to certain domains. The loss function is a layer of the generic type in Fig. 13.2, or specific to one of the particular output-to-rotation encoders, represented by the network in Fig. 13.3.

13.3.2 Network setup and training

All of our networks have the basic form of Fig. 13.1 with variants in the dimension of the output layer to match the encoders for each loss function. Each example encoding in Section 13.3.1 is paired with:

- **A matching network with an output vector or tensor v with specified dimensions.** The self-training network has a 3×3 rotation matrix output layer, all quaternions have a 4-vector output layer, etc.
- **A batch list generator.** In addition to the standard batch list generator that picks a random *initial* quaternion q from a distribution uniform on $\mathbf{S}^3$ and returns a matched pair of rotation matrices $R(q)$, one for the network input, and the other the identical target to be used by the loss function, we have an additional special class: instead of using random quaternions to generate each rotation, the quaternions are *restricted* in various ways to test the hypothesis that training to a subset that is a unique function is important. These classes are:

1. *all quaternion signs;*
2. *quaternions that generate all possible rotations, but only allowing half the quaternion space, with $q_0 > 0$;*
3. *quaternions that generate all possible rotations, but retain only a single-function-compatible 1/16 of the quaternion space, selected from $(\pm q_0, \pm q_1, \pm q_2, \pm q_3)$, assuming all the $q_i \geqslant 0$.*

- **A custom loss function.** Each output parameter set **v** has a transformation that maps it to a rotation matrix that can be compared to the target provided by the batch generator.

Procedure. Each candidate is set up in the Mathematica machine learning environment with its network and matched utilities. The network is specified to work with the ADAM optimizer, a training rate starting between 0.01 and 0.0001, and a default batch size of 64. All of the training data are computed on the fly by a batch generator and discarded. It was observed that the loss functions had individualized sensitivity to the learning rate and the total number of rounds required to achieve a stable final loss in the trained network. All experiments started with 5000 rounds and training rate 0.0001, and these were adjusted as needed to recover stable results in the asymptotic loss.

13.4 2D rotation training and results

Our first and most elementary check is to take our basic quaternion tests described above and reduce them to two dimensions. That is, we do the following:

- **Generate a random 2D quaternion distribution.** We create the 2D quaternions

$$(a = \cos(\theta/2),\ b = \sin(\theta/2))$$

with a uniform random distribution of angles θ, with the option of (1) permitting all signs, (2) permitting just $a > 0$, and (3) only allowing $(a > 0,\ b > 0)$, and producing training rotation matrices

$$R_{\text{init}}(a, b) = \begin{bmatrix} a^2 - b^2 & -2ab \\ 2ab & a^2 - b^2 \end{bmatrix}.$$

- **Create a network.** Then we create a network with 2×2 rotation matrix input and either that same output, or a 2-vector (a, b) as output. We show these networks in Fig. 13.4.

The number of rounds needed to train a network of the type we are using and the training rates are sensitive to the details of the problem because of lack of specificity in the network design, so we will report the parameters used in training to get the results quoted. We begin in this case with a very high-performance result for the 2D case: we used the ADAM optimizer, with 5000 rounds of training, at a specified initial training rate of 0.0001 for the 2×2 rotation matrix R output, and 0.00005 for the reduced quaternion (a, b) network; the rate is adjusted automatically to some extent by the system during the training rounds. All data are created on the fly rather than being stored and sampled, so each batch of 64 samples requested by the training supervisor has a new set of rotations created by uniformly random sampled quaternions.

In Fig. 13.5, we show the results of training our network to four cases, the quaternion network with all signs, with $a > 0$, and with $(a > 0, b > 0)$, and the 2D self-replicating rotation matrix. The results are remarkable for their predictability based on the arguments we have made in Chapter 12. The all-rotations 2D quaternions, with or without the restriction to the hemisphere $a > 0$, simply fail to respond to the network training in any reasonable way: they fluctuate randomly at a very high average loss. The self-replicating rotation matrix and the quaternion trained to only the positive quadrant (see Fig. 12.1 in

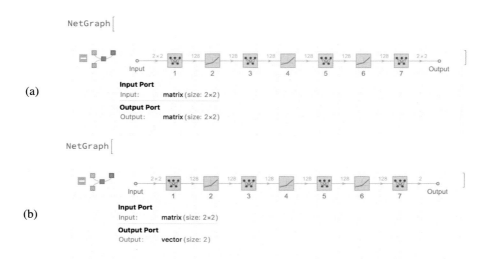

FIGURE 13.4

(a) The 2D network training to its own 2D rotation $R(a, b)$. (b) The 2D network training to a quaternion with possibly restricted training data.

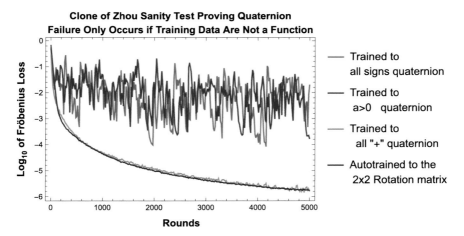

FIGURE 13.5

Sample 2D spatial rotation least-squares (Frobenius norm) losses for different reduced 2D (a, b) quaternion training choices. Training to a single 2D all "+" quadrant matches the performance of self-training to the rotation matrix itself. Training to quaternions producing all possible rotations prevents the learning procedure from converging because the entire manifold cannot be learned at once.

Chapter 12) follow identical learning paths and achieve a highly accurate result after only 5000 rounds of training.

13.5 3D rotation training and results

We now turn to the realistic case of 3D rotations and quaternions. Here we have distinct behaviors for different types of representations. Our data choices are:

- 3D rotation self-trained to the target rotation matrix;
- 3D rotation output generated from Euler angle triple data in the last layer;
- quaternion trained to rotation matrices generated by all signs of quaternion initialization data, 3D rotation generated from the quaternion in the last layer using Eq. (2.14);
- quaternion trained to rotation matrices generated by $q_0 > 0$ quaternion initialization data;
- quaternion trained to rotation matrices generated by all positive signs, the $(+, +, +, +)$ subset of quaternion initialization data;
- the Zhou et al. **R6** encoded rotation as output of the last layer, with the third row of the rotation matrix generated by the cross product of the two rows in **R6**;
- the quaternion **Adjugate** variable trained rotation, rotation generated by the quaternion of the maximal adjugate column.

From the results in Fig. 13.6, we see in particular that we can replicate the qualitative results of Fig. 13.5 with learning rate 0.0001 and only 5000 rounds of training for self-replicating 3×3 rotation matrices and quaternions trained to all rotations vs. quaternions trained to rotation matrices arising only from the $(+, +, +, +)$ quaternion sector. However, when we look at the **R6** and **Adjugate** training sectors, we get much slower convergence, and for just 5000 rounds they are far from reaching a trained equilibrium. We kept the learning rate at 0.0001, which was not necessarily the optimal choice, but in any case after 250,000 rounds, the **R6** and **Adjugate** results were essentially in line with all the others. We remark that Zhou et al. appear to have used as many as 500,000 rounds.

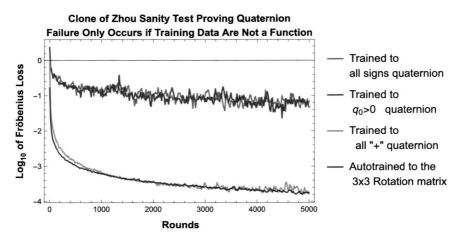

FIGURE 13.6

Sample 3D spatial rotation least-squares (Frobenius norm) losses for different $q = (q_0, q_1, q_2, q_3)$ quaternion training choices.

In Fig. 13.7, we show the results comparing all the 3D results. The rotation matrix self-training is the expected best-performing benchmark, and with 5000 rounds of training, it is matched by the $(+, +, +, +)$ quaternion. The Euler angle and all-signs quaternion training get lost in the noise without achieving a stable trained result as expected, but we see that the **R6** and **Adjugate** methods are still far from convergence if we plot just 5000 rounds of training. However, their training profiles after 250,000, rescaled to the graph of the 5000 round results, are virtually identical to the rotation and restricted quaternion outcomes. We can thus be confident that our analysis of how machine learning interacts with quaternion representations is valid.

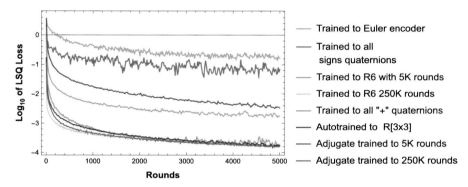

FIGURE 13.7

Comparison of training characteristics of various models applied in a network that trains different representations to match a given input rotation matrix. All data are perfect, no noise is added to the data.

Remarks. We note that the type of least-squares loss functions that we use typically are applied to each element of, say, the 3×3 rotation matrix, as a separate term. This turns the training process essentially into a dynamical system that is trying simultaneously to accommodate a list of Lagrange multipliers and reduce them all to zero simultaneously with the same overall weight coefficient. This is expected to be relatively easy for small numbers of variables, and to become increasingly erratic for larger numbers of effective Lagrange multipliers. We see exactly this behavior in the comparison of the Zhou six-variable **R6** system and our 10-variable **Adjugate** component systems with smaller systems such as the quaternion itself: the larger data output sizes can take orders of magnitude more rounds of training compared to smaller outputs to reach the same level of accuracy for the final stabilized loss.

13.6 Basic elements of applying a network to the cloud matching problem

In the beginning of this chapter, we introduced the RMSD 3D point-cloud matching task as a motivation for understanding the nature of quaternions as a method for encoding the rotation matrix needed to solve the problem. Now that we have some examples using an autoencoder network framework of what to beware of and how to solve the quaternion-related challenges pointed out in Chapter 12, let us examine a very elementary version of the network-based RMSD application.

The goal. Because of the existence of closed form solutions for the 3D point-cloud matching problem that we examined extensively starting in Chapter 8, we can choose to simplify our network considerably by avoiding entirely the expense of presenting the first layer with two long lists of 3D vectors as input. We know that, given a reference cloud $\mathbf{X}$ and a sample cloud $\mathbf{U}$ produced by applying a rotation R_{init} to $\mathbf{X}$, with the possible incorporation of noise, the entire information content of $\mathbf{X}$ and $\mathbf{U}$ needed to solve this problem is in the 3×3 cross-covariance matrix E_{ab} that we gave in Eq. (13.3). There is no more information – the value of the best possible rotation R_{opt} solving the least-squares optimization problem is entirely determined by E, with or without noise in the data. For the purpose of our academic exercises, we have the following resources to employ in various ways to feed, decode, and provide loss feedback to our cloud-matching (RMSD) machine learning environment:

- **Pre-input constraints.** The RMSD task begins with creating simulated data with a reference cloud and the choice of a rotation to apply to that cloud. We know from Section 13.6 that, if we attempt to use a single quaternion as an intermediate rotation representation, the success or failure of the training relies on whether the rotations used in the simulation correspond to samples of *all quaternions*, or a limited single-value-compatible sector such as quaternions with all plus signs. Therefore we want the option to apply a constraint filter on any quaternions employed to produce a rotation used in the RMSD cloud:cloud matching data construction.
- **Pre-input.** To supply data for the RMSD task, we can use random data methods for each element of a batch request on the fly to generate everything we need. Specifically, for each network input request, we require a new instance of the reference data $\mathbf{X}$, a possibly filtered random quaternion q_{init} (see Chapter 6), a corresponding rotation $R_{\text{init}} = R(q_{\text{init}})$ via Eq. (2.14), and the simulated sample cloud $\mathbf{U}$, possibly noisy. We do not need to save lists of these data because they are not reused, so we can generate each example in response to a batch request and discard it immediately. Nevertheless, if we *want* stored lists, e.g., for validation sets, we can construct those as well at the data preparation stage. From this information, we can construct the (Input:OutputToTraining) batch pair in various configurations.
- **Input: E-matrix.** Our goal in every case is to provide the input layer of our network with just the essential E-matrix, which we know is sufficient to solve the problem, so we will never submit $\mathbf{X}$ or $\mathbf{U}$ data to a network input layer. Thus we always have

$$\text{Input:} \; \rightarrow \mathbf{E}(3 \times 3) \, .$$

- **Target: $R_{\{\text{init},\text{opt}\}}$.** Typically, in tandem with the network input, we need to produce a paired desired Target output (typically called "Output" in the batch generator), which in this case is the rotation matrix that solves the RMSD best cloud matching task. However, this is a subtle question. One choice is to declare that "the creating rotation is the answer," in which case we supply

$$\text{Output:Target} \; \rightarrow \mathbf{R}_{\text{init}}(3 \times 3) \, .$$

However, as noted in our previous discussions, in the presence of noise, R_{init} cannot be computed from any available data, not from $\mathbf{X}$ and $\mathbf{U}$, and not from the cross-covariance E because the noise has broken any relationship that would allow R_{init} to be computed deterministically. However, even in the presence of noise, the rotation R_{opt}, or its corresponding quaternion q_{opt}, *can* be computed deterministically using the closed form solution minimizing the least-squares loss. If we want an answer with minimal loss (which we almost surely do), we should supply

$$\textbf{Output:Target} \rightarrow \mathbf{R}_{\text{opt}}(3 \times 3) .$$

Although there might be arguments for providing R_{init} as the target on occasion, one should really always provide R_{opt} as the train-to target if the objective is a network that deterministically returns the solution with the least possible loss. Having made that academic point, we will for simplicity be using noise-free data for our demonstrations here.

- **Target:** $q_{\{\text{init},\text{opt}\}}$. One of our points is of course that we do not necessarily need to deal with rotation matrices, as it can be challenging to enforce the constraints of a pure rotation matrix. Accordingly, we can alternatively provide either the quaternion q_{init} as our Target output, or the quaternion q_{opt}:

$$\textbf{Output:Target} \rightarrow \mathbf{q}_{\text{init}}(4) ,$$
$$\textbf{Output:Target} \rightarrow \mathbf{q}_{\text{opt}}(4) .$$

Of course we have the same issues just discussed regarding using R_{init} or R_{opt} as a rotation matrix output. Maintaining the quaternion $q \cdot q = 1$ unit length constraint is typically much easier than orthogonalizing a full rotation matrix, though one tactic could be to employ a Bar-Itzhack optimal-rotation extraction method. Nevertheless, using a quaternion as the output of the final layer potentially includes encountering all the problems mentioned already in Chapter 12. We really cannot train to a completely random quaternion, as the results are indeterminate, and the decoding map $q \rightarrow R(q)$ will simply reflect the impossibility of training reliably to the full quaternion manifold with a single quaternion 4-vector as an output.

- **Target: Adjugate$_{\{\text{init},\text{opt}\}}$.** The basic solution to the inability to train a network to a single quaternion from rotations based on random all-signs quaternion input data can be corrected, as described in detail in Chapter 12, by training to the entire multivalued quaternion adjugate matrix, which produces a valid quaternion for any possible sector of the manifold. From the initial cloud data, we can use q_{init} with, e.g., Eq. (12.46), to directly construct Adjugate$_{\text{init}}$, or q_{opt} to directly construct Adjugate$_{\text{opt}}$. As before, there are serious concerns about choosing to train to Adjugate$_{\text{init}}$. Rather than use the redundant 16 variables of the symmetric 4×4 adjugate matrix as the output layer data, we also typically configure the output layer to produce a vector of the 10 unique variables of the symmetric matrix. The batch-data-generating function can then provide the adjugate data as training data to be compared to the output layer in a Frobenius distance loss procedure. Since the target adjugate data will be constrained to a valid set of quaternions, the training loss will enforce those constraints on the network output, much like having a set of 10 Lagrange multipliers in a dynamical system.

- **Loss: Matrix least-squares norm.** For the two cases where our batch-data provider supplies a matrix, the cloud-matching 3D rotation matrix R_{init} or R_{opt} itself, with a rotation matrix output layer, we can just use the Frobenius norm or the sum of squared differences of the matrices to provide the loss function that drives the adjustment of weights to train the network. If the output layer is configured to provide the adjugate 10-vector, we can *decode* that into a rotation matrix $R(\text{Adjugate})$, and use that directly in the same loss function. Since the training data are valid rotation matrices, and the Frobenius style norm is effectively a collection of Lagrange multipliers for each matrix component, the network should be at least somewhat effective at producing an orthonormal matrix as the trained output.

- **Loss: Direct adjugate least-squares.** In the case where the batch-data provider supplies target data for the adjugate directly, that is, $\text{Adjugate}_{\text{init}}$ or $\text{Adjugate}_{\text{opt}}$, and the output layer is an adjugate symmetric-matrix-component 10-vector, the loss can be constructed directly as a sum of squared differences of the adjugate 10-vector output and target components. Again, this will be basically a collection of Lagrange multipliers for each adjugate component, and the target data being fully constrained by definition, the network should ultimately produce a constraint-satisfying quaternion adjugate output, from which the RMSD-solving optimal rotation matrix $R(\text{Adjugate}_{\text{opt}})$ can be constructed.

13.7 Experiments with training to the cross-covariance matrix $\mathbf{E}$

We ran machine learning experiments on a selected subset of the many possible configurations involving the 3D point-cloud matching problem. As already noted, the presence or absence of noise governs whether we would ordinarily take the extra step of computing the RMSD solution q_{opt} to provide more robust training, and it is possible that more sophisticated network approaches would be required to get good results for noisy data in any case: therefore, in this investigation, we restricted ourselves to noise-free cloud matching data.

The common thread for our experiments is that a reference 3D point cloud $\mathbf{X}$ is chosen, along with a uniform random quaternion that may be completely random over all of the $\mathbf{S}^3$ manifold, or may be filtered to allow only quaternions in certain submanifolds, including the unique-function sector q_{++++} with all elements in the positive hyperquadrant. The specified quaternion then generates a rotation matrix $R(q)$, and that in turn produces the sample data $R(q) \cdot \mathbf{X} = \mathbf{U}$, where we will omit the possibility of adding noise at this point. Based on this set of data, all networks will provide the cross-covariance matrix

$$\text{Input} \rightarrow E = \mathbf{X} \cdot \mathbf{U}^{\text{t}}$$

to the input layer: this is our universal input data for this study.

Then the output or training target data can be selected from the following pieces of cloud-related information that are available in addition to E to guide the learning of the network:

- q_{init}. The raw randomized quaternion rotation, correct for our noise-free data, but incomputable in the presence of noise, in which case we would provide q_{opt}.
- $R_{\text{init}} = R(q_{\text{init}})$. The 3D rotation computed from the quaternion.
- $A_{\text{init}} = \mathbf{Adjugate}(q_{\text{init}})$. The adjugate matrix $[q_i q_j]$ from Eq. (12.46), computed from the raw quaternion rotation.

Finally, we have the following selection of the corresponding networks:

- $E \rightarrow q : (3, 3) \rightarrow (4)$. Quaternion-compatible last layer output.
- $E \rightarrow R : (3, 3) \rightarrow (3, 3)$. Rotation-compatible last layer output.
- $E \rightarrow A : (3, 3) \rightarrow (10)$. The 4×4 symmetric adjugate matrix encoded in its 10 independent variables as a list for the last layer output.

Results. Our E-matrix results are summarized in Fig. 13.8. The R-matrix results shown in Fig. 13.7 involved defining a network whose only input was an isolated rotation matrix R, and then representing

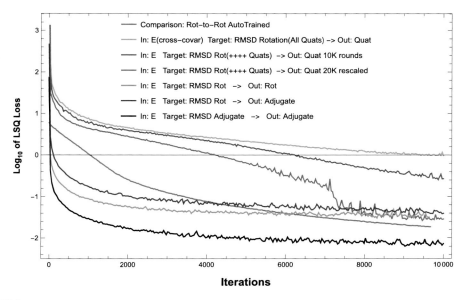

FIGURE 13.8

Cross-covariance matrix E input comparisons. Here we compare the training loss characteristics of various models applied to the task of determining the best-aligning rotation R_{opt} for the cloud matching problem given only the cross-covariance matrix of the cloud data, $E = \mathbf{X} \cdot \mathbf{U}^t$. We choose a variety of constraints on the quaternions used to generate the data, and a variety of output options data$_{out}$ that have to be decoded into R_{out} to be compared to the target R during training. We note in particular that the quaternion output trained to the all-positive quaternion hyperquadrant in bright magenta performs just as well as the expected high-quality rotation-to-rotation solution in green. Since the quaternion adjugate is one of our themes, we also implement a network that supplies the adjugate matrix as its target data, and trains the output layer adjugate directly to that by a Frobenius-style norm. That is by far the best-behaved training target, justifying some of our faith in the validity of the quaternion adjugate variables for these tasks. For our purposes here, all provided data are perfect, with no noise; working with noise may of course require more sophisticated network methods.

that matrix in various ways, both related and unrelated to quaternions. We found explicit confirmation of our concerns with trying to use full, topologically nontrivial, quaternions as training targets, and observed that our proposed methods for avoiding this issue worked exactly as expected. In this second experiment, using the E-matrix as input to the point-cloud matching problem, we have taken *exactly the same network*, but instead of supplying a pure rotation matrix, we created a cloud transformation with a rotation matrix and provided *the cross-covariance matrix E in place of the rotation matrix R* to the input layer of the network. So we are looking at discovering a rotation matrix now from a completely different input path.

As we see in Fig. 13.8, the pure quaternion output case in orange is the worst possible for all-sign quaternion origins in the data generation, and plateaus at a very large loss value. Training to the same all-sign E matrix data, the pure rotation output (training directly to the rotation matrix that produced E) shown in green is very good, confirming that the cloud-matching machine learning works well when trained directly to the rotation. As a cross-check comparing the level of loss values that we are getting, we superimpose the rotation-to-rotation loss evolution for the same parameters in red on our plot, and confirm that the behavior is quite similar.

Then we have our signature result for the only-plus-sign $(+, +, +, +)$ quaternion case. The initial network in dark magenta shows a slow start, clearly much better than the all-sign case, but after the standard 10,000 rounds, it appears to be still monotonically improving. So, as we did for some of the data in the rotation-rotation network training, we adjusted the quaternion $(+, +, +, +)$ training parameters, and found that after 20,000 rounds instead of 10,000, shown in bright magenta, it *completely matches* the performance of the corresponding pure rotation output network. This is again decisive verification of our model for the predicted behavior of a nontrivial quaternion manifold, and the limited circumstances under which it can be employed as the output data for a network learning a rotation.

Finally, with the quaternion adjugate parameterization, we get some additional interesting results. Using the adjugate's decoder to generate the actual rotation matrix candidate to be trained to the data-generating rotation, we get, at the end of the 10,000-round training session, almost exactly the same quality as the rotation matrix itself, which is expected. But if we actually compute the *adjugate matrix* from the input quaternion and use that as the training target, with a sum-of-squares loss function measuring the loss between the 10-parameter adjugate output layer and the training-value adjugate, we get an order of magnitude improvement in the loss curve performance. This is possibly an artifact of the distinct loss functions, but we note that the rotation loss function is itself just a sum of nine squared differences, so they are very similar. In any case, we see that the quaternion adjugate trains extremely well both to the solution of the pure rotation discovery task, shown in Fig. 13.7, and here for the E-matrix-based RMSD solution task in Fig. 13.8. And we remind ourselves that these are very different problems, one basically requiring introspective discovery of an output layer encoding that is trained to the input layer, and the case here where the input layer has only a very abstract relationship to the optimal rotation output that we seek.

Exact 2D rotations for cloud matching and pose estimation tasks[☆]

We now apply the quaternion adjugate variable representation to the elementary and completely solvable domain of two-dimensional point-cloud geometry, which is pedagogically instructive because, in its simplicity, it exposes hints of the more challenging structures we will encounter in 3D in Chapter 15. We first examine the task of matching rotation-related pairs of 2D clouds, illustrated in Fig. 14.1(a), then orthographic pose discovery given a 2D cloud and a 1D point-matched orthographic image projected from that cloud, illustrated in Fig. 14.1(b,c). In the final Section, we address the 2D perspective projection pose discovery problem, as shown in Fig. 14.2(a). Fig. 14.2(b) illustrates the alternative ways of computing errors for perspectively projected data: using image-space errors, requiring a nonlinear perspective division, and object-space errors (see Lu et al. (2000)), which are attractive because they are asymptotically the same as image-space errors, but escape the need for a division.

Least-squares loss function framework for 2D. We will work with the basic least-squares formulas comparing rotation-transformed reference data to (probably noisy) measurements derived from corresponding rotated data; we will assume that the rotation parameters being varied in the loss optimization search are expressed in quaternion variables q_i or quadratic quaternion adjugate variables q_{ij}. The traditional 2D-space loss functions whose optimum parameters correspond to finding the unknown rotation applied to the reference data (and thus solving the alignment problem) take the following forms:

$$\underset{q}{\text{argmin}} \quad \mathbf{S}_{(2D\ match)} = \sum_{k=1}^{K} \| R(q) \cdot \mathbf{x}_k - \mathbf{u}_k \|^2 \ , \tag{14.1}$$

$$\underset{q}{\text{argmax}} \quad \mathbf{\Delta}_{(2D\ match)} = \sum_{k=1}^{K} (\mathbf{u}_k \cdot R(q) \cdot \mathbf{x}_k) \ , \tag{14.2}$$

$$\underset{q}{\text{argmin}} \quad \mathbf{S}_{(2D\ ortho)} = \sum_{k=1}^{K} (P(q) \cdot \mathbf{x}_k - u_k)^2 \ , \tag{14.3}$$

$$\underset{q}{\text{argmin}} \quad \mathbf{S}_{(2D\ perspective\ image\ space)} = \sum_{k=1}^{K} \left(\frac{P(q) \cdot \mathbf{x}_k}{D(q) \cdot \mathbf{x}_k + \mathbf{C}} - u_k \right)^2 \ , \tag{14.4}$$

☆ Much of this chapter is inspired by Hanson and Hanson (2022).

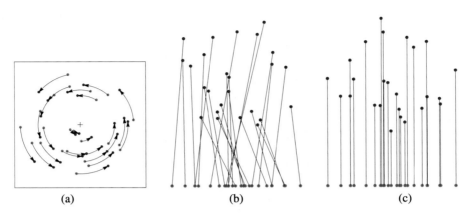

FIGURE 14.1

The cloud-to-cloud and orthographic pose matching problems in 2D. (a) Task of finding the 2D rotation aligning a 2D reference cloud with another 2D test cloud differing by a rotation. (b) Task of finding the 2D rotation locating the projection angle used to make a 1D image of a 2D point cloud. (c) The solution rotates the 2D cloud until its vertical projection aligns with the measured projected input points.

$$\underset{q}{\text{argmin}} \quad \mathbf{S}_{(2D\ Z\text{-prime Obj Space})} = \sum_{k=1}^{K} \|(R(q) \cdot \mathbf{x}_k + \mathbf{C}) - \mathbf{v}\,(D(q) \cdot \mathbf{x}_k + c)\|^2 \,, \tag{14.5}$$

$$\underset{q}{\text{argmin}} \quad \mathbf{S}_{(2D\ \text{LHM Obj space})} = \sum_{k=1}^{K} \|(R(q) \cdot \mathbf{x}_k + \mathbf{C}) - \hat{\mathbf{v}}_k\,(\hat{\mathbf{v}}_k \cdot (R(q) \cdot \mathbf{x}_k + \mathbf{C}))\|^2 \,, \tag{14.6}$$

$$\underset{q}{\text{argmin}} \quad \mathbf{S}_{(2D:\ \text{Radial Obj Space})} = \sum_{k=1}^{K} \|\hat{\mathbf{v}}_k\,(\hat{\mathbf{v}}_k \cdot (R(q) \cdot \mathbf{x}_k + \mathbf{C})) - \mathbf{v}\,(D(q) \cdot \mathbf{x}_k + c)\|^2 \,. \tag{14.7}$$

Here $\{\mathbf{x}_k\}$ is the set of K reference points describing a 2D cloud, with $\mathbf{X}$ denoting a $2 \times K$ matrix and $\{\mathbf{u}_k\}$ or $\mathbf{U}$ denoting our set of K test points. Other notation includes the projection matrices $P_a(q) = R_{1a}(q)$, $D_a(q) = R_{2a}(q)$, and the cloud-to-image-ray projector $\mathbf{v}_k = [u_k, 1]^t$, with $\hat{\mathbf{v}} = \mathbf{v}/\|\mathbf{v}\|$, and $\mathbf{C} = [0, c]^t$ the cloud's world center of mass. For the cloud-to-cloud matching problem, $\mathbf{X}$ and $\mathbf{U}$ denote 2D reference data $\{x_k, y_k\}$ and rotated data $\{u_k, v_k\}$. Eq. (14.1) permits the simplification $R^t \cdot R = \text{Identity Matrix}$, reducing the least squares minimization problem to the equivalent maximization problem Eq. (14.2) for the cross term, which can be written as

$$\mathbf{\Delta} = \text{tr}\,(R \cdot E) \tag{14.8}$$

in terms of the 3×3 cross-covariance

$$E_{ab} = \begin{bmatrix} \mathbf{X}_a \cdot \mathbf{U}_b{}^t \end{bmatrix}$$

after eliminating all constant terms. For the orthographic pose estimation problem, the measured data are constrained to a point-matched projected image line (see Fig. 14.1), and the necessarily incomplete

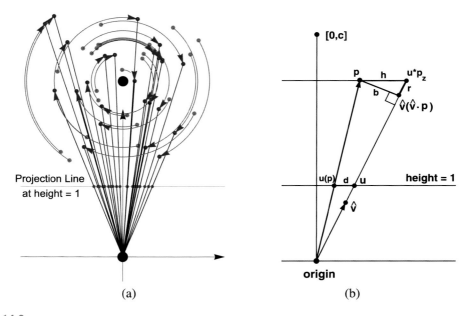

(a)   (b)

FIGURE 14.2

The perspective pose matching problem in 2D. (a) Representing the task of finding the 2D rotation applied to a point cloud (red) from the 1D perspective projected image at height = 1 of the rotated cloud (blue) . (b) The distinction between the image space error $d = (u - u(p))$ using the division-based projection loss and the division-free object space errors, **h** being the distance from **p** to its z-scaled image-space projection, and **b** derived from the Gram-Schmidt-style projection of each cloud point **p** onto the image point basis direction $\hat{\mathbf{v}}$, with **r** the radial distance along the ray between the ends of **h** and **b**. The loss-related vectors **b**, **h**, and **r** form a right triangle with three distinct weights obeying the Pythagorean theorem. All can be used interchangeably to study perspective imaging without division.

rotation matrices in the projection process no longer cancel out, so the full form of Eq. (14.3) must be retained. Perspective projection modifies the geometry to have the viewpoint through the origin and the image in the $y = 1$ line. That is, the projection line from a cloud point to the origin produces the corresponding image point u when it passes through $y = 1$, as illustrated in Fig. 14.2(a). Eq. (14.4) is the traditional image-space loss requiring a perspective division loss, with a difficult to handle rotation-containing denominator. The three object-space alternatives to the image-space loss are the object space depth-scaled loss Eq. (14.5), the 2D version Eq. (14.6) of the LHM (Lu et al., 2000) object-space loss function, and the radial loss Eq. (14.7). The geometry of this right-triangle set of scaled perspective loss quantities, shown in Fig. 14.2(b), has many advantages.

Applying the quaternion-like forms of the 2D rotation matrices. We will hereafter define the variables resulting from the quadratic quaternion forms as the *adjugate variables*, which we now define explicitly in terms of the quaternion-like 2D unit-length vector $[a, b]^{\mathrm{t}}$ as $\alpha = a^2$, $\beta = b^2$, and $\gamma = ab$.

Thus we have

$$\text{2D: Matching} \quad R(a, b) = \begin{bmatrix} a^2 - b^2 & -2ab \\ 2ab & a^2 - b^2 \end{bmatrix} \;\rightarrow\; R(\alpha, \beta, \gamma) = \begin{bmatrix} \alpha - \beta & -2\gamma \\ 2\gamma & \alpha - \beta \end{bmatrix}, \tag{14.9a}$$

$$\text{2D: Projection} \quad P(a, b) = \begin{bmatrix} a^2 - b^2 & -2ab \end{bmatrix} \;\rightarrow\; P(\alpha, \beta, \gamma) = \begin{bmatrix} \alpha - \beta & -2\gamma \end{bmatrix}, \tag{14.9b}$$

$$\text{2D: Divisor} \quad D(a, b) = \begin{bmatrix} 2ab & a^2 - b^2 \end{bmatrix} \;\rightarrow\; D(\alpha, \beta, \gamma) = \begin{bmatrix} 2\gamma & \alpha - \beta \end{bmatrix}. \tag{14.9c}$$

The 2D adjugate matrices themselves are thus given by

$$A(a, b) = \begin{bmatrix} a^2 & ab \\ ab & b^2 \end{bmatrix} \equiv \begin{bmatrix} \alpha & \gamma \\ \gamma & \beta \end{bmatrix}. \tag{14.10}$$

While the basic mathematics of studying the least-squares formula Eq. (14.1) without using the rotation matrix variables directly can be performed with either the quaternion (a, b) variables or the quaternion adjugate (α, β, γ) variables, the conversion from formulas that are quadratic or quartic in (a, b) to formulas that are linear or quadratic in (α, β, γ) can lead to some additional simplicity.

We need to consider two ways of applying Eq. (14.1). One is as the loss function for an **ArgMin** function, which takes as its input X and U, the reference and measured data points, and searches for the unknown rotation parameters, either the quaternions (a, b) with the constraint $a^2 + b^2 = 1$, or the adjugate variables (α, β, γ) with the constraints $\alpha + \beta = 1$ and $\alpha\beta = \gamma^2$. The output of this expensive numerical search is then generally sufficient to compute the originally unknown target rotation matrix from a loss based on a transformation such as Eq. (14.9a), though we could also use the projection-based losses. In principle, an equivalent result can be obtained by training a neural network to *discover* the optimal rotation and encode its formula in a high-speed trained network; that would evade repeating the expensive `argmin` search for every novel data set. However, this does not work smoothly unless one creates a network that outputs *multiple results* consisting of charts of functions covering different parts of the quaternion manifold: we have already seen in Chapter 12 that no single function suffices to describe the nontrivial quaternion manifold S^1 for 2D rotation tasks, much less S^3 for 3D rotation tasks.

However, here our focus will be on an alternate approach: discovering the algebraic function producing the optimal rotation from the inputs based on *direct analysis of the least-squares problem* defined by Eq. (14.1). Next, in Section 14.1, we will yet again work through this process for the exactly solvable 2D cloud-to-cloud matching problem that we have seen already in Chapters 7 and 12, this time as a lead-in to much more complex problems for which it will serve as a template. As in Chapter 12, we will first set up the idealized case of noise-free data, and then proceed to the richer and more nuanced case of noisy data. We repeat the process in Section 14.2 for the 2D orthographic pose estimation problem and in Section 14.3 for the perspective projection problem.

14.1 2D point-cloud orientation matching

We begin with the simplest example, which we will label as "2D matching," namely the problem of aligning a pair of 2D point clouds, X and U, where $X = \{x_k\} = \{[x, y]_k\}$ is a set of K 2D column vectors describing points in a reference set, and $\mathbf{U} = \{\mathbf{u}_k\} = \{[u, v]_k\}$ describes a set of 2D test points. Following our conventions in Eq. (14.1), the least-squares loss functions then take the forms

$$\underset{q}{\text{argmin}} \qquad \mathbf{S}_{(2D\,\text{match})} = \sum_{k=1}^{K} \|R(q) \cdot \mathbf{x}_k - \mathbf{u}_k\|^2 , \tag{14.11}$$

$$\underset{q}{\text{argmax}} \qquad \mathbf{\Delta}_{(2D\,\text{match})} = \sum_{k=1}^{K} (\mathbf{u}_k \cdot R(q) \cdot \mathbf{x}_k) \;\; = \;\; \text{tr}\,(R(q) \cdot E) \;\; = \;\; q \cdot M(E) \cdot q . \tag{14.12}$$

The corresponding optimization problem is easily solved in closed form in terms of the 2×2 cross-covariance matrix

$$E_{ab} = \sum_{k} (x^k)_a (u^k)_b = \begin{bmatrix} \text{ux} & \text{vx} \\ \text{uy} & \text{vy} \end{bmatrix} , \tag{14.13}$$

where with $[x_{a=1}, x_{a=2}] = [x, y]$, $[u_{a=1}, u_{a=2}] = [u, v]$, we adopt the abbreviation $\text{ux} = \sum_{k} u^k x^k$, etc. One can use a wide variety of methods, including quaternion-like forms exploiting the parameterization Eq. (14.9a) for $R(a, b)$; in this case, our optimal quaternion is the eigenvector of the maximum eigenvalue of the profile matrix of Eq. (14.13), which in this notation is

$$M = \begin{bmatrix} \text{ux} + \text{vy} & \text{vx} - \text{uy} \\ \text{vx} - \text{uy} & -(\text{ux} + \text{vy}) \end{bmatrix} \qquad \text{with } \lambda_{\text{opt}} = \sqrt{(\text{ux} + \text{vy})^2 + (\text{uy} - \text{vx})^2} . \tag{14.14}$$

The optimal eigenvector can be *either* column of the adjugate of the characteristic matrix

$$\chi(M) = [M - \lambda_{\text{opt}} I_2] , \tag{14.15}$$

where we can freely reverse the sign to get positive-definite diagonals for the unnormalized adjugate eigenvector matrix,

$$A(M) = -\text{Adj}(\chi(M)) = \begin{bmatrix} \lambda + (\text{ux} + \text{vy}) & \text{vx} - \text{uy} \\ \text{vx} - \text{uy} & \lambda - (\text{ux} + \text{vy}) \end{bmatrix} = \begin{bmatrix} \alpha & \gamma \\ \gamma & \beta \end{bmatrix} = \begin{bmatrix} a^2 & ab \\ ab & b^2 \end{bmatrix} . \tag{14.16}$$

From this, we can compute individual terms of the normalized rotation matrix: writing the unnormalized terms as $\alpha_0, \beta_0, \gamma_0$, the elements of the rotation matrix are constructed from

$$\alpha_0 - \beta_0 = 2(\text{ux} + \text{vy}) , \qquad\qquad 2\gamma_0 = 2(\text{vx} - \text{uy}) ,$$
$$(\text{norm})^2 = \lambda(x, y, u, v)^2 = 4 \left((\text{ux} + \text{vy})^2 + (\text{vx} - \text{uy})^2 \right) . \tag{14.17}$$

Dividing by the normalization of the columns and inserting them into the 2D rotation equation Eq. (14.9a), we find the same 2D rotation matrix that we have already encountered in different contexts, expressed explicitly in terms of the individual cross-covariances,

$$R_{\text{opt}}(a, b) = R_{\text{opt}}(\alpha, \beta, \gamma) = \begin{bmatrix} \dfrac{\text{ux} + \text{vy}}{\lambda(x, y, u, v)} & -\dfrac{\text{vx} - \text{uy}}{\lambda(x, y, u, v)} \\[2mm] \dfrac{\text{vx} - \text{uy}}{\lambda(x, y, u, v)} & \dfrac{\text{ux} + \text{vy}}{\lambda(x, y, u, v)} \end{bmatrix} . \tag{14.18}$$

Note that simply dividing the adjugate matrix by its trace $\alpha_0 + \beta_0 = 2\lambda$ immediately achieves the normalization that we did explicitly in Eq. (14.17).

Algebraic adjugate variable solution. We now introduce our first novel procedure that will lead to new insights as we proceed. Instead of using the quaternions themselves to represent the rotations in the loss function solvable using optimal eigensystems, we convert *entirely* to the adjugate variables that are *explicitly compatible* with being compared directly with rotation matrix variables in the optimization procedure. Here in our initial exercise, we verify that we get the same result from the alternative direct least-squares approach, parameterizing $R(q)$ in the loss function Eq. (14.11) not by the 2D quaternion itself, $q = (a, b)$, but now with the adjugate-inspired variables $\{\alpha, \beta, \gamma\}$ from Eq. (14.10) replacing $\{a^2, b^2, ab\}$, subject to the constraints $\alpha + \beta = 1$ and $\alpha\beta = \gamma^2$. The form $R(\alpha, \beta, \gamma)$ yields our explicit adjugate-based loss function

$$
\begin{aligned}
S_{(2D\ match)} &= uu + vv - 2\alpha\,ux + 2\beta\,ux - 4\gamma\,vx + \alpha^2\,xx + 4\gamma^2\,xx - 2\alpha\beta\,xx + \beta^2\,xx \\
&\quad + 4\gamma\,uy - 2\alpha\,vy + 2\beta\,vy + \alpha^2\,yy + 4\gamma^2\,yy - 2\alpha\beta\,yy + \beta^2\,yy \\
&= uu + vv - 2\alpha\,ux + 2\beta\,ux - 4\gamma\,vx + 4\gamma\,uy - 2\alpha\,vy + 2\beta\,vy + xx + yy ,
\end{aligned}
\tag{14.19}
$$

where we used both constraints to remove the quaternion presence in the $xx + yy$ terms (this is equivalent to exploiting $R \cdot R^t = I_2$ in Eq. (14.1)). The least-squares solution minimizing S in Eq. (14.19) is easy to find in the adjugate variable framework by using the constraints to eliminate β and γ in terms of α, requiring the derivative with respect to α to vanish, and solving the resulting quadratic equation for α.

The solution for the 2×2 (sign-adjusted) adjugate matrix, using the constraints $\alpha + \beta = 1$ and $\alpha\beta = \gamma^2$, can be written as

$$
A_{R\cdot x \to u} = \begin{bmatrix} \alpha & \gamma \\ \gamma & \beta \end{bmatrix} = \begin{bmatrix} a^2 & ab \\ ab & b^2 \end{bmatrix} = \begin{bmatrix} \frac{1}{2}\left(1 + \frac{ux+vy}{\lambda(x,y,u,v)}\right) & \frac{1}{2}\frac{vx-uy}{\lambda(x,y,u,v)} \\ \frac{1}{2}\frac{vx-uy}{\lambda(x,y,u,v)} & \frac{1}{2}\left(1 - \frac{ux+vy}{\lambda(x,y,u,v)}\right) \end{bmatrix} ,
\tag{14.20}
$$

where the independently derived $\lambda(x, y, u, v) = \sqrt{(ux+vy)^2 + (vx-uy)^2}$ is seen to be the maximal eigenvalue λ_{opt} of Eq. (14.14) appearing naturally in the quaternion profile matrix eigensystem approach. The optimal aligning 2D quaternion (a_{opt}, b_{opt}) is found as usual by identifying the maximum of α and β and normalizing its column, so the corresponding rotation matrix is $R_{opt} = R(a_{opt}, b_{opt})$. We can write the latter also in closed form as

$$
R_{opt}(\alpha, \beta, \gamma) = \begin{bmatrix} \alpha - \beta & -2\gamma \\ 2\gamma & \alpha - \beta \end{bmatrix} = \begin{bmatrix} \frac{ux+vy}{\lambda(x,y,u,v)} & -\frac{vx-uy}{\lambda(x,y,u,v)} \\ \frac{vx-uy}{\lambda(x,y,u,v)} & \frac{ux+vy}{\lambda(x,y,u,v)} \end{bmatrix} ,
\tag{14.21}
$$

in agreement with the alternative eigensystem-based derivation in Eq. (14.18). *Note:* One observes in practice that when extracting numerical values of the eigenvectors using the adjugate matrix method, there is typically an overall factor that must be removed before constructing a rotation matrix directly from an adjugate matrix as in Eq. (14.21). In many cases, simply dividing by the trace of the quaternion adjugate matrix restores the correct scale, though there may be exceptions, which can be avoided by first calculating a valid unit-length quaternion before proceeding.

Remark: There is one subtle point, which is that, because of the form of the constraints, the relative sign of a and b (the sign of γ) can be indeterminate. Both signs give an adjugate matrix that handles the possible singularities at $[a, b] = [0, 1]$ and $[a, b] = [1, 0]$, and in fact we can determine the appropriate sign from the data, yielding a result identical with the sign-resolved quadratic products of the results from the 2D quaternion eigensystem methods following from Eq. (14.14).

Determinant ratio solutions for error-free data. If the data are error-free, that is, if we can write exactly

$$
\begin{aligned}
\text{ux} &= r_{11}\,\text{xx} + r_{12}\,\text{xy}, & \text{vx} &= r_{21}\,\text{xx} + r_{22}\,\text{xy}, \\
\text{uy} &= r_{11}\,\text{xy} + r_{12}\,\text{yy}, & \text{vy} &= r_{21}\,\text{xy} + r_{22}\,\text{yy},
\end{aligned}
\tag{14.22}
$$

with r_{ij} the elements of a perfect rotation matrix, then there are in fact two more alternative ways to express the solution of the 2D matching problem. The first is to take the loss function Eq. (14.19) with free arguments (α, β, γ),

$$
\texttt{match2DLSQ} = \sum_k \| R(\alpha, \beta, \gamma) \cdot [x, y]^{\text{t}} - [u, v]^{\text{t}} \|^2,
$$

and use all three derivatives, but only the single constraint $\alpha + \beta = 1$, to get the values of the adjugate variables:

```
Module[{eqn = match2DLSQ},
    Solve[{D[eqn, alpha] == 0, D[eqn, beta] == 0, D[eqn, gamma] == 0,
        alpha + beta == 1},
        { alpha, beta, gamma}]].
```

The resulting values

$$
\alpha = \frac{\text{xx} + \text{yy} + (\text{ux} + \text{vy})}{2(\text{xx} + \text{yy})}, \qquad
\beta = \frac{\text{xx} + \text{yy} - (\text{ux} + \text{vy})}{2(\text{xx} + \text{yy})}, \qquad
\gamma = \frac{\text{vx} - \text{uy}}{2(\text{xx} + \text{yy})}
$$

produce a variant of Eq. (14.21) with a different denominator,

$$
R_{\text{2D error-free LSQ}}(X, U) =
\begin{bmatrix}
\dfrac{\text{ux} + \text{vy}}{\text{xx} + \text{yy}} & \dfrac{\text{uy} - \text{vx}}{\text{xx} + \text{yy}} \\[2ex]
\dfrac{\text{vx} - \text{uy}}{\text{xx} + \text{yy}} & \dfrac{\text{ux} + \text{vy}}{\text{xx} + \text{yy}}
\end{bmatrix}.
\tag{14.23}
$$

For error-free parameters obeying Eq. (14.22), we now show that this equation produces exactly the 2D rotation matrix elements r_{ij}. With the error-free property of pure 2D rotations that

$$
r_{11} = r_{22} \quad \text{and} \quad r_{12} = -r_{21},
$$

which is equivalent to observing that a pure 2D rotation matrix $R(\theta)$ has the form

$$
R = \begin{bmatrix}
\cos\theta & -\sin\theta \\
\sin\theta & \cos\theta
\end{bmatrix},
\tag{14.24}
$$

we see that

$$ux + vy = r_{11}xx + r_{22}yy = r_{11}(xx + yy),$$
$$vx - uy = r_{21}xx - r_{12}yy = r_{21}(xx + yy).$$

Thus with the error-free data constraint Eq. (14.22), our candidate formula Eq. (14.23) is a pure rotation, a solution of the least-squares equations for error-free data. Furthermore, the puzzle of the incompatibility of the denominators of Eq. (14.23) and Eq. (14.21) is resolved by substituting Eq. (14.22) into the eigenvalue expression and observing that it reduces exactly to the denominator of Eq. (14.23).

The Bar-Itzhack method in 2D corrects all error-induced flaws. There is one more important relationship between the solutions Eq. (14.23) and Eq. (14.21). We recall from Chapter 9 that the Bar-Itzhack method can be used to *correct* a defective rotation matrix, and that in fact Eq. (14.23) fails to maintain orthogonality as the data acquire errors. Applying the Bar-Itzhack correction to Eq. (14.21) has no effect, since by construction, its source remains a valid quaternion, and its rotation matrix remains orthogonal independent of the introduction of error. However, when we apply the Bar-Itzhack correction to Eq. (14.23), which definitely acquires defects unless the data are error-free, the result is *exactly* the perfect, noise-insensitive rotation of Eq. (14.21)! Bar-Itzhack permanently corrects the flaws if applied to Eq. (14.23) as a first approximation. Unfortunately, while we see substantial improvement after applying Bar-Itzhack in three dimensions, the results remain imperfect in that case.

And yet another method, determinant ratios. But we are not quite done. With foresight derived from solutions that will arise below in three dimensions, we can write down a third form that is valid for error-free data based on a linear algebra trick taking advantage of Eq. (14.22). If we consider the matrices of cross-covariances $(xx, \ldots, vy)$,

$$d_0 \rightarrow \det \begin{bmatrix} xx & xy \\ xy & yy \end{bmatrix},$$

$$d_{11} \rightarrow \det \begin{bmatrix} ux & xy \\ uy & yy \end{bmatrix}, \quad d_{12} \rightarrow \det \begin{bmatrix} xx & ux \\ xy & uy \end{bmatrix}, \tag{14.25}$$

$$d_{21} \rightarrow \det \begin{bmatrix} vx & xy \\ vy & yy \end{bmatrix}, \quad d_{22} \rightarrow \det \begin{bmatrix} xx & vx \\ xy & vy \end{bmatrix},$$

we can see that substituting Eq. (14.22) in each of these determinants and dividing by d_0 produces exactly the corresponding rotation r_{ij}. That is,

$$R_{\mathrm{opt}} = \frac{1}{d_0} \begin{bmatrix} d_{11} & d_{12} \\ d_{21} & d_{22} \end{bmatrix} = \begin{bmatrix} r_{11} & r_{12} \\ r_{21} & r_{22} \end{bmatrix}. \tag{14.26}$$

Comparison of losses on simulated cloud data. In Fig. 14.3, we assemble a simulated 100-element data set with unit cloud size and added normal error with $\sigma = 0.1$, which is a little large, but seems to be workable. The performance of the original rotation applied to the noisy data is in magenta, and the RMSD-style quaternion eigensystem results from Eq. (14.18) are in blue: since there is no way to *rediscover* the original rotation once noise is added, we expect the original rotation to behave worse than any other candidate set of rotations, and that is exactly what we see. To independently confirm the

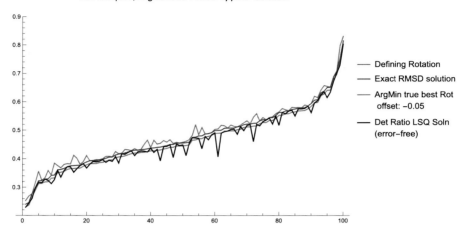

2D:2D Cloud:Target Matching Loss values for σ = 0.1:
100 samples, ArgMin and RMSD appear identical

FIGURE 14.3

The ranges of the 2D cloud matching problem losses based on noisy data from Eq. (14.13), with optimal matrices derived from the error-insensitive exact RMSD solutions Eq. (14.18) or Eq. (14.21), and Eq. (14.23), which is a valid rotation only for error-free data. For comparison, we add also the error-insensitive `argmin` solution using the adjugate variable loss Eq. (14.19); the exact RMSD solution and the `argmin` solution are essentially identical. [Magenta] is the original, simulation-creating, rotation, which is always worse in the presence of error. [Blue] is the exact RMSD solution, and [Red] is the `argmin` solution, displaced slightly since it is so close to the RMSD solution it is overwritten. [Black] shows the algebraic determinant-ratio solution Eq. (14.23) or Eq. (14.26), which deviates from a valid rotation for errorful data, but still solves the least-squares problem, and so has the smallest loss. Applying Bar-Itzhack correction to Eq. (14.23) yields *exactly* the perfect RMSD solution Eq. (14.18), so plotting it is irrelevant.

noise-insensitive RMSD solution, we can also apply the `argmin` numerical procedure directly to the loss function Eq. (14.11), e.g., to obtain an optimal quaternion adjugate matrix; the agreement between the RMSD solution and the `argmin` solution is so close we cannot sensibly display any difference. Finally, shown in black, we apply the noise-free-domain determinant-ratio solution given in Eq. (14.23) or Eq. (14.26), and see, interestingly, that it has the *smallest* loss of any candidate, but this is an illusion: while the resulting matrix is not a *rotation*, and so invalid for our task, it is still the *least-squares solution*, so it is expected to have the lowest possible values of the loss. We note that, in the absence of error, all methods converge to produce least-squares loss measures that basically vanish.

14.2 2D parallel projection pose estimation

Our next topic is the 2D pose estimation problem, "2D pose," using parallel projection. This is in principle a more complicated orientation problem than the 2D matching problem, even though its least-squares loss function is 1D and has only a single term. The problem studies the action of projections acting on 2D cloud points to obtain an image that is a line of matched points (see Fig. 14.1). The projection can be obtained by truncating the 2×2 rotation matrix to the first line as shown in Eq. (14.9b). Thus $P(a, b) = [a^2 - b^2, -2ab]$ or $P(\alpha, \beta, \gamma) = [\alpha - \beta, -2\gamma]$, with the adjugate matrix coordinates $\alpha = a^2, \beta = b^2, \gamma = ab$. In principle we should enforce the constraints $\alpha + \beta = 1$ and $\alpha\beta = \gamma^2$ to guarantee compatibility with the quaternion roots of our task, but it turns out that the reduced dimension of the pose estimation problem allows us some interesting additional freedom. Choosing the full least-squares loss expressed in adjugate form, our optimization problem takes the following explicit form:

$$\mathbf{S}_{(2D \text{ pose})} = \sum_{k=1}^{K} \left(P(\alpha, \beta, \gamma) \cdot [x_k, \ y_k]^t - u_k \right)^2 = \sum_{k=1}^{K} ((\alpha - \beta) x_k - 2\gamma \, y_k - u_k)^2$$

$$= (\alpha - \beta)^2 \, \mathrm{xx} - 4\gamma (\alpha - \beta) \, \mathrm{xy} + 4\gamma^2 \, \mathrm{yy} - 2(\alpha - \beta) \, \mathrm{ux} + 4\gamma \, \mathrm{uy} + \mathrm{uu} \ . \qquad (14.27)$$

Already we see something that might be interesting: while this equation is *quartic* in the quaternion (a, b) variables, making it unapproachable by the matrix methods applicable for the "2D match" problem, it is only *quadratic* in the adjugate (α, β, γ) variables. Might it be possible to complete the squares, and get an elegant system solvable as a transformed "2D match" problem? Unfortunately, this fails because the square completion transformation requires that the quadratic terms be represented by a symmetric nonsingular matrix (as its inverse is required), and the corresponding matrix following from Eq. (14.27) is

$$T = \begin{bmatrix} \mathrm{xx} & -\mathrm{xx} & -2\mathrm{xy} \\ -\mathrm{xx} & \mathrm{xx} & 2\mathrm{xy} \\ -2\mathrm{xy} & 2\mathrm{xy} & 4\mathrm{yy} \end{bmatrix} \Rightarrow \det T \equiv 0 \, ,$$

which obviously has *vanishing determinant*.

To exploit the reduced dimension of the adjugate parameters in the loss function Eq. (14.27), we proceed by requiring the vanishing of the derivatives of the loss function with respect to each variable, ignoring the constraints for the moment. We obtain three equations, but only the following two linear equations for our three variables are independent:

$$\left. \begin{aligned} \frac{dS}{d\alpha} &= \mathrm{ux} - \alpha \, \mathrm{xx} + \beta \, \mathrm{xx} + 2\gamma \, \mathrm{xy} = 0 \\ \frac{dS}{d\gamma} &= \mathrm{uy} - \alpha \, \mathrm{xy} + \beta \, \mathrm{xy} + 2\gamma \, \mathrm{yy} = 0 \end{aligned} \right\} . \qquad (14.28)$$

We find the partial solutions

$$\left. \begin{aligned} \alpha &= \beta + \frac{\mathrm{ux} \, \mathrm{yy} - \mathrm{uy} \, \mathrm{xy}}{\mathrm{xx} \, \mathrm{yy} - (\mathrm{xy})^2} \\ \gamma &= \frac{1}{2} \left(\frac{\mathrm{ux} \, \mathrm{xy} - \mathrm{uy} \, \mathrm{xx}}{\mathrm{xx} \, \mathrm{yy} - (\mathrm{xy})^2} \right) \end{aligned} \right\} . \qquad (14.29)$$

At this point we are ready to use one constraint, $\alpha = 1 - \beta$, inserted into the first line of Eq. (14.29) to solve for β; inserting *that* back into our equation for α, we have a complete solution in terms of only the usual cross-covariances (xx, xy, ..., ux), and we end up with

$$\alpha = \frac{1}{2}\left(1 + \frac{ux\,yy - uy\,xy}{xx\,yy - (xy)^2}\right),\tag{14.30}$$

$$\beta = \frac{1}{2}\left(1 - \frac{ux\,yy - uy\,xy}{xx\,yy - (xy)^2}\right),\tag{14.31}$$

$$\gamma = \frac{1}{2}\left(\frac{ux\,xy - uy\,xx}{xx\,yy - (xy)^2}\right).\tag{14.32}$$

Our algebraic solutions reduce to ratios of 2×2 determinants of the cross-covariances, which we choose to write using the notation

$$d_0 \rightarrow \begin{bmatrix} xx & xy \\ xy & yy \end{bmatrix}, \quad d_{11} = d_{22} \rightarrow \begin{bmatrix} ux & xy \\ uy & yy \end{bmatrix}, \quad d_{12} \rightarrow \begin{bmatrix} xx & ux \\ xy & uy \end{bmatrix}.\tag{14.33}$$

Thus

$$\alpha = \frac{1}{2}\left(1 + \frac{d_{11}}{d_0}\right), \quad \beta = \frac{1}{2}\left(1 - \frac{d_{11}}{d_0}\right), \quad \gamma = -\frac{1}{2}\left(\frac{d_{12}}{d_0}\right).\tag{14.34}$$

To complete our solution of the 2D pose problem, we note that while Eqs. (14.34) satisfy the constraints $\alpha + \beta = 1$, they do not satisfy our second constraint $\alpha\beta = \gamma^2$. However, it is important to remember that there is one less matrix element in the loss function Eq. (14.27) than a full rotation matrix loss. If we attempt to construct the projection, that is, the top line of the 2D rotation matrix, and insert the solutions Eq. (14.34), we find the following approximate first step, which in fact gives perfect solutions for noise-free test data:

$$\begin{aligned} \tilde{P}(\alpha, \beta, \gamma) &= [\alpha - \beta, \ -2\gamma] \\ &= \left[\frac{ux\,yy - uy\,xy}{xx\,yy - (xy)^2}, \ \frac{uy\,xx - ux\,xy}{xx\,yy - (xy)^2}\right] \\ &= \left[\frac{d_{11}}{d_0}, \ \frac{d_{12}}{d_0}\right] \end{aligned}.\tag{14.35}$$

We notice at once that this is identical to the first line of our second error-free 2D match solution Eq. (14.25). The second line of Eq. (14.25) is of course inaccessible to the pose problem because we do not have access to the (vx, vy) dimension of data. However, in 2D, the matrices are highly constrained for error-free data, and our next step is to apply those constraints to determine the full 2×2 orthonormal rotation matrix R implied by Eq. (14.35).

The cross product completion. Remarkably, without doing any further computation, simply taking the 2D cross product of Eq. (14.35) for the top row of R produces a legal second row of the rotation matrix element after normalization. Our final projection matrix, minimizing the noise-free least-squares loss for the 2D pose estimation problem, becomes

$$R(\alpha, \beta, \gamma) = \left. \begin{array}{c} \dfrac{\begin{bmatrix} (ux\ yy - uy\ xy) & (uy\ xx - ux\ xy) \\ (ux\ xy - uy\ xx) & (ux\ yy - uy\ xy) \end{bmatrix}}{\left((ux\ yy - uy\ xy)^2 + (uy\ xx - ux\ xy)^2\right)^{1/2}} \\[18pt] = \dfrac{1}{\sqrt{d_{11}^2 + d_{12}^2}} \begin{bmatrix} d_{11} & d_{12} \\ -d_{12} & d_{11} \end{bmatrix} \end{array} \right\}. \tag{14.36}$$

Evaluating this tentatively as the projection in the 2D pose loss function, Eq. (14.27), against an arbitrary list of pure or noisy data sets, we find that even though its scale varies through a range near unity, unlike a rotation matrix row, it still scores very well as a target for a minimizer of Eq. (14.27).

We have one final apparent puzzle to resolve: how can it be true that the numerators of Eq. (14.35) and Eq. (14.36) are the same, but their denominators differ? How can these expressions

$$\left. \begin{array}{c} d_0 = \det \begin{bmatrix} xx & xy \\ xy & yy \end{bmatrix} = xx\ yy - xy^2 \\[10pt] \left(d_{11}^2 + d_{12}^2\right) = (ux\ yy - uy\ xy)^2 + (uy\ xx - ux\ xy)^2 \\[10pt] = ux^2\left(xy^2 + yy^2\right) - 2ux\ uy\ xy(xx + yy) + uy^2\left(xx^2 + xy^2\right) \end{array} \right\} \tag{14.37}$$

be compatible? Once again we turn to the fact that for error-free data, we can explicitly write out the rotation that produced **U** from **X**, e.g., just use the top line of the 2D quaternion rotation $R(a, b)$ in ux and uy:

$$ux = xx\left(a^2 - b^2\right) - 2ab\ xy, \tag{14.38}$$

$$uy = xy\left(a^2 - b^2\right) - 2ab\ yy. \tag{14.39}$$

Inserting this into Eq. (14.37) we find

$$ux^2\left(xy^2 + yy^2\right) - 2\,ux\,uy\,xy\,(xx + yy) + uy^2\left(xx^2 + xy^2\right) = (a^2 + b^2)^2\left(xx\ yy - xy^2\right)^2 = (d_0)^2. \tag{14.40}$$

The same calculation works if we use (α, β, γ) with their constraints, and we see that Eq. (14.35) and Eq. (14.36) are indeed compatible.

Remark: We will see in the 3D pose problem that, to handle noisy data that move the least squares solution away from a pure rotation, we can take one more step: we can apply the Bar-Itzhack procedure to find the *optimal* pure rotation matrix given an approximate candidate. The 2D case has much less structure than 3D and typically can avoid that complication. For example, one can check that applying the Bar-Itzhack process to the numerator of Eq. (14.36) produces exactly the same rotation matrix.

In Fig. 14.4, we plot the values of the loss function for noisy data for 100 randomized data sets using four different projection matrix candidates. First, we use the rotation that was used to simulate the pose data; this should be very good, but it should be somewhat random and less deterministic than the least-squares solution, since it has *no way of knowing* what the least-squares formula is. Next, we use the rotation following from the single-constraint least-squares solution, Eq. (14.35): for noise-free

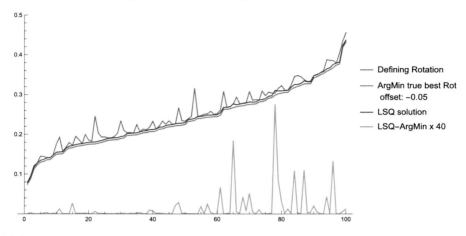

FIGURE 14.4

The ranges of the 2D pose problem losses from Eq. (14.27) using noisy data with projection matrices derived from alternate formulas. [Magenta] Original, simulation-creating, rotation. [Red] The "gold standard" highly accurate numerical `argmin` solution. [Blue] The bare least-squares solution Eq. (14.36) that is perfect for noise-free data, but veers slightly away from the perfect rotation for noisy data; it differs from the `argmin` solution only by the very small differences magnified in [Green] at the bottom of the graph. Applying the Bar-Itzhack optimal rotation procedure has no effect because the least-squares solution as we have constructed is orthonormal even in the presence of noise. This is unusual, and the 3D case does not have a comparable simple behavior.

test data, this produces losses that are consistently zero to machine accuracy, and in fact numerically obeys the constraint $\alpha\beta = \gamma^2$ that was not enforced, and it extends via the cross product to a perfect orthonormal rotation matrix for perfect data. For noisy data, it quickly loses its orthonormality, but *continues* to produce loss function values that appear to be smaller than can be achieved with a true rotation matrix; since we did not enforce orthonormality a priori, this appears to be a correct behavior – the least-squares solution need not be an orthonormal matrix. Then we use our true solution Eq. (14.36), which maintains its orthonormality for both noisy and perfect data, but may have loss function values that exceed those of Eq. (14.35). We apply the Bar-Itzhack process, which nominally finds the rotation that is the closest rotation matrix to Eq. (14.35), and find that up to machine precision, it is exactly the same as Eq. (14.36).

14.3 2D perspective projection

Finally, we study the most complicated pose estimation problem, which is to utilize perspectively projected data to determine the rotation applied to a 2D reference cloud before projection to the image.

The most obvious interpretation of our perspective image graphic in Fig. 14.2(a) is the conventional projective quotient form, with differences in the image space. Our task is to find the 2D quaternion or its adjugate form that minimizes this standard loss, assuming the camera displacement "c" along the viewing direction is sufficient to avoid division by small numbers:

$$S_{(2D \text{ perspective image space})} = \sum_{k=1}^{K} \left(\frac{P(q) \cdot \mathbf{x}_k}{D(q) \cdot \mathbf{x}_k + c} - u_k \right)^2. \tag{14.41}$$

(Notice that the object in parentheses is a scalar, not a vector.) However, we have no way to deal algebraically with the rotation elements appearing in the denominator of the quotient, though we can solve it in principle numerically with `argmin` optimization programs that support the imposition of the quaternion or adjugate constraints. We focus specifically on the quaternion-related object-space approach, which, in contrast to the traditional image space loss, involves no division and thus potentially allows more direct methods of solving the least squares problem. The three object-space geometries illustrated in Fig. 14.2(b) are essentially equivalent, with least-squares solutions that are asymptotically the same as the image-space formulation. These geometries are inspired by the "LHM" Gram–Schmidt projection onto the ray through the image point, labeled $\mathbf{b}$ in the figure and introduced by Lu et al. (2000), the "ZP" or z-prime measure that scales the $z = 1$ image data by the rotation-varying z scale of the associated object-space point, and the "Radial" vector loss given by the distance between the ends of the LHM and ZP vectors. (*Note:* We abuse notation here in 2D by referring to perspective depth as z, consistent with 3D depth notation when talking about depth, but denote spatial points as (x, y).) We add in the displacement to the cloud's world center of mass at $\mathbf{C} = [0, c]^t$, and write down the family of these loss functions that can be minimized over a quaternion or adjugate variable without needing a division. The object-space alternatives are

$$S_{(2D \text{ Z-prime perspective object space})}$$
$$= \sum_{k=1}^{K} \|(R(q) \cdot \mathbf{x}_k + \mathbf{C}) - \mathbf{v}(D(q) \cdot \mathbf{x}_k + c)\|^2$$
$$= \sum_{k=1}^{K} (N(q) \cdot \mathbf{x}_k - (D(q) \cdot \mathbf{x}_k + c) u)^2 \tag{14.42}$$

$$S_{(2D \text{ LHM perspective object space})}$$
$$= \sum_{k=1}^{K} \|(R(q) \cdot \mathbf{x}_k + \mathbf{C}) - \hat{\mathbf{v}}_k (\hat{\mathbf{v}}_k \cdot (R(q) \cdot \mathbf{x}_k + \mathbf{C}))\|^2, \tag{14.43}$$

$$S_{(2D \text{ Radial perspective object space})}$$
$$= \sum_{k=1}^{K} \|\hat{\mathbf{v}}_k (\hat{\mathbf{v}}_k \cdot (R(q) \cdot \mathbf{x}_k + \mathbf{C})) - \mathbf{v}(D(q) \cdot \mathbf{x}_k + c)\|^2, \tag{14.44}$$

where $\mathbf{v} = [u, 1]^t$, with $\hat{\mathbf{v}} = [u, 1]^t / \sqrt{1 + u^2}$. For simplicity, we decompose the 2D rotation into its first row, $N(q)$, and its bottom row, $D(q)$. Note that the vector in the initial definition of Eq. (14.42) reduces to a scalar due to the cancellation of identical second components, and that in contrast to Eq. (14.41) and (14.42), the arguments of Eq. (14.43) and (14.44) are actually 2D vectors. The object-space least-squares loss methods are very attractive measures, expressing the loss elements as differences between a rotated cloud point in space and a point on the ray through a noisy point in the image plane out into the cloud.

Object space solutions. We parameterize the rotation space in (2D) quaternion adjugate variables,

$$R(q) = \begin{bmatrix} \alpha - \beta & -2\gamma \\ 2\gamma & \alpha - \beta \end{bmatrix},$$

in order to examine our loss measure in potentially tractable symbolic terms, so our object-space least-squares symbolic loss measures become

$$S_{[\text{2D Z-prime perspective object space loss}]}$$
$$= \sum_{k=1}^{K} (c\,u_k - \alpha\,x_k + \alpha\,u_k\,y_k + \beta\,x_k - \beta\,u_k\,y_k + 2\,\gamma\,y_k + 2\,\gamma\,u_k x_k)^2 \tag{14.45}$$

$$S_{[\text{2D LHM perspective object space loss}]}$$
$$= \sum_{k=1}^{K} \frac{1}{(1 + u_k{}^2)} (c\,u_k - \alpha\,x_k + \alpha\,u_k\,y_k + \beta\,x_k - \beta\,u_k\,y_k + 2\,\gamma\,y_k + 2\,\gamma\,u_k x_k)^2 \tag{14.46}$$

$$S_{[\text{2D Radial perspective object space loss}]}$$
$$= \sum_{k=1}^{K} \frac{u_k{}^2}{(1 + u_k{}^2)} (c\,u_k - \alpha\,x_k + \alpha\,u_k\,y_k + \beta\,x_k - \beta\,u_k\,y_k + 2\,\gamma\,y_k + 2\,\gamma\,u_k x_k)^2 . \tag{14.47}$$

Note that the coefficients of the identical squared lengths in Eqs. (14.45), (14.46), and (14.47) sum to unity for each kth point, reflecting the Pythagorean-theorem requirements of the right-triangle geometry in Fig. 14.2(b) . In abstract generic form, the least squares forms of each object-space adjugate variable form are identical: the vector of possible adjugate polynomials is simply

$$\left\{1, \alpha, \beta, \gamma, \alpha^2, \beta^2, \gamma^2, \beta\gamma, \alpha\gamma, \alpha\beta\right\},$$

so the expanded forms of Eqs. (14.45), (14.46), and (14.47) can all be written in the form

$$S_{[\text{P}]}(a_k) = S_{[\text{Generic [P]erspective object space loss}]}$$
$$= a_0 + \alpha\,a_1 + \beta\,a_2 + \gamma\,a_3 + \alpha^2\,a_4 + \beta^2\,a_5 + \gamma^2\,a_6 + \beta\gamma\,a_7 + \alpha\gamma\,a_8 + \alpha\beta\,a_9 . \tag{14.48}$$

All the sums over the kth cloud point data elements reduce to individual numerical generalized cross-covariances. Comparing Eqs. (14.45), (14.46), and (14.47) to Eq. (14.48), we see that the abstract terms can be written as

$$
\begin{aligned}
a_0 &= c^2\text{uu} & a_1 &= 2\text{cuuyx} \\
a_2 &= -2\text{cuuyx} & a_3 &= 4\text{cuuxy} \\
a_4 &= \text{uyx2} & a_5 &= \text{uyx2} \\
a_6 &= 4\text{uxy2} & a_7 &= -4\text{uxyuyx} \\
a_8 &= 4\text{uxyuyx} & a_9 &= -2\text{uyx2} ,
\end{aligned} \tag{14.49}
$$

where the cross-covariance symbols uuyx, etc., are explicit sums of the form

$$uu = \sum_k w_k^i\, u_k^2\,, \qquad\qquad uyx = \sum_k w_k^i\, (u_k y_k - x_k)\,,$$

$$uxy = \sum_k w_k^i\, (u_k x_k + y_k)\,, \qquad uuyx = \sum_k w_k^i\, u_k(u_k y_k - x_k)\,,$$

$$uuxy = \sum_k w_k^i\, u_k(u_k x_k + y_k)\,, \qquad uxyuyx = \sum_k w_k^i\, (u_k x_k + y_k)(u_k y_k - x_k)\,,$$

$$uyx2 = \sum_k w_k^i\, (x_k - u_k y_k)^2\,, \qquad uxy2 = \sum_k w_k^i\, (u_k x_k + y_k)^2\,.$$

$$(14.50)$$

where

$$w_k^i = \left\{ 1,\ \frac{1}{(1+u_k^2)},\ \frac{u_k^2}{(1+u_k^2)} \right\},$$

and $i = \{1, 2, 3\}$ refers to the loss function of Eqs. (14.45), (14.46), and (14.47), respectively.
Thus we can alternatively rewrite the 2D object space perspective loss function as

$$\mathbf{S}_{[P]}(xy|u|c) = \mathbf{S}_{[2D\ [P]erspective\ object\ space\ loss]}$$
$$= c^2\, uu + 2c\alpha\, uuyx - 2c\beta\, uuyx + 4c\gamma\, uuxy + \alpha^2\, uyx2 + \beta^2\, uyx2$$
$$+ 4\gamma^2\, uxy2 - 4\beta\gamma\, uxyuyx + 4\alpha\gamma\, uxyuyx - 2\alpha\beta\, uyx2\,, \qquad (14.51)$$

where the cross-covariance coefficients include the $w_k^i(u_k)$ factors in Eq. (14.50). Since all the divisions by rotation matrix elements have disappeared, we can impose the 2D adjugate constraints $\alpha + \beta = 1$, $\alpha\beta = \gamma^2$, that is

$$\beta = 1 - \alpha$$
$$\gamma = s\sqrt{\alpha(1-\alpha)} \qquad (s = \pm 1) \qquad\qquad (14.52)$$

and attempt to solve directly for the form of $\alpha(a_k)$ (or $\alpha(xy|u|c)$) that minimizes the loss function and thus determines the adjugate quaternions for the optimal rotation.

Exact solutions for optimal perspective pose. In principle, solving for the vanishing derivative of $\mathbf{S}_{[P]}(\alpha; xy|u|c)$ or $\mathbf{S}_{[P]}(\alpha; a_k)$ with respect to α will produce the optimal perspective-projection rotation even in the presence of noisy data, but in practice it can be challenging: the first issue is the sign ambiguity, $s = \pm 1$ in Eq. (14.52), and the second is that the equation to be solved is quartic in α, so there is additional ambiguity in determining which of the four solutions, if any, actually describes the desired least squares optimal rotation. Another problem is that, with current technology, solving the explicit cross-covariance algebraic form Eq. (14.51) appears to be much harder for the computer algebra software than its abstraction Eq. (14.48), so we will generally employ the latter algebraic form. Let us examine how far we can take this approach for arbitrary data, and then we will look at an alternative more heuristic approach that works extremely well for error-free data, but is necessarily approximate for general data.

We take advantage of the Mathematica symbolic algebra system, as we have in other examples, and compute the derivative of Eq. (14.48) with the substitutions of Eq. (14.52) in $\mathbf{S}_{[P]}(\alpha; a_k)$. Taking the derivative $d\mathbf{S}_{[P]}(\alpha; a_k)/d\alpha$ with respect to α, and retaining only the numerator to solve directly for the vanishing points, gives us the following least squares minimization formula:

$$\Delta(\alpha) = 2s \sqrt{((1-\alpha)\alpha)} \, (a_1 - a_2 - 2a_5 + a_6 + a_9 + 2\alpha(a_4 + a_5 - a_6 - a_9))$$
$$+ \left(a_3 + a_7 + \alpha(-2a_3 - 5a_7 + 3a_8) + 4\alpha^2(a_7 - a_8) \right) . \tag{14.53}$$

Note that finding the zeroes of $\Delta(\alpha)$ with respect to α determines the solutions for *all* of the object-space loss functions, since for each case the a_k are simply numbers to be substituted in the formula. Solving for the zeroes in the remaining independent adjugate variable α using the solver utility

```
Solve[Delta[alpha] == 0, {alpha}]
```

yields four 400KB symbolic solutions for the variable $\alpha(a_k)$ (having *no dependence* on the sign s), which we must evaluate by examining the effects of the dependent variables. However, that step in the computation requires a sign choice, $s = \pm 1$, that can significantly change the resulting loss function, so in effect we have a total of 8 possible solutions for the adjugate rotation variables,

$$\left. \begin{array}{l} \alpha \quad [\text{choose one of four solutions, functions of } a_k] \\ \beta = 1 - \alpha \\ \gamma = s \sqrt{\alpha(1-\alpha)} \end{array} \right\} . \tag{14.54}$$

It remains to see whether there is a consistent single formula, or if, as can easily happen in least squares problems such as this, one needs a post-processing step to identify the lowest-loss alternative on a data-dependent basis. With 100 cloud data sets of 25 points and a distribution of rotations chosen from a uniformly random quaternion distribution, with no noise in the projected image, we found the following experimental results: three of the four solutions are consistently complex or unusable, but one or sometimes two generate a plausible rotation matrix, and furthermore the *location* of that solution appears completely random, distributed among the four returned suggestions. Therefore, we require in addition a data dependent choice of both the algebraic solution and the sign s (essentially the sign of $\gamma = ab$, the relative sign of the 2D quaternion components) to determine which option minimizes the standard loss function Eq. (14.41).

We have yet to find a deterministic algorithm for determining *which* of the symbolic algebraic solutions for the zeroes of Eq. (14.53) will be correct for a given data set, but in every example we have tried, a correct choice is always available. That is, in practical terms, we have verified that we can simply pick the instance of the possible eight selections that yields the smallest loss; it is possible that there is some algorithmic test yet to be determined that can make that choice deterministic.

Loss Comparisons. For error-free data, all methods produce essentially vanishing loss functions, so there is no opportunity for graphical comparison. Moving on to the case of noisy data, we illustrate in Fig. 14.5 the comparative loss distributions produced by our three sources of possible aligning rotations: the original data-defining rotation (inexact because noise is introduced in the image data), the "gold standard" rotation resulting from applying `argmin` to the noisy data, and the exact algebraic solutions to the least squares problem embodied in Eq. (14.53). Uniformly random quaternions were used to apply separate rotations to 100 unique unit-radius sampled clouds, which were then projected from their

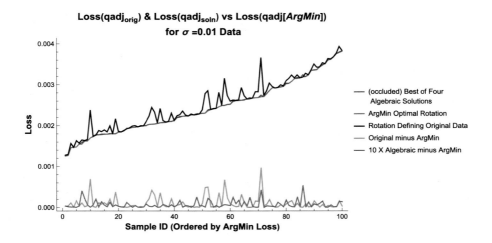

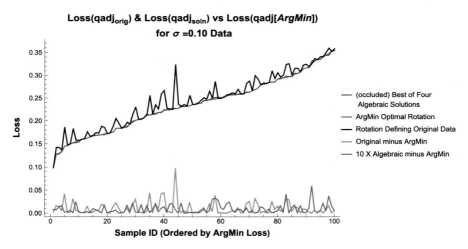

FIGURE 14.5

Distributions of 2D perspective losses using Eq. (14.41), with noisy data, $\sigma = 0.01$ (above) and $\sigma = 0.10$ (below). Red, blue, and black denote the `argmin` results, the best algebraic result of the four alternative solutions to the least squares measure Eq. (14.53), and the original rotation applied to the noisy data, respectively. Differences with respect to the `argmin` rotations' loss values are shown in green and magenta (scaled by 10 for visibility).

center at $\mathbf{C} = (0, c = 6)$ to the image line at $c = 1$, whose data are jittered with noise having $\sigma = 0.01$ in the top plot, and $\sigma = 0.10$ in the bottom plot. We remark that, since the error in the image is magnified by 6 in object space due to our choice of $\mathbf{C}$, $\sigma = 0.10$ is a fairly extreme error when viewed in object space. We identify the `argmin` optimal aligning rotation for each individual sample cloud-projection data set as our ground truth for comparison: that 2D rotation is inserted into Eq. (14.41) to generate loss values, sorted in ascending order and plotted in red. Loss values are then computed with original simulated-data-defining rotation matrices and displayed in black, and the loss values using the best of the four algebraic exact rotation solutions are displayed in blue. All are sorted to match the `argmin` loss order, and we see that the algebraic-solution based losses are so close to the ideal `argmin` losses that they are almost all occluded, so we display the actual differences magnified by 10 in magenta. The original versus `argmin` loss differences are shown in green. We remark that *none of the other rotations* produces a loss value below that of the `argmin` losses, confirming our assertion that the `argmin` results define the "gold standard" evaluation criteria for this problem.

Alternative heuristic method for error-free data. It is remarkable that there is an exact solution for the perspective problem in 2D that includes the errorful data case, matching the "gold standard" `argmin` numerical computation in every respect. However, that solution is also very complicated and takes hundreds of kilobytes of algebraic expressions to implement. We point out here that, for error-free data, there is an alternative approach that is interesting to consider: its advantage is that it is very simple, and its disadvantage is that it is highly unstable when we introduce noise in the data. We will find a similar situation in 3D in Chapter 15.

We consider an abstract algebraic form of the perspective least squares system that ignores the explicit interdependence of the quaternion adjugate variable system, and simply mixes together three heuristically chosen conditions, one adjugate constraint and two vanishing partial derivatives. Theoretically this should not work because it does not even consider the other constraints, and implicitly assumes the independence of the derivative conditions without justification. Nevertheless, suppose we start with our basic loss function for $S_{[P]}$ in Eq. (14.48) and investigate its behavior. We found that almost all of the obvious choices did not return usable answers with the Mathematica `Solve` utility. However, when we used only two vanishing partial derivatives instead of three, and replaced the third condition with the adjugate constraint $\alpha + \beta = 1$, that is, applying the following Mathematica system,

$$
\begin{aligned}
\texttt{Solve[\{ alpha + beta ==1,} \\
\texttt{D[Spersp[alpha, beta, gamma], beta] ==0,} \\
\texttt{D[Spersp[alpha, beta, gamma], gamma] ==0\},} \\
\texttt{\{alpha, beta, gamma\}]} \quad ,
\end{aligned}
\tag{14.55}
$$

we found a result that is correct when applied to input data *that are noise-free*. This formula, which is no longer an unwieldy and lengthy four-fold ambiguous solution as we found for the exact solution, is a *single unique* solution of the following form:

$$
\left.
\begin{aligned}
\alpha &\rightarrow \frac{2a_2a_6 - a_7(a_3 + a_7) + 4a_5a_6}{4a_5a_6 - 2a_6a_9 - a_7{}^2 + a_7a_8} \\[2mm]
\beta &\rightarrow \frac{-2a_2a_6 + a_3a_7 - 2a_6a_9 + a_7a_8}{4a_5a_6 - 2a_6a_9 - a_7{}^2 + a_7a_8} \\[2mm]
\gamma &\rightarrow \frac{a_2(a_7 - a_8) + a_3(a_9 - 2a_5) - 2a_5a_8 + a_7a_9}{4a_5a_6 - 2a_6a_9 - a_7{}^2 + a_7a_8}
\end{aligned}
\right\} .
\tag{14.56}
$$

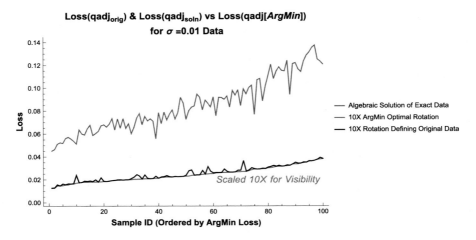

FIGURE 14.6

Distributions of 2D perspective losses using Eq. (14.41) with noise level $\sigma = 0.01$ in the image plane. Red, blue, and black denote the **argmin** results, the noise-free algebraic result of Eq. (14.58) that exactly solves the least squares measure Eq. (14.53), and the original rotation applied to the noisy data, respectively. For noise-free data, these are all perfect determinations of the pose rotation, and there is nothing to plot. The noise-free algebraic formula Eq. (14.58) for the adjugate rotation parameters is very simple and effective for data without errors, but we see here that even for the fairly small $\sigma = 0.01$ noise level, it immediately becomes very unstable.

In terms of the generalized cross-covariance expressions Eq. (14.50), this can also be written as

$$
\left.
\begin{aligned}
\alpha &\to \frac{-c\,\text{uuxy}\,\text{uxyuyx} + c\,\text{uuyx}\,\text{uxy2} - \text{uxy2}\,\text{uyx2} + \text{uxyuyx}^2}{2\,\text{uxyuyx}^2 - 2\,\text{uxy2}\,\text{uyx2}} \\[2mm]
\beta &\to \frac{-c\,\text{uuxy}\,\text{uxyuyx} + c\,\text{uuyx}\,\text{uxy2} + \text{uxy2}\,\text{uyx2} - \text{uxyuyx}^2}{2\,\text{uxy2}\,\text{uyx2} - 2\,\text{uxyuyx}^2} \\[2mm]
\gamma &\to \frac{c(\text{uuyx}\,\text{uxyuyx} - \text{uuxy}\,\text{uyx2})}{2\,\text{uxy2}\,\text{uyx2} - 2\,\text{uxyuyx}^2}
\end{aligned}
\right\}.
\tag{14.57}
$$

This in turn reduces to a simple form for the 2D rotation matrices for error-free perspective data,

$$
R(\mathbf{X}, U) =
\begin{bmatrix}
\dfrac{c(\text{uuyx}\,\text{uxy2} - \text{uuxy}\,\text{uxyuyx})}{\text{uxyuyx}^2 - \text{uxy2}\,\text{uyx2}} & \dfrac{c(\text{uuyx}\,\text{uxyuyx} - \text{uuxy}\,\text{uyx2})}{\text{uxyuyx}^2 - \text{uxy2}\,\text{uyx2}} \\[4mm]
\dfrac{c(\text{uuxy}\,\text{uyx2} - \text{uuyx}\,\text{uxyuyx})}{\text{uxyuyx}^2 - \text{uxy2}\,\text{uyx2}} & \dfrac{c(\text{uuyx}\,\text{uxy2} - \text{uuxy}\,\text{uxyuyx})}{\text{uxyuyx}^2 - \text{uxy2}\,\text{uyx2}}
\end{bmatrix}.
\tag{14.58}
$$

For error-free data, the algebraic solution Eq. (14.58) for the optimal rotation aligning any perspectively projected image agrees exactly with the **argmin** "gold standard" reference solution. Since our algebraic solution does not have all the constraints preserving its essential properties, it is expected, for example, that for noisy data the rotation matrix Eq. (14.58) will no longer even describe a valid rotation, and, indeed, that is the case. We exhibit this unstable sensitivity to noisy data in Fig. 14.6, where we applied our tools to an assembly of simulated data with unit cloud size and added normal error with $\sigma = 0.01$. The original rotation applied to the noisy data is in black, and the standard `argmin` numerically optimized solution is in red. Our heuristic algebraic solution to Eq. (14.55) in the form of the rotation Eq. (14.58), applied inappropriately to noisy data, is shown in blue. One can see that, although the algebraic results actually have a relatively low loss spectrum, for noisy data the `argmin` losses, and even the initial data-creation rotation value's losses, are over an order of magnitude better. Thus this object space least-squares solution, although perfect for error-free data, is very unstable with the introduction of noise, sometimes even becoming complex, and converges reliably to the correct best rotation only as the noise approaches zero. How to improve the behavior of the object-space algebraic solution while retaining the simplicity of its form is currently an open question.

Exact 3D rotations for cloud matching and pose estimation tasks☆

15

We have taken some time in Chapter 12 to get acquainted with the relationship between the *adjugate quaternion variables* and certain issues in converting back and forth between 2D and 3D rotation matrices and their quaternion equivalents, and then in Chapter 14 to see how the adjugate variables in 2D assist in exactly solving certain least-squares applications. Now we turn to 3D and the domain of significant practical applications. In particular, we find that some of our results extend from 2D applications to 3D applications: one of those discoveries is that we can exploit the adjugate framework to explicitly solve the symbolic *error-free* least-squares problem for the 3D → 3D cloud alignment ("matching") problem, the 3D → 2D orthographic projection alignment ("pose estimation") problem, and the 3D → 2D perspective projection alignment ("perspective") problem. We sketch the essence of these fundamental tasks in Fig. 15.1 and Fig. 15.2 to give an idea what these problems look like.

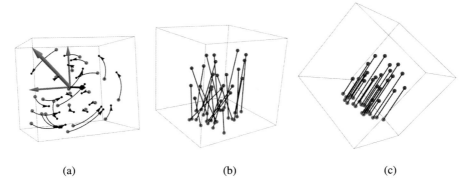

(a)　　　　　　　　　　(b)　　　　　　　　　　(c)

FIGURE 15.1

The fundamental 3D matching and orthographic pose estimation problems. (a) The task of finding the 3D rotation aligning a 3D reference cloud with another 3D test cloud differing by a rotation. The purple arrow is the axis of rotation, proportional to the vector part of the corresponding quaternion. (b) The task of finding the 3D rotation used to create the orthographically projected 2D image of a 3D point cloud. (c) The solution, rotating the cloud so the rays from the projected points align with the 3D virtual camera orientation frame.

☆ Much of this chapter is inspired by Hanson and Hanson (2022).

Visualizing More Quaternions. https://doi.org/10.1016/B978-0-32-399202-2.00025-3

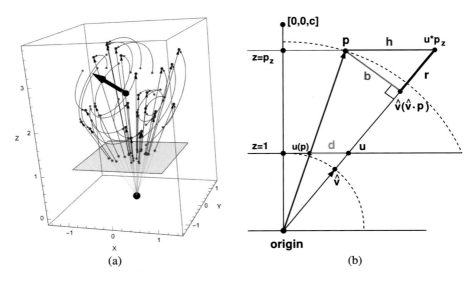

(a)   (b)

FIGURE 15.2

The fundamental 3D perspective projection problem. (a) Task of finding the 3D rotation aligning a 3D reference cloud with another 3D test cloud differing by a rotation, and then projected to the $z = 1$ plane. The black arrow is the axis of rotation, proportional to the vector part of the corresponding quaternion. The task is to find the axis and the amount of the rotation needed to move the known red template points to the set of blue points whose projections best align with the image in the image plane. (b) The contrast between the geometry of the standard image-plane based loss function, which uses the image-space distance $d = \|\mathbf{u} - \mathbf{u}(\mathbf{p})\|$, in green, compared to the LHM object-space loss $\mathbf{b} = \mathbf{p} - \hat{\mathbf{v}}(\hat{\mathbf{v}} \cdot \mathbf{p})$, in red, the Z-prime horizontal object-space loss $\mathbf{h} = \mathbf{p} - \mathbf{u}p_z$, in blue, and the Radial object space loss $\mathbf{r}$ in black. The object-space differences give basically the same results, as their differences are related by the Pythagorean theorem. Minimizing any of these types of loss over a rotation space, numerically or algebraically, solves the perspective pose problem in principle.

    The basic loss functions that we will need have essentially the same form as the 2D loss functions in Chapter 14, Eqs. (14.1), (14.2), (14.3), (14.4), (14.5), (14.6), and (14.7), and we will review their properties for specific applications as needed.

    When the rotated input data are no longer error-free, some of our exact solutions still solve the least-squares optimization problem, but are flawed because they then deviate from being pure orthonormal **SO**(3) rotation matrices. However, these least-squares solutions are still useful for the errorful case because they can serve as the very efficient approximate input for the *nearest pure rotation solution* that can be obtained using methods such as those of Bar-Itzhack (Bar-Itzhack, 2000, and Sections 12.3.4 and 12.4.2). As a bonus, we discover, as proven in Appendix I, that our exact least-squares solutions for error-free data can be extended to any dimension, for both the ND $\rightarrow$ ND matching task *and* the ND $\rightarrow$ (N $-$ 1)D orthographic pose estimation task. In parallel to the 3D errorful data method, the ND solutions can be used for errorful data by applying available nonquaternion methods such as SVD (Golub and van Loan, 1983) or the matrix-square-root construction (Horn et al., 1988) to find the nearest pure **SO**(N) rotation matrix. Finally, we study the object space method whose geometry is presented in Fig. 15.2; we will investigate features of two of them, the LHM (Lu et al., 2000) context and the superficially simpler alternative z-plane approach to the perspective projection problem. The significant details of the Radial object space loss were worked out in Chapter 14. We find that algebraic solutions exist that minimize the object space perspective projection losses in the error-free case; however, they are very complicated to apply, and seem again to be very unstable when noise is introduced.

15.1 Basics

The 3D quaternion-based matching problem will be the initial pedagogical context for our narrative, starting out very close to what we saw in Chapter 8. Thus we begin by reviewing some basic notation specifying the relation between unit quaternion 4-vectors and orthogonal 3×3 rotation matrices belonging to the special orthogonal group $\mathbf{SO}(3)$.

As we know, any unit-length 4D vector q, with $q \cdot q = 1$, is a quaternion point on the unit 3-sphere $\mathbf{S}^3$, and corresponds exactly to a *proper* 3D rotation matrix $R = R(q)$ through the equation

$$R(q) = \begin{bmatrix} q_0^2 + q_1^2 - q_2^2 - q_3^2 & 2q_1q_2 - 2q_0q_3 & 2q_1q_3 + 2q_0q_2 \\ 2q_1q_2 + 2q_0q_3 & q_0^2 - q_1^2 + q_2^2 - q_3^2 & 2q_2q_3 - 2q_0q_1 \\ 2q_1q_3 - 2q_0q_2 & 2q_2q_3 + 2q_0q_1 & q_0^2 - q_1^2 - q_2^2 + q_3^2 \end{bmatrix}. \tag{15.1}$$

Certain problems will be easier to solve when we recast our quaternion equations using the quaternion adjugate quadratic form variable transformation, $q_i q_j \to q_{ij}$, where we define the *quaternion adjugate matrix* as its own abstraction of the form

$$A(q_{ij}) = \begin{bmatrix} q_{00} & q_{01} & q_{02} & q_{03} \\ q_{01} & q_{11} & q_{12} & q_{13} \\ q_{02} & q_{12} & q_{22} & q_{23} \\ q_{03} & q_{13} & q_{23} & q_{33} \end{bmatrix}. \tag{15.2}$$

(From here on, we ignore the fact that various transformations on the actual adjugate forms for a given problem were needed to reach the equivalent form of Eq. (15.2).) The adjugate-valued rotation matrix is thus defined by replacing the quadratic quaternion transformation of Eq. (15.1) by a linear transformation in terms of the variables of Eq. (15.2):

$$R(q_{ij}) = \begin{bmatrix} q_{00} + q_{11} - q_{22} - q_{33} & 2q_{12} - 2q_{03} & 2q_{13} + 2q_{02} \\ 2q_{12} + 2q_{03} & q_{00} - q_{11} + q_{22} - q_{33} & 2q_{23} - 2q_{01} \\ 2q_{13} - 2q_{02} & 2q_{23} + 2q_{01} & q_{00} - q_{11} - q_{22} + q_{33} \end{bmatrix}. \tag{15.3}$$

When we use the adjugate quaternion variables, we must normally extend the single constraint $q \cdot q = 1$ imposed on a single quaternion to a family of seven constraints; this reduces the 10 adjugate variables to the required three independent degrees of 3D rotational freedom. Thus, for example, we may also require a set of constraints such as

$$\left. \begin{array}{ccc} & q_{00} + q_{11} + q_{22} + q_{33} = 1 & \\ q_{00}\, q_{11} = q_{01}^2 & q_{00}\, q_{22} = q_{02}^2 & q_{00}\, q_{33} = q_{03}^2 \\ q_{22}\, q_{33} = q_{23}^2 & q_{11}\, q_{33} = q_{13}^2 & q_{11}\, q_{22} = q_{12}^2 \end{array} \right\}. \tag{15.4}$$

However, we will discover that, in our applications, choosing exactly how to apply the quaternion adjugate constraints will be subtle.

15.2 3D point-cloud matching

We consider first the classic "3D match" problem, also known as the RMSD ("root mean square deviation") or "generalized orthogonal Procrustes" problem, whose task is to find the 3D rotation best aligning a reference cloud with a possibly noisy test cloud presumably derived by rotating the reference cloud. The classic methods for solving this problem have been presented in Chapter 8, and we outline them here to provide local context. We restrict ourselves initially to error-free data, as that is the domain where our formulas will be exact, and choose a least-squares loss function that compares a rotated reference cloud to a corresponding test cloud whose measured coordinates contain no added noise. We thus have two 3D point clouds of size K, with a reference set $\mathbf{X} = \{\mathbf{x}_k\} = \{[x_k, y_k, z_k]^t\}$ that we consider as a list of K columns of 3D points, and a test set of measured 3D points $\mathbf{U} = \{\mathbf{u}_k\} = \{[u_k, v_k, w_k]^t\}$ that is believed to be related, pointwise, to the reference set by a single unknown rotation whose computation is our goal. If we choose to express that rotation using Eq. (15.1) as $R(q)$, then we can write our loss function, the least-squares optimization target, as (see, e.g., Horn, 1987; Hanson, 2020)

$$\mathbf{S}_{\text{(3D match)}}(q) = \sum_{k=1}^{K} \| R(q) \cdot \mathbf{x}_k - \mathbf{u}_k \|^2 . \tag{15.5}$$

As we know from Chapter 8, there are well-known procedures using quaternion eigensystems to solve this problem in closed form either from the least-squares functional $\mathbf{S}(q)$ (see Faugeras and Hebert (1983, 1986)), or from the cross-term $\Delta(q)$, which reduces to a trace over the rotated 3×3 cross-covariance matrix E_{ab} (see, e.g., Horn, 1987),

$$\Delta(q) = \text{tr}\, R(q) \cdot \left(\mathbf{X} \cdot \mathbf{U}^t\right) = \text{tr}\, R(q) \cdot E . \tag{15.6}$$

Because the $R^t \cdot R$ factor that appears in the expansion of the least-squares loss function Eq. (15.5) is absent from the cross-term $\Delta(q)$, we can always express the problem of finding the optimal quaternion by maximizing the function $\Delta(q)$ *quadratic* in the quaternions, and the problem is solvable using standard linear algebra. (Nonquaternion methods such as SVD and the matrix-square-root construction are also widely used.) We emphasize that in the 3D→2D pose estimation problem to be dealt with in the next section, the $R^t \cdot R$ terms no longer cancel, the expression for $\mathbf{S}(q)$ becomes *quartic* in q, and Eq. (15.6) is no longer applicable.

In the standard quaternion solution of the optimal rotation problem, we rearrange the cross-term into a form that is optimized by the maximal eigenvector of the profile matrix $M(E)$ (see Chapter 8 and, e.g., Hanson (2020) for further references and a review). We find

$$\Delta(q) = \text{tr}\, R(q) \cdot E = [q_0, q_1, q_2, q_3] \cdot M(E) \cdot [q_0, q_1, q_2, q_3]^t \equiv q \cdot M(E) \cdot q , \tag{15.7}$$

where $M(E)$ is the traceless, symmetric 4×4 matrix that is composed of linear combinations of the elements of the 3×3 cross-covariance matrix:

$$E = \mathbf{X} \cdot \mathbf{U}^t = \begin{bmatrix} \text{ux} & \text{vx} & \text{wx} \\ \text{uy} & \text{vy} & \text{wy} \\ \text{uz} & \text{vz} & \text{wz} \end{bmatrix} , \tag{15.8}$$

where ux $= \sum_k u_k x_k$, vy $= \sum_k v_k y_k$, wz $= \sum_k w_k z_k$, etc. The profile matrix then can be written

$$M(E) = \begin{bmatrix} \text{ux} + \text{vy} + \text{wz} & \text{wy} - \text{vz} & \text{uz} - \text{wx} & \text{vx} - \text{uy} \\ \text{wy} - \text{vz} & \text{ux} - \text{vy} - \text{wz} & \text{vx} + \text{uy} & \text{uz} + \text{wx} \\ \text{uz} - \text{wx} & \text{vx} + \text{uy} & -\text{ux} + \text{vy} - \text{wz} & \text{wy} + \text{vz} \\ \text{vx} - \text{uy} & \text{uz} + \text{wx} & \text{wy} + \text{vz} & -\text{ux} - \text{vy} + \text{wz} \end{bmatrix}. \tag{15.9}$$

The adjugate form of M. Suppose for a moment that we relax our assumption that the measured rotated data $\mathbf{U}$ are error-free, and return to the original problem solved by the maximal quaternion eigenvalue method. Using whatever methods we have (e.g., a numerical library eigenvalue program or an explicit Cardano quartic symbolic solution package), we compute the maximal eigenvalue λ_{opt} of $M(E)$. Then we will find not one eigenvector, but a set of four equivalent eigenvectors of λ_{opt} by computing the characteristic matrix $\chi(E)$ along with its corresponding eigenvector list $A(E)$ from the profile matrix $M(E)$ as follows:

$$\chi(E) = \left[M(E) - \lambda_{\text{opt}} I_4 \right], \tag{15.10}$$

$$A(E) = -\text{Adjugate}\,(\chi(E)). \tag{15.11}$$

By definition the characteristic matrix $\chi(E)$ has vanishing determinant $\det(\chi(E)) = 0$, and since the adjugate multiplied into its source matrix is *by definition* its determinant times the identity matrix (determining the inverse if it exists), we have

$$\chi(E) \cdot A(E) = \left[M(E) - \lambda_{\text{opt}} I_4 \right] \cdot A(E) = \det(\chi(E))\, I_4 \equiv 0. \tag{15.12}$$

Each column of $A(E)$ is therefore *proportional* to the maximal eigenvector $q_{\text{opt}} = (q_0, q_1, q_2, q_3)$; thus each column is multiplied by some constant, and we can write

$$A(E) = \begin{bmatrix} c_0 q_0 & c_1 q_0 & c_2 q_0 & c_3 q_0 \\ c_0 q_1 & c_1 q_1 & c_2 q_1 & c_3 q_1 \\ c_0 q_2 & c_1 q_2 & c_2 q_2 & c_3 q_2 \\ c_0 q_3 & c_1 q_3 & c_2 q_3 & c_3 q_3 \end{bmatrix}. \tag{15.13}$$

However, we know also that, *by construction*, $M(E)$ and $\chi(E)$ are *symmetric*. Therefore, up to an overall constant that we can drop, the adjugate of the characteristic matrix is essentially

$$A(E) = \begin{bmatrix} (q_0)^2 & q_1 q_0 & q_2 q_0 & q_3 q_0 \\ q_0 q_1 & (q_1)^2 & q_2 q_1 & q_3 q_1 \\ q_0 q_2 & q_1 q_2 & (q_2)^2 & q_3 q_2 \\ q_0 q_3 & q_1 q_3 & q_2 q_3 & (q_3)^2 \end{bmatrix} = \begin{bmatrix} q_{00} & q_{01} & q_{02} & q_{03} \\ q_{01} & q_{11} & q_{12} & q_{13} \\ q_{02} & q_{12} & q_{22} & q_{23} \\ q_{03} & q_{13} & q_{23} & q_{33} \end{bmatrix}, \tag{15.14}$$

where we have converted the quaternion quadratic terms to the adjugate variables $q_{ij} = q_i q_j$. Thus the optimal rotation matrix can be calculated knowing only λ_{opt} and the numerical profile matrix $M(E) = M(\text{ux}, \ldots)$ from which to calculate the eigenvector list $A(E) = -\text{Adjugate}(\chi(E))$ in Eq. (15.14).

We know already from Chapter 12 that solving for the actual quaternion q_{opt} using Eq. (15.14) can be obstructed by 14 singular domains depending on various combinations of one, two, or three zeroes in the columnwise normalization factors, and requires checking to identify which solution best escapes the possible singularities. We immediately see that our optimal rotation matrix emerges from applying

Eq. (15.3) to the maximal norm column of $A(E)$; such a normalizable column always exists due to $q \cdot q = 1$, and thus none of the 14 singular domains can disturb the accuracy of the choice

$$R_{\text{opt}} = R(q_{\text{opt}}) = R(\text{normalized max-norm column of } A(E)) . \tag{15.15}$$

We note that in practice, the adjugate matrix emerging from the profile matrix can be quite large numerically, and that dividing by the trace $\text{tr}(A(E)) = \sum q_{ii}$ in effect normalizes $A(E)$ in both the row and column directions, potentially improving practical computation if the division itself does not create excessive numerical errors.

Note on the error-free case. The solution Eq. (15.15) for $R(\text{opt})$ is valid independently of whether there are errors in the measured $\mathbf{U}$ data or not, as long as we correctly calculate the maximal eigenvalue λ_{opt} of $M(E)$. However, as shown in Chapter 8, Section 8.4, a special simplification occurs in the error-free case: in the absence of errors, the coefficients of the characteristic equation $\det(\chi(E)) = 0$ are *independent* of the value of the unknown rotation R, and therefore the eigenvalues are also independent of R, and the maximal eigenvalue reduces to the trace of $E_0 = \mathbf{X} \cdot \mathbf{X}^{\text{t}}$, the self-covariance of the reference data. Thus

$$\lambda_{\text{opt}}(\text{error-free}) = \text{tr}(E_0) = \sum_k \left((x_k)^2 + (y_k)^2 + (z_k)^2 \right) = xx + yy + zz . \tag{15.16}$$

However, the *eigenvectors* must differ, depending on the profile matrix $M(E)$, even for identical eigenvalues. The optimal rotation for error-free data can thus be written down from the quaternion columns of the adjugate of $\chi = [M - \lambda_{\text{opt}} I_4]$ *with only simple matrix algebra.*

15.3 Direct solution of the error-free 3D match problem

The error-free case, for which a perfect rotation relates the reference and test data, has additional surprises in store for us. We now solve directly for the least-squares adjugate quaternion solution in the error-free case. Our successful strategy was to return to the richer structure of the full least-squares symbolic form of Eq. (15.5), written in Mathematica format expressing $\| R(q_{ij}) \cdot \mathbf{X} - \mathbf{U} \|^2$ as follows:

```
matchLSQ =
 Expand[({{q00 + q11 - q22 - q33, -2 q03 + 2 q12,  2 q02 + 2 q13},
    {2 q03 + 2 q12,  q00 - q11 + q22 - q33, -2 q01 + 2 q23},
    {-2 q02 + 2 q13, 2 q01 + 2 q23, q00 - q11 - q22 + q33}}.{x, y, z} - {u, v, w})
   . ({{q00 + q11 - q22 - q33, -2 q03 + 2 q12,  2 q02 + 2 q13},
    {2 q03 + 2 q12,  q00 - q11 + q22 - q33, -2 q01 + 2 q23},
    {-2 q02 + 2 q13, 2 q01 + 2 q23, q00 - q11 - q22 + q33}}.{x, y, z} - {u, v, w})]
  /. {x^2 -> xx, x y -> xy, x z -> xz, y x -> xy, y^2 -> yy,
     y z -> yz, z x -> xz, z y -> yz, z^2 -> zz, u x -> ux, u y -> uy,
     u z -> uz, v x -> vx, v y -> vy,v z -> vz, w x -> wx, w y -> wy,
     w z -> wz,  u^2 -> uu, v^2 -> vv, w^2 -> ww }  .
```

$$\tag{15.17}$$

Here we again employ the convenient notation $\{xx, xy, xz, \ldots, wz\}$ to denote the elements of the mixed types of point-summed cross-covariance elements such as $ux = u \cdot x = \sum_{k=1}^{K} u_k x_k$. This matrix conveniently represents all the componentwise data summations that we are referring to as cross-covariances:

$$
C = \begin{bmatrix}
xx & xy & xz & ux & vx & wx \\
xy & yy & yz & uy & vy & wy \\
xz & yz & zz & uz & vz & wz \\
ux & uy & uz & uu & uv & uw \\
vx & vy & vz & uv & vv & vw \\
wx & wy & wz & uw & vw & ww
\end{bmatrix} . \tag{15.18}
$$

Our loss function, **matchLSQ**, expanded out and written in terms of the measured data components summed over $k = 1, \ldots, K$, has both linear and quadratic terms in q_{ij}, and takes the form:

$$
\begin{aligned}
S_{\text{3D match}}(q_{ij}; x, y, z; u, v, w) = \\
& + q_{00}{}^2\, xx + q_{11}{}^2\, xx + q_{22}{}^2\, xx + q_{33}{}^2\, xx + q_{00}{}^2\, yy + q_{11}{}^2\, yy + q_{22}{}^2\, yy + q_{33}{}^2\, yy \\
& + q_{00}{}^2\, zz + q_{11}{}^2\, zz + q_{22}{}^2\, zz + q_{33}{}^2\, zz - 2q_{00}\, ux - 2q_{11}\, ux + 2q_{22}\, ux + 2q_{33}\, ux \\
& + 4q_{03}\, uy - 4q_{12}\, uy - 4q_{02}\, uz - 4q_{13}\, uz - 4q_{03}\, vx - 4q_{12}\, vx - 2q_{00}\, vy + 2q_{11}\, vy \\
& - 2q_{22}\, vy + 2q_{33}vy + 4q_{01}vz - 4q_{23}\, vz + 4q_{02}\, wx - 4q_{13}\, wx - 4q_{01}\, wy - 4q_{23}\, wy \\
& - 2q_{00}\, wz + 2q_{11}\, wz + 2q_{22}\, wz - 2q_{33}\, wz + 4q_{02}{}^2\, xx + 4q_{03}{}^2\, xx + 2q_{00}q_{11}\, xx \\
& + 8q_{03}q_{12}\, xx + 4q_{12}{}^2\, xx - 8q_{02}q_{13}\, xx + 4q_{13}{}^2\, xx - 2q_{00}q_{22}\, xx - 2q_{11}q_{22}\, xx \\
& - 2q_{00}q_{33}\, xx - 2q_{11}q_{33}\, xx + 2q_{22}q_{33}\, xx - 8q_{01}q_{02}\, xy - 8q_{03}q_{11}\, xy + 8q_{00}q_{12}\, xy \\
& + 8q_{01}q_{13}\, xy + 8q_{03}q_{22}\, xy - 8q_{02}q_{23}\, xy + 8q_{13}q_{23}\, xy - 8q_{12}q_{33}\, xy - 8q_{01}q_{03}\, xz \\
& + 8q_{02}q_{11}\, xz - 8q_{01}q_{12}\, xz + 8q_{00}q_{13}\, xz - 8q_{13}q_{22}\, xz + 8q_{03}q_{23}\, xz + 8q_{12}q_{23}\, xz \\
& - 8q_{02}q_{33}\, xz + 4q_{01}{}^2\, yy + 4q_{03}{}^2\, yy - 2q_{00}q_{11}\, yy - 8q_{03}q_{12}\, yy + 4q_{12}{}^2\, yy \\
& + 2q_{00}q_{22}\, yy - 2q_{11}q_{22}\, yy + 8q_{01}q_{23}\, yy + 4q_{23}{}^2\, yy - 2q_{00}q_{33}\, yy + 2q_{11}q_{33}\, yy \\
& - 2q_{22}q_{33}\, yy - 8q_{02}q_{03}\, yz + 8q_{02}q_{12}\, yz - 8q_{03}q_{13}\, yz + 8q_{12}q_{13}\, yz - 8q_{01}q_{22}\, yz \\
& + 8q_{00}q_{23}\, yz - 8q_{11}q_{23}\, yz + 8q_{01}q_{33}\, yz + 4q_{01}{}^2\, zz + 4q_{02}{}^2\, zz - 2q_{00}q_{11}\, zz \\
& + 8q_{02}q_{13}\, zz + 4q_{13}{}^2\, zz - 2q_{00}q_{22}\, zz + 2q_{11}q_{22}\, zz - 8q_{01}q_{23}\, zz + 4q_{23}{}^2\, zz \\
& + 2q_{00}q_{33}\, zz - 2q_{11}q_{33}\, zz - 2q_{22}q_{33}\, zz + uu + vv + ww . \tag{15.19}
\end{aligned}
$$

15.3.1 Solving the 3D match least-squares loss function algebraically

We now show how we can get algebraic solutions that minimize the least-squares loss function for the 3D match problem. We start with the **matchLSQ** form of the least-squares loss function for 3D matching given in Eq. (15.17) and its explicit form in Eq. (15.19) in terms of the quaternion adjugate variables q_{ij}. At this point there are many different approaches that might or might not allow us to obtain explicit solutions for the unknown q_{ij} that minimize Eq. (15.19). We might consider imposing all seven adjugate constraints in Eq. (15.4), or perhaps some subset, and perhaps picking some selection of the 10 q_{ij} variables to use in the search. Trying various apparently natural combinations of variables and constraints with the Mathematica Solve[ ] utility, we found that many seemingly reasonable attempts either did not return an answer after a very long wait, or returned an empty result within a second.

The successful method emerged from a sequence of closely related, basically equivalent approaches, one of which produced much more manageable results than the others. The first method was to take the 10 derivatives of Eq. (15.19) with respect to the q_{ij} set to zero, while using all 10 q_{ij} as free variables, with *no* imposed constraints: this produced an error saying only nine variables could be determined, and returned the nine q_{ij}'s *not* including q_{00}, with each a linear function of the form $q_{ij} = a_{ij} + b_{ij}q_{00}$. This was in principle solvable, but would have required processing a symbolic solution with a daunting 68,814,040 bytes of symbols. The second variant was to *delete* q_{00}'s derivative term and its appearance in the output list, and this yielded basically the same answer, but with no error flag, and half the size of the symbolic result, only 34,915,760 bytes.

Our final solution that led quickly to the closed form answer was to return to the form with *all* 10 free variables, and *all* 10 vanishing partial derivative equations and *in addition* imposing only the single stay-on-$\mathbf{S}^3$ constraint, $q_{00} + q_{11} + q_{22} + q_{33} = 1$, of the seven available adjugate variable constraints in Eq. (15.4). This may well be mathematically suspect, since presumably the constraints imply that the assumption that the partial derivative constraints can be applied independently is not necessarily correct. Nevertheless, the properties of the system framed in this fashion appears valid, as, with those parameters, the following solver runs for only 5 seconds or so, and returns a complete list of rules expressing all 10 q_{ij}'s in terms of only the measured quantities $\mathbf{X}$ and $\mathbf{U}$ as they appear in Eq. (15.18), comprising only 17,483,856 bytes:

```
the3D3DAdjMatchSolns =
    Module[{eqn = lossMatch3D3DAdj},
        Solve[ { D[eqn, q00] == 0,
            D[eqn, q11] == 0, D[eqn, q22] == 0, D[eqn, q33] == 0,
            D[eqn, q01] == 0, D[eqn, q02] == 0, D[eqn, q03] == 0,
            D[eqn, q23] == 0, D[eqn, q13] == 0, D[eqn, q12] == 0,
                q00 + q11 + q22 + q33 == 1 },
        {q00,  q11, q22, q33, q01, q02, q03, q23, q13, q12}]]   .
```

$$(15.20)$$

The resulting expressions respond to `Simplify[ ]` in seconds, and reduce to explicit compact equations for the q_{ij} in terms of the cross-covariances in Eq. (15.18), all of the general form

$$\begin{aligned}
q_{00} =&(\text{wz xy}^2 - \text{vz xy xz} - \text{wy xy xz} + \text{vy xz}^2 - \text{wz xx yy} + \text{uz xz yy}\\
&+\text{wx xz yy} + \text{xz}^2 \text{yy} + \text{vz xx yz} + \text{wy xx yz} - \text{uz xy yz} - \text{wx xy yz}\\
&-\text{uy xz yz} - \text{vx xz yz} - \text{2xy xz yz} + \text{ux yz}^2 + \text{xx yz}^2 - \text{vy xx zz}\\
&+\text{uy xy zz} + \text{vx xy zz} + \text{xy}^2 \text{zz} - \text{ux yy zz} - \text{xx yy zz})/\\
&(\text{4xz}^2 \text{yy} - \text{8xy xz yz} + \text{4xx yz}^2 + \text{4xy}^2 \text{zz} - \text{4xx yy zz}) .
\end{aligned}$$

$$(15.21)$$

However, this solution, which only uses one of the constraints Eq. (15.4), cannot guarantee *orthonormality* of the resulting rotation matrix, which depends on having all seven constraints obeyed. Remarkably, this simplified solution is *exact* as long as the rotated data are *error-free*. Furthermore, without ever quite solving for the known quaternion eigensystem solution Eq. (15.15), we will see below that we can easily get an orthonormal solution that is very close to optimal by finding the *nearest* exact orthonormal matrix.

Reassembling these solutions for the quaternion adjugate variables directly into a rotation matrix, we find

$$
R(x, y, z; u, v, w) =
\begin{bmatrix}
q_{00} + q_{11} - q_{22} - q_{33} & 2q_{12} - 2q_{03} & 2q_{13} + 2q_{02} \\
2q_{12} + 2q_{03} & q_{00} - q_{11} + q_{22} - q_{33} & 2q_{23} - 2q_{01} \\
2q_{13} - 2q_{02} & 2q_{23} + 2q_{01} & q_{00} - q_{11} - q_{22} + q_{33}
\end{bmatrix}
$$

$$
=
\begin{bmatrix}
\dfrac{d_{11}}{d_0} & \dfrac{d_{12}}{d_0} & \dfrac{d_{13}}{d_0} \\[2mm]
\dfrac{d_{21}}{d_0} & \dfrac{d_{22}}{d_0} & \dfrac{d_{23}}{d_0} \\[2mm]
\dfrac{d_{31}}{d_0} & \dfrac{d_{32}}{d_0} & \dfrac{d_{33}}{d_0}
\end{bmatrix} ,
\tag{15.22}
$$

which is orthonormal for error-free data. Here d_0 is the determinant of the self-covariance of $\mathbf{X}$ and the d_{ij} are essentially all possible $[\mathbf{x} \, \mathbf{x} \, \mathbf{u}]$, $[\mathbf{x} \, \mathbf{x} \, \mathbf{v}]$, and $[\mathbf{x} \, \mathbf{x} \, \mathbf{w}]$ determinants of the columns in the top three rows of Eq. (15.18); our solutions take the form of the following 10 determinants:

$$
d_0 \rightarrow
\begin{bmatrix}
xx & xy & xz \\
xy & yy & yz \\
xz & yz & zz
\end{bmatrix} ,
$$

$$
d_{11} \rightarrow
\begin{bmatrix}
xy & xz & ux \\
yy & yz & uy \\
zy & zz & uz
\end{bmatrix} , \quad
d_{12} \rightarrow
\begin{bmatrix}
xz & xx & ux \\
yz & yx & uy \\
zz & zx & uz
\end{bmatrix} , \quad
d_{13} \rightarrow
\begin{bmatrix}
xx & xy & ux \\
yx & yy & uy \\
zx & zy & uz
\end{bmatrix} ,
$$

$$
d_{21} \rightarrow
\begin{bmatrix}
xy & xz & vx \\
yy & yz & vy \\
zy & zz & vz
\end{bmatrix} , \quad
d_{22} \rightarrow
\begin{bmatrix}
xz & xx & vx \\
yz & yx & vy \\
zz & zx & vz
\end{bmatrix} , \quad
d_{23} \rightarrow
\begin{bmatrix}
xx & xy & vx \\
yx & yy & vy \\
zx & zy & vz
\end{bmatrix} ,
$$

$$
d_{31} \rightarrow
\begin{bmatrix}
xy & xz & wx \\
yy & yz & wy \\
zy & zz & wz
\end{bmatrix} , \quad
d_{32} \rightarrow
\begin{bmatrix}
xz & xx & wx \\
yz & yx & wy \\
zz & zx & wz
\end{bmatrix} , \quad
d_{33} \rightarrow
\begin{bmatrix}
xx & xy & wx \\
yx & yy & wy \\
zx & yz & wz
\end{bmatrix} .
\tag{15.23}
$$

Conversely, we can express the q_{ij} variables given by, e.g., Eq. (15.21), more compactly in terms of these same determinants, allowing quick access to the quaternion after zero-checking the adjugates:

$$
q_{00} = \frac{1}{4d_0}(d_0 + d_{11} + d_{22} + d_{33}) , \qquad
q_{11} = \frac{1}{4d_0}(d_0 + d_{11} - d_{22} - d_{33}) ,
$$

$$
q_{22} = \frac{1}{4d_0}(d_0 - d_{11} + d_{22} - d_{33}) , \qquad
q_{33} = \frac{1}{4d_0}(d_0 - d_{11} - d_{22} + d_{33}) ,
$$

$$
q_{01} = \frac{1}{4d_0}(d_{32} - d_{23}) , \qquad
q_{02} = \frac{1}{4d_0}(d_{13} - d_{31}) , \qquad
q_{03} = \frac{1}{4d_0}(d_{21} - d_{12}) ,
$$

$$
q_{23} = \frac{1}{4d_0}(d_{23} + d_{32}) , \qquad
q_{13} = \frac{1}{4d_0}(d_{13} + d_{31}) , \qquad
q_{12} = \frac{1}{4d_0}(d_{12} + d_{21}) .
$$

$$
\tag{15.24}
$$

Independent verification of the solution. We can now exploit the fact that we are assuming error-free measurements of the rotated cloud points to prove independently that Eq. (15.22) is correct. If we define the symbolic elements of the unknown rotation matrix as

$$R(r_{ij}) = \begin{bmatrix} r_{11} & r_{12} & r_{13} \\ r_{21} & r_{22} & r_{23} \\ r_{31} & r_{32} & r_{33} \end{bmatrix}, \tag{15.25}$$

we can write the individual components of each rotated element as

$$\begin{aligned} u &= r_{11}\, x + r_{12}\, y + r_{13}\, z, \\ v &= r_{21}\, x + r_{22}\, y + r_{23}\, z, \\ w &= r_{31}\, x + r_{32}\, y + r_{33}\, z. \end{aligned} \tag{15.26}$$

If we examine the d_{11} matrix element of R with $[u, v, w]$ replaced by their original source transformations Eq. (15.26), we see that two of the elements of the last column are simply multiples of the first two columns, and can be eliminated, leaving a pure multiple of the rotation matrix component r_{11}:

$$\begin{aligned} d_{11} &= \det \begin{bmatrix} xy & xz & ux \\ yy & yz & uy \\ zy & zz & uz \end{bmatrix} = \det \begin{bmatrix} xy & xz & r_{11}\,xx + r_{12}\,xy + r_{13}\,xz \\ yy & yz & r_{11}\,xy + r_{12}\,yy + r_{13}\,yz \\ zy & zz & r_{11}\,xz + r_{12}\,zy + r_{13}\,zz \end{bmatrix} \\[4pt] &= \det \begin{bmatrix} xy & xz & r_{11}\,xx \\ yy & yz & r_{11}\,xy \\ zy & zz & r_{11}\,xz \end{bmatrix} = r_{11}\, d_0. \end{aligned} \tag{15.27}$$

The same argument holds for all the d_{ij} determinants, so Eq. (15.22) is identical to Eq. (15.25), and our proof is complete:

$$R(x, y, z; u, v, w) \equiv R(r_{ij}). \tag{15.28}$$

Equivalence of the error-free profile matrix $M(E)$ solution. From the rotation independence of the optimal eigenvalue and Eq. (15.16), we realize that, for the error-free case to which Eq. (15.22) applies, the *exact* solution to which Eq. (15.15) applies also becomes much simpler. When we write down the error-free case for $\chi(\mathbf{X}, \mathbf{U})$, with the trivial eigenvalue Eq. (15.16), we find that the rotations do not seem to match. The coefficient that must be used as a divisor to obtain a rotation is

$$\begin{aligned} \text{[Adj coefficient]} = 8 \Big(&xx\,xy^2 + xx\,xz^2 - xx^2\,yy + xy^2\,yy - xx\,yy^2 + 2xy\,xz\,yz + yy\,yz^2 \\ &-xx^2\,zz + xz^2\,zz - 2xx\,yy\,zz - yy^2\,zz + yz^2\,zz - xx\,zz^2 - yy\,zz^2 \Big). \end{aligned} \tag{15.29}$$

In order to match the supposedly identical error-free rotation matrix, the corresponding divisor appears proportional to

$$\text{[Det matrix coefficient]} = \det(d_0) = xx\,yy\,zz - xz^2\,yy + 2xy\,xz\,yz - xx\,yz^2 - xy^2\,zz. \tag{15.30}$$

One might logically conclude that this can never be proportional to Eq. (15.29). In fact, for error-free orthonormal matrices acting on $[x, y, z]$ to produce the $[u, v, w]$ terms in the profile matrix Eq. (15.9), we find an expression only of (xx, xy, xz, yy, yz, zz) and a complicated combination of rotation matrix elements. However, it all works out in a strange way. Remember that we were able to use the $[u, v, w]^t \rightarrow R \cdot [x, y, z]^t$ substitution to show $\det(\text{matrix}_{ij}) = r_{ij}d_0$. Using this same trick, but parameterizing the rotation matrix elements using the orthonormal axis-angle trigonometric variables of Eq. (2.17), a massive cancelation appears, and we are left with exactly Eq. (15.29) multiplied by an ordinary rotation matrix in terms of axis-angle variables. Thus the *exact same* rotation matrices appear, except that in our earlier case, the simplification was easy in terms of the actual nine rotation elements r_{ij}, while in the $\text{Adj}(\chi)$ form, the simplification is accessible using the axis angle variables, each leaving behind a pure rotation matrix divided by Eq. (15.29) and Eq. (15.30), respectively.

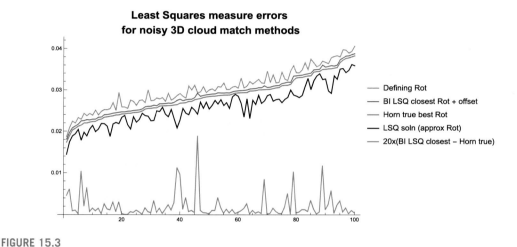

FIGURE 15.3

Comparison of solutions to the 3D point-cloud matching problem.

Comparison of match problem accuracy. To demonstrate the behavior of the various solutions and approximate solutions to the 3D matching problem, we generate sample data consisting of 100 clouds of 25 3D reference points **X** with ranges from $-1 \rightarrow +1$, rotate them each by a different random matrix, and apply a Gaussian noise distribution to each point with $\sigma = 0.1$. Our accuracy measure is the least-squares loss function Eq. (15.5). Using the exact best solution with the numerical eigenvalue of the (noisy) profile matrix as our standard, we sort all loss values in the order of this best solution, so these labels of the distinct clouds are the horizontal indices from 1 to 100. Then we take each set of cloud data, compute these four candidate optimally aligning rotation matrices, and plot the results in Fig. 15.3.

- **Original rotation.** In magenta, the rotation matrix used originally to generate the simulated data. Because we have added noise, this rotation matrix, unlike Eq. (15.15), is uncomputable, and must almost always have a larger error than the computable solutions.
- **Closest rotation to the least-squares solution.** In blue: using the Bar-Itzhack method, which is essentially the same as the process leading to Eq. (15.15) with the cross-covariance matrix E replaced by the transpose of the noisy least squares solution $R(x, y, z : u, v, w)$, we can find precisely

the closest possible rotation matrix approximating the least-squares solution. This is so close to the numerical optimum solution that we displace it upward slightly for visibility.

- **"Gold standard" solution.** In red, the "true" best possible optimal rotation solution from the Chapter 8 (Horn-style) optimal quaternion eigenvector of the profile matrix, using the adjugate of the characteristic equation and Eq. (15.15).
- **Least-squares solution.** In **black**, the least-squares solution Eq. (15.22) is an exact rotation matrix for noise-free input data; however, when noise is added, it *remains the least-squares solution*, and will have the smallest possible error measure, but it is *no longer a rotation matrix*.

The standard solution and the closest rotation solution are so close that we have added a plot of their *difference*, magnified 20× for visibility, in green; we see that while their difference can be very close to zero, it is *never negative*. This is exactly what we expect; our correction to the flawed least-squares solution is highly unlikely to match the optimal eigenvalue solution everywhere, but it can be very close. Thus Fig. 15.3 shows us a complete picture of the levels of accuracy expected for noisy data sets. Note that for *noise-free data*, all four graphs will be identical to machine accuracy.

15.4 3D to 2D orthographic pose estimation

We turn our attention now to the 3D orthographic projection pose estimation problem. This consists of a 3D cloud corresponding to a point-matched 2D image resulting from a rotation applied to the reference cloud before projection, and is analogous to the classic 3D cloud-cloud matching optimization problem we discussed in Sections 15.2 and 15.3. For this problem, we assume a 3D point-cloud reference set $\mathbf{X}$, a list of K columns of 3D points $\{\mathbf{x}_k\} = \{[x_k, y_k, z_k]^t\}$, and a paired test set of 2D points $\mathbf{U}$, with 2D image-plane components $\{\mathbf{u}_k\} = \{[u_k, v_k]^t\}$. Here we consider only error-free rotated data in the image plane, and parallel projection described by a 2×3 projection matrix $P(q)$ that is extracted from the top two rows of a 3D rotation matrix $R(q)$. Thus the third row of the rotation matrix, which we will call $D(q)$ indicating it handles the denominator, is absent and will not appear in the data or the loss function until Section 15.5, when we investigate perspective projection.

If we choose to express our rotation using Eq. (15.1) as $R(q)$, then we may write the projection as

$$
P(q) = \begin{bmatrix} q_0^2 + q_1^2 - q_2^2 - q_3^2 & 2q_1q_2 - 2q_0q_3 & 2q_1q_3 + 2q_0q_2 \\ 2q_1q_2 + 2q_0q_3 & q_0^2 - q_1^2 + q_2^2 - q_3^2 & 2q_2q_3 - 2q_0q_1 \end{bmatrix}, \tag{15.31}
$$

and the least-squares optimization target can be written

$$
\mathbf{S}_{\text{3D pose}} = \sum_{k=1}^{K} \| P(q) \cdot \mathbf{x}_k - \mathbf{u}_k \|^2 . \tag{15.32}
$$

While the 3D point-cloud matching loss function in Section 15.2 can be reduced to the quadratic cross-term Δ and solved using an optimal quaternion eigenvector, *this approach fails for pose estimation*. In the pose estimation problem we can no longer reduce the **x-x** term to a constant by the rotation matrix identity, thus failing to eliminate the quartic quaternion part of the optimization, and the problem becomes potentially much more complex. Therefore we must typically solve Eq. (15.32) by numerical methods such as applying an `argmin` utility to the least-squares expression Eq. (15.32) while simultaneously imposing the constraint $q \cdot q = 1$, or by an algebraic method applied directly to the full least-squares minimization problem of Eq. (15.32).

Our next step is to actually attempt such an algebraic solution. Our approach again is to reduce the order of the rotation variable polynomial in Eq. (15.32) by replacing the individual quaternions as they appear in Eq. (15.31) by their adjugate quadratic forms, $q_i q_j \rightarrow q_{ij}$, so our adjugate-valued projection matrix becomes

$$P(q_{ij}) = \begin{bmatrix} q_{00} + q_{11} - q_{22} - q_{33} & 2q_{12} - 2q_{03} & 2q_{13} + 2q_{02} \\ 2q_{12} + 2q_{03} & q_{00} - q_{11} + q_{22} - q_{33} & 2q_{23} - 2q_{01} \end{bmatrix}. \quad (15.33)$$

We note the projection matrix is lacking these matrix elements, $q_{13} - q_{02}$, $q_{23} + q_{01}$, and $q_{00} - q_{11} - q_{22} + q_{33}$, so the number of constraints needed can in principle be reduced. Secondly, those variables are in a specific sense recoverable because, since $P(q_{ij})$ is part of an orthonormal 3×3 matrix, the missing bottom row can be recovered if the first two rows have been determined accurately: we simply take the cross product of the two rows of $P(q_{ij})_{\text{opt}}$ and normalize the result (if necessary) to get the missing last row of a full orthonormal rotation matrix. The loss expression Eq. (15.32) has now become a quadratic function in the adjugate variables q_{ij} that can be solved, in principle, using least-squares algebraic methods, possibly resulting in a computable adjugate matrix from which a guaranteed nonsingular quaternion could be extracted. To turn our least-squares formula Eq. (15.32) into a symbolic expression for the next step, we expand its form in Mathematica:

lossOrtho3D2DAdj =
```
 Expand[({{q00 + q11 - q22 - q33, -2 q03 + 2 q12,  2 q02 + 2 q13},
   {2 q03 + 2 q12,  q00 - q11 + q22 - q33, -2 q01 + 2 q23}} . {x, y, z} - {u, v}})
  . ({{q00 + q11 - q22 - q33, -2 q03 + 2 q12,  2 q02 + 2 q13},
   {2 q03 + 2 q12,  q00 - q11 + q22 - q33, -2 q01 + 2 q23}}  .{x, y, z} - {u, v})]
 /. {x^2 -> xx, x y -> xy, x z -> xz, y x -> xy, y^2 -> yy,
    y z -> yz, z x -> xz, z y -> yz, z^2 -> zz, u x -> ux, u y -> uy,
    u z -> uz, v x -> vx, v y -> vy, v z -> vz, u^2 -> uu, v^2 -> vv }   ,
```
$$\quad (15.34)$$

where, as before, we use the notation of Eq. (15.18) for the cross-covariance data sums, with $\text{ux} = \sum_k u_k x_k$, etc. The projection task loss expression is complicated by the appearance of both linear and quadratic terms in q_{ij}, as we see clearly in the full symbolic loss expansion:

$$S_{\text{3D:2D ortho loss}}(q_{ij}; x, y, z; u, v) =$$
$$q_{00}^2 \text{xx} + q_{11}^2 \text{xx} + q_{22}^2 \text{xx} + q_{33}^2 \text{xx} + q_{00}^2 \text{yy} + q_{11}^2 \text{yy} + q_{22}^2 \text{yy} + q_{33}^2 \text{yy}$$
$$- 4q_{00}q_{01} \text{yz} + 4q_{00}q_{02} \text{xz} + 2q_{00}q_{11} \text{xx} - 2q_{00}q_{11} \text{yy} + 8q_{00}q_{12} \text{xy}$$
$$+ 4q_{00}q_{13} \text{xz} - 2q_{00}q_{22} \text{xx} + 2q_{00}q_{22} \text{yy} + 4q_{00}q_{23} \text{yz} - 2q_{00}q_{33} \text{xx} - 2q_{00}q_{33} \text{yy}$$
$$- 2q_{00} \text{ux} - 2q_{00} \text{vy} + 4q_{01}^2 \text{zz} - 8q_{01}q_{03} \text{xz} + 4q_{01}q_{11} \text{yz} - 8q_{01}q_{12} \text{xz}$$
$$- 4q_{01}q_{22} \text{yz} - 8q_{01}q_{23} \text{zz} + 4q_{01}q_{33} \text{yz} + 4q_{01} \text{vz} + 4q_{02}^2 \text{zz} - 8q_{02}q_{03} \text{yz}$$
$$+ 4q_{02}q_{11} \text{xz} + 8q_{02}q_{12} \text{yz} + 8q_{02}q_{13} \text{zz} - 4q_{02}q_{22} \text{xz} - 4q_{02}q_{33} \text{xz} - 4q_{02} \text{uz}$$
$$+ 4q_{03}^2 \text{xx} + 4q_{03}^2 \text{yy} - 8q_{03}q_{11} \text{xy} + 8q_{03}q_{12} \text{xx} - 8q_{03}q_{12} \text{yy} - 8q_{03}q_{13} \text{yz}$$

$$+ 8q_{03}q_{22}\,\text{xy} + 8q_{03}q_{23}\,\text{xz} + 4q_{03}\,\text{uy} - 4q_{03}\,\text{vx} + 4q_{11}q_{13}\,\text{xz} - 2q_{11}q_{22}\,\text{xx}$$
$$- 2q_{11}q_{22}\,\text{yy} - 4q_{11}q_{23}\,\text{yz} - 2q_{11}q_{33}\,\text{xx} + 2q_{11}q_{33}\,\text{yy} - 2q_{11}\,\text{ux} + 2q_{11}\,\text{vy}$$
$$+ 4q_{12}{}^2\text{xx} + 4q_{12}{}^2\text{yy} + 8q_{12}q_{13}\,\text{yz} + 8q_{12}q_{23}\,\text{xz} - 8q_{12}q_{33}\,\text{xy} - 4q_{12}\,\text{uy}$$
$$- 4q_{12}\,\text{vx} + 4q_{13}{}^2\text{zz} - 4q_{13}q_{22}\,\text{xz} - 4q_{13}q_{33}\,\text{xz} - 4q_{13}\,\text{uz} + 4q_{22}q_{23}\,\text{yz}$$
$$+ 2q_{22}q_{33}\,\text{xx} - 2q_{22}q_{33}\,\text{yy} + 2q_{22}\,\text{ux} - 2q_{22}\,\text{vy} + 4q_{23}{}^2\text{zz} - 4q_{23}q_{33}\,\text{yz}$$
$$- 4q_{23}\,\text{vz} + 2q_{33}\,\text{ux} + 2q_{33}\,\text{vy} + \text{uu} + \text{vv} \,. \tag{15.35}$$

(Note that it is the sum of q_{ii} that adds up to unity, not the sum of $q_{ii}{}^2$, so there is no simplification in the first line.) Parallel to the 2D case, one cannot complete the squares to recover a simpler quadratic form in a transformed variable set because the 10×10 matrix incorporating the quadratic products of the adjugate variables is singular. As usual, the redundancy of the adjugate variables has to be reduced by the imposition of constraints such as those in Eq. (15.4). However, we are potentially lacking some degrees of freedom in the projection-matrix adjugate variables, so if we try to constrain all the variables, we may come up with no solutions. Thus we will find that we do not need (or cannot utilize) all seven adjugate constraints in order to obtain the canonical three rotational degrees of freedom in a rotation.

15.4.1 Least-squares solution to error-free orthographic pose problem

We now show that we can get full solutions to the error-free least-squares problem defined by the general form of Eq. (15.31) using a specific choice of the adjugate constraints in Eq. (15.4). When we impose the four constraints containing the adjugate variable q_{00}, with **lossOrtho3D2DAdj** denoting the algebraic expression Eq. (15.35) and symbols (xx, xy, ...) for the cross-covariance sums, this Mathematica program yields a list of eight candidate solutions:

theOrtho3D2DAdjSolns =
```
    Module[{eqn = lossOrtho3D2DAdj },
        Solve[ { D[eqn, q00] == 0,
          D[eqn, q11] == 0, D[eqn, q22] == 0, D[eqn, q33] == 0,
          D[eqn, q01] == 0, D[eqn, q02] == 0,  D[eqn, q03] == 0,
          D[eqn, q23] == 0, D[eqn, q13] == 0,  D[eqn, q12] == 0,
             q00 + q11 + q22 + q33 == 1,
             q00 q11 == q01 q01, q00 q22 == q02 q02, q00 q33 == q03 q03
           (* q22 q33==q23 q23,q11 q33==q13 q13,q11 q22==q12 q12 *)   },
        {q00,  q11, q22, q33, q01, q02, q03, q23, q13, q12}]]      .
```
$$\tag{15.36}$$

(The three unused constraints are commented out, retained for reference.) The resulting set of eight algebraic expressions for the q_{ij} in terms of the data sums can be tested with randomly generated projected-cloud data sets to see, first, whether they produce an adjugate matrix with a valid solution for the initial rotation, and also which of the seven constraints in Eq. (15.4) are satisfied by the resulting solutions. For exact data, four of the solutions were usually complex, and thus unusable. Four of the solutions were always real, and, strangely, exactly *one* of them always produced the quaternion (via the adjugate procedure) that was used to generate the data. For pure data, all the constraints were in fact obeyed, while for errorful data, the constraint identities that were *enforced* were always maintained, while those that were not enforced (commented out in the **theOrtho3D2DAdjSolns** expression) were no longer valid. Effectively this means that, as expected, the solutions can deviate from producing a pure orthogonal rotation matrix when the data are noisy.

Form of the candidate orthographic projection solutions. We were able to find a useful symbolic representation of the four promising solutions, drastically reducing the original several megabytes of the initial symbolic expressions. All the algebraic solutions reduce to functions of these determinants of cross-covariances of Eq. (15.18) that we used in the loss function Eq. (15.35):

$$d_0 \rightarrow \begin{bmatrix} xx & xy & xz \\ xy & yy & yz \\ xz & yz & zz \end{bmatrix},$$

$$d_{11} \rightarrow \begin{bmatrix} xy & xz & ux \\ yy & yz & uy \\ zy & zz & uz \end{bmatrix}, \quad d_{12} \rightarrow \begin{bmatrix} xz & xx & ux \\ yz & yx & uy \\ zz & zx & uz \end{bmatrix}, \quad d_{13} \rightarrow \begin{bmatrix} xx & xy & ux \\ yx & yy & uy \\ zx & zy & uz \end{bmatrix},$$

$$d_{21} \rightarrow \begin{bmatrix} xy & xz & vx \\ yy & yz & vy \\ zy & zz & vz \end{bmatrix}, \quad d_{22} \rightarrow \begin{bmatrix} xz & xx & vx \\ yz & yx & vy \\ zz & zx & vz \end{bmatrix}, \quad d_{23} \rightarrow \begin{bmatrix} xx & xy & vx \\ yx & yy & vy \\ zx & zy & vz \end{bmatrix},$$

$$\tilde{d}_{31} \rightarrow \begin{bmatrix} xx & ux & vx \\ yx & uy & vy \\ zx & uz & vz \end{bmatrix}, \quad \tilde{d}_{32} \rightarrow \begin{bmatrix} xy & ux & vx \\ yy & uy & vy \\ zy & uz & vz \end{bmatrix}, \quad \tilde{d}_{33} \rightarrow \begin{bmatrix} xz & ux & vx \\ yz & uy & vy \\ zz & uz & vz \end{bmatrix}.$$

(15.37)

Here, as before, the *self-covariance* d_0 will play a special role in overall normalization. The last row, $\tilde{d}_{3i}$, *differs in form* from the last row of Eq. (15.23), as it contains no (wx, wy, wz) terms (the [w] data element does not even appear in the problem). The explanation, as we will see explicitly in a moment, is that for error-free data, the cross product of the top two rows, which involve no w terms, must be numerically the same as the last row of Eq. (15.23) to maintain orthonormality. None of this is true if the data contain noise, but it is remarkable that for our noise-free case, the last row appears automatically as the cross product of the first two.

The four usable versions of the least-squares solutions differ by pairs of signs of square roots in the expressions for $(q_{01}, q_{02}, q_{23}, q_{13})$, so we can write the 3D pose least-squares solutions as a function of their corresponding square root signs, which we denote by s_{ij}. We can then express the four solutions in terms of the combinations of signs that distinguish one from another as $\omega(s_{01}, s_{02}, s_{23}, s_{13})$, where

$$\left.\begin{aligned} \text{soln}(1) &= \omega(+1, +1; \ +1, +1) \\ \text{soln}(2) &= \omega(+1, -1; \ +1, -1) \\ \text{soln}(3) &= \omega(-1, +1; \ -1, +1) \\ \text{soln}(4) &= \omega(-1, -1; \ -1, -1) \end{aligned}\right\}. \tag{15.38}$$

Then the explicit form of the solutions for q_{ij} in terms of the cross-covariance determinants of Eq. (15.37) can be written

$$\omega(s_{01}, s_{02}, s_{23}, s_{13}) =$$

$$
\begin{cases}
q_{00} \rightarrow \dfrac{\sqrt{d_0^2\left(d_{12}+(-d_{22}-d_{11})^2-d_{21}\right)}+d_0(d_{11}+d_{22})}{4d_0^2}, \\[2ex]
q_{11} \rightarrow \dfrac{d_0(2d_0-d_{22}+d_{11})-\sqrt{d_0^2\left(d_{12}+(d_{11}+d_{22})^2-d_{21}\right)}}{4d_0^2}, \\[2ex]
q_{22} \rightarrow -\dfrac{\sqrt{d_0^2\left(d_{12}+(d_{11}+d_{22})^2-d_{21}\right)}+d_0(-2d_0+-d_{22}+d_{11})}{4d_0^2}, \\[2ex]
q_{33} \rightarrow \dfrac{\sqrt{d_0^2\left(d_{12}+(d_{11}+d_{22})^2-d_{21}\right)}-d_0(d_{11}+d_{22})}{4d_0^2}, \\[2ex]
q_{01} \rightarrow -\dfrac{s_{01}\sqrt{2(d_0-d_{22})\left(\sqrt{d_0^2\left(d_{12}+(d_{11}+d_{22})^2-d_{21}\right)}+d_0(d_{11}+d_{22})\right)-d_0(d_{12}-d_{21})^2}}{4\sqrt{d_0^3}}, \\[2ex]
q_{02} \rightarrow -\dfrac{s_{02}\sqrt{2(d_0-d_{11})\left(\sqrt{d_0^2\left(d_{12}+(d_{11}+d_{22})^2-d_{21}\right)}+d_0(d_{11}+d_{22})\right)-d_0(d_{12}-d_{21})^2}}{4\sqrt{d_0^3}}, \\[2ex]
q_{03} \rightarrow \dfrac{-d_{12}+d_{21}}{4d_0}, \\[2ex]
q_{23} \rightarrow \dfrac{d_{21}}{2d_0} - \dfrac{s_{23}\sqrt{2(d_0-d_{22})\left(\sqrt{d_0^2\left(d_{12}+(d_{11}+d_{22})^2-d_{21}\right)}+d_0(d_{11}+d_{22})\right)-d_0(d_{12}-d_{21})^2}}{4\sqrt{d_0^3}}, \\[2ex]
q_{13} \rightarrow \dfrac{d_{13}}{2d_0} + \dfrac{s_{13}\sqrt{2(d_0-d_{11})\left(\sqrt{d_0^2\left(d_{12}+(d_{22}+d_{11})^2-d_{21}\right)}+d_0(d_{11}+d_{22})\right)+d_0(d_{12}-d_{21})^2}}{4\sqrt{d_0^3}}, \\[2ex]
q_{12} \rightarrow \dfrac{d_{12}+d_{21}}{4d_0}.
\end{cases}
\tag{15.39}
$$

With careful examination and experimentation, the daunting form of Eq. (15.39) reveals some remarkable structure. If one takes a list of error-free data sets and evaluates all four functions in Eq. (15.39) *against the pose loss function* Eq. (15.35), all four least-squares solutions produce a *perfect match*. This seems impossible until one carefully checks the steps, and discovers that, although the elements $(q_{01}, q_{02}, q_{23}, q_{13})$ differ among the four functions, when all are combined together to form the 2×3 *projection matrix* $P(q_{ij})$, *the differences cancel and all four produce the same projection with no square roots*.

Since only the top two lines enter into the least-squares loss formula, this is completely logical: our solution only asks to minimize those two lines, and the third line does not appear at all, and in fact if we substitute the four versions of q_{ij} solutions in Eq. (15.39) into the third line of the adjugate-parameterized rotation matrix Eq. (15.3), they are *all different*. Now we come to a procedure that remarkably takes us full circle back to the calculations for optimal matches to noisy rotation matrices in Section 12.4.3. First, since we get a perfect first two lines of the rotation matrix for all four solutions, we can simply construct the third line by taking the cross product of the two projection-matrix components. The second step is to observe that for noisy test data, the perfect success of Eq. (15.39) is deformed and, in general, we do not know exactly what process is going on. However, we soon realize that, when errors are present, what we truly need is the *ideal* orthonormal approximation to our error-free data solution. We already know how to do that from our work in previous sections on extracting adjugate vectors, and hence quaternions, using Bar-Itzhack methods.

The pose projection matrix solution and its full rotation matrix. We can now write the projection-matrix solution in terms of the cross-covariance determinants in Eq. (15.37) as follows:

$$\widetilde{P}(x, y, z; u, v) = \begin{bmatrix} \dfrac{d_{11}}{d_0} & \dfrac{d_{12}}{d_0} & \dfrac{d_{13}}{d_0} \\[2ex] \dfrac{d_{21}}{d_0} & \dfrac{d_{22}}{d_0} & \dfrac{d_{23}}{d_0} \end{bmatrix}. \tag{15.40}$$

On any error-free data set, this projection remarkably is a perfect least-squares solution, is orthonormal, and produces a vanishing loss function to 30 orders of magnitude accuracy. We ignore the disagreements with the form of the third rotation matrix line coming from the four solutions for q_{ij}, and simply take the cross product of the two lines of the projection matrix, that is, $\widetilde{P}_1 \times \widetilde{P}_2$, and that gives the unique answer for the third row of Eq. (15.37), $(\tilde{d}_{31}, \tilde{d}_{32}, \tilde{d}_{33})$:

$$\tilde{d}_{31} \rightarrow \begin{bmatrix} xx & ux & vx \\ yx & uy & vy \\ zx & uz & vz \end{bmatrix}, \quad \tilde{d}_{32} \rightarrow \begin{bmatrix} xy & ux & vx \\ yy & uy & vy \\ zy & uz & vz \end{bmatrix}, \quad \tilde{d}_{33} \rightarrow \begin{bmatrix} xz & ux & vx \\ yz & uy & vy \\ zz & uz & vz \end{bmatrix}. \tag{15.41}$$

With this third row, our initial form for the full 3D pose rotation matrix solution thus takes the form of nine ratios of determinants as did Eq. (15.22), except the bottom line of Eq. (15.37) replaces the bottom line of Eq. (15.23):

$$\widetilde{R}(x, y, z; u, v) = \begin{bmatrix} \dfrac{d_{11}}{d_0} & \dfrac{d_{12}}{d_0} & \dfrac{d_{13}}{d_0} \\[2ex] \dfrac{d_{21}}{d_0} & \dfrac{d_{22}}{d_0} & \dfrac{d_{23}}{d_0} \\[2ex] \dfrac{\tilde{d}_{31}}{d_0} & \dfrac{\tilde{d}_{32}}{d_0} & \dfrac{\tilde{d}_{33}}{d_0} \end{bmatrix}. \tag{15.42}$$

We can also *reverse-engineer* a new version of the adjugate variables solutions found by hand-solving the least-squares optimization. We know that Eq. (15.42) corresponds with the adjugate variables via Eq. (15.3), so if we simply solve that for the q_{ij}, we find the adjugate variables directly in terms of the cross-covariance determinants that produce Eq. (15.42):

$$q_{00} = \frac{1}{4d_0}(d_0 + d_{11} + d_{22} + \tilde{d}_{33}), \qquad q_{11} = \frac{1}{4d_0}(d_0 + d_{11} - d_{22} - \tilde{d}_{33}),$$

$$q_{22} = \frac{1}{4d_0}(d_0 - d_{11} + d_{22} - \tilde{d}_{33}), \qquad q_{33} = \frac{1}{4d_0}(d_0 - d_{11} - d_{22} + \tilde{d}_{33}),$$

$$q_{01} = \frac{1}{4d_0}(-d_{23} + \tilde{d}_{32}), \qquad q_{02} = \frac{1}{4d_0}(d_{13} - \tilde{d}_{31}), \qquad q_{03} = \frac{1}{4d_0}(-d_{12} + d_{21}),$$

$$q_{23} = \frac{1}{4d_0}(d_{23} + \tilde{d}_{32}), \qquad q_{13} = \frac{1}{4d_0}(d_{13} + \tilde{d}_{31}), \qquad q_{12} = \frac{1}{4d_0}(d_{12} + d_{21}).$$

$$\tag{15.43}$$

We note that the expressions for q_{03} and q_{12} in Eq. (15.43) were already present in the original symbolic solution Eq. (15.39), and thus we conclude that, for error-free data, there are constraints similar to several we have already seen that reduce the elaborate expressions in the other terms of Eq. (15.39) to coincide with Eq. (15.43).

Proof of exact rotation for error-free data. We claim the elements of Eq. (15.42) must be identical to those of Eq. (15.22), even though they obviously do not contain the variables $[w]$. In fact, for error-free data and an orthonormal rotation matrix, the $[w]$ components of the matched test data set are superfluous; we can establish this by a variant of the proof of Eq. (15.27) as follows:

$$\tilde{d}_{31} = \det \begin{bmatrix} xx & ux & vx \\ yx & uy & vy \\ zx & uz & vz \end{bmatrix} = \det \begin{bmatrix} xx & r_{11}\,xx + r_{12}\,xy + r_{13}\,xz & r_{21}\,xx + r_{22}\,xy + r_{23}\,xz \\ yx & r_{11}\,yx + r_{12}\,yy + r_{13}\,yz & r_{21}\,yx + r_{22}\,yy + r_{23}\,yz \\ zx & r_{11}\,zx + r_{12}\,zy + r_{13}\,zz & r_{21}\,zx + r_{22}\,zy + r_{23}\,zz \end{bmatrix}$$

$$= \det \begin{bmatrix} xx & xy & xz \\ yx & yy & yz \\ zx & zy & zz \end{bmatrix} \cdot \begin{bmatrix} 1 & r_{11} & r_{21} \\ 0 & r_{12} & r_{22} \\ 0 & r_{13} & r_{23} \end{bmatrix} = (r_{12}r_{23} - r_{22}r_{13}) \det \begin{bmatrix} xx & xy & xz \\ yx & yy & yz \\ zx & zy & zz \end{bmatrix} \qquad (15.44)$$

$$= (r_{12}r_{23} - r_{13}r_{22})\, d_0 = r_{31}\, d_0 .$$

Here we used the explicit form of the cross product of the top two rows of an arbitrary orthonormal rotation matrix in the form of Eq. (15.25),

$$\begin{bmatrix} r_{11} & r_{12} & r_{13} \end{bmatrix} \times \begin{bmatrix} r_{21} & r_{22} & r_{23} \end{bmatrix} = \begin{bmatrix} r_{12}r_{23} - r_{13}r_{22} & r_{13}r_{21} - r_{11}r_{23} & r_{11}r_{22} - r_{12}r_{21} \end{bmatrix}$$
$$= \begin{bmatrix} r_{31} & r_{32} & r_{33} \end{bmatrix} , \qquad (15.45)$$

from which we see that the analogous identities hold for all the elements of the third row of Eq. (15.42). Therefore the 3D pose match solution Eq. (15.42) is identical in form to the 3D match solution Eq. (15.22). Thus, for error-free measurements, the optimal rotation can be computed with either the full 3D-cloud-to-3D-cloud match, or the 3D-cloud-to-projected-2D-image match: the two superficially different computations using different elements of the cross-covariance array C defined in Eq. (15.18) *give the same answer*:

$$R(x, y, z; u, v, w) \equiv R(x, y, z; u, v) . \qquad (15.46)$$

Remark: Since we know the 3D points in the reference data set **X**, and we know the rotation matrix $R(x, y, z; u, v)$ that produced the 2D projected points $[u_k,\ v_k]$ embodied in the image **U**, we can *recover* all the missing 3D points $[w_k]$ simply by multiplying by the rotation:

$$[u_k,\ v_k,\ w_k]^t = R(x, y, z; u, v) \cdot [x_k,\ y_k,\ z_k]^t . \qquad (15.47)$$

This enables all kinds of interesting iterative algorithms even with perspective distortion and measurement errors in the data. Our approximate but plausible starting estimate for $R(x, y, z; u, v)$ converts the next stage of the process to a full 3D-to-3D matching problem, for which very accurate closed form answers are known even for noisy data.

Correcting for errorful data. We call Eq. (15.42) our "initial form" instead of our final solution because it is a true rotation only for *error-free* data. When we use error-containing data for the pose problem, the perfect match of Eq. (15.40) and its extension to an actual 3×3 camera model matrix in Eq. (15.42) breaks down. When one allows data with measurement errors, the components of $\widetilde{R}(x, y, z; u, v)$ in Eq. (15.42) are *not even normalized to unity*, much less orthogonal. This cannot be the optimal answer for a rotation matrix orienting a noisy 2D point image into a corresponding 3D cloud scene. The matrix will still be a least-squares solution minimizing the cost function Eq. (15.32), but since it does not preserve the properties of a rotation matrix, it will not actually correspond to an optimal *rotation*, which is what we require of the pose estimation problem. Our issue here is virtually identical to the distinctions we found in Sections 12.4.1, 12.4.2, and 12.4.3 in the treatment of exact rotation matrices vs. error-containing measurements of rotation matrices. Our proposed solution is the same: we can use the approximate matrix that is given by Eq. (15.42) as a starting point in the presence of noisy data, and transform it into its closest pure rotation using the Bar-Itzhack optimization method introduced in Section 12.4.3.

Final steps of adjugate solution to the 3D:2D orthographic pose estimation problem. For completeness, we review the additional step that so far as we know produces the best available closed form pure rotation solving the 3D:2D orthographic pose estimation task. First, we examine Eq. (15.42), which in the errorful-data case effectively comes from the cross-covariance matrix appearing in the 3D cloud matching task, Eq. (15.6), and transform it into its profile matrix corresponding to Eq. (15.9), with the result

$$
M(x, y, z; u, v) =
\begin{bmatrix}
d_{11} + d_{22} + \tilde{d}_{33} & \tilde{d}_{32} - d_{23} & d_{13} - \tilde{d}_{31} & d_{21} - d_{12} \\
\tilde{d}_{32} - d_{23} & d_{11} - d_{22} - \tilde{d}_{33} & d_{21} + d_{12} & \tilde{d}_{31} + d_{13} \\
d_{13} - \tilde{d}_{31} & d_{21} + d_{12} & -d_{11} + d_{22} - \tilde{d}_{33} & d_{23} + \tilde{d}_{32} \\
d_{21} - d_{12} & \tilde{d}_{31} + d_{13} & d_{23} + \tilde{d}_{32} & -d_{11} - d_{22} + \tilde{d}_{33}
\end{bmatrix} . \quad (15.48)
$$

We then compute the maximal eigenvalue using any method we like, but we note that it is not hard to compute the analytic algebraic formula using the Cardano equations (see Appendix J). Given that eigenvalue

$$
\lambda_{\max} = \textbf{Maximal Eigenvalue } (M(x, y, z; u, v)) , \quad (15.49)
$$

we form the characteristic matrix χ by subtracting $\lambda_{\max}$,

$$
\chi(x, y, z; u, v) = [M(x, y, z; uv) - \lambda_{\max} I_4] . \quad (15.50)
$$

Recall that the critical feature is the maximal eigenvector, whose normalized value is the quaternion giving the optimal solution for the sought-for rotation matrix. As usual, we now just compute the adjugate, which, up to a normalization, will now always be four copies of the needed optimal quaternion,

$$
A(x, y, z; u, v) = -\text{Adjugate}(\chi(x, y, z; u, v)) =
\begin{bmatrix}
q_0^2 & q_0 q_1 & q_0 q_2 & q_0 q_3 \\
q_0 q_1 & q_1^2 & q_1 q_2 & q_1 q_3 \\
q_0 q_2 & q_1 q_2 & q_2^2 & q_2 q_3 \\
q_0 q_3 & q_1 q_3 & q_2 q_3 & q_3^2
\end{bmatrix} . \quad (15.51)
$$

The final answer is found by choosing a nonsingular row from the adjugate $A(x, y, z; u, v)$ for normalization to determine q_{opt}:

$$q_{opt} = (\textbf{normalize row with largest diagonal})\, (A(x, y, z; u, v)) \,. \tag{15.52}$$

We noted in previous simpler examples that, while q_{opt} can be very complicated, reassembling the quaternions into the rotation matrix itself, $R_{opt} = R(q_{opt})$, can result in a simplified final form. For this case, as in Eq. (15.15), we can skip the quaternion step entirely and go straight from the adjugate matrix to the rotation matrix

$$R_{opt}(x, y, z; u, v) = [\text{normalized map of Eq. (15.3)}](A(x, y, z; u, v)) \tag{15.53}$$

$$= R\left(q_{ij}\right) \,. \tag{15.54}$$

Comparison of accuracy. To demonstrate the behavior of the various solutions and approximate solutions to the 3D pose estimation problem, we generate sample data consisting of 100 clouds of 25 3D reference points $\mathbf{X}$ with ranges from $-1 \Rightarrow +1$, choose a random rotation matrix for each data set, delete the bottom line to produce a 3×2 projection matrix, and apply a Gaussian noise distribution to each projected 2D point with $\sigma = 0.1$. Our accuracy measure is the least-squares projection loss function Eq. (15.32). For the pose estimation case, there is no exact solution for the best rotation analogous to Eq. (15.15). In place of the latter, we use a numerical method based on `argmin` and Eq. (15.32) with quaternion arguments. The results appear to have exactly the same properties as the exact solution for the 3D matching problem, so we believe they are reliable. Again, we sort all loss values in the order of this best numerical pose solution. Then we compute the same four rotation matrices for each data set as we did for Fig. 15.3 and plot the results in Fig. 15.4. The `argmin` solution and the closest rotation solution are very close, so the closest rotation is displaced slightly upward for visibility. Plotting the *difference* of the closest rotation minus the best numerical solution (magnified $20\times$ for visibility) in green, we see that they are quite close, but that, as before, their difference can be very close to zero, but *never negative*. Thus Fig. 15.4 shows us a complete picture of the levels of accuracy expected for noisy data sets. Again note that, for *noise-free data*, all four graphs will be identical to machine accuracy.

15.4.2 Remarks on variational approaches to solving matching problems

The algebraic solution methods we have outlined here are not necessarily the most effective approach. In particular, numerical approximation methods involving `argmin` numerical searches for the numerical optimization of the loss functions that we have considered here can be quite accurate, and can efficiently enforce constraints on either the quaternion variables themselves, or on the adjugate variables. The latter have the advantage of course that the singular domains of the quaternion determination can be explicitly avoided, and the constraints essentially function as Lagrange multipliers if the optimization is considered as a dynamical system. Similarly, trainable neural networks can function more or less equivalently to numerical search optimization, with the particular advantage that, if successfully trained, the expensive search process in an `argmin` implementation, which is repeated in every new data instance, is *skipped in every later application* of a pretrained neural network. However, there is a major *caveat*: as argued in Chapter 12, quaternions *cannot be represented as single functions*, but must be encoded as a set of charts of an atlas covering the nontrivial quaternion manifold $\mathbf{S}^3$. That encoding is most effectively represented as the *adjugate matrix*, and neural nets with adjugate output encoding

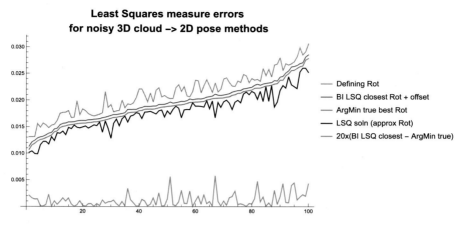

FIGURE 15.4

Comparison of solutions to the 3D orthographic projection pose estimation problem. Here we use 100 data sets lying in a $-1 \leqslant x \leqslant +1$ cube with error standard deviation $\sigma = 0.1$. The Bar-Itzhack correction to the algebraic error-free LSQ solution (in black) is shown in blue, displaced slightly, because it is almost indistinguishable from the "gold standard" best-practice `argmin` solution in red.

the $\mathbf{S}^3$ atlas can then function, in principle, to the same level of accuracy as an `argmin` implementation in which the variation of the quaternion parameters is constrained to move only on the manifold itself, without requiring an atlas of separate charts.

15.5 3D pose estimation with perspective projection

To conclude our tour through the tasks of extraction of rotation data from matched clouds, we turn now to the most challenging problem, the full-perspective projection case. We take as usual a set of K 3D reference points $\mathbf{X}$ centered at the origin, but now assumed to be rotated by some rotation matrix $R(q)$ and translated to a displaced center $\mathbf{C} = [0, 0, c]^t$, with $c \gg 1$. Finally, these rotated points are projected to the origin through the $z = 1$ plane, where they form a possibly noisy image of measured 2D data $\mathbf{U}$, as shown in Fig. 15.5.

We take as our standard perspective projection loss function

$$S_{\text{3D perspective}} = \sum_{k=1}^{K} \left\| \frac{P(q) \cdot \mathbf{x}_k}{D(q) \cdot \mathbf{x}_k + c} - \mathbf{u}_k \right\|^2, \tag{15.55}$$

where q is a quaternion or a quaternion adjugate, $P(q)$ is the 2×3 top two rows of the target rotation matrix, and $D(q)$ is the bottom row performing the perspective division to the $z = 1$ plane along the line of sight with cloud-center displacement c. The optimization problem defined by the loss Eq. (15.55), even with introduced error, can in principle be solved to high accuracy using available numerical `argmin:minimize` utilities. We have also seen that, for any data, using the closed form

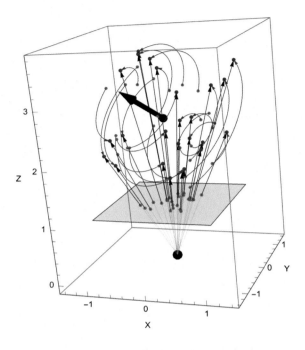

FIGURE 15.5

Geometry of a standard 3D perspective projection. The reference cloud input points $\{\mathbf{X}_k = [x_k, y_k, z_k]\}$ are shown in red. The axis of rotation $\hat{\mathbf{n}}$ of the unknown rotation $R(\theta, \hat{\mathbf{n}})$ is represented by a black arrow. This rotation, which is to be discovered by computation, transforms the red points to the blue points traveling along the arcs. The other input data are the image points $\{\mathbf{u}_k = [u_k, v_k]\}$, which are projected from a pinhole camera at the origin to the plane at $z = 1$.

determinant-ratio rotation matrix for noise-free orthogonal projection data, plus the correction to the closest pure rotation, compares very favorably with the standard numerical `argmin` solution. The open question is therefore whether the additional insights we gained from the adjugate variable approach to the 3D:3D least-squares loss matching problem and the least-squares loss 3D:2D orthographic pose estimation problem will be applicable to the perspective projection task.

The first thing that is clear is that although the classic perspective division-based expression in the *image space* loss function of Eq. (15.55) is apparently the only reasonable way to construct simulated pinhole camera image data as symbolized in Fig. 15.5, it is also apparently computationally intractable. The division effectively forces an infinite series upon us that cannot be reasonably solved in the same fashion that we used in Sections 15.3 and 15.4.

However, remarkably, we have another option: there are division-free loss functions based on *the same perspective input data* that coincide at the solution point with the optimal solution of the loss of Eq. (15.55). An elegant example is the loss introduced by Lu et al. (2000), which we will refer to hereafter as the "LHM" method, that replaces the image-space loss function of Eq. (15.55) and its division with an *object space* difference expression using equivalent penalty distances in the loss function that are obtained *in the cloud* instead of in the image plane, and avoid divisions altogether.

LHM object space projection geometry. To construct the LHM object space loss, we start by reviewing the essence of the classic image space loss. We start as usual with a unit-size reference cloud $\mathbf{X} = \{\mathbf{x}_k = [x_k, y_k, z_k]\}$ and a rotation R. However, here already we have a new feature: in order to get an unambiguous image in the $z = 1$ film plane, one traditionally takes the cloud reference data to have the center of mass at the origin of some coordinate system, applies the rotation, and then sends the cloud out to a distance $c \gg z = 1$. This provides a clear projection path along the lines to the points in the cloud centered at $C = [0, 0, c]^t$, to which we can add measurement-like errors to the points, either in 3D space or in the image plane. Then the result in the $z = 1$ image plane produces our measured 2D data set $\mathbf{U} = \{\mathbf{u}_k = [u_k, v_k]\}$. A division by the z value of each 3D point in the rotated reference cloud implements the projection to the $z = 1$ plane producing a planar 3D image cloud that we understand as follows:

$$\mathbf{v}_k = \frac{R(q) \cdot \mathbf{x}_k + \mathbf{C}}{D(q) \cdot \mathbf{x}_k + c} + \langle 2D\ noise \rangle = [u_k, v_k, 1]\ . \tag{15.56}$$

That is, we extend the concept of the 2D projected image $\mathbf{u}$ to a specific embedding $\mathbf{v}$ in 3D space. Here $D_i(q) = R_{3i}(q)$ is the third row of the rotation matrix, and we examine the rays from the origin through the target point $\mathbf{x}'_k = R \cdot \mathbf{x}_k + \mathbf{C}$ compared to the rays through the (typically noise-affected) measured image points $\mathbf{u}_k$ in the $z = 1$ image plane, as illustrated in Fig. 15.6. We see that if we construct a unit vector

$$\hat{\mathbf{v}}_k = \frac{[u_k, v_k, 1]}{\sqrt{1 + u_k^2 + v_k^2}} \tag{15.57}$$

from the origin through the 3D projected data point $\mathbf{v} = [u, v, 1]$ in the $z = 1$ plane, we can compare the sample ray $\mathbf{x}' = R \cdot \mathbf{x} + \mathbf{C}$ with its corresponding ray projected onto $\hat{\mathbf{v}}$ (remember, we assume the reference cloud and the image points have predetermined correspondences for k in an ordered list of K such pairs). The formula for this decomposition is just the 3D extension of the projection we already studied in the 2D LHM projection problem:

$$\mathbf{x} = \mathbf{x}_\| + \mathbf{x}_\perp = \hat{\mathbf{v}}(\hat{\mathbf{v}} \cdot \mathbf{x}) + \mathbf{x}_\perp \tag{15.58}$$

so that $\mathbf{x}_\perp$, which is effectively a Gram–Schmidt decomposition, takes the form

$$\mathbf{x}_\perp = \mathbf{x} - \hat{\mathbf{v}}(\hat{\mathbf{v}} \cdot \mathbf{x}) \tag{15.59}$$

and is explicitly perpendicular to $\hat{\mathbf{v}}$,

$$\hat{\mathbf{v}} \cdot \mathbf{x}_\perp = 0\ .$$

Fig. 15.6 illustrates schematically the object space approach to the coordinate system of the cloud projection, in contrast to the image space method. Adopting this coordinate system to analyze the loss measure of the projected image, we have a 3D way rather than a 2D way of measuring the closeness of the measured image to the ideal image using the loss function

$$S_{3D\ LHM\ perspective}(q) = \sum_{k=1}^{K} \left\| (R(q) \cdot \mathbf{x}_k + \mathbf{C}) - \hat{\mathbf{v}}_k \left(\hat{\mathbf{v}}_k \cdot (R(q) \cdot \mathbf{x}_k + \mathbf{C}) \right) \right\|^2\ . \tag{15.60}$$

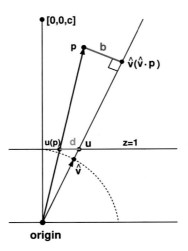

FIGURE 15.6

Geometry of the LHM perspective loss function. We select a single reference cloud input point $\mathbf{X} = [x, y, z]$ and an unknown rotation $R(\theta, \hat{\mathbf{n}})$ acting to give the rotated point $\mathbf{p} = [x', y', z'] = R(\theta, \hat{\mathbf{n}}) \cdot [x, y, z]$. For simplicity, we display the geometry in 2D graphics without loss of generality: the rotated scene point $\mathbf{p}$ produces a projected image $\mathbf{u}(p)$ in the $z = 1$ plane, varying with the rotation parameters being applied to the corresponding reference point. The usual perspective loss function compares the search varying point $\mathbf{u}(p)$ with the noisy measured image point $\mathbf{u} = [u, v, 1]$ in the $z = 1$ plane, and performs the sum of their squared differences as projected to that plane. The LHM object space loss measure, in contrast, performs a projection from $\mathbf{p}$ onto the ray $\hat{\mathbf{v}} = [u, v, 1]/\sqrt{1 + u^2 + v^2}$ through the measured image point, resulting in the *object space measure* given by the perpendicular distance $\|\mathbf{b}\|$ from the tip of $\mathbf{p}$ (embodying the unknown rotation acting on $\mathbf{X}$) to the $\hat{\mathbf{v}}$ ray. Minimizing the LHM object space distances is effectively indistinguishable from minimizing the $z = 1$ plane $\mathbf{u}$ distances.

One can use either the quaternion parameterization or the corresponding quaternion adjugate variable form to parameterize the rotation $R(q)$ in Eq. (15.60). Both forms provide a successful argmin:minimize solution when all the appropriate constraints are included in the problem definition, though increasing the working precision may be required to achieve the same accuracy when using the adjugate variables due to the computational stress of the extra constraints. However, the quaternion adjugate form has the advantage of reducing the degree of the least-squares loss function, so our next step in attempting an algebraic solution is to expand the loss explicitly in the adjugate variables. After expansion and collection of coefficients of the q_{ij} variables, we find this abstract form:

$$S_{\text{3D LHM perspective}}(q_{ij}) =$$

$$a_{01} + a_{02}\,q_{12} + a_{03}\,q_{12}{}^2 + a_{04}\,q_{13} + a_{05}\,q_{12}\,q_{13} + a_{06}\,q_{13}{}^2 + a_{07}\,q_{23} + a_{08}\,q_{12}\,q_{23} +$$

$$a_{09}\,q_{13}\,q_{23} + a_{10}\,q_{23}{}^2 + a_{11}\,q_{03} + a_{12}\,q_{03}\,q_{12} + a_{13}\,q_{03}\,q_{13} + a_{14}\,q_{03}\,q_{23} + a_{15}\,q_{03}{}^2 + a_{16}\,q_{02} +$$

$$a_{17}\,q_{02}\,q_{12} + a_{18}\,q_{02}\,q_{13} + a_{19}\,q_{02}\,q_{23} + a_{20}\,q_{02}\,q_{03} + a_{21}\,q_{02}{}^2 + a_{22}\,q_{01} + a_{23}\,q_{01}\,q_{12} + a_{24}\,q_{01}\,q_{13} +$$

$$a_{25}\,q_{01}\,q_{23} + a_{26}\,q_{01}\,q_{03} + a_{27}\,q_{01}\,q_{02} + a_{28}\,q_{01}{}^2 + a_{29}\,q_{33} + a_{30}\,q_{12}\,q_{33} + a_{31}\,q_{13}\,q_{33} + a_{32}\,q_{23}\,q_{33} +$$

$$a_{33}\,q_{03}\,q_{33} + a_{34}\,q_{02}\,q_{33} + a_{35}\,q_{01}\,q_{33} + a_{36}\,q_{33}{}^2 + a_{37}\,q_{22} + a_{38}\,q_{12}\,q_{22} + a_{39}\,q_{13}\,q_{22} + a_{40}\,q_{22}\,q_{23} +$$
$$a_{41}\,q_{03}\,q_{22} + a_{42}\,q_{02}\,q_{22} + a_{43}\,q_{01}\,q_{22} + a_{44}\,q_{22}\,q_{33} + a_{45}\,q_{22}{}^2 + a_{46}\,q_{11} + a_{47}\,q_{11}\,q_{12} + a_{48}\,q_{11}\,q_{13} +$$
$$a_{49}\,q_{11}\,q_{23} + a_{50}\,q_{03}\,q_{11} + a_{51}\,q_{02}\,q_{11} + a_{52}\,q_{01}\,q_{11} + a_{53}\,q_{11}\,q_{33} + a_{54}\,q_{11}\,q_{22} + a_{55}\,q_{11}{}^2 + a_{56}\,q_{00} +$$
$$a_{57}\,q_{00}\,q_{12} + a_{58}\,q_{00}\,q_{13} + a_{59}\,q_{00}\,q_{23} + a_{60}\,q_{00}\,q_{03} + a_{61}\,q_{00}\,q_{02} + a_{62}\,q_{00}\,q_{01} +$$
$$a_{63}\,q_{00}\,q_{33} + a_{64}\,q_{00}\,q_{22} + a_{65}\,q_{00}\,q_{11} + a_{66}\,q_{00}{}^2 . \tag{15.61}$$

The 66 coefficients a_{ij} in Eq. (15.61) correspond to the counts of one constant, 10 single powers of q_{ij}, and $10*11/2 = 55$ double powers, $1 + 10 + 55 = 66$. Each coefficient is a function of a sum over all the cloud points of certain functions of $(x, y, z; u, v; c)$. The first eight, for example, are, in terms of individual k-th elements,

$$a_{01} = \frac{c^2\left(u^2 + v^2\right)}{u^2 + v^2 + 1} , \qquad\qquad a_{02} = -\frac{4c(uy + vx)}{u^2 + v^2 + 1} ,$$
$$a_{03} = \frac{4\left(\left(u^2 + 1\right)x^2 - 2uvxy + \left(v^2 + 1\right)y^2\right)}{u^2 + v^2 + 1} , \qquad a_{04} = \frac{4c\left(u^2x - uz + v^2x\right)}{u^2 + v^2 + 1} ,$$
$$a_{05} = -\frac{8\left(vx(uz + x) + y(ux - z) + v^2(-y)z\right)}{u^2 + v^2 + 1} , \qquad a_{06} = \frac{4\left(u^2x^2 - 2uxz + v^2\left(x^2 + z^2\right) + z^2\right)}{u^2 + v^2 + 1} , \tag{15.62}$$
$$a_{07} = \frac{4c\left(u^2y + v(vy - z)\right)}{u^2 + v^2 + 1} , \qquad\qquad a_{08} = \frac{8\left(u^2xz - vy(uz + x) - uy^2 + xz\right)}{u^2 + v^2 + 1} ,$$
$$\cdots = \cdots , \qquad\qquad\qquad\qquad \cdots = \cdots .$$

More useful for implementation, we convert these into specific sums over k that appear in the full loss function, where we redefine the notation for the LHM context as

$$\mathrm{uy} = \sum_{k=1}^{K} \frac{u_k y_k}{u_k^2 + v_k^2 + 1} , \qquad \mathrm{uxy} = \sum_{k=1}^{K} \frac{u_k x_k y_k}{u_k^2 + v_k^2 + 1} , \qquad \mathrm{uvxy} = \sum_{k=1}^{K} \frac{u_k v_k x_k y_k}{u_k^2 + v_k^2 + 1} , \tag{15.63}$$

and so on, to symbolize the independent extended cross-covariances that appear in the object space perspective projection loss. With this notation, we find expressions such as these that appear, e.g., in a typical program implementation:

$$a_{01} = c**2 * (uu + vv) ,$$
$$a_{02} = -4 * c * (uy + vx) ,$$
$$a_{03} = 4 * (uuxx - 2 * uvxy + vvyy + xx + yy) ,$$
$$a_{04} = 4 * c * (uux - uz + vvx) ,$$
$$a_{05} = -8 * (uvxz + uxy - vvyz + vxx - yz) ,$$
$$a_{06} = 4 * (uuxx - 2 * uxz + vvxx + vvzz + zz) , \tag{15.64}$$
$$a_{07} = 4 * c * (uuy + vvy - vz) ,$$
$$a_{08} = 8 * (uuxz - uvyz - uyy - vxy + xz) ,$$
$$a_{09} = 8 * (uuxy - uvzz - uyz + vvxy - vxz) ,$$
$$a_{10} = 4 * (uuyy + uuzz + vvyy - 2 * vyz + zz) ,$$
$$\cdots = \cdots .$$

A full list is also supplied in Appendix H. The individual numerical a_k coefficients can also be computed from their expressions in terms of $(x, y, z; u, v; c)$ to convert Eq. (15.62) to Eq. (15.64) using a program such as this in Mathematica:

```
getNumbersFromCoeffListExprs [coeffExprList66_, cloud_, proj_, cc_] :=
    Module[{npts = Length[cloud], xx, yy, zz, uu, vv },
        Sum[{xx, yy, zz} = cloud[[k]]; {uu, vv} = proj[[k]];
            coeffExprList66 /. {x -> xx, y -> yy, z -> zz, u -> uu, v -> vv, c -> cc},
        {k, 1, npts}] ] .
```

The LHM object space loss function Eq. (15.60) produces basically the same equilibrium point for the minimizing rotation R_{opt} as Eq. (15.55), except that now there is *no division* involving the rotation matrix in the loss function, so the computation becomes potentially much more tractable, as we shall see shortly, when we actually work on the algebraic solution to the minimization of the generic object-space loss function Eq. (15.61).

The alternative object-space variants of the LHM perspective geometry. As we already explored in Chapter 14, the geometry of Fig. 15.2 implies that we can approach the object-space loss method in alternative ways. The $\boxed{\textit{Z-prime loss}}$ function simply moves the denominator of the standard perspective loss function Eq. (15.55) over to multiply the $(z = 1)$-plane image point $\mathbf{v} = [u, v, 1]^{\text{t}}$, producing a perfectly acceptable alternative loss function, with argument linear in the rotation matrix. This approach measures the perspective loss in the plane from which the data were projected, as shown in Fig. 15.7, and takes this explicit alternative form to consider alongside Eq. (15.60):

$$\left. \begin{aligned} S_{\text{3D Z-prime perspective}}(q) &= \sum_{k=1}^{K} \| (R(q) \cdot \mathbf{x}_k + \mathbf{C}) - \mathbf{v}\,(D(q) \cdot \mathbf{x}_k + c) \|^2 \\ &= \sum_{k=1}^{K} \| (P(q) \cdot \mathbf{x}_k) - \mathbf{u}\,(D(q) \cdot \mathbf{x}_k + c) \|^2 \end{aligned} \right\}, \tag{15.65}$$

where $\mathbf{v} = [u, v, 1]^{\text{t}}$ and $\mathbf{u} = [u, v]^{\text{t}}$. As in the 2D case, the image-plane coordinate cancels out, reducing the dimension of the loss expression by one. This results in some simplifications in the cross-covariance coefficients embedded in $a_{ij}(xyz|uv|c)$. The corresponding loss formula still contains all of the multiple powers of the measured variables occurring in the LHM variant, and its abstract form in terms of a_{ij} is identical to Eq. (15.61). The two expressions therefore *share* the same algebraic solutions for the adjugate quaternions $q_{mn}(a_{ij})$. Thus the algebraic solutions for both the LHM loss and the Z-prime loss function are *identical functions* of a_{ij}, and differ only in the map to the generalized cross-covariance expressions $a_{ij}(xyz|uv|c)$.

The alternative $\boxed{\textit{Radial loss}}$ with measure labeled by $\mathbf{r}$ in Fig. 15.2 is also possible, but in fact adds complexity to the LHM Gram-Schmidt loss measure, while the Z-prime loss reduces the complexity, so we will focus here on the basic properties and commonalities of the LHM loss and the Z-prime loss.

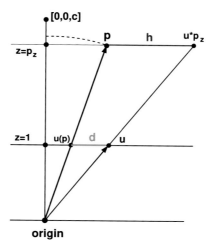

FIGURE 15.7

Geometry of the Z-prime perspective loss function. We select a single reference cloud input point $\mathbf{X} = [x, y, z]$ and an unknown rotation $R(\theta, \hat{\mathbf{n}})$ acting to give the rotated point $\mathbf{p} = [x', y', z'] = R(\theta, \hat{\mathbf{n}}) \cdot [x, y, z]$. Then the Z-prime horizontal object-space loss $\mathbf{h} = \mathbf{p} - \mathbf{u}p_z$, in blue, is a scaled version of the standard image-plane based loss $d = \|\mathbf{u} - \mathbf{u}(\mathbf{p})\|$, in green.

The $a_{ij}(x, y, z|u, v|c)$ cross-covariance expressions for the Z-prime loss are *different* from the LHM expressions, shorter in many cases, and lacking the factor $1/(1 + u^2 + v^2)$ in the cross-covariance sums. In particular,

$$
\begin{aligned}
a_{01} &= c^2\left(u^2 + v^2\right), & a_{02} &= -4c(uy + vx), \\
a_{03} &= 4\left(x^2 + y^2\right), & a_{04} &= 4c\left(u^2x - uz + v^2x\right), \\
a_{05} &= -8\left(uxy + vx^2 - yz\right), & a_{06} &= 4\left(u^2x^2 - 2uxz + v^2x^2 + z^2\right), \\
a_{07} &= 4c\left(u^2y + v^2y - vz\right), & a_{08} &= -8\left(uy^2 + vxy - xz\right), \\
\cdots &= \cdots, & \cdots &= \cdots.
\end{aligned}
\tag{15.66}
$$

These expressions correspond to specific sums over k that appear in the full loss function, which we redefine for the Z-prime optimization problem as in Eq. (15.63) but lacking the denominator:

$$
\mathrm{uy} = \sum_{k=1}^{K} u_k y_k, \qquad \mathrm{uxy} = \sum_{k=1}^{K} u_k x_k y_k, \qquad \mathrm{uvxy} = \sum_{k=1}^{K} u_k v_k x_k y_k,
\tag{15.67}
$$

and so on.

With this notation, we find expressions such as these that appear in actual computation:

$$a_{01} = c ** 2 * (uu + vv) ,$$
$$a_{02} = -4 * c * (uy + vx) ,$$
$$a_{03} = 4 * (xx + yy) ,$$
$$a_{04} = 4 * c * (uux - uz + vvx) ,$$
$$a_{05} = -8 * (uxy + vxx - yz) ,$$
$$a_{06} = 4 * (uuxx - 2 * uxz + vvxx + zz) , \qquad (15.68)$$
$$a_{07} = 4 * c * (uuy + vvy - vz) ,$$
$$a_{08} = -8 * (uyy + vxy - xz) ,$$
$$a_{09} = 8 * (uuxy - uyz + vvxy - vxz) ,$$
$$a_{10} = 4 * (uuyy + vvyy - 2 * vyz + zz) ,$$
$$\cdots = \cdots .$$

A full list is also supplied in Appendix H.

15.5.1 The exact perspective solution

Starting from the abstract form of the full object space least-squares loss function for perspective projection Eq. (15.61), we now claim that we can employ the quaternion adjugate representation to *exactly solve* for the desired rotation that aligns the reference points to the error-free perspective projection data. The starting point for any such solution process is to compute the derivatives of Eq. (15.61) with respect to the adjugate data $(q_{00}, q_{11}, q_{22}, q_{33}, q_{01}, q_{02}, q_{03}, q_{23}, q_{13}, q_{12})$, and impose appropriate constraints from the set Eq. (15.4). This involves subtle choices and some luck: the obvious choice is to require all seven of the constraints listed in Eq. (15.4) in addition to the vanishing of the 10 derivatives of Eq. (15.61):

$$\frac{\partial S(q_{ij})}{\partial q_{00}} = 0 , \quad \frac{\partial S(q_{ij})}{\partial q_{11}} = 0 , \quad \frac{\partial S(q_{ij})}{\partial q_{22}} = 0 , \quad \frac{\partial S(q_{ij})}{\partial q_{33}} = 0 , \quad \frac{\partial S(q_{ij})}{\partial q_{01}} = 0 ,$$

$$(15.69)$$

$$\frac{\partial S(q_{ij})}{\partial q_{02}} = 0 , \quad \frac{\partial S(q_{ij})}{\partial q_{03}} = 0 , \quad \frac{\partial S(q_{ij})}{\partial q_{23}} = 0 , \quad \frac{\partial S(q_{ij})}{\partial q_{13}} = 0 , \quad \frac{\partial S(q_{ij})}{\partial q_{12}} = 0 .$$

This fails to return an answer of any kind in our current Mathematica machine algebra implementation. Other reasonable guesses such as using the first four constraints, which worked in the orthographic projection case, also fail to return any solution. However, experimenting fearlessly with all different kinds of combinations of vanishing derivatives and constraints, a pattern of partial solutions emerges, indicating that the polynomial solution algorithms have a potential path to success. The one that actually works is to use exactly 10 equations in the solution argument, simply substituting the fundamental *single quaternion constraint* $q_{00} + q_{11} + q_{22} + q_{33} = 1$, which restricts the solution to a sort of cross-section with an $\mathbf{S}^3$ subspace, for one of the vanishing derivatives. We warn ourselves as usual that this set of partial derivative conditions is highly suspect mathematically due to the seven constraints on the adjugate quaternion manifold, and yet it seems to be sufficient to imply the correct constraint space in the special case of error-free data. The simplest working choice is just to omit the $\partial S/\partial q_{00} = 0$

derivative. The successful calculation takes the form of the Mathematica function (with `lhmLossQadj` being the loss function $S(q_{ij})$ of Eq. (15.61))

```
TheLHMSoln  =
  Solve[{ (* OMIT: D[lhmLossQadj, q00] == 0,*)
    q00 + q11 + q22 + q33 == 1,
    D[lhmLossQadj, q11] == 0,  D[lhmLossQadj, q22] == 0,  D[lhmLossQadj, q33] == 0,
    D[lhmLossQadj, q01] == 0,  D[lhmLossQadj, q02] == 0,  D[lhmLossQadj, q03] == 0,
    D[lhmLossQadj, q23] == 0,  D[lhmLossQadj, q13] == 0,  D[lhmLossQadj, q12] == 0},
        { q00, q11, q22, q33, q01, q02, q03, q23, q13, q12} ] [[1]] .
```

In about a minute, this returns a list of rules for all 10 q_{ij} quaternion adjugate variables in terms of complicated but computable data-determined coefficients $a_k(x, y, z|u, v|c)$ of the form

```
{q00 -> -((-(((-((-(((-((((-((((-((-(a38*a46) + a37*a47)*(a24*a47 - a23*a48)) +
              (-(a23*a46) + a22*a47)*(a39*a47 - a38*a48))*
              (a32*a39*a47^2 - a31*a40*a47^2 - a32*a38*a47*a48 +
               a30*a40*a47*a48 + a31*a38*a47*a49 - a30*a39*a47*a49) -
               (-((-(a38*a46) + a37*a47)*(a31*a47 - a30*a48)) +
               (-(a30*a46) + a29*a47)*(a39*a47 - a38*a48))*
               ((a39*a47 - a38*a48)*(a25*a47 - a23*a49) ...  )))))))))),
        q11 ->  ... ,   q22 -> ... ,   q33 -> ... ,   ...      ... , q12 -> ... } .
```

From these, we can construct the exact rotation solving the error-free perspective projection case,

$$
R_{\text{opt}}\left(q_{ij}(a_{mn})\right) =
\begin{bmatrix}
q_{00} + q_{11} - q_{22} - q_{33} & 2q_{12} - 2q_{03} & 2q_{13} + 2q_{02} \\
2q_{12} + 2q_{03} & q_{00} - q_{11} + q_{22} - q_{33} & 2q_{23} - 2q_{01} \\
2q_{13} - 2q_{02} & 2q_{23} + 2q_{01} & q_{00} - q_{11} - q_{22} + q_{33}
\end{bmatrix} . \tag{15.70}
$$

Disappointingly, if there is a condensed form like the determinant-ratio expressions Eq. (15.22) and Eq. (15.42) that we found for the 3D:3D and 3D:2D:orthographic pose estimation problems, it is very hard to find, and we do not yet have a candidate. The full expressions are so large when written as symbolic digital data that we will not attempt to present them here, with the q_{ij} sizes

ByteCount[{q00, q11, q22, q33, q01, q02, q03, q23, q13, q12}] ⇒
{15,824K, 23,735K, 27,690K, 27,690K, 29,668K, 30,656K, 31,149K, 30,955K, 30,955K, 30,955K} .

However, the actual solution expression is easily evaluated, and for error-free data, reproduces the exact initial rotation matrices used to simulate each cloud data set to machine accuracy. One presumes that the system is able to find efficient transformations to facilitate this numerical evaluation that are not obvious from the direct symbolic representation without additional insights, although we have some evidence of numerical instability that might be improved by algebraic reduction combined with customized compilation. We suspect these are issues that will eventually be resolved, but we should feel fortunate to have any algebraic form at all for the perspective projection pose estimation problem.

Implications of the coincidence of the LHM and Z-prime solutions. When we insert the two very distinct pairs of cross-covariance expressions Eq. (15.64) and Eq. (15.68) into the algebraic solution of the loss minimization problem for Eq. (15.61), we get *identical numerical answers* for the adjugate quaternions defining the pose for any reasonable numerical perspective projection data. This signifies that something very special is happening, since the expressions for $a_{ij}(xyz|uv|c)$ are extremely different. The same will be true of the object-space Radial loss. We do not entirely understand, in the first place, how the constraints are implicitly contained in the equations we are solving, but only for error-free data, nor how distinct values of the numerical a_{ij} coefficients yield identical numbers for the adjugate $q_{ij}(a_{mn})$ values. Among the things that could give additional insights would be to discover the relationship between the infinite focal length limit of our solutions and our existing very simple determinant ratio form Eq. (15.42) solving the orthographic projection problem.

Iterative solutions. Adapting our methods to the LHM iterative algorithm proposed by Lu et al. (2000) is a further option. That procedure is based on picking an initial explicit rotation matrix R(numerical), substituting it into the right-hand ($\hat{\mathbf{v}}_k$) element of Eq. (15.60), and solving that system using the RMSD methods of Chapter 8. With the resources we have developed here, we can suggest incremental improvements on that algorithm. We have seen experimentally, for example, that the orthogonal determinant-ratio solution can be a very good approximation for projected data alignment, especially when corrected to the closest true rotation via the Bar-Itzhack procedure. So the latter should be already very close to a solution to use as the initial state, and from there the iteration should converge very quickly. Alternatively, the closed form LHM solution to the noise-free case of Eq. (15.61), just described in this section, could be an even better choice for the initial guess in the LHM iterative solution procedure. It appears that which of these closed form choices for the initial LHM rotation works best can depend on the exact nature of the noise in the data; with no noise, our full-perspective solution of Eq. (15.61) makes the iteration procedure unnecessary, but for higher noise levels, the orthogonal projection solution may behave better for the initial rotation in the iteration.

Comparisons of accuracy. To demonstrate the behavior of the object space error-free perspective solution and our available alternative approaches when error is introduced into the perspective projection alignment problem, we show a series of comparative loss plots for image space errors $\sigma = \{0.001, 0.01, \text{and } 0.05\}$ in Figs. 15.8, 15.9, and 15.10, respectively. We generated simulated sample data consisting of 100 clouds of 50 3D reference points $\mathbf{X}$ with ranges in a cube from $-1 \Rightarrow +1$. Following the pattern of Eq. (15.55), we chose a random rotation matrix for each data set, deleted the bottom line to produce a 3×2 projection matrix, divided by a translation combined with the deleted bottom line of the rotation, and applied a Gaussian noise distribution to each projected 2D point with spatial error σ to produce the target points $\mathbf{U}$. We examined several different σ values to exhibit properties dependent on the data error magnitude. We use the least-squares perspective projection loss function Eq. (15.55) defined in image space, though we would get basically the same results with the object space loss Eq. (15.61). The "gold standard" best solution must be obtained from the numerical `argmin` algorithm, based on minimizing the chosen loss function, as we have no other source of a guaranteed optimal answer. Note that the `argmin` algorithm is in principle insensitive to the multi-chart quaternion problem because it is constrained to search within an *always valid* quaternion manifold. In practice, using the adjugate variables with their additional constraint terms in the `argmin` calculation may require additional working precision compared to using pure quaternions due to the extra complexity of the constraint manifold searched by the algorithm.

The different methods shown in the figures and their properties are the following:

- **Original rotation.** This is the rotation applied to the cloud before projection and incorporation of noise. It may be close to the correct rotation, but it is not determinable from the measured data and is thus uncomputable. It should generally have the largest error, but is unexpectedly quite good here.
- **"Gold standard"** `argmin` **solution.** The best solution we know of follows in this case from the application of a quaternion `argmin` procedure applied to the image space perspective projection loss function Eq. (15.55), incorporating the $q \cdot q = 1$ constraint restricting the quaternion search to legal normalized values. No 0/0 singular situations can occur when this constraint is maintained.
- **Orthographic matrix ratio solution.** Since the perspective projection data, especially near the center of the projection, are closely related to the orthographic projection data, it is entirely possible that the orthographic closed form is a viable approximation. In its pure form, for noisy data, this ratio of determinants is not actually a rotation matrix, so the loss values do not actually correspond to a solution to the optimization problem. In fact, the errors for this version are off-scale on the first two plots, and only appear in the third.
- **Corrected orthographic solution.** By applying the Bar-Itzhack nearest-pure-rotation algorithm, we can map each orthographic closed form determinant-ratio solution to a best candidate for the optimal rotation. This is then a legitimate candidate for a set of optimal real rotations, and it is astonishingly good, so it might be appealing for practical applications because it is so simple and appears to have behavior more stable than our actual LHM solution. We do not understand exactly why this seems to be the best perspective solution.
- **Exact LHM object space solution.** Our main result in this section was the computation of an exact solution to the LHM object space least-squares loss function, assuming noise-free image data. This again is not a true solution when noise is introduced, even though, as clearly shown on each graph, it is guaranteed to have the smallest possible error; this low loss is an illusion since the matrix is not even a valid rotation for noisy data.
- **Corrected LHM object space solution.** Applying the Bar-Itzhack nearest-pure-rotation algorithm, we should be able to find the closest valid rotation to each object space closed form approximate solution. We would expect this to be our best final legitimate candidate for the optimal perspective projection rotation in the presence of noisy measurement data. However, in the figures, we see that this is not true. Numerical instabilities disturb the accuracy of the correction algorithm in the presence of noise for this solution, and the Bar-Itzhack corrected *orthographic approximation* is actually a better performer. There could be many reasons for this, including the extreme complexity of the representation of the symbols needed to implement a numerical computer calculation.

The comparison step involves computing the candidate rotation matrices for the list of perspective projection data sets using each of several methods to derive corresponding lists of proposed optimal rotations aligning the **X** and **U** data sets. We sort all the collections of candidate rotations based on the `argmin` "gold standard" and plot the resulting curves. The plots can be challenging to read because there are several solutions that are so close they are hard to distinguish. The results are shown in Fig. 15.8 for very small $\sigma = 0.001$, and again in Fig. 15.9 for $\sigma = 0.01$ and Fig. 15.10 for $\sigma = 0.05$. The results vary dramatically with the imposed error, and we see that while our *exact* least-squares algebraic solution for the perspective projection problem indeed has the smallest error in all cases, this is of little use because the resulting candidate rotation matrix, unlike the `argmin` result, is not a valid rotation matrix. One would expect the Bar-Itzhack correction procedure to generate an accurate corrected matrix, but in fact the least-squares solution appears unstable under that procedure unless the error is very close to zero. Somewhat surprisingly, the *nominally deficient* closed form orthographic

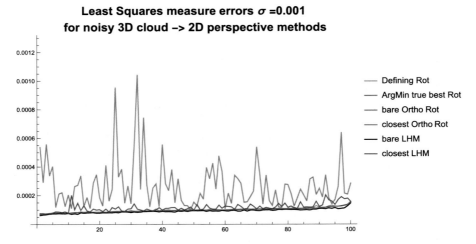

FIGURE 15.8

$\sigma=0.001$. Relative accuracy of the perspective projection pose estimation solutions for low-noise $\sigma = 0.001$ projection data. As the noise approaches zero, the solution errors approach zero. The measure here is the least-squares loss, sorted relative to the perfect `argmin` solution values, in red, obscured when other values are close; the bare orthographic losses are too large for the plot scale. In this plot, the corrected LHM solution in blue is better than the corrected orthographic approximation in green, but surprisingly worse than the original data-defining rotation in cyan.

solution applied to the *incompatible perspective data* produces a Bar-Itzhack-corrected result that is very close to the `argmin` standard. This erratic behavior of the exact object space LHM result is likely due to numerical instabilities in the closed form solution when error is introduced, as, with no error, its loss vanishes to machine precision.

15.6 Remarks on handling data with measurement errors

For the special case of error-free target data in the problem of finding the optimal rotation to align a 3D reference point cloud with (a) a rotated cloud, (b) an orthographic projected image of the rotated cloud, or (c) a perspectively projected image of the rotated cloud, we have exploited the quaternion adjugate variables to discover algebraic solutions for the desired rotations. With modest regularity assumptions, these algebraic formulas reproduce the rotation matrices that best align the reference cloud to the error-free measured target data in each case.

However, it is easy to check that, as soon as errors are introduced in the data, the solutions of Eq. (15.22), Eq. (15.42), and Eq. (15.70) *remain* the solutions of the least-squares minimization problem, but they *are no longer orthonormal rotation matrices*. The actual problem we wish to solve is find the optimal *pure rotation*, and we only know how to solve that problem analytically for the 3D:3D RMSD matching task. Nevertheless, we have some options for the more realistic errorful cases, in par-

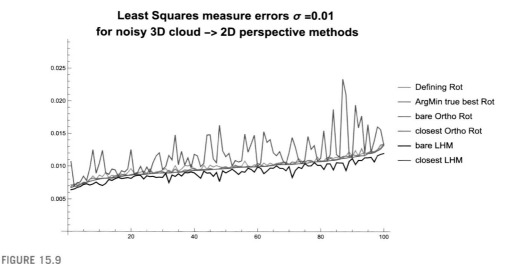

FIGURE 15.9

σ=0.01. Relative accuracy of the perspective projection pose estimation solutions for noisier $\sigma = 0.01$ projection data. When this much noise is added to the range ± 1 of the cloud data, computing instabilities start to overwhelm the very complex exact LHM solution, and the corrected orthographic approximation in green is much more accurate numerically than the algebraic LHM solution corrected by the Bar-Itzhack closest-rotation method, in blue. While the bare orthographic solution continues to be off-scale, the defining rotation is still quite good relative to the `argmin` and closest orthographic solutions. The bare LHM solution shown in black, which is not a legitimate rotation, is now clearly still the least-squares solution, lying below the lowest possible valid `argmin` solution in red.

ticular, the literature is well populated with a wide variety of numerical approximation schemes. We point out some musings related to our philosophy.

Numerically minimizing the loss functions. Existing `argmin` software such as Mathematica's `ArgMin` and Python's `scipy.optimize.minimize()` applications transparently incorporate constraints such as those we place on unit quaternions or the 10 quaternion adjugate variables and do extremely well in practice at determining the optimal quaternions or optimal adjugates relative to a given loss function, although adjustments in the working precision may be required to get matching results as the number of constraints increases. The advantage of the constrained `argmin` process is that it is immune to the multi-chart manifold problem because it is a *search constrained to a valid manifold*. On the other hand, a major drawback of the `argmin` method is that, unlike an algebraic function, the algorithm must repeatedly perform an exhaustive search through the parameter space without ever learning anything about the structure of the problem. All the detailed work of finding the parameters for each novel data set has to be redone, and the brute force search must be repeated in full, each and every time. For large problems, or problems depending on speed, the `argmin` method is often not feasible.

Trained neural networks. Some elementary demonstrations of neural network applications to quaternion-related rotation extraction problems were presented in Chapter 13. There we tested and

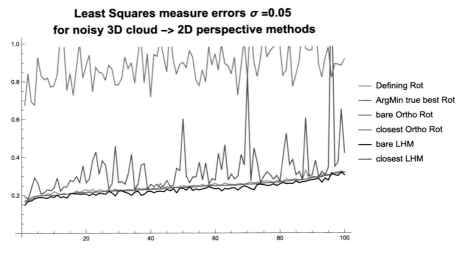

Least Squares measure errors σ =0.05 for noisy 3D cloud –> 2D perspective methods

— Defining Rot
— ArgMin true best Rot
— bare Ortho Rot
— closest Ortho Rot
— bare LHM
— closest LHM

FIGURE 15.10

σ=0.05. Relative accuracy of the perspective projection pose estimation solutions for quite noisy $\sigma = 0.05$ projection data relative to cloud range ± 1. The badly fitting raw orthographic approximation is now visible, but its corrected matrix remains better than the corrected symbolic LHM least-squares solution, probably due to numerical instabilities. In fact, the closest orthographic solution now overwrites the `argmin` solution at this scale, and the bare LHM least-squares solution again lies below the optimal solutions, as expected.

verified the applicability of the result of Hornik et al. (1989), that neural nets are intrinsically capable of modeling well-defined functions to any desired accuracy, but serious problems arise when one tries to train a single output to an object like a quaternion that lives on a nontrivial manifold. We confirmed that training to objects consistent with the required description of an output manifold was possible and effective. Clearly applying machine learning techniques to a variety of problems, when done with proper attention to avoiding confusion by training simultaneously to too many regions of a complicated manifold, can be of significant interest. Unlike the `argmin` approach, which obtains accurate answers but never learns how to make its parameter search faster, the neural net essentially "compiles" all the parameters of the abstract functional solution into a compact set of weights in what can be a fairly simple trained network. Data come into the trained network and are quickly routed through a path that simulates the desired output function. Once trained, in contrast to `argmin`, a network can find its result very quickly and never has to repeat unnecessary work.

Improving exact optimal solutions. For the 3D matching problem, we know the exact numerical solution, and can use SVD, the matrix-square-root method, or the optimal eigenvalue method computed either numerically or with the Cardano-based algebraic solution. For all three cases, however, we have found solutions that are exact for error-free data generation and measurement. In the absence of extreme errors, these solutions, though no longer true rotation matrices, can still be quite close to the desired optimal result. What we have advocated throughout is that these approximate rotation matrices can be refined into *pure* rotation matrices by finding the *closest pure rotation matrix* to the flawed exact solution formula. The closest-rotation algorithm falls into our domain serendipitously as essentially a corollary of *any* of the methods that solve the 3D:3D RMSD matching problem. If one simply replaces the cross-covariance matrix E in the RMSD problem by the transpose of the attempted exact rotation, $E \to \widetilde{R}^{\mathrm{t}}$, the exact rotation R_{opt} that emerges has the smallest Frobenius difference of any possible pure rotation, and thus R_{opt} is the closest pure rotation to the approximation $\widetilde{R}$.

Quaternion proteomics

4

Quaternion methods of representing and analyzing data are applicable particularly to representing, analyzing, and displaying 3D orientation frames and their interrelations. The chapter and appendix related to this part focus on studying protein geometry from a quaternion-based viewpoint. The fundamental fact of proteomics is that the protein backbone structure is completely described by the 3D positions and orientations of the component residues; the additional features of side-chain interactions are also important, but we will not attempt to deal with that here. Our goal is to describe and encode the fundamental spatial orientation parameters of the C_α-based residue orientations for entire proteins: understanding this geometry using quaternions is potentially a fertile ground for developing novel insights into protein structure. In Chapter 16, we describe a thorough investigation of the characterization of a protein as a list of residue orientations, exploring the diverse ways in which these orientations can be studied using quaternions (Hanson and Thakur, 2012). In Appendix K, we include a detailed study of the relationship between our quaternion-based methods for encoding protein residue orientation features and the traditional Ramachandran plot approach.

Quaternion protein frame maps[☆]

16.1 Quaternion maps of global protein structure

We now explore a family of global visualization methods for exploiting quaternion maps of intrinsic orientation frames assigned to amino acids of a protein. The advantage of quaternion maps is that a single quaternion point embodies the full three-degree-of-freedom transformation from the identity frame triad in 3D to an arbitrary frame triad; therefore, a quaternion frame representation is much simpler than the usual frame representation using a triple of orthogonal 3D vectors, and simultaneously it is much richer than the Ramachandran plot, which in addition to having only two degrees of freedom, can only represent the relative orientations of immediately neighboring residues. Quaternions naturally expose global similarities among all residues in a protein complex, no matter how near or how distant, and can easily be extended across component proteins in multi-part structures.

Our canonical methods for visually representing quaternion values as geometric points in space superficially resemble geographic maps of the world globe, but the distinction is profound: while similar problems are addressed by the relation between a flat globe map such as a Mercator projection and the actual spherical surface of the Earth, the Mercator-to-Globe relationship is identical in dimension and produces identical local representations in a sufficiently small neighborhood (your town's map is flat for all practical purposes). There is no local correspondence between a quaternion and a frame triad: a quaternion is a point in four Euclidean dimensions, constrained to move inside a particular three-degree-of-freedom spherical space, while a frame triad contains nine separate components constrained to have the properties of a 3×3 orthogonal matrix, a completely distinct embodiment of the three-degree-of-freedom 3D orientation system. The correspondence also has a deep mathematical context: one of the greatest geometric achievements of the 19th century was the discovery of the quadratic form, based on quaternions, that embodies an exact map from a quaternion point to the nine elements of an arbitrary 3×3 orthogonal rotation matrix (an arbitrary 3D frame triad in our context), together with the reverse two-fold ambiguous mapping from any such frame matrix to a quaternion point.

Here we investigate the details of these mappings as they can be applied to reveal properties of the spatial orientations of protein systems. Section 16.2 reviews previous work in this area. Section 16.3 outlines the mathematical and geometrical properties of quaternions that we will be exploiting. Two classes of quaternion visualization methods are provided, one based on a visual geometric context ("geometric view"), the other based on parallel coordinates and some innovative quaternion-driven variants ("coordinate view"). Section 16.6 provides numerous intuition-building examples of quaternion frame methods applied first to ideal mathematical curves and then to idealized spline curves used

---

☆ This chapter follows closely the treatment in Hanson and Thakur (2012).

Visualizing More Quaternions. https://doi.org/10.1016/B978-0-32-399202-2.00027-7

traditionally to represent a high-level protein structure. Finally, in Section 16.7, we illustrate applications to discrete frames given by the atomic positions of residue components in a protein's PDB file. Section 16.8 expands our scope to a variety of protein data domains and applications, including in particular a treatment of the orientation variations present in the statistical distributions of NMR data. Section 16.9 summarizes the spectrum of tools that can be applied to studies of quaternion maps, while Appendix K is devoted to a detailed pedagogical study of the relationship between Ramachandran plots and our quaternion maps. In summary, quaternion maps have the potential to expose novel properties and features of protein geometry, with particular applicability to questions of global overall structure.

16.2 Related work

While quaternions have been employed extensively to encode molecular orientations (see Karney (2007); Kearsley (1989); Coutsias et al. (2004); Mackay (1984); Horn (1987)), and have also been applied to RNA (see Magarshak (1993)), applications of quaternions to protein structures have been limited in scope (see, for example, Albrecht et al. (1996); Dunker et al. (1998); Coutsias et al. (2004); Siminovitch (1997a,b); Quine (1999); Srinivasan et al. (1993); Kneller and Calligari (2006)). The most widely used approach to analyzing orientations of amino acid residues is the classic work of Ramachandran et al. (1963) and Branden and Tooze (1999), which encodes only local information about orientation angles, although alternative orientation visualization methods have been proposed, e.g., by Bojovic et al. (2011). Topics such as quaternion derivatives have been explored by Robert Hanson et al. (2011). Other treatments, such as Morris et al. (1992), use both local and global structures to ascertain the stereochemical nature of proteins, but their visualizations of protein stereochemistry are limited to 2D plots and histograms. Our treatment here is somewhat complementary to these, focusing on visualizing global residue orientation properties directly in quaternion space, inspired by the "quaternion Gauss map" treated in Hanson and Ma (1994, 1995), Hanson (1998), and Hanson (2006).

16.3 Introduction to quaternion maps

As is our custom for chapters involving a deep exploitation of a particular quaternion-based data treatment, we review for the reader's convenience the most important quaternion formulas needed to follow the arguments, replicating intentionally some material that appears first in Chapter 2. This section begins by introducing orientation frames in quaternion form, the geometric view of quaternions, and the coordinate view of quaternions, working with displays that show a related series of quaternion points. In particular we introduce new approaches for looking at collections of quaternion points, including parallel coordinates and "star" displays, that are unique to the applications in this chapter.

16.3.1 Quaternion orientation frames

We begin by reviewing this chapter's most essential quaternion feature, the *quaternion orientation frame*. A 3D frame $\mathbf{F}$ can be specified as a triple of mutually orthogonal normalized 3-vectors, where the identity frame consists of the three columns representing the x-axis, the y-axis, and the z-axis. Any frame whatsoever can, by a theorem of Euler, be expressed as a rotation $\mathbf{R}(\theta, \hat{\mathbf{n}})$ that acts on the identity frame and rotates it about a fixed direction $\hat{\mathbf{n}}$ by some angle θ, where $\hat{\mathbf{n}}$ is the unique real eigenvector of $\mathbf{R}(\theta, \hat{\mathbf{n}})$. The columns of the matrix $\mathbf{R}$ are *exactly* the three vectors describing the corresponding frame $\mathbf{F}$.

   Rotation matrices and the actions of rotations in three dimensions, and hence orientation frames, can alternatively be represented by *unit-length quaternions* (see, e.g., Chapter 2 or Hanson (2006)). Just as a unit-length complex number $\cos\theta + i\sin\theta = \exp(i\theta)$ with $i^2 = -1$ can be represented by a pair of real numbers (x, y) satisfying $x^2 + y^2 = 1$, a unit-length quaternion $(q_0 + \mathbf{I}q_x + \mathbf{J}q_y + \mathbf{K}q_z) = \exp(\mathbf{\Xi} \cdot \hat{\mathbf{n}}\,(\theta/2))$ with $\mathbf{\Xi} = (\mathbf{I}, \mathbf{J}, \mathbf{K})$ and $\mathbf{I}^2 = \mathbf{J}^2 = \mathbf{K}^2 = \mathbf{IJK} = -1$ can be represented as a quadruple of real numbers

$$
\begin{aligned}
q(\theta, \hat{\mathbf{n}}) &= (q_0, q_x, q_y, q_z)\\
&= (q_0, \mathbf{q})\\
&= (w, \mathbf{x})\\
&= \left(\cos\left(\frac{\theta}{2}\right), \hat{\mathbf{n}}\sin\left(\frac{\theta}{2}\right)\right),
\end{aligned}
\tag{16.1}
$$

where it is sometimes convenient to define $w = q_0$ and $\mathbf{x} = (x, y, z) = (q_x, q_y, q_z)$. Here the rotation axis $\hat{\mathbf{n}}$ and the angle θ correspond precisely to those introduced already in $\mathbf{R}(\theta, \hat{\mathbf{n}})$; it is easy to verify that this parameterization has unit length, $q \cdot q = q_0^2 + q_x^2 + q_y^2 + q_z^2 = 1$, and has only the obligatory three free rotation parameters since $\hat{\mathbf{n}} \cdot \hat{\mathbf{n}} = 1$ as well. The solutions of $q \cdot q = 1$ (which define the 3D topological space of unit quaternions) define $\mathbf{S}^3$, or the *3-sphere*.

   Quaternions as represented in Eq. (16.1) have some additional properties of particular interest to us here:

- There exists a quadratic formula that defines a two-to-one mapping from a quaternion q to a frame represented as a 3×3 rotation matrix $\mathbf{R}$, as follows:

$$
\begin{bmatrix}
q_0^2 + q_x^2 - q_y^2 - q_z^2 & 2q_xq_y - 2q_0q_z & 2q_xq_z + 2q_0q_y\\
2q_xq_y + 2q_0q_z & q_0^2 - q_x^2 + q_y^2 - q_z^2 & 2q_yq_z - 2q_0q_x\\
2q_xq_z - 2q_0q_y & 2q_yq_z + 2q_0q_x & q_0^2 - q_x^2 - q_y^2 + q_z^2
\end{bmatrix}.
\tag{16.2}
$$

Furthermore, given a rotation matrix, one can find the two unique corresponding diametrically opposed quaternions (see Chapter 9).
- The identity frame corresponds both to the quaternion $q_{\text{ID}} = (1, 0, 0, 0)$ and to $-q_{\text{ID}} = (-1, 0, 0, 0)$.
- If we require two quaternions q_1 and q_2 to multiply together using the following order-dependent (noncommutative) rule originally discovered by Hamilton,

$$
Q = q_1 \star q_2 = (w_1 w_2 - \mathbf{x}_1 \cdot \mathbf{x}_2,\ w_1\mathbf{x}_2 + w_2\mathbf{x}_1 + \mathbf{x}_1 \times \mathbf{x}_2),
$$

where $\star$ is quaternion multiplication, the resulting quaternion Q *remains embedded* in $\mathbf{S}^3$ and generates the composite 3×3 rotation matrix $\mathbf{R}(Q) = \mathbf{R}_1(q_1) \cdot \mathbf{R}_2(q_2)$. We reiterate that the order matters: neither quaternions nor 3D rotation matrices commute in general.
- The inverse of $q = q(\theta, \hat{\mathbf{n}})$ is just $q^{-1} = q(\theta, -\hat{\mathbf{n}}) = q(-\theta, \hat{\mathbf{n}})$, or $(q_0, -q_x, -q_y, -q_z)$, and corresponds to the inverse 3D rotation matrix.

- There exists a *quaternion distance formula*, $d_{12} = \theta_{12} = 2\cos^{-1}|q_1 \cdot q_2|$, that gives a precise and rigorous definition of the similarity between frames. This corresponds intuitively to a great circle or geodesic minimal-length arc connecting two points on an ordinary sphere. Smooth spline-like curves, based on the properties of the distance formula and embedded in the quaternion sphere $\mathbf{S}^3$, can be constructed that smoothly connect sequences of quaternion points. More details are given next, in Section 16.4.

16.4 Quaternion distance

Distances between orientation frames are properly computed as the shortest paths lying within the quaternion 3-sphere $\mathbf{S}^3$. This distance between two frames in isolation can be computed either from axis-angle rotation matrix methods or using quaternion methods. Placing these distances in a global context, however, requires quaternions. We can get a good estimate of the spherical distance from our 3D quaternion visualization methods, but there is typically some spherical distortion in the projections that requires compensation, as discussed in detail in Chapter 4. (This is similar to the problem of trying to make a distance-preserving projection of the Earth onto a flat piece of paper; distances between cities on a globe are perfectly correct, but we cannot take a satellite picture that allows accurate measurement of these distances with a simple ruler.) The distance of the frame defined by $\mathbf{R}(\theta, \hat{\mathbf{n}})$ from the identity frame may be understood in practice as the angle θ itself. We can express this distance in invariant form by noting that the 4D inner product (dot product) of the identity frame $q_{\text{ID}} = (1, 0, 0, 0)$ with $q = (\cos(\theta/2), \hat{\mathbf{n}}\sin(\theta/2))$ is exactly $q \cdot q_{\text{ID}} = \cos(\theta/2)$. Taking into account the sign ambiguities of quaternions, the distance between any arbitrary pair of frames is thus seen to be

$$d_{12} = \theta_{12} = 2\cos^{-1}|q_1 \cdot q_2|.$$

Since $q_1 \cdot q_2 = (q_1 \star q_2^{-1}) \cdot q_{\text{ID}}$, some prefer to define this distance in terms of the first element of $q_1 \star q_2^{-1}$.

A practical representation of this minimal distance between two arbitrary quaternions q_1 and q_2 in $\mathbf{S}^3$ is the so-called `slerp` (Shoemake, 1985). This smooth minimal-length geodesic curve (which also projects to a smooth curve in any of our geometric views) is given by

$$\texttt{slerp}(q_1, q_2; t) = q_1 \frac{\sin((1-t)\phi)}{\sin\phi} + q_2 \frac{\sin(t\phi)}{\sin\phi}, \tag{16.3}$$

where $\cos\phi = q_1 \cdot q_2$.

16.5 Protein frame geometry

In order to apply quaternion maps to a protein, we must identify full 3D coordinate frames that can be constructed uniquely from the atomic positions in a protein structure file, typically obtained from the Protein Data Bank (PDB, 2011), and derived from crystallographic or NMR data.

Each amino acid by itself defines a tetrahedral framework containing five atoms centered on the alpha-carbon atom. Ignoring the upward-pointing hydrogen, we have the alpha-carbon C_α at the apex

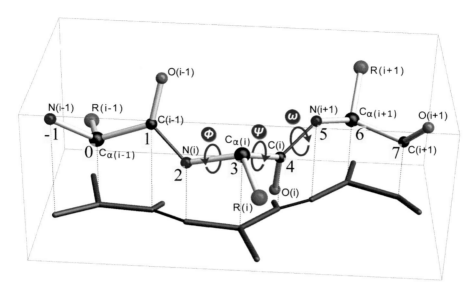

FIGURE 16.1

Amino acid neighboring structure. Triples of atoms $(i-1)$, (i), and $(i+1)$ correspond to a single amino acid residue. The group of six atoms, $C:N:C_\alpha:C:N:C_\alpha$, starting at label "1" in the figure for the first C, defines the Ramachandran angles as "hinge" angles of the three groups of four atoms in the sequence of six. The planes of the peptide bonds connecting to adjacent amino acids define the ψ and ϕ dihedral angles. The angle ω describes the normally negligible torsion of the peptide bond, which is relatively rigid. The central tetrahedron has the alpha-carbon at its top center, and we note that the orientation is the dominant L-form: the implicit hydrogen points upwards and the [CO]-R-N triangle goes clockwise.

of a tetrahedron whose triangular base is formed by the carbon atom of the COOH carboxyl group, the side-chain R, and the nitrogen atom from the original amino group, as shown in Fig. 16.1. Note that amino acids come in two isomers, D and L, and that most biological systems involve the L isomeric form shown in Fig. 16.1, where the left-handed [CO]RN sequence points the thumb in the direction of the C_α and its upward H at the apex of the tetrahedron.

We can choose any three atoms to define a frame with one atom at the origin, and vectors from that central atom to the other two serving to define the rest of the frame. Although we generally choose the C_α frame, with the triplet $N:C_\alpha:C$ forming the basis, we can choose any other local frame containing only atoms in a single residue, and it will be related to the C_α frame (or any other local frame) by a body frame rotation that is universal for all amino acids (with the exception of a two-fold ambiguity for glycine, for which the side-chain (R) = H). We can also choose frames based on triples of atoms that cross residue boundaries; such frames form the basis of the Ramachandran angles.

Alpha-carbon frame: For a local, single-residue frame, we therefore only need to define the C_α frame. Starting from the position vectors for each atom of the $N:C_\alpha:C$ triplet, we define the canonical C_α frame

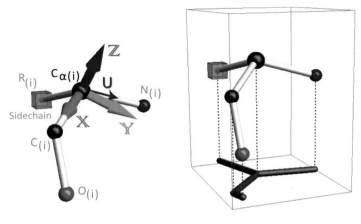

FIGURE 16.2

Amino acid geometry of the C_α frame, alongside the corresponding tetrahedron with the left-handed [CO]-R-N triangle as its base and C_α as the apex. The computation of our default discrete orthogonal frame is based on the directions from the C_α to the neighboring C and N atoms, whose cross product is the frame's **Z**-axis. The frame vectors **X** (red), **Y** (green), and **Z** (blue) are superimposed on the basic amino acid unit structure.

as follows:

$$\mathbf{X} = \frac{\mathbf{C} - \mathbf{C}_\alpha}{|\mathbf{C} - \mathbf{C}_\alpha|},$$

$$\mathbf{U} = \frac{\mathbf{N} - \mathbf{C}_\alpha}{|\mathbf{N} - \mathbf{C}_\alpha|},$$

$$\mathbf{Z} = \frac{\mathbf{X} \times \mathbf{U}}{|\mathbf{X} \times \mathbf{U}|},$$

$$\mathbf{Y} = \mathbf{Z} \times \mathbf{X},$$

as shown in Fig. 16.2.

Any such construction gives us a *frame* constructed from the fixed atomic vertices of an amino acid residue in a protein. The frame itself is representable as the 3×3 orthonormal matrix

$$\mathbf{F} = \begin{bmatrix} \mathbf{X} & \mathbf{Y} & \mathbf{Z} \end{bmatrix}$$

and the corresponding quaternion $q(\mathbf{F})$ can be constructed (up to a sign) by the algorithms in Chapter 9 and Appendix G.

Peptide bond frame: The peptide bond frame, or "P frame," uses atomic positions from two neighboring residues sharing a peptide bond. Starting from the position vectors for each atom of the $N{:}C_\alpha{:}C$

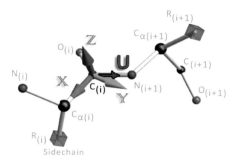

FIGURE 16.3

The coordinates of the P frame definition; the frame is centered on the carbon atom, and based on the connection to its own C_α and then the nitrogen N_{i+1} of the neighboring residue. This cross product defines the P frame's **Z**-axis.

triplet along with its following neighbor, $N':C'_\alpha:C'$, we define the canonical P frame as follows:

$$\mathbf{X} = \frac{C_\alpha - C}{|C_\alpha - C|},$$

$$\mathbf{U} = \frac{N' - C}{|N' - C|},$$

$$\mathbf{Z} = \frac{\mathbf{X} \times \mathbf{U}}{|\mathbf{X} \times \mathbf{U}|},$$

$$\mathbf{Y} = \mathbf{Z} \times \mathbf{X}.$$

The schematic image corresponding to the P frame construction is shown in Fig. 16.3.

Side-chain frame: In addition to the main protein backbone coordinates, the PDB data files contain information on the positions of the atoms in the residue side-chains that can also be studied. Since the side-chain geometry and composition varies considerably, starting with the essentially structureless side-chain of glycine, which contains only a single hydrogen, one might need to customize the quaternion frame description on a case-by-case basis. Typically one would start with a framework such as the $C:C_\alpha:C_\beta$ triplet where C_β (if it exists) is the carbon atom on the side-chain group connected to C_α. A prototype side-chain frame might then look something like

$$\mathbf{X} = \frac{C_\beta - C_\alpha}{|C_\beta - C_\alpha|},$$

$$\mathbf{U} = \frac{C - C_\alpha}{|C - C_\alpha|},$$

$$\mathbf{Z} = \frac{\mathbf{X} \times \mathbf{U}}{|\mathbf{X} \times \mathbf{U}|},$$

$$\mathbf{Y} = \mathbf{Z} \times \mathbf{X}.$$

Among other options, once could interchange C and C_β, or substitute N for C in the triad.

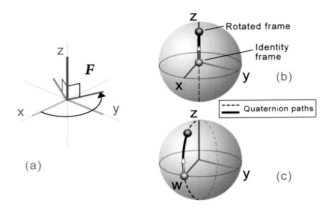

FIGURE 16.4

(a) This shows a simple frame **F** obtained by rotating the identity frame by a 160-degree angle about the direction $\hat{z}$. The smaller white spheres in panels (b) and (c) show the location of the identity frame and the larger black spheres show the quaternion points representing the outcome of the rotation. The thick black lines show the quaternion path starting at the identity, and the dashed lines show the path of a continued z-axis rotation. (b) Here we show the canonical (x, y, z) projection, which produces a cyclic line up and down the z-axis, and (c) shows the (w, y, z) projection, which produces a circle in the wz-plane. Note that the total angle of 160 degrees in 3D space appears as an angle of 80 degrees in quaternion space.

16.5.1 Visualizing quaternions as geometry

A simple example of a frame **F** resulting from applying a rotation about the $\hat{z}$ direction to the identity frame is shown in Fig. 16.4(a). Our next task is to relate this frame to its quaternion representation and to convert the standard 3D display of this frame to a quaternion display. In this subsection, we will explore explicit geometric views of the frame quaternion, and in the following subsection, we will examine alternative coordinate views.

For a positive (counterclockwise) rotation about the z-axis, the matrix **F** becomes

$$\mathbf{F} = \mathbf{R}(\theta, \hat{z}) = \begin{bmatrix} \cos\theta & -\sin\theta & 0 \\ \sin\theta & \cos\theta & 0 \\ 0 & 0 & 1 \end{bmatrix},$$

and the columns are easily seen to be the components of **F**. Since we know that the fixed rotation axis is $\hat{n} = (0, 0, 1)$, we can also write down the corresponding quaternion from Eq. (16.1) as

$$q_{\mathbf{F}}(\theta, \hat{z}) = (\cos(\theta/2), 0, 0, \sin(\theta/2)) .$$

How do we use this information to make a geometric view of $q_{\mathbf{F}}$? We have already become accustomed to the fact that any unit-length quaternion 4-vector q corresponds to a point on the 3-manifold $\mathbf{S}^3$, a 3-sphere embedded in four Euclidean dimensions, and the fact that since quaternions have unit length, the fourth component w is redundant and is just the solution of the equation $w = \pm\sqrt{1 - x^2 - y^2 - z^2}$. Hence, up to a sign specifying the hemisphere of the two halves of $\mathbf{S}^3$, all the orientation frame information embodied in a quaternion can be represented by a point (x, y, z) in 3D Euclidean space, plotted as a point $\boxed{\hat{\mathbf{n}} \sin(\theta/2)}$ inside a unit-radius solid sphere. In protein geometry, we rarely need to dwell on the two hemispheres because, in practice, we study sequences of linked frames that do not frequently cross between them; nevertheless, it is important to be aware of the possibility when it happens, and to be prepared, e.g., to use different colors to encode in which hemisphere a point lies. Another variant of the geometric view is to choose alternative projections, picking, say, $x = \pm\sqrt{1 - w^2 - y^2 - z^2}$ instead of w as the "extra" variable, and plotting the point (w, y, z) inside a distinct unit-radius solid ball. These two choices are represented in quaternion coordinates in Fig. 16.4(b,c) for rotations about the $\hat{\mathbf{z}}$-axis leading from the identity quaternion to $q_{\mathbf{F}}$, the quaternion representation of our sample frame **F**.

16.5.2 Visualizing quaternions as coordinates

In our experience, most situations involving frame comparisons are most effectively represented using the geometric view of quaternion coordinates. Nevertheless, in some cases one may prefer a very explicit (if less visually intuitive) representation showing a list of quaternion coordinate values. The conventional representation of this type is the *parallel coordinate representation* (Inselberg, 2009). This representation in our case would consist of taking a 4D quaternion vector representing an orientation frame (in some fixed, arbitrary order in 4D such as (q_0, q_x, q_y, q_z)) and drawing three lines connecting a graph of those four numbers, giving a display for one point like that in Fig. 16.5(a) corresponding exactly to Fig. 16.4(b,c).

Another useful coordinate view for multidimensional data is the *star plot* (see, e.g., Chambers et al. (1983) or Fanea et al. (2005)). In this approach, the real line in Fig. 16.5(a) is essentially deformed to a point and the graph connecting the four coordinate values becomes a piecewise-linear circle bounded by a diamond-shaped polygon. However, because quaternions have the special property of unit length, we again can pick three independent coordinates instead of four constrained coordinates and construct a three-axis star plot instead of a four-axis star plot.

There are several variants to the star plot. The four-axis quaternion star plot is sufficiently similar to the parallel coordinate plot (basically "wrapped around" parallel coordinates) that we will omit detailed discussion. The three-axis star plot is based on the 2×3 projection matrix

$$\sigma = \begin{bmatrix} -\frac{\sqrt{3}}{2} & \frac{\sqrt{3}}{2} & 0 \\ -\frac{1}{2} & -\frac{1}{2} & 1 \end{bmatrix}. \tag{16.4}$$

From this matrix we can construct several variants:

- Plot the triangle formed by $(\sigma \cdot [x, 0, 0]^{\mathrm{t}}, \sigma \cdot [0, y, 0]^{\mathrm{t}}, \sigma \cdot [0, 0, z]^{\mathrm{t}})$. This has the advantage of placing the identity frame uniquely at the origin, but negative values of z appear in the same region as positive values of x and y.
- Modifying the above by displacing the (x, y, z)-coordinates to be always positive, i.e., $[x + 1, 0, 0]^{\mathrm{t}}$, etc., effectively makes the graphing areas unique. However, the identity frame is now an odd finite triangle.
- A novel variation is to map the identity frame uniquely to the origin by using the absolute values or the squares of the coordinates in the map. Although we lose some specific information, we prefer

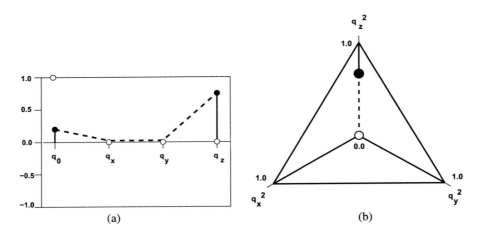

FIGURE 16.5

(a) A classic parallel coordinate map for a single quaternion point such as the frame in Fig. 16.4. The four quaternion coordinates are represented in this case by the four values placed side-by-side and connected by three line segments to denote a single point. (b) The quaternion star-squared map (see Eq. (16.4)), showing the quaternion frame of $\mathbf{F}$ as a line from the q_x^2 and q_y^2 points at the origin to the q_z^2 value on the vertical axis. White dots denote zeroes or identity frame values.

the star plot of the *squared* values, that is, the triangle connecting the three points

$$\text{plot of } x = x^2 \begin{bmatrix} -\frac{\sqrt{3}}{2} \\ -\frac{1}{2} \end{bmatrix}, \qquad \text{plot of } y = y^2 \begin{bmatrix} +\frac{\sqrt{3}}{2} \\ -\frac{1}{2} \end{bmatrix}, \qquad \text{plot of } z = z^2 \begin{bmatrix} 0 \\ 1 \end{bmatrix}, \qquad (16.5)$$

etc., which shows clearly how close the quaternion value is to the identity frame (the center of the star plot).

In Fig. 16.5(b) we show the star-squared quaternion plot corresponding to Fig. 16.4(b,c). For our simple z-axis rotation example, all one sees is a line to a point on the vertical axis. More general frames would lie inside an equilateral triangle, with frames near the identity frame converging on the center. Remark: One could also produce Euclidean 2D star-like plots with each frame defined by a 2D point

$$[u, v]^t = \sigma \cdot [x^2, y^2, z^2]^t,$$

instead of combining the nonrectangular triangular axis points into a triangle; however, one can easily see from Eq. (16.4) that the corresponding summations formed by any set of coordinates with equal values of $(|x|, |y|, |z|)$ are confused with the identity.

16.5.3 Collections of frames via quaternion maps

A significant property of the quaternion views just described is that they provide a visual image of an orientation frame as a *quaternion map*, either as a single point in space or as a graph. We now show how quaternion maps can be used to expose the absolute similarity of two 3D orientation frames (arbitrarily separated in 3D distance) using the proximity of the two quaternion points in the plot (see Section 16.4). Selected groups of dozens of orientation frames occurring at widely different spatial positions may correspond to quaternion points falling close to one another or in a revealing pattern in the quaternion map. Dissimilarities can similarly be exposed. While the geometric view has the most powerful tools for exposing global similarities, the parallel coordinate or star plot approaches to representing frames with the coordinate view can also suggest interesting relationships among frames.

For a *collection of frames* (in our case, a set of frames corresponding to a sequence of residues in a protein), each frame is then represented as a distinct point or graph of some sort, and the *ordered sequence* of frames can be represented by a collection of these. A special technique is typically used for sequences of quaternion frames to enforce continuity of the quaternion value: since any frame can be represented by *either q* or $-q$, we must eventually choose one. We therefore compare the inner product $q_k \cdot q_{k+1}$ for each neighboring pair of frames $(k, k + 1)$, and replace $q_{k+1} \to -q_{k+1}$ if the inner product is negative. This is the $\boxed{\text{force-close-frames}}$ method.

Our choices of representations that *embody the intrinsic quaternion distances* include the following:

- **S^3 map.** Using the projection directly from the 4D quaternion value in S^3 to a 3D subspace such as xyz produces a spherically deformed map of the actual quaternion distances (like looking at a country on a globe of the Earth from an oblique angle). However, the deformation is completely predictable (see Chapter 4), and distances for pairs near the center, for example, are reliable. Interactive 4D rotations (see Hanson (2006); Yan et al. (2012)) can place pairs anywhere one would like in the projection. The metrically most accurate distance in the projection is found by transforming the scene so one of the desired pair of quaternion frames is at the origin in the (x, y, z) projection. As a matter of practice, the curves connecting collections of quaternion points in the spherical projection are typically drawn as geodesics (shortest-distance paths constrained to the 3-sphere), though for simplicity one may also draw them as piecewise linear paths. The advantage of this representation is that no matter how distant one object is from another in 3D space or along the sequence, objects that have similar orientation frames can always be aligned to expose the fact that their frames have nearby quaternion points.

- **Parallel coordinate map.** A typical parallel coordinate plot for a collection has all the points superimposed on a single 2D plot. Our case is different from the usual case because we have the *additional* relative quaternion distance information available, which we can use to *displace* each set of plotted 4D coordinates from its neighbors in a meaningful way. That is, we take each individual parallel coordinate plot and displace it in a perpendicular direction by the value of the quaternion distance to its next neighbor. This approach has the advantage that *all* absolute orientation differences from neighboring frames around the curve are represented as completely accurate Euclidean distances. It has the disadvantage that it is hard to get an intuitive feeling of whether spatially distant objects (far apart in space and/or the parallel coordinate plot) have similar quaternion frame values or not.

- **Star-squared map.** Sequences of the star maps that we have defined using the projection Eq. (16.4) can also be displaced perpendicularly relative to each neighbor by the amount of the quaternion distance. Again, this gives a metrically accurate neighboring frame distance representation for large numbers of frames, and exposes patterns that have visual advantages similar to the geometric projection, but relating distant frames to one another is a challenge.

First example with real protein data. We now go to the main source of the data we will proceed to examine exhaustively, the "PDB" or Protein Data Bank that is maintained for public use at https:// www.wwpdb.org/. From these data, we can construct lists of quaternion orientation frames using the techniques that we will describe shortly. Here, our purpose is to illustrate how, once we have such quaternion data, basic quaternion displays can be constructed using the techniques just explored in this section. In Figs. 16.6, 16.7, 16.8, and 16.9 we illustrate our display methods for a generic sequence of frames taken from the 1AIE PDB data.

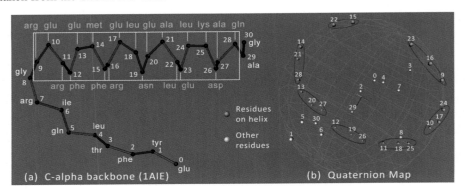

(a) C-alpha backbone (1AIE)   (b) Quaternion Map

FIGURE 16.6

(a) The PDB file geometry of 1AIE containing an alpha-helix. (b) The quaternionic geometric view of the 1AIE quaternion frame coordinates in the (w, y, z) projection.

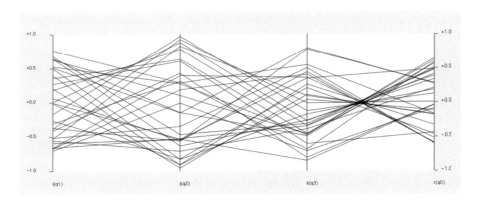

FIGURE 16.7

Standard parallel coordinate plot of the 1AIE standard quaternion frame 4-vectors, with no means of distinguishing neighboring residues.

Summary. This completes our treatment of some of the ways that, once we have *single* instances of quaternion frames, we can start to keep track of *sequences* of quaternion frames in various contexts.

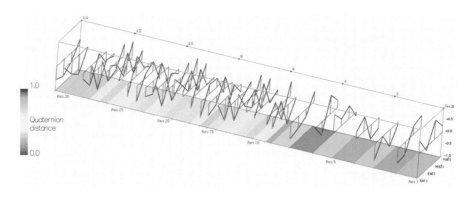

FIGURE 16.8

Parallel coordinate plot of the **1AIE** standard quaternion frame 4-vectors, spaced by quaternion distance.

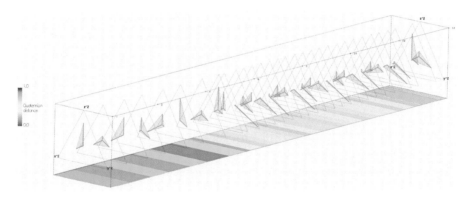

FIGURE 16.9

Star-squared parallel coordinate plot of the **1A05** quaternion frame sequence, spaced by quaternion distance.

Each method has specific domains of utility. Our own preference is for representations with clear geometric properties as opposed to coordinate-value properties, and thus we will for the most part make use of the geometric quaternion point projections in (x, y, z)- or (w, y, z)-coordinates.

16.6 Studies in quaternion frame maps

We now explore some specific examples of frame maps for idealized mathematical objects that correspond closely to the behavior of real protein data, providing additional intuitive grounding. The following section will show an illustrative sequence of examples taken directly from PDB file data.

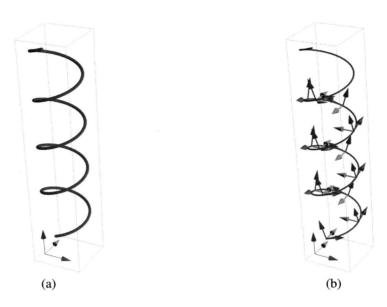

(a)   (b)

FIGURE 16.10

(a) A helix defined by the parametric equation $(r\cos(t),\, r\sin(t),\, pt)$. (b) A set of frames on the helical curve defined by the Frenet–Serret equation. Note the relation of the identity frame at bottom left to the first actual helix frame.

16.6.1 Alpha-helix model: quaternion frames of idealized curves

We first turn to an elementary application of quaternion Frenet–Serret frames (Hanson and Ma, 1995; Hanson, 2006) to the study of helical curves, which correspond to alpha-helices in proteins. Due to the double-valued nature of the relation between quaternions and rotations, two full turns of the helix correspond to exactly one closed circuit in quaternion space. The quaternion map in this case is a circular closed loop that has an elliptical projection into the (x, y, z)-coordinates, determined by the axis direction of the helix and its pitch. Fig. 16.10 shows an ideal mathematical helix and a sampling of the continuous Frenet–Serret frames determined by the local curve derivatives; Fig. 16.11 presents the corresponding xyz and wyz quaternion maps of the orientation frames in Fig. 16.10. Note that the Frenet–Serret frame may not be suitable for certain classes of curves; if there is a straight section or an inflection point (typical of cubic curves, for example), the second derivative vanishes and the Frenet–Serret frame becomes undefined.

A quick outline of how one actually does a quaternion calculation for a helix may prove useful for understanding the quaternion frames of an alpha-helix structure. We start with the equation of a helix of radius r and pitch p, along with its first and second derivatives:

$$\left. \begin{aligned} \mathbf{x}(t) &= (a\cos t,\, b\sin t,\, pt) \\ \mathbf{x}'(t) &= (-a\sin t,\, b\cos t,\, p) \\ \mathbf{x}''(t) &= (-a\cos t,\, -b\sin t,\, 0) \end{aligned} \right\} . \tag{16.6}$$

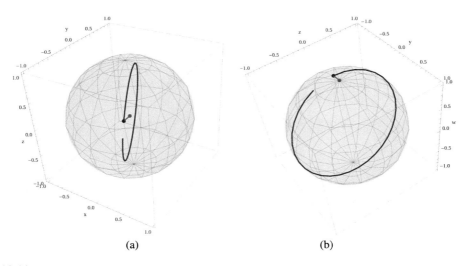

(a)   (b)

FIGURE 16.11

The quaternion maps for a helix defined by the parametric equation $(r\cos(t),\ r\sin(t),\ pt)$. (a) The xyz quaternion map and (b) the wyz quaternion map of the continuous frames attached to the helix. The red line is the path from the identity frame (at the red dot) to the first actual helix frame.

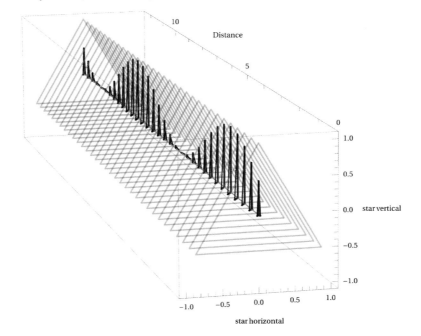

FIGURE 16.12

Parallel xyz star-squared coordinates for the frames of a helix.

We would take $a = b = r$ to produce a circular helix, and $a \gg b$ to make a flattened elliptical helix. We use the value of the tangent (first derivative) to determine the direction of the first frame axis, which we label $\mathbf{X}$. Typically, the next frame axis direction is computed from the cross product $\mathbf{x}'(t) \times \mathbf{x}''(t)$, whose direction we label $\mathbf{Z}$; then the remaining frame axis direction is $\mathbf{Y} = \mathbf{Z} \times \mathbf{X}$. Normalizing to unit length, we obtain the result for the frame triad of orthonormal vectors for any point t on the helix:

$$\mathbf{X}(t) = \left(-\frac{r \sin(t)}{\sqrt{p^2 + r^2}}, \frac{r \cos(t)}{\sqrt{p^2 + r^2}}, \frac{p}{\sqrt{p^2 + r^2}} \right) , \tag{16.7}$$

$$\mathbf{Y}(t) = (-\cos(t), -\sin(t), 0) , \tag{16.8}$$

$$\mathbf{Z}(t) = \left(\frac{p \sin(t)}{\sqrt{p^2 + r^2}}, -\frac{p \cos(t)}{\sqrt{p^2 + r^2}}, \frac{r}{\sqrt{p^2 + r^2}} \right) . \tag{16.9}$$

Do not forget that these are the *column* vectors for the frame matrix, not the *row* vectors.

The quaternion frame can then be computed as a rotation about the z-axis acting on the initial frame at $t = 0$, which reduces after a bit of algebra to the form

q helix at 0 $=$

$$\left(\frac{1}{2}\sqrt{\frac{\sqrt{p^2 + r^2} + r}{\sqrt{p^2 + r^2}}}, \frac{p}{2\sqrt{r\sqrt{p^2 + r^2} + p^2 + r^2}}, \frac{-p}{2\sqrt{r\sqrt{p^2 + r^2} + p^2 + r^2}}, \frac{1}{2}\sqrt{\frac{\sqrt{p^2 + r^2} + r}{\sqrt{p^2 + r^2}}} \right) . \tag{16.10}$$

Multiplying to the left by the quaternion $q_{z\,\mathrm{rot}}(t) = (\cos(t/2), 0, 0, \sin(t/2))$ rotating about the z-axis, the full quaternion frame for the helix is then

$$q_{\mathrm{helix}}(t) = q_{z\,\mathrm{rot}}(t) \star q_{\mathrm{helix\ at\ 0}} . \tag{16.11}$$

Plugging these values into the quadratic form in Eq. (16.2), one finds the matrix whose columns are the vectors in Eq. (16.9).

Fig. 16.11 shows the explicit helical quaternion maps in spherical geometric coordinates, in contrast to Fig. 16.12, which shows the star-squared parallel coordinate representation. The geometric forms in Fig. 16.11 are pure circles in the 4D geometry, and also in the (w, y, z) projection, but must be flattened ellipses in any other projection. The periodic circular path of the quaternion frame in Fig. 16.11 is reflected in the perfectly periodic pattern along the z-projection axis in the star-squared plot in Fig. 16.12.

16.6.2 Beta-sheet model: extreme quaternion frames

A crude mathematical approximation to a beta-sheet can be constructed by using a flattened helix such as Eq. (16.6) with $a \gg b$. Sampling the Frenet–Serret frames at $t = n\pi + \epsilon$ for a small random ϵ produces the alternating pairs of frame orientations characteristic of beta-sheets. In real data, the beta-sheet also twists systematically, which we could include in the model by a slow rotation in the xy-plane. We show in Figs. 16.13 and 16.14 the beta-sheet analogs of the alpha-helix model features shown in Figs. 16.10 and 16.11. The parallel-coordinate-based star-squared map of the beta-sheet model shown in Fig. 16.15 corresponds to Fig. 16.12 for the helix.

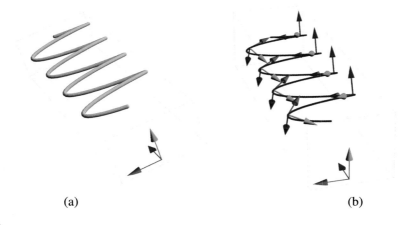

(a)   (b)

FIGURE 16.13

(a) A beta-sheet modeled by the parametric equation $(\cos(t), 0.1\sin(t), 0.5t)$. (b) A set of Frenet–Serret frames at roughly the expected places on the equation of the curve. Note the relation of the identity frame in the foreground to the first actual sampled frame.

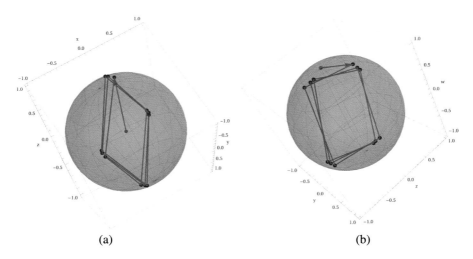

(a)   (b)

FIGURE 16.14

A beta-sheet modeled by the parametric equation $(\cos(t), 0.1\sin(t), 0.5t)$. (a) The corresponding xyz quaternion map of the continuous frames attached to the helix. The red arrow is the path from the identity frame (the red dot) to the first actual helix frame. (b) The wyz quaternion map. (Note: The discontinuous nature of the beta-sheet frames is reflected in our choice of straight lines to connect neighboring frames here; in most cases, we will prefer to use smooth quaternion geodesics reflecting the shortest rotation path from one frame to the next.)

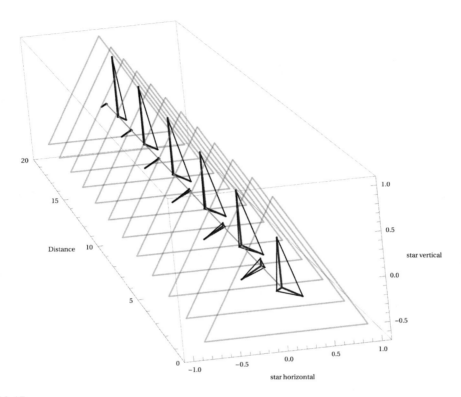

FIGURE 16.15

Parallel xyz star-squared coordinates for the frames of an idealized beta-sheet model.

For idealized beta-sheets, we observe clusters at intervals of roughly 90 degrees in the quaternion plots, corresponding to the approximate 180-degree flips between neighboring residue orientations in a beta-sheet. In practice, the real-world noisiness of the data will tend to interrupt the regular pattern of the mathematical model.

16.6.3 Quaternion frames from spline curves of PDB backbones

We next examine smooth frame sequences that can be associated directly with measured helical protein structures. Fig. 16.16(a) shows the structure of a helix-containing subsequence of a protein, the leucine zipper from the PDB file 1C94, whose dominant element is a single helical structure consisting of approximately seven loops. The idealized curve is defined by a smooth B-spline approximation to the path of the C_α atoms making up the backbone. This curve is continuously differentiable and is suitable for defining continuous moving frames along the curve such as the Frenet–Serret frames, samples of which are shown in Fig. 16.16(b). Fig. 16.16(c) is the quaternion xyz map of the Frenet–Serret frames for 1C94, showing the quaternion form of the sequence of orientations and their global relationship for the whole protein. Comparison with the pure mathematical helix in Fig. 16.10 and Fig. 16.11 clearly shows the close resemblance.

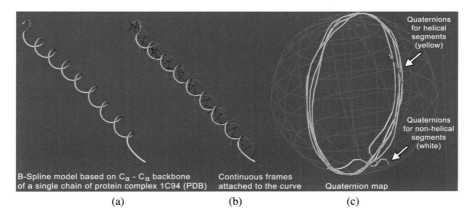

FIGURE 16.16

A simple protein section, the leucine zipper in the 1C94 data set. (a) The standard B-spline curve derived from the underlying C_α backbone of the protein; note that this curve passes near, but not through, each C_α in order to achieve a slightly artificial degree of smoothness. (b) The Frenet–Serret frames determined by this curve. (c) The xyz quaternion map of the continuous Frenet–Serret frames. Compare these maps with those of the ideal helix in Fig. 16.10 and Fig. 16.11.

16.7 Quaternion frames from discrete PDB data

We now turn our attention to quaternion descriptions of discrete 3D frames determined by exact atomic positions, rather than idealized curves. This will allow us to explore applications involving sequences of amino acid residue orientations.

There are many possible frame choices that can be assigned to components of a protein. We find it most natural to study those defined *within a residue*. Thus our prototypical frame will be the one anchored at the C_α carbon (C_α frame), shown in Fig. 16.2. Another useful but very distinct frame is the "P frame" (discussed below), which includes the direction of the peptide bond connecting a pair of residues, and thus utilizes atomic positions from both. The geometry of these frames is defined in detail in Section 16.5.

There is one potential deficiency of the C_α frame, as pointed out in Robert Hanson et al. (2011): it is possible to fix both the absolute and relative orientation of two adjacent residues via their C_α frames, and *still* have a potentially significant ambiguity in the local geometric structure due to the so-called "bicycle motion" (see Fig. 16.17). The bond between the C_i atom and the N_{i+1} atom could possibly serve as the spinning crank joining two C_α frame "bicycle pedals," though that action is severely limited by the rigidity of the peptide bond. This is of course true for *any* adjacent sets of three atoms in the protein backbone used to define independent frames.

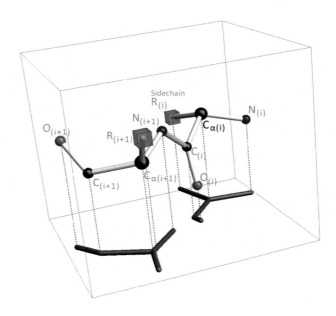

FIGURE 16.17

Drop shadow representation of the geometry for two adjacent residues. The tinted C-N' bond is central to the peptide bond, and embodies the "bicycle" ambiguity of the two neighboring C_α frames. The P frame incorporates this peptide bond instead of being isolated in a single residue.

In order to construct the protein geometry completely (up to whatever effect might arise from local distortions of the bond features), we would need at least one more intermediate frame such as the "P frame" relating two adjacent residues. The P frame is shown schematically in Fig. 16.3 and defined in Section 16.5. (Note that the Ramachandran angles, described in Appendix K, do not actually fully describe the transformation between adjacent frames.)

When we pass to sequences of discrete frames, remember that we must resolve the sign ambiguity between adjacent quaternion frame values by choosing the minimum quaternion distance to the preceding quaternion frame (the *force-close-frames* procedure). For an ordered sequence of frames such as those produced by protein residues, the resulting map is a sequence of points in S^3 that can be connected by piecewise-continuous minimal-length quaternion curves (Shoemake, 1985) contained in S^3, embedded in 4D Euclidean space, and projected according to the methods detailed above.

Our first example, the 1AIE structure, was introduced in Fig. 16.6 as a prototypical alpha-helix. Applying the C_α frame map and the P frame map side-by-side, we find the results in Fig. 16.18, showing similar but not identical elliptical helical structures in the spherical (x, y, z) quaternion projection.

A more complicated configuration, the leucine zipper 1C94 double helix, is shown in Fig. 16.19, along with its C_α frame quaternion maps. Each frame is represented by a single quaternion point in the map, and the ordered sequence of amino acids produces an ordered sequence of quaternion points. Any two points in this sequence, whether adjacent or not, can be connected pairwise by quaternion curves that correspond to the smallest rotation transforming one orientation frame to the other. The lengths of

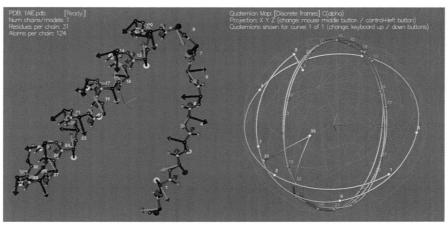

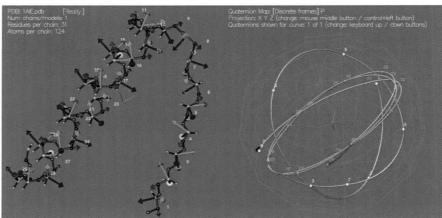

FIGURE 16.18

Protein structure of **1AIE** with its C_α frames and its P frames, and the corresponding quaternion frames joined as a sequence of spherical arcs.

these minimal curves provide a precise measure of the similarity of the orientation frames. Amino acid residue frames that are close in quaternion space, whether nearby or distant in the ordered sequence, have similar global orientations.

Beta-sheet example. We next examine the signature of beta-sheets, which form widely spaced clusters of similar orientations in the quaternion maps, as shown in Fig. 16.20 for 2HC5, and later in Fig. 16.26. Our conventional coercion of neighboring quaternion frames to have positive inner products is not always effective for widely spaced frames such as those in beta-sheets. In Fig. 16.21 we show an alternative method, plotting *both* signs for frames suspected to form beta-sheets, and clearly exhibiting the theoretically expected four-fold clustering.

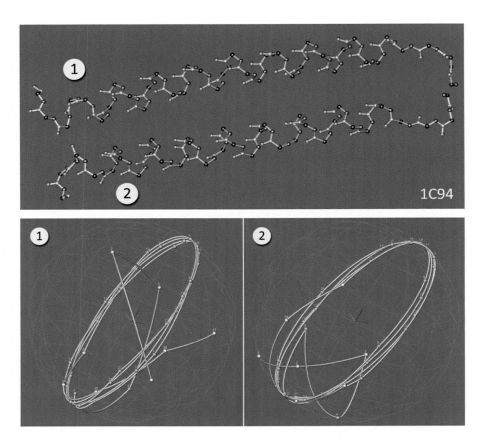

FIGURE 16.19

(above) Geometry of the double-stranded protein structure 1C94 (the leucine zipper). Segments that are part of helices in the two strands are depicted in yellow. (below) Discrete quaternion maps for C_α frames of the two strands of 1C94.

Another type of example is shown in Fig. 16.22, which includes a B-spline model of the protein 1A05, based on the C_α backbone. Orientation frames for residues that correspond to a beta-sheet are drawn over the model shown in Fig. 16.22(a). The quaternion map in Fig. 16.22(b) shows quaternion points corresponding to the orientation frames. The quaternion map reveals that pairs of alternating residues have similar orientations. Some of the pairs of similar orientations are highlighted (a–d in the figure).

Observations. Examining Fig. 16.6 and Fig. 16.18, we see that alpha-helices also produce clusters of similar orientations, and that every seventh amino acid frame falls close to its predecessor. For the particular case of 1AIE (see Fig. 16.6), the number of residues in the helix is small enough that we can single out seven distinct groups of two or three (marked by oval outlines in the figure) that

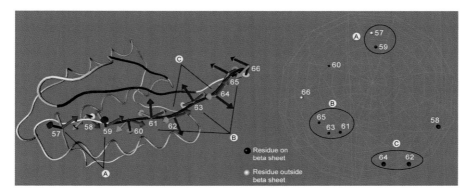

FIGURE 16.20

Protein structure of **2HC5** and a quaternion map of its beta-sheet structure. Neighboring frames are given matching quaternion signs in this map.

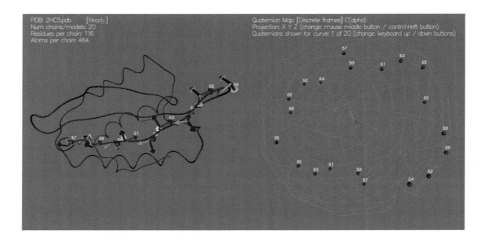

FIGURE 16.21

Beta-sheet structure with each quaternion frame displayed twice, with both possible overall signs. One can see that the beta-sheet does not lie exactly in a plane, but twists slowly, causing the four expected 90-degree-quaternion-spaced groups (180 degrees in physical space) to spread out across quaternion space. Locate, for example, residues 57, 58, 59, and 60.

are spaced seven apart in the sequence making up the helix. This is an example of the application of the quaternion map to highlight global orientation patterns that may be difficult to extract by other methods. In contrast, beta-sheets will produce isolated clusters for short sequences, and more highly spread patterns for longer sequences. We can thus exploit the quaternion map to extract similarities in orientation patterns.

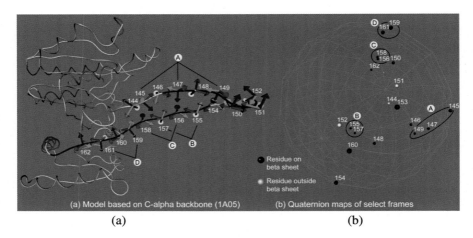

(a)   (b)

FIGURE 16.22

(a) A model of protein **1A05** constructed using a B-spline curve, which is based on backbone C_α atoms of the protein. The region with frames corresponds to the beta-sheet structure. The frames are labeled by the sequence number of the amino acids they belong to. (b) Quaternion map of the select frames associated with discrete amino acids making up the protein **1AIE**. Numbered points represent the quaternions corresponding to a single frame defined for each amino acid.

16.8 Example applications of discrete global quaternion frames

Applications of quaternion maps to the analysis of orientation frames fall into several categories:

- **Single or composite rigid protein frame groupings.** The available data sets are dominated by explicit atomic locations for one single protein or a few closely associated proteins. The most useful information for such data sets is probably the set of discrete global frames based on a single residue, such as the C_α frame. Incorporating information from neighboring residues to form alternate sets of frames is possible, and can produce quaternion alternatives to the Ramachandran plot (see Appendix K). The backbone atoms can also be exploited to generate approximate polynomial curves representing protein structure; the analysis of such curves is exhaustively detailed elsewhere (Hanson, 1998, 2006).

  We will focus on the residue-local C_α frame in our examples. Such discrete frames are particularly appropriate for identifying clusters of globally similar frames, which may be near one another physically or farther away but belonging to a regular geometric structure. Such clusters expose the natural relationships among groups of frames with diverse spatial relationships.

- **Patterns and straightness.** Proteins arrange themselves into secondary geometric groups such as alpha-helices and beta-sheets. Quaternions can be used for detailed analysis of the global orientations into patterns identified with secondary structures, and approaches have been found that use quaternion-based "straightness measures" to effectively identify structural patterns (Robert Hanson et al., 2011).

- **Nonrigid class groupings.** The quaternion analysis does not depend on the rigidity assumptions underlying the X-ray crystallography data for atom locations in the PDB database. We can examine instead the nonrigid groupings of NMR data, which produce clusters of similar geometries for the same sequence of amino acid residues. These sheaf-like groupings of protein strands that present themselves in the NMR data provide an entirely different opportunity: here each individual amino acid appears multiple times, and quaternion measures provide essentially the only rigorous metric for quantifying the similarities of the orientations of the multiple instances of an amino acid in each of the strands. We employ both the spherical mean and the standard deviation (Buss and Fillmore, 2001) to evaluate overall qualitative features of the cluster, and utilize outlier-excluding convex hulls for more robust descriptions of the low-rank statistics of these clusters. Examples are shown in Figs. 16.23, 16.24, 16.25, 16.26, 16.27, and 16.28.
- **Functional activation groupings.** Current research on enzyme functionality seeks to identify groups of active residues with side-chains that arrange themselves geometrically to facilitate biological functions. Given a hypothesis about, say, a triplet of side-chains, we can compute quaternion representations not only for the basic C_α frame, but also for the orientations of relevant side-chains with respect to the C_α frame. Quaternion frames provide a relatively straightforward method for surveying proposed activation groupings for matching orientation patterns. See Fig. 16.29 and Fig. 16.30.

16.8.1 Quaternion frames of rigid proteins

Example: 1C94 (PDB), the leucine zipper. An elementary example is provided by a protein fragment known as the "leucine zipper," which consists of two alpha-helices that align with one another in a tertiary structure (i.e., two or more associated proteins). The top of Fig. 16.19 shows the two strands and C_α-C_α backbones.

We compare the maps corresponding to the two different strands making up the protein complex of 1C94. The bottom of Fig. 16.19 shows quaternion maps for the two strands. While the maps for the two strands are expected to be similar because of the similarity of two strands, the maps can be used along with other metrics to uncover subtle differences, either within the same complex or across protein complexes.

16.8.2 Statistical quaternion groupings of NMR data

Example: NMR data for 1T50 (PDB): a water-borne pheromone from the mollusk *Aplysia* attractin. A more complex protein, the 1T50 pheromone, is depicted in Fig. 16.23(a). The structures in the figure correspond to a selection of 20 NMR data sets for the C_α-C_α backbone of the protein complex 1T50. Fig. 16.23(b) singles out one of these for reference. Fig. 16.23(c) shows the single model color coded by quaternion distances between neighboring residues. The quaternion map for the reference segment (b) is displayed in Fig. 16.23(d).

Example: NMR data for 2HC5 (PDB). Twenty different geometric configurations of protein YvyC from *Bacillus subtilis*, listed on the PDB as 2HC5, are shown in Fig. 16.24. In this data set, the variations in the orientation of each amino acid can be clearly seen in the quaternion map, along with clusters of similar and dissimilar orientations. The spatial displacements in the data have only minimal correlation with the orientation displacements observed in the quaternion map; however, their cluster centers and

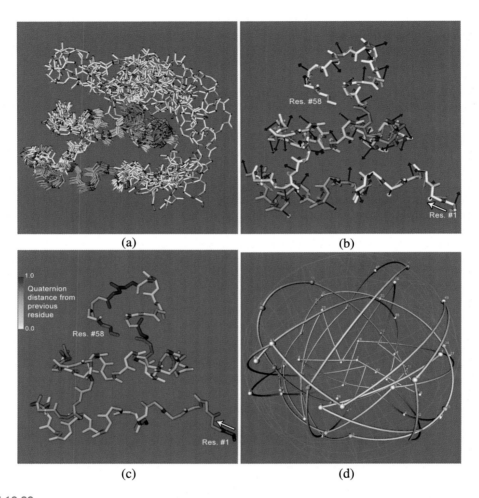

FIGURE 16.23

Structure and quaternion map of the protein **1T50** (PDB), a water-borne pheromone. (a) Twenty models of the protein **1T50** determined using NMR. (b) Backbone of a single model of the NMR data set showing C_α frames. (c) The model shown in (b) color coded by quaternion distance of a residue relative to its preceding residue on the backbone. (d) Quaternion map of the frames shown in (b).

statistical characteristics can be clearly identified, with very "floppy" arms of the protein generating large orientation variances, and relatively rigid branches keeping close to one another in quaternion space.

In Fig. 16.25, we interactively select a particular helical region of the protein to study the orientation distribution of its elements. Small quaternion regions correspond to fairly rigid configurations, and

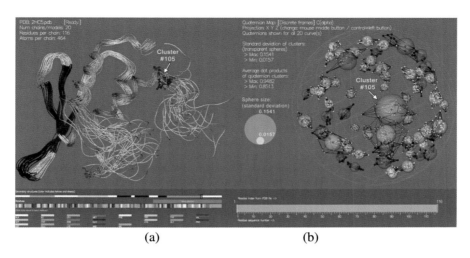

(a)   (b)

FIGURE 16.24

Quaternion maps for NMR data describing 10 different observed geometries for the protein **yvyC** from *Bacillus subtilis*, recorded as **2HC5** on PDB. (a) The collection of alternative geometries. (b) Quaternion maps showing the *orientation space* geometry spreads for each individual amino acid.

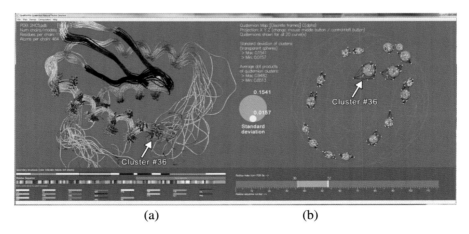

(a)   (b)

FIGURE 16.25

Isolating a selected section of the protein **YvyC** from *Bacillus subtilis*, using the **2HC5** data. (a) The selected region. (b) Quaternion maps showing the *orientation space* geometry spreads for each individual amino acid in this region.

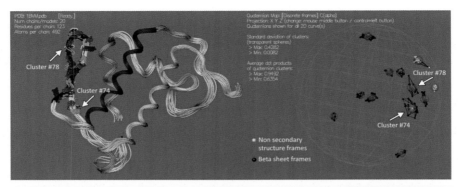

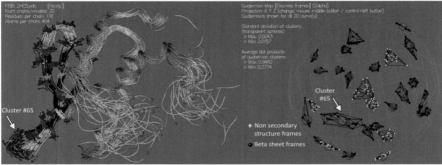

FIGURE 16.26

NMR data for **1BVM** and **2HC5** beta sheet examples.

large regions have large quaternion distance spreads around the spherical mean, indicating nonrigid behavior. Since the NMR data are selected on a relatively qualitative basis by the contributors, the precise meaning of some components of the differences are elusive, and it may well be possible to perform further quaternion-based analysis to further refine the apparent deviations in the data.

Two particularly interesting examples of NMR analysis with quaternion visualization of the statistical distance distributions are given in Fig. 16.27, which shows a very "tight" distribution in the protein shape flexibility, and a contrasting situation in Fig. 16.28, which shows significant variation in the spatial structure, but retains relatively close distributions in the orientation frames (quaternion dot products within the 0.9 range).

16.8.3 Enzyme functional structures

Comparing His-Tyr-Arg structures. It is known that catalytic residues can exhibit characteristic geometric structures (Xin et al., 2010). Among the groups of structures that could have similar structure and behavior, the proteins **1CB8** and **1QAZ** form an interesting pair of examples, with similar physical locations of His, Tyr, and Arg, with quaternion frames noted in the table.

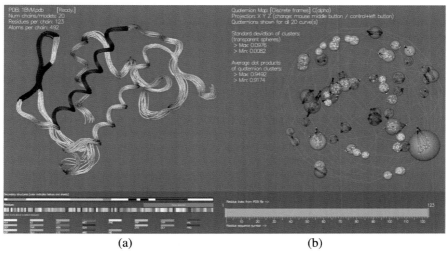

(a)   (b)

FIGURE 16.27

Quaternion maps for NMR data describing 20 different geometries for the protein Bovine pancreatic phospholipase A2 derived from **1BVM** PDB data. (a) The collection of alternative geometries. (b) Quaternion maps showing the *orientation space* geometry spreads for secondary structures of each of the predicted chains. This example has very close spatial similarity and quaternion frame similarity in the collection of alternative structures.

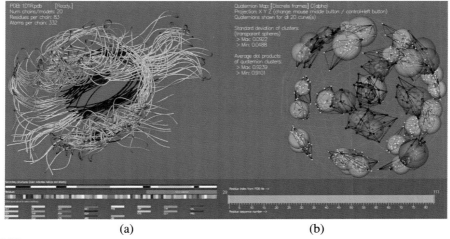

(a)   (b)

FIGURE 16.28

Quaternion maps for NMR data describing 20 different geometries for the protein obtained from **1D1R** data; the protein is derived from genetic information in the **YciH** gene of the *E. coli* bacterium. (a) The collection of alternative geometries. (b) Quaternion maps showing the *orientation space* geometry spreads for secondary structures of each of the predicted chains. Note that even though the predicted structures in (a) are widely displaced in space, the error in orientations among corresponding residues is relatively low. It would be very hard to see this using any method except the quaternion plot.

ID	Residue	(q_0, q_x, q_y, q_z)
1CB8	His225	$(0.861990, 0.491292, -0.124869, -0.003658)$
	Tyr234	$(0.146887, 0.799738, 0.297092, 0.500579)$
	Arg288	$(-0.637504, 0.412136, -0.383576, 0.525929)$
1QAZ	His192	$(0.385861, -0.419153, -0.585444, 0.576781)$
	Arg239	$(0.117073, 0.047674, 0.638640, 0.759052)$
	Tyr246	$(-0.565738, 0.038947, 0.598576, -0.565801)$
1FMI	Glu330	$(0.443820, -0.519370, -0.290678, -0.669915)$
	Asp463	$(-0.200536, 0.850733, 0.058078, 0.482355)$
	Glu599	$(0.082620, -0.582741, -0.708285, 0.389768)$
1RFN	His57	$(-0.154496, 0.404741, -0.857509, 0.277477)$
	Asp102	$(-0.116981, 0.070199, -0.983358, -0.119979)$
	Ser195	$(-0.699076, -0.128476, 0.354901, -0.607316)$

In Fig. 16.29, we show how these structures appear in 3D space. A different type of detail is shown in the quaternion plots: the objective in Fig. 16.30 is to successively align parts of the enzyme orientation in sequence to single out similarities and differences in these very distinct structures. Fig. 16.30(a) begins the process by showing the identified active sites listed above for the four enzymes 1CB8, 1QAZ, 1FMI, and 1RFN, with the quaternion plots of their orientations all transformed to have the reference frame of the first residue as the origin (the identity frame). Technically, that means that all orientations have been multiplied by the inverse of the frame matrix of the first residue. Fig. 16.30(b) is the result of identifying the axis of rotation characterizing the rotation of the second residue from the identity frame, and applying a global rotation on the entire enzyme to align that axis with the q_x-axis. At this point, there is still one more degree of freedom, since the quaternion curve denoting the rotation from the second residue's frame to the third residue's frame can still be realigned.

16.9 Tools for exploring and comparing quaternion maps

Visual analysis of the quaternion maps can reveal interesting global information about protein structures. However, there are in addition several useful techniques at our disposal that can enhance the visualizations or provide supplementary information, and can therefore aid in the exploration and comparison of the quaternion maps.

16.9.1 Aids for exploring quaternion maps

Our standard method relies on the fact that if we plot just the vector element $\mathbf{q}$ of the full unit quaternion $q = (q_0, \mathbf{q})$ obeying $q \cdot q = 1$, then we have in principle a complete picture, since the fourth component $q_0 = \pm\sqrt{1 - \mathbf{q} \cdot \mathbf{q}}$ is redundant up to a sign (see Chapter 4). Curves, surfaces, and even volumes can be plotted in this way to show the global features of the quaternion orientation families and to represent available degrees of freedom. Several specialized techniques can further aid the visualizations:

- **Double frames.** The quaternion $q_{\mathbf{F}}$ corresponding to a given 3D frame $\mathbf{F}$ is ambiguous up to a sign. In principle, we should not be able to distinguish anything about the same family of frames if we assigned the sign at random. One approach to this feature is simply to place each frame twice in the

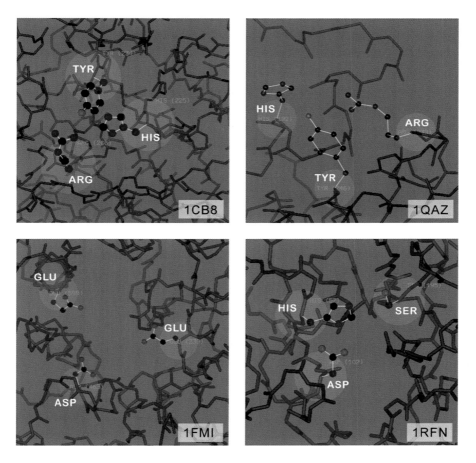

FIGURE 16.29

The 3D locations of the active catalytic features of the proteins 1CB8, 1QAZ, 1FMI, and 1RFN.

plot, with both signs always present. For statistical clustering, and particularly for beta-sheets, this can have some advantages; for applications dependent upon spatial sequencing of residues, this is less useful since any piecewise linear connections become very cluttered.

- **Force close frames.** One useful technique for studying sequences of quaternion frames is to assume that neighboring pairs are not wildly different in their orientations. Each pair of initially assigned neighboring quaternion frames can be characterized by the sign of their mutual dot product; the "force-close-frames" algorithm makes a smallest-rotation assumption, and changes the sign of the *next* quaternion frame if its dot product with its predecessor is initially negative. The result is a unique single sequence of quaternion frames, with no sign ambiguity or duplicated points, in which each neighboring pair of quaternion frames has a nonnegative dot product, i.e., $q_{(i)} \cdot q_{(i+1)} \geqslant 0$.

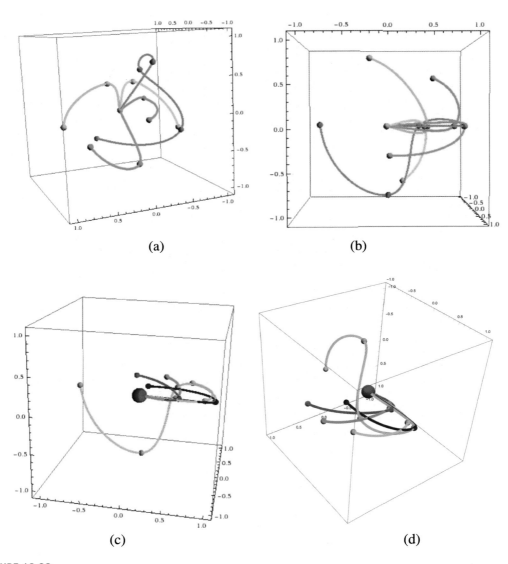

FIGURE 16.30

(a) The quaternion maps of the listed active sites for **1IA6** (red), **1FMI** (yellow), **1CB8** (green), **1QAZ** (cyan), and **1RFN** (blue), transformed to the same quaternion origin (first residue is the reference identity frame). (b) Result when the quaternion paths for all five enzymes from the first to the second residue are rotated to lie on the same axis. (c) Quaternion map that results when we perform the maximum possible alignment, with the frames of the third residue rotated to lie in a common xy-plane. (d) Oblique view of the maximal quaternion space alignment.

- **Color coding.** While the value of q_0 is in principle superfluous, it can be useful to supply redundant information about its value, particularly in complicated long sequences, where q_0 may change sign several times. A simple way to do this is to color code the value of the unseen q_0 component at each plotted point of **q**.
- **Cycle through displayed quaternion components.** While our default xyz quaternion projection displays **q** and omits q_0, it may be useful to display q_0 as one of the three visible components and omit one of the q_k, e.g., the wyz projection. This is particularly useful for exposing certain types of circular or cyclic structures.
- **Grouping by skipping.** Many proteins have global orientation patterns that are not exactly sequential, but that may be exposed by sampling the protein at intervals. Thus if we group quaternion points corresponding to interval samplings, we can sometimes see the global orientation structure more clearly.

16.9.2 Metric for comparing quaternion maps

The quaternion maps for complex and large proteins can be dense, and discerning the relevant structure visually may become difficult. In such cases, we can use selection tools that rely on quaternion space distances to pull out various subsequences or similar regions.

Similar protein frames are characterized by quaternions that have larger mutual dot products (cosine of the angle between 4-vectors closer to one) and so are closer in quaternion space. We can select locations on the protein whose orientations are similar to any given prototype point by thresholding the dot product.

Another measurement of the global properties of protein frames is the total turning along the helices. This is incrementally measured by the angle that takes one frame and rotates it into the other. This turning angle is given as usual by the quaternion-based measure computed from the dot product.

16.9.3 Quaternion rings: orientation freedom spaces

Further global information can be obtained from the full space of the possible orientations of families of frames (for more detailed discussion of spaces of frames, see Hanson (2006)). If an object can have a continuous cyclic family of related orientations, for example, the rotations resulting from spinning a single frame about its fixed axis, that family can be drawn as a circle or closed curve in quaternion space. We now apply this concept to include the appearance of such frames of a given protein, and describe the resulting global structure in the quaternion space.

Given a vector fixed by the molecular geometry, one can describe all possible orientations formed by spinning about this vector using a quaternion plot. This gives rise to families of orientation frames expressed locally as rings in the corresponding quaternion map, as studied in Hanson (1998).

Examples of such ring structures are displayed in Fig. 16.31, which shows the quaternion ring structure exposing the axial degrees of freedom for one of the strands of the protein 1C94. Based on the B-spline attached to the simplest possible skeleton, the direct C_α-C_α curve, the figure exhibits the global ring structure under the (x, y, z) and (w, x, z) projections.

The ring structures for the simpler protein 1AIE are also shown in Fig. 16.31 for comparison with those of the protein 1C94.

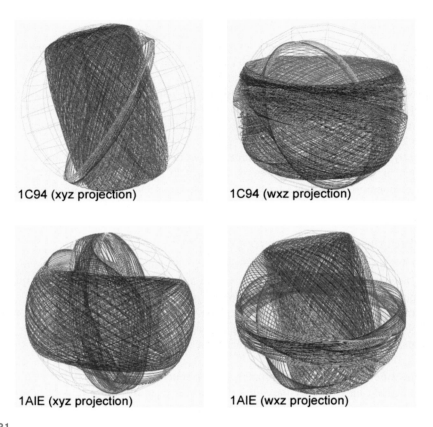

1C94 (xyz projection)   1C94 (wxz projection)

1AIE (xyz projection)   1AIE (wxz projection)

FIGURE 16.31

Quaternion ring structures for two helical proteins. Top row: rings for the protein **1C94** shown in Fig. 16.16. Bottom row: rings for the protein **1AIE** shown in Fig. 16.6. The maps in the first column are under (x, y, z) projection, and those in the second column are under (w, x, z) projection.

Significance of quaternion ring structures. The quaternion rings precisely describe the one degree of freedom allowed for rotation around a bond or the freedom of rotations in the plane of a triangle used to constrain a frame choice. These are closely related to the type of degree of freedom expressed in a Ramachandran plot, except that quaternions possess fully valid metric properties relating neighboring values. The full space of orientations for all the bonds thus forms an *orientation landscape*, points on which correspond to allowable configurations of protein chains.

16.10 Conclusion

We have attacked the problem of defining global frames appropriate to the sequences of amino acids that make up proteins. Traditional methods for analyzing protein orientations such as the Ramachandran plot are useful for local relationships but say nothing about global orientation patterns or statistical distributions of absolute orientations. Quaternion maps embody a powerful technology for revealing global orientation patterns and similarities, and we have exhibited explicit examples of methods that create quaternion maps of both discrete and continuous orientation sequences derived directly from the PDB file structure for any given protein with crystallographic or NMR-based geometry. Quaternion maps thus provide a unique bridge between sequence and structure, and establish a framework that enables the asking of novel classes of questions.

Spherical geometry and quaternions

In this part, our focus is on the fact that quaternion geometry is intimately related to the properties of spheres. Ordinary Euclidean geometry is relatively easy to understand, and is the source of many common and useful concepts such as averaging a set of points, finding centers of mass, barycentric coordinates, and linear interpolation. Therefore we begin in Chapter 17 with a brief review of these standard properties of Euclidean geometry in real N-dimensional space.

One might think at first that all these familiar Euclidean constructs could be easily extended to spheres, and in particular to the 3-sphere $\mathbf{S}^3$ that defines the space of quaternions. This is not entirely true, and although there are a number of useful properties shared by Euclidean and spherical geometry, there are a number of simple applications that are much more difficult for spheres, or even impossible. We explore these properties in the other two chapters. In Chapter 18, we look into the question of finding the best quaternion to represent a *collection* of quaternions, which, in our context, can of course be considered as the best rotation to represent a collection of rotations. In Euclidean geometry, this is just the center of mass; the quaternion or rotation counterpart is quaternion averaging, that is, how do you find a spherical center of mass. Finally, in Chapter 19, we examine the next logical step, which is how to encode the position of a spherical point in terms of a meaningful weighted basis; we would like to know the spherical counterpart of the Euclidean barycentric coordinates using a simplicial basis. Both of these spherical analogs of simple constructs of Euclidean linear algebra turn out to be very challenging.

Euclidean center of mass and barycentric coordinates[☆]

Our task here is to introduce some basic concepts in Euclidean geometry that frequently appear as desirable in the context of spherical geometry, and, in particular, the quaternion 3-sphere $\mathbf{S}^3$. This chapter thus serves as a reference context for what follows in Chapter 18 and Chapter 19, where we will see that extending all the features of Euclidean geometry operations exactly to spherical geometry is surprisingly challenging and in some respects impossible.

Two of the concepts in question, the Euclidean center of mass, also sometimes referred to as the barycenter, and the barycentric coordinates often employed for interpolation of an intermediate point among a set of anchor points, have confusingly similar names in much of the literature but are quite distinct.

Definition of the center of mass. The *center of mass* or *barycenter* is a point that in some sense is the average of a collection of points. The objective of computing the center of mass is to take as input a collection of *many* (possibly weighted) points to output a *single* unique point best representing the entire collection of input points. In classical mechanics, this would be the point in an object that describes the trajectory of the object as a whole, even if its mass distribution around that point is very complicated.

Definition of barycentric coordinates. The *barycentric coordinates* of a point, in contrast, cannot be defined in isolation, but are referred to a chosen framework of points that serve essentially as a coordinate system, but, unlike the standard Cartesian system, have their own meaning in some geometric context. While there is a substantial literature both for Euclidean and spherical barycentric coordinates relative to arbitrary numbers of points (Wachspress, 1975; Warren, 1996; Floater, 2003; Floater et al., 2006; Hormann and Floater, 2006; Ju et al., 2005a,b; Langer et al., 2006), we will devote our attention only to the *minimal* framework, which in N dimensions is a set of $N + 1$ nondegenerate points (with nonvanishing N-volume and a well-defined orientation), known as the *simplex*. The relationship between the barycenter and barycentric coordinates is that the barycenter of the anchor points can be defined as the point at which all the barycentric coordinates are equal and sum to unity. We will see that this apparently essential relationship becomes complicated in spherical geometry.

17.1 The 1D simplex: $N = 1$

The very simplest thing we can think of in the world of geometric objects and their centers is a line segment in $\mathbb{R}^1$. If we consider the two points $\mathbf{x}_1$ and $\mathbf{x}_2$ as shown in Fig. 17.1, the line segment between

---

[☆] Much of this chapter is based on material from Hanson (1994).

Visualizing More Quaternions. https://doi.org/10.1016/B978-0-32-399202-2.00029-0
273

them is a 1D simplex. It is obvious that two points are required to define the simplex, "two points determine a line," and this generalizes to any Euclidean dimension $\mathbb{R}^N$: we need $(N+1)$ points with N-dimensional coordinates to determine an N-simplex.

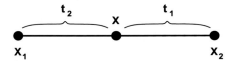

FIGURE 17.1

A point lying on a 1-simplex (a line segment).

The barycentric coordinates. The *barycentric coordinates* of an arbitrary point $\mathbf{x}$ are defined in this framework *relative to the coordinate system* determined by the simplex vertices, in this case just the two points $\mathbf{x}_1$ and $\mathbf{x}_2$. The barycentric coordinates are *not* defined just by the distances $\|\mathbf{x}-\mathbf{x}_1\|$ and $\|\mathbf{x}_2-\mathbf{x}\|$ but are in fact a pair of *signed* parameters, only one of which is independent. We define the barycentric coordinates of $\mathbf{x}$ relative to the chosen coordinate system with *complementary* labels, so t_1 is opposite the vertex $\mathbf{x}_1$. As t_1 approaches unity, we can think of it as "pushing" the variable point $\mathbf{x}$ along the line to the left until it runs into $\mathbf{x}_1$, and similarly t_2 pushes $\mathbf{x}$ to the right towards $\mathbf{x}_2$. The labels and signs of our barycentric coordinates work out as follows:

$$\mathbf{x} = t_1\mathbf{x}_1 + t_2\mathbf{x}_2 ,$$

$$t_1 = \frac{(x_2 - x)}{(x_2 - x_1)} , \qquad\qquad t_2 = \frac{(x - x_1)}{(x_2 - x_1)} , \qquad (17.1)$$

where we introduce *signed lengths*, so, e.g., the denominator

$$d = x_2 - x_1 \qquad (17.2)$$

is just the signed length of the whole line segment. We confirm that the signs are arranged so that $\mathbf{x}(t_1 = 1, t_2 = 0) = \mathbf{x}_1$ and $\mathbf{x}(t_1 = 0, t_2 = 1) = \mathbf{x}_2$. Of course only one t_k is independent because

$$t_1 + t_2 = \frac{1}{d}(x_2 - x + x - x_1) = 1 . \qquad (17.3)$$

The barycenter. If we choose equally distributed weights,

$$t_1 = t_2 = \frac{1}{2} ,$$

we define the actual *barycenter* of $\mathbf{x}_1$ and $\mathbf{x}_2$ to be the corresponding point

$$\mathbf{x}_{\text{barycenter}} = \frac{1}{2}(\mathbf{x}_1 + \mathbf{x}_2) \qquad (17.4)$$

corresponding to the center of mass. While these equations spell out the essence of barycenters, there are further important properties that do not start to emerge clearly until we examine some higher-dimensional cases. Thus we next look at two and three dimensions and their explicit properties.

17.2 Triangular simplex: $N = 2$

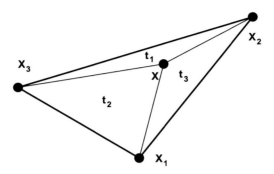

FIGURE 17.2

Simplicial barycentric coordinates in two dimensions.

For our next set of barycentric coordinates, consider three points $\{\mathbf{x}_k\}$, $k = (1, 2, 3)$, in $\mathbb{R}^2$ defining a triangle in a plane. If we consider a fourth point $\mathbf{x}$ inside the triangle, we see, as in Fig. 17.2, that this point divides the triangular area into the union of three triangles, each having $\mathbf{x}$ as one vertex. The *ratios of these areas* define 2D barycentric coordinates using three dimensionless ratios of triangle areas, summing to unity, which in turn uniquely define the location of the point $\mathbf{x} = [x, y]$ without actually referring to its location in Euclidean coordinates. That is, we can describe a point in $\mathbb{R}^2$ relative to a coordinate system defined by the three points $\{\mathbf{x}_k\}$ using the *barycentric coordinates* (Möbius, 1827) that are ratios of triangle areas. However, to express the idea of triangle areas to something that extends easily to higher dimensions, we introduce first the determinant formula for the area of a parallelogram defined by the rays $(\mathbf{x}_1 - \mathbf{x}_3)$ and $(\mathbf{x}_2 - \mathbf{x}_3)$,

$$\textbf{2D parallelogram area} = 2 \times \textbf{triangle area} = \det \begin{bmatrix} x_1 - x_3 & x_2 - x_3 \\ y_1 - y_3 & y_2 - y_3 \end{bmatrix} = \det \begin{bmatrix} x_1 & x_2 & x_3 \\ y_1 & y_2 & y_3 \\ 1 & 1 & 1 \end{bmatrix},$$

which is precisely twice the area of the triangle $\{\mathbf{x}_1, \mathbf{x}_2, \mathbf{x}_3\}$ cutting across the diagonal of the parallelogram. We see that if $\mathbf{x}_3 = [0, 0]^t$ is at the origin, then this is a positive area if the vertices are in right-handed order. Therefore we now define the area of any triangle using a generalizable volume formula:

$$\textbf{Triangle area} = V(\mathbf{x}_1, \mathbf{x}_2, \mathbf{x}_3) = \frac{1}{2} \det \begin{bmatrix} x_1 & x_2 & x_3 \\ y_1 & y_2 & y_3 \\ 1 & 1 & 1 \end{bmatrix}. \tag{17.5}$$

The barycentric coordinates. We now can define the barycentric coordinates suggested by Fig. 17.2 explicitly in terms of ratios of the three subdividing inner triangle areas to their obvious enclosing outer triangle area $d = V(\mathbf{x}_1, \mathbf{x}_2, \mathbf{x}_3)$ as

$$t_1 = \frac{1}{d} V(\mathbf{x}, \mathbf{x}_2, \mathbf{x}_3), \quad t_2 = \frac{1}{d} V(\mathbf{x}_1, \mathbf{x}, \mathbf{x}_3), \quad t_3 = \frac{1}{d} V(\mathbf{x}_1, \mathbf{x}_2, \mathbf{x}), \tag{17.6}$$

obeying

$$t_1 + t_2 + t_3 = 1 \ . \tag{17.7}$$

With these coefficients, we can now write any inner subdividing point $\mathbf{x}$ in terms of its barycentric coordinates t_i as

$$\mathbf{x} = t_1 \mathbf{x}_1 + t_2 \mathbf{x}_2 + t_3 \mathbf{x}_3 \ . \tag{17.8}$$

Due to the constraint Eq. (17.7), the barycentric coordinates t_i describe the two independent degrees of freedom of $\mathbf{x}$ in a symmetric notation via a partition of unity. This extremely attractive notion will fail for spherical geometry.

The barycenter. Once again, we find the center of mass of the three points $\{\mathbf{x}_k\}$ by setting all the parameters t_k equal to one another,

$$t_1 = t_2 = t_3 = \frac{1}{3} \tag{17.9}$$

and thus

$$\mathbf{x}_{\text{barycenter}} = \frac{1}{3} (\mathbf{x}_1 + \mathbf{x}_2 + \mathbf{x}_3) \ . \tag{17.10}$$

17.3 Tetrahedral simplex barycenter: $N = 3$

We now study the 3D case for the tetrahedral simplex for reasons that may not be so obvious at first: we want to have this Euclidean structure worked out explicitly because it is the closest thing we have in Euclidean space to the analogous structures relevant to quaternions. Quaternions live intrinsically on the 3-sphere $\mathbf{S}^3$ that in very small portions looks very much like 3D Euclidean space, so the fundamental nontrivial small-scale structure on $\mathbf{S}^3$ must therefore be a solid 3D object. This object is the *spherical tetrahedron*, an apparently simple structure with enormous hidden complexity.

Therefore we will now work out for completeness the details of the tetrahedron and tetrahedral barycentric coordinates based on the fundamental simplex in $\mathbb{R}^3$, Euclidean 3D space. We start with four (assumed noncoplanar) points $\{\mathbf{x}_k\}$, $k = (1, 2, 3, 4)$, in $\mathbb{R}^3$, defining a tetrahedron in 3D space, as shown in Fig. 17.3. If we consider a fifth point $\mathbf{x}$ inside the tetrahedron, we see, as in Fig. 17.3, that this point divides the tetrahedral volume into the union of four smaller tetrahedra, each having $\mathbf{x}$ as one vertex. The pattern that emerges of course is that we scratch out each of $N + 1$ vertices in turn and replace it with the interior point $\mathbf{x}$, creating $N + 1$ smaller simplices with the same number of vertices subdividing *exactly* the entire original simplex.

The barycentric coordinates. The *dimensionless ratios of volumes* are again barycentric coordinates for the point $\mathbf{x} = [x, y, z]^t$ relative to the *tetrahedral* minimal coordinate basis defined by the $\{\mathbf{x}_k\}$. Just to be pedantic, we write them out to expose the general pattern. Taking the volume of the entire

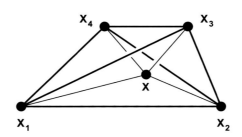

Tetrahedral barycentric coordinates.

tetrahedron as

$$d = \textbf{tetrahedron volume} = V(\mathbf{x}_1, \mathbf{x}_2, \mathbf{x}_3, \mathbf{x}_4) = \frac{1}{6} \det \begin{bmatrix} x_1 & x_2 & x_3 & x_4 \\ y_1 & y_2 & y_3 & y_4 \\ z_1 & z_2 & z_3 & z_4 \\ 1 & 1 & 1 & 1 \end{bmatrix}, \tag{17.11}$$

then we can define

$$\mathbf{x} = t_1 \mathbf{x}_1 + t_2 \mathbf{x}_2 + t_3 \mathbf{x}_3 + t_4 \mathbf{x}_4, \tag{17.12}$$

$$t_1 = \frac{1}{d} V(\mathbf{x}, \mathbf{x}_2, \mathbf{x}_3, \mathbf{x}_4), \quad t_2 = \frac{1}{d} V(\mathbf{x}_1, \mathbf{x}, \mathbf{x}_3, \mathbf{x}_4), \quad t_3 = \frac{1}{d} V(\mathbf{x}_1, \mathbf{x}_2, \mathbf{x}, \mathbf{x}_4), \quad t_4 = \frac{1}{d} V(\mathbf{x}_1, \mathbf{x}_2, \mathbf{x}_3, \mathbf{x}).$$
$$\tag{17.13}$$

The four values of t_k are of course a linearly constrained partition of unity, obeying

$$t_1 + t_2 + t_3 + t_4 = 1, \tag{17.14}$$

which symmetrically parameterizes the three actual degrees of freedom in the placement of $\mathbf{x}$ in terms of the tetrahedral simplicial basis $(\mathbf{x}_1, \mathbf{x}_2, \mathbf{x}_3, \mathbf{x}_4)$.

The barycenter. As before, if we set all the parameters to the same value,

$$t_k = \frac{1}{N+1} = \frac{1}{4},$$

the relations of Eq. (17.12) are satisfied and $\mathbf{x}$ is the center of mass,

$$\mathbf{x}_{\text{barycenter}} = \frac{1}{4}(\mathbf{x}_1 + \mathbf{x}_2 + \mathbf{x}_3 + \mathbf{x}_4). \tag{17.15}$$

17.4 The *N*-simplex barycenter

The generalizations of the simplicial barycenter equations can now be deduced easily from the formulas we have written down for 1D, 2D, and 3D.

In N-dimensional Euclidean space $\mathbb{R}^N$, we must have $(N + 1)$ nondegenerate (not lying in the same $(N - 1)$-dimensional hyperplane) points to define a coordinate system on $\mathbb{R}^N$, and we denote the set of those points by $\{\mathbf{x}_k\}$, with distinct points labeled by the index $k = (1, \ldots, N + 1)$. Since the dimensions are arbitrary, instead of denoting the components of each point as $\mathbf{x} = [x, y, z, \ldots]$, we choose the admittedly cumbersome, but more flexible, notation that represents one N-dimensional vector as $\mathbf{x} = x^{(a)}$, $a = 1, \ldots, N$. The points $\mathbf{x}_k$ define a line segment in 1D, a triangle in 2D, a tetrahedron in 3D, etc., as we have seen in Figs. 17.1, 17.2, and 17.3, that is, they explicitly define the geometry of an N-simplex, a polytope enclosing a volume in N-dimensional space using the smallest possible number of points. The determinant of the $(N + 1) \times (N + 1)$-dimensional square matrix defined by the homogeneous forms of the vectors $\mathbf{x}$, which is to say the vector with an extra unit scale dimension appended, is the *volume of the (hyper)parallelepiped* with the N defining edges radiating from the first vertex $\mathbf{x}_1$ (or any other) to the remaining vertices. The volume of the corresponding N-simplex that is the convex hull of the $(N + 1)$ coordinate-frame-defining vertices is $1/N!$ times the (hyper)parallelepiped volume, or

$$
V_0 = \text{simplex volume} = \frac{1}{N!} \det \begin{bmatrix}
x_1^{(1)} & x_2^{(1)} & \cdots & x_{N+1}^{(1)} \\
x_1^{(2)} & x_2^{(2)} & \cdots & x_{N+1}^{(2)} \\
\vdots & \vdots & \ddots & \vdots \\
x_1^{(N)} & x_2^{(N)} & \cdots & x_{N+1}^{(N)} \\
1 & 1 & \cdots & 1
\end{bmatrix}. \tag{17.16}
$$

Now, as we can see by analogy to Fig. 17.3, if one places an arbitrary test point $\mathbf{x}$ probing the interior of the simplex, one can immediately define $(N + 1)$ *interior simplices* by simply excluding each original point $\mathbf{x}_k$ in turn and replacing it by the test point $\mathbf{x}$. The volumes V_k of each of those interior simplices are given by Eq. (17.16) with the column containing the omitted $\mathbf{x}_k$ replaced by the test point $\mathbf{x}$. The barycentric coordinates of any arbitrary test point with respect to the simplicial coordinate system $\{\mathbf{x}_k\}$ are just the dimensionless ratios of those volumes relative to the total volume V_0 in Eq. (17.16):

$$
t_1 = \frac{V_1}{V_0}, \quad t_2 = \frac{V_2}{V_0}, \quad \cdots \quad t_N = \frac{V_N}{V_0}, \quad t_{N+1} = \frac{V_{N+1}}{V_0}, \tag{17.17}
$$

with $\mathbf{x}$ in general therefore taking the form

$$
\mathbf{x} = t_1 \mathbf{x}_1 + t_2 \mathbf{x}_2 + \cdots + t_N \mathbf{x}_N + t_{N+1} \mathbf{x}_{N+1}. \tag{17.18}
$$

Since the $(N + 1)$ partial volumes exactly fill the entire N-simplex, the barycentric coordinates t_k sum to unity, and only N of them are actually independent coordinates in $\mathbb{R}^N$, as required:

$$
t_1 + t_2 + \cdots + t_N + t_{N+1} = 1. \tag{17.19}
$$

Finally, the actual *simplicial barycenter* is defined by setting all the parameters equal,

$$
t_k = \frac{1}{N + 1} \mid \forall k,
$$

which corresponds to having all the partial volumes exactly equal to one another, a concept going back at least as far as Möbius (1827).

17.5 Variational approach to the Euclidean center of mass

Suppose we consider again an arbitrary N-dimensional space with coordinates $\mathbf{x} = (x^{(1)}, x^{(2)}, \ldots, x^{(N)})$ and a list of K such coordinates $\mathbf{X} = \{\mathbf{x}_1, \mathbf{x}_2, \ldots, \mathbf{x}_K\}$; we can choose K to be anything we want, including $K = N + 1$ if we wish to describe the N-simplex itself. We now consider a completely different approach to finding a point $\mathbf{x}$ that describes our whole collection of coordinates: we define a loss function penalizing the squared distance of each point from $\mathbf{x}$ as follows:

$$F(\mathbf{x}, \{\mathbf{x}_k\}) = \sum_{k=1}^{K} \|\mathbf{x} - \mathbf{x}_k\|^2 \; . \tag{17.20}$$

A *least-squares* optimization problem searches for the solution minimizing the *derivative* of $F(\mathbf{x})$ with respect to each component of $\mathbf{x}$, which gives the equation

$$\frac{\partial (F((\mathbf{x}, \{\mathbf{x}_k\}))}{\partial \mathbf{x}} = 2 \sum_{k=1}^{K} (\mathbf{x} - \mathbf{x}_k) \tag{17.21}$$

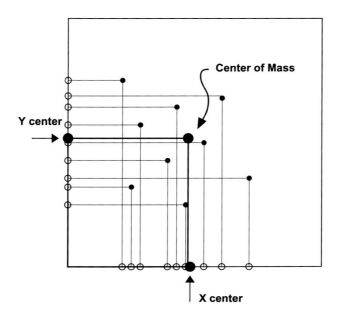

FIGURE 17.4

The Euclidean barycenter is an independent average in each Cartesian direction.

$$= 2 \left(K\mathbf{x} - \sum_{k=1}^{K} \mathbf{x}_k \right) \tag{17.22}$$

$$= 0 . \tag{17.23}$$

This is easily solved, and we recover exactly the usual barycenter equation

$$\mathbf{x}_{\text{barycenter}} = \frac{1}{K} \sum_{k=1}^{K} \mathbf{x}_k \tag{17.24}$$

as the solution minimizing the sum of the squared distances $\|\mathbf{x} - \mathbf{x}_k\|^2$ to the test point $\mathbf{x}$. The weighting factor $1/K$ of the solution is of course a partition of unity,

$$\sum_{k=1}^{K} \frac{1}{K} = 1 ,$$

which connects the classic center of mass formula to the generalization of the simplicial barycentric coordinates. In the simplified diagram of Fig. 17.4, we illustrate that, in Euclidean space, the coordinates projected onto one or more of the axes of any orthogonal coordinate system are also centers of mass in any of those subspaces.

Quaternion averaging and the barycenter

Our purpose here is to investigate how far we can go in generalizing the definition and applications of the Euclidean center of mass in Chapter 17 to the quaternion sphere $\mathbf{S}^3$. The basic result is that the theoretically ideal concept of an average quaternion representing a cluster of points on $\mathbf{S}^3$ requires geodesic distances whose averages can be computed numerically but not algebraically. Alternative methods, e.g., using chordal distances, admit algebraic forms but do not minimize exact spherical measures. This task of defining a center of mass on the quaternion sphere appears frequently in the literature related to the "rotation averaging" or the "quaternion averaging" problem.

18.1 Introduction

The generalized non-Euclidean center of mass or *Riemannian barycenter* is generally associated with the work of Grove et al. (1974), and may also be referred to as the *Karcher mean* (Karcher, 1977), defined as the point that minimizes the sum of squared geodesic distances from the elements of a collection of fixed points on a manifold. The general class of such optimization problems has also been studied, e.g., by Manton (2004). We are interested here in the barycenter of a collection of quaternions, which we know are points $q \in \mathbb{R}^4$ on the spherical 3-manifold $\mathbf{S}^3$ defined by the unit-quaternion subspace restricted to $q \cdot q = 1$. This subject has been investigated by a number of authors. For example, Buss and Fillmore (2001) proposed a solution applicable to computer graphics 3D orientation interpolation problems in 2001, inspired to some extent by Shoemake's 1985 introduction of the quaternion `slerp` of Eq. (2.19) as a way to perform geodesic orientation interpolations in 3D using Eq. (2.14) for $R(q)$. There are a variety of methods and studies related to the quaternion barycenter problem. In 2002, Moahker published a rigorous account on averaging in the group of rotations (Moakher, 2002), while subsequent treatments include the work of Markley et al. (2007), focusing on aerospace and astronomy applications, and the comprehensive review by Hartley et al. (2013), aimed in particular at the machine vision and robotics community, with additional attention to conjugate rotation averaging (the "hand-eye calibration" problem in robotics) and multiple rotation averaging. While we will focus on L_2 norms, measures starting from sums of squares that lead to closed form optimization problems, Hartley et al. (2011) have carefully studied the utility of the corresponding L_1 norm and the iterative Weiszfeld algorithm for finding its optimal solution numerically.

There are many reasons why we might need to explore spherical geometry and try to extend features of the Euclidean center of mass to the spherical context in the process of studying quaternion applications. For example, given a distribution of frames, whether biased, localized, or uniform, we may often want to describe the entire collection by the *average rotation* or, equivalently, the *spherical center of mass* of a cluster or neighborhood of quaternions in $\mathbf{S}^3$.

Visualizing More Quaternions. https://doi.org/10.1016/B978-0-32-399202-2.00030-7

We know from Chapter 17 the basic machinery of sets of points in N-dimensional Euclidean space: there are closed form solutions of least-squares methods that produce exact centers of mass that are equivalent to Euclidean point averaging. Since our applications of quaternion geometry rely heavily on the hyperspherical geometry of the 3-sphere $\mathbf{S}^3$ and its equivalence to the space of unit quaternions, we are therefore motivated to investigate spherical analogs of Euclidean geometry and Euclidean barycentric coordinates that might help us understand and exploit quaternions.

A thorough review of orientation representations can be found in, e.g., Huynh (2009), and many features of the rotation averaging nongeodesic approximations that we will explore appear in the literature, including in particular Moakher (2002); Markley et al. (2007); Hartley et al. (2011, 2013); Huggins (2014).

18.2 The geometry of inter-quaternion distances

As a brief introduction, we review the geometry of the way we need to think about distance relationships between two quaternions, which we will write as q and p. The terms in these relationships can be counterintuitive because an individual quaternion is associated to 3D axis-angle rotations in physical space by the half-angle: if the rotation in space is by an angle θ, the scalar quaternion component is expressed with half that angle, $q_0 = \cos(\theta/2)$. Since the scalar product of two quaternions is invariant under any 4D orthogonal rotation, one quaternion in a dot product can be rotated to the identity $(1, 0, 0, 0)$, identifying the full dot product as equivalent to a rotation to some angle relative to the identity appearing in the transformed q_0. Thus the dot product has dual properties: defining $q \cdot p = \cos\alpha$, then α is the angle between the two quaternions considered as natural unit 4-vectors in Euclidean $\mathbb{R}^4$. But that is also related to the actual 3D relative rotation angle θ via the equation

$$q \cdot p = q_0 p_0 + \mathbf{q} \cdot \mathbf{p} = \cos\alpha = \cos(\theta/2) . \tag{18.1}$$

Here, as noted, we can imagine having transformed to a coordinate system where p is an identity quaternion, so θ is the angle of the 3D *rotation* that would transform $R(p)$ into $R(q)$ (or if $R(p) = I_3$, just the parameter of $R(q) = R(\theta, \hat{\mathbf{n}})$). So we have two perfectly geometrical ways to interpret the geodesic minimum-spherical-distance arc length relating q and p. This arc length, $\alpha = \arccos(q \cdot p)$, is our "gold standard" for the most correct quaternion-quaternion distance measure, the measure corresponding most closely to the Euclidean distance in Euclidean geometry.

Note on rotation proximity: We must of course keep in mind that, since $R(q) = R(-q)$, as the distance between actual quaternions passes from $q \cdot p = 1$ (arc length 0) to $q \cdot p = 0$ (arc length $\pi/2$) to $q \cdot p = -1$ (maximum arc length π), the relative proximity of the corresponding *rotations* goes from closest to farthest and back to closest. Thus if we want the quaternion-valued calculation of *rotation distance*, we must use the sign-insensitive alternative geodesic distance based on

$$\textbf{Rotation proximity} = |q \cdot p| = |q_0 p_0 + \mathbf{q} \cdot \mathbf{p}|$$
$$= \max(q \cdot p, -q \cdot p) . \tag{18.2}$$

However, we are going to be looking at modifying the gold standard, to what we might call the "silver standard," exploiting the simplicity of Euclidean geometry to define measures that, while not

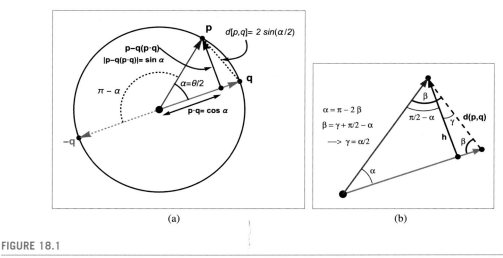

(a)   (b)

FIGURE 18.1

(a) The geometry of quaternion distances relating two (unit-length) quaternions q and p. Their geodesic distance is the arc length $\alpha = \arccos(q \cdot p)$ in radians, the perpendicular distance is $\|p - q(q \cdot p)\| = \sin \alpha$. (b) Additional detail on the Euclidean "chord length" distance between the pair of quaternions, $d(q, p) = 2\sin(\alpha/2)$ (up to signs).

exact, are good approximations to the geodesic arc length measure, and are *far easier* to use in computational applications. In Fig. 18.1, we illustrate typical quantities that we will consider for the task of quantifying inter-quaternion distances. The geodesic arc length is of course the angle between the directions from the origin to the unit-radius points q and p. Since we do not care about the signs of the quaternions, we would typically replace any $q \cdot p < 0$ case by its positive-sign counterpart, e.g., taking $\alpha' = \pi - \alpha \to \alpha$, or we could simply use $|q \cdot p|$ before taking the inverse cosine. The important quantities that we see in Fig. 18.1 are

dot product:	$= q \cdot p = \cos \alpha = \cos(\theta/2)$	(18.3)
dropped perpendicular to q:	$= p - q(q \cdot p) = p - q\cos\alpha$	(18.4)
squared length of dropped perpendicular:	$= \|p - q(q \cdot p)\|^2 = 1 - (q \cdot p)^2 = 1 - \cos^2\alpha = \sin^2\alpha$	(18.5)
length of dropped perpendicular:	$= h = \|p - q(q \cdot p)\| = \sin\alpha$	(18.6)
chordal square distance:	$= d(q, p)^2 \equiv \|q - p\|^2 = 2 - 2(q \cdot p) = 2(1 - \cos\alpha)$	(18.7)

$$= 4\left(\sin^2\frac{\alpha}{2}\right) = 4\left(\sin^2\frac{\theta}{4}\right) \tag{18.8}$$

chordal distance:
$$= d(q, p) = \frac{\|p - q(q \cdot p)\|}{\cos(\alpha/2)} = \frac{\sin\alpha}{\cos(\alpha/2)} = 2\sin\left(\frac{\alpha}{2}\right), \tag{18.9}$$

where we used the fact that $h/d(q, p) = \cos\gamma$ and, from Fig. 18.1(b),

$$\alpha + 2\beta = \pi, \qquad \beta - \gamma = \left(\frac{\pi}{2} - \alpha\right) \ \to \ \gamma = \frac{\alpha}{2}.$$

18.3 Analysis of the quaternion averaging problem

We begin with the definition of quaternion averaging as the minimization of a loss measure or cost function. This is necessary because the Euclidean definition of an average based on a simple sum breaks down on curved spaces like a sphere, with the important exception of the circle $\mathbf{S}^1$, for which the values of angles in a hemicircle behave as a Euclidean line. Thus we make the crucial assumption that our task is to find the best ways to generalize the Euclidean cost function, $S(\mathbf{x}) = \sum_k \|\mathbf{x} - \mathbf{x}_k\|^2$, whose least-squares minimization we exploited as an alternative method for finding the Euclidean center of mass.

The geodesic spherical loss function. We start with a data set of K rotation matrices $R(p_k)$ that are represented by the quaternions p_k, and we want to find a quaternion q that gives a rotation $R(q)$ as close as possible to the set of $R(p_k)$. To obtain that rotation matrix, or its associated quaternion point, we choose the analog of the Euclidean distance in our Euclidean barycenter loss measure for a set of Euclidean points to be the quaternion sign-insensitive *geodesic arc length* on the quaternion sphere. We have the option of choosing any norm power-parameter p, giving us in general

$$\mathbf{S}_{p[\text{geodesic}]}(q) = \sum_{k=1}^{K} \arccos |q \cdot p_k|^p \ . \tag{18.10}$$

For $p = 1$, we have the L_1 norm preferred in some cases, and for $p = 2$, we have the standard squared error convention with an L_2 norm. Following Buss and Fillmore (2001), we emulate the Euclidean case and use the squared error $L_{p=2}$ loss function to start our exploration; we also define $\cos\alpha_k = |q \cdot p_k|$, sidestepping the possible occurrence of $\pi - \alpha_k$ when $q \cdot p_k < 0$. Thus we can write

$$q_{\text{opt}} = \underset{q}{\text{argmin}} \left(\sum_{k=1}^{K} \arccos |q \cdot p_k|)^2 \right) = \underset{q}{\text{argmin}} \left(\sum_{k=1}^{K} (\alpha_k)^2 \right) \ . \tag{18.11}$$

This is essentially the method explored by Buss and Fillmore (2001) in the context of trying to produce more general rotation spline techniques. But to the best of anyone's knowledge, there is no way to apply linear algebra to find the corresponding exact "average quaternion" q_{opt} and its implied optimal barycentric rotation solution $R(q_{\text{opt}})$. Nevertheless, numerically solving Eq. (18.11) is what we should do if our application *requires* an extremely accurate answer.

However, once we relax the rigorous requirements of the perfect geodesic loss function Eq. (18.10), we can find a number of connections between the rotation matrix Frobenius norms and the quaternion *chord measure* approximations, as well as our investigations of the Bar-Itzhack closest-rotation methods in Chapter 9, matching orientation frame clouds in Chapter 10, and quaternion adjugate matrix exploitation in Chapter 12.

Rotation-matrix-based measures. One method that has the advantage of being insensitive to the (usually irrelevant) quaternion sign is to consider complete rotation matrices $R(q)$ compared using the Frobenius measure $\|M\|^2 = \text{tr}(M \cdot M^t)$. This leads us to the following starting point for an attempt to describe a set of K rotations R_k generated by quaternions p_k in terms of a closest rotation R described by its own quaternion q. We want the value of $q = q_{\text{opt}}$ that minimizes this sum of Frobenius distances:

$$
\begin{aligned}
\mathbf{S}_{\text{barycenter}}(q) &= \sum_{k=1}^{K} \| R(q) - R(p_k) \|^2 \\
&= \sum_{k=1}^{K} \text{tr}\left([R(q) - R(p_k)] \cdot [R^{\text{t}}(q) - R^{\text{t}}(p_k)] \right) \\
&= \sum_{k=1}^{K} \text{tr}\left(2\, I_4 - 2 R(q) \cdot R^{\text{t}}(p_k) \right) \\
&= \sum_{k=1}^{K} (8 - 2\, \text{tr}\, R(q) \cdot R(\overline{p}_k)) \\
&= \sum_{k=1}^{K} (8 - 2\, \text{tr}\, R(q \star \overline{p}_k))
\end{aligned}
\qquad (18.12)
$$

where we have used $R^{\text{t}}(q) = R(\overline{q})$ and $R(q) \cdot R(p) = R(q \star p)$ from Eq. (2.15). Now we need a few algebraic manipulations, starting by defining the composite quaternion Q as follows:

$$
q \star \overline{p} = (q_0 p_0 + \mathbf{q} \cdot \mathbf{p}, \; -q_0 \mathbf{p} + p_0 \mathbf{q} - \mathbf{q} \times \mathbf{p} = (Q_0, Q_1, Q_2, Q_3) \,, \qquad (18.13)
$$
$$
Q_0 = q \cdot p = q_0 p_0 + \mathbf{q} \cdot \mathbf{p} = \cos\alpha \,. \qquad (18.14)
$$

Then if we drop the constants and change the sign in Eq. (18.12), we get the familiar cross-term function $\Delta(q)$ that needs to be maximized instead of minimized:

$$
\begin{aligned}
\Delta_{\text{barycenter}}(q) &= \sum_{k=1}^{K} (\text{tr}\, R(q) \cdot R(\overline{p}_k)) \\
&= \sum_{k=1}^{K} (\text{tr}\, R(q \star \overline{p}_k)) \\
&= \sum_{k=1}^{K} (\text{tr}\, R(Q_k)) \\
&= \sum_{k=1}^{K} \left(4 \left(Q_{0k} \right)^2 - 1 \right) \\
&= \sum_{k=1}^{K} \left(4 \cos^2 \alpha_k - 1 \right)
\end{aligned}
\qquad (18.15)
$$

Remove constants and simplify:

$$
\Delta_{\text{barycenter}}(q) = \sum_{k=1}^{K} \cos^2 \alpha_k
$$

The Frobenius norm approach thus eliminates dependencies on the sign of the quaternion and produces a maximal L_2 cost function in terms of the quaternion-quaternion cosine, $q \cdot p_k$, and implies the existence of an L_1–L_2 family of justifiable choices for our optimization function,

$$\boldsymbol{\Delta}_{\text{max}:L_1}(q) = \sum_{k=1}^{K} |\cos \alpha_k| ,$$

$$\boldsymbol{\Delta}_{\text{max}:L_2}(q) = \sum_{k=1}^{K} \cos^2 \alpha_k = \sum_{k=1}^{K} \left(1 - \sin^2 \alpha_k\right) ,$$

$$\boldsymbol{\Delta}_{\text{min}:L_2}(q) = \sum_{k=1}^{K} \sin^2 \alpha_k .$$

These appear difficult to optimize analytically, but in fact retaining the rotation matrix form

$$\boldsymbol{\Delta}_{\text{barycenter}}(q) = \sum_{k=1}^{K} (\text{tr} \, R(q) \cdot R(\bar{p}_k))$$

for the L_2 maximization version of the problem allows us immediately to apply the quaternion eigensystem method that we studied in Chapter 8.

Quaternion eigensystem optimization. If we proceed as usual and drop the constants in Eq. (18.12) and change the sign to get Eq. (18.15) cross-term $\boldsymbol{\Delta}(q)$ that is to be maximized, we see that we have precisely the chord-distance-squared loss function proposed by Markley et al. (2007) and Hartley et al. (2013):

$$\left.\begin{aligned}
\boldsymbol{\Delta}_{\text{trial barycenter}}(q) &= \frac{1}{4} \sum_{k=1}^{K} \text{tr} \left(R(q) \cdot R(\bar{p}_k)\right) \\
&= q \cdot \left[\sum_{k=1}^{K} F_k(p)\right] \cdot q
\end{aligned}\right\} . \tag{18.16}$$

We now proceed as in Chapter 8, beginning with the k-th element of an initial profile matrix,

$$F_k(p)_{\text{initial}} =$$
$$\frac{1}{4} \begin{bmatrix} 3p_0{}^2 - p_1{}^2 - p_2{}^2 - p_3{}^2 & 4p_0 p_1 & 4p_0 p_2 & 4p_0 p_3 \\ 4p_0 p_1 & -p_0{}^2 + 3p_1{}^2 - p_2{}^2 - p_3{}^2 & 4p_1 p_2 & 4p_1 p_3 \\ 4p_0 p_2 & 4p_1 p_2 & -p_0{}^2 - p_1{}^2 + 3p_2{}^2 - p_3{}^2 & 4p_2 p_3 \\ 4p_0 p_3 & 4p_1 p_3 & 4p_2 p_3 & -p_0{}^2 - p_1{}^2 - p_2{}^2 + 3p_3{}^2 \end{bmatrix}_k . \tag{18.17}$$

Next, as pointed out, e.g., by Markley et al. (2007), we can add one copy of the identity matrix in the form $(1/4)I_4 = (1/4)(p_0{}^2 + p_1{}^2 + p_2{}^2 + p_3{}^2)I_4$ to each of the matrices $F_k(p)$, and the result is simply an additive constant that has no effect on the optimization process. We then get a much simpler

k-th partial profile matrix:

$$
F_k(p) = \begin{bmatrix}
p_0{}^2 & p_0 p_1 & p_0 p_2 & p_0 p_3 \\
p_0 p_1 & p_1{}^2 & p_1 p_2 & p_1 p_3 \\
p_0 p_2 & p_1 p_2 & p_2{}^2 & p_2 p_3 \\
p_0 p_3 & p_1 p_3 & p_2 p_3 & p_3{}^2
\end{bmatrix}_k .
\tag{18.18}
$$

Here we see again that the adjusted profile matrix for finding optimal rotations using the Frobenius norm Eq. (18.12) leads to a version of the profile matrix that is *precisely* in the form of a quaternion adjugate matrix, though the actual adjugate matrix is the adjugate of the corresponding characteristic matrix $\chi(F(p))$ with unit eigenvalue. However, here we are going one step further, taking more than a single Bar-Itzhack term. To complete our task, we take the *sum* of the terms in Eq. (18.18), and that is *not* going to correspond directly to a rotation matrix. In effect, we are computing the Bar-Itzhack solution from Chapter 9 to obtain an optimal exact rotation approximating an inexact rotation that is a sum of contributions from a whole list of rotation matrices.

After taking the sum over k, the profile matrix in terms of the quaternion columns $[p_k]$ becomes essentially the self-covariance matrix of $\mathbf{P} = [p_k]_a$,

$$
F(\mathbf{P})_{ab} = \sum_{k=1}^{N} [p_k]_a [p_k]_b = \left[\mathbf{P}^t \cdot \mathbf{P} \right]_{ab} ,
\tag{18.19}
$$

with quaternion indices (a, b) ranging from 0 to 3. Finally, we can write the expression for the chord-based barycentric measure that is maximized by q_{opt} as

$$
\begin{aligned}
\Delta_{\text{barycenter}}(q) &= q \cdot [F(\mathbf{P})] \cdot q \\
&= \sum_{k=1}^{N} (q \cdot p_k)(q \cdot p_k) \\
&= \sum_{k=1}^{N} (\cos \alpha_k)^2
\end{aligned}
\tag{18.20}
$$

confirming an alternate path to our previous Eq. (18.15) in terms of the Euclidean projection of p_k onto q, the standard Euclidean alignment measure, and not a geodesic distance.

Optimization of the approximate chord-measure Eq. (18.20) for the L_2 norm has the advantage that optimization to find the "average rotation," the "quaternion average," or the spherical barycenter now just reduces, as before, to finding the (normalized) eigenvector corresponding to the largest eigenvalue of F. It is also significant that the initial F_{trial} matrix in Eq. (18.17) is *traceless*, so the traceless algebraic eigenvalue methods would apply, while the simpler F matrix in Eq. (18.19) is *not traceless*, and thus, in order to apply the algebraic Cardano eigenvalue method, we would have to use the generalization presented in Appendix J that includes an arbitrary trace term. Our solution, using the approximate L_2 chord measure and the eigensystem of the profile matrix Eq. (18.19), then becomes

$$
\left.
\begin{aligned}
\lambda_{\text{opt}} &= \text{max eigenvalue: } [F(\mathbf{P})] \\
q_{\text{opt}} &= \text{max eigenvector: } q_{\text{opt}} = \text{max norm column of: } \mathbf{Adjugate}\left(\left[F - \lambda_{\text{opt}} I_4\right]\right) \\
R_{\text{opt}} &= R(q_{\text{opt}}) \\
\mathbf{\Delta}_{\text{opt}} &= q_{\text{opt}} \cdot F(\mathbf{P}) \cdot q_{\text{opt}} = \lambda_{\text{opt}}
\end{aligned}
\right\}.
\tag{18.21}
$$

Remark: Matrix-only data. If, for some reason, we only have numerical 3×3 rotation matrices R_k and not their associated quaternions p_k, we can recast Eq. (18.16) in terms of a single target matrix formed by summing over each matrix R_k,

$$
R_{\text{total}}^{\text{t}} = \frac{1}{K} \sum_{k=1}^{K} R_k^{\text{t}},
$$

and applying the Bar-Itzhack method through Eq. (9.7) to get the closest pure rotation to the cluster of R_k matrices. We note that, as usual, inverse matrices need to appear so that, at the optimum, the product of $R(q)$ with the target is approaching the identity matrix. The observation of this possibility is nontrivial because the simple eigensystem form of the profile matrix F for the Bar-Itzhack task was originally constructed only for *one* rotation data matrix, and as soon as we start summing over additional matrices, all of that simplicity disappears, though the eigensystem problem remains intact.

Relation to the quaternion chordal measure and the L_1 norm. Our other option is to work directly with the quaternion chordal measure, though in fact the Frobenius measure and the chordal measure are related, as noted by Markley et al. (2007) and elaborated by Hartley et al. (2013). The key is that Eq. (18.12) allows one more step of simplification to create a relationship to the quaternion chordal distance.

If we replace the geodesic arc length $\arccos(q \cdot p)$ comparing the locations of two unit quaternions on the sphere $\mathbf{S}^3$ by their *Euclidean distance*, we see that once we work out the identities, the explicit chordal distance as shown in Fig. 18.1 reduces to the L_1 norm, even though it starts out as quadratic:

$$
\left.
\begin{aligned}
D(q\{p_k\}) = \textit{(Minimize): } & \sum_k \min\left(|q - p_k|^2, |q + p_k|^2\right) \\
= \textit{(Maximize): } & \sum_k \max\left(q \cdot p_k, -q \cdot p_k\right) \\
= \textit{(Maximize): } & \sum_k \max\left(\cos\alpha_k, \cos(\pi - \alpha_k)\right)
\end{aligned}
\right\},
\tag{18.22}
$$

where $q \cdot p_k = \cos\alpha_k$ is again the Euclidean projection of one quaternion, as a Euclidean vector, on the other. The challenge of making a clean loss equation is that both options are required because the Euclidean distance between the tips of the 4D quaternion unit vectors,

$$
d(q, p)^2 = (q_0 - p_0)^2 + (q_1 - p_1)^2 + (q_2 - p_2)^2 + (q_3 - p_3)^2 = 2 - 2(q \cdot p) = 2(1 - \cos\alpha) = 4\cos^2\left(\frac{\alpha}{2}\right),
$$

cannot distinguish between q and $-q$. So at $\alpha = \pi$ we have $q \cdot p \approx -1$ and $\cos(\alpha/2) \approx 0$, and we must still make the branched choice, the maximum of $d(q, p)$ for α vs. $\pi - \alpha$.

The quaternion barycentric coordinate problem

In this chapter, we continue our examination of possible quaternion extensions of the Euclidean geometry introduced in Chapter 17, focusing now on the properties of *barycentric coordinates*, which we reiterate are a collection of coefficients, as opposed to the *barycenter* or quaternion average examined in Chapter 18, which is a single point representing a collection of measured data. In the abstract, barycentric coordinates are generally taken to have the property of being a partition of unity that defines a point as a combination of a set of anchor points that establish a reference coordinate system. We will see that the partition of unity property that we found for Euclidean geometry in Chapter 17 will not have an exact parallel for spherical geometry.

19.1 Introduction

While arbitrary numbers of reference coordinates may be employed, for example, when modeling sophisticated spline techniques, we will not attempt to deal with multipoint splines here: we will restrict ourselves to *simplexes*, which in both Euclidean and spherical N-dimensional contexts are sets of $N + 1$ nondegenerate points that can be used to define a finite N-volume as well as the basis for a set of reference coordinates. We will study mostly spherical dimensions 1, 2, and 3, with the corresponding spherical simplexes being:

- **N = 1**: Two points defining a circular line segment; a geodesic arc on a circle $\mathbf{S}^1$ possibly embedded in a higher-dimensional sphere.
- **N = 2**: Three points defining a spherical triangle embedded in $\mathbf{S}^2$, possibly embedded in turn as a spherical "face" in a higher-dimensional sphere.
- **N = 3**: Four points defining a spherical tetrahedron embedded in $\mathbf{S}^3$.

Any of these objects can appear with a quaternion context.

For Euclidean geometry, where the coordinate system is linear, a point and its barycentric coordinates can always be written with a simple partition of unity in the form

$$\mathbf{p}(t_i) = \sum_{i=1}^{N+1} t_i \mathbf{x}_i \;\Big|\; \sum_{i=1}^{N+1} t_i = 1 \,. \tag{19.1}$$

We noted the relation to the "barycenter" itself, or the center of mass, in Chapter 17 that defines the center of mass of the $N + 1$ points $\mathbf{x}_i$ by choosing the t_i to all be equal, $t_i = \frac{1}{N+1}$, so

$$\mathbf{p}_{\mathrm{cm}}(\mathbf{x}_i) = \frac{1}{N+1} \sum_{i=1}^{N+1} \mathbf{x}_i \,. \tag{19.2}$$

Visualizing More Quaternions. https://doi.org/10.1016/B978-0-32-399202-2.00031-9

Unfortunately, we cannot immediately extend Eq. (19.1) or Eq. (19.2) to our target spherical domains for $N > 1$ because linear summation of bare points on a sphere does not even produce a point on a sphere.

1D spherical barycentric coordinates. The circle, or 1-sphere $\mathbf{S}^1$, has the unique advantage that, when points are parameterized by their angular location θ (in a way that does not cause problems by crossing over the equatorial $0:\pi$ border), some Euclidean geometry can still be used. For example, suppose we take K points $p_k = [\cos\theta_k, \sin\theta_k]$ with, say, $-\pi/2 \leqslant \theta_k \leqslant \pi/2$, and form the Euclidean average

$$\theta_{\mathrm{cm}} = \sum t_k \theta_k \overset{t_k = 1/K}{=} \frac{1}{K}\sum_{k=1}^{K}\theta_k .$$

Then the spherical average barycenter following from this partition of unity with all the same weights $t_k = 1/K$ is just

$$\mathbf{p}_{\mathrm{cm}} = [\cos\theta_{\mathrm{cm}}, \sin\theta_{\mathrm{cm}}]^{\mathrm{t}} .$$

What is particularly striking is that the analog of the Euclidean partition of unity is effectively *inside* the trigonometric function map. We might hope that for spherical triangles, tetrahedra, etc., in the absence of a linear exterior partition of unity as found in the Euclidean center of mass, we could find an *interior* partition of unity in the argument of something like a trigonometric function that would produce an exactly correct spherical barycenter. We will suggest that no solution analogous to the Euclidean case exists, but that we can propose an intuitively satisfying alternative for spherical geometry.

The slerp. We now argue that some desirable properties can be retained on higher-dimensional spheres if we focus on *geodesic arcs* between two points on a sphere. Suppose we have a pair of $(N+1)$-dimensional unit vectors $\mathbf{p}_1$ and $\mathbf{p}_2$ on $\mathbf{S}^N$ embedded in $\mathbb{R}^{N+1}$. These points lie completely in a 2D subspace of $\mathbb{R}^{N+1}$ defined by $\mathbf{p}_1, \mathbf{p}_2$, and the origin, so we can perform a *spherical linear interpolation* (slerp) (Shoemake, 1985) that smoothly parameterizes any point $\mathbf{p}(t)$ on the geodesic in $\mathbf{S}^N$ connecting $\mathbf{p}_1$ and $\mathbf{p}_2$. Defining the angle θ between the vectors by $\cos\theta = \mathbf{p}_1 \cdot \mathbf{p}_2$ (remember that on any unit sphere, including the circle, $\mathbf{p}_1 \cdot \mathbf{p}_1 = \mathbf{p}_2 \cdot \mathbf{p}_2 = 1$), the desired slerp can be written

$$\mathbf{p}(t) = \mathrm{slerp}(\mathbf{p}_1, \mathbf{p}_2; t) = \frac{\sin(t_1\theta)}{\sin\theta}\mathbf{p}_1 + \frac{\sin(t_2\theta)}{\sin\theta}\mathbf{p}_2 , \tag{19.3}$$

where $t_1 + t_2 = 1$ is our partition of unity, so we can take $t_1 = 1 - t$ and $t_2 = t$. This gives us $\mathbf{p}(0) = \mathbf{p}_1$ and $\mathbf{p}(1) = \mathbf{p}_2$, matching the linear interpolation convention $\mathbf{p}(t) = (1-t)\mathbf{p}_1 + t\mathbf{p}_2$, which is also the $\theta \to 0$ limit of Eq. (19.3).

Counterintuitively, we see that t_1 is *not* the distance from the point $\mathbf{p}_1$ to $\mathbf{p}(t)$, but the *complementary distance*. In Fig. 19.1, we show geometrically how, with $t_1 + t_2 = 1$ or $t_1 = 1 - t$ and $t_2 = t$, taking

$$\theta = \theta_1 + \theta_2 = t_1\theta + t_2\theta$$

gives $\theta = \theta_1$ if $t = 0$ ($t_1 = 1$), and $\theta = \theta_2$ if $t = 1$ ($t_2 = 1$). This is a standard feature of barycentric coordinates: they take on their label from the subsimplex *opposite* to the vertex in question, applying virtual pressure that pushes the interior point towards the coordinate's namesake. These properties are

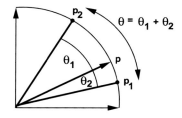

FIGURE 19.1

Illustrating the subdivision of spherical arcs $\theta_1 + \theta_2 = \theta$ in the weighted points derivation of the `slerp` equations. Note that θ_1 is the angle *opposite* the vertex $\mathbf{v}_1$, and similarly for θ_2 and $\mathbf{v}_2$. Effectively, as $\theta_1 \to \theta$, it *pushes* $\mathbf{p}$ towards $\mathbf{p}_1$.

repeated naturally in the notation chosen for the `slerp`: the simplest possible connection between geodesic arcs and barycentric coordinates is

$$t_1 = 1, \qquad t = 0, \qquad \mathrm{slerp}(\mathbf{p}_1, \mathbf{p}_2; 0) = \mathbf{p}_1,$$
$$t_2 = 1, \qquad t = 1, \qquad \mathrm{slerp}(\mathbf{p}_1, \mathbf{p}_2; 1) = \mathbf{p}_2,$$

while the midpoint, the spherical average of $\mathbf{p}_1$ and $\mathbf{p}_2$, is just

$$t_1 = t_2 = \frac{1}{2}, \qquad \mathrm{slerp}\left(\mathbf{p}_1, \mathbf{p}_2; \frac{1}{2}\right) = \frac{1}{\sqrt{2(1+\cos\theta)}}\mathbf{p}_1 + \frac{1}{\sqrt{2(1+\cos\theta)}}\mathbf{p}_2.$$

It is crucial that, in addition, the unit norm is automatically preserved by the `slerp` for all t,

$$\mathrm{slerp}(\mathbf{p}_1, \mathbf{p}_2; t) \cdot \mathrm{slerp}(\mathbf{p}_1, \mathbf{p}_2; t) = 1. \tag{19.4}$$

In the `slerp`, as well as in the $\mathbf{S}^1$ barycenter, the partition of unity that we can rigorously employ for a simple circular arc appears *inside* a transcendental function, not in the weight of the linear combination of the points themselves, as we saw in Euclidean geometry.

19.2 Review of spherical trigonometry

Three unit-length points $\mathbf{v}_1$, $\mathbf{v}_2$, and $\mathbf{v}_3$ define a spherical triangle. Remember that the `slerp` is a valid point-to-point geodesic interpolation in any dimension, and thus we can use it for any *pair* of points on $\mathbf{S}^2$. Thus we can create a parameterization of the spherical triangle simply by first defining a pair of `slerp`'s,

$$\left.\begin{aligned}
\mathbf{v}_{12}(t) &= \mathrm{slerp}(\mathbf{v}_1, \mathbf{v}_2; t), \\
\mathbf{v}_{13}(t) &= \mathrm{slerp}(\mathbf{v}_1, \mathbf{v}_3; t), \\
\text{and then combining these into another } &\texttt{slerp} \\
\text{between the two moving points:} \\
\mathbf{p}_{123}(t_1, t_2) &= \mathrm{slerp}\left(\mathbf{v}_{12}(t_1), \mathbf{v}_{23}(t_1); t_2\right)
\end{aligned}\right\}. \tag{19.5}$$

This construction is actually surprisingly effective: it creates a perfect spherical triangle submanifold in any dimension, and extends to spherical K-simplexes in any appropriate dimension. While the algebraic expression is inelegant and very much dependent on the order of the vertices, it can even in principle be integrated to get the spherical simplex volumes. However, it is a *rectangular* map from the parameter space (t_1, t_2) to the spherical triangle, and surely, at the very least, we would hope to find a more symmetric *partition of unity* map, something like a space (t_1, t_2, t_3) with $t_1 + t_2 + t_3 = 1$.

Area formulas for spherical triangles. In the case of the 1D parameterized function slerp(t), we eventually can see via such representations as Fig. 19.1 that the ultimate source of the partition of unity in the two-point `slerp` is the arc length and ratios of arc lengths. To extend this idea, we should logically attempt to study the areas of spherical triangles in $\mathbf{S}^2$ and their partition into subtriangles via an inserted point: if everything works right (it will not), a partition of unity based on ratios of spherical triangle areas should be the ideal way to study and exploit spherical geometry, in parallel to what we did with the geometry of higher-dimensional Euclidean simplexes in Chapter 17.

There are many approaches to computing the properties of $\mathbf{S}^2$ spherical triangles from a triple of points on a sphere, dating back hundreds of years. Some are founded on basic properties of spherical trigonometry, which supplements the Euclidean angles between edges with the information on angles between arcs passing through the spherical vertices themselves, and some are based on determinants in the Euclidean $\mathbb{R}^3$ in which the sphere is embedded. The classic formula for the spherical triangle area is *Girard's theorem* using the *angular excess* of the vertex angles (A, B, C) measured in the tangent plane of the sphere at the vertices $(\mathbf{v}_1, \mathbf{v}_2, \mathbf{v}_3)$, as shown in Fig. 19.2:

$$E(1\,2\,3) = A + B + C - \pi . \tag{19.6}$$

Eq. (19.6) is essentially a measure of the curvature embodied in the triangle, with the excess vanishing as the spherical triangle approaches an infinitesimal Euclidean triangle, with vanishing area.

Constructing the 3D dual basis. The vertex tangent plane angles (A, B, C) can be computed directly from geometric principles using normals to the planes through the origin defined by $\mathbf{v}_i$ and $\mathbf{v}_j$. These give us the 3D version of the spherical triangle's *dual basis*, which we will now use repeatedly. Most features of the dual basis, attributed to Möbius (Möbius, 1846; Alfeld et al., 1996), extend to N-dimensional spheres $\mathbf{S}^N$ (see Appendix M). We start with a 2-sphere $\mathbf{S}^2$ embedded in $\mathbb{R}^3$, along with the unit-length spherical vertices $\mathbf{v}_1$, $\mathbf{v}_2$, and $\mathbf{v}_3$, as shown in Fig. 19.2; most of our calculations can be extended to larger portions of the sphere than the first octant shown in Fig. 19.2, but we will assume the properties of restriction to the first octant to avoid getting bogged down handling special cases. Our basic object is the normalized set of three vertex-pair cross products,

$$\left.\begin{array}{l} \mathbf{n}_1 = \dfrac{\mathbf{v}_2 \times \mathbf{v}_3}{\|\mathbf{v}_2 \times \mathbf{v}_3\|} = \dfrac{\mathbf{v}_2 \times \mathbf{v}_3}{\sqrt{1 - |\mathbf{v}_2 \cdot \mathbf{v}_3|^2}} = \dfrac{\mathbf{v}_2 \times \mathbf{v}_3}{\sqrt{1 - c_{23}^2}} = \dfrac{\mathbf{v}_2 \times \mathbf{v}_3}{s_{23}} \\[2ex] \mathbf{n}_2 = \dfrac{\mathbf{v}_3 \times \mathbf{v}_1}{\|\mathbf{v}_3 \times \mathbf{v}_1\|} = \dfrac{\mathbf{v}_3 \times \mathbf{v}_1}{\sqrt{1 - |\mathbf{v}_3 \cdot \mathbf{v}_1|^2}} = \dfrac{\mathbf{v}_3 \times \mathbf{v}_1}{\sqrt{1 - c_{13}^2}} = \dfrac{\mathbf{v}_3 \times \mathbf{v}_1}{s_{13}} \\[2ex] \mathbf{n}_3 = \dfrac{\mathbf{v}_1 \times \mathbf{v}_2}{\|\mathbf{v}_1 \times \mathbf{v}_2\|} = \dfrac{\mathbf{v}_1 \times \mathbf{v}_2}{\sqrt{1 - |\mathbf{v}_1 \cdot \mathbf{v}_2|^2}} = \dfrac{\mathbf{v}_1 \times \mathbf{v}_2}{\sqrt{1 - c_{12}^2}} = \dfrac{\mathbf{v}_1 \times \mathbf{v}_2}{s_{12}} \end{array}\right\} , \tag{19.7}$$

where we use the index for $\mathbf{n}$ corresponding to the unused opposite vertex, and introduce the notation $\mathbf{v}_i \cdot \mathbf{v}_j = \cos\theta_{ij} = c_{ij}$ and $\sin\theta_{ij} = s_{ij}$ to simplify our notation for upcoming large expressions. The first

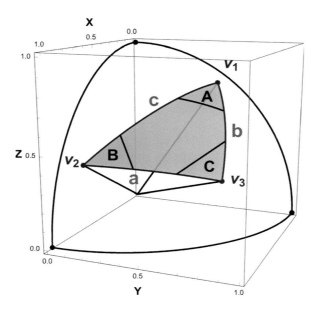

FIGURE 19.2

Illustrating the geometry of Girard's theorem, identifying the arc lengths a, b, c between each pair of surface points $\mathbf{v}_i$ and the opposing interior surface angles (A, B, C) that allow a simple calculation of the area of a spherical triangle in terms of a curvature measure, the angular defect $E = A + B + C - \pi$.

application of the dual basis is the fact that the elements are normals to the planes enclosing each pair of the surface vertices, so their inner products produce the cosines of the angles (A, B, C), and their cross products align with the intersections of their perpendicular planes, which are just the lines through the origin and the vertices $\mathbf{v}_1$, $\mathbf{v}_2$, and $\mathbf{v}_3$ themselves. The surface angle definitions at each vertex and the corresponding reconstructions of the vertices are thus

$$\left.\begin{array}{l} \cos A = \mathbf{n}_2 \cdot \mathbf{n}_3 \\[4pt] \mathbf{v}_1 = \dfrac{\mathbf{n}_2 \times \mathbf{n}_3}{\sqrt{1 - (\cos A)^2}} = \dfrac{\mathbf{n}_2 \times \mathbf{n}_3}{\sin A} \\[10pt] \cos B = \mathbf{n}_3 \cdot \mathbf{n}_1 \\[4pt] \mathbf{v}_2 = \dfrac{\mathbf{n}_3 \times \mathbf{n}_1}{\sqrt{1 - (\cos B)^2}} = \dfrac{\mathbf{n}_3 \times \mathbf{n}_1}{\sin B} \\[10pt] \cos C = \mathbf{n}_1 \cdot \mathbf{n}_2 \\[4pt] \mathbf{v}_3 = \dfrac{\mathbf{n}_1 \times \mathbf{n}_2}{\sqrt{1 - (\cos C)^2}} = \dfrac{\mathbf{n}_1 \times \mathbf{n}_2}{\sin C} \end{array}\right\} . \tag{19.8}$$

Note that these equations each guarantee that $\|\mathbf{v}_i\| = 1$, so the vertices are on the unit sphere as required. We will assume in general that the vertices are in right-handed counterclockwise order, as indicated in

Fig. 19.2, and that the signed directions of the $\mathbf{n}_i$ are also in the natural cyclic order so that the signs in Eq. (19.8) are correct.

The area of an S^2 spherical triangle in terms of Euclidean geometry. Finally, there are a number of formulas that permit the computation of trigonometric functions of the spherical triangle area from geometric constructions involving determinants and inner products (see, for example, Eriksson, 1990; Tuynman, 2013; Fillmore and Fillmore, 2018). These formulas generally involve the scaled determinant form for the volume of the parallelepiped defined by the three vertices of the spherical triangle, which can be written conveniently by exploiting the product theorem, $\det M \cdot M^t = (\det M)^2$ for square matrices M. If we take M to be the 3×3 matrix whose columns are the unit vectors $\mathbf{v}_1$, $\mathbf{v}_2$, and $\mathbf{v}_3$, we see that, up to a sign determined by the first line's determinant,

$$
\left.
\begin{aligned}
V(123) \; [\text{parallelepiped}] &= \det \begin{bmatrix} \mathbf{v}_1 & \mathbf{v}_2 & \mathbf{v}_3 \end{bmatrix} \\
&= \left(\det \begin{bmatrix} \mathbf{v}_1 \cdot \mathbf{v}_1 & \mathbf{v}_1 \cdot \mathbf{v}_2 & \mathbf{v}_1 \cdot \mathbf{v}_3 \\ \mathbf{v}_2 \cdot \mathbf{v}_1 & \mathbf{v}_2 \cdot \mathbf{v}_2 & \mathbf{v}_2 \cdot \mathbf{v}_3 \\ \mathbf{v}_3 \cdot \mathbf{v}_1 & \mathbf{v}_3 \cdot \mathbf{v}_2 & \mathbf{v}_3 \cdot \mathbf{v}_3 \end{bmatrix} \right)^{1/2} \\
&= \left(\det \begin{bmatrix} 1 & \cos\theta_{12} & \cos\theta_{13} \\ \cos\theta_{12} & 1 & \cos\theta_{23} \\ \cos\theta_{13} & \cos\theta_{23} & 1 \end{bmatrix} \right)^{1/2} \\
&= \sqrt{1 - c_{23}{}^2 - c_{13}{}^2 - c_{12}{}^2 + 2\,c_{23}\,c_{13}\,c_{12}}
\end{aligned}
\right\} .
\tag{19.9}
$$

The intuitive meaning of the spherical parallelepiped volume Eq. (19.9) is a sort of "generalized sine function," related to the *polar sine*, *polsine*, or "polsin" notation analogous to the usual "sin." This interpretation naturally follows from the upper left 2×2 matrix, whose determinant corresponds to the usual "sin" function:

$$
\left(\det \begin{bmatrix} \mathbf{v}_1 & \mathbf{v}_2 \end{bmatrix} \right)^2 = \det \begin{bmatrix} \mathbf{v}_1 \cdot \mathbf{v}_1 & \mathbf{v}_1 \cdot \mathbf{v}_2 \\ \mathbf{v}_1 \cdot \mathbf{v}_2 & \mathbf{v}_2 \cdot \mathbf{v}_2 \end{bmatrix} = \|\mathbf{v}_1 \times \mathbf{v}_2\|^2 = 1 - \cos^2\theta_{12} = \sin^2\theta_{12} .
\tag{19.10}
$$

From Eq. (19.9) one can construct trigonometric functions of the spherical triangle areas (see, e.g., Eriksson, 1990)

$$
\sin \frac{E(123)}{2} = \frac{V(123)}{\sqrt{2(1 + c_{23})(1 + c_{13})(1 + c_{12})}} ,
\tag{19.11}
$$

$$
\tan \frac{E(123)}{2} = \frac{V(123)}{1 + c_{23} + c_{13} + c_{12}} ,
\tag{19.12}
$$

where the identity $\tan\theta = \sin\theta / \sqrt{1 - \sin^2\theta}$ creates the interestingly close correspondence between the two trigonometric functions. In fact, we can use the half-angle formula to obtain a formula for the sphere that is closer to the sine ratios appearing in the circular arc `slerp`. We have

$$
\sin E(123) = 2 \sin \frac{E(123)}{2} \cos \frac{E(123)}{2} = 2 \frac{\sin^2(E(123)/2)}{\tan(E(123)/2)} = V(123) \frac{1 + c_{23} + c_{13} + c_{12}}{(1 + c_{23})(1 + c_{13})(1 + c_{12})} .
\tag{19.13}
$$

Yet another transformation results from examining the *midpoints* of the arcs between each pair of vertices (see, e.g., Fillmore and Fillmore, 2018),

$$\sin \frac{E(123)}{2} = \det \left[\frac{\mathbf{v}_2 + \mathbf{v}_3}{\sqrt{2(1 + c_{23})}} \quad \frac{\mathbf{v}_3 + \mathbf{v}_1}{\sqrt{2(1 + c_{13})}} \quad \frac{\mathbf{v}_1 + \mathbf{v}_2}{\sqrt{2(1 + c_{12})}} \right], \tag{19.14}$$

where determinant rearrangement identities transform Eq. (19.14) back to the form of Eq. (19.11). It is interesting to note identities of this sort because simplex edge-midpoint calculations may in some cases be more generalizable to difficult spherical problems than vertex-based calculations. An important example of this is the relation to the `polsin`, which has the property

$$\sin \frac{E(123)}{2} = \text{polsin (midpoints of edges of } E(123)) . \tag{19.15}$$

As the midpoints of the edges of spherical triangles subdivide the original triangle into four triangles, with the interior instance being the midpoint-defined spherical triangle, this is a fundamental relation (see, e.g., Eriksson, 1990, and historical references cited therein).

Extracting spherical triangle vertices from edge midpoints. As a practical matter, given the relation between the sine of the half-area of a spherical triangle and the `polsin` of the midpoints of that same triangle, we could have some interest in converting a set of midpoints into the spherical triangle in which they are embedded. This can be done strictly in terms of analytic geometry, depicted schematically in Fig. 19.3, and which we now describe. First, the only data we actually have are the midpoint unit vectors, call them $(\mathbf{m}_1, \mathbf{m}_2, \mathbf{m}_3)$, and the only structures we have to work with in 3D based on these are their mutual dot products and cross products. If we look at the normalized cross products of the midpoint pairs,

$$\left. \begin{aligned} \mathbf{x}_1 &= \frac{\mathbf{m}_2 \times \mathbf{m}_3}{\sqrt{1 - \cos^2 \phi_{23}}} \\ \mathbf{x}_2 &= \frac{\mathbf{m}_3 \times \mathbf{m}_1}{\sqrt{1 - \cos^2 \phi_{31}}} \\ \mathbf{x}_3 &= \frac{\mathbf{m}_1 \times \mathbf{m}_2}{\sqrt{1 - \cos^2 \phi_{12}}} \end{aligned} \right\}, \tag{19.16}$$

where $\cos \phi_{ij} = \mathbf{m}_i \cdot \mathbf{m}_j$, we can infer that they define the normals to the planes containing the geodesic arcs between each pair of midpoints. These planes themselves are not useful, as they are not related to the vertices we seek. However, they are based on the midpoints, which are joined by arcs exactly parallel to the arcs through pairs of vertices including the midpoint associated with each pair. That is, for each of these normals computed from two midpoints, there is an omitted midpoint $\mathbf{m}$ that, by definition, shares a circular arc through its two vertices; we can find the *plane* of that arc by applying a Gram–Schmidt process to extract just the portion $\mathbf{u}$ of the normal $\mathbf{x}$ that is perpendicular to the omitted $\mathbf{m}$, that is,

$$\mathbf{u} = \frac{\mathbf{x} - \mathbf{m} (\mathbf{m} \cdot \mathbf{x})}{\sqrt{1 - (\mathbf{m} \cdot \mathbf{x})^2}} . \tag{19.17}$$

In fact it is obvious that these are just the opposite directions of our previous dual basis vectors $\mathbf{n}_i$ from Eq. (19.7),

$$\mathbf{u}_i = -\mathbf{n}_i ,$$

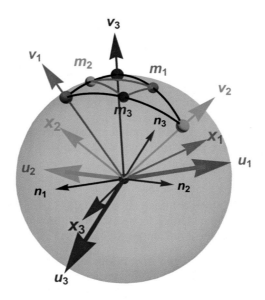

FIGURE 19.3

The geometry of finding the spherical triangle vertices, $\mathbf{v}_1$, $\mathbf{v}_2$, and $\mathbf{v}_3$ shown as red, green, and blue dots, respectively, given only the midpoints of the opposite edges, $(\mathbf{m}_1, \mathbf{m}_2, \mathbf{m}_3)$, shown also as red, green, and blue dots at the centers of the connecting arcs. Cross products of the midpoints, shown as bright red, green, and blue short arrows, produce the normals to the planes of the midpoint pairs. We tilt that plane to pass through the unused midpoint, and the normals $\mathbf{u}_i$ to those planes, shown in darker corresponding colors, *must* pass through a pair of vertices, so in fact they are just the opposite directions of the dual basis vectors $\mathbf{n}_i$ of Eq. (19.7), noted as black arrows, which are derived from entirely different input data. Cross products of those $\mathbf{u}_i$ pairs, shown as red, green, and blue arrows, locate the intersections of pairs of planes, defining the desired vertices $\mathbf{v}_1$, $\mathbf{v}_2$, and $\mathbf{v}_3$, using *only* the midpoint data, $\mathbf{m}_1$, $\mathbf{m}_2$, and $\mathbf{m}_3$.

which are noted as small black arrows in Fig. 19.3 to complete the picture. Given any pair of these normals, their perpendicular planes will intersect in the line passing through their common vertex. For example, defining the plane of the arcs containing the midpoints of the edges joining the spherical triangle vertices $(\mathbf{v}_1, \mathbf{v}_3)$ and $(\mathbf{v}_2, \mathbf{v}_3)$, both arcs must pass through $\mathbf{v}_3$ and the origin. Thus the cross product of *that* pair of normals to those planes must coincide with the direction of the line through the origin to $\mathbf{v}_3$. All three vertices therefore correspond to the normalized cross products,

$$\mathbf{v}_i = \frac{\mathbf{u}_j \times \mathbf{u}_k}{\sqrt{1 - \cos^2 \psi_{jk}}} = \frac{\mathbf{n}_j \times \mathbf{n}_k}{\sqrt{1 - \cos^2 \psi_{jk}}} \,, \tag{19.18}$$

where $\cos \psi_{jk} = \mathbf{u}_j \cdot \mathbf{u}_k = \mathbf{n}_j \cdot \mathbf{n}_k$ and (i, j, k) are cyclic, as shown by the red, green, and blue arrows in the outline of this construction shown in Fig. 19.3. This would be the process we would need to follow if, for example, we wanted to compute the `polsin` of spherical triangle vertices having knowledge only of the midpoints.

19.3 Spherical triangles as coordinates

The success of the `slerp` in interpolating a point following a geodesic curve between two spherical points in any dimension, maintaining its unit length, can be interpreted as a method of separating the total arc length $\theta = \arccos(\mathbf{v}_1 \cdot \mathbf{v}_2)$ between the unit vectors $\mathbf{v}_1$ and $\mathbf{v}_2$ on the sphere into two parts, as illustrated earlier in Fig. 19.1:

$$\left.\begin{array}{l} \theta_1 = t_1\,\theta = (1-t)\,\theta \\ \theta_2 = t_2\,\theta = t\,\theta \\ \theta = \theta_1 + \theta_2 \end{array}\right\} . \tag{19.19}$$

We now attempt to generalize this partition to the sphere $\mathbf{S}^2$, the next dimension higher, and eventually conclude that an exactly parallel generalization of the Euclidean external partition of unity cannot succeed; a more complex mechanism must be utilized.

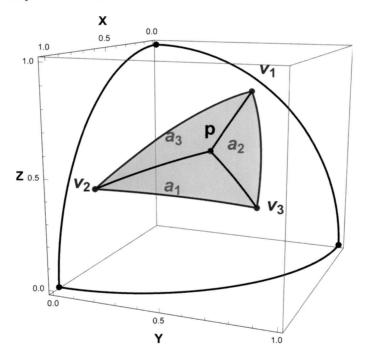

FIGURE 19.4

The basic geometry of a spherical triangle, with vertices $(\mathbf{v}_1, \mathbf{v}_2, \mathbf{v}_3)$, and interior point $\mathbf{p}$ defining a partition of unity with the areas (a_1, a_2, a_3) that add up to $a_{(total)}$, the total spherical triangle area.

In our quest to generalize the success of partitions of unity based on ratios of lengths, areas, or volumes to $\mathbf{S}^2$, we would naturally look at a spherical triangle as a simplex defined by three points, as in the Euclidean case. As shown in Fig. 19.4, if we start with three points $(\mathbf{v}_1, \mathbf{v}_2, \mathbf{v}_3)$, when we put another spherical point $\mathbf{p}$ inside the spherical triangle $(\mathbf{v}_1, \mathbf{v}_2, \mathbf{v}_3)$, it immediately partitions the original triangle into three smaller spherical triangles with surface areas (a_1, a_2, a_3) computable from Eq. (19.11) with sequential vertices replaced by $\mathbf{p}$. Since these are close analogs of the Euclidean partition forms, except that the vector components are unit vectors on the sphere, we might expect to be able to use these areas as a partition of unity

$$
\left.\begin{aligned}
a_{(total)} &= 2\arcsin\left(\frac{\det[\mathbf{v}_1\,\mathbf{v}_2\,\mathbf{v}_3]}{\sqrt{2(1+\cos\theta_{23})(1+\cos\theta_{13})(1+\cos\theta_{12})}}\right) \\
a_1 &= 2\arcsin\left(\frac{\det[\mathbf{p}\,\mathbf{v}_2\,\mathbf{v}_3]}{\sqrt{2(1+\cos\theta_{23})(1+\cos\theta_{p3})(1+\cos\theta_{p2})}}\right) \\
a_2 &= 2\arcsin\left(\frac{\det[\mathbf{v}_1\,\mathbf{p}\,\mathbf{v}_3]}{\sqrt{2(1+\cos\theta_{13})(1+\cos\theta_{p1})(1+\cos\theta_{p3})}}\right) \\
a_3 &= 2\arcsin\left(\frac{\det[\mathbf{v}_1\,\mathbf{v}_2\,\mathbf{p}]}{\sqrt{2(1+\cos\theta_{12})(1+\cos\theta_{p1})(1+\cos\theta_{p2})}}\right)
\end{aligned}\right\}. \tag{19.20}
$$

Observe that we label the triangles by their *omitted* point, as already suggested by Fig. 19.1, and illustrated for $\mathbf{S}^2$ in Fig. 19.4; this is natural because $\mathbf{p} \to \mathbf{v}_i$ as the size of $a_i \to a_{(total)}$. We can now *attempt* to form a useful partition of unity by taking ratios of these areas with the total area $a_{(total)}(\mathbf{v}_1, \mathbf{v}_2, \mathbf{v}_3)$ of the main triangle:

$$
t_1 = \frac{a_1}{a_{(total)}}, \qquad t_2 = \frac{a_2}{a_{(total)}} \qquad t_3 = \frac{a_3}{a_{(total)}}. \tag{19.21}
$$

Thus we have a candidate for a partition of unity and a set of coordinates for the spherical point $\mathbf{p}$ with

$$
\sum_{i=1}^{3} \frac{a_i}{a} = \sum_{i=1}^{3} t_i = 1. \tag{19.22}
$$

Spherical triangle barycentric coordinates are incompatible but slerps work. Attempting to use spherical triangle ratios as barycentric coordinates to define a point $\mathbf{p}$ on the sphere $\mathbf{S}^2$ in terms of the three spherical triangle vertices $(\mathbf{v}_1, \mathbf{v}_2, \mathbf{v}_3)$ has a problem: this algebraic formula analogous to the Euclidean case

$$
\mathbf{p} = t_1\mathbf{v}_1 + t_2\mathbf{v}_2 + t_3\mathbf{v}_3 \tag{19.23}
$$

does not preserve the unit norm of $\mathbf{p}$. Thus we fail to express a unit vector $\mathbf{p}$ as a linear combination of the triangle vertices and pure area ratios $t_i = a_i/a$ as a partition of unity, and there is no exact analog of the Euclidean equations Eq. (17.18) or Eq. (17.24) because the required identity $\mathbf{p} \cdot \mathbf{p} = 1$ is not satisfied for non-Euclidean partitions of unity. Of course this is not really a surprise, since even the 2D geodesic arc `slerp` has the two-point partition of unity hidden away *inside* transcendental functions. And yet, as we will see, there does exist a formula, analogous to the Euclidean coefficients Eq. (17.17), that is *not a partition of unity* for vertices on a spherical simplex but, thanks to the properties of the dual basis, *does satisfy* $\mathbf{p} \cdot \mathbf{p} = 1$.

What we really need, and would expect in the best of possible worlds, is a generalization of the `slerp` of Eq. (19.3) that concealed a partition of unity inside a sine-like function that locally approached a Euclidean partition of unity in the limit of large sphere radius. We can in fact accomplish that goal in a correct but seemingly arbitrary fashion by nesting `slerps` with vertex-asymmetric rectangular parameters, as opposed to vertex-symmetric partition-of-unity parameters, in any order:

$$\mathbf{p}(t_1, t_2) = \text{slerp}\,(\text{slerp}(\mathbf{v}_1, \mathbf{v}_2; t_1), (\text{slerp}(\mathbf{v}_1, \mathbf{v}_3; t_1); t_2)) \tag{19.24}$$

$$\overset{\text{limit}}{\rightarrow} (1 - t_2)\,((1 - t_1)\mathbf{v}_1 + t_1\mathbf{v}_2) + t_2\,((1 - t_1)\mathbf{v}_1 + t_1\mathbf{v}_3)$$

$$= \mathbf{v}_1 - t_1\mathbf{v}_1 + t_1\mathbf{v}_2 - t_1 t_2\mathbf{v}_2 + t_1 t_2\mathbf{v}_3\,,$$

$$\mathbf{p}(t_1, t_2) = \text{slerp}\,(\text{slerp}(\mathbf{v}_1, \text{slerp}(\mathbf{v}_2, \mathbf{v}_3; t_2); t_1)) \tag{19.25}$$

$$\overset{\text{limit}}{\rightarrow} (1 - t_1)\mathbf{v}_1 + t_1\,((1 - t_2)\mathbf{v}_2 + t_2\mathbf{v}_3)$$

$$= \mathbf{v}_1 - t_1\mathbf{v}_1 + t_1\mathbf{v}_2 - t_1 t_2\mathbf{v}_2 + t_1 t_2\mathbf{v}_3 = \text{\textbf{the same flat limit}}\,.$$

We see that in the infinitesimal angle flat limit, these nested `slerps` reduce to the *same* bilinear interpolation. These parameterizations guarantee that each edge of the spherical triangle is a geodesic, and that every interior point lies on the sphere, but, unlike the Euclidean partition of unity, there is no edge-permutation symmetry, except of course in the flat limit, where the coefficients of $\mathbf{v}_1$, $\mathbf{v}_2$, and $\mathbf{v}_3$ sum to unity.

Numerical spherical barycenter approach. Barycentric coordinates must support two operations: (1) given a basis and a point $\mathbf{p}$, compute the barycentric coordinates uniquely describing the point; (2) given the basis and the barycentric coordinates, compute the location of the point $\mathbf{p}$. One *can* in fact do this if one does not ask for an equation for $\mathbf{p}$ of the form of Eq. (19.23), and that approach was explored, e.g., by Buss and Fillmore (2001). For the first requirement, we already know from Eq. (19.20) that, given $\mathbf{p}$, it is easy to compute the corresponding spherical subtriangles and their areas. However, the second requirement has no analytic solution: *given* a three-subarea partition of unity of some spherical triangle, we cannot find an analytic expression for $\mathbf{p}$. However, we have the straightforward capability of performing a computer-based *numerical search*. Heuristic search methods exist that can locate the corresponding parameterized unit vector $\mathbf{p}$ given a partition of a spherical triangle's area, and the numerical method can be very effectively initialized to a starting value that is obtained simply by normalizing the right-hand side of Eq. (19.23). We cannot use this normalization to find $\mathbf{p}$ because it is highly nonlinear, but it works quite well as an approximation to start an exact search. Obviously our other goal, finding the actual rigorous *spherical center of mass*, can be found in this way by setting the partition coefficients equal to one another. This is essentially the philosophy of Buss and Fillmore (2001), and it is one method, though inelegant, to deal with the quaternion barycentric coordinate and spherical barycenter problems by exploiting the partition of a spherical triangle. We remark that for $\mathbf{S}^3$ and higher-dimensional spheres, even Eq. (19.20) for the spherical N-simplex volumes has no known analog, so the numerical approach has to be extended a step further to compute the spherical simplex volumes themselves.

19.4 Brown–Worsey no-go theorem

In fact there is even more to say about the spherical triangle partition of unity. Despite the fact that the spherical triangle partition of unity t_i only implicitly specifies its corresponding point $\mathbf{p}(t_1, t_2, t_3)$ on $\mathbf{S}^2$, and the fact that we have no area-ratio partition-of-unity-based formula as the weighted sum of

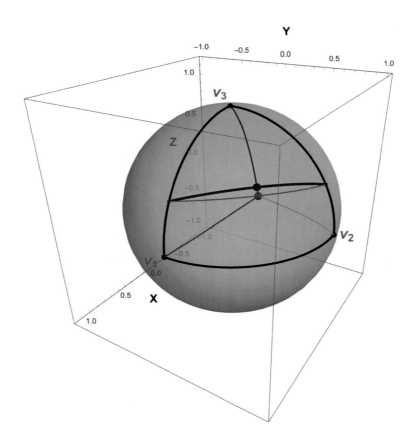

FIGURE 19.5

Visualizing the Brown–Worsey (1992) no-go theorem, showing that the geodesic (in blue) between two points on a spherical triangle does not in general correspond to the path (in red) interpolated by a ratio of spherical subtriangle areas if one subtriangle area is held fixed.

three points on $\mathbf{S}^2$ analogous to the Euclidean case, we might expect our $\mathbf{S}^2$ area-ratio partition of unity to serve as an appropriate interpolation scheme *extending* a `slerp`-style interpolation to $\mathbf{S}^2$. This hope fails, as shown by Brown and Worsey (1992).

The basic argument that Brown and Worsey use to prove their no-go theorem is illustrated in Fig. 19.5. We define a spherical triangle with vertices $\mathbf{v}_1 = \hat{\mathbf{x}}$, $\mathbf{v}_2 = \hat{\mathbf{y}}$, $\mathbf{v}_3 = \hat{\mathbf{z}}$ aligning with the orthogonal axes of the first octant of $\mathbf{S}^2$, with edges that are the geodesic arcs between each pair of vertex points. Given any point $\mathbf{p}$ in this spherical triangle, there are three spherical triangles partitioning the whole area, which we label as usual by their *missing* vertex replaced by $\mathbf{p}$,

$$T_1(\text{triangle: } \mathbf{p}, \mathbf{v}_2, \mathbf{v}_3) , \quad T_2(\text{triangle: } \mathbf{v}_1, \mathbf{p}, \mathbf{v}_3) , \quad T_3(\text{triangle: } \mathbf{v}_1, \mathbf{v}_2, \mathbf{p}) .$$

If any one triangle's area *vanishes*, the remaining triangles' areas serve exactly as a two-part partition of unity for two end points related by an angle θ (actually $\pi/2$ in this example) creating a two-point `slerp`,

$$\theta = \theta_1 + \theta_2 = t_1\theta + t_2\theta \quad \text{with } t_1 + t_2 = 1 ,$$

thus providing a perfect inside-the-sine barycentric coordinate system *confined to the geodesic edge.*

Now suppose we pick a point parameterized by $\mathbf{u}_1 = (z, 0, (1-z))$ on the $\widehat{\mathbf{v}_1\mathbf{v}_3}$ edge, and another point equidistant from $\mathbf{v}_3 = \hat{\mathbf{z}} = [0, 0, 1]$, at $\mathbf{u}_2 = (0, z, (1-z))$ on the $\widehat{\mathbf{v}_2\mathbf{v}_3}$ edge. A partition of unity interpolator in Euclidean space must follow a straight path between two points interpolated linearly in the partition of unity, so the form

$$[z(1-t), zt, 1-z] \tag{19.26}$$

should describe the shortest path between $\mathbf{u}_1$ at $(z, 0, (1-z))$ and $\mathbf{u}_2$ at $(0, z, (1-z))$. In Fig. 19.5, we plot in red the sequence of points whose spherical partitions of unity are parameterized by Eq. (19.26). Effectively this means that we *fix the area* of the bottom triangle, opposite the vertex $\mathbf{v}_3$: as the red dot moves, the spherical triangle area defined by $\mathbf{v}_1$, $\mathbf{v}_2$, and the red dot must be kept at $(1-z) \times a_{\text{(total)}}$. In blue, we plot the actual geodesic from $\mathbf{u}_1$ to $\mathbf{u}_2$ using the standard `slerp` on $\mathbf{S}^2$. The spherical area-ratio partition interpolator and the `slerp` interpolator agree at the end points $\mathbf{u}_1$ and $\mathbf{u}_2$ on the edges $\widehat{\mathbf{v}_1\mathbf{v}_3}$ and $\widehat{\mathbf{v}_1\mathbf{v}_3}$ by construction, but they *differ* at every point on $0 < t < 1$. Choosing $z = 1/2$ and looking at the midpoint $t = 1/2$, we find

$$\texttt{slerp}(\mathbf{u}_1, \mathbf{u}_2; \tfrac{1}{2}) = (0.408, 0.408, 0.816) ,$$

$$\text{Solve } (t_i = \frac{a_i}{a} = \left(\frac{1}{4}, \frac{1}{4}, \frac{1}{2}\right) \text{ for } \mathbf{p}) = (0.442, 0.442, 0.780) .$$

The spherical triangle ratio thus fails even the simplest requirement for a consistent interpolator. Note that to perform this test, we implemented a numerical gradient-descent solver to locate the spherical point $\mathbf{p}$ described by any given spherical-triangle area-ratio partition of unity. Brown and Worsey utilize more general algebraic techniques in their proof, for which we refer the interested reader to their paper (Brown and Worsey, 1992).

19.5 Möbius dual coordinates for points in a spherical simplex

Having now seen a wide range of challenges in the construction of parameterizations of points on a sphere in terms of a nonorthogonal basis provided by the vertices of an ordered spherical triangle, we conclude that no simple partition of unity parameterization for geometry or interpolation exists based on analogs to the Euclidean case. We therefore turn finally to examine in more detail the properties of the *dual basis* in the classic work of Möbius (1846), and its revival by, e.g., Alfeld et al. (1996), that appear to supply the best available generalization of the ideal single-arc solution of Shoemake (1985). We will focus here on 2D (coordinates on the ordinary sphere $\mathbf{S}^2$) and 3D (coordinates on the quaternion sphere $\mathbf{S}^3$), but refer the interested reader to the general abstract treatment of the N-dimensional spherical dual coordinate basis in Appendix M.

The dual coordinates of a spherical triangle. We start with $\mathbf{S}^2$ and our usual three vertices $(\mathbf{v}_1, \mathbf{v}_2, \mathbf{v}_3)$ of a spherical triangle, assumed for simplicity to be oriented in a right-handed order, so their determinant is positive, indicating that the parallelepiped that they determine has a positive volume that is equal to that determinant, as schematized in Fig. 19.2. There are only three interesting quantities we can calculate from our vertices, the **dot products**,

$$\mathbf{v}_i \cdot \mathbf{v}_j = \cos\theta_{ij} = c_{ij},$$

the **cross products**, which we leave unnormalized, while noting the relation of their magnitudes to the sines of θ_{ij},

$$\begin{aligned}
\mathbf{n}_1 &= \mathbf{v}_2 \times \mathbf{v}_3 = \det\begin{bmatrix} \mathbf{e} & \mathbf{v}_2 & \mathbf{v}_3 \end{bmatrix}, & \|\mathbf{n}_1\|^2 &= 1 - \cos^2\theta_{23} = \sin^2\theta_{23} \\
\mathbf{n}_2 &= \mathbf{v}_3 \times \mathbf{v}_1 = \det\begin{bmatrix} \mathbf{v}_1 & \mathbf{e} & \mathbf{v}_3 \end{bmatrix}, & \|\mathbf{n}_2\|^2 &= 1 - \cos^2\theta_{13} = \sin^2\theta_{13} \\
\mathbf{n}_3 &= \mathbf{v}_1 \times \mathbf{v}_2 = \det\begin{bmatrix} \mathbf{v}_1 & \mathbf{v}_2 & \mathbf{e} \end{bmatrix}, & \|\mathbf{n}_3\|^2 &= 1 - \cos^2\theta_{12} = \sin^2\theta_{12}
\end{aligned} \Bigg\}, \tag{19.27}$$

where $\mathbf{e} = [\mathbf{e}_x, \mathbf{e}_y, \mathbf{e}_z]^t$ are the Cartesian basis elements, and the **parallelepiped volume**, which is

$$\mathbf{V}(123) = \det\begin{bmatrix} \mathbf{v}_1 & \mathbf{v}_2 & \mathbf{v}_3 \end{bmatrix} \equiv \det\mathbf{V}. \tag{19.28}$$

We add the crucial observation, with clear higher-dimensional analogs, that if we add a point $\mathbf{p}$ on the sphere inside a spherical triangle, it creates three new parallelepiped volumes that are determinants of the form

$$\begin{aligned}
d_1 &= \det[\mathbf{p}\ \mathbf{v}_2\ \mathbf{v}_3] = \mathbf{p} \cdot \mathbf{n}_1 \\
d_2 &= \det[\mathbf{v}_1\ \mathbf{p}\ \mathbf{v}_3] = \mathbf{p} \cdot \mathbf{n}_2 \\
d_3 &= \det[\mathbf{v}_1\ \mathbf{v}_2\ \mathbf{p}] = \mathbf{p} \cdot \mathbf{n}_3
\end{aligned} \Bigg\}. \tag{19.29}$$

Now suppose we postulate an expression for the expansion of the unit-length point $\mathbf{p}$ on a sphere, with unknown coefficients $b_i(p; v)$, in the form

$$\mathbf{p} = b_1(p; v)\mathbf{v}_1 + b_2(p; v)\mathbf{v}_2 + b_3(p; v)\mathbf{v}_3. \tag{19.30}$$

We next observe that the dual basis elements $\mathbf{n}_i$ in Eq. (19.27) obey the relation

$$\mathbf{n}_i \cdot \mathbf{v}_j = \delta_{ij} \det\mathbf{V}, \tag{19.31}$$

as well as its dual form

$$\sum_{i=1}^{3} \left(v_i{}^a\, n_i{}^b\right) = \delta_{ab} \det\mathbf{V}. \tag{19.32}$$

Thus we can use the dual basis to project out each coefficient $b_i(p; v)$ individually to find

$$\mathbf{p} \cdot \mathbf{n}_i = b_i(p; v)\, \mathbf{v}_i \cdot \mathbf{n}_i = b_i(p; v)\det\mathbf{V} \tag{19.33}$$

$$\Rightarrow b_i(p; v) = \frac{\mathbf{p} \cdot \mathbf{n}_i}{\det\mathbf{V}}. \tag{19.34}$$

Our expansion Eq. (19.30) of $\mathbf{p}$ in terms of the dual spherical triangle basis becomes

$$\mathbf{p} = \sum_{i=1}^{3} \mathbf{v}_i \left(\frac{\mathbf{n}_i \cdot \mathbf{p}}{\det \mathbf{V}} \right). \tag{19.35}$$

In addition, we can verify, due to Eq. (19.32), that this choice for an expansion of an arbitrary unit vector $\mathbf{p}$ in terms of the nonorthogonal spherical triangle basis $\mathbf{V}$ *preserves the unit-norm property* that we require:

$$1 = \mathbf{p} \cdot \mathbf{p} = \sum_i (\mathbf{p} \cdot \mathbf{v}_i) \left(\frac{\mathbf{n}_i \cdot \mathbf{p}}{\det \mathbf{V}} \right) = \frac{1}{\det \mathbf{V}} \sum_{a,b} p_a p_b \sum_i \left(v_i{}^a n_i{}^b \right) \equiv \mathbf{p} \cdot \mathbf{p}. \tag{19.36}$$

Spherical barycentric coordinates as polsine ratios. The determinant defining the parallelepiped volume for edges at the base vertex in the unit sphere $\mathbf{S}^N$ in any dimension is known as the "polar sine" or `polsin` (see, e.g., Eriksson, 1978). We can thus recast our expansion Eq. (19.30) of the dual basis coefficients of an arbitrary spherical point $\mathbf{p}$ identifying the determinant forms as ratios of polsines:

$$b_1 = \frac{\mathbf{p} \cdot \mathbf{n}_1}{\mathbf{v}_1 \cdot \mathbf{n}_1} = \frac{\det[\mathbf{p}\,\mathbf{v}_2\,\mathbf{v}_3]}{\det[\mathbf{v}_1\,\mathbf{v}_2\,\mathbf{v}_3]}, \quad b_2 = \frac{\mathbf{p} \cdot \mathbf{n}_2}{\mathbf{v}_2 \cdot \mathbf{n}_2} = \frac{\det[\mathbf{v}_1\,\mathbf{p}\,\mathbf{v}_3]}{\det[\mathbf{v}_1\,\mathbf{v}_2\,\mathbf{v}_3]}, \quad b_3 = \frac{\mathbf{p} \cdot \mathbf{n}_3}{\mathbf{v}_3 \cdot \mathbf{n}_3} = \frac{\det[\mathbf{v}_1\,\mathbf{v}_2\,\mathbf{p}]}{\det[\mathbf{v}_1\,\mathbf{v}_2\,\mathbf{v}_3]}. \tag{19.37}$$

Remarkably, this collection of determinant weights is an expansion of a classic identity, described also for arbitrary dimensions in Appendix M, that guarantees that $\mathbf{p}$ has unit length,

$$\mathbf{p} \cdot \mathbf{p} \equiv 1, \tag{19.38}$$

for unit-length nonorthonormal basis vectors $\|\mathbf{v}_i\| = 1$. Our fundamental identities for representing a unit vector $\mathbf{p}$ in our dimensions of interest, $N = 2, 3, 4$, can now be written as

2D: $\quad \mathbf{p} \det[\mathbf{v}_1\,\mathbf{v}_2] = \mathbf{v}_1 \det[\mathbf{p}\,\mathbf{v}_2] + \mathbf{v}_2 \det[\mathbf{v}_1\,\mathbf{p}],$ \hfill (19.39)

3D: $\quad \mathbf{p} \det[\mathbf{v}_1\,\mathbf{v}_2\,\mathbf{v}_3] = \mathbf{v}_1 \det[\mathbf{p}\,\mathbf{v}_2\,\mathbf{v}_3] + \mathbf{v}_2 \det[\mathbf{v}_1\,\mathbf{p}\,\mathbf{v}_3] + \mathbf{v}_3 \det[\mathbf{v}_1\,\mathbf{v}_2\,\mathbf{p}]],$ \hfill (19.40)

$$\text{4D:} \quad \mathbf{p} \det[\mathbf{v}_1\,\mathbf{v}_2\,\mathbf{v}_3\,\mathbf{v}_4] = \left\{ \begin{array}{l} \mathbf{v}_1 \det[\mathbf{p}\,\mathbf{v}_2\,\mathbf{v}_3\,\mathbf{v}_4] + \mathbf{v}_2 \det[\mathbf{v}_1\,\mathbf{p}\,\mathbf{v}_3\,\mathbf{v}_4] \\ +\mathbf{v}_3 \det[\mathbf{v}_1\,\mathbf{v}_2\,\mathbf{p}\,\mathbf{v}_4] + \mathbf{v}_4 \det[\mathbf{v}_1\,\mathbf{v}_2\,\mathbf{v}_3\,\mathbf{p}] \end{array} \right. . \tag{19.41}$$

Note that, unlike the spherical subtriangle partition, we satisfy both basic requirements: given a unit vector $\mathbf{p}$, we know its weights in any spherical simplex, and, conversely, if we know its weights, we can find $\mathbf{p}$. The 2D case is especially revealing: referring back to the `slerp` geometry in Fig. 19.1, we find the correspondences

$$\begin{array}{l} \det[\mathbf{v}_1\,\mathbf{v}_2]^2 = \det \begin{bmatrix} 1 & \cos\theta \\ \cos\theta & 1 \end{bmatrix} = \sin^2\theta \quad \Rightarrow \quad \det[\mathbf{v}_1\,\mathbf{v}_2] = \texttt{slerp denominator} \\[2ex] \det[\mathbf{p}\,\mathbf{v}_2]^2 = \det \begin{bmatrix} 1 & \cos\theta_1 \\ \cos\theta_1 & 1 \end{bmatrix} = \sin^2\theta_1 \quad \Rightarrow \quad \det[\mathbf{p}\,\mathbf{v}_2] = \texttt{slerp first numerator} \\[2ex] \det[\mathbf{v}_1\,\mathbf{p}]^2 = \det \begin{bmatrix} 1 & \cos\theta_2 \\ \cos\theta_2 & 1 \end{bmatrix} = \sin^2\theta_2 \quad \Rightarrow \quad \det[\mathbf{v}_1\,\mathbf{p}] = \texttt{slerp second numerator} \end{array} \Bigg\} . \tag{19.42}$$

Thus, for 2D arcs, the dual basis expansion is *exactly* the `slerp` sine-ratio form for a point $\mathbf{p}$ having an angle $\cos\theta_1 = \cos(t_1\,\theta)$ relative to the basis point $\mathbf{v}_1$ and $\cos\theta_2 = \cos(t_2\,\theta)$ relative to the basis point $\mathbf{v}_2$.

So when **p** is a linear combination of only two vertices, then its dual basis representation is exactly a perfect `slerp`.

The properties of these coefficients, originally studied by Möbius, have many parallels to the barycentric coordinates of a Euclidean point relative to Euclidean simplex vertices, perfectly describe *any* unit-length point in a spherical triangle in terms of a basis consisting of the triangle vertices alone, and reduce to the individual $(\mathbf{v}_1, \mathbf{v}_2, \mathbf{v}_3)$ values as **p** approaches each of those vertices. The most significant difference is that these coefficients are *not a partition of unity*. We conclude, in agreement with Alfeld et al. (1996), that the notion that spherical generalizations of barycentric coordinates must be partitions of unity is *false*, and that the properties we have in Eq. (19.30), Eq. (19.37), and Eq. (19.38) are, instead, the essential properties required.

More geometry of the spherical dual basis coordinates. In fact, we have barely begun to explore the properties of Eq. (19.30). Next, following in part the treatment of Alfeld et al. (1996), we will explore the actual angles in the spatial geometry that are described by the ratio-of-determinants coefficients b_i that we have introduced. There is significant geometric structure hidden in the inner products between the spherical triangle vertices $(\mathbf{v}_1, \mathbf{v}_2, \mathbf{v}_3)$ and the *normalized* dual basis elements

$$\hat{\mathbf{n}}_i = \frac{\mathbf{v}_j \times \mathbf{v}_k}{\|\mathbf{v}_j \times \mathbf{v}_k\|} = \frac{\mathbf{v}_j \times \mathbf{v}_k}{\sin \theta_{jk}} , \tag{19.43}$$

where we have assumed cyclic index order for (i, j, k). First, we take the $\mathbf{S}^2$ equation Eq. (19.40) and note that each term is the numerator of a dot product projection onto a dual basis axis. Consider, for example, the projections of $\mathbf{v}_1$ and **p** onto the dual basis unit vector $\hat{\mathbf{n}}_1$ as shown in Fig. 19.6:

$$\mathbf{p} \cdot \hat{\mathbf{n}}_1 = \frac{\det[\mathbf{p}\, \mathbf{v}_2\, \mathbf{v}_3]}{\|\mathbf{n}_1\|} = \frac{\det[\mathbf{p}\, \mathbf{v}_2\, \mathbf{v}_3]}{\sin\theta_{23}} = \cos\psi_1 = \cos\left(\frac{\pi}{2} - \alpha_1\right) = \sin\alpha_1 , \tag{19.44}$$

$$\mathbf{v}_1 \cdot \hat{\mathbf{n}}_1 = \frac{\det[\mathbf{v}_1\, \mathbf{v}_2\, \mathbf{v}_3]}{\|\mathbf{n}_1\|} = \frac{\det[\mathbf{v}_1\, \mathbf{v}_2\, \mathbf{v}_3]}{\sin\theta_{23}} = \cos\phi_1 = \cos\left(\frac{\pi}{2} - \beta_1\right) = \sin\beta_1 . \tag{19.45}$$

Note that the normalizations of the cross products *cancel* when we take the ratio, so we can also consider the list of the b_i to be these ratios of projections of the four vectors, $(\mathbf{p}, \mathbf{v}_1, \mathbf{v}_2, \mathbf{v}_3)$, onto the dual basis cross products. Because of the normalization of the $\hat{\mathbf{n}}_i$ and the unit length of all the vectors involved, these will be valid trigonometric functions. We summarize these observations in the following set of relationships:

$$b_1 = \frac{\det[\mathbf{p}\, \mathbf{v}_2\, \mathbf{v}_3]}{\det[\mathbf{v}_1\, \mathbf{v}_2\, \mathbf{v}_3]} = \frac{\det[\mathbf{p}\, \mathbf{v}_2\, \mathbf{v}_3]}{\sin\theta_{23}} \times \frac{\sin\theta_{23}}{\det[\mathbf{v}_1\, \mathbf{v}_2\, \mathbf{v}_3]} = \frac{\mathbf{p} \cdot \hat{\mathbf{n}}_1}{\mathbf{v}_1 \cdot \hat{\mathbf{n}}_1} = \frac{\sin\alpha_1}{\sin\beta_1} , \tag{19.46}$$

$$b_2 = \frac{\det[\mathbf{v}_1\, \mathbf{p}\, \mathbf{v}_3]}{\det[\mathbf{v}_1\, \mathbf{v}_2\, \mathbf{v}_3]} = \frac{\det[\mathbf{p}\, \mathbf{v}_2\, \mathbf{v}_3]}{\sin\theta_{13}} \times \frac{\sin\theta_{13}}{\det[\mathbf{v}_1\, \mathbf{v}_2\, \mathbf{v}_3]} = \frac{\mathbf{p} \cdot \hat{\mathbf{n}}_2}{\mathbf{v}_1 \cdot \hat{\mathbf{n}}_2} = \frac{\sin\alpha_2}{\sin\beta_2} , \tag{19.47}$$

$$b_3 = \frac{\det[\mathbf{v}_1\, \mathbf{v}_2\, \mathbf{p}]}{\det[\mathbf{v}_1\, \mathbf{v}_2\, \mathbf{v}_3]} = \frac{\det[\mathbf{p}\, \mathbf{v}_2\, \mathbf{v}_3]}{\sin\theta_{12}} \times \frac{\sin\theta_{12}}{\det[\mathbf{v}_1\, \mathbf{v}_2\, \mathbf{v}_3]} = \frac{\mathbf{p} \cdot \hat{\mathbf{n}}_3}{\mathbf{v}_1 \cdot \hat{\mathbf{n}}_3} = \frac{\sin\alpha_3}{\sin\beta_3} , \tag{19.48}$$

for which we can see precisely the geometric meaning of the b_i if we study Fig. 19.6.

There is even more geometric meaning hidden in the dual coefficients. We have already noted that for the 2D `slerp` case, the b_1 and b_2 correspond precisely to the `slerp` ratio of sines of the angles

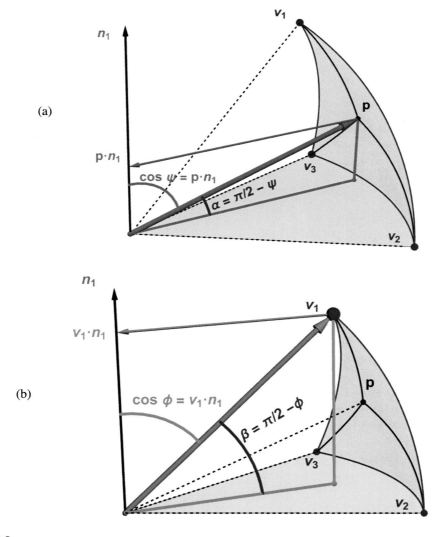

FIGURE 19.6

Geometry based on projections to the dual basis vector $\hat{\mathbf{n}}_1 = \mathbf{v}_2 \times \mathbf{v}_3 / \sin\theta_{23}$. (a) Projecting $\mathbf{p}$ to $\hat{\mathbf{n}}_1$ defines two angles, $\cos\psi = \mathbf{p} \cdot \hat{\mathbf{n}}_1$ and its 90-degree complement $\alpha = \pi/2 - \psi$, with $\cos\psi = \sin\alpha$ and $\cos\alpha = \|\mathbf{p} - \hat{\mathbf{n}}_1 \cdot \mathbf{p}\| = \sin\psi$. (b) Projecting $\mathbf{v}_1$ to $\hat{\mathbf{n}}_1$ defines two angles, $\cos\phi = \mathbf{v}_1 \cdot \hat{\mathbf{n}}_1 = \det \mathbf{v}_1\mathbf{v}_2\mathbf{v}_3 = $ parallelehedron volume and its 90-degree complement $\beta = \pi/2 - \phi$, with $\cos\phi = \sin\beta$ and $\cos\beta = \|\mathbf{v}_1 - \hat{\mathbf{n}}_1 \cdot \mathbf{v}_1\| = \sin\phi$.

embodying our only working spherical partition of unity as shown in Fig. 19.1. What about our spherical triangle area ratios, for which we saw the answers for the $\mathbf{S}^2$ case in Eq. (19.11), Eq. (19.12), Eq. (19.13), and Eq. (19.20)? Here again we can factor out the $\sin E(\mathbf{v}_1 \mathbf{v}_2 \mathbf{v}_3) = E(123)$ spherical triangle area term along with the subareas $(E(p12), E(1p3), E(12p))$ partitioned by the point $\mathbf{p}$ using Eq. (19.13) as follows:

$$
\begin{aligned}
b_1 &= \frac{\det[\mathbf{p}\,\mathbf{v}_2\,\mathbf{v}_3]}{\det[\mathbf{v}_1\,\mathbf{v}_2\,\mathbf{v}_3]} = \frac{\sin E(p23)(1+c_{23})(1+c_{p3})(1+c_{p2})}{\sin E(123)(1+c_{23})(1+c_{13})(1+c_{12})} \frac{(1+c_{23}+c_{13}+c_{12})}{(1+c_{23}+c_{p3}+c_{p2})} \\
&= \frac{\sin E(p23)}{\sin E(123)} \, f_1(p23) \,,
\end{aligned}
\tag{19.49}
$$

$$
\begin{aligned}
b_2 &= \frac{\det[\mathbf{v}_1\,\mathbf{p}\,\mathbf{v}_3]}{\det[\mathbf{v}_1\,\mathbf{v}_2\,\mathbf{v}_3]} = \frac{\sin E(1p3)(1+c_{p3})(1+c_{13})(1+c_{1p})}{\sin E(123)(1+c_{23})(1+c_{13})(1+c_{12})} \frac{(1+c_{23}+c_{13}+c_{12})}{(1+c_{p3}+c_{13}+c_{1p})} \\
&= \frac{\sin E(1p3)}{\sin E(123)} \, f_2(1p3) \,,
\end{aligned}
\tag{19.50}
$$

$$
\begin{aligned}
b_3 &= \frac{\det[\mathbf{v}_1\,\mathbf{v}_2\,\mathbf{p}]}{\det[\mathbf{v}_1\,\mathbf{v}_2\,\mathbf{v}_3]} = \frac{\sin E(12p)(1+c_{2p})(1+c_{1p})(1+c_{12})}{\sin E(123)(1+c_{23})(1+c_{13})(1+c_{12})} \frac{(1+c_{23}+c_{13}+c_{12})}{(1+c_{2p}+c_{1p}+c_{12})} \\
&= \frac{\sin E(12p)}{\sin E(123)} \, f_3(12p) \,.
\end{aligned}
\tag{19.51}
$$

Clearly we have evidence of the influence of a `slerp`-like partition, since the spherical area partitions add up by definition,

$$
E(123) = E(p23) + E(1p3) + E(12p) \,,
\tag{19.52}
$$

and this is just a different notation for Eq. (19.22), after invoking the half-angle trigonometric identities. We also see *exactly* what goes wrong: for arcs in a circle in a 2D plane, the case where the standard `slerp` is applicable, the extra weighting coefficients $f_i(p:123)$ eliminate all the cosines containing a component "3," and cancel out, leaving only the ratios of sines. We will also argue in a moment that for the quaternion $\mathbf{S}^3$ case, where spherical tetrahedra replace spherical triangles in our geometry, we lose even the relative simplicity of Eq. (19.49), as there *do not exist* equivalent equations in terms of easily computable quantities like determinants for the volumes of spherical tetrahedra. Treating the extra dimension appearing in quaternion spherical geometry introduces *significant* extra complexity, much of which is an area of current research.

Exploring the spherical mid-triangle arc. We have one more interesting property of $\mathbf{S}^2$ geometry to explore. The geometric methods we have been exploiting all basically ask the question "I have this list of known variables in the spherical triangle problem: what quantities can I calculate?" From that question, we realize that we have not fully written down all of the quantities in the geometry that we can compute from our three triangle vertices $(\mathbf{v}_1\,\mathbf{v}_2\,\mathbf{v}_3)$ and our sample point $\mathbf{p}$ inside that triangle. What we have left out is the fact that every pair $(\mathbf{v}_1, \mathbf{p})$, $(\mathbf{v}_2, \mathbf{p})$, and $(\mathbf{v}_3, \mathbf{p})$ defines its own plane, and an arc in that plane from $\mathbf{v}_i$ through the interior point $\mathbf{p}$ to intersect with *the opposite plane* defined by the remaining two points $(\mathbf{v}_j\,\mathbf{v}_k)$, cyclic, with the arc through $\mathbf{p}$ intersecting the opposite plane in a new, and interesting, point $\mathbf{y}_i$ (Alfeld et al., 1996).

The location of $\mathbf{y}_i$ is determined by computing first our familiar dual basis normal defined by $\mathbf{v}_j \times \mathbf{v}_k$, combined with the normal $\mathbf{p} \times \mathbf{v}_i$ to the plane containing the arc to $\mathbf{p}$. Normalization is irrelevant

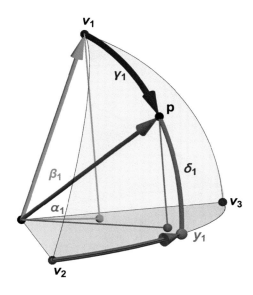

FIGURE 19.7

Showing both $\mathbf{p}$ and $\mathbf{v}_1$ together, we see the angle γ_1 that relates them and the other angles, with $\cos \gamma_1 = (\mathbf{p} \cdot \mathbf{v}_1)$.

for this purpose, so we define

$$
\left.
\begin{aligned}
\mathbf{n}_1 &= \mathbf{v}_2 \times \mathbf{v}_3 = \det [\mathbf{e}\ \mathbf{v}_2\ \mathbf{v}_3] & \mathbf{m}_1 &= \mathbf{p} \times \mathbf{v}_1 = \det [\mathbf{e}\ \mathbf{p}\ \mathbf{v}_1] \\
\mathbf{n}_2 &= \mathbf{v}_3 \times \mathbf{v}_1 = \det [\mathbf{v}_1\ \mathbf{e}\ \mathbf{v}_3] & \mathbf{m}_2 &= \mathbf{p} \times \mathbf{v}_2 = \det [\mathbf{e}\ \mathbf{p}\ \mathbf{v}_2] \\
\mathbf{n}_3 &= \mathbf{v}_1 \times \mathbf{v}_2 = \det [\mathbf{v}_1\ \mathbf{v}_2\ \mathbf{e}] & \mathbf{m}_3 &= \mathbf{p} \times \mathbf{v}_3 = \det [\mathbf{e}\ \mathbf{p}\ \mathbf{v}_3]
\end{aligned}
\right\} .
\tag{19.53}
$$

The remainder of the calculation is straightforward: as illustrated in Fig. 19.7, since each pair of planes must contain its corresponding point $\mathbf{y}_i$, which is a unit vector through the origin, the normals to *both* planes must be perpendicular to $\mathbf{y}_i$. The only piece of geometry with that property is the cross product of the normals to each plane, normalized to unity. Thus we have

$$
\left.
\begin{aligned}
\mathbf{y}_1 &= \frac{\mathbf{n}_1 \times \mathbf{m}_1}{\|\mathbf{n}_1 \times \mathbf{m}_1\|} = \frac{\det [\mathbf{e}\ \mathbf{n}_1\ \mathbf{m}_1]}{\sin \lambda_1} \\
\mathbf{y}_2 &= \frac{\mathbf{n}_2 \times \mathbf{m}_2}{\|\mathbf{n}_2 \times \mathbf{m}_2\|} = \frac{\det [\mathbf{e}\ \mathbf{n}_2\ \mathbf{m}_2]}{\sin \lambda_2} \\
\mathbf{y}_3 &= \frac{\mathbf{n}_3 \times \mathbf{m}_3}{\|\mathbf{n}_3 \times \mathbf{m}_3\|} = \frac{\det [\mathbf{e}\ \mathbf{n}_3\ \mathbf{m}_3]}{\sin \lambda_3}
\end{aligned}
\right\} ,
\tag{19.54}
$$

where as usual $\mathbf{e}$ is the Cartesian basis, and the angle λ_i is defined by $\cos \lambda_i = \mathbf{n}_i \cdot \mathbf{m}_i$.

The geodesic arc from $\mathbf{v}_i$ through $\mathbf{p}$ lands on the point $\mathbf{y}_i$, which is itself on the geodesic arc $\widehat{\mathbf{v}_j \mathbf{v}_k}$, (i, j, k) cyclic. Therefore this pair of arcs can be interpreted on its own as a nested `slerp`, and, as such, can in principle be generalized to applications in higher dimensions, such as the quaternion $\mathbf{S}^3$, without actually knowing anything specific about the volume element formulas for the higher-dimensional

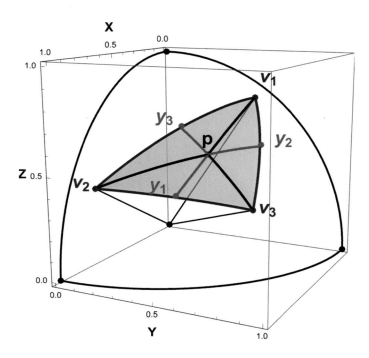

FIGURE 19.8

The key structure of the geometry of the spherical triangle is that, from any vertex $\mathbf{v}_i$, we can define a plane with $\mathbf{p}$ containing a geodesic arc intersecting the opposite geodesic edge at a point $\mathbf{y}_i$, which can be directly computed from the cross product of plane normals. The $\widehat{\mathbf{v}_i\mathbf{p}}$ curve implicitly implements a `slerp` starting at $\mathbf{v}_i$ arcing towards $\mathbf{y}_i$ and stopping at $\mathbf{p}$ with its chosen barycentric coordinate weight based on the dual basis of $(\mathbf{v}_1, \mathbf{v}_2, \mathbf{v}_3)$.

spherical simplexes. In Fig. 19.7 and more generally in Fig. 19.8, we see that we can define three useful angles,

$$\left. \begin{aligned} \cos\eta_1 &= \mathbf{v}_1 \cdot \mathbf{y}_1 \\ \cos\gamma_1 &= \mathbf{v}_1 \cdot \mathbf{p} \\ \cos\delta_1 &= \mathbf{p} \cdot \mathbf{y}_1 \end{aligned} \right\}, \tag{19.55}$$

where obviously $\eta_1 = \gamma_1 + \delta_1$, so we can immediately express $\mathbf{p}$ in terms of an internal pair of `slerp` interpolations with easily computable parameters due to $\gamma = t_y\eta$ and $\delta = t_1\eta$ being a partition of unity with $t_y + t_1 = 1$, so as $t_y \to 1$, $\mathbf{p} \to \mathbf{y}$, and as $t_1 \to 1$, $\mathbf{p} \to \mathbf{v}_1$. Similar arguments obviously hold for other permutations of arcs through $\mathbf{p}$ that are shown in Fig. 19.8. We can conclude this argument by going back to Eq. (19.24), and writing it in terms of the angles and geometry of Fig. 19.7. We can see that the first "inner" step is to move $\mathbf{v}_2$ along the arc $\widehat{\mathbf{v}_2 : \mathbf{v}_3}$ to the point $\mathbf{y}_1$, which is a `slerp` by an angle with $\cos\mu_1 = \mathbf{v}_2 \cdot \mathbf{y}_1$, which is a certain fraction $u = \mu_1/\theta_{23}$ of the angle between $\mathbf{v}_2$ and $\mathbf{v}_3$. The "outer" step is also straightforward, moving $\mathbf{v}_1$ along the arc $\widehat{\mathbf{v}_1 : \mathbf{y}_1}$ to the point $\mathbf{p}$, which is a `slerp` by an angle with $\cos\beta = \mathbf{v}_1 \cdot \mathbf{p}$, which is a certain fraction $s = \beta/\delta$ of the angle between $\mathbf{v}_1$ and $\mathbf{y}_1$, where we computed $\mathbf{y}_1$ in the first step. Our modified version of Eq. (19.24), adapting the double `slerp` for the full surface of a spherical triangle to the task of finding a single point by taking advantage of the dual basis geometry, is then

$$\mathbf{p} = \text{slerp}\,(\mathbf{v}_1, \text{slerp}(\mathbf{v}_2, \mathbf{v}_3; u); s) ,\qquad (19.56)$$

$$u = \frac{\arccos(\mathbf{v}_2 \cdot \mathbf{y}_1)}{\arccos(\mathbf{v}_2 \cdot \mathbf{v}_3)} = \frac{\mu_1}{\theta_{23}} ,\qquad\qquad s = \frac{\arccos(\mathbf{v}_1 \cdot \mathbf{p})}{\arccos(\mathbf{v}_1 \cdot \mathbf{y}_1)} = \frac{\gamma_1}{\eta_1} .$$

19.6 Quaternion barycentric coordinates

We conclude with our main theme, understanding the geometry of quaternions and the mental images that help us to see how they fit into their place. Remarkably, now that we know the role of the dual coordinate basis and the determinant identity that guarantees preservation of unit length, we can extend some of our $\mathbf{S}^2$ arguments to the quaternion space $\mathbf{S}^3$. We now propose an expansion of a quaternion q in terms of a nonorthogonal, nonpartition-of-unity set of spherical simplex weights that preserve $q \cdot q = 1$. In fact, the Möbius spherical barycentric coordinate framework itself can easily be extended to any dimension for $\mathbf{S}^N$, and that is the subject of Appendix M.

We begin with a nondegenerate spherical tetrahedron on $\mathbf{S}^3$ embedded in $\mathbb{R}^4$, and specified by four unit vectors (v_1, v_2, v_3, v_4) with a finite positive parallelepiped volume $V_{1234} = \det[v_1, v_2, v_3, v_4]$, adopting our shorthand for the volume determinant as before. Next we choose a unit-length quaternion point q inside the spherical tetrahedral volume and define

$$d_1 = \det[q, v_2, v_3, v_4] , \quad d_2 = \det[v_1, q, v_3, v_4] , \quad d_3 = \det[v_1, v_2, q, v_4] , \quad d_4 = \det[v_1, v_2, v_3, q] ,$$
$$(19.57)$$

all of which we assume are positive. By induction from our treatment of the $\mathbf{S}^2$ spherical triangle case, we hypothesize that we can define `slerp`-like barycentric coordinates of the form

$$b_1 = \frac{\sin\alpha_1}{\sin\gamma_1} = \frac{q \cdot \hat{n}_1(v_2, v_3, v_4)}{v_1 \cdot \hat{n}_1(v_2, v_3, v_4)} = \frac{\det[q\, v_2\, v_3\, v_4]/\|v_2\, v_3\, v_4\|}{\det[v_1\, v_2\, v_3\, v_4]/\|v_2\, v_3\, v_4\|} = \frac{\det[q\, v_2\, v_3\, v_4]}{\det[v_1\, v_2\, v_3\, v_4]\|} \qquad (19.58)$$

and the cyclic partners (b_2, b_3, b_4). Here we have introduced the 4D cross product and the mechanism for creating a unit normal dual basis coordinate $\hat{n}_i$ that is perpendicular to the hyperplane spanned by (v_2, v_3, v_4) and its cyclic partners using

$$\hat{n}_1 = \frac{\mathbf{n}_1}{\|\mathbf{n}_1\|} = \frac{\times(v_2\, v_3\, v_4)}{\sqrt{\times(v_2\, v_3\, v_4) \cdot \times(v_2\, v_3\, v_4)}} = \frac{\det(e\, v_2\, v_3\, v_4)}{\sqrt{\times(v_2\, v_3\, v_4) \cdot \times(v_2\, v_3\, v_4)}} , \qquad (19.59)$$

where the e_i are the Euclidean basis 4-vectors in $\mathbb{R}^4$, and $\det[e\, v_2\, v_3\, v_4] = \times(v_2\, v_3\, v_4)$, etc., are the generalized cross products in 4D.

To obtain our normal vectors for the dual basis in 4D, we need to compute the normalizing denominator. The trick is to write the expansion in terms of the 4D Levi-Civita symbol from Appendix L, as follows:

$$\|\mathbf{n}_1\|^2 = \times(v_2\, v_3\, v_4) \cdot \times(v_2\, v_3\, v_4) = \sum_n \epsilon_{nijk}\epsilon_{ni'j'k'}v_2{}^i v_3{}^j v_4{}^k v_2{}^{i'} v_3{}^{j'} v_4{}^{k'}$$

$$= \det \begin{bmatrix} 1 & v_2 \cdot v_3 & v_2 \cdot v_4 \\ v_2 \cdot v_3 & 1 & v_3 \cdot v_4 \\ v_2 \cdot v_4 & v_3 \cdot v_4 & 1 \end{bmatrix}, \tag{19.60}$$

where we used our unit-vector constraints $v_i \cdot v_i = 1$. We recognize this form, easily generalizable to any dimension, as the `polsin` appearing earlier in Eq. (19.9), that generalizes the 2D cross product sine relationship

$$\|\mathbf{v}_1 \times \mathbf{v}_2\|^2 = 1 - (\mathbf{v}_1 \cdot \mathbf{v}_2)^2 = 1 - \cos^2 \theta = \sin^2 \theta \tag{19.61}$$

to the quaternion $\mathbf{S}^3$ case.

Our task is now to work out the dual coordinate construction starting with $q \cdot \hat{n}_1 = q \cdot d_1 / \|d_1\|$ to produce a cosine of the $\widehat{v_1 : \hat{n}_1}$ angle that is the *sine* of the angle α_1 between q and the intersection point y_1 of the $\widehat{v_1 : q}$ arc with the opposing spherical triangle "face" (v_2, v_3, v_4) of the $\mathbf{S}^3$ spherical tetrahedron.

The point y_1 in the spherical triangle $T(v_2, v_3, v_4)$ is itself a point defined by an $\mathbf{S}^2$ double `slerp` parameterization of the type already studied in the $\mathbf{S}^2$ case. Again, with

$$b_1 = \frac{\sin \alpha_1}{\sin \gamma_1} = \frac{\det[q\, v_2\, v_3\, v_4]}{\det[v_1\, v_2\, v_3\, v_4]}$$

and the parameterization

$$q = \frac{\sin \alpha_1}{\sin \gamma_1} v_1 + \frac{\sin \beta_1}{\sin \gamma_1} y_1(v_2, v_3, v_4) \mid \gamma_1 = \alpha_1 + \beta_1, \tag{19.62}$$

we can argue by symmetry and the absence of any v_1 component in the spherical triangle intersection point y_1 that q must be parameterized by the coefficients

$$b_1 = \frac{\det[q\, v_2\, v_3\, v_4]}{\det[v_1\, v_2\, v_3\, v_4]} \mid \text{(cyclic)}, \tag{19.63}$$

so

$$\begin{aligned} q &= b_1 v_1 + b_2 v_2 + b_3 v_3 + b_4 v_4 \\ &= \frac{\det[q\, v_2\, v_3\, v_4]}{\det[v_1\, v_2\, v_3\, v_4]} v_1 + \frac{\det[v_1\, q\, v_3\, v_4]}{\det[v_1\, v_2\, v_3\, v_4]} v_2 + \frac{\det[v_1\, v_2\, q\, v_4]}{\det[v_1\, v_2\, v_3\, v_4]} v_3 + \frac{\det[v_1\, v_2\, v_3\, q]}{\det[v_1\, v_2\, v_3\, v_4]} v_4 \end{aligned} \left. \right\}, \tag{19.64}$$

where one can verify directly that for unit vectors (v_1, v_2, v_3, v_4), which we of course take to be unit quaternions, defining a spherical tetrahedron in the quaternion 3-sphere $\mathbf{S}^3$, q *remains* on $\mathbf{S}^3$, with $q \cdot q = 1$, and thus is always a valid quaternion. Choosing any particular vertex as our base, we can write down explicit nested three-part `slerps` for any quaternion inside the $\mathbf{S}^3$ tetrahedron in the form

$$\begin{aligned} q(v_1, v_2, v_3, v_4; t_1, t_2, t_3) &= \text{slerp}_3(v_1, v_2, v_3, v_4; t_1, t_2, t_3) \\ &= \text{slerp}(v_1, y_1 = \text{slerp}(v_2, y_2 = \text{slerp}(v_3, v_4; t_3); t_2); t_1). \end{aligned} \tag{19.65}$$

Fig. 19.9 illustrates the basic geometry of this quaternion-preserving representation of a quaternion point relative to `slerp`-related Möbius spherical barycentric coordinates defining a spherical tetrahedron.

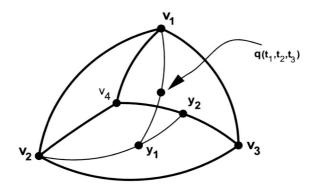

FIGURE 19.9

One of the most fundamental structures of the full quaternion space $\mathbf{S}^3$ is the solid spherical tetrahedron $T(v_1, v_2, v_3, v_4)$, whose interior points $q(t_1, t_2, t_3)$ can be described using the 4D spherical barycentric coordinates of Eq. (19.64), and parameterized, as indicated schematically here, by a three-parameter nested `slerp` such as Eq. (19.65).

19.7 Remark on intractability of spherical volume ratios for $N > 2$

One may be curious what happens if we investigate the use of tetrahedral volume ratios in the quaternion space $\mathbf{S}^3$ as partitions of unity, paralleling our discussion for the $\mathbf{S}^2$ case. One might expect the elegant, though ultimately flawed, area ratios, e.g., as in Eq. (19.21), for interpolation in $\mathbf{S}^2$ using spherical triangles, to extend straightforwardly to $\mathbf{S}^3$ and spherical tetrahedra. *This is definitively not the case.* In fact, for the 3-sphere $\mathbf{S}^3$ the situation for tetrahedral patches defined by four points becomes extraordinarily complex (see, e.g., Schläfli, 1860; Coxeter, 1935; Milnor, 1994; Cho and Kim, 1999; Mohanty, 2002; Doyle and Leibon, 2003 & 2018; Murakami and Yano, 2005; Murakami, 2012), which is why we have not attempted to study this here.

Computing spherical tetrahedral volumes (and their generalizations to spherical N-simplexes in $\mathbf{S}^N$) turns out to be incredibly complicated, involving enormous algebraic expressions, dilogarithms, and issues that have challenged leading mathematicians, going back as far as 1860 with the work of Schläfli (1860). An excellent introduction to the subject, not available as a publication elsewhere, can be found in John Milnor's complete works (Milnor, 1994). Amid an assortment of treatments with widely varying notations, we found the most readable, though still very complicated, description of how to actually carry out the relevant volume computations to be that of Murakami (2012). The recent date of publications such as Murakami's is a sign that these questions are still under active investigation. Encountering the body of classic mathematical literature on this subject, one is led to wonder: *why* does this question get so complicated for spheres and spherical simplexes in spheres of dimension greater than $\mathbf{S}^2$? Are there even more exotic facts about quaternions that are hinted at by this complexity?

Extending quaternions to **4D** and **SE(3)**

6

The standard Hamilton quaternions that we have dealt with correspond exactly to 3D rotations, providing alternative ways of exploiting the structure inherent in those rotations. However, there are a number of relatively straightforward modifications that preserve all the actions that can be done with the Hamilton quaternions and the three-parameter rotation group **SO**(3), but contain *three additional free parameters* that have very interesting consequences. There are in fact even more extensions of quaternions, from long ago (see, e.g., Clifford, 1873, 1882; Study, 1891) and more recent times (see, e.g., Frenkel and Libine, 2008, 2011, 2021; Antonuccio, 2013, 2015), that we will omit from our treatment because their related applications are not as clearly of practical use as the two variants that we now investigate. In Part 7, we will explore yet another fundamental extension of quaternions to incorporate in a very natural way the additional three parameters of Einstein's theory of special relativity using complexified quaternions.

The particular quaternion extensions that we will study in the chapters in this part are these:

- **4D rotations: the rotation group SO**(4) **and 4D coordinate frames.** The standard method of deriving a rotation action in ordinary space from quaternions is the bilinear conjugation, acting on the left and the right by a quaternion and that quaternion's conjugate. Each of those quaternions contain the *same* three parameters. However, if we use *different* quaternions for the left and right actions, there are *six* free parameters, and those correspond precisely to the six parameters of the 4D Euclidean rotation group **SO**(4) and the 4D coordinate frames encoded in the columns of a 4D rotation matrix. Following essentially the same procedures as we used in 3D in Chapters 8, 9, and 10, this allows us to extend our entire treatment of point-cloud matching and frame-cloud matching from 3D to 4D by exploiting the double-quaternion form of the rotation group **SO**(4).

- **Dual quaternions: including translations to support the group SE(3).** While quaternions are extremely useful for representing 3D rotations, they permit no explicit reference to the location of an object in space or the center of the rotation being applied. Many practical problems in areas like robotics or simulated body part and skin element motion in animation require integration of spatial location with orientation. In this chapter, we introduce *dual quaternions*, which combine the actions of translation and quaternion rotations in a unified six-degree-of-freedom quaternion-based framework that has become popular in applications like robotics. The corresponding group is the *special Euclidean group* **SE**(3) that integrates the three 3D translation parameters with the three 3D rotation parameters. An important aspect of any attempt like dual quaternions to combine translations with rotations is that the units are *incompatible*, so, somewhere, a dimensional constant of arbitrary magnitude will be involved, and that will complicate any attempt to create a completely transparent unification.

The 4D quaternion-based coordinate and orientation frame alignment problems[☆]

20

This chapter starts from the basic ideas of 3D matching presented in Chapter 8 and breaks new ground by extending them to 4D. The key fact is that 4D rotation matrices can be represented by pairs of distinct ordinary quaternions to treat almost any application in 4D that can be addressed by a single quaternion in 3D. Our main topic here is thus to show how quaternions can be applied to 4D point-cloud alignment problems, along with 4D orientation frame description and alignment. In addition we will extend the methods of Chapter 9 to the determination of the quaternion *pairs* that correspond to a perfect or noisy measured 4D rotation matrix.

20.1 Foundations of quaternions for 4D problems

Starting from the quaternion properties from Chapter 2 that we have used up until now in our 3D analysis, we now systematically add the extensions that can be exploited to handle the 4D case.

Double quaternions and 4D rotations. We begin by extending Eq. (2.14) from three Euclidean dimensions to four Euclidean dimensions by choosing two *distinct* quaternions and generalizing Eq. (2.14) to 4D points $\mathbf{x}_4 = (w, x, y, z)$ as follows:

$$p \star (w, x, y, z) \star \bar{q} = R_4(p, q) \cdot \mathbf{x}_4 . \tag{20.1}$$

Here R_4 turns out to be an orthonormal 4D rotation matrix that is quadratic in the *pair* (p, q) of unit quaternion elements, which together have exactly the six degrees of freedom required for the most general 4D Euclidean rotation produced by the special orthogonal group $\mathbf{SO}(4)$. The algebraic form of this 4D rotation matrix is

$$R_4(p, q) = \begin{bmatrix} p_0 q_0 + p_1 q_1 + p_2 q_2 + p_3 q_3 & -p_1 q_0 + p_0 q_1 + p_3 q_2 - p_2 q_3 \\ p_1 q_0 - p_0 q_1 + p_3 q_2 - p_2 q_3 & p_0 q_0 + p_1 q_1 - p_2 q_2 - p_3 q_3 \\ p_2 q_0 - p_3 q_1 - p_0 q_2 + p_1 q_3 & p_3 q_0 + p_2 q_1 + p_1 q_2 + p_0 q_3 \\ p_3 q_0 + p_2 q_1 - p_1 q_2 - p_0 q_3 & -p_2 q_0 + p_3 q_1 - p_0 q_2 + p_1 q_3 \end{bmatrix}$$

☆ This chapter closely follows the Supporting Information associated with Hanson (2020).

Visualizing More Quaternions. https://doi.org/10.1016/B978-0-32-399202-2.00033-2

$$\left.\begin{array}{cc} -p_2q_0 - p_3q_1 + p_0q_2 + p_1q_3 & -p_3q_0 + p_2q_1 - p_1q_2 + p_0q_3 \\ -p_3q_0 + p_2q_1 + p_1q_2 - p_0q_3 & p_2q_0 + p_3q_1 + p_0q_2 + p_1q_3 \\ p_0q_0 - p_1q_1 + p_2q_2 - p_3q_3 & -p_1q_0 - p_0q_1 + p_3q_2 + p_2q_3 \\ p_1q_0 + p_0q_1 + p_3q_2 + p_2q_3 & p_0q_0 - p_1q_1 - p_2q_2 + p_3q_3 \end{array}\right], \tag{20.2}$$

where $\det R_4(p, q) = (p \cdot p)^2 (q \cdot q)^2$ and $\operatorname{tr} R_4(p, q) = 4p_0q_0$. Since this is a quadratic form in p and q, the rotation is unchanged under $(p, q) \to (-p, -q)$, and the quaternions are again a double covering. If we set $p = q$, we recover a matrix that leaves the w component invariant, and is just the rotation Eq. (2.14) for the $\mathbf{x}_3 = (x, y, z)$ component. If we set p and q in turn to the identity quaternion q_{ID}, we find the interesting result that $R_4(p, q_{\mathrm{ID}}) = Q(p)$ from Eq. (2.6), and $R_4(q_{\mathrm{ID}}, \bar{p}) = \widetilde{Q}(p)$ from Eq. (2.7).

Rotations in 4D can be composed in quaternion form parallel to the 3D case, with

$$R_4(p, q) \cdot R_4(p', q') = R_4(p \star p', q \star q'). \tag{20.3}$$

This behavior means that there is a group action of 4D Euclidean rotations that is representable as a pair of quaternion group actions, which we write as $\mathbf{SU(2)} \times \mathbf{SU(2)}$, the covering group of $\mathbf{SO(4)}$. As in 3D, any number of repeated 4D rotations can be collapsed into a single rotation whose quaternion arguments are corresponding quaternion products in the same sequence.

We also observe that the 4D columns of Eq. (20.2) can be used to define 4D Euclidean orientation frames in the same fashion as the 3D columns of Eq. (2.14), and we will exploit this to treat the 4D orientation frame alignment problem below.

Remark: *Eigensystem and properties of R_4:* In Appendix C, we worked out explicit forms for the eigensystems of 3D rotation matrices. We can also compute the eigenvalues of our 4D rotation matrix $R_4(p, q)$ from Eq. (20.2). The 3D form of $R_3(q)$ in terms of explicit fixed axes does not have an exact analog in 4D because 4D rotations leave a *plane* invariant, not an axis. Nevertheless, we can still find a very compact form for the 4D eigenvalues. Our exact 4D analog of Eq. (C.1) in Appendix C, after applying the transformations $q_1^2 + q_2^2 + q_3^2 \to 1 - q_0^2$ for q and p to simplify the expression, is just

$$\left\{ \begin{array}{l} p_0q_0 - \operatorname{sign}(p_0q_0)\left(+\sqrt{(1 - p_0^2)(1 - q_0^2)} + i\sqrt{(1 - p_0^2)q_0^2} + i\sqrt{p_0^2(1 - q_0^2)} \right) \\[2ex] p_0q_0 - \operatorname{sign}(p_0q_0)\left(+\sqrt{(1 - p_0^2)(1 - q_0^2)} - i\sqrt{(1 - p_0^2)q_0^2} - i\sqrt{p_0^2(1 - q_0^2)} \right) \\[2ex] p_0q_0 - \operatorname{sign}(p_0q_0)\left(-\sqrt{(1 - p_0^2)(1 - q_0^2)} + i\sqrt{(1 - p_0^2)q_0^2} - i\sqrt{p_0^2(1 - q_0^2)} \right) \\[2ex] p_0q_0 - \operatorname{sign}(p_0q_0)\left(-\sqrt{(1 - p_0^2)(1 - q_0^2)} - i\sqrt{(1 - p_0^2)q_0^2} + i\sqrt{p_0^2(1 - q_0^2)} \right) \end{array} \right\}. \tag{20.4}$$

In these 4D eigenvalue expressions, the overall sign in the right-hand terms depends on the sign of $p_0q_0 = (1/4)\operatorname{tr} R_4(p, q)$. This feature is subtle, and arises in the process of removing a spurious apparent asymmetry between p_0 and q_0 in the eigenvalue expressions associated with the appearance of $\sqrt{q_0^2}$ and $\sqrt{p_0^2}$; incorrect signs arise in removing the square roots without the sign term, which is required to make the determinant equal to the products of the eigenvalues. The eigenvectors can be computed in the usual way, but we know of no informative simple algebraic form. Interestingly, the eigensystem of the *profile matrix* of $R_4(p, q)$, discussed below in Section 20.4, is much simpler.

20.2 Double-quaternion approach to the 4D RMSD problem

Here and in the following sections, we present the nontrivial steps needed to understand and solve the 4D spatial and orientation frame RMSD optimization problems in the quaternion framework, as well as how to extract the quaternion representation of a 4D rotation matrix. A particularly important step is to extend our understanding of the eigensystems solving the RMSD problem for 3D Euclidean data to 4D: unlike the 3D profile matrix M_3, which is symmetric and traceless, the 4D profile matrix M_4 is unconstrained, and that complicates the problem.

While we might expect the quaternion eigensystem of the 4D profile matrix to allow us to solve the 4D RMSD problem in exactly the same fashion as in 3D, this is, interestingly, false. We will need several stages of analysis to actually find the correct way to exploit quaternions in the 4D RMSD optimization context. In this section, we study the 4D coordinate matching problem by itself and discover a method that can be easily solved using a quaternion approach that parallels the algorithms traditional in the 3D problem. We note that Appendix J gives detailed treatment of the *algebraic* solutions to the eigensystems of the 4×4 symmetric real matrices that are relevant to our quaternion-based spatial and orientation frame alignment problems in both 3D and 4D.

‡ *Advanced: Notes on 4D and mirrored geometry.* While applications of 4D methods are typically theoretical, we note that some applications have been considered, e.g., by Immel et al. (2018, 2019). Another issue is that mirrored structures, which we mentioned in Chapter 8, have distinct properties in 4D. In 3D, rotations admit no central inversion, so mirrored geometries cannot be reached by standard rotations. In even dimensions like 4D, the rotation group contains the central inversion.

20.2.1 Starting point for the 4D RMSD problem

The 4D double quaternion matrix Eq. (20.2) provides the most general quaternion context that we know of for expressing a 4D RMSD cloud alignment task. We start with the RMSD minimization problem for aligning 4D Euclidean point data, with a reference set $\mathbf{X}$ of N columns of 4-vectors, rotated by an unknown matrix R_4 to generate the corresponding target data $\mathbf{U}$. The optimization task can be expressed as the maximization problem for the by-now-familiar cross-term expression

$$\Delta_4 = \sum_{k=1}^{N} (R_4 \cdot x_k) \cdot u_k = \sum_{a=0,b=0}^{3} R_4{}^{ba} E_{4:ab} = \operatorname{tr} R_4 \cdot E_4 \,, \tag{20.5}$$

where

$$E_{4:ab} = \sum_{k=1}^{N} [x^k]_a \, [u^k]_b = \left[\mathbf{X} \cdot \mathbf{U}^{\mathrm{t}} \right]_{ab} \tag{20.6}$$

is the cross-covariance matrix whose (a, b) indices we will usually write as (w, x, y, z) in the manner of Eq. (8.9). Using Eq. (20.2) in Eq. (20.5) to perform the 4D version of the rearrangement of the optimization function, we can rewrite our measure as

$$\Delta_4 = \operatorname{tr} R_4(p, q) \cdot E_4 = [p_0, p_1, p_2, p_3] \cdot M_4(E_4) \cdot [q_0, q_1, q_2, q_3]^{\mathrm{t}} \equiv p \cdot M_4(E_4) \cdot q , \qquad (20.7)$$

where we have p and q, not p and $\bar{q}$, in Eq. (20.7), and the profile matrix for the 4D data now becomes

$$M_4(E_4) = \begin{bmatrix} E_{ww} + E_{xx} + E_{yy} + E_{zz} & +E_{yz} - E_{zy} - E_{wx} + E_{xw} \\ +E_{yz} - E_{zy} + E_{wx} - E_{xw} & E_{ww} + E_{xx} - E_{yy} - E_{zz} \\ +E_{zx} - E_{xz} + E_{wy} - E_{yw} & +E_{xy} + E_{yx} + E_{wz} + E_{zw} \\ +E_{xy} - E_{yx} + E_{wz} - E_{zw} & +E_{zx} + E_{xz} - E_{wy} - E_{yw} \end{bmatrix}$$
$$\begin{bmatrix} +E_{zx} - E_{xz} - E_{wy} + E_{yw} & +E_{xy} - E_{yx} - E_{wz} + E_{zw} \\ +E_{xy} + E_{yx} - E_{wz} - E_{zw} & +E_{zx} + E_{xz} + E_{wy} + E_{yw} \\ E_{ww} - E_{xx} + E_{yy} - E_{zz} & +E_{yz} + E_{zy} - E_{wx} - E_{xw} \\ +E_{yz} + E_{zy} + E_{wx} + E_{xw} & E_{ww} - E_{xx} - E_{yy} + E_{zz} \end{bmatrix} . \qquad (20.8)$$

We note that, in contrast to the traceless, symmetric 3D profile matrix $M_3(E_3)$ from Eq. (8.9),

$$M_3(E_3) = \begin{bmatrix} E_{xx} + E_{yy} + E_{zz} & E_{yz} - E_{zy} & E_{zx} - E_{xz} & E_{xy} - E_{yx} \\ E_{yz} - E_{zy} & E_{xx} - E_{yy} - E_{zz} & E_{xy} + E_{yx} & E_{zx} + E_{xz} \\ E_{zx} - E_{xz} & E_{xy} + E_{yx} & -E_{xx} + E_{yy} - E_{zz} & E_{yz} + E_{zy} \\ E_{xy} - E_{yx} & E_{zx} + E_{xz} & E_{yz} + E_{zy} & -E_{xx} - E_{yy} + E_{zz} \end{bmatrix} , \qquad (20.9)$$

our 4D profile matrix $M_4(E_4)$ in Eq. (20.8) is *neither traceless nor symmetric*. The nonzero trace is not a problem, and can be treated easily with the quartic eigenvalue formulas in Appendix J; however, our nonsymmetric matrix, even though it is real, adds substantial additional complexity to the work of optimizing Eq. (20.7) using the profile matrix Eq. (20.8). We now turn to that problem.

20.2.2 A tentative 4D eigensystem

Our task is now to find an algorithm that allows us to successfully compute the quaternion pair $(p_{\mathrm{opt}}, q_{\mathrm{opt}})$, or, equivalently, the global rotation $R_4(p_{\mathrm{opt}}, q_{\mathrm{opt}})$, that maximizes the measure

$$\Delta_4 = \operatorname{tr} R_4(p, q) \cdot E_4 = p \cdot M_4(E_4) \cdot q , \qquad (20.10)$$

with $M_4(E_4)$ a general real matrix with a generic trace and no symmetry conditions. Note that now we can have *both* left and right eigenvectors p and q for a single eigenvalue of the profile matrix M_4. Thus q would correspond to the eigenvectors of M_4, and p would correspond to the eigenvectors of the transpose M_4^{t}. *Warning:* The eigensystem of M_4 typically has some complex eigenvalues and is furthermore *insufficient* by itself to solve the 4D RMSD optimization problem, so additional refinements will be necessary. We now explore a path to an optimal solution amenable to quaternion-based numerical evaluation, with corresponding algebraic approaches based on Cardano's method for solving generic quartic eigensystems elaborated in Appendix J.

For some types of calculations, we may find it useful to decompose M_4 in a way that isolates particular features using the form

$$M_4(w, x, y, z, \ldots) = \begin{bmatrix} w+x+y+z & a-a_w & b-b_w & c-c_w \\ a+a_w & w+x-y-z & C-C_w & B+B_w \\ b+b_w & C+C_w & w-x+y-z & A-A_w \\ c+c_w & B-B_w & A+A_w & w-x-y+z \end{bmatrix}, \quad (20.11)$$

where $(w, x, y, z) = (E_{ww}, E_{xx}, E_{yy}, E_{zz})$, $a = E_{yz} - E_{zy}$, cyclic, $A = E_{yz} + E_{zy}$, cyclic, $a_w = E_{wx} - E_{xw}$, cyclic, $A_w = E_{wx} + E_{xw}$, cyclic, and $\mathrm{tr}(M_4) = 4w$. This effectively exposes the structural symmetries of M_4.

We next review the properties of the eigenvalue equation $\det[M_4 - eI_4] = 0$, where e is the variable for which we solve to obtain the four eigenvalues ϵ_k, and I_4 denotes the 4D identity matrix; transposing M_4 does not change the eigenvalues but does interchange the distinct left and right eigenvectors. While M_4 itself has new properties, the corresponding expressions in terms of e and ϵ_k, along with the outcome of eliminating e (Abramowitz and Stegun, 1970), are by now familiar:

$$\det[M_4 - eI_4] = e^4 + e^3 p_1 + e^2 p_2 + ep_3 + p_4 = 0, \quad (20.12)$$

$$(e - \epsilon_1)(e - \epsilon_2)(e - \epsilon_3)(e - \epsilon_4) = 0, \quad (20.13)$$

$$\left.\begin{aligned} p_1 &= (-\epsilon_1 - \epsilon_2 - \epsilon_3 - \epsilon_4) \\ p_2 &= (\epsilon_1\epsilon_2 + \epsilon_1\epsilon_3 + \epsilon_2\epsilon_3 + \epsilon_1\epsilon_4 + \epsilon_2\epsilon_4 + \epsilon_3\epsilon_4) \\ p_3 &= (-\epsilon_1\epsilon_2\epsilon_3 - \epsilon_1\epsilon_2\epsilon_4 - \epsilon_1\epsilon_3\epsilon_4 - \epsilon_2\epsilon_3\epsilon_4) \\ p_4 &= \epsilon_1\epsilon_2\epsilon_3\epsilon_4 \end{aligned}\right\}. \quad (20.14)$$

We make no assumptions about M_4, so its structure includes a trace term $4w = -p_1$ as well as the possible antisymmetric components shown in Eq. (20.11), yielding the following expressions for the $p_k(E_4)$ following from the expansion of $\det[M_4 - eI_4]$:

$$p_1(E_4) = -\mathrm{tr}[M_4] = -4w, \quad (20.15)$$

$$\begin{aligned} p_2(E_4) &= \frac{1}{2}(\mathrm{tr}[M_4])^2 - \frac{1}{2}\mathrm{tr}[M_4 \cdot M_4] \\ &= 6w^2 - 2(x^2 + y^2 + z^2) - A^2 - a^2 - B^2 - b^2 - C^2 - c^2 \\ &\quad + A_w{}^2 + a_w{}^2 + B_w{}^2 + b_w{}^2 + C_w{}^2 + c_w{}^2, \end{aligned} \quad (20.16)$$

$$\begin{aligned} p_3(E_4) &= -\frac{1}{6}(\mathrm{tr}[M_4])^3 + \frac{1}{2}\mathrm{tr}[M_4 \cdot M_4]\,\mathrm{tr}[M_4] - \frac{1}{3}\mathrm{tr}[M_4 \cdot M_4 \cdot M_4] \\ &= -8xyz + 4w(x^2 + y^2 + z^2) \\ &\quad - 2ABC - 2Abc - 2aBc - 2abC \\ &\quad + 2A^2x - 2a^2x + 2B^2y - 2b^2y + 2C^2z - 2c^2z \\ &\quad - 2AB_wC_w + 2Ab_wc_w - 2aB_wc_w + 2ab_wC_w \\ &\quad\quad - 2A_wBC_w + 2a_wBc_w - 2a_wbC_w + 2A_wbc_w \\ &\quad\quad - 2A_wB_wC + 2a_wb_wC - 2A_wb_wc + 2a_wB_wc \\ &\quad + 2a^2w + 2A^2w - 2A_w^2w - 2A_w^2x - 2a_w^2w + 2a_w^2x \\ &\quad\quad + 2b^2w + 2B^2w - 2B_w^2w - 2B_w^2y - 2b_w^2w + 2b_w^2y \end{aligned}$$

$$+ 2c^2 w + 2C^2 w - 2C_w^2 w - 2C_w^2 z - 2c_w^2 w + 2c_w^2 z \,, \tag{20.17}$$

$$p_4(E_4) = \det[M_4] \,. \tag{20.18}$$

20.2.3 Issues with the naive 4D approach

We previously found that we could maximize $\Delta_3 = \operatorname{tr}(R_3 \cdot E_3)$ over the 3D rotation matrices R_3 by mapping E_3 to the profile matrix M_3, with $\Delta_3 = q \cdot M_3 \cdot q$, solving for the maximal eigenvalue ϵ_{opt} of the symmetric matrix M_3, and choosing $R_{\text{opt}} = R_3(q_{\text{opt}})$ with q_{opt} the normalized quaternion eigenvector corresponding to $\Delta_3(\text{opt}) = \epsilon_{\text{opt}}$. The obvious 4D extension of the 3D quaternion RMSD problem would be to examine $\Delta_4 = \operatorname{tr}(R_4 \cdot E_4) = q_\lambda \cdot M_4 \cdot q_\rho$. This is defined over the 4D rotation matrices R_4, where M_4 in Eq. (20.8) turns out no longer to be symmetric, so we must split the eigenvector space into a separate left quaternion q_λ and right quaternion q_ρ. We might guess that, as in the 3D case, M_4 would have a maximal eigenvalue ϵ_{opt} (already a problem – it may be complex), and we could use the "optimal" left and right eigenvectors $q_{\lambda:\text{opt}}$ and $q_{\rho:\text{opt}}$ that could be obtained as the corresponding eigenvectors of M_4 and $M_4{}^{\text{t}}$. Then the solution to the 4D optimization problem would look like this:

$$\Delta_4(\text{opt}) \overset{?}{=} q_{\lambda:\text{opt}} \cdot M_4 \cdot q_{\rho:\text{opt}} = (q_{\lambda:\text{opt}} \cdot q_{\rho:\text{opt}})\epsilon_{\text{opt}} \,. \tag{20.19}$$

Unfortunately, this is wrong. First, even when this result is real, Eq. (20.19) is typically smaller than the actual maximum of $\operatorname{tr}(R_4(q_\lambda, q_\rho) \cdot E_4)$ over the space of 4D rotation matrices (or their equivalent representations in terms of a search through q_λ and q_ρ). Even a simple `slerp` through q_{ID} and just beyond the apparent optimal eigenvectors $q_{\lambda:\text{opt}}$ and $q_{\rho:\text{opt}}$ from an eigenvalue of M_4 can yield *larger* values of Δ_4! And, to add insult to injury, starting with those eigenvectors $q_{\lambda:\text{opt}}$ and $q_{\rho:\text{opt}}$, one does not in general even find a *basis* for some normalized linear combination that yields the true optimal result. What is going wrong, and what is the path to our hoped-for quaternionic solution to the 4D RMSD problem, which seems so close to the 3D RMSD problem, but then fails so spectacularly to correspond to the obvious hypothesis?

20.2.4 Insights from the singular value decomposition

We know that the 3D version of Eq. (20.19) is certainly correct with ϵ_{opt} the maximal eigenvalue of $M_3(E_3)$, and we know also that there is *some* rotation matrix $R_4(q_\lambda, q_\rho)$ that maximizes $\operatorname{tr}(R_4(q_\lambda, q_\rho) \cdot E_4)$, and therefore the 4D expression Eq. (20.19) must describe $\Delta_4(\text{opt})$ for *some* nontrivial pair of quaternions (q_λ, q_ρ). The crucial issue is that the 3D RMSD problem and the 4D RMSD problem differ, with 3D being a special case due to the symmetry of the 4×4 profile matrix. We know also that the SVD form of the optimal rotation matrix is valid in *any* dimension, so we conjecture that the key is to look at the commonality of the SVD solutions in 3D and 4D, and work backwards to see how those nonquaternion-driven equations might relate to what we know is *in principle* a quaternion approach to the 4D problem that looks like Eq. (20.19). Details of the relationship between the quaternion and SVD solutions for the 3D problem are addressed in Appendix N; some aspects of that treatment that were not essential in the 3D case will be found helpful here in 4D.

Therefore, we first look at the general SVD for the spatial alignment problem (Schönemann, 1966; Golub and van Loan, 1983) and then analyze the 3D and 4D problems to understand how we can recover a quaternion-based construction of the 4D spatial RMSD solution. For 3D and 4D, the basic

SVD construction of the optimal rotation for a cross-covariance matrix E takes the form

$$\{U, S, V\} = \text{Singular Value Decomposition}(E)\,, \tag{20.20}$$

where

$$E(U, S, V) = U \cdot S \cdot V^{\mathrm{t}}\,, \tag{20.21}$$

$$R_{\text{opt}}(U, D, V) = V \cdot D \cdot U^{\mathrm{t}}\,, \tag{20.22}$$

$$D_3 = \text{Diagonal}\left(1, 1, \text{sign det}(V \cdot U^{\mathrm{t}})\right)\,, \tag{20.23}$$

$$D_4 = \text{Diagonal}\left(1, 1, 1, \text{sign det}(V \cdot U^{\mathrm{t}})\right)\,. \tag{20.24}$$

Here U and V are orthogonal matrices that are usually ordinary rotations, while D is usually the identity matrix but can be nontrivial in more situations than one might think. A critical component for this analysis is the diagonal matrix S, whose elements are the all-positive square roots of the eigenvalues of the symmetric matrix $E_4^{\mathrm{t}} \cdot E_4$ (the trace of this matrix is the squared Frobenius norm of E). The first key fact is that in any dimension the RMSD cross-term obeys the following sequence of transformations following from the SVD relations of Eqs. (20.21)–(20.24):

$$
\left.
\begin{aligned}
\Delta(\text{opt}) &= \text{tr}(R_{\text{opt}} \cdot E) = \text{tr}(R_{\text{opt}} \cdot [U \cdot S \cdot V^{\mathrm{t}}]) \\
&= \text{tr}\left([V \cdot D \cdot U^{\mathrm{t}}] \cdot [U \cdot S \cdot V^{\mathrm{t}}]\right) \\
&= \text{tr}(D \cdot S)
\end{aligned}
\right\} \,. \tag{20.25}
$$

Note the appearance of D in the SVD formula for the optimal measure; we found in numerical experiments that including this term is absolutely essential to guaranteeing agreement with brute force verification of the optimization results, particularly in 4D.

3D context. Thus an alternative to considering the 3D optimization of $\text{tr}(R \cdot E)$ in the context of E alone is to look at the 3×3 matrices

$$
\left.
\begin{aligned}
F &= E^{\mathrm{t}} \cdot E \\
F' &= E \cdot E^{\mathrm{t}}
\end{aligned}
\right\} \tag{20.26}
$$

and to note that, although E itself will not in general be symmetric, F and F' are intrinsically symmetric. Thus they have the same eigenvalues, and like all nonsingular matrices of this form, and unlike E itself, will have real positive eigenvalues (Golub and van Loan, 1983) that we can write as $(\gamma_1, \gamma_2, \gamma_3)$. From Eq. (20.21), we can show that $\text{tr}\,F = \text{tr}\,F' = \text{tr}(S \cdot S)$, and since the trace is the sum of the eigenvalues, the eigensystem of F or F' determines S. The diagonal elements that enter naturally into the SVD are therefore just the square roots

$$S(E) = \text{Diagonal}\left(\sqrt{\gamma_1}, \sqrt{\gamma_2}, \sqrt{\gamma_3}\right)\,. \tag{20.27}$$

So far, this has no obvious connection to the quaternion system. For our next step, let us now examine how the 3D SVD system relates to the profile matrix $M_3(E_3)$ derived from the quaternion decomposition to give the form in Eq. (20.9) above.

We define the analogs of Eq. (20.26) for a profile matrix as

$$
\left.
\begin{aligned}
G &= M^{\mathrm{t}} \cdot M \\
G' &= M \cdot M^{\mathrm{t}}
\end{aligned}
\right\}\,, \tag{20.28}
$$

where we recall that in 3D, ϵ_{opt} is just the maximal eigenvalue of $M_3(E)$. Thus if we arrange the eigenvalues of $M_3(E)$ in descending order as $(\epsilon_1, \epsilon_2, \epsilon_3, \epsilon_4)$, we obviously have

$$\text{Eigenvalues}(G) = \text{Eigenvalues}(G') = (\alpha_1, \alpha_2, \alpha_3, \alpha_4) = (\epsilon_1{}^2, \epsilon_2{}^2, \epsilon_3{}^2, \epsilon_4{}^2). \tag{20.29}$$

Therefore, since we already know that $\epsilon_1(M) = \Delta(\text{opt})$, we have precisely the sought-for connection,

$$\sqrt{\text{Max eigenvalue}(G)} = \sqrt{\alpha_1} = \text{tr}(D \cdot S) = \Delta(\text{opt}) = \epsilon_1(M). \tag{20.30}$$

That is, given E, compute $M(E)$ from the quaternion decomposition, and, instead of examining the eigensystem of $M(E)$ itself, take the square root of the maximal eigenvalue of the manifestly symmetric, positive-definite real matrix $G = M^{\text{t}} \cdot M$. This is the quaternion-based translation of the 3D application of the SVD method to obtaining the optimal rotation: numerical methods in particular do not care whether you are computing the maximal eigenvalue of a symmetric quaternion-motivated matrix M_3 or of the associated symmetric matrix $M_3{}^{\text{t}} \cdot M_3$.

Remark: In 3D, we can compute *all four* of the eigenvalues of G from the *three* elements of S (Coutsias et al., 2004): defining

$$\text{Diagonal}(D \cdot S) = (\lambda_1, \lambda_2, \lambda_3), \tag{20.31}$$

then we can write

$$\begin{bmatrix} \alpha_1 \\ \alpha_2 \\ \alpha_3 \\ \alpha_4 \end{bmatrix} = \begin{bmatrix} (+\lambda_1 + \lambda_2 + \lambda_3)^2 \\ (-\lambda_1 - \lambda_2 + \lambda_3)^2 \\ (-\lambda_1 + \lambda_2 - \lambda_3)^2 \\ (+\lambda_1 - \lambda_2 - \lambda_3)^2 \end{bmatrix}, \tag{20.32}$$

where obviously $\sqrt{\alpha_1} = \text{tr}(D \cdot S)$ is the maximal eigenvalue determining the value of the optimal quaternion eigenvector.

The final step is to connect $R_3(\text{opt})$ to a quaternion via $R_3(q_{\text{opt}})$ without requiring prior knowledge of the SVD solution Eq. (20.22). We know that the square root of the maximal eigenvalue of $G = M^{\text{t}} \cdot M$, which depends only on the quaternion decomposition, gives us $\text{tr}(D \cdot S) = \Delta(\text{opt})$ without using the SVD, and we know that in 3D the profile matrix M is symmetric, so G and G' share a single maximal eigenvector v corresponding to $\alpha_1 = (\text{tr}(D \cdot S))^2 = (\Delta(\text{opt}))^2$. Using this eigenvector we thus have

$$v \cdot G \cdot v = (M \cdot v)^{\text{t}} \cdot (M \cdot v) = v \cdot \left((\text{tr}(D \cdot S))^2 \cdot v \right) = (\Delta(\text{opt}))^2,$$

so in this case $v = q_{\text{opt}}$ is itself the optimal eigenvector determining $R_3(q_{\text{opt}})$.

4D context. The 4D case, as we are now aware, cannot be solved using the nonsymmetric profile matrix $M_4(E_4)$ directly. But now we can see a more general way to exploit the 4D quaternion decomposition of Eq. (20.8) by constructing the *manifestly symmetric products*

$$\left. \begin{array}{l} G = M_4{}^{\text{t}} \cdot M_4 \\ G' = M_4 \cdot M_4{}^{\text{t}} \end{array} \right\}. \tag{20.33}$$

Although this superficially extends Eq. (20.28) to 4D, it is quite different because M_4 is not itself symmetric (as M_3 was), so, while G and G' have the same eigenvalues, they have *distinct eigenvectors* q_ρ and q_λ, respectively. If we use the maximal eigenvalue α_1 to solve for q_ρ and q_λ as follows, these in fact will produce the optimal quaternion system. First we solve these equations using the maximal eigenvalue α_1 of G,

$$\left. \begin{array}{l} G \cdot q_\rho = \alpha_1 \, q_\rho = (\operatorname{tr}(D \cdot S))^2 \, q_\rho \\ G' \cdot q_\lambda = \alpha_1 \, q_\lambda = (\operatorname{tr}(D \cdot S))^2 \, q_\lambda \end{array} \right\} . \tag{20.34}$$

At this point, the *signs* of the eigenvectors have to be checked for a correction, since the eigenvector is still correct whatever its sign or scale. But we know that the value of $q_\lambda \cdot M_4(E_4) \cdot q_\rho$ must be positive, so we simply check that sign, and change, say, $q_\lambda \to -q_\lambda$ if needed to make the sign positive. There is still an *overall* sign ambiguity, but that is natural and an intrinsic part of the rotation $R_4(q_\lambda, q_\rho)$, so now we can use these eigenvectors to generate the optimal measure for the 4D RMSD matching problem using *only* the quaternion-based data, giving finally the whole spectrum of ways to write $\Delta_4(\text{opt})$:

$$\Delta_4(\text{opt}) = \operatorname{tr}(R_{4:\text{opt}}(q_\lambda, q_\rho) \cdot E_4) = q_\lambda \cdot M_4(E_4) \cdot q_\rho = \sqrt{\alpha_1} . \tag{20.35}$$

Remark: In 4D, we can compute all the eigenvalues of G from the *four* elements of S: defining

$$\text{Diagonal}(D \cdot S) = (\lambda_1, \lambda_2, \lambda_3, \lambda_4) , \tag{20.36}$$

then we can write

$$\begin{bmatrix} \alpha_1 \\ \alpha_2 \\ \alpha_3 \\ \alpha_4 \end{bmatrix} = \begin{bmatrix} (+\lambda_1 + \lambda_2 + \lambda_3 + \lambda_4)^2 \\ (+\lambda_1 + \lambda_2 - \lambda_3 - \lambda_4)^2 \\ (+\lambda_1 - \lambda_2 + \lambda_3 - \lambda_4)^2 \\ (+\lambda_1 - \lambda_2 - \lambda_3 + \lambda_4)^2 \end{bmatrix} , \tag{20.37}$$

where again $\sqrt{\alpha_1} = \operatorname{tr}(D \cdot S)$ is maximal.

Summary: Now we have the entire algorithm for solving the RMSD spatial alignment problem in 4D by exploiting the quaternion decomposition of Eq. (20.7) and Eq. (20.8), based on Eq. (20.2), inspired by the SVD solution to the problem:

- **Compute the profile matrix.** Using the quaternion decomposition Eq. (20.2) of the general 4D rotation matrix $R_4(p, q)$, extract the 4D profile matrix $M_4(E_4)$ of Eq. (20.8) from the initial optimization measure

$$\Delta_4 = \operatorname{tr}(R_4(p, q) \cdot E_4) = p \cdot M_4(E_4) \cdot q . \tag{20.38}$$

So far all we know is the numerical value of M_4 and the fact the Δ_4 can be maximized by exploring the entire space of the quaternion pair (p, q).

- **Construct the symmetric matrices and extract the optimal eigenvalue.** The maximal eigenvalue α_1 of the 4×4 symmetric matrix $G = M_4{}^{\text{t}} \cdot M_4$ is itself easily obtained by numerical means, just as has traditionally been done for M_3. If all we need is the optimal value of the measure for data comparison tasks, we are done:

$$\Delta_4(\text{opt}) = \sqrt{\text{Max eigenvalue}\,(G = M_4{}^t \cdot M_4)} = \sqrt{\alpha_1} \,. \tag{20.39}$$

The algebraic methods for computing the eigenvalues are discussed in detail in Appendix J.

- **If needed, compute the left and right eigenvectors of G.** Our two distinct symmetric matrices, $G = M_4{}^t \cdot M_4$ and $G' = M_4 \cdot M_4{}^t$, have their own distinct maximal eigenvectors, both corresponding to the maximal eigenvalue α_1 shared by G and G', so we can easily use this common maximal numerical eigenvalue to solve

$$\left. \begin{aligned} (G - \alpha_1 I_4) \cdot q_{\text{opt}:\rho} &= 0 \\ (G' - \alpha_1 I_4) \cdot q_{\text{opt}:\lambda} &= 0 \end{aligned} \right\} \tag{20.40}$$

for the numerical values of $q_{\text{opt}:\lambda}$ and $q_{\text{opt}:\rho}$. Remember that there may be singularities, but these are taken care of by the adjugate method studied in Chapter 12. We correct the signs so that $q_{\text{opt}:\lambda} \cdot M_4(E_4) \cdot q_{\text{opt}:\rho} > 0$, and then these in turn yield the required 4D rotation matrix

$$R_{4:\text{opt}}\left(q_{\text{opt}:\lambda},\, q_{\text{opt}:\rho}\right)$$

from Eq. (20.2).

If everything is in order, all of the following ways of expressing $\Delta_4(\text{opt})$ should now be equivalent:

$$\Delta_4(\text{opt}) = \text{tr}(R_{4:\text{opt}}\left(q_{\text{opt}:\lambda},\, q_{\text{opt}:\rho}\right) \cdot E_4) = q_{\text{opt}:\lambda} \cdot M_4(E_4) \cdot q_{\text{opt}:\rho} = \sqrt{\alpha_1} \,, \tag{20.41}$$

independently of the fact that one knows from the SVD decomposition of E_4 that

$$\Delta_4(\text{opt}) = \text{tr}(D \cdot S) = \sqrt{\alpha_1} \,.$$

20.3 4D orientation frame alignment

In this section, we extend the treatment of 3D orientation frame alignment presented in Chapter 10 to handle the case of 4D orientation frame alignment, completing the picture we started in Section 20.2 on the 4D spatial coordinate alignment problem. As the 3D treatment of orientation frame alignment in Chapter 10 was extremely detailed, here we can be somewhat more compact in our description.

20.3.1 The 4D orientation frame alignment problem

Orientation frames in four dimensions have axes that are the columns of a 4D rotation matrix taking the identity frame to the new orientation frame. Therefore, in parallel with the 3D case, such frames can be represented either as 4D rotation matrices R_4 (the action on a 4D identity frame to get columns that are a new set of four orthogonal axes), or as the pair of quaternions (q, q') used in Eq. (20.2) to define $R_4(q, q')$. As in the 3D frame case, we will take advantage of the chord distance linearization of the ideal but intractable geodesic angular measure, and we shall present two alternative approaches to the optimization measure.

Quadratic form. In 3D, with Eq. (10.15) having a single quaternion involved in the rotation, we were able to write down Δ_{chord} in terms of a simple expression linear in the quaternion q and the cumulative data V, and we observed that a quadratic expression $(q \cdot V)^2$ would also produce the same optimal eigenvector $q = V/\|V\|$. The optimal frame problem in 4D, in contrast, already requires a pair of quaternions, and one strategy is to split the analogs of the 3D quadratic expression into two parts, yielding

$$\Delta_{4:\text{chord-sq}}(q, q') = (q \cdot V)(q' \cdot V') = q_a \left(V_a V_b'\right) q_b' = q \cdot \Omega_4 \cdot q' \tag{20.42}$$

as the generalization from 3D to 4D. Here, each 4D test frame consists of frames denoted by the quaternion pair (p, p'), and each reference frame employs a pair (r, r'), so we build the data coefficients starting from

$$\left. \begin{array}{l} V = \displaystyle\sum_{k=1}^{N} (r_k \star \bar{p}_k) = \sum_{k=1}^{N} t_k \\[2em] V' = \displaystyle\sum_{k=1}^{N} (r'_k \star \bar{p}'_k) = \sum_{k=1}^{N} t'_k \end{array} \right\} \tag{20.43}$$

and then apply the transformation

$$\left. \begin{array}{l} t_k \to \widetilde{t}_k = t_k \, \text{sign}(q \cdot t_k) \\[1em] t'_k \to \widetilde{t}'_k = t'_k \, \text{sign}(q \cdot t'_k) \end{array} \right\} \tag{20.44}$$

to achieve consistent (local) signs. According to Eq. (10.21), V could also be constructed from $W_{ab} = \sum_{k=1}^{N} [\bar{p}_k]_a \, [r_k]_b$, and V' from $W'_{ab} = \sum_{k=1}^{N} [\bar{p}'_k]_a \, [r'_k]_b$, noting that here p is transformed by the "tilde" of Eq. (20.44). Now, for the 4D frame pairs, the solution for the optimal quaternions must achieve the maximum for *both* elements of the pair, so we obtain as a solution maximizing Eq. (20.42)

$$\left. \begin{array}{l} q_{\text{opt}} = \dfrac{V}{\|V\|} \\[1.5em] q'_{\text{opt}} = \dfrac{V'}{\|V'\|} \\[1.5em] \Delta_{4:\text{chord-sq}}(\text{opt}) = \|V\| \|V'\| \end{array} \right\} . \tag{20.45}$$

Remark: There is a particular reason to prefer Eq. (20.42) for the 4D orientation frame problem: in the next section, we will see that the separate *pre-summation* arguments for V and V', gathered together, are *exactly* equal to the joint summand of the 4D triple rotation pre-summation arguments, following the pattern seen in Eq. (10.28) for the 3D orientation frame analysis.

Quartic triple rotation form. One can also eliminate the sign choice step altogether by defining a 4D frame similarity measure that is the exact analog of Eq. (10.27) in 3D as follows:

$$\Delta_{\text{RRR4}} = \sum_{k=1}^{N} \text{tr}\left[R(q,q') \cdot R(p_k, p'_k) \cdot R^{-1}(r_k, r'_k) \right] \tag{20.46}$$

$$= \sum_{k=1}^{N} \text{tr}\left[R(q,q') \cdot R(p_k \star \bar{r}_k,\ p'_k \star \bar{r}'_k) \right] \tag{20.47}$$

$$= \sum_{k=1}^{N} \text{tr}\left[R(q,q') \cdot R^{-1}(t_k, t'_k) \right] \tag{20.48}$$

$$= q \cdot U(p, p';\, r, r') \cdot q' . \tag{20.49}$$

Remarkably, there is a 4D version of the 3D identity Eq. (10.28) relating the triple rotation measure to the quadratic realizations of the linear quaternion rotation measures, namely

$$\left.\begin{aligned}
\frac{1}{4}\sum_{k=1}^{N} \text{tr}\left[R(q,q') \cdot R(p_k, p'_k) \cdot R(\bar{r}_k, \bar{r}'_k) \right] &= \sum_{k=1}^{N} \left((q \star p_k) \cdot r_k\right)\left((q' \star p'_k) \cdot r'_k\right) \\
&= \sum_{k=1}^{N} \left(q \cdot (r_k \star \bar{p}_k)\right)\left(q' \cdot (r'_k \star \bar{p}'_k)\right) \\
&= \sum_{k=1}^{N} (q \cdot t_k)(q' \cdot t'_k) \\
&= \sum_{a,b} q_a \left(\sum_{k=1}^{N} [t_k]_a\,[t'_k]_b\right) q'_b \\
&= q \cdot A\left(t = r \star \bar{p},\ t' = r' \star \bar{p}'\right) \cdot q'
\end{aligned}\right. .$$

Thus the pre-summation version of the arguments in the $(q \cdot V)(q' \cdot V')$ version of the 4D chord measure turns out to be *exactly* the same as the triple-matrix product measure summand without the additional trace term that is present in 3D. Furthermore, as long as one follows the rules of changing *both* the primed and unprimed signs together (the condition for $R_4(q,q')$'s invariance), this measure is sign-independent. The 4×4 matrix $A(t,t')$ is the 4D profile matrix equivalent to that of Markley et al. (2007); Hartley et al. (2013) for the 3D chord-based quaternion averaging problem. We can therefore use either the measure Δ_{RRR4} or

$$\Delta_{\text{A4}} = q \cdot A(t,t') \cdot q' \tag{20.50}$$

with $A(t,t')_{ab} = \sum_{k=1}^{N} [t_k]_a\,[t'_k]_b$ as our rotation-matrix-based sign-insensitive chord distance optimization measure.

To get an expression in terms of R, we now use Eq. (20.2) for $R(q,q')$ to decompose the measure Eq. (20.47) into the rotation averaging form

$$\Delta_{\text{RRR4}} = \text{tr}\left[R(q,q') \cdot T(p,p';r,r') \right] \tag{20.51}$$

$$= q \cdot U(T) \cdot q' , \tag{20.52}$$

where $T(p, p'; r, r') = \sum_{k=1}^{N} R^{-1}(t_k, t'_k)$ and $U(T)$ has the same relationship to T as the 4D profile matrix $M(E)$ in Eq. (20.8) does to the cross-covariance matrix E. In the next section, we will see that the singleton version of this map is unusually degenerate, with rank 1, though that feature does not persist for data sets with $N > 1$.

Now, as in the 4D spatial RMSD analysis, we might naturally assume that we could follow the 3D case by determining the maximal eigenvalue ϵ_0 of U and its left and right eigenvectors q_λ and q_ρ, which would give

$$\Delta_{\text{RRR4}} \overset{?}{=} q_\lambda \cdot U \cdot q_\rho = (q_\lambda \cdot q_\rho)\epsilon_0 \,.$$

As before, this is not a maximal value for the measure Δ_{RRR4} over the possible range of $R(q, q')$. To solve the optimization correctly, we must again be very careful, and work with the maximal eigenvalue $\alpha(\text{RRR4:opt})$ of $G = U^{\text{t}} \cdot U$ and $G' = U \cdot U^{\text{t}}$, which we can get numerically as usual, or algebraically from the quartic solution for the eigenvalues for symmetric 4×4 matrices with a trace, yielding

$$\Delta_{\text{RRR4}}(\text{opt}) = \sqrt{\text{max eigenvalue } (U^{\text{t}} \cdot U)} = \sqrt{\alpha(\text{RRR4:opt})} \,.$$

If we need the actual optimal rotation matrix solving

$$\Delta_{\text{RRR4}}(\text{opt}) = \text{tr}\left(R_4(q_{\text{opt}}, q'_{\text{opt}}) \cdot S \right) = q_{\text{opt}} \cdot U \cdot q'_{\text{opt}} = \sqrt{\alpha(\text{RRR4:opt})} \,,$$

then we just use our optimal eigenvalue to solve

$$(G - \alpha(\text{RRR4:opt})I_4) \cdot q = 0 \,,$$
$$(G' - \alpha(\text{RRR4:opt})I_4) \cdot q' = 0$$

for q_{opt} and q'_{opt}, or use the equivalent adjugate column method to extract the eigenvectors. That gives the desired 4D rotation matrix $R_4(q_{\text{opt}}, q'_{\text{opt}})$ explicitly via Eq. (20.2). The same approach applies to the solution of $\Delta_{\text{A4}} = q \cdot A(t, t') \cdot q'$. Note that this can all be accomplished numerically, directly as above, with SVD, or using the quaternion eigenvalue decomposition on the symmetric matrices either numerically or algebraically, using the quartic solutions of Appendix J.

20.4 Extracting quaternion pairs from 4D rotation matrices

Our task in this section is to show how the methods of Chapter 9 introduced by Bar-Itzhack (2000) can be adapted to the 4D case, enabling us to efficiently obtain a unique optimal quaternion pair from a possibly noisy and imperfectly measured 4D rotation matrix. The basic method was given in Hanson (2020), rediscovered with a different derivation in (Sarabandi and Thomas, 2022). Two fundamental steps are needed: the first is to determine what changes in the 3D eigenvalue method are needed for the 4D nearest-pure-rotation extraction, and the second is to determine how to use that information to determine the actual quaternion pair corresponding to the 4D rotation.

First, we recall from Eq. (20.2) that any 4D orthogonal matrix $R_4(p, q)$ can be expressed as a quadratic form in two independent unit quaternions. This is a consequence of the fact that the six-parameter orthogonal group **SO**(4) is double covered by the composition of two smaller three-parameter unitary groups, that is, **SU**(2) × **SU**(2); the group **SU**(2) has essentially the same properties as a single quaternion, so it is not surprising that **SO**(4) should be related to a pair of quaternions.

We begin our treatment of the 4D case by extending Eq. (9.5) to 4D with a numerical **SO**(4) matrix R_4, giving us a Bar-Itzhack measure to maximize over a rotation matrix target $S(\ell, r)$ of the form

$$\Delta_{4:\mathrm{BI}} = \operatorname{tr} S(\ell, r) \cdot R_4^{\mathrm{t}} = \ell \cdot K_4(R_4) \cdot r = \ell \cdot K_4(p, q) \cdot r . \tag{20.53}$$

Here (ℓ, r) are the left and right quaternions over which we are varying the measure to be maximized, and $K_4(R_4)$ is the 4D generalization of Eq. (9.7). To compute $K_4(R_4)$, we define a general 4D rotation matrix candidate with columns $\mathbf{W} = (w_0, w_1, w_2, w_3)$, etc., so the matrix takes the form $R_4 = [\mathbf{W}|\mathbf{X}|\mathbf{Y}|\mathbf{Z}]$, producing a numerical profile matrix of the form (taking into account the transpose in Eq. (20.53))

$$K_4(R_4) =$$
$$\frac{1}{4} \begin{bmatrix} w_0 + x_1 + y_2 + z_3 & -w_1 + x_0 + y_3 - z_2 & -w_2 - x_3 + y_0 + z_1 & -w_3 + x_2 - y_1 + z_0 \\ w_1 - x_0 + y_3 - z_2 & w_0 + x_1 - y_2 - z_3 & -w_3 + x_2 + y_1 - z_0 & w_2 + x_3 + y_0 + z_1 \\ w_2 - x_3 - y_0 + z_1 & w_3 + x_2 + y_1 + z_0 & w_0 - x_1 + y_2 - z_3 & -w_1 - x_0 + y_3 + z_2 \\ w_3 + x_2 - y_1 - z_0 & -w_2 + x_3 - y_0 + z_1 & w_1 + x_0 + y_3 + z_2 & w_0 - x_1 - y_2 + z_3 \end{bmatrix} . \tag{20.54}$$

Now, from Eq. (20.2), we know that we also have an analog to Eq. (9.8) for exact data, and with $R_4(p, q)$ in place of the numerical $R_4(w, x, y, z)$ in $K_4(R_4)$, this takes the remarkably compact algebraic form

$$K_4(p, q) = \begin{bmatrix} p_0 q_0 & p_0 q_1 & p_0 q_2 & p_0 q_3 \\ p_1 q_0 & p_1 q_1 & p_1 q_2 & p_1 q_3 \\ p_2 q_0 & p_2 q_1 & p_2 q_2 & p_2 q_3 \\ p_3 q_0 & p_3 q_1 & p_3 q_2 & p_3 q_3 \end{bmatrix} . \tag{20.55}$$

We see that Eq. (20.55) is exactly the outer product of p and q, with vanishing determinant, rank 1, and a lone eigenvalue equal to its trace $(p \cdot q)$. Its deceptively beautiful simple eigensystem is

$$\epsilon = \{p \cdot q, \ 0, \ 0, \ 0\} , \tag{20.56}$$

$$r_{\mathrm{right}} = \left\{ \begin{bmatrix} p_0 \\ p_1 \\ p_2 \\ p_3 \end{bmatrix}, \begin{bmatrix} -q_1 \\ q_0 \\ 0 \\ 0 \end{bmatrix}, \begin{bmatrix} -q_2 \\ 0 \\ q_0 \\ 0 \end{bmatrix}, \begin{bmatrix} -q_3 \\ 0 \\ 0 \\ q_0 \end{bmatrix} \right\}, \tag{20.57}$$

$$\ell_{\mathrm{left}} = \left\{ \begin{bmatrix} q_0 \\ q_1 \\ q_2 \\ q_3 \end{bmatrix}, \begin{bmatrix} -p_1 \\ p_0 \\ 0 \\ 0 \end{bmatrix}, \begin{bmatrix} -p_2 \\ 0 \\ p_0 \\ 0 \end{bmatrix}, \begin{bmatrix} -p_3 \\ 0 \\ 0 \\ p_0 \end{bmatrix} \right\}. \tag{20.58}$$

However, we have seen this before in Section 20.2: the maximal eigensystems of *nonsymmetric* 4D matrices are *not optimal*. We can in fact see that Eqs. (20.57) and (20.58) give $\Delta_{4:\mathrm{BI}} = (p \cdot q)^2$, whereas we know that ideally we must have a value that corresponds to the identity matrix, or eigenvalue unity. That is, the right and left eigenvectors in Eqs. (20.57) and (20.58) appear to be *reversed*.

The key to fixing this is now familiar: we must construct the *symmetric* matrices

$$G(p,q) = K_4^t \cdot K_4 = \begin{bmatrix} q_0^2 & q_0q_1 & q_0q_2 & q_0q_3 \\ q_0q_1 & q_1^2 & q_1q_2 & q_1q_3 \\ q_0q_2 & q_1q_2 & q_2^2 & q_2q_3 \\ q_0q_3 & q_1q_3 & q_2q_3 & q_3^2 \end{bmatrix}, \tag{20.59}$$

$$G'(p,q) = K_4 \cdot K_4^t = \begin{bmatrix} p_0^2 & p_0p_1 & p_0p_2 & p_0p_3 \\ p_0p_1 & p_1^2 & p_1p_2 & p_1p_3 \\ p_0p_2 & p_1p_2 & p_2^2 & p_2p_3 \\ p_0p_3 & p_1p_3 & p_2p_3 & p_3^2 \end{bmatrix}. \tag{20.60}$$

These both have a lone eigenvalue equal to one, that is, $\epsilon = \operatorname{tr} G = \operatorname{tr} G' = 1$, and eigenvectors that are the *reverse* of Eqs. (20.57) and (20.58), that is,

$$\left. \begin{array}{l} G(p,q) \to r_{\text{right}} = q \\ G'(p,q) \to \ell_{\text{left}} = p \end{array} \right\}. \tag{20.61}$$

Note: One can easily see the subtle emergence of the eigensystems of G and G' from the following calculation:

$$G \cdot q = K_4^t \cdot K_4 \cdot q = K_4^t \cdot (q \cdot q)p = (p \cdot p)(q \cdot q)q = q,$$
$$G' \cdot p = K_4 \cdot K_4^t \cdot p = K_4 \cdot (p \cdot p)q = (q \cdot q)(p \cdot p)p = p.$$

So q and p are clearly the right and left quaternion vectors that maximize $\Delta_{4:BI}$, with eigenvalue unity, always greater than or equal to the false candidate eigenvalue $(p \cdot q)$.

To solve the problem, we thus use our numerical data, e.g., the 4D rotation matrix data in Eq. (20.54), and compute the right and left normalized numerical eigenvectors q and p from G and G', respectively, with eigenvalue equal to one for perfect data, and use those to optimally describe $R_4(p,q)$. Again, if a statistical distribution in the double quaternion space is desired, the signs can be chosen randomly, consistent with the sign of $\operatorname{tr} K_4(R_4)$. Explicitly, we can either use any (normalized) adjugate column or equivalent algorithm to solve directly for the eigenvectors. We prefer to obtain the eigenvectors from the optimal adjugate columns of the following characteristic equations. *No further computation is required.*

$$\chi(G) = \left[G(R) - \epsilon_{\text{opt}} \times I_4 \right] \qquad r_{\text{opt}} = q = (\textit{normalize best adjugate})(\chi(G)), \tag{20.62}$$

$$\chi(G') = \left[G'(R) - \epsilon_{\text{opt}} \times I_4 \right] \qquad \ell_{\text{opt}} = p = (\textit{normalize best adjugate})(\chi(G')). \tag{20.63}$$

The solution to our problem is thus $R_4(p,q) = R_4(\ell_{\text{opt}}, r_{\text{opt}})$. For perfect data, we will have $\epsilon_{\text{opt}} = 1$. As in 3D, if the numerical matrix R_4 has some moderate errors and the numerical maximum eigenvalues of (G, G') *differ* significantly from unity, we can solve for the actual maximal eigenvalue ϵ_{opt} and use that in Eqs. (20.62) and (20.63) to find the left and right eigenvectors numerically.

Nonideal cases. It is important to note, as emphasized by Bar-Itzhack, that if there are *significant errors* in the numerical matrix R, then the actual nonunit maximal eigenvalue of $G(R)$ can be computed numerically or algebraically as usual, and then the corresponding eigenvector pair comprises the *closest exact* quaternion pair describing the errorful rotation matrix R_4, which can be very useful since such quaternions always produce a valid rotation matrix.

In any case, *up to an overall sign*, (r_{opt}, ℓ_{opt}) is the desired numerical quaternion pair (q, p) corresponding to the target numerical rotation matrix $R_4(opt) = R_4(q, p)$. In some circumstances, one is looking for a uniform statistical distribution of quaternions, in which case the overall signs should be chosen randomly, consistent with $\operatorname{tr} K_4(R_4)$.

The Bar-Itzhack approach solves the problem of extracting the quaternion of an arbitrary numerical 4D rotation matrix or its quaternion pair in a fashion that involves no singularities and only trivial testing for special cases.

We emphasize an important caveat: the 3D quaternion rotation $R_3(q)$ does not care what the sign of q is, but the 4D quaternion rotation $R_4(p, q)$ is only invariant under *both* $p \rightarrow -p$ and $q \rightarrow -q$ in tandem. To ensure that $R_4(p, q)$ is the correct matrix, the signs of the quaternions may need to be adjusted after the initial computation so that the sign of $(\ell \cdot r)$ matches the sign of the numerical input value of $R_{4(1,1)} = \operatorname{tr} K_4(R_4) = p \cdot q$. That guarantees that the solution describes the same matrix that we used as input, and not its negative.

Dual quaternions and $\mathbf{SE}(3)$

Quaternions, as we have seen, are extremely useful for representing 3D, and, on occasion, 4D, rotations in space, with no explicit reference to the location of an object in space or the center of the rotation being applied. Many practical problems in areas like robotics or simulated body part and skin element motion in animation require integration of spatial location with orientation. In this chapter, we will look at the question of combining translations with rotations, first reviewing traditional methods in 2D and 3D to introduce the problem. We then focus on *dual quaternions*, which combine the actions of translation and quaternion rotations in a unified quaternion-based framework that has become popular in applications such as robotics and character animation. Here we meet a new algebraic symbol "ϵ" that in a very special way generalizes the imaginary unit $i^2 = -1$. This new symbol obeys $\epsilon^2 = 0$, has dimensions of *inverse length*, and creates the ability to effortlessly represent spheres of infinite radius, which do not roll, but are able instead to represent *translations* when we try to roll them. We note that, while 3D rotations are associated with the *special orthogonal group*, written $\mathbf{SO}(3)$, the combination of rotations with 3D translations is a group written as $\mathbf{SE}(3)$, referring to the *special Euclidean group*, preserving the spatial relationships among collections of points in 3D Euclidean space. In physics, with the inclusion of time, this type of transformation in nonrelativistic Euclidean space is part of the *Galilean group*, distinguished from the *Poincaré group*, which refers to the inclusion of 4D spacetime translations with relativistic Lorentz transformations.

Remark: *Incompatibility of dimensional units.* The reader should bear in mind that regardless of the methods used to combine translations and rotations, this process involves incompatible units, and thus any attempt to unify translations, which have dimensions of length, and rotations, which have dimensionless parameters, into a single formula encounters subtle difficulties. In the end, all combinations of translations with rotations must either separate the parts of the transformations, or, if unified, e.g., as a single quaternion-like object such as dual quaternions, the unification must somehow involve an extra parameter with dimensions of length. Dual quaternions depend critically on the dual parameter "ϵ" having dimensions of inverse length, corresponding heuristically to the inverse radius of an infinite sphere.

21.1 Background

Our basic problem is that to place objects arbitrarily in 3D space, rotations are insufficient. We must also translate. There are several ways that translation can be incorporated with rotation in a computational context. Perhaps the most traditional method is to extend the representation of a vector by one dimension into what are effectively projective coordinates, and perform transformations with a matrix whose extra dimension implements Euclidean translations. Some of the elements of these transformation matrices have dimensionless coordinates and some have dimensions of length.

2D homogeneous coordinates and translations. Starting with the simplest 2D case, we write a 2D point as a 3D *extended* vector

$$\begin{bmatrix} x \\ y \\ 1 \end{bmatrix} \tag{21.1}$$

and the transformation of a rotated 2D point to an arbitrary location (t_x, t_y) in space as multiplication by a 3×3 matrix:

$$\begin{bmatrix} x' \\ y' \\ 1 \end{bmatrix} = \begin{bmatrix} \cos\theta & -\sin\theta & t_x \\ \sin\theta & \cos\theta & t_y \\ 0 & 0 & 1 \end{bmatrix} \cdot \begin{bmatrix} x \\ y \\ 1 \end{bmatrix} = \begin{bmatrix} x\cos\theta - y\sin\theta + t_x \\ x\sin\theta + y\cos\theta + t_y \\ 1 \end{bmatrix} . \tag{21.2}$$

The vector $[x, y, 1]^t$ is actually the *inhomogeneous* form of the *homogeneous* projective vector $[u, v, w]^t$, which is formally invariant under scaling,

$$[\lambda u, \lambda v, \lambda w]^t \cong [u, v, w]^t .$$

For actual calculations, we can therefore set $\lambda = 1/w$ for any finite point and use the inhomogeneous coordinates

$$[x = u/w, \ y = v/w, \ 1]^t .$$

The one context where this becomes especially useful is when we need to do mathematics with points at infinity, which is very awkward with inhomogeneous coordinates: here we can fluidly deal with infinity by setting $w = 0$, so the *direction* in 2D space $[u, v, 0]^t$ describes a unique point on the line at infinity, which we can normalize and treat as a topological circle.

3D homogeneous coordinates and translations. Turning to the 3D case, the standard approach employs 3D extended coordinates and 4×4 matrix multiplication to implement translations. We write a 3D point as an extended 4D vector

$$\begin{bmatrix} x \\ y \\ z \\ 1 \end{bmatrix} \tag{21.3}$$

and the transformation of a 3D point or collection of points to an arbitrarily situated point in space as multiplication by the 4×4 matrix:

$$\begin{bmatrix} x' \\ y' \\ z' \\ 1 \end{bmatrix} = \begin{bmatrix} r_{11} & r_{12} & r_{13} & t_x \\ r_{21} & r_{22} & r_{23} & t_y \\ r_{31} & r_{32} & r_{33} & t_z \\ 0 & 0 & 0 & 1 \end{bmatrix} \cdot \begin{bmatrix} x \\ y \\ z \\ 1 \end{bmatrix} = \begin{bmatrix} r_{11}x + r_{12}y + r_{13}z + t_x \\ r_{21}x + r_{22}y + r_{23}z + t_y \\ r_{31}x + r_{32}y + r_{13}z + t_z \\ 1 \end{bmatrix} . \tag{21.4}$$

Again the vector $[x, y, z, 1]^t$ is actually the *inhomogeneous* form of the *homogeneous* projective vector $[u, v, w, s]^t$, which is formally invariant under scaling,

$$[\lambda u, \lambda v, \lambda w, \lambda s]^t \cong [u, v, w, s]^t .$$

When we now set $\lambda = 1/s$ for finite distances and remove the dependence on s, we can use the inhomogeneous coordinates

$$[x = u/s, \ y = v/s, \ z = w/s, \ 1]^t .$$

We can again deal transparently with the sphere at infinity by setting $s = 0$, so the *direction* in 3D space $[u, v, w, 0]^t$ describes a unique point on the space at infinity that is topologically the celestial sphere $\mathbf{S}^2$.

In computer graphics, the freedom to set $s = 0$ is exploited to use the standard $\mathbf{SE}(3)$ pipeline of Eq. (21.4) to handle objects like *light rays*, which have no location in physical space. It is easy to see that with $s = 0$, adding *any finite translation* to the corresponding ray produces no change: this is a manifestation of the definition of infinity: $\infty +$ any number $\equiv \infty$.

21.2 Introduction to dual quaternions

A brief history of dual quaternions. The adoption of dual quaternions, which include the six parameters of the 3D special Euclidean group $\mathbf{SE}(3)$, three rotational and three translational, has been significant in several application areas. A coherent approach to incorporating translations in a framework applying the style of quaternion multiplication dates back at least to Clifford (Clifford, 1873, 1882), who called them *biquaternions*. This was a logical part of his path that led to the Clifford algebra as a generalization of quaternion properties that forms a fundamental element of modern mathematics. Clifford's method was elaborated by Study (1891), then by Ball (1900) and others, and eventually came to be called by its modern name *dual quaternions*. Various treatments related to mechanics, robotics, and computer graphics elaborate on applications, and the books by Blaschke (1960), Müller (1962), McCarthy (1990), and Roth and Bottema (2012), for example, supply more modern treatments of applications of dual quaternions to mechanics. An extensive literature on "geometric algebra," for example, Dorst et al. (2007); Doran and Lasenby (2003); Dorst et al. (2002); Hestenes (2000); Hestenes and Sobczyk (1984), expands on the extensions of Clifford's biquaternion methods to many areas of modern mathematical physics. The robotics "hand-eye" problem has been studied extensively in the dual quaternion context, nicely reviewed by, e.g., Daniilidis (1999), and a comprehensive approach to the "skinning problem" in computer graphics can be found, for example, in Kavan et al. (2008).

Motivation. The motivation for modifying the quaternion framework to contain a coherent representation for Euclidean vectors goes back to the beginning of quaternions themselves, and many, many years of argument and controversy. The confusion between a Euclidean vector $\mathbf{x}$ and the pure imaginary quaternion $(0, \mathbf{x})$, which is simply a quaternion rotation by 180 degrees, is inevitable. It is a fact that, under quaternion conjugation, the transformation of the fixed rotation axis of a given 180-degree quaternion rotation to a new axis of rotation is indistinguishable from the transformation of a 3D vector under the same action. This calculation looks like this: given a vector $\mathbf{x}$ in radial form as a unit direction $\hat{\mathbf{n}}$, with $\hat{\mathbf{n}} \cdot \hat{\mathbf{n}} = 1$, and radius r, so we can write $\mathbf{x} = r\hat{\mathbf{n}}$, we have

$$\left.\begin{array}{rll}\text{Vector rotation:} \quad \mathbf{x}' = & r\, R(q) \cdot \hat{\mathbf{n}} & = R(q) \cdot \mathbf{x} \\ \text{Quaternion conjugation:} \quad (0, \mathbf{x}') = q \star (0, \mathbf{x}) \star q^{-1} = & r\, q \star (0, \hat{\mathbf{n}}) \star q^{-1} \\ = & r\, (0, R(q) \cdot \hat{\mathbf{n}}) & = (0, R(q) \cdot \mathbf{x})\end{array}\right\} . \tag{21.5}$$

In Appendices D and E, we present alternative, explicitly group-theoretical, derivations of 3D vector rotations related to the group action of **SU(2)** and the equivalence of its algebra to the quaternion algebra. Furthermore, in Chapter 24, we will discover a strong argument that, in relativistic physics, the spatial part of a spacetime 4-vector has very important properties that justify considering the rotation transformation subset of the Lorentz group to be part of a quaternion-like conjugation in the context of the group **SL(2, C)**. However, whether or not there is a strict justification for the derivation of fundamental equations such as Eq. (2.14) via conjugation of a 3D spatial vector written as a pure imaginary quaternion, that does not mean we would not be interested in exploiting a related formalism that merges translation with rotation. It might very well be useful to have in hand a quaternion-like *square-root* form of the full Euclidean group **SE(3)** and an embedded way to treat translations in the context of that square root. Dual quaternions provide this opportunity.

Remark on notation: The context of dual quaternions repurposes some of the standard mathematical symbols we use for its own notation. Beware that in this chapter, *complex conjugation* changes from $\bar{z} = (x - \mathrm{i}y) \to (x + \mathrm{i}y)^* = (x - \mathrm{i}y)$ and $\bar{q} \to q^* = (q_0, -\mathbf{q})$, while certain *unit-length vectors* no longer are written with the "hat" symbol, so $\hat{\mathbf{n}} \to \mathbf{n}$. That change is necessary because, throughout this chapter, we use the notation $\hat{q}$ to stand for the extended *dual* quaternion to distinguish it from the ordinary quaternion q.

21.3 The fundamental idea of quaternion-like translations

The basic idea is that if rotations can exploit exponentials of $\mathrm{i} = \sqrt{-1}$ to represent 2D rotations and the generalization to quaternions to represent 3D rotations, we might look first for a similar exponential power series that would generate translations. Consider first 2D rotations and their multiple-rotation composition done in this way:

$$\text{Usual: } \mathrm{i}^2 = -1: \quad e^{\mathrm{i}\theta} = 1 + \mathrm{i}\theta - \frac{1}{2}\theta^2 - \frac{1}{3!}\mathrm{i}\theta^3 + \cdots$$
$$= \cos\theta + \mathrm{i}\sin\theta ,$$
$$e^{\mathrm{i}\theta} e^{\mathrm{i}\phi} = \cos(\theta + \phi) + \mathrm{i}\sin(\theta + \phi)$$
$$= 1 + \mathrm{i}(\theta + \phi) + O(\text{higher powers}) .$$

Clearly the first term in the power series is linear, so in the $\theta \to 0$ limit, we already have something formally resembling a translation. Alternatively, suppose θ is considered as a dimensionless ratio of two lengths,

$$\theta = \frac{x}{R} \to e^{\mathrm{i}\theta} \approx 1 + \mathrm{i}\frac{x}{R} + O\left(\frac{1}{R^2}\right) .$$

We know in fact that for a very large circle with radius R, this is a good way to look at the approximate behavior of θ. The way that translations are introduced into dual quaternions follows this idea by introducing a new symbol, analogous in some ways to $i^2 = -1$ except now we treat it as an abstraction of $1/R$ for which we are not allowed to retain higher powers of $1/R$, simulating the large R limit of a rotation by the angle x/R. Defining the symbol ϵ with dimensions of *inverse length*,

$$\text{New dual factor} = \epsilon \ , \quad \text{where } \epsilon^2 = 0 \ , \tag{21.6}$$

we see that when we examine its exponentials,

$$e^{\epsilon t} = 1 + \epsilon t + 0 \ ,$$
$$e^{\epsilon x} e^{\epsilon t} = 1 + \epsilon(x + t) \ ,$$

it behaves exactly like a translation operator; or alternatively it looks like a rotation over a very large circle whose usual exponential power series is truncated to replace trigonometric functions by linear addition.

The dual translation form and its square root. It is obvious that we can now replace the controversial *pure quaternion* form $(0, \mathbf{x})$ representing a 3D point by a new form that makes *all 3D points into dual numbers*. That is, we represent locations by the *sum* of a *unit quaternion* $(1, 0, 0, 0)$ and a *dual pure quaternion* using the traditional form *multiplied by* ϵ:

$$\text{Spatial coordinate:} \quad \mathbf{x} \equiv (1, 0, 0, 0) + \epsilon(0, x, y, z)$$
$$= (1, \epsilon\mathbf{x}) \ . \tag{21.7}$$

This can now serve to denote a *position in space*, but it is slightly odd: quaternions do not represent 3D rotations, but are basically *square roots* of rotations, with 3D rotations being the "spin one" representation of the quaternion group. Even if we allow Eq. (21.7) to stand for a spatial vector upon which we can act by rotations and translations, we need another step to produce a quaternion-like *square-root translation action*.

The solution is to *encode the act of translation* separately from the vector in physical space upon which we are acting, and, following what we know about quaternions naturally containing the half-angle in $q_0 = \cos(\theta/2)$, we realize that we can convert the translation process (not the location itself) into a conjugation by *half-translations* in quaternion space, exactly mimicking the appearance of the half rotation angle in the standard formulas for the action of quaternions to produce a rotation in physical space. Defining

$$\tau(\mathbf{x}) = \left(1, \epsilon\frac{\mathbf{x}}{2}\right)$$

we see that to implement the act of translation as quaternion-like conjugation

$$\mathbf{x} = (1, \ \epsilon\mathbf{x}) = \tau(\mathbf{x}) \star (1, \ \epsilon\mathbf{0}) \star \tau(\mathbf{x}) \ , \tag{21.8}$$

we require a subtle change in the definition of conjugation, with the *dual conjugation* taking the form

$$\overline{a + b\epsilon} = a - b\epsilon \ ,$$

while redefining our usual complex-like conjugation to use the alternative notation $q^* = (q_0, -\mathbf{q})$ in the context of this chapter. Thus, in order to obtain the correct addition of both half-translations in Eq. (21.8), we must conjugate the dual vector part in tandem,

$$\overline{(a, \epsilon\mathbf{b})^*} = (a, -\epsilon\mathbf{b})^* = (a, +\epsilon\mathbf{b}) .$$

This definition combining regular and dual conjugations guarantees that the sandwiching of translations adds up in the final displacement from a vector $(1, \epsilon\mathbf{0})$ *at the origin* to give

$$\mathbf{x} = (1, \epsilon\mathbf{x}) = \tau(\mathbf{x}) \star (1, \epsilon\mathbf{0}) \star \overline{\tau(\mathbf{x})^*} = \tau(\mathbf{x}) \star (1, \epsilon\mathbf{0}) \star \tau(\mathbf{x}) .$$

Obviously any translation in space just adds two vectors, as we can see by checking the composite translation:

$$\mathbf{x}' = \tau(\mathbf{t}) \star (1, \epsilon\mathbf{x}) \star \overline{\tau(\mathbf{t})^*} = \mathbf{x} + \mathbf{t} .$$

Folding in quaternion rotations. Rotations acting on a given vector are now computed using the dual quaternion representation for the arbitrary vector, but the usual $q \star (0, \mathbf{x}) \star q^*$ is performed on a *dual* imaginary quaternion instead of the *pure* imaginary quaternion $(0, \mathbf{x})$:

$$(q(\theta, \mathbf{n}) \star (1, \epsilon\mathbf{x}) \star q(\theta, \mathbf{n})^* = (1, \epsilon R(\theta, \mathbf{n}) \cdot \mathbf{x})$$
$$= (1, \epsilon\mathbf{x}') .$$

In this form, the action of rotation by conjugation is no longer confounded with a rotation by 180 degrees, since the vector that is rotated is a *translatable point in space*. One can check that under the translation action of Eq. (21.8) the rotation-fixed axis $\mathbf{n}$ of a regular quaternion is unchanged.

Full SE(3) transform: composing translations and rotations. In order to produce combined translations and rotations in dual quaternion form, we find that we must combine the quaternion rotation directly with the dual quaternion term in the translation. To accomplish the needed full **SE(3)** transformation from Eq. (21.4),

$$\mathbf{x}' = R(\theta, \mathbf{n}) \cdot \mathbf{x} + \mathbf{t} ,$$

we first conjugate by $q(\theta, \mathbf{n})$ to rotate about the object's origin, and then translate it to a new Euclidean position using an outer conjugation by $\tau(\mathbf{t})$, as follows:

$$\mathbf{x}' \equiv (1, \epsilon\mathbf{x}') = \tau(\mathbf{t}) \star q(\theta, \mathbf{n}) \star (1, \epsilon\mathbf{x}) \star q(\theta, \mathbf{n})^* \star \overline{\tau(\mathbf{t})^*}$$
$$= (1, \epsilon(R(\theta, \mathbf{n}) \cdot \mathbf{x} + \mathbf{t})) . \tag{21.9}$$

The dual quaternion. We now introduce the "hat" notation $\hat{q}$ to denote a *dual quaternion* that contains both rotation and translation actions when it is used to transform a vector $\mathbf{x}$ in 3D Euclidean space. We can then write the general rigid-body motion, including translation, in terms of the dual quaternion

$$\hat{q}(\theta, \mathbf{n}, \mathbf{t}) = \tau(\mathbf{t}) \star q(\theta, \mathbf{n}) = (1, \epsilon\mathbf{t}/2) \star (\cos(\theta/2), \mathbf{n}\sin(\theta/2)) ,$$

and write the combined transformation Eq. (21.9) as

$$(1, \epsilon\mathbf{x}') = \hat{q} \star (1, \epsilon\mathbf{x}) \star \overline{\hat{q}^*} . \tag{21.10}$$

21.4 Refining the dual quaternion to use dual trigonometry

With Eq. (21.10), we only have half the story. First, we expect from our initial observations about power series involving dual numbers that we could exploit those properties to reformulate our transformations using some kind of trigonometry in which angles become dual numbers, and can represent both rotations and translations. We will see that this is true, but complicated, and involves advanced concepts using the *Plücker coordinates* to represent arbitrary fixed lines in space about which we perform rotations. This will lead us to a very old, but somewhat bizarre, concept known as *Chasles' theorem*, which expresses any motion as a "screw motion" combining a rotation about a translated axis and a displacement along that axis. In addition, we surely must have a condition that is the analog for $\hat{q}$ to the unit-quaternion constraint $q \cdot q = 1$ for ordinary quaternions. This is required to maintain the topological features of the dual quaternion under large numbers of repeated transformations: enforcing that condition is what maintains the orthonormality of the rotation matrix under repeated quaternion transformations. This requirement adds one more constraint to our implementation of **SE**(3) transformations using dual quaternions, but happily that constraint is enforced if we correctly encode the screw motion.

Intuition from the fixed-point rule. The standard computer graphics approach to performing a rotation of an object that has already been translated to a point $\mathbf{r}$ in its fundamental orientation is to apply the *fixed-point rule* $T(\mathbf{r}) \cdot R \cdot T^{-1}(\mathbf{r})$. (Here we are using the 4×4 matrices of Eq. (21.4) with R a pure rotation and T a pure translation.) In dual quaternion form, this comes out as

$$\text{Quaternion fixed-point rule:} \quad = \tau(\mathbf{r}) \star \left(\cos \frac{\theta}{2}, \mathbf{n} \sin \frac{\theta}{2} \right) \star \tau(-\mathbf{r})$$

$$= \left(\cos \frac{\theta}{2}, (\mathbf{n} + \epsilon \, \mathbf{r} \times \mathbf{n}) \sin \frac{\theta}{2} \right).$$

We see that the fixed point $\mathbf{r}$ is automatically incorporated into a vector that one might call a "dual rotation axis"

$$\text{Dual axis} = \epsilon \mathbf{r} \times \mathbf{n} \,.$$

This dual vector defines the normal to the plane containing the vectors $\mathbf{r}$ and $\mathbf{n}$, and thus, e.g., displacing $\mathbf{r}$ in the direction of $\mathbf{n}$ leaves the dual axis and the plane unchanged. So this is not enough to handle an arbitrary translation; next we will exploit dual angles to make a translation to an arbitrary location as a sum of a vector in the plane through the origin defined by $\mathbf{n}$, and an additional multiple of $\mathbf{n}$.

Dual trigonometry. Our next step is to investigate dual angles, which will lead us to an elegant solution for achieving a screw motion. First, suppose we consider our object to be at the origin, but we look at the possibility of making the *angle* dual, so $\theta \to \theta + \epsilon \lambda$. Then expanding in a Taylor series is very simple due to $\epsilon^2 = 0$, and we find

$$\begin{aligned} \cos(a + \epsilon \, b) &= \cos a - \epsilon \, b \sin a \,, \\ \sin(a + \epsilon \, b) &= \sin a + \epsilon \, b \cos a \,. \end{aligned} \tag{21.11}$$

Then, with only the dual trigonometry expansion, but still with the dual vector part vanishing, we see

$$q(\theta + \epsilon \lambda, \mathbf{n}) = \left(\cos \frac{\theta}{2} - \epsilon \frac{\lambda}{2} \sin \frac{\theta}{2}, \mathbf{n} \left(\sin \frac{\theta}{2} + \epsilon \frac{\lambda}{2} \cos \frac{\theta}{2} \right) \right) \,. \tag{21.12}$$

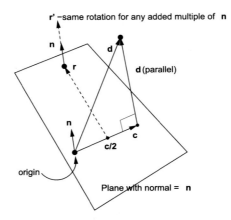

FIGURE 21.1

Steps starting from the fixed-point rule to achieving an arbitrary displacement **d** in two perpendicular steps, $\mathbf{d} = \mathbf{d}_\perp + \mathbf{d}_\| = \mathbf{c} + \lambda\mathbf{n}$, where **c** is perpendicular to the plane normal **n** and $\lambda = \mathbf{n} \cdot \mathbf{d}$ is the parallel displacement along **n**.

Curiously, in the $\theta \to 0$ limit, we find that this translates the origin along **n**, exactly the behavior we are looking for:

$$q(\epsilon\lambda, \mathbf{n}) = (1, \epsilon\,\mathbf{n}\frac{\lambda}{2}) .$$

If we now define our desired displacement to be **d**, using $\lambda = \mathbf{n} \cdot \mathbf{d}$ to decompose it into $\mathbf{d}_\perp$ and $\mathbf{d}_\|$, we can write

$$\mathbf{d} = \mathbf{d}_\perp + \mathbf{d}_\| = \mathbf{c} + \lambda\mathbf{n} . \tag{21.13}$$

Now we see the introduction of $\mathbf{c} = \mathbf{d} - \lambda\mathbf{n}$, which appears in Fig. 21.1 as the shortest distance from the origin to the line dropped from **d** to the plane defined by the origin and rotation axis **n**. *Any line* can be defined by a unit-length direction **n** and the vector **c** to its closest point on the plane through the origin perpendicular to **n**:

n and **c** are the *Plücker coordinates* of this line,

so $\mathbf{P}(t) = \mathbf{c} + t\,\mathbf{n}$ parameterizes points on the line.

These will be the natural quantities defining our screw motion using dual quaternions.

Displacing the rotation to start the screw motion. Next, we need to show how to obtain a combined rotation of the right size from a certain fixed point **r** that must be computed.

In Fig. 21.2 we show a top-down view of the (x, y)-plane with the dark vertical arrow at the origin denoting our initial orientation. Our target rotation by $R(\theta, \mathbf{n})$ in the plane perpendicular to the rotation axis **n** is applied to that orientation to get the desired new orientation located at the point **c**. What the screw motion must do, in this top-down projected view, is to rotate the arrow about the *fixed point* **r**

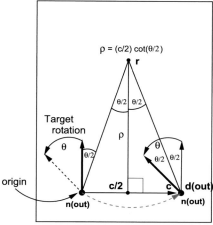

$$\rho = (c/2)\cot(\theta/2)$$

FIGURE 21.2

The geometry of the partial screw motion for a rotation by $R(\theta, \mathbf{n})$ restricted to the plane perpendicular to the rotation axis $\mathbf{n}$. (Think of this as the xy-plane, and $\mathbf{n}$ as the z direction **(out)** pointing out of the plane at the viewer, but the actual orientation in physical space is arbitrary.) Using the point $\mathbf{r}$ along the line perpendicular to the vector $\mathbf{c}$ as the fixed point of our rotation, with $\rho = \|\mathbf{r} - \mathbf{c}/2\|$, we can calculate the position of $\mathbf{r}$ as proportional to $\cot(\theta/2)$ from elementary trigonometry and the requirement that a line at the origin be rotated by angle θ after it swings over to its new position at $\mathbf{c}$.

to simultaneously displace the origin to the point $\mathbf{c}$, the projected portion of the displacement $\mathbf{d}$ in the plane perpendicular to $\mathbf{n}$, while *rotating* the initial vector by exactly the angle θ.

Clearly if we place the fixed point $\mathbf{r}$ at $\mathbf{c}/2$, a 180-degree rotation will move the origin to the point $\mathbf{c}$ while applying a 180-degree rotation to our reference direction, the dark vertical arrow in Fig. 21.2. If we extend $\mathbf{r}$ to infinity, the displacement to $\mathbf{c}$ still works, but the rotation angle essentially vanishes. At some intermediate distance $\rho = \|\mathbf{r} - \mathbf{c}/2\|$ along the line from $\mathbf{c}/2$ in the direction $\mathbf{n} \times \mathbf{c}$, we will be able to achieve a rotation by θ. As represented in Fig. 21.2, we can find the necessary point

$$\mathbf{r} = \frac{\mathbf{c}}{2} + \rho \frac{\mathbf{n} \times \mathbf{c}}{c}, \tag{21.14}$$

where we note the normalization of the direction $\|\mathbf{n} \times \mathbf{c}\|^2 = \|\mathbf{n}\|^2 \|\mathbf{c}\|^2 - (\mathbf{n} \cdot \mathbf{c})^2 = c^2$. Since the total known angle θ at the vertex $\mathbf{r}$ is divided into two fractions $\theta/2$ by the perpendicular dropped to $\mathbf{c}/2$ (the midpoint of the part of the total displacement $\mathbf{d}$ perpendicular to $\mathbf{n}$), we see at once that

$$\frac{\rho}{c/2} = \cot\frac{\theta}{2} \quad \rightarrow \quad \mathbf{r} = \frac{\mathbf{c}}{2} + \frac{1}{2}\mathbf{n} \times \mathbf{c} \cot\frac{\theta}{2}. \tag{21.15}$$

This gives us all the geometric data we need to complete the dual quaternion form of the screw motion displacing an object at the origin and its identity frame to an arbitrary point and orientation corresponding to

$$\mathbf{x}' = R(\theta, \mathbf{n}) \cdot \mathbf{x} + \mathbf{d} . \tag{21.16}$$

Unit length condition for dual quaternions. Finally, we can see how to represent $T(\mathbf{d}) \cdot R$ as a *unit dual trigonometric quaternion* $\hat{q}$. The general trigonometric form for a dual quaternion $\hat{q}$ can be split into a "normal" and a dual part:

$$
\begin{aligned}
\hat{q}(\hat{\theta}, \hat{\mathbf{n}}) &= q_n + \epsilon q_\epsilon \\
&= (q_0 + \epsilon q_{0\epsilon}, \, \mathbf{q} + \epsilon \mathbf{q}_\epsilon) \\
&= \left(\cos \frac{\theta + \epsilon \lambda}{2}, \, (\mathbf{n} + \epsilon \mathbf{r} \times \mathbf{n}) \sin \frac{\theta + \epsilon \lambda}{2} \right) .
\end{aligned} \tag{21.17}
$$

Now, having the ability to use a dual quaternion to create an arbitrary rotation and displacement of a 3D point represented as a dual pure quaternion vector, we can check the requirements that we actually have a quaternion-like "square-root" representation of **SE**(3) that obeys the fundamental requirements:

- *Unit length:* $q \cdot q = 1 \;\rightarrow\; \hat{q} \cdot \hat{q} = 1.$
- *Multiplicative norm:* $\|q \star p\| = \|q\| \|p\| \;\rightarrow\; \|\hat{q} \star \hat{p}\| = \|\hat{q}\| \|\hat{p}\| = 1.$

Separating out the normal and dual components as $\hat{q} = q_n + \epsilon q_\epsilon$, the norm can be decomposed as

$$\|\hat{q}\|^2 = \|q_n\|^2 + 2\epsilon q_n \cdot q_\epsilon + 0 = 1 + 2\epsilon q_n \cdot q_\epsilon .$$

Thus a *unit* dual quaternion must have $q_n \cdot q_\epsilon = 0$. That is satisfied by the trigonometric form because

$$\mathbf{n} \cdot (\mathbf{r} \times \mathbf{n}) \equiv 0 .$$

We can always parameterize the 4×4 matrix $T(\mathbf{d}) \cdot R(\theta, \mathbf{n})$, equivalent to $((1, \epsilon(R(\theta, \mathbf{n}) \cdot \mathbf{x} + \mathbf{d}))$, using the unit dual quaternion

$$\hat{q}(\hat{\theta}, \hat{\mathbf{n}}) = \hat{q}(\theta + \epsilon(\mathbf{d} \cdot \mathbf{n}), \mathbf{n} + \epsilon \mathbf{r} \times \mathbf{n}) , \tag{21.18}$$

where the rotation center $\mathbf{r}$ given in Eq. (21.15) is the rotation center, computable from θ, $\mathbf{n}$, and $\mathbf{c}$ (or $\mathbf{d}$). The entire dual screw motion action is shown in Fig. 21.3. Note that in the exceptional case $\theta = 0$, we just take the action to be a pure translation. For completeness, we note the following parameters used in the context of this transformation:

$$
\begin{aligned}
[\textit{the Plücker coordinate } \mathbf{c}]:& \quad \mathbf{c} = \mathbf{d} - \mathbf{n}(\mathbf{n} \cdot \mathbf{d}) , \\
[\textit{the part of } \mathbf{d} \textit{ parallel to } \mathbf{n}]:& \quad \mathbf{d}_\| = \mathbf{n}(\mathbf{d} \cdot \mathbf{n}) = \lambda \mathbf{n} , \\
[\textit{direction of the fixed point}]:& \quad \hat{\mathbf{r}} = \frac{\mathbf{n} \times \mathbf{c}}{\|\mathbf{n} \times \mathbf{c}\|} = \frac{\mathbf{n} \times \mathbf{c}}{c} .
\end{aligned}
$$

We can confirm the normed division algebra property (see Chapter 3) by working out the norm of the product of dual quaternions. One finds

$$\|\hat{p} \star \hat{q}\|^2 = \hat{p} \star \hat{q} \star \hat{q}^* \star \hat{p}^* = \hat{p} \star (\|\hat{q}\|^2) q_{\text{ID}} \star \hat{p}^* = \|\hat{p}\|^2 \|\hat{q}\|^2 .$$

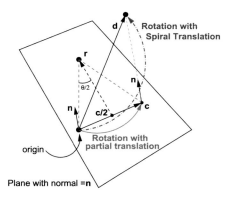

FIGURE 21.3

Rotation can produce displacement, but to get a full displacement, you need both the rotation in the plane about the **n**-axis at **r** (causing a tandem displacement from the origin to **c**) *and* a displacement at **c** in the direction of **n**.

Interpolation. The screw motion realization of the **SE**(3) rotation plus translation transformation of a 3D coordinate frame gets us where we want to be, from one frame to another via Eq. (21.18), as illustrated schematically by the dashed path in Fig. 21.3. Furthermore, we can extend the simplest interpolation formula from the identity frame to a new orientation and location directly when we use the trigonometric form

$$\hat{q}(t) = \cos t \, \frac{\hat{\theta}}{2} + \hat{n} \sin t \, \frac{\hat{\theta}}{2} \,, \tag{21.19}$$

where we define our dual parameters as $\hat{\theta} = \theta + \epsilon \, \mathbf{d} \cdot \mathbf{n}$, and $\hat{n} = \mathbf{n} + \epsilon \, \mathbf{r} \times \mathbf{n}$. However, the "rotation-induced" translation along $\mathbf{c} \perp \mathbf{n}$ has *a different speed* from the translation along $\mathbf{n}$ (the $\lambda = (\mathbf{d} \cdot \mathbf{n})$ part). Furthermore, the translation is *not straight*, as we can see in the more realistic depiction in Fig. 21.4, which follows a shape from an identity frame at the origin, with $t = 0$ in Eq. (21.19) to $t = 1$. *This is the cost of elegance in the dual trigonometric form.*

Some useful properties. We conclude this section with a handful of properties of dual quaternions that are conceptually parallel to properties of normal quaternions that we have found useful for certain applications.

- **Exponential and log.** Using power series, one can extend the usual quaternion exponential and log formulas:

$$\hat{q} = \exp\left(\hat{s} \, \frac{\hat{\theta}}{2}\right) = \cos \frac{\hat{\theta}}{2} + \hat{s} \sin \frac{\hat{\theta}}{2} \,.$$

Thus we also have $\log \hat{q} = \hat{s} \, \hat{\theta} / 2$.

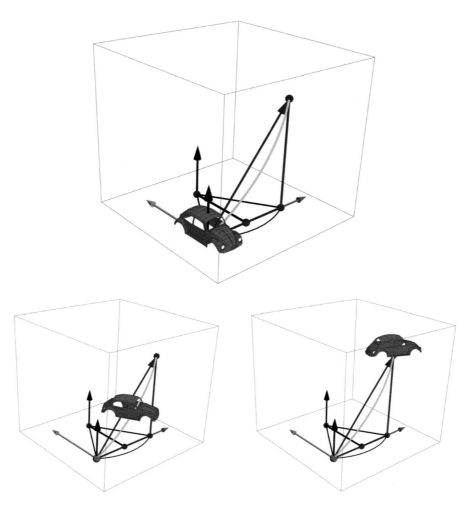

FIGURE 21.4

Interpolation in 3D space taking an identity frame to a rotated and translated frame via the dual quaternion "screw motion." The 3D path is not straight, and therefore not generally suitable for applications like animation modeling.

- **Inverse:** The inverse of a dual object is defined only when the normal quaternion scalar term is nonvanishing, $a \neq 0$, in which case we can write

$$(a + \epsilon b)^{-1} = \frac{1}{a + \epsilon b} = \frac{1}{a} - \epsilon \frac{b}{a^2} \,.$$

This can easily be confirmed from the dual identity

$$(a + \epsilon b)(c + \epsilon d) = ac + \epsilon (ad + bc) \,.$$

- **Square root:** For ordinary quaternions, we know from Appendix B that the square root can be written

$$p = \sqrt{q} = \pm \frac{1+q}{\sqrt{2(1+q_0)}} = \pm \left(\sqrt{\frac{1+q_0}{2}}, \, \hat{\mathbf{n}} \sqrt{\frac{1-q_0}{2}} \right) ,$$

giving $p \star p = q$ if $q \cdot q = 1$ and $q_0 \neq -1$ for the first form, and the second form shows that for any $\hat{\mathbf{n}} \cdot \hat{\mathbf{n}} = 1$, $p = (0, \pm\hat{\mathbf{n}})$ is a solution for $q_0 = -1$. A similar formula works for the square root of any dual number,

$$\sqrt{a + \epsilon b} = \sqrt{a} + \epsilon \frac{b}{2\sqrt{a}} , \tag{21.20}$$

provided $a \neq 0$.

21.5 The motion of a dual quaternion

We can express the interpolating rotation between two pure orientation frames at the origin using a `slerp`. However, the essence of this interpolation is embodied simply in terms of a minimal rotation from the identity frame to the frame $q_1^{-1} \star q_2 = q(\theta, \hat{\mathbf{n}})$ using $q(\lambda\theta, \hat{\mathbf{n}})$ as the parameter λ varies from $0 \leqslant \lambda \leqslant 1$. If we do the same thing for a general **SE**(3) motion expressed as a dual quaternion, we obtain a surprising result: first, it can be shown that the most general unit-length dual quaternion can be written in terms of a dual angle

$$\hat{\theta} = \theta + \epsilon \mathbf{t} \cdot \mathbf{n} ,$$

where $\mathbf{t}$ is the translation vector and $\mathbf{n}$ is the rotation axis. (We normally write $\mathbf{n}$ as $\hat{\mathbf{n}}$, but we remove that "hat" in this chapter because in the dual quaternion context, "hat" is conventionally used to represent mixed quaternion and dual objects). Secondly, we can replace the unit vector $\mathbf{n}$ in the regular quaternion $q = \left(\cos \frac{\theta}{2}, \, \mathbf{n} \sin \frac{\theta}{2} \right)$ by

$$\mathbf{s} = \mathbf{n} + \epsilon \mathbf{r} ,$$

where $\mathbf{r}$ is the *axis of screw motion* following from Chasles' theorem, and takes the form

$$\mathbf{r} = \frac{1}{2} \left(\mathbf{t} + \cot \frac{\theta}{2} (\mathbf{n} \times \mathbf{t}) \right) . \tag{21.21}$$

Thus the entire dual quaternion can be written in terms of the *fixed* values of the displacement $\mathbf{t}$ and the rotation axis $\mathbf{n}$, with *interpolation* only in θ, using the interpolation $\lambda\theta$ in the following form:

$$\hat{q}(\lambda) = \left(\cos \frac{\lambda\theta + \mathbf{n} \cdot \mathbf{t}}{2}, \, (\mathbf{n} + \epsilon \mathbf{r}) \sin \frac{\lambda\theta + \mathbf{n} \cdot \mathbf{t}}{2} \right) . \tag{21.22}$$

This expression contains all the necessary parameters, three rotation parameters in θ and $\mathbf{n}$ (which is unit-length) and three translation parameters $\mathbf{t} = (t_x, t_y, t_z)$, with the screw-axis $\mathbf{r}$ expressed via Eq. (21.21) in terms of the other variables, keeping the $\cot \frac{\theta}{2}$ term fixed since it would introduce a singularity in the interpolation.

The result is that the smooth interpolation from an identity frame to an arbitrary frame, as shown in Fig. 21.4, follows a curved path in space, as though being acted upon automatically by a force, whereas we normally want paths for which we specify all such details in a highly constrained physical model. Thus, unfortunately, this seemingly natural interpolation is not suitable for approximating, e.g., free-body motion in the way that the regular quaternion is suitable for interpolating orientations.

21.6 Brief discussion of applications

A proper discussion of the applications of dual quaternions could easily fill its own book. We will not attempt that level of treatment here, but we will outline a couple of examples that we believe are good examples of successful exploitation of the basic idea behind the invention of dual quaternions:

> *Create a consistent system for the 3D special Euclidean group* **SE**(3) *that incorporates translations into a context with the same square-root-like properties that quaternions provide for the pure 3D rotations* **SO**(3).

Blending and skinning. Dual quaternions permit an unusually smooth static blending of weighted vertices associated to two or more skeletal elements in character animation. A typical application is getting realistic skin attached to the elements of a complicated underlying skeleton. The most rigorous methods are essentially dual quaternion extensions of the spherical center-of-mass methods of Buss and Fillmore (2001). Faster, but less accurate methods use an extension of the concept of Phong shading, renormalizing a linear combination of data sets (see, e.g., Kavan et al. (2008)).

Interpolation. Blending is a static process, and needs to be done to combine character body elements such as skin vertices at each moment. Interpolation for simulating moving object kinematics and controlling camera motion can also be accomplished by extending standard quaternion interpolation techniques to dual quaternions, though challenging issues such as how to control dual parameters and how to match rotational and translational speeds in a single interpolation introduce additional complexity and possible artifacts.

Robotics control systems. One of the more widely used applications of dual quaternions is to reduce the complexity of the coordination of the parts of robotics systems. A typical application discussed extensively in, e.g., Daniilidis (1999), is the understanding of the six-degree-of-freedom transformations relating a physical robotic gripper to the view of an associated camera. This problem, known as the "hand-eye calibration" problem, is an essential part of many robotics applications, and the ability of dual quaternions to coordinate frame orientations with relative positions of the system elements has been effectively exploited.

Quaternion applications in physics

7

Our quaternion applications have so far been heavily concentrated on the role of quaternions in 3D, and sometimes 4D, rotations or orientation-related tasks. We have intensely exploited the geometric identification of quaternions with the rotation group $\mathbf{SO}(3)$ and the connection of the quaternion group $\mathbf{SU}(2)$ with the topological 3-sphere $\mathbf{S}^3$. In this next part of our story, we explore the ways that the complexification of quaternions, implemented by extending the $\mathbf{SU}(2)$ to $\mathbf{SL}(2, \mathbf{C})$, plays an important role in physics, including quantum mechanics, Lorentz transformations, and special relativity, and thus spacetime itself. The physical problems that we will look at include the appearance of quaternions in the description of quantum computing, an introduction to special relativity, and a comprehensive treatment of the Lorentz group in terms of complexified quaternions. Even more physical applications will appear in Part 8, where we will see the appearance of quaternions in Einstein's theory of general relativity.

Quantum computing and quaternion qubits[☆]

22.1 Context of quantum computing

Quantum computing addresses the task of exploiting the distinction between information contained in quantum states and the classical information accessible to an ordinary computer program. Unlike typical computation, quantum computing is intrinsically probabilistic, involves complex variables that are essentially the square roots of traditional probabilities, is reversible, and can exploit entanglement, which inseparably mixes the properties of states in ways that can be exploited. All this combines to produce a computational framework that, for certain problems, allows massive acceleration of the rate of solving a problem relative to the times typical for solving the same problem using a classical computer.

The most basic objects in quantum computing are complex vectors whose absolute value is unity, which signifies that all the probabilities combine to cover all possibilities, that is, conserving probability. What this means is that for certain important classes of states, the entirety of a given probability-conserving system is constrained to a (possibly high-dimensional) sphere. In the case of a single elementary state, described by two complex variables, or four real variables, this means that the state data are constrained to lie on the 3-sphere $\mathbf{S}^3$, and thus can be viewed as a quaternion. (In fact, it is more complicated than that, but the probability-conserving constraint to a sphere is fundamental.)

At first glance, this relationship to quaternions is *geometrical*, related specifically to $\mathbf{S}^3$, and not *algebraic*, and thus not related to the Hamilton quaternion algebra for combining rotations. However, there are two relevant contexts where quaternions assume additional significance in quantum mechanics. First, since a vast spectrum of quantum mechanical systems involve states that have at least part of their description in the context of 3D space, rotations play an important role. In fact, the discrete states of any atom are explicitly dependent on representations of the rotation group that involve quaternion polynomials (see, e.g., Appendix E). Furthermore, the electron itself cannot be described by ordinary rotations, but requires such constructs as the Pauli matrices and the Dirac matrices, which are all keyed to the fact that the electron has spin 1/2, meaning that it has states that return to their original form only after *two* full rotations in space, exactly the behavior of a quaternion as it rotates in space due to the appearance of $\theta/2$ in Eq. (2.16). Finally, there exists a substantial body of literature beyond the scope of this book that *does* treat quaternion quantum mechanics using the quaternion algebra to extend the complex arithmetic of ordinary quantum mechanics; see, e.g., Adler (1995).

☆ Selected material in this chapter is derived from Hanson et al. (2013).

22.2 The qubit

The most fundamental structure in the mathematics of quantum computing is the single quantum bit or "qubit." This is a 2D complex vector that represents something very different from any other kind of vector, as it is effectively the complex square root of a probability, a *probability amplitude*. Represented in vector form, a qubit looks something like this:

$$\text{one qubit} = \begin{bmatrix} a_0 \\ a_1 \end{bmatrix} = a_0 \begin{bmatrix} 1 \\ 0 \end{bmatrix} + a_1 \begin{bmatrix} 0 \\ 1 \end{bmatrix}. \tag{22.1}$$

In traditional quantum mechanics notation, this 2D vector space is usually written in the slightly more abstract "bra-ket" notation, attributed to Dirac, with $[1, 0]^t$ denoted by the "ket" $|0\rangle$ and $[0, 1]^t$ by the "ket" $|1\rangle$.[1] The kets are matched by their corresponding "bra" symbols $\langle 0|$ and $\langle 1|$, representing the conjugate basis used for multiplying on the left to produce a projection onto an expectation value. By definition, the product of a bra and a ket is $\langle i|j\rangle = \delta_{ij}$. The qubit and its complex conjugate then take the general form for any possible state

$$|\psi\rangle = a_0|0\rangle + a_1|1\rangle\,, \tag{22.2a}$$

$$\langle\psi| = \bar{a}_0\langle 0| + \bar{a}_1\langle 1|\,. \tag{22.2b}$$

This is, however, not yet enough to define a unique, consistent qubit. In quantum mechanics, any wave function must conserve probability, which means in practice the absolute square of the wave function must be the real number "one," signifying that the totality of all possible probabilistic outcomes must be conserved. Mathematically, we express this as

$$\begin{aligned} \|\psi\|^2 = \langle\psi|\psi\rangle &= (\bar{a}_0\langle 0| + \bar{a}_1\langle 1|)(a_0|0\rangle + a_1|1\rangle) \\ &= |a_0|^2 + |a_1|^2 \\ &= 1\,. \end{aligned} \tag{22.3}$$

Note that we distinguish the *norm* $\|\cdot\|$ of a complex vector from the *modulus* $|\cdot|$ of a complex number. This is immediately recognizable as a description of the 3-sphere $\mathbf{S}^3$, making the real coordinates of the one-qubit amplitude geometrically indistinguishable from a quaternion, that is,

$$\left. \begin{array}{l} a_0 = x_0 + iy_0 \qquad a_1 = x_1 + iy_1 \\ \quad = q_0 + iq_1 \qquad\quad = q_2 + iq_3 \\ \qquad\quad \text{giving the constraints} \\ \qquad\qquad |a_0|^2 + |a_1|^2 = 1 \\ \Rightarrow \quad q_0{}^2 + q_1{}^2 + q_2{}^2 + q_3{}^2 = 1 \end{array} \right\}. \tag{22.4}$$

We observe that, in this context, only the *geometry* of the quaternion's $\mathbf{S}^3$ space is evident: there is no sign of the quaternion *algebra* so far. We will return to this issue in a moment. Starting from this vector space notation, we now discover and impose various constraints on the single-qubit system.

---

[1] In some applications, $|0\rangle$ and $|1\rangle$ are replaced by $|+\frac{1}{2}\rangle$ and $|-\frac{1}{2}\rangle$, without changing the basic framework.

22.3 The Bloch sphere

The basic qubit $|\psi\rangle$ in Eqs. (22.2a) and (22.2b) with norm 1 from Eq. (22.3) satisfies the basic principles of quantum mechanics, distributing its probability over the alternative possibilities $|0\rangle$ and $|1\rangle$, but unfortunately one can easily see that it is ambiguous: since $\|\psi\|^2$ is invariant under any global phase transformation $a_k \rightarrow a_k \exp(i\phi)$, we cannot uniquely specify the state of a qubit without eliminating this extra degree of freedom. If we look at the free parameters of a qubit, we see that we started with four real variables, the two complex variables a_0 and a_1, imposed probability conservation via the unit norm to get three real degrees of freedom, and now we need to eliminate one more degree of freedom to remove the phase ambiguity, which will leave us with two. We can accomplish this using exactly the quadratic forms we exploited to remove the quaternion sign ambiguity in the map to the 3D rotation $R(q)$.

To see this, we look back at Eq. (2.14), and extract all *six* possible rows and columns,

$$
\begin{bmatrix} q_0^2 + q_1^2 - q_2^2 - q_3^2 \\ 2q_1q_2 + 2q_0q_3 \\ 2q_1q_3 - 2q_0q_2 \end{bmatrix} , \quad
\begin{bmatrix} 2q_1q_2 - 2q_0q_3 \\ q_0^2 - q_1^2 + q_2^2 - q_3^2 \\ 2q_2q_3 + 2q_0q_1 \end{bmatrix} , \quad
\begin{bmatrix} 2q_1q_3 + 2q_0q_2 \\ 2q_2q_3 - 2q_0q_1 \\ q_0^2 - q_1^2 - q_2^2 + q_3^2 \end{bmatrix} ,
$$
$$
\begin{bmatrix} q_0^2 + q_1^2 - q_2^2 - q_3^2 \\ 2q_1q_2 - 2q_0q_3 \\ 2q_1q_3 + 2q_0q_2 \end{bmatrix} , \quad
\begin{bmatrix} 2q_1q_2 + 2q_0q_3 \\ q_0^2 - q_1^2 + q_2^2 - q_3^2 \\ 2q_2q_3 - 2q_0q_1 \end{bmatrix} , \quad
\begin{bmatrix} 2q_1q_3 - 2q_0q_2 \\ 2q_2q_3 + 2q_0q_1 \\ q_0^2 - q_1^2 - q_2^2 + q_3^2 \end{bmatrix} .
\tag{22.5}
$$

We recognize this as a family of six equivalence classes of quadratic maps that take the remaining three degrees of freedom in Eq. (22.4) and reduce them to two degrees of freedom by effectively removing $e^{i\phi}$ ("fibering out by the circle $\mathbf{S}^1$"). The standard form of this class of maps, "the Hopf map" or "the Hopf fibration," is

$$
\begin{aligned}
X &= 2\,\mathrm{Re}\; a_0\, a_1^* = 2x_0x_1 + 2y_0y_1 \, , \\
Y &= 2\,\mathrm{Im}\; a_0\, a_1^* = 2x_1y_0 - 2x_0y_1 \, , \\
Z &= |a_0|^2 - |a_1|^2 = x_0^2 + y_0^2 - x_1^2 - y_1^2 \, ,
\end{aligned}
\tag{22.6}
$$

and all six maps in Eq. (22.5) can be re-expressed in this form. These transformed coordinates are in effect just the normalization conditions for the unit-length 3-vector rows or columns of a rotation matrix, and therefore the vector $\mathbf{X} = [X, Y, Z]^t$ lies on a sphere $\mathbf{S}^2$ referred to as the *Bloch sphere*:

$$
\|\mathbf{X}\|^2 = X^2 + Y^2 + Z^2 = \left(|a_0|^2 + |a_1|^2 \right)^2 = 1 \, .
\tag{22.7}
$$

Thus $\mathbf{X}$ has only two remaining degrees of freedom describing all possible distinct one-qubit quantum states. In Fig. 22.1 we illustrate schematically the family of circles *each one of which is collapsed to a point* on the Bloch sphere $X^2 + Y^2 + Z^2 = 1$ by the Hopf map.

The resulting manifold is the 2-sphere $\mathbf{S}^2$ (a 2-manifold) embedded in $\mathbb{R}^3$. Suppose we choose one of many possible coordinate systems obeying Eq. (22.4) and thus describing $\mathbf{S}^3$ such as

$$
\begin{aligned}
(x_0, y_0, x_1, y_1) = \\
\left(\cos\left(\frac{\theta+\phi}{2}\right)\cos\frac{\psi}{2},\; \sin\left(\frac{\theta+\phi}{2}\right)\cos\frac{\psi}{2},\; \cos\left(\frac{\theta-\phi}{2}\right)\sin\frac{\psi}{2},\; \sin\left(\frac{\theta-\phi}{2}\right)\sin\frac{\psi}{2} \right),
\end{aligned}
\tag{22.8}
$$

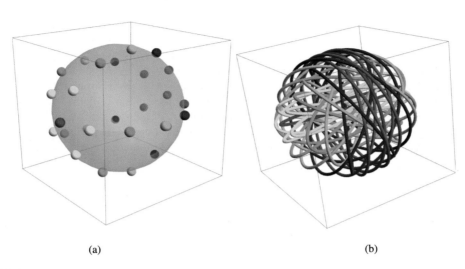

(a)   (b)

FIGURE 22.1

(a) The Bloch sphere represented by Eq. (22.7), which is the irreducible space of one-qubit states, along with a representative set of points on the sphere. (b) Representation of the Hopf fibration as a family of circles in $\mathbf{S}^3$, each collapsing to a single point on the sphere $\mathbf{S}^2$ in (a). Points in (a) are color coded corresponding to circles in (b), e.g., one red point represents the red elliptical circle, and the diametrically opposite (faded) red point corresponds to the large perfectly round red circle at the vertical equator in (b).

where $x_0{}^2 + y_0{}^2 + x_1{}^2 + y_1{}^2 = 1$ and $0 \leqslant \psi \leqslant \pi$, with $0 \leqslant \dfrac{\theta + \phi}{2} < 2\pi$ and $0 \leqslant \dfrac{\theta - \phi}{2} < 2\pi$. (This basically rotates the (θ, ϕ) coordinate system by $-\pi/4 = -45$ degrees so that for $0 \leqslant \theta \leqslant \pi$, ϕ is constrained to $-\theta \leqslant \phi \leqslant \theta$, and for $\pi \leqslant \theta \leqslant 2\pi$, ϕ is constrained to $\theta - 2\pi \leqslant \phi \leqslant 2\pi - \theta$, as summarized schematically in Fig. 22.2.) Inserting these variables into the quadratic expressions Eq. (22.6) for (X, Y, Z) defining the Hopf map, we see that the variable θ simply disappears from the map, leaving a pure two-parameter $\mathbf{S}^2$,

$$(X, Y, Z) = (\cos\phi \sin\psi, \ \sin\phi \sin\psi, \ \cos\psi) \ . \tag{22.9}$$

Thus the single-qubit state described by the sphere (X, Y, Z) is independent of θ, and we can choose $\theta = \phi$ without loss of generality, reducing the form of the unique one-qubit states to

$$|\phi, \psi\rangle = e^{i\phi} \cos\frac{\psi}{2} |0\rangle + \sin\frac{\psi}{2} |1\rangle \ , \tag{22.10}$$

and these irreducible states can be represented as points on the sphere, as shown in Fig. 22.3(a).

    Thus the geometry of a single qubit reduces to transformations among points on $\mathbf{S}^2$, which can be parameterized in an infinite one-parameter family of transformations, one of which is the geodesic or minimal-length transformation between any given pair of points. Explicitly, given two one-qubit states denoted by points $\mathbf{a}$ and $\mathbf{b}$ on $\mathbf{S}^2$, the shortest rotation carrying the state represented by the unit vector $\mathbf{a}$ to the state represented by the unit vector $\mathbf{b}$ on $\mathbf{S}^2$ is the `slerp` (spherical linear interpolation)

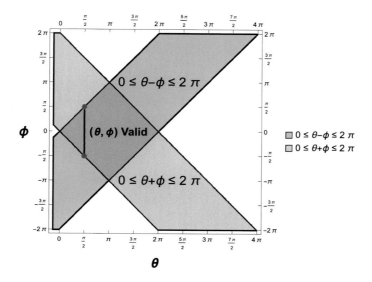

FIGURE 22.2

Showing the nature of the constrained parameter space in θ and ϕ used in Eq. (22.8) for the quaternion. This combination allows us to explicitly project out all dependence on the variable θ and reduce our qubit description to the unique Bloch sphere parameters ϕ and ψ. A legal range of ϕ for a given θ is marked with red dots.

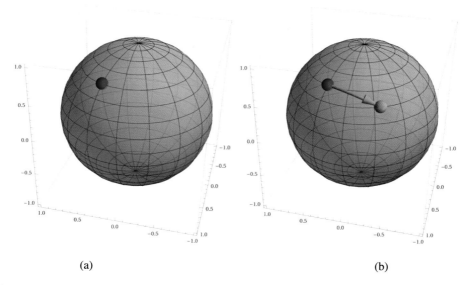

(a)   (b)

FIGURE 22.3

(a) The conventional Bloch sphere with a unique state represented by the point at the red sphere. (b) The geodesic shortest-distance arc connecting two one-qubit quantum states, implementable as a `slerp`.

introduced in Eq. (2.19),

$$\text{slerp}(\mathbf{a}, \mathbf{b}; t) = \mathbf{a}\,\frac{\sin((1 - t)\,\theta)}{\sin\theta} + \mathbf{b}\,\frac{\sin(t\,\theta)}{\sin\theta}. \tag{22.11}$$

Here $\mathbf{a} \cdot \mathbf{b} = \cos\theta$, and remember that, although the `slerp` was introduced in the context of quaternions, it works on spheres of any dimension. Fig. 22.3(b) illustrates the minimal path traced by a `slerp` between two such irreducible one-qubit states on the Bloch sphere.

22.4 Quaternions as quantum computing operators

The single qubit represents a state that has a 2D basis, and thus is interpretable as an arbitrary mixture of two pure states, which we can label in various ways, e.g., $\{|0\rangle, |1\rangle\}$ for bits of information, or, say, $\{|+1/2\rangle, |-1/2\rangle\}$ for the up and down states of a spin $1/2$ elementary particle such as an electron.

However, quantum mechanical states like a qubit have a very special property: not only must they have unit norm to conserve probability, but *they can only be transformed into one another by unitary operators*, which preserve that unit norm. What that means is that the only operators that can evolve a single qubit to a new state are the elements of the unitary matrix group $\mathbf{U(2)}$, which is a *quaternion-valued* $\mathbf{SU(2)}$ matrix with an arbitrary multiplicative *phase* $(\exp i\psi)$. Written out, this type of qubit transformation takes the 2-vector form

$$|\psi'\rangle = a'|0\rangle + b'|1\rangle$$

$$= \begin{bmatrix} a' \\ b' \end{bmatrix} = [U(\psi; q_0, q_x, q_y, q_z)] \cdot \begin{bmatrix} a \\ b \end{bmatrix} = e^{i\psi} \times \begin{bmatrix} q_0 - iq_z & -i(q_x - iq_y) \\ -i(q_x + iq_y) & q_0 + iq_z \end{bmatrix} \cdot \begin{bmatrix} a \\ b \end{bmatrix}. \tag{22.12}$$

The result is that basically any unitary phase-multiplied quaternion matrix $U(\psi; q)$ can be used to move a qubit around the Bloch sphere, preserving the probability-conserving nature of any transformed state. In particular, one frequently sees $\mathbf{SU(2)}(q_x = 1) = -i\sigma_x$ and $\mathbf{SU(2)}(q_y = 1) = -i\sigma_y$ used in quantum computing manipulations. Other important operators incorporate a nontrivial phase, e.g., the Hadamard operator H, and the phase operator Φ can be written in terms of quaternions as

$$H = \frac{1}{\sqrt{2}}\begin{bmatrix} 1 & 1 \\ 1 & -1 \end{bmatrix} = i \times \mathbf{SU(2)}\left(0, \frac{1}{\sqrt{2}}, 0, \frac{1}{\sqrt{2}}\right) = \frac{i}{\sqrt{2}}\begin{bmatrix} -i & -i \\ -i & +i \end{bmatrix}, \tag{22.13}$$

$$\Phi(\phi) = \begin{bmatrix} 1 & 0 \\ 0 & e^{i\phi} \end{bmatrix} = e^{i\phi/2} \times \mathbf{SU(2)}\left(\cos(\phi/2), 0, 0, \sin(\phi/2)\right) = e^{i\phi/2} \times \begin{bmatrix} e^{-i\phi/2} & 0 \\ 0 & e^{i\phi/2} \end{bmatrix}. \tag{22.14}$$

We now see that single qubit states themselves follow directly from the Hopf fibration of the quaternion sphere $\mathbf{S}^3$, and all possible transformations on those states are quaternion-based 2×2 unitary operators in $\mathbf{U(2)}$ that move states around the Bloch sphere. Thus everything upon which single qubit quantum computing is based is fundamentally related to quaternions.

22.5 Higher-dimensional qubit geometry

Many of the concepts involved in the construction and manipulation of a single qubit using a quaternion can be extended to qubit applications in much more complicated quantum computing circuits. In this section, we follow a diversion to examine the ways in which complex geometry is used to exploit the properties of higher-dimensional qubits beyond the quaternion-based single qubit. While not as closely connected to quaternions as the single qubit, this material explores extensions of related geometric arguments that can have valuable applications.

Quantum computation for n-qubit states has the following basic properties:

1. Orthonormal basis vectors of an n-qubit state have dimension $D = 2^n$. Think of these simply as a set of 2^n states labeled by sequential n-bit binary numbers,

$$(|00\ldots00\rangle, |00\ldots01\rangle, |00\ldots10\rangle, |00\ldots11\rangle, \ldots, |11\ldots10\rangle, |11\ldots11\rangle) \ .$$

2. $D = 2^n$ normalized complex probability amplitude coefficients describe the contribution of each basis vector consistent with conservation of probability.

3. There exists a set of probability-conserving $n \times n$ unitary matrix operators that suffice to describe all required state transformations of an n-qubit quantum circuit.

4. There exists a measurement framework that converts complex amplitudes describing a quantum state into probabilities by summing the absolute squares of the amplitudes.

We remark that there are many things that are assumed in ordinary quantum computing, such as the absence of zero-norm states for nonzero vectors, and the decomposition of any state's complex amplitudes into a pair of ordinary real numbers. One also typically assumes the existence of a Hilbert space with an orthonormal basis, allowing us to write n-qubit *pure* states in general as Hilbert space vectors with a Hermitian inner product:

$$|\Psi\rangle = \sum_{i=0}^{D-1} a_i |i\rangle \ , \qquad\qquad \langle\Psi| = \sum_{i=0}^{D-1} \bar{a}_i \langle i| \ . \tag{22.15}$$

Here the $a_i \in \mathbb{C}$ are complex probability amplitudes and $\bar{z} = \overline{x + \mathrm{i}y} = x - \mathrm{i}y$ is complex conjugation. The variables are complex vectors, $\vec{a} \in \mathbb{C}^D$, and the $\{|i\rangle\}$ is an orthonormal basis of D states obeying

$$\langle i | k \rangle = \delta_{ik} \ . \tag{22.16}$$

This means that any state $|\Phi\rangle = \sum_{i=0}^{D-1} \beta_i |i\rangle$ can be projected onto another state $|\Psi\rangle$ by writing

$$\langle\Phi|\Psi\rangle = \sum_{i=0}^{D-1} \bar{\beta}_i a_i \ , \tag{22.17}$$

thus quantifying the proximity of the two states. This is one of many properties we take for granted in standard quantum mechanics.

Details of n-qubit geometry. For n qubits, with state vector dimension $D = 2^n$, the irreducible states are encoded in a family of geometric structures known formally as the complex projective space

$\mathbf{CP}^{D-1}$. For one qubit, $D = 2$, the manifold $\mathbf{CP}^1$ is in fact the Bloch sphere $\mathbf{S}^2$. To derive this fundamental generalization of the Bloch sphere, we start with the $D = 2^n$ initially unnormalized complex coefficients of the n-qubit state basis

$$|\Psi\rangle = \sum_{i=0}^{D-1} a_i |i\rangle . \tag{22.18}$$

Following the first step of the one-qubit procedure, we enforce the conservation of probability, requiring that the norm of the complex vector $\mathbf{a}$ be normalized to unity:

$$\langle\Psi|\Psi\rangle = \|\mathbf{a}\|^2 = \sum_{i=0}^{D-1} |a_i|^2 = 1 . \tag{22.19}$$

Considering all the real and imaginary parts of $\mathbf{a}$ together, with $a_k = x_k + iy_k$, they define a real space $\mathbb{R}^{2D}$ of dimension $2 \times 2^n = 2^{n+1}$, and therefore the *real* components of Eq. (22.19) describe a very large sphere, $\mathbf{S}^{2D-1} = \mathbf{S}^{2^{n+1}-1}$, in explicit form as

$$[n\text{-qubit sphere of dimension } (2^{n+1} - 1)] \Rightarrow \sum_{i=0}^{D-1} |a_i|^2 = \sum_{i=0}^{D-1} \left(x_i{}^2 + y_i{}^2\right) = 1 . \tag{22.20}$$

This sphere in turn is ambiguous up to the usual overall phase, inducing an $\mathbf{S}^1$ or $\mathbf{U}(1)$ symmetry action, and identifying $\mathbf{S}^{2D-1}$ as an $\mathbf{S}^1$ bundle, whose base space must have dimension one less after removing the $\mathbf{S}^1$, or $2D - 2$ real dimensions. This manifold is a complex space with complex dimensions $(2D - 2)/2 = D - 1$, and in particular can be shown to be the $(D - 1)$-complex-dimensional projective space $\mathbf{CP}^{D-1}$. This is the generalization of the Bloch sphere to the space of irreducible degrees of freedom $((D - 1)$ complex, $(2D - 2)$ real) for an n-qubit quantum state with a $D = 2^n$-dimensional basis, $\{|i\rangle \,|\, i = 0, \ldots, D - 1\}$. In mathematical notation, the process we have just performed to go from the initial unnormalized n-qubit quantum state to its irreducible topological manifold $\mathbf{CP}^{D-1}$ by removing a circle from the normalization sphere looks like

$$\mathbf{S}^1 \hookrightarrow \mathbf{S}^{2D-1} \to \mathbf{CP}^{D-1} , \tag{22.21}$$

with $D = 2^n$ as usual. For $n = 1$, the single qubit, we have $D - 1 = 2^n - 1 = 2 - 1 = 1$, and the base space of the circle bundle is $\mathbf{CP}^1 = \mathbf{S}^2$, the usual Bloch sphere. Note that only for $n = 1$ is this actually a sphere-like geometry due to an accident of low-dimensional topology.

Explicit n-qubit Hopf map. Just as we used Eq. (22.6) to implement the single-qubit Hopf map to the irreducible Bloch sphere, there is a parallel approach to go from the abstraction of Eq. (22.21) to an explicit embedding. However, it is as challenging to exploit as it is simple to write down. We start with the $D \times D$ *density matrix*, $\rho = \left[a_i \bar{a}_j\right]$, for an n-qubit system,

$$\rho = \begin{bmatrix} |a_0|^2 & a_0\bar{a}_1 & \cdots & a_0\bar{a}_{D-1} \\ a_1\bar{a}_0 & |a_1|^2 & \cdots & a_1\bar{a}_{D-1} \\ \vdots & \vdots & \ddots & \vdots \\ a_{D-1}\bar{a}_0 & \cdots & a_{D-1}\bar{a}_{D-2} & |a_{D-1}|^2 \end{bmatrix} . \tag{22.22}$$

Note that, like Eq. (22.6), every term in Eq. (22.22) is *invariant* under the global phase transformation $a_k \rightarrow e^{i\phi} a_k$.

We can now take advantage of the complex generalization of the classical Veronese coordinate system for embedding projective geometry. Our approach is to convert the density matrix into a real unit vector $\mathbf{X}(\mathbf{a})$ describing a $(D^2 - 1)$-dimensional sphere embedded in a huge phase invariant real space of dimension D^2. If we define

$$\mathbf{X} = \left[|a_i|^2, \ldots, \sqrt{2}\, \text{Re}\, a_i \bar{a}_j, \ldots, \sqrt{2}\, \text{Im}\, a_i \bar{a}_j, \ldots \right] , \tag{22.23}$$

the norm of this construction reduces to the square of the probability normalization condition,

$$\mathbf{X} \cdot \mathbf{X} = \left(\sum_{i=0}^{D-1} |a_i|^2 \right)^2 = 1 , \tag{22.24}$$

and thus is indeed an embedded sphere. While each point of $\mathbf{X}$ in its description of $\mathbf{CP}^{D-1}$ is a unique irreducible point of the n-qubit quantum state, it is of course ridiculously redundant, with many more dimensions than are necessary. Nevertheless, this is precisely a realization of the desired n-qubit Hopf fibration Eq. (22.21).

22.6 Conclusion

Quaternions and quaternion operators in the form of 2×2 **SU(2)** matrices and their phase-multiplied extensions to **U(2)** appear in a wide variety of roles throughout quantum mechanics, but their geometry is particularly evident in the single qubit. The Bloch sphere is a critical concept at the very foundations of quantum computing through its appearance in the reduction of the degrees of freedom in a qubit to a unique two-variable parameterization on $\mathbf{S}^2$. For the single qubit case we focused on here, our study of quaternions sheds new light revealing the intimate relation of the Hopf fibration to the construction of the columns and rows of the 3D rotation matrix $R(q)$ in Eq. (22.5) that came directly and inevitably from the Hamilton quaternion. Given any such single qubit state, the phase-enhanced 2×2 matrix realization of quaternions provides us with the entire library of possible one-qubit unitary transformations. Both quaternion *geometry* and quaternion *algebra* are thus fundamental core concepts in quantum computing.

Introduction to quaternions and special relativity

We have thus far used quaternions to create powerful ways of dealing with and gaining insights into 3D Euclidean space and 3D orientation frames. However, there is much more that we can do if we think of quaternions themselves as the special real-valued case of an object that can be extended to complex variables. Here we explore the fact that complexified quaternions can take us smoothly into the domain of Einstein's theory of special relativity: that will allow us to expand our scope from the symmetries and transformations of 3D Euclidean space to those of 4D Minkowski space. Our familiar Euclidean rotations will have their cosines and sines change to hyperbolic cosines and hyperbolic sines, turning rotations into Lorentz transformations; the extra three complex degrees of freedom will be seen to parameterize the velocity direction of a Lorentz frame. As we have done previously, we will initially take advantage of the simplicity of 2D space to establish our basic mathematical tool set, and we will see that there is really nothing that unusual about relativistic transformations once we have mastered ordinary rotations.

23.1 Rotations in 2D tell us a lot about 2D relativity

Review of 2D rotations. The simplest 2D rotation, as we know well, can be written as the quaternion

$$[a, 0, 0, b] \,, \tag{23.1}$$

where $a^2 + b^2 = 1$ and, with $a = \cos(\theta/2)$ and $b = \sin(\theta/2)$, we can use Eq. (23.1) to reduce the quaternion quadratic form of a 3D rotation to an arbitrary 2D rotation matrix of the form

$$R_2(a, b) = \begin{bmatrix} a^2 - b^2 & -2ab \\ 2ab & a^2 - b^2 \end{bmatrix} = \begin{bmatrix} \cos\theta & -\sin\theta \\ \sin\theta & \cos\theta \end{bmatrix} . \tag{23.2}$$

This matrix has unit determinant, $\det(R) = (a^2 + b^2)^2 = \cos^2\theta + \sin^2\theta = 1$, and is *orthogonal*, that is, its transpose is equal to its inverse, and so it *preserves the 2D identity matrix*:

$$R_2 \cdot \begin{bmatrix} 1 & 0 \\ 0 & 1 \end{bmatrix} \cdot R_2{}^{\mathrm{t}} = \begin{bmatrix} 1 & 0 \\ 0 & 1 \end{bmatrix} . \tag{23.3}$$

The action of the 2D rotation and the invariance of the distance from the origin are illustrated in Fig. 23.1.

Visualizing More Quaternions. https://doi.org/10.1016/B978-0-32-399202-2.00037-X

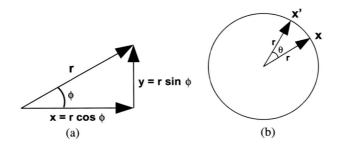

FIGURE 23.1

(a) The action of a 2D rotation by an angle ϕ on the point $\mathbf{x} = (r, 0)$ gives $\mathbf{x}' = (x = r \cos \phi, y = r \sin \phi)$. (b) Rotating a completely arbitrary point $\mathbf{x} = (x, y)$ by some angle θ gives a new point $x' = x \cos \theta - y \sin \theta$ and $y' = x \sin \theta + y \cos \theta$. Any such rotation leaves the distance between two points invariant. The radius $r = \sqrt{x^2 + y^2}$ giving the distance between the origin and any point on the circle is an obvious example.

Introduction to 2D Lorentz transformations. Eq. (23.3) seems pretty trivial, but a very small change will lead us directly to Einstein's theory of special relativity: all we do is to require our transformation to preserve a matrix in which the sign of one diagonal is changed.

This is accomplished by redefining (a, b) to be hyperbolic instead of trigonometric, that is, $a = \cosh(\xi/2)$ and $b = \sinh(\xi/2)$, and changing the form of Eq. (23.2) to

$$B_2(a, b) = \begin{bmatrix} a^2 + b^2 & +2ab \\ +2ab & a^2 + b^2 \end{bmatrix} = \begin{bmatrix} \cosh \xi & \sinh \xi \\ \sinh \xi & \cosh \xi \end{bmatrix}. \tag{23.4}$$

The (a, b) form in Eq. (23.4) is essentially the complex quaternion parameterization of a 2D Lorentz transformation, and the (cosh, sinh) form is equivalent to a standard Lorentz transformation with velocity $v = \tanh \xi$, logically analogous to the 2D (cos, sin) rotation form in Eq. (23.2). The resulting matrix B_2 still has $\det(B_2) = \cosh^2 \xi - \sinh^2 \xi = +1$ but, instead of preserving the Euclidean identity matrix, it preserves the *Lorentz metric* matrix $\eta_2 = \text{diag}(1, -1)$ as follows:

$$B_2 \cdot \eta_2 \cdot B_2 = B_2 \cdot \begin{bmatrix} 1 & 0 \\ 0 & -1 \end{bmatrix} \cdot B_2$$

$$= \begin{bmatrix} \cosh^2 \xi - \sinh^2 \xi & 0 \\ 0 & -\cosh^2 \xi + \sinh^2 \xi \end{bmatrix} = \begin{bmatrix} 1 & 0 \\ 0 & -1 \end{bmatrix} = \eta_2. \tag{23.5}$$

The action of a 2D Lorentz transform is shown in Fig. 23.2.

We note that, since B_2 is real symmetric, it is self-adjoint, and we can omit the "transpose" for the inverse that is needed on the right-hand side of Eq. (23.3) for Euclidean rotations. The significant difference is that the inverse of B_2 is *not* defined by contracting B_2 with its inverse using an identity matrix, but by contracting with the *Lorentz metric* $\eta_2 = \text{diag}(1, -1)$. The mathematical notation for rotations of the form R_2 is **SO**(2), the *special orthogonal group*, and for B_2 we write **SO**(1, 1), the 2D spacetime *Lorentz group*, denoting that the diagonal matrix η_2 of Eq. (23.5) is preserved.

We will say more about the meaning of the parameters later, but it is useful to make the explicit connection here between 2D spacetime and our parameters. First, we examine a typical 2D Lorentz

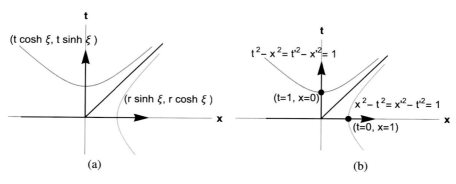

FIGURE 23.2

(a) The action of a 2D boost by a rapidity ξ (velocity $\tanh \xi$) on the point $\mathbf{p} = (t, 0)$ gives the upward hyperbola $\mathbf{p}' = (t \cosh \xi, t \sinh \xi)$, and the boost on the point $\mathbf{p} = (0, x)$ gives the rightward hyperbola $\mathbf{p}' = (x \sinh \xi, x \cosh \xi)$. (b) Boosting a completely arbitrary point $\mathbf{p} = (t, x)$ by a range of velocity parameters ξ gives new points that preserve the invariant Minkowski distances $t^2 - x^2 > 0$ for time-like points, and $x^2 - t^2 > 0$ for space-like points. The set of disjoint invariant hyperbolic curves replaces the constant-radius invariant circle in Fig. 23.1.

transform in our notation, and differentiate with respect to t and x to get infinitesimals representing a kind of tangent space:

$$t' = t \cosh \xi + x \sinh \xi , \qquad\qquad x' = t \sinh \xi + x \cosh \xi ,$$

$$\frac{dt'}{dt} = \cosh \xi , \qquad\qquad \frac{dx'}{dt} = \sinh \xi , \qquad (23.6)$$

$$v = \frac{dx'}{dt'} = \frac{\sinh \xi}{\cosh \xi} = \tanh \xi .$$

Here we have set the velocity of light to unity, so v is the velocity of the Lorentz frame in units of a fraction of the speed of light "c." With full units, the velocity parameter of Eq. (23.6) is often written as $\beta = v/c$. The rest of the conventional parameters follow from expressing our hyperbolic trigonometric functions in terms of $v = \tanh \xi$. Defining

$$\gamma = \frac{1}{\sqrt{1 - v^2}} = \frac{1}{\sqrt{1 - \beta^2}} , \qquad (23.7)$$

we have

$$\gamma = \cosh \xi \quad = \frac{\cosh \xi}{\sqrt{\cosh \xi^2 - \sinh \xi^2}} = \frac{1}{\sqrt{1 - \tanh \xi^2}} = \frac{1}{\sqrt{1 - v^2}} , \qquad (23.8)$$

$$\beta \gamma = \sinh \xi \quad = \tanh \xi \cosh \xi = \frac{v}{\sqrt{1 - v^2}} . \qquad (23.9)$$

Finally, we recall that the hyperbolic "angle" parameter ξ

$$\xi = \tanh^{-1}(v) \qquad (23.10)$$

is referred to as the *rapidity* of the Lorentz transformation, since at $\xi = 0$ the velocity vanishes, the sign of ξ tells the direction of the velocity, and as $\xi \to \infty$, $v \to 1$, that is, v approaches the velocity of light.

There are "a few more details" along the way, but that is basically all there is to special relativity. We can visualize what is happening by replacing the rotation action of $R_2(\theta)$ on a Euclidean point (x, y) by acting with $B_2(\xi)$ on a 2D spacetime point (t, x); this comparison is the objective of Fig. 23.2. One important distinction is that a pure "time" point, say $(1, 0)$, is always stuck to be mostly time, and cannot be connected to a "space" point such as $(0, 1)$ by applying a $B_2(\xi)$ matrix, and vice versa. In Fig. 23.3, we look at this another way, interpolating with $-1 \leqslant \xi \leqslant +1$ in tandem with a pair of orthogonal directions, one starting at $(t, 0) = (1, 0)$ (time-like), and the other at $(0, x) = (0, 1)$ (space-like). We see in the figure that instead of having the point simply move in a circle at fixed unit radius from the origin, for B_2, the time-like point $(1, 0)$ moves in the vertical direction as it gets farther from the origin in the horizontal dimension, tracing a vertical hyperbola; the space-like point $(0, 1)$ moves to the right as it gets farther from the origin in the vertical direction, tracing one branch of a completely distinct horizontal hyperbola. What is special about this plot is that the starting pair of points $\mathbf{t} = (1, 0)$, $\mathbf{x} = (0, 1)$ are orthogonal, $\mathbf{t} \cdot \eta_2 \cdot \mathbf{x} = 0$, and, under the action of B_2, boosted line pairs mirrored across the $x = t$ diagonal *remain* perpendicular, $\mathbf{t}' \cdot \eta_2 \cdot \mathbf{x}' = 0$, even though the *plot* using Cartesian coordinates appears very unlike our more familiar pairs of Euclidean perpendicular lines. Another way to look at this is to observe that any dot product of spacetime vectors using η_2 is *Lorentz-invariant* and cannot change under the action of $B_2(\xi)$, so obviously $\mathbf{t} \cdot \eta_2 \cdot \mathbf{x} = 0$ cannot change.

What is happening with B_2 in Fig. 23.3? Let us go back to $\eta_2 = \mathrm{diag}\,(1, -1)$, where we are writing *time* as having the positive metric sign, the first component of the vector (t, x), but, by ancient convention, we will plot the time on the *vertical* axis, and the horizontal (space-like) x-axis is plotted on the *horizontal* axis even though it appears in the second position, which in Euclidean coordinates would be the y- or vertical axis. Confusing at first, but one almost always sees these coordinates plotted this way, so we need to stay consistent with other images the reader might see on this subject.

Back to the point, what is happening is that the moving points have constant squared hyperbolic differences, which we call *Minkowski differences*, in contrast to constant *Euclidean differences*. That is actually why the hyperbolic cosines and sines that we are using are named as they are. So, if $\mathbf{t}_0 = (1, 0)$, so $\mathbf{t}_0 \cdot \eta_2 \cdot \mathbf{t}_0 = +1$, and $\mathbf{t} = B_2 \cdot \mathbf{t}_0 = (\cosh \xi, \sinh \xi)$, then

$$\mathbf{t} \cdot \eta_2 \cdot \mathbf{t}^{\mathbf{t}} = (\cosh \xi)^2 - (\sinh \xi)^2 = 1\,,$$

and the Minkowski length of the vector $\mathbf{t}$ is independent of ξ. Likewise, if $\mathbf{x}_0 = (0, 1)$, so $\mathbf{x}_0 \cdot \eta_2 \cdot \mathbf{x}_0 = -1$, and $\mathbf{x} = B_2 \cdot \mathbf{x}_0 = (\sinh \xi, \cosh \xi)$, then

$$\mathbf{x} \cdot \eta_2 \cdot \mathbf{x}^{\mathbf{t}} = (\sinh \xi)^2 - (\cosh \xi)^2 = -1\,,$$

so the length of $\mathbf{x}$ is also independent of ξ. Think of these features as a generalization of the Euclidean constant unit radius of $(x, y) = (\cos \theta, \sin \theta)$. Finally, if we compute the *Minkowski differences* between two general spacetime points $\mathbf{p}_1 = [t_1, x_1]$ and $\mathbf{p}_2 = [t_2, x_2]$ before and after transformation by $B_2(\xi)$,

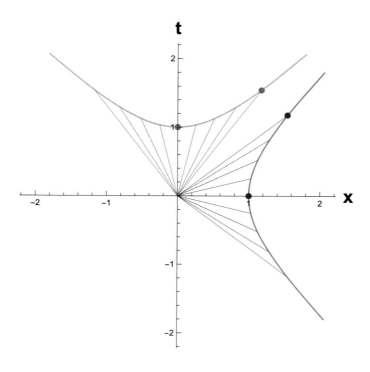

FIGURE 23.3

We plot 2D spacetime points $[t, x]$ with the time coordinate t on the vertical axis, and the space coordinate x on the horizontal axis. The motion of a time-like point $t = [1, 0]$ is shown in red under the action of the relativistic boost transformation $B_2(\xi)$ in Eq. (23.4) for $-1 \leqslant \xi \leqslant +1$. The path of the space-like point $x = [1, 0]$ under the same 2D Lorentz transformation is shown in blue. The pairs of vectors ending in mirror image points relative to the $t = x$ diagonal line are in fact Minkowski space *orthogonal pairs*.

$$(d_{12})^2 = (\mathbf{p}_1 - \mathbf{p}_2) \cdot \begin{bmatrix} 1 & 0 \\ 0 & -1 \end{bmatrix} \cdot (\mathbf{p}_1 - \mathbf{p}_2)$$

$$= (t_1 - t_2)^2 - (x_1 - x_2)^2,$$

$$(d'_{12})^2 = (B_2(\xi) \cdot \mathbf{p}_1 - B_2(\xi) \cdot \mathbf{p}_2) \cdot \begin{bmatrix} 1 & 0 \\ 0 & -1 \end{bmatrix} \cdot (B_2(\xi) \cdot \mathbf{p}_1 - B_2(\xi) \cdot \mathbf{p}_2)$$

$$= (\mathbf{p}_1 - \mathbf{p}_2)_j \, B_{2(ij)} \begin{bmatrix} 1 & 0 \\ 0 & -1 \end{bmatrix}_{ik} B_{2(kl)} \, (\mathbf{p}_1 - \mathbf{p}_2)_l$$

$$= (\mathbf{p}_1 - \mathbf{p}_2) \cdot \begin{bmatrix} 1 & 0 \\ 0 & -1 \end{bmatrix} \cdot (\mathbf{p}_1 - \mathbf{p}_2)$$

$$= (t_1 - t_2)^2 - (x_1 - x_2)^2,$$

we see that $d_{12} = d'_{12}$, and the Minkowski distances are unchanged. This has a bizarre consequence when we plot a sequence of Lorentz-transformed points in Fig. 23.3 as a function of ξ: since we start with two spacetime points $\mathbf{p}_1 = [1, 0]$ and $\mathbf{p}_2 = [0, 1]$ that are orthogonal, $\mathbf{p}_1 \cdot \eta_2 \cdot \mathbf{p}_2 = 0$, at $\xi = 0$, and Lorentz transform them in tandem, their vectors from the origin become closer and closer together along the upper right diagonal, *but they remain orthogonal throughout*. The curves swept out by the $B_2(\xi)$ transformation are plotted in Fig. 23.3 as a function of ξ, with the mirrored sets of curve points being *orthogonal pairs*. All this may *look* strange, but it is no different conceptually from the rotation invariance of the radius of a circle or the preservation of a right angle under rigid rotation in ordinary Euclidean space. Furthermore, it describes a fundamental property of the physical world in which we live!

23.2 Relativity with one time and two space dimensions

While a world with one time and one space dimension can tell us a lot of fundamental properties of Lorentz transformations and special relativity, when we add another spatial dimension, so we have one time and *two* space dimensions, even more interesting things begin to happen, and these are worth a brief look before we go on to full $3 + 1$ spacetime in Chapter 24. To begin our intuitive framework for 2 space + 1 time, we note that the corresponding toy world has these properties:

- *Objects* are polygons (at one time).
- Polygon *vertices* sweep out spacetime lines that are dominated by time.
- A whole *spacetime object* is a time-like tube for which the paths of its points remain inside the cone $t^2 > x^2 + y^2$.
- *Cameras* can be thought of as tubes that are intersected by *cones* of object-emitted light rays having paths converging on a vertex that is the spacetime path of the focal point.

$2 + 1$ **Spacetime boost matrices.** To make the transition from $1 + 1$, one space and one time dimension, to $2 + 1$, two space dimensions plus one time dimension, we need first to see what happens to our now-familiar $1 + 1$ dimension boost. Our fundamental transformation from a point (t, x) to a different velocity frame (t', x') is implemented by the hyperbolic variant $B_2(\xi)$ of the 2D rotation,

$$B_2(\xi) = \begin{bmatrix} \cosh \xi & \sinh \xi \\ \sinh \xi & \cosh \xi \end{bmatrix},$$

where ξ is the rapidity and the 1D velocity parameter is $v = \sinh \xi / \cosh \xi = \tanh \xi$.

Putting off a derivation until later, when we derive the full $3 + 1$, 3 space + 1 time, Lorentz transformations, we declare that the full formula for an $SO(2, 1)$ boost to a new frame with 2D velocity $\mathbf{v} = [v_x, v_y]$ takes the form, with matrix order (t, x, y),

$$B_3(\mathbf{v}) = \begin{bmatrix} \cosh \xi & \hat{v}_x \sinh \xi & \hat{v}_y \sinh \xi \\ \hat{v}_x \sinh \xi & 1 + \hat{v}_x^2(\cosh \xi - 1) & \hat{v}_x \hat{v}_y(\cosh \xi - 1) \\ \hat{v}_y \sinh \xi & \hat{v}_x \hat{v}_y(\cosh \xi - 1) & 1 + \hat{v}_y^2(\cosh \xi - 1) \end{bmatrix}. \tag{23.11}$$

Here $\det B_3 = 1$, and we can verify that this has the appropriate $1 + 1$ limits when $\hat{v}_x = 0$ or $\hat{v}_y = 0$. The "hat" notation means, as usual, that $\hat{\mathbf{v}} \cdot \hat{\mathbf{v}} = \hat{v}_x \hat{v}_x + \hat{v}_y \hat{v}_y = 1$ and we define the 2D velocity as $\mathbf{v} = \hat{\mathbf{v}} \tanh \xi$, with the units chosen so the velocity of light is 1. Given the definition that in any dimension the magnitude of the velocity is $\|\mathbf{v}\| = v = \tanh\xi$, we see that this is the obvious way to transition from $1+1$ to $2+1$ spacetime. We also note that Eq. (23.5) extends, as it must, to leave invariant $\eta_3 = \mathrm{diag}(1, -1, -1)$, the $2+1$ version of the Minkowski metric:

$$
B_3 \cdot \eta_3 \cdot B_3 = B_3 \cdot \begin{bmatrix} 1 & 0 & 0 \\ 0 & -1 & 0 \\ 0 & 0 & -1 \end{bmatrix} \cdot B_3
$$

$$
= \begin{bmatrix} \cosh^2\xi - \sinh^2\xi & 0 & 0 \\ 0 & -\cosh^2\xi + \sinh^2\xi & 0 \\ 0 & 0 & -\cosh^2\xi + \sinh^2\xi \end{bmatrix} = \begin{bmatrix} 1 & 0 & 0 \\ 0 & -1 & 0 \\ 0 & 0 & -1 \end{bmatrix} = \eta_3 .
$$

(23.12)

Note that Eq. (23.11) for B_3 will give very strange results if one forgets to insert the minus signs of η_3 when projecting boosted vectors onto each other to get the Lorentz version of a Euclidean dot product!

23.3 Sequential noncommuting transformations

We next explore the implications of noncommutativity in two space and one time dimension, and their interesting parallels to what we already know about the behaviors following from the noncommutativity of ordinary 3D rotations.

Euclidean 3D transformations. We recall that an important feature of 3D rotations is that they are *order-dependent*. We learned in Chapter 5 that, as a consequence, performing a sequence of small-angle rotations such as $R(\epsilon, \hat{\mathbf{x}}) \cdot R(\epsilon, \hat{\mathbf{y}})$ differs from the opposite-order action $R(\epsilon, \hat{\mathbf{y}}) \cdot R(\epsilon, \hat{\mathbf{x}})$, and we can *quantify* that difference. As illustrated in Fig. 23.4, this clockwise circuit, starting with a small positive fixed-$\hat{\mathbf{y}}$-axis rotation, generates an infinitesimal positive $\hat{\mathbf{z}}$-*axis rotation*, acting to rotate vectors in the xy-plane:

$$
R(\epsilon, -\hat{\mathbf{x}}) \cdot R(\epsilon, -\hat{\mathbf{y}}) \cdot R(\epsilon, \hat{\mathbf{x}}) \cdot R(\epsilon, \hat{\mathbf{y}}) = \begin{bmatrix} 1 & 0 & 0 \\ 0 & 1 & 0 \\ 0 & 0 & 1 \end{bmatrix} + \epsilon^2 \begin{bmatrix} 0 & -1 & 0 \\ 1 & 0 & 0 \\ 0 & 0 & 0 \end{bmatrix} + O(\epsilon^3) \approx R(\epsilon, \hat{\mathbf{z}}) . \quad (23.13)
$$

Thus in 3D Euclidean space, sequences of rotations *counter-rotate*: This reversed-direction rotation relative to the action direction for infinitesimal rotation sequences is the "rolling ball" effect (Hanson, 1992). Note, for comparison below with the two-spatial-dimension boost, that this rotation is in the xy-plane, and does not actually act in the third dimension. The rolling ball phenomenon follows from the noncommutativity of 3D rotations, and is true for any set of actions in a plane.

$2+1$ **Spacetime transformations.** If we now perform an analogous sequence of 2 space + 1 time *boosts* in the xy-plane, we find an astonishing result: as illustrated in Fig. 23.5, a clockwise set of infinitesimal boosts *also* does not commute, but the resulting final action is not another pure boost, but

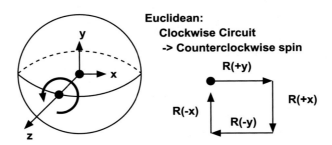

FIGURE 23.4

This is the *rolling ball* effect in 3D rotations. The effect can be observed by moving the palm of one's hand, resting on the top of a ball, in a small circular motion.

a *clockwise rotation* in the xy-plane. The action is the following (recall our index order is (t, x, y)):

$$B(\epsilon, -\hat{\mathbf{x}}) \cdot B(\epsilon, -\hat{\mathbf{y}}) \cdot B(\epsilon, \hat{\mathbf{x}}) \cdot B(\epsilon, \hat{\mathbf{y}}) = \begin{bmatrix} 1 & 0 & 0 \\ 0 & 1 & 0 \\ 0 & 0 & 1 \end{bmatrix} + \epsilon^2 \begin{bmatrix} 0 & 0 & 0 \\ 0 & 0 & 1 \\ 0 & -1 & 0 \end{bmatrix} + O(\epsilon^3) . \tag{23.14}$$

The first term is an infinitesimal clockwise rotation in the xy-plane, in the opposite direction from the corresponding Euclidean rotation sequence!

That is, the rotation is in the *same* direction as the ordered path of the boost actions, not the *opposite* of the path as we in saw in Fig. 23.4 for 3D Euclidean rotations. Since in 3D, we can restrict ourselves to exactly this same path of boosts in the xy-plane subspace, we can alternatively look at this phenomenon as corresponding to a clockwise rotation (right-handed about the negative $\hat{\mathbf{z}}$-axis) in full 3D space + time. This observed rotation following from a sequence of noncollinear boosts is known in 3D relativity theory as the *Thomas precession*.

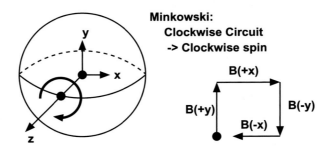

FIGURE 23.5

This is the *relativistic rolling ball* effect, showing the origin of the *Thomas precession* for 3D circular boost sequences in the xy-plane.

The Thomas precession is the *exact analog* of the Euclidean 3D "rolling ball" effect. Its physical significance is that it effectively *modifies the magnetic coupling* of atomic electrons in accelerated circular motion by causing an angular velocity

$$\omega = -(\cosh\xi - 1)\frac{v \times \dot{v}}{v^2} \approx -\frac{1}{2}v \times \dot{v}$$

to be applied to the rest frame of an orbiting electron. (One can check that $(\cosh\xi - 1) \approx \frac{1}{2}\xi^2$, while by definition $\|\mathbf{v}\|^2 = (\hat{\mathbf{v}} \cdot \hat{\mathbf{v}})(\tanh\xi)^2 \approx \xi^2$, giving the indicated approximate formula.)

23.4 Quaternion framework for 2 + 1 relativity

We recall that every 3D rotation $R(\theta, \hat{\mathbf{n}})$ has a corresponding quaternion quadratic form $R(q)$,

$$\begin{bmatrix} q_0^2 + q_1^2 - q_2^2 - q_3^2 & 2q_1q_2 - 2q_0q_3 & 2q_1q_3 + 2q_0q_2 \\ 2q_1q_2 + 2q_0q_3 & q_0^2 - q_1^2 + q_2^2 - q_3^2 & 2q_2q_3 - 2q_0q_1 \\ 2q_1q_3 - 2q_0q_2 & 2q_2q_3 + 2q_0q_1 & q_0^2 - q_1^2 - q_2^2 + q_3^2 \end{bmatrix}.$$

Choosing to parameterize the unit quaternion, with $q \cdot q = 1$, as $q = (\cos\frac{\theta}{2}, \hat{\mathbf{n}}\sin\frac{\theta}{2})$, with $\hat{\mathbf{n}}$ a unit 3-vector, $\hat{\mathbf{n}} \cdot \hat{\mathbf{n}} = 1$, then $R(q(\theta, \hat{\mathbf{n}})) = R(\theta, \hat{\mathbf{n}})$ is the usual 3D rotation by θ in the plane perpendicular to $\hat{\mathbf{n}}$.

In 2 space + 1 time, which also requires a 3×3 matrix, we can construct exactly the same type of quadratic form for the *boost*. We start by postulating a close analog to $q(\theta, \hat{\mathbf{n}})$ that is 3D instead of 4D, but contains the boost parameters in similar places, taking the form

$$\mathbf{h} = (h_0, h_x, h_y) = (\cosh(\xi/2), \hat{\mathbf{v}}\sinh(\xi/2)), \tag{23.15}$$

with $v = \sinh\xi/\cosh\xi$ and $\|\hat{\mathbf{v}}\| = 1$. By inspection of Eq. (23.11), we can identify the appropriate parts of the quaternion-like object $\mathbf{h}$, and show that $B_3(\mathbf{v})$ can indeed be written in the form

$$B_3(\mathbf{h}) = \begin{bmatrix} h_0^2 + h_x^2 + h_y^2 & 2h_0h_x & 2h_0h_y \\ 2h_0h_x & h_0^2 + h_x^2 - h_y^2 & 2h_xh_y \\ 2h_0h_y & 2h_xh_y & h_0^2 + h_y^2 - h_x^2 \end{bmatrix}. \tag{23.16}$$

The required relativistic constraint on $B_3(\mathbf{h})$, namely $B_3 \cdot \eta_3 \cdot B_3 = \eta_3$, is obeyed *provided* $\mathbf{h}$ has Minkowski space unit length in parallel with $q \cdot q = 1$ for Euclidean rotations:

$$\mathbf{h} \cdot \eta_3 \cdot \mathbf{h} = h_0^2 - h_x^2 - h_y^2 = \cosh^2\xi - \sinh^2\xi = 1.$$

Note that the usual unit-determinant requirement we applied to 3D rotations, with $\det[R(q)] = (q \cdot q)^3 = 1$, is satisfied, now taking on the form $\det[B_3(\mathbf{h})] = (\cosh^2\xi - \sinh^2\xi)^3 \equiv 1$.

Eq. (23.16) is exactly the quaternion form of the $\mathbf{SO}(2, 1)$ Lorentz transformation, and we will see below, in our treatment of 3D space + 1 time and its $\mathbf{SO}(3, 1)$ Lorentz transformation, that all this basically follows from adding extra complex degrees of freedom to the standard quaternion. In fact, we can see that if we attempt to form a quaternion product of two distinct copies of our three-component boost parameters, with complex spatial parts to convert a rotation to a boost using $\mathbf{h} \rightarrow (h_0, ih_x, ih_y)$, we find this formula:

$$\mathbf{g} \star \mathbf{h} = \left(g_0 h_0 + g_x h_x + g_y h_y, \; ig_0 h_x + ig_x h_0, \; ig_0 h_y + ig_y h_0, \; -(g_x h_y - g_y h_x)\right) . \tag{23.17}$$

The attempted quaternion product *departs* from the three-component subspace of Eq. (23.15) and introduces a fourth component; that last component in fact exhibits a negative sign multiplying what would be a rotation about the z component of the cross product of the spatial parts of $\mathbf{g}$ and $\mathbf{h}$. This is what we should in fact expect when we try to make a quaternion algebra out of a pure boost.

Remark: Because of the Thomas precession, even though $\mathbf{h} = (\cosh(\xi/2), \hat{\mathbf{v}} \sinh(\xi/2))$ generates $B(\mathbf{v})$, the boosts alone are *incomplete* and do not in themselves form a group. The full group of $\mathbf{SO}(2, 1)$ includes the missing third component, rotations in the xy-plane that emerge from sequences of boosts.

23.5 Features of light in lower dimensions of spacetime

It is challenging to draw informative graphic illustrations of many full $3 + 1$ spacetime phenomena, and we will not get too distracted from our main quaternion-related topics by attempting to study all of special relativity as well. However, restricting ourselves to $1 + 1$ and $2 + 1$ subspaces of physical spacetime allows us to see a few interesting things that can build a little bit of intuition. Here we will look at a few examples of how Lorentz-transforming a light ray can *change its direction*. First we look at a simple $\mathbf{SO}(1, 1)$ spacetime transformation that can also be looked at as an $\mathbf{SO}(2, 1)$ transformation in the $y = 0$ subspace:

$$t' = x \sinh \xi + t \cosh \xi ,$$
$$x' = x \cosh \xi + t \sinh \xi .$$

We note that even if $x < 0$, we can have $x' > 0$ if $t \sinh \xi > x \cosh \xi$. Now let θ parameterize an isotropic distribution of *light-like vectors* with $(t, x, y) = (1, \cos\theta, \sin\theta)$, so $t^2 - x^2 - y^2 = 0$ for any θ. Then a boost with $\hat{\mathbf{v}}$ in the $\hat{\mathbf{x}}$ direction gives us

$$t' = \cos\theta \sinh \xi + \cosh \xi ,$$
$$x' = \cos\theta \cosh \xi + \sinh \xi ,$$
$$y' = \sin\theta .$$

The time slice t sits in the *observer frame*, so the observed geometry is

$$\tan\theta' = \frac{y'}{x'} = \frac{\sin\theta}{\cosh \xi (\cos\theta + \tanh \xi)} = \frac{\sin\theta}{\cosh \xi (\cos\theta + v)} ,$$

where as usual $v = \tanh \xi$. Thus our isometric distribution

$$(t, x, y) = (1, \cos\theta, \sin\theta)$$

deforms using formulas like $\cos\theta = 1/\sqrt{1 + \tan^2\theta}$ and so we find

$$\sin\theta' = \frac{\sin\theta}{(1 + v\cos\theta)\cosh\xi}, \qquad \cos\theta' = \frac{v + \cos\theta}{1 + v\cos\theta},$$

$$\tan\theta' = \frac{\sin\theta}{(v + \cos\theta)\cosh\xi}.$$

(23.18)

For the purposes of seeing the qualitative features of this transformation on light-like vectors, $\tan\theta'$ probably has the clearest properties, as it is dominated by the $1/\cosh\xi = \sqrt{1 - v^2}$ term, and thus *shrinks* dramatically as $v \to 1$.

Observations on relativistic light distortion.

- $\tan\theta' \propto 1/\cosh\xi = \sqrt{1 - v^2}$.
- So, as $v = \sinh/\cosh \to 1$ ...
- ... the *aberration of light* (resembling a *searchlight*) swings all the rays to the *forward direction!*

Visualizing aberration: light cones. Looking down on rightward boosted spacetime cones representing symmetric light ray distributions:

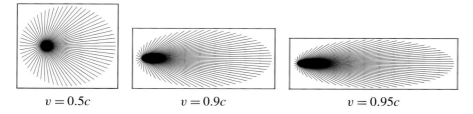

$v = 0.5c$ $v = 0.9c$ $v = 0.95c$

Visualizing aberration: circular distributions. Now we look down on rightward boosted 2D symmetric light ray distributions in a local circle. The density of rays moves intensely forward, but we see only distortion, never crossing rays that would change the occlusion geometry:

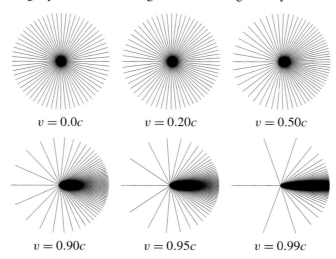

$v = 0.0c$ $v = 0.20c$ $v = 0.50c$

$v = 0.90c$ $v = 0.95c$ $v = 0.99c$

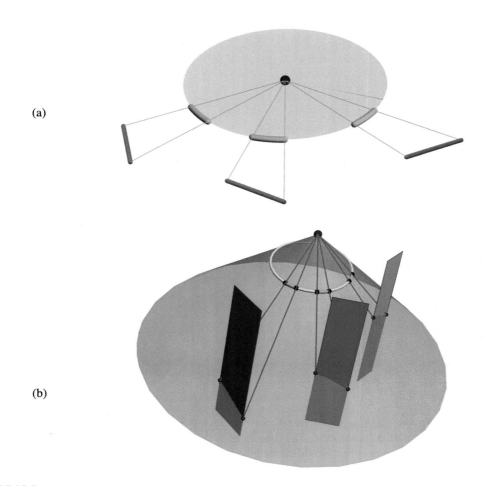

FIGURE 23.6

This is a simple model of imaging without and with finite speed of light. (a) A nonrelativistic camera geometry with infinite speed of light. Light passes from the linear objects in the 2D world, focused on the central focal point, and forms an image on what is in effect a circular camera, looking at the entire world. (b) A 2 + 1 space-time world, visualizing the finite-time paths of light rays from different scene objects as they arrive at the rest frame camera. The spacetime paths of a line-segment 2D object travel up the light cone to the focal point, and create images as they hit the circular camera that has a view of the entire 2D world. The light rays arriving at one time point in the camera film have departed from their world objects at many distinct time points in history.

23.6 Some simple 2 + 1 relativistic camera effects

The way in which the recording of a photograph must be represented in relativistic physics is a challenge. For example, perpendicular lines as drawn in various velocity frames do not appear perpendicular, and simultaneous events in one frame are not simultaneous in another. So when we try to

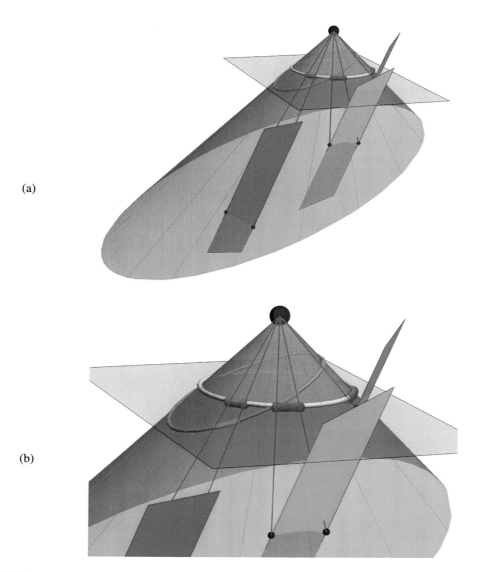

(a)

(b)

FIGURE 23.7

The boosted camera model. (a) A 2 + 1 spacetime world that has been boosted to a different Lorentz frame, which squeezes the formerly isotropic light rays in the boost direction, and makes the previous circular camera "lens" asynchronous in time, so we replace it by a new camera, in yellow, that is at a constant moment of time in the new frame. We leave the old camera to show how the rays from the objects intersect the images at different time points. (b) A closeup view of the original rest frame camera and the boosted frame camera.

see how a camera works in a relativistic setting, we need to remember that, because the speed of light is finite but also a constant, the camera effectively looks *backwards in time* to acquire light rays that may have been emitted at very different times, and yet arrive at the film surface simultaneously. Finally, the *light cone*, defined as the surface $t^2 = x^2 + y^2$, has its shape preserved under Lorentz transformations: the path of simultaneous image construction is different for the light-cone camera ray paths in different frames, but the shape of the light cone is always the same.

We now draw some schematic images, showing a simplistic scene with three objects, at different distances, emitting light rays that can be detected on a piece of circular film with focal point at the center. Fig. 23.6(a) shows what such a camera's detection process looks like in the absence of relativity, with light rays passing instantly, with no delay, between the scene and our circular camera film. In Fig. 23.6(b), with a camera and scene at rest, the 1D image pixels we *detect at the same time* in the camera, depicted as a yellow circle, can be intuitively understood as light rays from the night sky, looking back into the early Universe: the light rays hitting the camera at one moment in its own frame could be emitted at any time, and we would have to do extensive analysis to actually discover the history of the origin of each distinct light ray. The final sample in Fig. 23.7(a,b) shows our simple scene boosted to a moving frame at about 1/2 the velocity of light. The light cone containing the camera remains a cone no matter what, but the parts of the scene objects that are seen simultaneously differ depending on the frame. Since the camera film itself loses simultaneity under the boost, we establish a new simultaneously recording camera in the fixed-time yellow line; again, since we use a circular film, this camera looks at the entire scene of simultaneously arriving light rays from diverse temporal origins. In addition to departing their scene objects at different times, all the light rays may change direction, but, since their *relative* paths vary smoothly, light rays seen by the camera never cross each other – there are no *new* occlusions that result from Lorentz transformations, and thus a picture of a scene at rest or moving at relativistic velocities can both be constructed from the exact same light rays; taking advantage of this fact is called *relativistic image-based rendering* (Weiskopf et al., 2000).

23.7 What is next?

In the next chapter, we build on the 2 space + 1 time framework we have studied as simplified model system in this chapter. In this context, we have shown how complexified quaternion-like quadratic forms rigorously correspond to boosts in 1 space + 1 time and 2 space + 1 time dimensions. We will extend these methods to the 3 space + 1 time real world of special relativity by considering the full complexification of our usual 4D real quaternions. Retaining the constraint $q \cdot q = 1$ with complex values of q gives us eight total variables, but *two* constraints, one real and one complex, leaving us with exactly the six required free parameters of the full Lorentz group $\mathbf{SO}(3, 1)$. We will see that fundamental insights come from extending the $\mathbf{SU}(2)$ three-parameter unitary matrix form of the real quaternion, and the associated quaternion algebra, to the six-parameter complex 2×2 matrices with unit determinant comprising the matrix group $\mathbf{SL}(2, \mathbf{C})$.

Complex quaternions and special relativity

24.1 Introducing 3 space and 1 time

In the previous chapter, our purpose was to get familiar with adding a time dimension and looking at the intuitive features of special relativity in lower dimensions, observing that relativistic boosts are closely related to the rotations, and that they admit quaternion-like forms analogous to those that we have already extensively explored to represent rotations.

We now transition to the real world from our lower-dimensional mathematical exercises, beginning with a summary of what we know about the real relativistic world with three spatial dimensions plus one temporal dimension:

- *Coordinate frame transformations:* The spacetime transformations of coordinate frames (excluding translations) in dimensions $3 + 1$ include a total of *six* parameters defining the group $\mathbf{SO}(3, 1)$. We will derive these from complex quaternion techniques, noting that the six relativistic parameters are closely related to the existence of six parameters in the quaternion approach to $\mathbf{SO}(4)$ rotations in four Euclidean dimensions that we saw in Chapter 20. Three of these $\mathbf{SO}(3, 1)$ parameters are our familiar three Euclidean 3D spatial *rotations*, e.g., the three Euler angles that can be parameterized by quaternion-related axis-angle variables such as $(\theta, \hat{\mathbf{n}})$. The other three are the *boosts* that mix space and time, parameterized by a 3D velocity $\mathbf{v}$, with a unit direction $\hat{\mathbf{v}}$ and rapidity ξ, $\mathbf{v} = \hat{\mathbf{v}} \tanh \xi$. We studied many significant features of the boosts already in dimensions $1 + 1$ and $2 + 1$, noting many parallels to rotations.
- *Invariance:* Every possible Lorentz transformation in the six-parameter group $\mathbf{SO}(3, 1)$ leaves invariant the Minkowski metric matrix $\eta = \text{diag}(1, -1, -1, -1)$. Lorentz-invariant quantities are constructed from contraction with η, e.g., for two spacetime 4-vectors a and b, transformed to a' and b' by an element of $\mathbf{SO}(3, 1)$, their Lorentz inner product is invariant, $a \cdot \eta \cdot b = a' \cdot \eta \cdot b'$.
- *Aberration:* The restructuring of the directions of light rays under the boost transformations has a similar basic form in all dimensions, aligned with and concentrated around the velocity $\mathbf{v}$ of the boost axis. This occurs naturally due to motions of points on the Lorentz-invariant hyperbolic curve collecting along the $x = t$ diagonal at high velocity parameters.
- *Light rays and their relation to imaging:* Imaging is based on light rays emitted along a light cone, and we have a full $\mathbf{S}^2$ "celestial sphere" of possible ray directions in the $(3 + 1)$-dimensional spacetime context; we can think of the real-life light cone at any fixed time as a growing sphere surrounding an instantaneous point light source. Since boosts only change the directional distribution of the light rays, but never cross them, all occlusions are rigorously preserved, and image-based rendering is always valid and straightforward even after applying a boost (Weiskopf et al., 2000).

Visualizing More Quaternions. https://doi.org/10.1016/B978-0-32-399202-2.00038-1

Unfortunately, $(3+1)$-dimensional imaging is much more challenging to draw and visualize than $1+1$ and $2+1$, and a realistic treatment is beyond our scope here.

- *Terrell effect:* In $3+1$ dimensions, transformations of spheres leave the overall visible shape unchanged (Penrose, 1959), and cube-like objects appear to rotate (Terrell, 1959). These are phenomena that are now well-known counterintuitive properties of the observation of relativistic objects, but went unnoticed for half a century.

- *Relation to complexified quaternions:* This will be the focus of our study of $3+1$ dimensions, and the approaches we introduce are closely related to modern physics. Quaternion methods and Minkowski space *combine forces* in a way that has been well known in physics since the work of Wigner (1939). The remarkable fact is that if we simply complexify quaternions, then the real part corresponds to our usual 3D space quaternion rotations, the three imaginary extensions produce a 3D boost to a new coordinate frame with the velocity parameter **v**, and the related algebra is isomorphic to the properties of the 2×2 complex matrix group $\mathbf{SL(2,C)}$. We will build on the extended quaternion language arising from the $\mathbf{SL(2,C)}$ group to explore the foundations of relativistic physics.

24.2 Exploring complexified quaternions

We are going to choose an amusing pedagogical route through the path from quaternions to a proper incorporation of quaternions in Einstein's theory of special relativity. In this section, we will follow a procedure invoking the simplest possible ways to get relativistic transformations in real $(3+1)$-dimensional spacetime; the method is related to a number of treatments appearing in the literature. We will see how, knowing what the answers should be, we can impose some *ad hoc* assumptions on the equations and get results that appear to derive all relativistic transformations from complex quaternions. However, it turns out that there are superfluous assumptions embedded in that process, and so we will then move forward to the $\mathbf{SL(2,C)}$ group-theoretical origins of the special relativistic transformations. Our ultimate and most appealing method effectively exploits a completely rigorous approach based on very old principles dating back to the origins of relativistic quantum field theory, e.g., in the work of Dirac (1930) and Wigner (1939).

As usual, we begin with the basic idea of a real quaternion $q = [q_0, q_1, q_2, q_3] = [q_0, \mathbf{q}]$ being a unit-length 4-vector, with $q \cdot q = 1$ or $\mathrm{Re}\,(q \star \bar{q}) = 1$, thus defining a point on the 3-sphere $\mathbf{S}^3$, and the extraction of a rotation matrix from the conjugation operation via the quaternion multiplication rule $q \star [0, \mathbf{x}] \star \bar{q}$. However, this time we will make a slight change by emphasizing the first component, replacing the "0" by a constant identified with the *time* of an event in spacetime, and examining the results more carefully. First, we confirm that

$$q \star [t, \mathbf{x}] \star \bar{q} = [t, R_3(q) \cdot \mathbf{x}]^{\mathrm{t}} = R(q) \cdot [t, \mathbf{x}]^{\mathrm{t}}, \tag{24.1}$$

so our usual quaternion rotations *do not affect the time coordinate.* Here $R(q)$ is actually the 4×4 matrix

$$R(q) = \begin{bmatrix} q \cdot q & 0 & 0 & 0 \\ 0 & q_0^2 + q_1^2 - q_2^2 - q_3^2 & 2q_1q_2 - 2q_0q_3 & 2q_1q_3 + 2q_0q_2 \\ 0 & 2q_1q_2 + 2q_0q_3 & q_0^2 - q_1^2 + q_2^2 - q_3^2 & 2q_2q_3 - 2q_0q_1 \\ 0 & 2q_1q_3 - 2q_0q_2 & 2q_2q_3 + 2q_0q_1 & q_0^2 - q_1^2 - q_2^2 + q_3^2 \end{bmatrix}. \tag{24.2}$$

So we see the neglected fact that conjugation by the usual "rotation quaternion" is not actually computing an $SO(3)$ group element, but a 4×4 matrix in the relativistic group $SO(3, 1)$, of which $SO(3)$ is the lower right 3×3 subgroup in our notation. It is ambiguous whether this is actually $SO(3, 1)$ or $SO(4)$: as the relevant matrix elements vanish, this 4×4 matrix $R(q)$ preserves *both* $\mathrm{diag}(1, 1, 1, 1)$ and $\mathrm{diag}(1, -1, -1, -1)$. In any case, we can start by recognizing that the matrix $R(q)$ in Eq. (24.2) does preserve $\eta_{\mu\nu} = \mathrm{diag}(1, -1, -1, -1)$, in parallel with what we found for $1+1$ and $2+1$ Lorentz transformations in the previous section:

$$R(q) \cdot \eta \cdot R(q)^{\mathrm{t}} = \eta = \begin{bmatrix} 1 & 0 & 0 & 0 \\ 0 & -1 & 0 & 0 \\ 0 & 0 & -1 & 0 \\ 0 & 0 & 0 & -1 \end{bmatrix} .$$

Complexification. Now we are at last ready to make a very intuitive, but ultimately dubious, extension from our usual real quaternion q to complex quaternions. The obvious thing to do is just imagine that the vector components of the complexified quaternion coefficients are of the form $\mathbf{z} = \mathbf{q} + \mathrm{i}\mathbf{h}$, set $\mathbf{q} = 0$, and replace q_0 by h_0. We note that the independent parameters are analogous to the three components in the exponential form $q = \exp(-\mathrm{i}(\hat{\mathbf{n}} \cdot \boldsymbol{\sigma})\theta/2)$, so q_0, and by extension, h_0, should technically be considered dependent variables derived from the vector part in the power series expansion. Adding this degree of freedom to the Pauli matrices effectively extends the 2×2 matrix group $SU(2)$ that we have been using for pure rotations to the related group $SL(2, C)$, the complex 2D matrices with unit determinant, leaving three complex degrees of freedom.

To observe the behavior of a pure imaginary complex quaternion, we thus make the replacement

$$q = [q_0, q_1, q_2, q_3] \ \rightarrow \ h = [h_0, \mathrm{i}h_1, \mathrm{i}h_2, \mathrm{i}h_3] . \tag{24.3}$$

We can confirm that this complexified form can be constrained to be a unit quaternion with respect to the same inner product used for q, using $\mathrm{i} = \sqrt{-1}$ to simulate the Minkowski metric $\eta_{\mu\nu}$, that is,

$$h \cdot h = h_0{}^2 - h_1{}^2 - h_2{}^2 - h_3{}^2 = 1 . \tag{24.4}$$

We can thus consider h to be a Minkowski space unit vector, a hyperbolic constant instead of a spherical constant, in order to reduce the number of degrees of freedom from four to the three physical boost velocity parameters. Then we perform the usual bilinear map, but with a very important additional assumption, which is that our use of complex conjugation for rotations, $\bar{q} = [q_0, -q_1, -q_2, -q_3]$, is extended for *complex* quaternions to *Hermitian* conjugation, which in the context of the 2×2 matrix group $SL(2, C)$ means that h is Hermitian, $h = h^{\dagger}$, that is, unchanged by the complex conjugation combined with an interchange consistent with $SL(2, C)$ transposition. We can thus propose a boost construction equation in parallel with Eq. (24.1) as

$$h \star [t, \mathbf{x}] \star h^{\dagger} = h \star [t, \mathbf{x}] \star h = B_a(h) \cdot [t, \mathbf{x}]^{\mathrm{t}} , \tag{24.5}$$

where $B_a(h)$ (think of "a" as "attempt") is the following 4×4 *complex* matrix:

$$B_a(h) =$$

$$
\begin{bmatrix}
h_0^2 + h_1^2 + h_2^2 + h_3^2 & -2ih_0h_1 & -2ih_0h_2 & -2ih_0h_3 \\
2ih_0h_1 & h_0^2 + h_1^2 - h_2^2 - h_3^2 & 2h_1h_2 & 2h_1h_3 \\
2ih_0h_2 & 2h_1h_2 & h_0^2 - h_1^2 + h_2^2 - h_3^2 & 2h_2h_3 \\
2ih_0h_3 & 2h_1h_3 & 2h_2h_3 & h_0^2 - h_1^2 - h_2^2 + h_3^2
\end{bmatrix}.
$$

$$(24.6)$$

We see immediately that the matrix $B_a(h)$ has some problems due to the appearance of $i = \sqrt{-1}$ compared to the boost matrices we used in our $(1 + 1)$- and $(2 + 1)$-dimensional spacetimes: the elements of the matrix $B(h)$ should in principle be *real*. However, the matrix in Eq. (24.6) is in fact Hermitian and does have the desirable property,

$$B_a(h) \cdot \eta \cdot B_a(h)^\dagger = B_a(h) \cdot \eta \cdot B_a(h) = \eta \,,$$

guaranteeing that the length of spacetime vectors is preserved under transformation by $B_a(h)$. Thus we are on the right track: simple complexification of the raw quaternion has produced a candidate for the 3D boost of $\mathbf{SO}(3, 1)$ that has basically the right global structure and the right numbers, but apparently with troublesome coefficients.

The complexified spacetime fix. At this point, it is tempting to look at $h = [h_0, ih_1, ih_2, ih_3]$ and our candidate spacetime operand $X = [t, x, y, z]$ in Eq. (24.6) and wonder why we might not redefine X in the same form as h: if we took $X = [t, ix, iy, iz]$, then we would have $X \cdot X = t^2 - \mathbf{x} \cdot \mathbf{x} = [t, x, y, z] \cdot \eta \cdot [t, x, y, z]^t$, and our Euclidean vector length is transformed to a Minkowski vector length. This method is used in a number of sources such as (Berry and Visser, 2021; Greiter and Schuricht, 2003) that must distinguish $i = \sqrt{-1}$ from the Hamilton $(\mathbf{I}, \mathbf{J}, \mathbf{K})$ to be consistent. We shall soon see that the resulting mixtures of notation will be unnecessary when classic quantum field theory methods exploiting $\mathbf{SL}(2, \mathbf{C})$ group theory are used instead.

However, let us proceed to explore for ourselves the attractive *ad hoc* method adapted from Berry and Visser (2021) to fix up the issues in Eq. (24.6). Changing X to look like the complexified quaternion h, with an intrinsic Minkowski length, we find

$$h \star [t, i\mathbf{x}] \star h^\dagger = h \star [t, i\mathbf{x}] \star h = B_b(h) \cdot [t, \mathbf{x}]^t \,,$$

$$(24.7)$$

where

$$B_b(h) =$$

$$
\begin{bmatrix}
h_0^2 + h_1^2 + h_2^2 + h_3^2 & 2h_0h_1 & 2h_0h_2 & 2h_0h_3 \\
2ih_0h_1 & i(h_0^2 + h_1^2 - h_2^2 - h_3^2) & 2ih_1h_2 & 2ih_1h_3 \\
2ih_0h_2 & 2ih_1h_2 & i(h_0^2 - h_1^2 + h_2^2 - h_3^2) & 2ih_2h_3 \\
2ih_0h_3 & 2ih_1h_3 & 2ih_2h_3 & i(h_0^2 - h_1^2 - h_2^2 + h_3^2)
\end{bmatrix}.
$$

$$(24.8)$$

(Think of "b" as "better.") This also satisfies our basic requirement, preserving the Lorentz metric, if we are careful to use the Hermitian conjugate, since this is no longer invariant under Hermitian conjugation,

$$B_b(h) \cdot \eta \cdot B_b(h)^\dagger = \eta \,.$$

So this form also preserves the length of spacetime vectors under conjugation by $B_b(h)$ and $B_b(h)^\dagger$. At the vector level, this does basically work, since if we multiply $B_b(h)$ by $[t, x, y, z]^t$, we get a new vector with factors of i multiplying each vector component of the complexified quaternion in precisely the form of our input:

$$\begin{bmatrix} t \\ \mathrm{i}\,x \\ \mathrm{i}\,y \\ \mathrm{i}\,z \end{bmatrix} \rightarrow \begin{bmatrix} t' \\ \mathrm{i}\,x' \\ \mathrm{i}\,y' \\ \mathrm{i}\,z' \end{bmatrix} = B_b(h) \cdot \begin{bmatrix} t \\ x \\ y \\ z \end{bmatrix}. \tag{24.9}$$

What about rotations? Furthermore, since multiplying the vector part of a quaternion by a constant coefficient, in this case "i," simply changes the transformed result by the same factor, we can now *repeat* our quaternion rotation calculation of Eq. (24.1),

$$q \star [t, \mathrm{i}\,\mathbf{x}] \star \bar{q} = [t', \mathrm{i}\,\mathbf{x}']^t = X', \tag{24.10}$$

where the resulting quaternion correctly has its vector part multiplied by i:

$$X' = \begin{bmatrix} t\left(q_0{}^2 + q_1{}^2 + q_2{}^2 + q_3{}^2\right) \\ \mathrm{i}\left(x\left(q_0{}^2 + q_1{}^2 - q_2{}^2 - q_3{}^2\right) + 2q_0q_2\,z - 2q_0q_3\,y + 2q_1q_2\,y + 2q_1q_3\,z\right) \\ \mathrm{i}\left(y\left(q_0{}^2 - q_1{}^2 + q_2{}^2 - q_3{}^2\right) - 2q_0q_1\,z + 2q_0q_3\,x + 2q_1q_2\,x + 2q_2q_3\,z\right) \\ \mathrm{i}\left(z\left(q_0{}^2 - q_1{}^2 - q_2{}^2 + q_3{}^2\right) + 2q_0q_1\,y - 2q_0q_2\,x + 2q_1q_3\,x + 2q_2q_3\,y\right) \end{bmatrix}. \tag{24.11}$$

In addition, the transformation matrix does not care if we use $[t, \mathbf{x}]$ or $[t, \mathrm{i}\,\mathbf{x}]$; it leaves both $\eta_{\mu\nu}$ and $-\eta_{\mu\nu}$ invariant. However, using the complexified spatial component is somewhat disturbing, as we had thought that we were dealing with quaternion-based applications related only to rotations in real Euclidean 3D space. Apparently, buried in all of the machinery of the classical Hamilton quaternions, we can change our interpretation of the context of 3D rotations to have imaginary space coordinates, and merge them seamlessly into Minkowski space, Lorentz transformations, and Einstein's theory of special relativity!

24.3 Relativity from $\mathbf{SL(2,C)}$ and complexified quaternions

We now rederive the general Lorentz transformations using the properties of the matrix group $\mathbf{SL(2,C)}$, which is the set of complex matrices of determinant 1 that are precisely equivalent to complexified quaternions. However, the methods we use do not actually use the quaternions themselves to find how a point transforms under a given real or imaginary parameter in the quaternion: instead we consider the spacetime 4-vector to be embodied in a *basis for a representation* of the action of the matrix group $\mathbf{SL(2,C)}$.

Our fundamental framework is based on the 2D identity matrix I_2 and the three Pauli matrices $\sigma = \{\sigma^1, \sigma^2, \sigma^3\}$, where we use upper indices here rather than the more conventional lower indices because when we get to the boost transformation, distinguishing upper and lower indices will be essential. The standard form of the Pauli matrices is

$$\sigma^1 = \begin{bmatrix} 0 & 1 \\ 1 & 0 \end{bmatrix}, \qquad \sigma^2 = \begin{bmatrix} 0 & -i \\ i & 0 \end{bmatrix}, \qquad \sigma^3 = \begin{bmatrix} 1 & 0 \\ 0 & -1 \end{bmatrix}. \tag{24.12}$$

They obey many fundamental identities. Among those of importance, we note the following:

$$\sigma^1 \cdot \sigma^1 = \sigma^2 \cdot \sigma^2 = \sigma^3 \cdot \sigma^3 = I_2 ,$$
$$\sigma^i \cdot \sigma^j = \delta_{ij} I_2 + i \epsilon_{ijk} \sigma^k ,$$
from which one can deduce the following:
$$\sigma^i \cdot \sigma^j + \sigma^j \cdot \sigma^i = 2\delta_{ij} I_2 , \tag{24.13}$$
$$\sigma^i \cdot \sigma^j - \sigma^j \cdot \sigma^i = 2 i \epsilon_{ijk} \sigma^k ,$$
$$\sigma^i = \frac{1}{2 i} \epsilon_{ijk} \, \sigma^j \cdot \sigma^k .$$

Here we used the Kronecker delta, δ_{ij}, which is essentially the components of an identity matrix, the totally antisymmetric 3D Levi-Civita symbol ϵ_{ijk}, which is described in detail in Appendix L, and the Einstein summation convention in which repeated indices are assumed to be summed over their range.

The rotation: unitary quaternions with three real parameters. The $\mathbf{SL}(2, \mathbf{C})$ matrices correspond-ing to our usual rotation-parameterizing quaternions belong to the three-parameter subgroup $\mathbf{SU}(2)$ of $\mathbf{SL}(2, \mathbf{C})$, so the corresponding matrices are additionally restricted to be unitary, and, in this new context, we repeat our previous finding that we can write q as a 2×2 unitary *matrix* $[q]$:

$$[q] = q_0 \sigma^0 - i q_1 \sigma^1 - i q_2 \sigma^2 - i q_3 \sigma^3 = \begin{bmatrix} q_0 - i q_3 & -i q_1 - q_2 \\ -i q_1 + q_2 & q_0 + i q_3 \end{bmatrix}. \tag{24.14}$$

Remark: One often sees this equation written with plus signs, $+i$, but this is wrong: the minus signs, $-i$, are required to reproduce the Hamilton quaternion algebra in matrix form. With the plus signs, the resulting rotation is instead the inverse, $R(q) \to R^{-1}(q) = R^t(q)$, or alternatively, the *leftward acting* rotation instead of our convention that rotation matrices multiply forward, acting to the right.

The unitary matrix $[q]$ obeys its own identities due to the unit quaternion constraint,

$$\text{(vector form): } q \cdot q = 1 \;\leftrightarrow\; \text{(matrix form): } [q] \cdot [q]^\dagger = I_2 \, (q \cdot q) = I_2 ,$$
$$\det[q] = (q \cdot q) = 1 . \tag{24.15}$$

The boost: self-adjoint quaternions with three imaginary parameters. The $\mathbf{SL}(2, \mathbf{C})$ matrices corresponding to the boost, or Lorentz-transform-parameterizing quaternions, belong to the three-parameter self-adjoint sector of $\mathbf{SL}(2, \mathbf{C})$, but they do *not* form a subgroup because sequences of the self-adjoint generators do not stay self-adjoint, but spill over into rotations. So in fact the closure of the boost transformations is the *entire* $\mathbf{SO}(3, 1)$ group; the rotations that result from (noncollinear) se-quences of boosts are a famous phenomenon known as the *Thomas precession* (Thomas, 1926; Fisher, 1972), which we investigated in a simplified $\mathbf{SO}(2, 1)$ case in Chapter 23. Denoting our self-adjoint (or Hermitian) group elements of the $\mathbf{SL}(2,\mathbf{C})$ matrices by the coefficients $\{h_0, h_1, h_2, h_3\}$, we can integrate these real parameters into a Hermitian $\mathbf{SL}(2,\mathbf{C})$ matrix that we denote by $[h]$, constructed in the following way:

$$[h] = h_0\,\sigma^0 + h_1\,\sigma^1 + h_2\,\sigma^2 + h_3\,\sigma^3 = \left[\begin{array}{cc} h_0 + h_3 & h_1 - i\,h_2 \\ h_1 + i\,h_2 & h_0 - h_3 \end{array}\right]. \tag{24.16}$$

In our conventional quaternion form, this would be denoted using a purely imaginary vector part as $q(h) = [h_0, i\,h_1, i\,h_2, i\,h_3]$, and this would be the expression inserted into our quaternion multiplication rule Eq. (2.4) to get results exactly equivalent to $\mathbf{SL}(2,\mathbf{C})$ matrix multiplication by $[h]$ in Eq. (24.16).

The matrix $[h]$ obeys its own identities, and produces a matrix that is a unit quaternion provided the real vector h is a *Minkowski unit vector*, meaning that instead of the $\mathbf{S}^3$ spherical unit-length constraint

$$q \cdot q = q \cdot I_4 \cdot q = q_0{}^2 + q_1{}^2 + q_2{}^2 + q_3{}^2 = 1 \tag{24.17}$$

we have the hyperbolic unit-length constraint on the real component variables

$$h \cdot \eta \cdot h = h_0{}^2 - h_1{}^2 - h_2{}^2 - h_3{}^2 = 1 \tag{24.18}$$

along with the unit determinant (required by the definition of $\mathbf{SL}(2,\mathbf{C})$)

$$\det[h] = h^\mu \eta_{\mu\nu} h^\nu = h \cdot \eta \cdot h = 1\,. \tag{24.19}$$

Composite complex quaternions. Every possible Lorentz transformation, including the self-contained subgroup of rotations in 3D space (which leaves the time component of the spacetime 4-vector untouched), can be composed from products of the two matrices, Eq. (24.14) and Eq. (24.16). The simplest example of a composite action is the "rolling ball" rotation phenomenon introduced in Chapter 5, Section 5.1, and revisited in Chapter 23. The origin of the rolling ball rotation (in the opposite direction of a sequence of rotations) is the sign of the commutator of the infinitesimal generators of $\mathbf{SO}(3)$, which we can see clearly using the quaternion rotation form. Now we define $r_1 = [q_0, q_1, q_2, q_3]$ as the first (rightmost in the sense of the composite matrix) of two quaternion rotations to act on the world and $r_2 = [p_0, p_1, p_2, p_3]$ as the second, leftmost acting, and examine the results: the rotation keeping the $\mathbf{q}$-axis fixed is our first, rightmost acting, matrix, and the second, leftmost acting, matrix keeps the $\mathbf{p}$-axis fixed, so in the quaternion product, $r_2(p) \star r_1(q)$, the quaternion product will produce a new composite axis of rotation in the direction $\mathbf{p} \times \mathbf{q}$. However, the order of the guided right-handed directions in space, with $\mathbf{q}$ up, rolling first to the right, and with $\mathbf{p}$ pointing right, with the second rolling motion downward, produces a cyclic motion with the opposite positive axis, in the direction $\mathbf{q} \times \mathbf{p}$. As shown also using $\mathbf{SO}(3)$ rotations in Chapter 23, Fig. 23.4, this four-step circuit in the counterclockwise direction of sequential quaternion rolls produces an opposing clockwise rotation about $\mathbf{p} \times \mathbf{q}$:

$$\left.\begin{array}{l}\text{Counterclockwise}\\ \text{circuit of rotations:}\end{array}\right\} = r_2(\bar{p}) \star r_1(\bar{q}) \star r_2(p) \star r_1(q)$$

$$= [q_0', q_1', q_2', q_3']$$

$$\stackrel{\text{infinitesimal}}{\underset{\text{limit}}{\equiv}} [q_0',\ 2(-q_2 p_3 + p_2 q_3),\ 2(-q_3 p_1 + p_3 q_1),\ 2(-q_1 p_2 + p_1 q_2)]$$

$$= [q_0',\ +2\,\mathbf{p} \times \mathbf{q}]\,,$$

where

$$\mathbf{p} \times \mathbf{q} = [p_1, p_2, p_3] \times [q_1, q_2, q_3] = [p_2 q_3 - q_2 p_3, \ p_3 q_1 - q_3 p_1, \ p_1 q_2 - q_1 p_2],$$

so *the rolling ball quaternion action counter-rotates.*

The Thomas precession, the rotation of a coordinate frame following the application of a sequence of noncollinear boosts, is also easy to compute in quaternion form. If we take $b_1 = [h_0, ih_1, ih_2, ih_3]$ and $b_2 = [g_0, ig_1, ig_2, ig_3]$ and simply do two boosts, with $b_1(h)$ applied first, and $b_2(g)$ next, we find, with no approximations, that the result splits into an imaginary part corresponding to the composite boost result, and a real vector part specifying a rotation axis, with the **h** component being the *first* component of the cross product, and the **g** component acting *later*, but now the direction of the composite-action rotation axis is $\mathbf{h} \times \mathbf{g}$. As shown also using **SO(2, 1)** rotations in Chapter 23, Fig. 23.5, this composite complex quaternion boosting action *coincides* with the direction of the resulting rotation:

$$\left.\begin{array}{l} \text{Real part of} \\ \text{counterclockwise} \\ \text{circuit of boosts:} \end{array}\right\} = \mathrm{Re}\left(b_2(g)^{-1} \star b_1(h)^{-1} \star b_2(g) \star b_1(h)\right)$$

$$\underset{\text{infinitesimal}}{\overset{}{\underset{\text{limit}}{=}}} \ [\mathrm{Re}\,(h_0'), \ 2(h_2 g_3 - g_2 h_3), \ 2(h_3 g_1 - g_3 h_1), \ 2(h_1 g_2 - g_1 h_2)]$$

$$= [\mathrm{Re}\,(h_0'), \ +2\mathbf{h} \times \mathbf{g}],$$

where

$$\mathbf{h} \times \mathbf{g} = [h_1, h_2, h_3] \times [g_1, g_2, g_3] = [h_2 g_3 - g_2 h_3, h_3 g_1 - g_3 h_1, h_1 g_2 - g_1 h_2],$$

so *the Thomas precession quaternion action co-rotates.*

Finally, we can combine rotations and boosts to make a composite controllable arbitrary Lorentz transformation in the full group **SO(3, 1)**. The usual form would apply first a rotation to orient the object at the origin in the desired aspect, followed by a boost to a new Lorentz frame, though of course one could reverse that for some reason. For this case, we write out the entire quaternion to see what it looks like. We take the initial rotation to be the quaternion $q = [q_0, q_1, q_2, q_3]$ and the subsequent boost to be $h = [h_0, ih_1, ih_2, ih_3]$. The composite quaternion action is

$$\text{Composite Lorentz transform}: w = h \star q$$

$$= [h_0 q_0, h_0 q_1, h_0 q_2, h_0 q_3]$$
$$+ i[-h_1 q_1 - h_2 q_2 - h_3 q_3,$$
$$h_1 q_0 + h_2 q_3 - h_3 q_2,$$
$$h_2 q_0 + h_3 q_1 - h_1 q_3,$$
$$h_3 q_0 + h_1 q_2 - h_2 q_1]. \tag{24.20}$$

We can check that this is still a quaternion because, even with the complexification,

$$w \cdot w = \left(q_0{}^2 + q_1{}^2 + q_2{}^2 + q_3{}^2\right)\left(h_0{}^2 - h_1{}^2 - h_2{}^2 - h_3{}^2\right) = 1.$$

This is of course guaranteed because our abstract quaternion multiplication is identical to **SL(2, C)** matrix multiplication of unit-determinant matrices, and the determinant of a product is the product of the determinants. Reversing the order to $q \star h$ simply changes the signs of the vector terms $h_i q_j$ for $(i, j) = \{(2, 3), (3, 2), (3, 1), (1, 3), 1, 2), (2, 1)\}$.

24.4 Four-vector Pauli matrices and the spacetime Lorentz group

We now have a pretty good idea how the complexified quaternions and their alternate expression in terms of $\mathbf{SL(2, C)}$ matrices can be exploited to embody all the properties of the six-parameter Lorentz group $\mathbf{SO}(3, 1)$. Our final step is to use the methods of relativistic quantum field theory to complete the picture of the construction of the $\mathbf{SO}(3, 1)$ Lorentz transformation matrices from $\mathbf{SL(2, C)}$ without getting confused about whether a spacetime 4-vector is itself a quaternion or a complex quaternion. In fact, we will be able to deduce an answer to that question indirectly after we build our Lorentz matrices and their explicit parameterizations in terms of world-embodied variables without referring directly to the form of the vector space upon which the matrices act.

We begin by going back to the $\mathbf{SL(2, C)}$ matrix quaternion form, and a method of identifying those matrices in a nontrivial way with a mapping from $\mathbf{SL(2, C)}$ to 4×4 representations of $\mathbf{SO}(3, 1)$, of which $\mathbf{SL(2, C)}$ is the double covering. We follow methods originating with Wigner (1939) and worked out in more detail in numerous places in the physics literature, with our treatment adapted from Muller-Kirsten and Wiedemann (1987).

We recall that $\mathbf{SL(2, C)}$ matrices are built from coefficients starting with four complex variables or eight real variables, but are restricted to have unit determinant, which imposes two real constraints and leaves us with six real variables, the real and imaginary parts of three complex variables. However, maintaining the unit determinant criterion imposes extra constraints on the relationship between the real, rotational quaternion parameters, and pure imaginary, Lorentz boost quaternion parameters. Accordingly, we will treat those two cases separately, and combine them as separate unit-determinant entities since, if each matrix has unit determinant, any arbitrary product of the two types will continue to have unit determinant without imposing any constraints between sets of variables.

Our fundamental tool will be the Minkowski metric η for flat $(3, 1)$ spacetime, in (t, x, y, z) order,

$$\eta_{\mu\nu} = \eta^{\mu\nu} = \begin{bmatrix} 1 & 0 & 0 & 0 \\ 0 & -1 & 0 & 0 \\ 0 & 0 & -1 & 0 \\ 0 & 0 & 0 & -1 \end{bmatrix}, \tag{24.21}$$

where $(\mu, \nu) \in \{0, 1, 2, 3\}$ and we chose the plus sign for the η_{00} element in the upper left to match our quaternion notation conventions. We then define the fundamental relativistic $\mathbf{SL(2, C)}$ matrices as they are used to adapt Hamilton quaternion notation to relativistic Lorentz transform notation. In the approach common in relativistic quantum field theory (Muller-Kirsten and Wiedemann, 1987), one defines two pairs of complex 2×2 matrices related to the Pauli matrices:

$$\sigma_\mu = \tilde{\sigma}^\mu = \left\{ \begin{bmatrix} 1 & 0 \\ 0 & 1 \end{bmatrix} \begin{bmatrix} 0 & 1 \\ 1 & 0 \end{bmatrix} \begin{bmatrix} 0 & -i \\ i & 0 \end{bmatrix} \begin{bmatrix} 1 & 0 \\ 0 & -1 \end{bmatrix} \right\}, \tag{24.22}$$

$$\sigma^\mu = \tilde{\sigma}_\mu = \left\{ \begin{bmatrix} 1 & 0 \\ 0 & 1 \end{bmatrix} \begin{bmatrix} 0 & -1 \\ -1 & 0 \end{bmatrix} \begin{bmatrix} 0 & i \\ -i & 0 \end{bmatrix} \begin{bmatrix} -1 & 0 \\ 0 & 1 \end{bmatrix} \right\}. \tag{24.23}$$

Note that the indices are raised and lowered by multiplying by the Minkowski metric η in Eq. (24.21).

The spacetime 4-vector. We are now well aware that in the "standard quaternion approach," the quaternion algebra can be looked at as a noncommutative algebra of an abstract 4-vector, as a generalized complex number using the Hamilton imaginaries $(\mathbf{I}, \mathbf{J}, \mathbf{K})$, or as a 2×2 complex matrix algebra with the Hamilton imaginaries replaced by $-i\,\boldsymbol{\sigma}$ (Pauli). We also know that bilinear conjugation of a quaternion of the form $(0, \hat{\mathbf{x}})$ denoting a 180-degree rotation about an axis $\mathbf{x} = r\hat{\mathbf{x}}$ can produce a standard 3D rotation of the vector $\mathbf{x}$, but that there are some misgivings about calling that 180-degree quaternion rotation a "3-vector" in a rigorous sense (Altmann, 1986, 1989).

Now we will see that when spacetime is introduced, the *quaternion-like spacetime 4-vector* (t, x, y, z) associated with the extended Pauli matrices appears as the legitimate target of $\mathbf{SL(2, C)}$ transformations, which are themselves complex quaternions represented as 2×2 matrices.

The rest of the algorithms in this section will compute the Lorentz transformations in the group $\mathbf{SO(3, 1)}$ in their normal, completely real form, without the factors of $i = \sqrt{-1}$ encountered above in $B_a(h)$ and $B_b(h)$. Consistency in this context now requires that we implicitly or explicitly assume that a spacetime coordinate is represented by the matrix

$$[X] = t\,I_2 + \mathbf{x} \cdot \boldsymbol{\sigma} = \begin{bmatrix} t+z & x-iy \\ x+iy & t-z \end{bmatrix}. \tag{24.24}$$

This coordinate has the essential property that its determinant is its Lorentz-invariant length:

$$\det[X] = t^2 - x^2 - y^2 - z^2 = [t, x, y, z] \cdot \eta \cdot [t, x, y, z]^{\mathrm{t}}. \tag{24.25}$$

We are thus inclined to soften the criticism of using the quaternion $(0, \mathbf{x})$ as a vector to construct $R(q)$. In fact the approach to spacetime symmetries in field theory, as given, e.g., in the book on supersymmetry by (Muller-Kirsten and Wiedemann, 1987), uses $[p] = t\,I_2 + \mathbf{x} \cdot \boldsymbol{\sigma}$ with $\det[p] = t^2 - x^2 - y^2 - z^2$ throughout as the basis for a spacetime point p being acted upon by Lorentz group transformations.

We will see in just a moment that this leads to a quaternion-matrix-based Minkowskian inner product as well. In fact, we can define Minkowski 4-vector versions of the Pauli matrices, removing the actual spacetime coordinates from $[X]$, and using the matrices themselves as *free indices* to define the real $\mathbf{SO(3, 1)}$ matrices *immediately* with no extra steps involving imaginary numbers. Thus some of the doubts about the standard-quaternion conjugation construction of the $\mathbf{SO(3)}$ matrix $R(q)$ appear to become irrelevant when $\mathbf{SO(3)}$ emerges as a subgroup of the process used to define representations of $\mathbf{SO(3, 1)}$ from $\mathbf{SL(2, C)}$. We will now work through this in detail.

Identities of the relativistic Pauli matrices. The 2×2 matrices $(\sigma^\mu, \tilde{\sigma}^\mu)$ with spacetime indices $\mu \in (0, 1, 2, 3)$, raised and lowered with the Minkowski metric η of Eq. (24.21), have a special meaning: they are a matrix representation of the $(3, 1)$ *Clifford algebra* (see, e.g., Clifford, 1873; Doran and Lasenby, 2003; Dorst et al., 2007). In particular, their symmetric sum (anticommutator) produces the metric

$$\tilde{\sigma}_\mu \cdot \sigma_\nu + \tilde{\sigma}_\nu \cdot \sigma_\mu = 2\eta_{\mu\nu}\,I_2, \tag{24.26}$$

while their antisymmetric sum (commutator) produces a Levi-Civita symbol expression

$$\tilde{\sigma}_\mu \cdot \sigma_\nu - \tilde{\sigma}_\nu \cdot \sigma_\mu = i\,\epsilon_{\mu\nu}{}^{\alpha\beta}\,\tilde{\sigma}_\alpha \cdot \sigma_\beta, \tag{24.27}$$

and these combine to give

$$\tilde{\sigma}_\mu \cdot \sigma_\nu = \eta_{\mu\nu} I_2 + \frac{i}{2} \epsilon_{\mu\nu}{}^{\alpha\beta} \tilde{\sigma}_\alpha \cdot \sigma_\beta . \tag{24.28}$$

As a consequence of these, one can find many more identities, often expressed in terms of the traces of these matrices; we will only need one of these, the metric trace identity

$$\begin{aligned} \mathrm{tr}\left(\tilde{\sigma}_\mu \cdot \sigma_\nu\right) &= 2\eta_{\mu\nu} , \\ \mathrm{tr}\left(\tilde{\sigma}^\mu \cdot \sigma^\nu\right) &= 2\eta^{\mu\nu} . \end{aligned} \tag{24.29}$$

In this notation, our usual quaternion takes the form

$$q \to q_0\sigma^0 - iq_1\sigma^1 - iq_2\sigma^2 - iq_3\sigma^3 = \begin{bmatrix} q_0 - iq_3 & -iq_1 - q_2 \\ -iq_1 + q_2 & q_0 + iq_3 \end{bmatrix},$$

while the boost is

$$h \to h_0\sigma^0 + h_1\sigma^1 + h_2\sigma^2 + h_3\sigma^3 = \begin{bmatrix} h_0 + h_3 & h_1 - ih_2 \\ h_1 + ih_2 & h_0 - h_3 \end{bmatrix},$$

and

$$\tilde{q} \to q_0\tilde{\sigma}^0 - iq_1\tilde{\sigma}^1 - iq_2\tilde{\sigma}^2 - iq_3\tilde{\sigma}^3 = \begin{bmatrix} q_0 + iq_3 & +iq_1 + q_2 \\ +iq_1 - q_2 & q_0 - iq_3 \end{bmatrix},$$

while the boost form becomes

$$\tilde{h} \to h_0\tilde{\sigma}^0 + h_1\tilde{\sigma}^1 + h_2\tilde{\sigma}^2 + h_3\tilde{\sigma}^3 = \begin{bmatrix} h_0 - h_3 & -h_1 + ih_2 \\ -h_1 - ih_2 & h_0 + h_3 \end{bmatrix} .$$

These matrices belong to the group $\mathbf{SL}(2, \mathbf{C})$, the double cover of $\mathbf{SO}(3, 1)$, and we note the following properties:

$$\left.\begin{aligned} \det q &= \det \tilde{q} = 1 \\ \det h &= \det \tilde{h} = 1 \\ \textit{(inner product)}\ \mathrm{tr}\, q \cdot \tilde{p} &= q_0 p_0 + q_1 p_1 + q_2 p_2 + q_3 p_3 \\ \textit{(inner product)}\ \mathrm{tr}\, h \cdot \tilde{g} &= h_0 g_0 - h_1 g_1 - h_2 g_2 - h_3 g_3 \end{aligned}\right\} . \tag{24.30}$$

The use of $\mathrm{tr}\, \tilde{\sigma}^\mu \sigma^\nu = 2\eta^{\mu\nu}$ defines our Minkowski space inner product with the $\mathbf{SO}(3, 1)$ metric $\eta^{\mu\nu} = \eta_{\mu\nu} = \mathrm{diag}(1, -1, -1, -1)$, and with the "$-i$" coefficients in $[q]$ and $[\tilde{q}]$, the diagonal minus signs change to plus signs as required to make $q \cdot q = 1$ for ordinary quaternions. This is the full "time-plus-3D space" environment of the relativistic world, featuring the requirement that Lorentz transformations leave $\eta^{\mu\nu}$ invariant.

The double covering of SO(3, 1) from the SL(2, C) Clifford algebra. The Clifford algebra embodied in our $\mathbf{SL}(2, \mathbf{C})$ matrices is a double covering of the 4×4 Lorentz transform matrices, and hence the Lorentz matrices can be derived from double products directly, without specifically choosing a set of spacetime coordinates, as the Clifford algebra indices themselves form such a basis. Accordingly, an arbitrary element $\Lambda^\mu{}_\nu$ of $\mathbf{SO}(3,1)$ can be written in terms of an arbitrary element A of $\mathbf{SL}(2,\mathbf{C})$ as

$$\Lambda^\mu{}_\nu = \frac{1}{2}\,\mathrm{tr}\left(\tilde{\sigma}^\mu\,A\,\sigma_\nu\,A^\dagger\right), \tag{24.31}$$

where $A^\dagger$ denotes the Hermitian conjugate or complex conjugate transpose. Our claim that the $\mathbf{SL}(2,\mathbf{C})$ matrices A doubly cover $\mathbf{SO}(3,1)$ is now clear from the fact both A and $-A$ produce the same $\mathbf{SO}(3,1)$ matrix.

The 3D rotation subgroup of $\mathbf{SO}(3,1)$. Basically all of our relations connecting quaternions to 3D rotations follow from applying Eq. (24.31) to the three-parameter unitary subgroup $\mathbf{SU}(2)$ of $\mathbf{SL}(2,C)$

$$\pm A(q) \equiv \pm[q] = q_0\sigma_0 - i\mathbf{q}\cdot\boldsymbol{\sigma}, \tag{24.32}$$

where of course we must have $\det A(q) = q \cdot q = 1$ for $A(q)$ to be an $\mathbf{SL}(2,\mathbf{C})$ matrix. From this follows the form for the symbolic matrix, without any extra manipulations of factors of i that were required in the previous section:

$$R^\mu{}_\nu = \frac{1}{2}\,\mathrm{tr}\left(\tilde{\sigma}^\mu\,[q]\sigma_\nu\,[q]^\dagger\right),$$

$$R(q) = \begin{bmatrix} q\cdot q & 0 & 0 & 0 \\ 0 & q_0{}^2+q_1{}^2-q_2{}^2-q_3{}^2 & 2q_1q_2-2q_0q_3 & 2q_1q_3+2q_0q_2 \\ 0 & 2q_1q_2+2q_0q_3 & q_0{}^2-q_1{}^2+q_2{}^2-q_3{}^2 & 2q_2q_3-2q_0q_1 \\ 0 & 2q_1q_3-2q_0q_2 & 2q_2q_3+2q_0q_1 & q_0{}^2-q_1{}^2-q_2{}^2+q_3{}^2 \end{bmatrix}. \tag{24.33}$$

Alternatively, we can express $A(q)$ directly in terms of the rotation angle θ and the unit-length direction $\hat{\mathbf{n}}$ that is fixed by the rotation,

$$\pm A(q(\theta,\hat{\mathbf{n}})) = \exp\left(-i\frac{\theta}{2}\hat{\mathbf{n}}\cdot\boldsymbol{\sigma}\right)$$

$$= \sigma_0\cos\frac{\theta}{2} - i\hat{\mathbf{n}}\cdot\boldsymbol{\sigma}\sin\frac{\theta}{2}. \tag{24.34}$$

As we have already seen, $q(\theta,\hat{\mathbf{n}})$ explicitly generates the axis-angle form of the 3D rotation matrix in the lower 3×3 submatrix of the 4×4 Lorentz matrix,

$$R^\mu{}_\nu(\theta,\hat{\mathbf{n}}) = \frac{1}{2}\,\mathrm{tr}\left(\tilde{\sigma}^\mu\,[q(\theta,\hat{\mathbf{n}})]\sigma_\nu\,[q(\theta,-\hat{\mathbf{n}})]\right)$$

$$= \begin{bmatrix} 1 & 0 & 0 & 0 \\ 0 & \cos\theta+(1-\cos\theta)\hat{n}_1{}^2 & (1-\cos\theta)\hat{n}_1\hat{n}_2-\sin\theta\,\hat{n}_3 & (1-\cos\theta)\hat{n}_1\hat{n}_3+\sin\theta\,\hat{n}_2 \\ 0 & (1-\cos\theta)\hat{n}_1\hat{n}_2+\sin\theta\,\hat{n}_3 & \cos\theta+(1-\cos\theta)\hat{n}_2{}^2 & (1-\cos\theta)\hat{n}_2\hat{n}_3-\sin\theta\,\hat{n}_1 \\ 0 & (1-\cos\theta)\hat{n}_1\hat{n}_3-\sin\theta\,\hat{n}_2 & (1-\cos\theta)\hat{n}_2\hat{n}_3+\sin\theta\,\hat{n}_1 & \cos\theta+(1-\cos\theta)\hat{n}_3{}^2 \end{bmatrix}. \tag{24.35}$$

The boost matrix of $\mathbf{SO}(3,1)$. Similarly, we can apply Eq. (24.31) to the three-parameter self-adjoint matrices in $\mathbf{SL}(2,\mathbf{C})$ that generate pure Lorentz transformation boosts. Remember that this is not a subgroup of $\mathbf{SL}(2,\mathbf{C})$ because its algebra *does not close* without including all rotations, e.g., the Thomas precession, in addition to the boosts. The $\mathbf{SL(2, C)}$ Clifford algebra form for the boost, equivalent to the complexified quaternion $[h_0, i\mathbf{h}]$, is

$$\pm A(h) \equiv \pm[h] = h_0\sigma_0 + \mathbf{h} \cdot \boldsymbol{\sigma} \,, \tag{24.36}$$

where again $\det A(h) = h_0{}^2 - \mathbf{h} \cdot \mathbf{h} = 1$ for $A(h)$ to be an $\mathbf{SL(2, C)}$ matrix. The symbolic boost matrix that we have already discovered now appears without any extra manipulations of factors of $i = \sqrt{-1}$ if we use the bilinear Clifford algebra construction from the $\mathbf{SL(2, C)}$ matrix of Eq. (24.36), noting that $[h]$ is *not* unitary like $[q]$, but is instead a Hermitian (self-adjoint) matrix:

$$
\begin{aligned}
B^\mu{}_\nu(h) &= \frac{1}{2}\operatorname{tr}\left(\tilde{\sigma}^\mu [h]\sigma_\nu [h]\right) \\
&= \begin{bmatrix}
h_0^2 + h_x^2 + h_y^2 + h_z^2 & 2h_0h_x & 2h_0h_y & 2h_0h_z \\
2h_0h_x & h_0^2 + h_x^2 - h_y^2 - h_z^2 & 2h_xh_y & 2h_xh_z \\
2h_0h_y & 2h_xh_y & h_0^2 + h_y^2 - h_x^2 - h_z^2 & 2h_yh_z \\
2h_0h_z & 2h_xh_z & 2h_yh_z & h_0^2 + h_z^2 - h_x^2 - h_y^2
\end{bmatrix} .
\end{aligned}
\tag{24.37}
$$

Note that the $\mathbf{SO(2, 1)}$ boost explored in Eq. (23.16) in Chapter 23 is obtained by setting $h_z = 0$. The explicit form for the boost in terms of the velocity direction $\hat{\mathbf{v}}$ and magnitude $v = \tanh\xi$ in $\mathbf{SL(2, C)}$ form is

$$\pm A(h(\xi, \hat{\mathbf{v}})) = \exp\left(\frac{\xi}{2}\hat{\mathbf{v}} \cdot \boldsymbol{\sigma}\right) = \sigma_0 \cosh\frac{\xi}{2} + \hat{\mathbf{v}} \cdot \boldsymbol{\sigma} \sinh\frac{\xi}{2} \,. \tag{24.38}$$

This matrix is equivalent to the complexified quaternion $h(\xi, \hat{\mathbf{v}}) = (\cosh\xi/2, i\hat{\mathbf{v}}\sinh\xi/2)$, where $\hat{\mathbf{v}} \cdot \hat{\mathbf{v}} = 1$. We can now generate the 4×4 Lorentz boost matrix using the $\mathbf{SL(2, C)}$ quadratic matrix construction

$$
\begin{aligned}
B^\mu{}_\nu(\xi, \hat{\mathbf{v}}) &= \frac{1}{2}\operatorname{tr}\left(\tilde{\sigma}^\mu \cdot [h(\xi, \hat{\mathbf{v}})] \cdot \sigma_\nu \cdot [h(\xi, \hat{\mathbf{v}})]\right) \\
&= \begin{bmatrix}
\cosh\xi & v_x\sinh\xi & v_y\sinh\xi & v_z\sinh\xi \\
v_x\sinh\xi & 1 + v_x^2(\cosh\xi - 1) & v_xv_y(\cosh\xi - 1) & v_xv_z(\cosh\xi - 1) \\
v_y\sinh\xi & v_xv_y(\cosh\xi - 1) & 1 + v_y^2(\cosh\xi - 1) & v_yv_z(\cosh\xi - 1) \\
v_z\sinh\xi & v_xv_z(\cosh\xi - 1) & v_yv_z(\cosh\xi - 1) & 1 + v_z^2(\cosh\xi - 1)
\end{bmatrix} .
\end{aligned}
\tag{24.39}
$$

Here $\det[B] = (\cosh^2\xi - \sinh^2\xi)^4 \equiv 1$ and transformations leave the Minkowski metric matrix $\eta_{\mu\nu} = \operatorname{diag}(1, -1, -1, -1)$ invariant. Thus $B^\mu{}_\nu$ is precisely an orthogonal transformation in Minkowski space, denoted as $\mathbf{SO(3, 1)}$, to distinguish it from the 4D Euclidean transformations $\mathbf{SO(4)}$.

Remark: The boost quaternion $\mathbf{h} = (\cosh\xi/2, \hat{\mathbf{v}}\sinh\xi/2)$ generates the most general pure 4×4 boost matrix, $B(\mathbf{v})$, but this transformation alone leaves our full intuitive picture of 4D spacetime incomplete, since rotations must be folded in with boosts to complete the symmetries of $3 + 1$ spacetime. The *Thomas precession* resolves this issue via the relativistic version of the "rolling ball" method from Chapter 5 by producing all possible 3D rotations in the context of multiple pure boosts:

> *One can generate all three missing* **SO**(3 , 1) *rotations by circular rolling-ball-style boost circuits in the yz-, xz-, and xy-planes.*

We have also seen that the full Lorentz group **SO**(3 , 1) has a quadratic form corresponding to its complexified quaternion "double covering group," namely **SL**(**2** , **C**). This group is directly derivable from Clifford algebra methods, and is also written as **Spin**(3, 1). It corresponds exactly to our six-parameter group realized in terms of the complex 2 × 2 matrices **SL**(**2** , **C**), and eventually leads to the Dirac equation for the relativistic spin 1/2 electron.

24.5 Miscellaneous properties

There are several interesting facts and opportunities that present themselves in the process of investigating the Lorentz group in quaternion form, or, more precisely, the **SL**(**2** , **C**) form. This is not by any means a complete list, but simply a few pieces of information that caught our attention.

- **The symbolic exponential form of the 3D rotation.** If we look at the quaternion analog of the "$i\theta$" logarithm for an ordinary complex 2D rotation written as $e^{i\theta}$, we have

$$[q] = \exp\left(-i\frac{\theta}{2}\hat{\mathbf{n}}\cdot\boldsymbol{\sigma}\right) .$$

Normally, we impose $\hat{\mathbf{n}}\cdot\hat{\mathbf{n}} = n_x{}^2 + n_y{}^2 + n_z{}^2 = 1$ to specify the unit direction of the fixed axis of the rotation. However, we see something interesting when we leave that normalization unspecified in the expansion of the exponential, and we get the components of the quaternion rotation as

$$\left[\cos\left(\frac{\theta}{2}\sqrt{n_x{}^2+n_y{}^2+n_z{}^2}\right), \frac{n_x\sin\left(\frac{\theta}{2}\sqrt{n_x{}^2+n_y{}^2+n_z{}^2}\right)}{\sqrt{n_x{}^2+n_y{}^2+n_z{}^2}},\right.$$
$$\left.\frac{n_y\sin\left(\frac{\theta}{2}\sqrt{n_x{}^2+n_y{}^2+n_z{}^2}\right)}{\sqrt{n_x{}^2+n_y{}^2+n_z{}^2}}, \frac{n_z\sin\left(\frac{\theta}{2}\sqrt{n_x{}^2+n_y{}^2+n_z{}^2}\right)}{\sqrt{n_x{}^2+n_y{}^2+n_z{}^2}}\right] . \tag{24.40}$$

Thus we see that $\sqrt{\hat{\mathbf{n}}\cdot\hat{\mathbf{n}}}$ appears throughout essentially as the *scale* for the angle θ, and that setting that scale to unity recovers our usual

$$q = \left[\cos\frac{\theta}{2}, n_x\sin\frac{\theta}{2}, n_y\sin\frac{\theta}{2}, n_z\sin\frac{\theta}{2}\right] .$$

- **The symbolic exponential form of the 4D boost.** If we look at the complex quaternion exponential with the "ξ" logarithm, we have

$$[h] = \exp\left(\frac{\xi}{2}\hat{\mathbf{v}}\cdot\boldsymbol{\sigma}\right) .$$

Normally, we impose $\hat{\mathbf{v}} \cdot \hat{\mathbf{v}} = v_x^2 + v_y^2 + v_z^2 = 1$ to specify the unit direction of the direction of the boost velocity. Again we see something interesting when we leave that normalization unspecified in the expansion of the exponential, and we get the components of the complexified quaternion boost as

$$
\left[\cosh\left(\frac{\xi}{2}\sqrt{v_x^2 + v_y^2 + v_z^2}\right), \ \mathrm{i}\, \frac{v_x \sinh\left(\frac{\xi}{2}\sqrt{v_x^2 + v_y^2 + v_z^2}\right)}{\sqrt{v_x^2 + v_y^2 + v_z^2}}, \right.
$$
$$
\left. \mathrm{i}\, \frac{v_y \sinh\left(\frac{\xi}{2}\sqrt{v_x^2 + v_y^2 + v_z^2}\right)}{\sqrt{v_x^2 + v_y^2 + v_z^2}}, \ \mathrm{i}\, \frac{v_z \sinh\left(\frac{\xi}{2}\sqrt{v_x^2 + v_y^2 + v_z^2}\right)}{\sqrt{v_x^2 + v_y^2 + v_z^2}} \right]. \tag{24.41}
$$

Thus again we see that $\sqrt{\hat{\mathbf{v}} \cdot \hat{\mathbf{v}}}$ appears as the *scale* for the boost rapidity ξ, and that setting that scale to unity and pulling out the coefficients recovers our usual quaternion form of the boost operation (the double cover of the Lorentz boost):

$$
h = \left[\cosh\frac{\xi}{2}, \ \mathrm{i}\,v_x \sinh\frac{\xi}{2}, \ \mathrm{i}\,v_y \sinh\frac{\xi}{2}, \ \mathrm{i}\,v_z \sinh\frac{\xi}{2} \right].
$$

- **The general SL(2, C) matrix from its exponential.** We have examined the full **SO(3, 1)** Lorentz transformation group so far by separating out the Hermitian boost sector from the unitary rotation sector and multiplying together the two unit-determinant matrices to get a composite matrix that satisfies the **SL(2, C)** conditions by preserving the unit determinant. Guessing how to do that in a way that does not involve separate boost and rotation components for composite transformations is difficult because, among other things such as noncommutativity, the zeroth, time-like, components are mixed together in a way that is difficult to parse. However, the *exponential form*, which basically derives the Lie *group* from a finite form of the infinitesimal generators of the Lie *algebra*, computes the zeroth component automatically. So we can get a pretty good idea of what the most general **SL(2, C)** matrix looks like in quaternion form if we exponentiate with a complex six-degree-of-freedom parameter as follows:

$$
[s] = \exp\left(\frac{1}{2}\left(\xi\hat{\mathbf{v}} - \mathrm{i}\theta\,\hat{\mathbf{n}}\right) \cdot \boldsymbol{\sigma} \right).
$$

Normally, we impose $\hat{\mathbf{n}} \cdot \hat{\mathbf{n}} = n_x^2 + n_y^2 + n_z^2 = 1$ and $\hat{\mathbf{v}} \cdot \hat{\mathbf{v}} = v_x^2 + v_y^2 + v_z^2 = 1$ to specify unit lengths of the axis of the rotation and velocity direction. However, we see something interesting when we leave that normalization unspecified in the expansion of the exponential, and we get the components of the quaternion rotation as

$$
[s] = \left[\cos\left(\frac{\sqrt{\alpha^2}}{2}\right), \ \frac{\sin\left(\frac{\sqrt{\alpha^2}}{2}\right)(n_x\theta + \mathrm{i}v_x\xi)}{\sqrt{\alpha^2}}, \ \frac{\sin\left(\frac{\sqrt{\alpha^2}}{2}\right)(n_y\theta + \mathrm{i}v_y\xi)}{\sqrt{\alpha^2}}, \ \frac{\sin\left(\frac{\sqrt{\alpha^2}}{2}\right)(n_z th + \mathrm{i}v_z\xi)}{\sqrt{\alpha^2}} \right],
$$
$$
\tag{24.42}
$$

where the complex angle "α" effectively normalizes the three components of the complex exponential *vector*, that is,

$$\alpha^2 = \left\| \theta\,\hat{\mathbf{n}} + \mathrm{i}\,\xi\,\hat{\mathbf{v}} \right\|^2 = \theta^2 + 2\,\mathrm{i}\,\hat{\mathbf{n}}\cdot\hat{\mathbf{v}}\,\theta\,\xi - \xi^2\;.$$

The quantity α describes what we might call a Lorentz group complex angle, mixing the concepts of a 3D rotation angle and a 4D rapidity. This mapping also *preserves* $\det[s] = 1$ *without any constraints*, and so the six free variables in $\{\theta, \xi, \hat{\mathbf{n}}, \hat{\mathbf{v}}\}$ are all independent parameters of a legal $\mathbf{SL}(2, \mathbf{C})$ matrix; we do not know exactly what this combination means, but it is the best we can do for a general legal form, guaranteed by the exponentiation-based derivation. The full 4×4 matrix $\Lambda^\mu{}_\nu$ can be computed, but handling the mixture of degrees of freedom in α and $(\theta\,\hat{\mathbf{n}} + \mathrm{i}\,\xi\,\hat{\mathbf{v}})$ is complicated and not very informative.

- **Aberration of light distributions in $3 + 1$ spacetime.** Finally, we briefly examine ways of representing 3D *spatial* light ray distributions for symmetric point sources, much as we did for 1D and 2D spatial distributions of light sources. The following figure shows the distribution of light ray paths for three different upward-moving velocities of a point source as seen by a stationary observer:

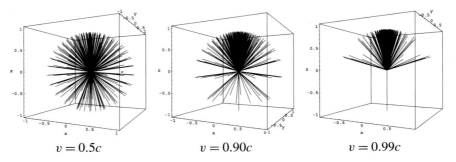

$$v = 0.5c \qquad\qquad v = 0.90c \qquad\qquad v = 0.99c$$

As an alternative approach to the 3D visualization, here we show a solid sphere plot of 3D space light ray distributions for an upward-moving symmetric point source:

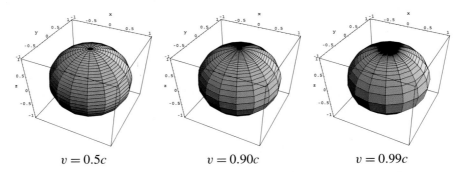

$$v = 0.5c \qquad\qquad v = 0.90c \qquad\qquad v = 0.99c$$

In relativistic rendering applications, *texture maps* on these distorted spheres provide an implementation of *relativistic image-based rendering* (IBR) (Weiskopf et al., 2000). This property of the Lorentz transformation on light rays basically permits a spherical image surrounding a central scene in one Lorentz frame to be represented in any Lorentz frame with the correct image distortion given only the image texture map in one Lorentz frame.

24.6 Summary of (3 + 1)-dimensional relativity and complexified quaternions

- The boost $B(\mathbf{v})$ is an orthogonal 4×4 matrix that preserves the Minkowski metric η instead of a Euclidean metric, so $B \cdot \eta \cdot B = \eta$. Boosts are based on having hyperbolic functions, cosh and sinh, replacing the trigonometric functions, cos and sin, that appear in ordinary (quaternion-derived) 3D rotations. Boost matrices leave *Minkowski-metric dot products* unchanged, just as rotations leave Euclidean distances unchanged.

- Quaternion-like forms exist, rigorously corresponding to the representations and algebra of **SL(2, C)**, with $\cosh(\xi/2)$ and $\sinh(\xi/2)$ supplementing the familiar $\cos(\theta/2)$ and $\sin(\theta/2)$ of rotational quaternions.

- Occlusion invariance and the computability of light aberration allow relativistic IBR to be implemented naturally. The occlusion of light rays by polygons is *relativistically invariant*.

- Objects are made up of vertices tracing *world lines*, linked into edges, polygons, and polyhedra to form 3D spatial structures that include their evolution in time, and their shapes distort as the velocity of the observer's frame approaches the speed of light. Camera images can be formed by tracing light rays backward in time on negative light cones until they hit scene objects that emitted light at an earlier time point; or, conversely, all light rays emitted at all time points from light sources can be traced forward on their light cones to a pixel on a camera, though this is typically more costly.

- Cycles of 3D rotations in one plane, say the xy-plane, generate an incremental *counter-rotating z-axis spin* or rolling ball effect. By essentially the same mechanism of cycles of noncommuting matrices, cycles of Lorentz transformations in a plane generate the *co-rotating* Thomas precession.

- By means of the Thomas precession, if one possesses only access to the three boosts, the remaining missing three rotation group elements can be generated in an interactive exploration environment using commutators of boosts in the plane of any given rotation.

Quaternion discrete symmetries and gravitons

In this final part of our journey we will look at another part of mathematical physics related to quaternions that goes beyond even special relativity into the realm of Einstein's theory of general relativity. The initial trigger leading to interest in our subject was the discovery in 1975 of a family of so-called *instanton* solutions to classical Yang–Mills field theories for the group $\mathbf{SU(2)}$ in 4D Euclidean space (Belavin et al., 1975). The Yang–Mills result stimulated a successful search for a gravitational analog by Eguchi and Hanson (1978) corresponding to the A_1 Kleinian discrete group of the sphere, and the subsequent generalization of that solution by Gibbons and Hawking (1978) to the infinite A_k family of discrete Kleinian groups. In very short order, Hitchin (1979) recognized that the entire so-called "ADE" family of Kleinian groups (Klein, 1884) might classify *all possible gravitational instantons* of this type, continuing a contemporary legendary merger of the interests of the physics and mathematics research communities, and inspiring our own narrative in Eguchi, Gilkey, and Hanson (1980). Hitchin's conjecture was eventually proven to be correct by Kronheimer (1989a; 1989b), stimulating further interest in the ADE groups. The chapters in this part provide a modest introduction to the features of this family of discrete groups, known since the 19th century, that attracted renewed attention after the discovery of the first gravitational instantons. Both the physics and the mathematics related to the Kleinian groups are now vast fields of investigation: our goals here will be limited to their most basic aspects along with a study of how quaternions can be exploited to help develop a coherent and intuitive understanding of these groups. We begin with the fundamental descriptions of the geometry of the regular spherical polyhedra, divisions of the 2-sphere S^2 into identical polygons. Next, we show how every possible such regular tessellation admits discrete symmetric actions of the rotation group, and how these are realized for both ordinary 3D $SO(3)$ rotations and for their corresponding quaternions: these are the ADE discrete symmetry groups of S^2, namely the A_k cyclic groups, the D_k dihedral groups, and the E_6, E_7, and E_8 groups corresponding to actions on the tetrahedron, octahedron, and icosahedron, respectively. The last chapter devoted to the ADE symmetries examines the construction of a set of invariant polynomials studied by Klein, and presents visualizations of some of their exotic properties. In conclusion, a final chapter describes how quaternion methods facilitate the discovery of a visualizable geometric object embedded in 11 dimensions (Eguchi and Hanson, 1978; Hanson and Sha, 2017) whose induced metric is exactly the A_1 instanton solution of Einstein's equations in 4D Euclidean space.

Geometry of the ADE symmetries of the 2-sphere

25.1 Introduction to discrete groups on the sphere

In this chapter, we will study the nature of discrete rotational symmetry groups operating on the ordinary sphere $\mathbf{S}^2$, with the eventual goal of investigating their quaternion representations. There are important applications of these groups throughout both early and modern mathematics. Felix Klein in 1884 wrote an exhaustive treatment of these discrete groups in his book *Vorlesungen über das Ikosaeder und die Auflösung der Gleichungen vom fünften Grade* (*Lectures on the icosahedron and the solution [of equations] of the fifth degree*) (Klein, 1884). Remarkably, a century later, these same groups came to play a fundamental role in the classification of certain solutions of Einstein's theory of general relativity in four Euclidean dimensions (see, e.g., Eguchi and Hanson, 1978; Gibbons and Hawking, 1978; Eguchi and Hanson, 1979; Hitchin, 1979; Eguchi et al., 1980; Kronheimer, 1989a,b; Lindström et al., 2000). The groups themselves are referred to as the ADE groups, and are labeled A_k, D_k, E_6, E_7, and E_8, corresponding to the cyclic, dihedral, tetrahedral, octahedral, and icosahedral subdivisions of the sphere $\mathbf{S}^2$ into identical spherical polygons (Coxeter and Moser, 1980). In general relativity, these are associated with solutions of the 4D open Euclidean Einstein manifolds whose asymptotic shapes are locally flat 3-spaces that are quotients of $\mathbf{S}^3$ by each of the ADE group actions; these Einstein manifolds are known as the *asymptotically locally Euclidean* (ALE) manifolds. It is believed that at the core of each of these Einstein manifolds is one of the resolved singularities that we study in Chapter 28. At this writing, explicit solutions are known only for the A_k class of the ALE Einstein spaces.

The appearance of the ADE classification has long been known in geometry in the context of a fundamental set of 3D *lens spaces* of the 3-sphere $\mathbf{S}^3$, possible ways by which a new smooth space can be built from $\mathbf{S}^3$ by identifications under the action of the ADE discrete rotations (see, e.g, Adams, 1994; Rolfson, 1976). These spaces are of course intimately related to quaternions since $\mathbf{S}^3$ is precisely the space of all quaternions. The ADE groups, the discrete actions on the symmetric tessellations of $\mathbf{S}^2$, and their correspondence with lens spaces of $\mathbf{S}^3$ appear in a wide variety of contexts, and there are many ways of looking at and studying the ADE groups and their associated geometry. We will begin with the elementary geometry of $\mathbf{S}^2$ embedded in 3D Euclidean space, and systematically work our way up to examining increasingly exotic properties of the ADE discrete groups.

Underlying geometry of the Kleinian ADE groups. We are now ready to begin examining additional details of how the sphere $\mathbf{S}^2$ can be decomposed into sets of identical geometric shapes, leading into the properties of the associated symmetry groups. We can think of these decompositions either as tessellations of the sphere into spherical polygons or as corresponding planar-faced polygons embedded in a sphere; the one exception is the A_k digon or spherical lune, basically the surface of an orange slice, which has a sensible definition as a spherical polygon with two vertices, two edges, and one face, but

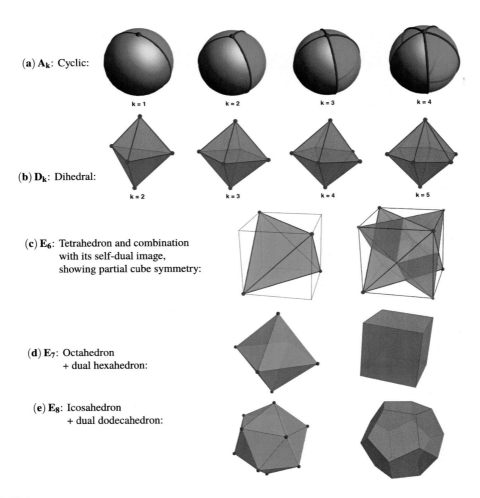

(a) A_k: Cyclic:

k = 1   k = 2   k = 3   k = 4

(b) D_k: Dihedral:

k = 2   k = 3   k = 4   k = 5

(c) E_6: Tetrahedron and combination with its self-dual image, showing partial cube symmetry:

(d) E_7: Octahedron + dual hexahedron:

(e) E_8: Icosahedron + dual dodecahedron:

FIGURE 25.1

(a) Samples of the A_k digon or lune "orange-slice-shaped" partitions of the sphere for shapes labeled by $k = 1, 2, 3, 4$. (b) Samples of the D_k dihedral fundamental shapes for $k = 2, 3, 4, 5$. (c) The tetrahedron and its self-dual tetrahedron, which have most of the symmetries of a cube, with actions in the E_6 group. (d) The octahedron and its dual hexahedron (the cube), which have symmetries in the E_7 group. (e) The icosahedron and its dual dodecahedron, which have symmetries in the E_8 group.

cannot be represented as a flat finite polygon embedded in the sphere, though some treatments depict it as a regular $(k + 1)$-sided polygon extruded into a faceted cylinder of indefinite length. Initially, we will think of these as shapes that are acted upon by rotating the sphere around one of its axes using basic $\mathbf{SO}(3)$ 3D axis-angle parameterized rotation matrices, keeping in mind that we will return eventually to quaternion versions of these rotations.

There are five basic classes of the geometric objects tessellating the sphere. Two of these, the hoso-hedron, dividing the sphere into equal digons, and the dihedron, dividing the sphere into equal isosceles triangles with bases around the equator extending up or down to one of the poles, are infinite and labeled by integers k corresponding to the number of stepped rotations in one full cycle about the vertical polar axis, excluding the identity. Three more are unique, corresponding to the polyhedral boundaries of the Platonic solids merged with their duals, the tetrahedron (self-dual), the octahedron (dual to the cube or hexahedron), and the icosahedron (dual to the dodecahedron). For each of the Platonic polyhedra, we choose the one with equilateral triangular faces to obtain uniform edge and face group actions. A visual summary of the five types of geometric partitions of the sphere S^2 is shown in Fig. 25.1. The ADE group actions corresponding to our list of geometric objects in Fig. 25.1(a,b,c,d,e) are the following:

(a)	Hosohedron	A_k: the cyclic group of digons.
(b)	Dihedron	D_k: the dihedral group of isosceles triangles.
(c)	Tetrahedron	E_6: the tetrahedral (self-dual) group of equilateral triangles.
(d)	Octahedron	E_7: the octahedral (dual: hexahedral) group of equilateral triangles.
(e)	Icosahedron	E_8: the icosahedral (dual: dodecahedral) group of equilateral triangles.

25.2 Coordinates of the A_k cyclic geometry

The digon subdivision in Fig. 25.1(a) acted upon by the A_k cyclic group in fact has only two vertices, the North and South poles, which we can label as "n" and "s":

$$v_n = \begin{bmatrix} 0 & 0 & +1 \end{bmatrix},$$
$$v_s = \begin{bmatrix} 0 & 0 & -1 \end{bmatrix}. \tag{25.1}$$

Lacking further vertices, we can specify the rest of the tessellation using, e.g., the equatorial midpoints of the $(k+1)$ half-circular arcs of the digon edges passing from the North pole to the South pole, in right-handed order relative to the North pole,

$$
\begin{aligned}
e_0 &= \begin{bmatrix} 1 & 0 & 0 \end{bmatrix}, \\
e_1 &= \begin{bmatrix} \cos\left(\dfrac{2\pi}{k+1}\right) & \sin\left(\dfrac{2\pi}{k+1}\right) & 0 \end{bmatrix}, \\
\vdots &= \begin{bmatrix} \vdots & \vdots & \vdots \end{bmatrix}, \\
e_{k-1} &= \begin{bmatrix} \cos\left(\dfrac{2\pi(k-1)}{k+1}\right) & \sin\left(\dfrac{2\pi(k-1)}{k+1}\right) & 0 \end{bmatrix}, \\
e_k &= \begin{bmatrix} \cos\left(\dfrac{2\pi k}{k+1}\right) & \sin\left(\dfrac{2\pi k}{k+1}\right) & 0 \end{bmatrix}.
\end{aligned}
\tag{25.2}
$$

The face centers (the centers of the digon spherical slices) are obviously also in the equatorial plane, displaced by $d\theta = \pi/(k+1)$ from the edge centers.

25.3 Coordinates of the D_k dihedral geometry

We choose the coordinates for the dihedron geometry shown in Fig. 25.1(b) that form the basis of the D_k group action starting with the same pair of polar vertices as the A_k, labeling them with "n" and "s" for North and South polar coordinates. However, for the dihedra, the $k + 1$ equatorial coordinates are actually included in the geometry, forming the successive bottom pairs of vertices for isosceles triangles spanning two pyramids with $(k + 1)$-gon bases; one polygon-based pyramid has a North pole peak vertex, and its mirror image shares the polygonal base, but has an apex corresponding to the South pole peak vertex. We will see later that there is a more group-theory-oriented view of the South pole triangles, interpreted as the result of performing a 180-degree "flip" about an equatorial axis, which we will choose here to be the $\hat{x}$-axis, so the North pole swings to the South pole in the plane perpendicular to the $\hat{x}$-axis before we apply the mirror sequence of rotations. In terms of the geometry, the dihedron vertex coordinates for each k are the same as the A_k digon polar vertex and edge center coordinates, except that now instead of digons built from semicircular arcs, we have triangle vertices defining $2(k + 1)$ isosceles triangles. We choose the position of the initial vertex to constrain the *first edge center* to be aligned with the $\hat{x}$-axis, as that will be our natural flip axis corresponding to the two-fold edge symmetries throughout the ADE groups; choosing a vertex for that action would be less universally consistent. Our vertices for the poles and the equatorial sampling are thus

$$v_n = \begin{bmatrix} 0 & 0 & +1 \end{bmatrix},$$

$$v_s = \begin{bmatrix} 0 & 0 & -1 \end{bmatrix},$$

$$v_0 = \begin{bmatrix} \cos\left(0 - \dfrac{\pi}{k+1}\right) & \sin\left(0 - \dfrac{\pi}{k+1}\right) & 0 \end{bmatrix},$$

$$v_1 = \begin{bmatrix} \cos\left(\dfrac{2\pi}{k+1} - \dfrac{\pi}{k+1}\right) & \sin\left(\dfrac{2\pi}{k+1} - \dfrac{\pi}{k+1}\right) & 0 \end{bmatrix},$$

$$v_2 = \begin{bmatrix} \cos\left(\dfrac{2\pi \times 2}{k+1} - \dfrac{\pi}{k+1}\right) & \sin\left(\dfrac{2\pi \times 2}{k+1} - \dfrac{\pi}{k+1}\right) & 0 \end{bmatrix}, \tag{25.3}$$

$$\vdots = \begin{bmatrix} \vdots & \vdots & \vdots \end{bmatrix},$$

$$v_{k-1} = \begin{bmatrix} \cos\left(\dfrac{2\pi(k-1)}{k+1} - \dfrac{\pi}{k+1}\right) & \sin\left(\dfrac{2\pi(k-1)}{k+1} - \dfrac{\pi}{k+1}\right) & 0 \end{bmatrix},$$

$$v_k = \begin{bmatrix} \cos\left(\dfrac{2\pi k}{k+1} - \dfrac{\pi}{k+1}\right) & \sin\left(\dfrac{2\pi k}{k+1} - \dfrac{\pi}{k+1}\right) & 0 \end{bmatrix}.$$

As we can tell from Fig. 25.1(b), the edge-center directions of the dihedral structure are

$$e(n; i) = \{v_n, v_i\} \ (i = 0, \dots, k), \tag{25.4}$$

$$e(z = 0; i) = \{v_i, v_{i+1|k}\} \ (i = 0, \dots, k), \tag{25.5}$$

$$e(s; i) = \{v_s, v_i\} \ (i = 0, \dots, k), \tag{25.6}$$

where $\{\_,\_\}$ denotes the normalized vector average, and the first equatorial edge center for any k is the normalized sum $v_0 + v_1$, or

$$e(z = 0; 0) = \{v_0, v_1\} = [1, 0, 0],$$

that is, lying in the $\hat{x}$ direction as desired. The faces have vertices from the $(k + 1)$-face North pole collection and the $(k + 1)$-face South pole collection:

$$f(n; i) = \{v_n, v_i, v_{i+1|k}\} \ (i = 0, \ldots, k), \tag{25.7}$$
$$f(s; i) = \{v_s, v_{i+1|k}, v_i\} \ (i = 0, \ldots, k). \tag{25.8}$$

The characterizing features of the dihedron geometry are that only the single edge flip is required to generate all symmetries, and that there are no additional face-center symmetries. We will assume that one single North-to-South edge flip plus a second set of A_k-like polar rotations include all supplementary D_k symmetries, although technically the four-fold partition case with $k = 3$ has isosceles triangles that are accidentally equilateral.

25.4 Coordinates of the E_6 tetrahedral geometry

We choose the set of tetrahedral coordinates that form the basis of the E_6 group action by aligning the tetrahedral edges with the diagonals of a cube, as shown in Fig. 25.1(c). This has the advantage of aligning the directions of the edge centers with the Cartesian axes. The four vertices are then simply one diagonal pair at the top of a cube, and an orthogonal diagonal pair at the bottom, with the values chosen to make the first vertex correspond to a certain group element that we will introduce in Chapter 27, and the first three chosen in right-handed order to define the face-center direction $(+1, +1, +1)$. Our E_6 tetrahedral vertices are then

$$
\begin{aligned}
v_1 &= \left[+\frac{1}{\sqrt{3}} \quad -\frac{1}{\sqrt{3}} \quad +\frac{1}{\sqrt{3}} \right], \\
v_2 &= \left[+\frac{1}{\sqrt{3}} \quad +\frac{1}{\sqrt{3}} \quad -\frac{1}{\sqrt{3}} \right], \\
v_3 &= \left[-\frac{1}{\sqrt{3}} \quad +\frac{1}{\sqrt{3}} \quad +\frac{1}{\sqrt{3}} \right], \\
v_4 &= \left[-\frac{1}{\sqrt{3}} \quad -\frac{1}{\sqrt{3}} \quad -\frac{1}{\sqrt{3}} \right],
\end{aligned}
\tag{25.9}
$$

and the edge centers are

$$
\begin{aligned}
e_1 &= [\ +1 \quad 0 \quad 0 \ ], \\
e_2 &= [\ -1 \quad 0 \quad 0 \ ], \\
e_3 &= [\ 0 \quad +1 \quad 0 \ ], \\
e_4 &= [\ 0 \quad -1 \quad 0 \ ], \\
e_5 &= [\ 0 \quad 0 \quad +1 \ ], \\
e_6 &= [\ 0 \quad 0 \quad -1 \ ].
\end{aligned}
\tag{25.10}
$$

The face centers in each of the four cases are simply the opposites of the vertex coordinates, and thus mirror the tetrahedral vertex symmetry,

$$
\left.
\begin{aligned}
f_1 &= \left[-\frac{1}{\sqrt{3}} \quad +\frac{1}{\sqrt{3}} \quad -\frac{1}{\sqrt{3}} \right] \\
f_2 &= \left[-\frac{1}{\sqrt{3}} \quad -\frac{1}{\sqrt{3}} \quad +\frac{1}{\sqrt{3}} \right] \\
f_3 &= \left[+\frac{1}{\sqrt{3}} \quad -\frac{1}{\sqrt{3}} \quad -\frac{1}{\sqrt{3}} \right] \\
f_4 &= \left[+\frac{1}{\sqrt{3}} \quad +\frac{1}{\sqrt{3}} \quad +\frac{1}{\sqrt{3}} \right]
\end{aligned}
\right\} .
\tag{25.11}
$$

In Fig. 25.2 we show the directions of the possible fixed axes of rotation symmetry for the tetrahedron: four vertex directions, six edge-center directions, and four face-center directions (which coincide with the vertex axes).

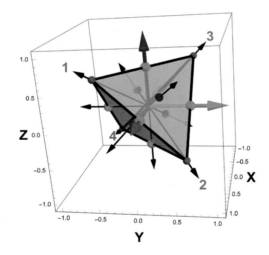

FIGURE 25.2

The E_6 tetrahedral group vertex, edge, and face rotation centers, indicating the axes that correspond to elements of the E_6 Kleinian 3D rotation group actions. Considering mirror-image rotations to be the same, we see that in this picture every vertex-based rotation is matched by a face-center rotation, and every edge-center rotation is matched by an opposite edge-center rotation. To choose unique group elements, we pick only one of each of these pairs.

25.5 Coordinates of the E_7 octahedral geometry

We choose the set of octahedral coordinates that form the basis of the E_7 group action by aligning the six vertices of the octahedron with the Cartesian coordinate axes as shown in Fig. 25.1(d), essentially the same directions as the edge centers we chose for the tetrahedral coordinate system:

$$v_1 = [\quad +1 \quad 0 \quad 0 \quad] , \qquad v_4 = [\quad -1 \quad 0 \quad 0 \quad] ,$$
$$v_2 = [\quad 0 \quad +1 \quad 0 \quad] , \qquad v_5 = [\quad 0 \quad -1 \quad 0 \quad] , \qquad (25.12)$$
$$v_3 = [\quad 0 \quad 0 \quad +1 \quad] , \qquad v_6 = [\quad 0 \quad 0 \quad -1 \quad] .$$

These vertices generate a set of 12 edge centers corresponding to the 12 90-degree-rotated edge centers of the cube that is the dual of our octahedron. These become, normalized to unit vectors, this list in which every direction has an opposite-sign pair:

$$e_1 = [-\tfrac{1}{\sqrt{2}} \quad \tfrac{1}{\sqrt{2}} \quad 0 \quad] , \qquad e_7 = [\tfrac{1}{\sqrt{2}} \quad \tfrac{1}{\sqrt{2}} \quad 0 \quad] ,$$
$$e_2 = [-\tfrac{1}{\sqrt{2}} \quad 0 \quad -\tfrac{1}{\sqrt{2}} \quad] , \qquad e_8 = [\quad 0 \quad -\tfrac{1}{\sqrt{2}} \quad -\tfrac{1}{\sqrt{2}} \quad] ,$$
$$e_3 = [-\tfrac{1}{\sqrt{2}} \quad 0 \quad \tfrac{1}{\sqrt{2}} \quad] , \qquad e_9 = [\tfrac{1}{\sqrt{2}} \quad 0 \quad -\tfrac{1}{\sqrt{2}} \quad] ,$$
$$e_4 = [-\tfrac{1}{\sqrt{2}} \quad -\tfrac{1}{\sqrt{2}} \quad 0 \quad] , \qquad e_{10} = [\quad 0 \quad -\tfrac{1}{\sqrt{2}} \quad \tfrac{1}{\sqrt{2}} \quad] , \qquad (25.13)$$
$$e_5 = [\quad 0 \quad \tfrac{1}{\sqrt{2}} \quad -\tfrac{1}{\sqrt{2}} \quad] , \qquad e_{11} = [\tfrac{1}{\sqrt{2}} \quad 0 \quad \tfrac{1}{\sqrt{2}} \quad] ,$$
$$e_6 = [\quad 0 \quad \tfrac{1}{\sqrt{2}} \quad \tfrac{1}{\sqrt{2}} \quad] , \qquad e_{12} = [\tfrac{1}{\sqrt{2}} \quad -\tfrac{1}{\sqrt{2}} \quad 0 \quad] .$$

The eight triangular faces, in opposite-direction redundant pairs, have centers corresponding to these normalized vectors:

$$f_1 = [\quad \tfrac{1}{\sqrt{3}} \quad -\tfrac{1}{\sqrt{3}} \quad \tfrac{1}{\sqrt{3}} \quad] ,$$
$$f_2 = [\quad \tfrac{1}{\sqrt{3}} \quad \tfrac{1}{\sqrt{3}} \quad \tfrac{1}{\sqrt{3}} \quad] ,$$
$$f_3 = [-\tfrac{1}{\sqrt{3}} \quad \tfrac{1}{\sqrt{3}} \quad \tfrac{1}{\sqrt{3}} \quad] ,$$
$$f_4 = [-\tfrac{1}{\sqrt{3}} \quad -\tfrac{1}{\sqrt{3}} \quad \tfrac{1}{\sqrt{3}} \quad] ,$$
$$f_5 = [-\tfrac{1}{\sqrt{3}} \quad -\tfrac{1}{\sqrt{3}} \quad -\tfrac{1}{\sqrt{3}} \quad] , \qquad (25.14)$$
$$f_6 = [\quad \tfrac{1}{\sqrt{3}} \quad -\tfrac{1}{\sqrt{3}} \quad -\tfrac{1}{\sqrt{3}} \quad] ,$$
$$f_7 = [-\tfrac{1}{\sqrt{3}} \quad \tfrac{1}{\sqrt{3}} \quad -\tfrac{1}{\sqrt{3}} \quad] ,$$
$$f_8 = [\quad \tfrac{1}{\sqrt{3}} \quad \tfrac{1}{\sqrt{3}} \quad -\tfrac{1}{\sqrt{3}} \quad] .$$

We show the geometry of the octahedral vertices, edge centers, and face centers in Fig. 25.3, annotated with the axes of ordinary 3D rotations about the vertex, edge-center, and face-center directions. For the octahedron, it is particularly easy to see the exact correspondence between the vertex, edge, and face rotational symmetries of the octahedron and the face, edge, and vertex rotational symmetries of its dual cube.

25.6 Coordinates of the E_8 icosahedral geometry

We choose the set of 12 icosahedral vertex coordinates that form the basis of a certain traditional form of the E_8 group action, which is based on the fact that the 12 icosahedral vertices can be grouped into

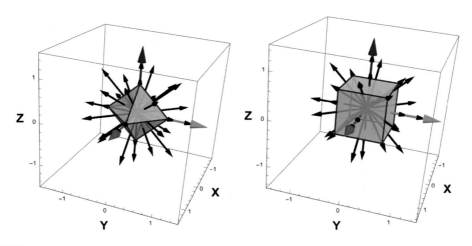

FIGURE 25.3

The E_7 octahedral group vertex, edge, and face rotation centers, for both the octahedral group and its dual, the cubic or hexahedral group.

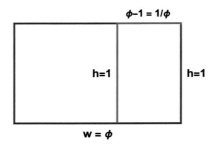

FIGURE 25.4

The geometry of the golden ratio, $w/h = \phi = h/(w-h) = 1/(\phi - 1)$, so $(\phi - 1) = 1/\phi$ and $\phi = \frac{1}{2}\left(1 + \sqrt{5}\right)$.

three sets of four. Each set of four vertices lying in the planes $x = 0$, $y = 0$, or $z = 0$ outlines a *golden rectangle* embedded in that plane, as shown in Fig. 25.4 with width w and height h obeying

$$\frac{w}{h} = \frac{h}{w-h} \quad \rightarrow \quad \frac{w}{h} = \phi = \frac{1}{2}\left(1 + \sqrt{5}\right) = 1.61803 = \textit{the golden ratio}.$$

The 12 vertices of the icosahadron, as well as the 12 face centers of the dodecahedron, fall into these three groups of orthogonal golden rectangles. With the golden ratio $\phi = (1 + \sqrt{5})/2$, our chosen explicit form of the icosahedral vertices embedded in the unit sphere $\mathbf{S}^2$ can be written

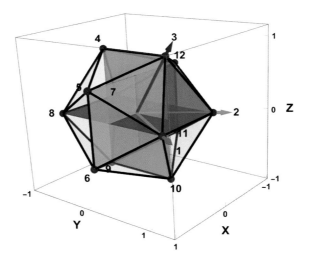

FIGURE 25.5

The E_8 icosahedral geometry with the vertices of Eq. (25.15), emphasizing the subtle $\hat{\mathbf{x}} = 0$ plane, $\hat{\mathbf{y}} = 0$ plane, $\hat{\mathbf{z}} = 0$ plane triple golden rectangle structure hidden within the icosahedron, shown in red, green, and blue, respectively. The fundamental vertex direction used in the traditional E_8 group generators is the first vertex indicated by the red arrow. The first three vertices form the first triangular face shown in magenta.

$$v_1 = \frac{1}{\sqrt{\phi^2 + 1}} [\ 1 \quad \phi \quad 0\ ], \qquad v_2 = \frac{1}{\sqrt{\phi^2 + 1}} [\ -1 \quad \phi \quad 0\ ],$$

$$v_3 = \frac{1}{\sqrt{\phi^2 + 1}} [\ 0 \quad 1 \quad \phi\ ], \qquad v_4 = \frac{1}{\sqrt{\phi^2 + 1}} [\ 0 \quad -1 \quad \phi\ ],$$

$$v_5 = \frac{1}{\sqrt{\phi^2 + 1}} [\ \phi \quad 0 \quad 1\ ], \qquad v_6 = \frac{1}{\sqrt{\phi^2 + 1}} [\ \phi \quad 0 \quad -1\ ],$$

$$v_7 = \frac{1}{\sqrt{\phi^2 + 1}} [\ -1 \quad -\phi \quad 0\ ], \qquad v_8 = \frac{1}{\sqrt{\phi^2 + 1}} [\ 1 \quad -\phi \quad 0\ ], \qquad (25.15)$$

$$v_9 = \frac{1}{\sqrt{\phi^2 + 1}} [\ 0 \quad -1 \quad -\phi\ ], \qquad v_{10} = \frac{1}{\sqrt{\phi^2 + 1}} [\ 0 \quad 1 \quad -\phi\ ],$$

$$v_{11} = \frac{1}{\sqrt{\phi^2 + 1}} [\ -\phi \quad 0 \quad -1\ ], \qquad v_{12} = \frac{1}{\sqrt{\phi^2 + 1}} [\ -\phi \quad 0 \quad 1\ ].$$

(Note that, although it does not greatly change the computation, golden ratio identities such as $(\phi^2 + 1) = \sqrt{5}\,\phi = \phi + 2$ and $1 + \phi = \phi^2$ can be used to simplify various icosahedron-related equations.) In Fig. 25.5, we show the fundamental structure behind this coordinate system, which is that, in remarkable parallel to the tetrahedron and the octahedron, the icosahedron also has a basic underlying shape corresponding to a Cartesian cube. Each of the vertex coordinates in Eq. (25.15) lies in either the $\hat{\mathbf{x}} = 0$ plane, the $\hat{\mathbf{y}} = 0$ plane, or the $\hat{\mathbf{z}} = 0$ plane, and the four coordinates in each plane outline an embedded golden rectangle as shown in Fig. 25.5. We can easily see from Eq. (25.15) that

$$\frac{v_1 - v_8}{v_1 - v_2} = \frac{2\phi}{2} = \phi\ .$$

The first vertex, in the direction $(1, \phi, 0)$, is the vertex direction traditionally chosen for the fifth-order generator of the E_8 group "Presentation" (Wikipedia, 2023d), which we will explore shortly.

The icosahedral edge and face centers. To complete the picture of the geometry defining the actions of our most complicated group, the icosahedral E_8 discrete group, we need the edge centers and the face centers in addition to the vertices. We easily calculate those just from averaging the vertices at each end of an edge, and averaging all the vertices of a polygonal face. The result of applying the edge center extraction to the vertices of Eq. (25.15) is

$$
\begin{aligned}
e_1 &= \frac{1}{2\phi}[\quad 0 \qquad 2\phi \qquad 0 \quad], & e_2 &= \frac{1}{2\phi}[\quad -1 \qquad \phi+1 \qquad \phi \quad], \\
e_3 &= \frac{1}{2\phi}[\quad 1 \qquad \phi+1 \qquad \phi \quad], & e_4 &= \frac{1}{2\phi}[\quad 0 \qquad 0 \qquad 2\phi \quad], \\
e_5 &= \frac{1}{2\phi}[\quad \phi \qquad -1 \qquad \phi+1 \quad], & e_6 &= \frac{1}{2\phi}[\quad \phi \qquad 1 \qquad \phi+1 \quad], \\
e_7 &= \frac{1}{2\phi}[\quad -1 \qquad -\phi-1 \qquad \phi \quad], & e_8 &= \frac{1}{2\phi}[\quad 2\phi \qquad 0 \qquad 0 \quad], \\
e_9 &= \frac{1}{2\phi}[\quad \phi+1 \qquad -\phi \qquad -1 \quad], & e_{10} &= \frac{1}{2\phi}[\quad \phi+1 \qquad -\phi \qquad 1 \quad], \\
e_{11} &= \frac{1}{2\phi}[\quad 1 \qquad -\phi-1 \qquad \phi \quad], & e_{12} &= \frac{1}{2\phi}[\quad \phi+1 \qquad \phi \qquad 1 \quad], \\
e_{13} &= \frac{1}{2\phi}[\quad \phi+1 \qquad \phi \qquad -1 \quad], & e_{14} &= \frac{1}{2\phi}[\quad 1 \qquad \phi+1 \qquad -\phi \quad], \\
e_{15} &= \frac{1}{2\phi}[\quad \phi \qquad 1 \qquad -\phi-1], & e_{16} &= \frac{1}{2\phi}[\quad 0 \qquad 0 \qquad -2\phi \quad], \quad\quad (25.16) \\
e_{17} &= \frac{1}{2\phi}[\quad -\phi \qquad 1 \qquad -\phi-1], & e_{18} &= \frac{1}{2\phi}[\quad -1 \qquad \phi+1 \qquad -\phi \quad], \\
e_{19} &= \frac{1}{2\phi}[\quad \phi \qquad -1 \qquad -\phi-1], & e_{20} &= \frac{1}{2\phi}[\quad -\phi \qquad -1 \qquad -\phi-1], \\
e_{21} &= \frac{1}{2\phi}[-\phi-1 \qquad \phi \qquad -1 \quad], & e_{22} &= \frac{1}{2\phi}[\quad 1 \qquad -\phi-1 \qquad -\phi \quad], \\
e_{23} &= \frac{1}{2\phi}[\quad -1 \qquad -\phi-1 \qquad -\phi \quad], & e_{24} &= \frac{1}{2\phi}[\quad 0 \qquad -2\phi \qquad 0 \quad], \\
e_{25} &= \frac{1}{2\phi}[-\phi-1 \qquad -\phi \qquad -1 \quad], & e_{26} &= \frac{1}{2\phi}[-\phi-1 \qquad \phi \qquad 1 \quad], \\
e_{27} &= \frac{1}{2\phi}[\quad -\phi \qquad 1 \qquad \phi+1 \quad], & e_{28} &= \frac{1}{2\phi}[\quad -\phi \qquad -1 \qquad \phi+1 \quad], \\
e_{29} &= \frac{1}{2\phi}[-\phi-1 \qquad -\phi \qquad 1 \quad], & e_{30} &= \frac{1}{2\phi}[\quad -2\phi \qquad 0 \qquad 0 \quad].
\end{aligned}
$$

Note that the denominator 2ϕ normalizes every edge vector to unity; the golden ratio has many hidden identities that make this possible, as unlikely as it looks; also observe that several edge centers, when divided by 2ϕ, simply become one of the six Cartesian signed axis directions. All of the available symmetry axes, including reflection redundancy, are shown for both the icosahedron and its dual do-

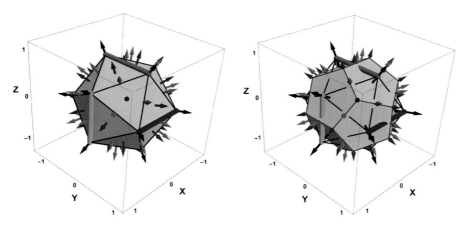

FIGURE 25.6

The E_8 icosahedral geometry with the vertices of Eq. (25.15) marked in black arrows, the edge centers in red arrows, and the face centers in blue arrows. These are the fundamental fixed directions of the ordinary 3D **SO**(3) rotation versions of the E_8 discrete group actions on the icosahedron. The RGB-colored bars accentuate the appearance of the Cartesian axes embedded in the icosahedron. The exact same symmetries are shown in the dual dodecahedron on the right. Mirror reflection symmetry produces duplication of the group actions that need to be taken into account.

decahedron in Fig. 25.6; the red, green, and blue bars accent the edges whose centers are the Cartesian axis directions.

Performing the face-center calculation, we find

$$f_1 = \tfrac{1}{\sqrt{3}(1+\phi)} [\quad 0 \quad\quad 2\phi+1 \quad\quad \phi \quad], \qquad f_2 = \tfrac{1}{\sqrt{3}(1+\phi)} [\ \ \phi+1 \quad\quad \phi+1 \quad\quad \phi+1 \ ],$$

$$f_3 = \tfrac{1}{\sqrt{3}(1+\phi)} [\ 2\phi+1 \quad\quad \phi \quad\quad 0 \quad], \qquad f_4 = \tfrac{1}{\sqrt{3}(1+\phi)} [\ \ \phi+1 \quad\quad \phi+1 \quad\quad -\phi-1 \ ],$$

$$f_5 = \tfrac{1}{\sqrt{3}(1+\phi)} [\quad 0 \quad\quad 2\phi+1 \quad\quad -\phi \quad], \qquad f_6 = \tfrac{1}{\sqrt{3}(1+\phi)} [\ -\phi-1 \quad\quad -\phi-1 \quad\quad \phi+1 \ ],$$

$$f_7 = \tfrac{1}{\sqrt{3}(1+\phi)} [\ -2\phi-1 \quad\quad -\phi \quad\quad 0 \quad], \qquad f_8 = \tfrac{1}{\sqrt{3}(1+\phi)} [\ -\phi-1 \quad\quad -\phi-1 \quad\quad -\phi-1 \ ],$$

$$f_9 = \tfrac{1}{\sqrt{3}(1+\phi)} [\quad 0 \quad\quad -2\phi-1 \quad\quad -\phi \quad], \qquad f_{10} = \tfrac{1}{\sqrt{3}(1+\phi)} [\quad 0 \quad\quad -2\phi-1 \quad\quad \phi \quad],$$

$$f_{11} = \tfrac{1}{\sqrt{3}(1+\phi)} [\ -\phi-1 \quad\quad \phi+1 \quad\quad \phi+1 \ ], \qquad f_{12} = \tfrac{1}{\sqrt{3}(1+\phi)} [\quad \phi \quad\quad 0 \quad\quad 2\phi+1 \ ],$$

$$f_{13} = \tfrac{1}{\sqrt{3}(1+\phi)} [\ 2\phi+1 \quad\quad -\phi \quad\quad 0 \quad], \qquad f_{14} = \tfrac{1}{\sqrt{3}(1+\phi)} [\quad \phi \quad\quad 0 \quad\quad -2\phi-1 \ ],$$

$$f_{15} = \tfrac{1}{\sqrt{3}(1+\phi)} [\ -\phi-1 \quad\quad \phi+1 \quad\quad -\phi-1 \ ], \qquad f_{16} = \tfrac{1}{\sqrt{3}(1+\phi)} [\quad -\phi \quad\quad 0 \quad\quad 2\phi+1 \ ],$$

$$f_{17} = \tfrac{1}{\sqrt{3}(1+\phi)} [\ \ \phi+1 \quad\quad -\phi-1 \quad\quad \phi+1 \ ], \qquad f_{18} = \tfrac{1}{\sqrt{3}(1+\phi)} [\ \ \phi+1 \quad\quad -\phi-1 \quad\quad -\phi-1 \ ],$$

$$f_{19} = \tfrac{1}{\sqrt{3}(1+\phi)} [\quad -\phi \quad\quad 0 \quad\quad -2\phi-1 \ ], \qquad f_{20} = \tfrac{1}{\sqrt{3}(1+\phi)} [\ -2\phi-1 \quad\quad \phi \quad\quad 0 \quad],$$

$$(25.17)$$

where now the universal normalization factor is $\sqrt{3}(1 + \phi)$, which again seems improbable, but in fact the numerator of every face-center vector possesses exactly that length. Fig. 25.6 shows all the rotation axes, the icosahedral vertices, edge centers, and face centers, on which the symmetry group elements of E_8 that we are investigating can act, along with their dual appearances in the dodecahedron.

The discrete ADE rotation groups and their quaternions

26

26.1 Introduction

In this chapter, we will build on our knowledge from Chapter 25 about the nature of the symmetric tessellations of the 2-sphere $\mathbf{S}^2$. From the geometric properties of the observed tessellations, we can identify the geometry of the *symmetric rotations*, as each such rotation depends on writing down the direction in 3D space kept fixed by a given rotational symmetry action. For both an axis-angle-parameterized 3D rotation in $\mathbf{SO}(3)$, and its corresponding doubled quaternion, all we need is the unit normal in 3D space $\hat{\mathbf{n}}$ that is kept fixed, and the fraction of 360 degrees that is the smallest symmetry-preserving rotation. Accordingly, in this chapter, we will first review the ways in which our geometric knowledge from Chapter 25 tells us how to define the appropriate 3D rotations $R(\theta, \hat{\mathbf{n}})$ from Eq. (2.17), and then move finally to exploiting the same data to define the quaternion groups, the so-called binary groups, using quaternion axis-angle rotations $q(\theta, \hat{\mathbf{n}})$ from Eq. (2.16). We then display the quaternions of the ADE group actions in their entirety, distinguishing each group element as a point in quaternion space using various methods from Chapter 4. In the following chapter, we introduce the concept of a subset of three group actions comprising a "Presentation," from which the whole group can be generated, and a simplified diagrammatic method inspired by basic group theory to encode the properties of each ADE group.

26.2 Actions of the $\mathbf{SO}(3)$ ADE rotations in Euclidean 3D space

We begin with the classical 3D rotation group, and how it fits into the way symmetric rotations act on the five collections of tessellations of $\mathbf{S}^2$ into identical spherical polygons. The actual ADE symmetries are simply the collections of all rotational actions that return any given symmetrically tessellated $\mathbf{S}^2$ to an identically appearing object. If we study each of the polyhedra in Fig. 25.1, we can see by inspection what we have been implying all along: that the symmetry actions can be classified entirely by the available repetitive actions on the vertices, edge centers, and face centers.

Now let us work out in detail *what form* those 3D rotations take, and *how many* repeated similar actions there are among the classes for each ADE symmetry group. We have already noted that there are cases where a given rotation has an identical mirrored partner acting elsewhere on the spherical tessellation, and we need to take such redundancies into account for situations in which we want each action to be unique. Our basic philosophy will be to choose only one of any paired action when we speak of or list $\mathbf{SO}(3)$ rotations, and often to *include* duplicate pairs when we deal with quaternion representations of rotations: the latter constructions are known technically as *binary groups* (Wikipedia,

2023c,h,f,e,d; Coxeter and Moser, 1980). We note that when we work with quaternions as an equivalent context for matrix-based group theory, the unit quaternion 4-vector $[q_0, q_1, q_2, q_3]$ denoting a point on the 3-sphere $\mathbf{S}^3$ is equivalent to the $\mathbf{SU(2)}$ matrix group element constructed from Pauli matrices, which we write as

$$[q] = q_0 I_2 - i\mathbf{q} \cdot \boldsymbol{\sigma} = \left[\begin{array}{cc} q_0 - iq_3 & -iq_1 - q_2 \\ -iq_1 + q_2 & q_0 + iq_3 \end{array} \right]. \tag{26.1}$$

We emphasize once again that the "$-i$" in this matrix expression is not a choice, but is required to conform to the Hamilton quaternion algebra.

Review of the properties of polygonal vertices, edges, and faces. For completeness, let us gather together in one place the relevant properties of vertices, edges, and faces in the tessellation of our polygons, and how rotations about those features are embodied in the five ADE classes of $\mathbf{S}^2$ tessellations whose geometry we are examining:

Vertices: Vertex-based rotations fix an axis from the sphere's center to a vertex, rotating by exactly the amount needed to move one edge ending on the vertex to the position of another edge ending on the vertex. If there are n edges terminating on a given vertex, there are $(n-1)$ nontrivial rotation elements, since the identity element executes no motion, and is a shared common element with all the transformations of one tessellation. The A_k and D_k groups have specific k-fold rotations about the $\hat{\mathbf{z}}$-axis (excluding the identity), doubled by a reflection for D_k, while the E_6, E_7, and E_8 groups have three, four, and five edges, respectively, radiating from each vertex, and thus have two, three, and four nonidentity vertex-aligned actions.

Edges: Edge-based rotations fix an axis from the origin to the *midpoint of an edge*, flipping the edge end for end in a 180-degree rotation. The edge transformation always has only a *single* nontrivial element. The A_k group has no edge-flip actions, and the D_k group needs only a single edge flip combined with its $\hat{\mathbf{z}}$-axis vertex spins.

Faces: Face-based rotations fix an axis from the center to the *midpoint of a face*, rotating a given vertex belonging to the face into the equivalent position of another vertex. For the A_k and D_k groups, there are no independent face-centered transformations. For the E_6, E_7, and E_8 groups, our face-based transformations always act on the center of an *equilateral triangle*, as we have chosen the set of equivalent dual Platonic solids all to be represented by those with triangular faces. Thus there are exactly *two* nontrivial face actions in the E_6, E_7, and E_8 discrete groups besides the identity, namely the two nontrivial possible rotations acting to rotate one equilateral triangle into another.

The transformations that rotate each tessellated $\mathbf{S}^2$ to a copy of itself are now easy to list, remembering that one member of the elements in the following totals is always the *identity transformation*:

Geometric element	How many rotations (including the shared identity)
Vertices	Number of edges (or faces) sharing the vertex
Edges	Always two when applicable
Faces	Always three for E_7 and E_8, not used for A_k and D_k, duplicated by vertices for E_6

The Euler characteristic, the 2-sphere, and the ADE tessellations. Each class of geometries in Fig. 25.1 specifically divides the sphere $\mathbf{S}^2$ into identical spherical surface patches (or equivalently identical flat-faced polygons, except for the tricky case of A_k, which needs digons). Each ADE tessellation has a certain number V of vertices, E of edges, and F of faces. This leads to the following properties to keep in mind:

- **All of the ADE geometries are spheres with Euler characteristic two.** One of the most important theorems of the elementary geometry of orientable 2D surfaces is that the Euler characteristic χ is an invariant determined by the numbers V, E, and F, and for the 2-sphere $\mathbf{S}^2$, that invariant is

$$\chi = V - E + F = 2 \,. \tag{26.2}$$

- **The vertex-face interchanging duality transform leaves the Euler characteristic invariant.** A fundamental feature of the Euler characteristic is that $V - E + F = F - E + V$, so there is a basic "duality" equivalence between pairs of surfaces related by interchanging their vertices with their faces. Fundamentally, duality amounts to rotating every edge by 90 degrees, and leaving it otherwise intact, and replacing every face by its center, calling that the new set of vertices, and expanding every vertex in a radiating spider-web-like line until it meets the new vertices and becomes a face. This transformation has important implications for the five Platonic solids, since we have to pick inequivalent members of that set to choose sensible unique discrete group actions (and our solution is to pick the triangle-based alternatives).
- **Collecting the V, E, and F counts for the ADE geometries.** The following table enumerates the vertex, edge, and face features and dualities of the cyclic and dihedral partitions, as well as the Platonic solids, that are associated with the elementary members of the ADE classification:

The cyclic and dihedral spaces					
Name	**V**	**E**	**F**	**Faces**	
A_k = Cyclic	2	$k+1$	$k+1$	Digons or "lunes"	
D_k = Dihedron	$2 + (k+1)$	$3(k+1)$	$2(k+1)$	Isosceles triangles	
The Platonic solids					
Name	**V**	**E**	**F**	**Faces**	**Dual**
E_6 = Tetrahedron	4	6	4	Equilateral triangles	*Self-dual*
E_7 = Octahedron	6	12	8	Equilateral triangles	*Hexahedron*
Hexahedron	8	12	6	Squares	*Octahedron*
E_8 = Icosahedron	12	30	20	Equilateral triangles	*Dodecahedron*
Dodecahedron	20	30	12	Pentagons	*Icosahedron*

Despite their increasing complexity, it is easy to check that the Euler characteristic is $\chi = V - E + F = 2$ for all these ways of symmetrically subdividing the sphere, and to deduce that there is no reason to have both a Platonic solid figure and its dual in the list of groups, because the pairs will have exactly the same discrete rotations due to the face-vertex symmetry of the duality transform.

- **The group actions may be redundant and require culling.** One of the other features we can extract from Fig. 25.1, is that there are many examples of situations where there are vertices exactly opposite vertices, edges opposite edges, and faces opposite faces; in fact for the tetrahedron, each three-fold face is opposite a three-fold vertex. In each of these cases, a group action that we choose to realize on one of these pairs will be *exactly the same action* as that on the mirror-image opposing partner.

Thus only one of these pairs will correspond to a unique group action. This becomes somewhat more interesting when we pass to the quaternion form of the discrete rotations, and realize that in the so-called "binary group" ADE representation, the doubling of the quaternions and the doubling of rotations corresponding to the mirror image elements can be viewed as equivalent: the quaternions could be perceived as "pulling apart" redundant rotations and providing them with an independent existence in quaternion space.

26.3 Identifying the 3D symmetric ADE transformations

Now that we have an idea of the sets of identical polyhedron tessellations of the topological sphere we are dealing with to define the ADE discrete group actions, we can define the transformations themselves. We again restrict ourselves for the time being to the ordinary $SO(3)$ 3D rotation matrix forms for the ADE actions, exploiting the axis-angle form $R(\theta, \hat{\mathbf{n}})$ as our fundamental tool. We have already argued that the vertices, edge centers, and face centers (when present as an independent source of symmetry) are the basic places at which we expect to be able to define a symmetric action. Since those three collections of points remain fixed under the corresponding rotations, the corresponding rotation matrices have those collections of points as their fixed axes $\hat{\mathbf{n}}$. We present the details of the entire set of such constructions graphically in Figs. 26.1, 26.2, 26.3, 26.4, and 26.5, with individual details and discussion in the following paragraphs:

$\mathbf{A}_k$ **symmetries.** As one can see from Fig. 25.1 and Fig. 26.1, the A_k symmetry transformation of $\mathbf{S}^2$ simply rotates the sphere around the vertical axis in $k + 1$ equal steps, with $n = (0, 1, \ldots, k - 1, k)$ labeling possible actions that rotate the $k + 1$ segment "orange slice" tessellation into itself. $n = 0$ (and also $n = k + 1$) are the identity action and execute no motion, so only k elements of the group have nontrivial actions, and each of the actions can be achieved by repeating the smallest rotation by angle $2\pi/(k + 1)$ up to k times. There are no actions about edge or face centers for this basic transformation.

$\mathbf{D}_k$ **symmetries.** From Fig. 25.1 and Fig. 26.2, we can see that the dihedral symmetry transformation group of $\mathbf{S}^2$ includes the A_k polar rotations both without and with a 180-degree edge flip fixing the $\hat{\mathbf{x}}$-axis that interchanges the North and South poles of the sphere. In addition to the flip, the remaining actions can be regarded as rotating the sphere around the South vertical axis in $k + 1$ equal steps, with $n = (0, 1, \ldots, k - 1, k)$ possible actions that rotate the $k + 1$ segment "isosceles triangle mirror pyramid" tessellation into itself. In this second set of $\hat{\mathbf{z}}$-axis rotations, $n = 0$ is no longer the identity, but the nontrivial result of flipping the identity orientation matrix, while $n = k + 1$ repeats the $n = 0$ case and produces no new action. Thus the entire group consists of $2k + 2$ distinct actions, one of which is the identity. There are no additional independent actions, e.g., for face centers, in this group because the triangles are never equilateral except for the special case of $k = 3$, which has the same shape as the basis for the $\mathbf{E}_7$ octahedral group, but retains its dual identity as a basis element of each set, acted on by distinct groups.

$\mathbf{E}_6$, $\mathbf{E}_7$, and $\mathbf{E}_8$ are Platonic solid symmetries. The remaining rotations in the Kleinian groups take place as rotations that transform the five Platonic solids into themselves. The basic structure of the symmetric rotations on these polyhedra can be seen in Fig. 25.1, along with Figs. 25.2, 25.3, and 25.6.

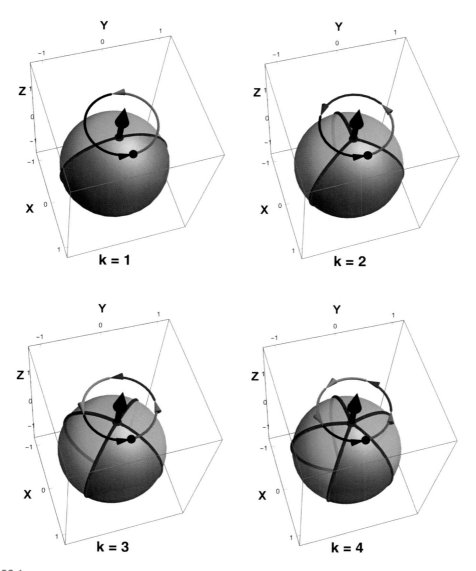

FIGURE 26.1

The A_k cyclic group vertex rotation geometry, with k individual rotations by angle $2\pi/(k+1)$. The unique actions of the rotation group are given by the colored circular arcs, with the final black arc denoting the $(k+1)$-st rotation returning to the identity at the black dot starting point.

We repeat these Platonic solid figures in Figs. 26.3, 26.4, and 26.5, now with an eye towards recognizing their group-theoretical significance separately from their geometric construction. We recall first that the reason there are only three such groups instead of five is because the Platonic solids are related by the

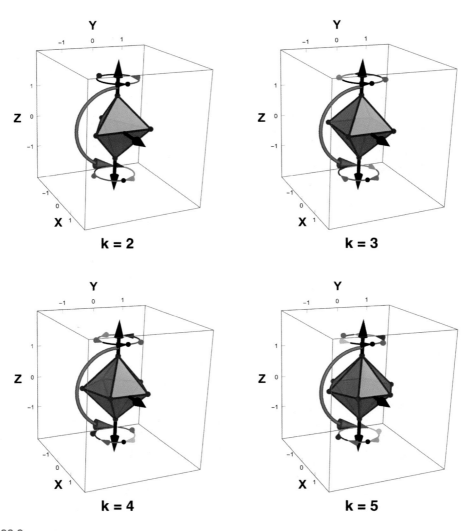

FIGURE 26.2

The D_k dihedral group rotation geometry, combining polar vertex axis rotations denoted by circular arcs, with the final black arc in the upper group denoting the return to the identity orientation. The D_k equatorial edge flip around the marked edge-centered $\hat{\mathbf{x}}$-axis is shown as a blue 180-degree arc.

duality transform, which converts any such polygon into another polygon by shrinking each face to a point, and expanding each point to a face, leaving the same number of lines rotated by 90 degrees in place about their midpoint. When we count the vertices, edges, and faces of the Platonic solids, we discover, following the table in Section 26.2, that we only need to consider three of them, the tetrahedron, the octahedron, and the icosahedron, all with only equilateral triangular faces.

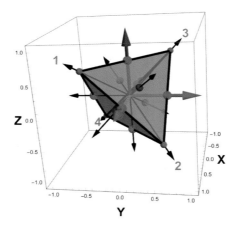

FIGURE 26.3

The tetrahedral group vertex, edge, and face rotation centers, as red, green, and blue dots, indicating the axes that correspond to elements of the E_6 tetrahedral 3D rotation group actions. Considering mirror-image rotations to be the same, we see that in this picture every vertex-based rotation is matched by a face-center rotation, and every edge-center rotation is matched by an opposite edge-center rotation. To choose unique group elements, we only pick one of each of these pairs.

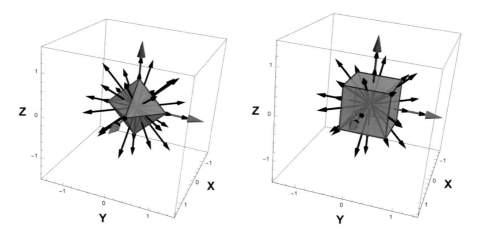

FIGURE 26.4

The octahedral group vertex, edge, and face rotation centers, indicating the axes that correspond to elements of the E_7 Kleinian 3D rotation group actions. We first observe the duality relation between the octahedron and the cube: all the possible rotation axes occur in both geometries, so we can focus on the octahedron. Then, in order to exclude mirror-image copies of the same group in the octahedron, we notice that every vertex, every edge center, and every face center has a mirror-image copy. Therefore, to identify the minimal group actions, we only need consider the action of one of each of these mirror pairs.

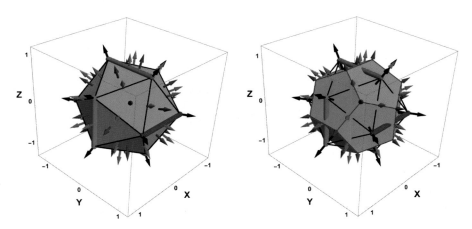

FIGURE 26.5

The icosahedral group vertex, edge, and face rotation centers, indicating the axes about which the rotations of E_8 act. Our final Kleinian group acts symmetrically on both the icosahedron and its dual, the dodecahedron. The two objects obviously have the same set of possible axes that can be fixed by a symmetry, so we need consider only the icosahedron. The icosahedral actions fix the vertices, marked in black arrows, the edge centers in red arrows, and the face centers corresponding to blue arrows. These are the fundamental fixed directions of 3D **SO**(3) rotation versions of the E_8 discrete group actions. Mirror reflection symmetry occurs for every vertex, every edge center, and every face center, so the family of unique E_8 group actions requires only one of each of these pairs.

    As we now count the sets of unique rotations E_6, E_7, and E_8 acting on the Platonic solid surface elements, we recall first that each vertex, edge, and face has a group that starts first with its own identity element: since these are *all identical* we must take care to count only one of them. In addition, there are mirror-image geometry elements for each fixed axis of rotation, so, in one way or another, we may only count one each, leading to a division by two when we tally the numbers in most cases. We thus get the following table, where we include the doubled tally for equivalent quaternions for future reference (remember that indeed even the *identity* is doubled for quaternions):

Platonic solid symmetries					
Name	**V**	**E**	**F**	**+ Identity = Total**	**q:Total**
E_6 = Tetrahedron	$4(3-1)=8$	$\dfrac{6(2-1)}{2}=3$	zero: faces = vertices	$8+3+0+1=12$	$2 \times 12 = 24$
E_7 = Octahedron	$\dfrac{6(4-1)}{2}=9$	$\dfrac{12(2-1)}{2}=6$	$\dfrac{8(3-1)}{2}=8$	$9+6+8+1=24$	$2 \times 24 = 48$
E_8 = Icosahedron	$\dfrac{12(5-1)}{2}=24$	$\dfrac{30(2-1)}{2}=15$	$\dfrac{20(3-1)}{2}=20$	$24+15+20+1=60$	$2 \times 60 = 120$

    The exceptions in this table are the following: first, the tetrahedron has a face exactly opposite each vertex due to its self-duality, so the vertex-centered and face-centered rotations are identical and can only be counted once. However, *all edges* in the Platonic solids have a paired edge exactly opposite, so only *half* of the edge-centered rotations are independent, and furthermore, for the octahedron and icosahedron, while the face-centered and edge-centered rotations are independent, they *all* have a paired opposite, so only half are independent. Finally, we have subtracted one from each list of possible rotations at a vertex, edge, or face to account for the identity element: at the end, we add back *one* identity element to the total number of rotations to account for those subtractions.

26.4 The quaternion actions of the Klein groups

We now turn to the quaternion forms of the ADE list of discrete rotations, which reappear throughout mathematics and physics with surprising frequency. We have already shown in Section 26.3 using 3D geometric arguments how the rotations that produce symmetries can be carried out in principle with ordinary $\mathbf{SO}(3)$ transformations. In order to be completely explicit about the forms of the rotations, the next step would be to take the coordinates for the vertices, edges, and faces in the ADE symmetric sphere tessellations and transform them into their corresponding axis-angle parameters. However, this is exactly the same calculation that is needed to compute the corresponding quaternions, and so we move immediately from the $\mathbf{SO}(3)$ ADE group actions to the quaternion binary groups. In addition, there are obstacles to making good graphical representations of the properties of the $\mathbf{SO}(3)$ transformations compared to very powerful tools that we can use to represent all the transformations, including all signs of the binary groups, using quaternions. In this section, we will work from the beginning with the full quaternion treatment, making reference to the $\mathbf{SO}(3)$ transformations as consequences of the more fundamental, and *visualizable*, quaternion actions.

26.4.1 The A_k cyclic quaternion group

In Fig. 25.1, we saw the $(k+1)$-fold sliced sphere (digon partitioned) geometry that embodies the fundamental structure of the A_k symmetry, and in Fig. 26.1 selected explicit examples of the group action. The basic action of A_k is defined as the cyclic group, all the rotations leaving the z-axis fixed (that is, rotations in the xy-plane) and returning to the identity after $k+1$ repetitions. First we note that the relevant *continuous* group action is just a right-handed rotation about the $\hat{\mathbf{z}}$-axis given by the quaternion

$$q(\theta, \hat{\mathbf{z}}) = \left[\cos\frac{\theta}{2}, \, 0, \, 0, \, \sin\frac{\theta}{2} \right],$$

(26.3)

or the $\mathbf{SU}(2)$ matrix

$$[q] = \begin{bmatrix} \exp\left(-i\frac{\theta}{2}\right) & 0 \\ 0 & \exp\left(+i\frac{\theta}{2}\right) \end{bmatrix}.$$

(26.4)

Note the reversal of the sign of i, required to preserve our standard $(\mathbf{I}, \mathbf{J}, \mathbf{K})$ orientation with $\mathbf{IJK} = -1$. The discrete version of the continuous fixed-z-axis rotation that implements a cyclic group rotation of order k has $k+1$ elements, labeled by $n = (0, 1, \ldots, k)$, which are rotations in physical 3D space by the angles

$$\theta_{k,n} = \frac{2\pi n}{(k+1)}.$$

(26.5)

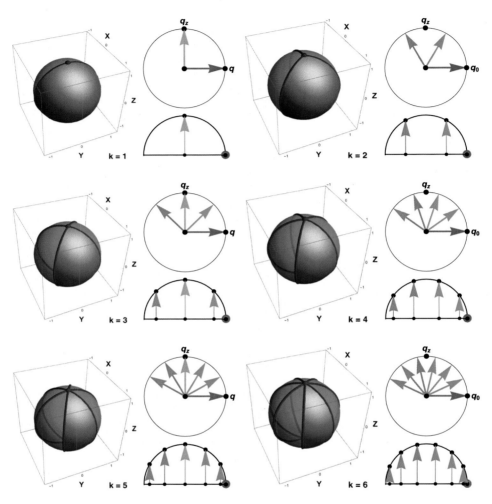

FIGURE 26.6

We show the geometry of A_k for $k = 1, \ldots, 6$ together with the plots of each quaternion group element on the circle to which q_0 and q_z are constrained. The fundamental "slice" domain in blue is rotated by the cyclic group into the other k identical slices to cover the entire $\mathbf{S}^2$ exactly once. Beneath the fundamental $q_0 q_z$-plane quaternion plots of each A_k symmetry group, we separate out the vector q_z components plotted in a half-disk to show how they mirror one another. Adopting the color coding of Chapter 4, we draw the $q_0 < 0$ components in a reddish color, the $q_0 = 0$ unit-length components on the surface of the equatorial $\mathbf{S}^2$ in orange, and the $q_0 > 0$ components in a bluish color; the zero-length black dot embedded in red is the identity.

Note that this *includes* the identity, with $n = 0$, as one of the $k + 1$ group elements, and *omits* $n = k + 1$, since that just returns full circle to the identity. To obtain the quaternion form whose geometric properties in $\mathbf{S}^3$ we will be exploiting for visualization purposes, we must divide θ by two, leading to a quaternion, with two copies of every group element, of the following form:

$$q_A(k, n) = \left[\cos \frac{\theta_{k,n}}{2}, \ 0, \ 0, \ \sin \frac{\theta_{k,n}}{2} \right] = \left[\cos \left(\frac{n\pi}{(k+1)} \right), \ 0, \ 0, \ \sin \left(\frac{n\pi}{(k+1)} \right) \right]. \tag{26.6}$$

The **SU(2)** matrix is just Eq. (26.4) with $\theta = \theta_{k,n}$ given by Eq. (26.5), and technically including the double-covering range $n = (0, 1, \dots, 2k + 1)$.

Our primary task now is to visualize the quaternion corresponding to the discrete A_k cyclic group actions given by Eq. (26.6). One can see that there are in general overlapping mirror-image pairs in the sine terms of the quaternion vector part, in addition to having overlapping q_3 vectors on the z-axis if we try to draw them using $\mathbf{q}$ alone and dropping the q_0 component. Thus drawing these using the vector-only 3D display suggested in Chapter 4 is problematic. However, since there are only two nonvanishing quaternion components, we can just use a circular plot in the 2D $q_0 q_z$-plane shown in Fig. 26.6 to easily distinguish the A_k group components. Since the distinct sine terms have a nice structure on their own, we add a supplementary plot, underneath the $q_0 q_z$ plot, displaying the q_z-axis terms individually, spaced according to their q_0 values.

All of the distinct group elements for each example of A_k are shown in Fig. 26.6, with the identity quaternion as a lone horizontal vector. The unique q_z elements are color coded to show their associated q_0 sign, with red, orange, and blue corresponding to $q_0 < 0, = 0,$ and > 0. When we move on to the dihedral group D_k, all of the A_k elements will be repeated, and then doubled in an interesting way to fill an independent sector of quaternion space.

Remark: The quaternion elements of the ADE groups are all doubled, with opposite-sign copies, realizing the *binary A_k* group. Thus a more exhaustive version of Fig. 26.6 would have a $q_0 q_z$ plot that continued around the full circle with negatives of the primary quaternions, starting with the $q = (-1, 0, 0, 0)$ alternative identity quaternion on the left horizontal axis. In the following, we will add the doubled version when it seems useful, e.g., when it helps us see the more complete symmetries of the binary groups.

26.4.2 The D_k dihedral quaternion group

The dihedral group D_k can be thought of as containing *two copies* of the cyclic group, with the second having a North-pole-to-South-pole flip imposed before each A_k cyclic group element acts. Examples of the spherical tessellations for which the D_k group preserves the geometry are shown in Fig. 25.1, and examples of the actions of the D_k group elements are shown in Fig. 26.2.

The first cyclic quaternion actions of the D_k dihedral group consist of the $k + 1$ A_k elements with rotation angle $\theta_{k,n}$ from Eq. (26.5),

$$q = \left[\cos \frac{\theta_{k,n}}{2}, \ 0, \ 0, \ \sin \frac{\theta_{k,n}}{2} \right] = \left[\cos \left(\frac{n\pi}{(k+1)} \right), \ 0, \ 0, \ \sin \left(\frac{n\pi}{(k+1)} \right) \right]. \tag{26.7}$$

These correspond to the 3D rotation $R(\theta_{k,n}, \hat{\mathbf{z}})$ for $n = 0, 1, 2, \dots, k - 1, k$, with $n = 0$ being the "no-action" identity q_{ID}. It is important to include q_{ID} because the second set of these elements in the D_k group includes a "flipping action" that changes q_{ID} so it is no longer trivial, but $(0, 1, 0, 0)$, a 180-degree rotation about the $\hat{\mathbf{x}}$-axis.

The pole-interchanging flip, assuming the poles are aligned with the $\hat{z}$-axis, can be chosen to leave *any* equatorial vertex or edge-center point fixed, so the list of options lies in a circle in the xy-plane. The quaternion representation of this flip for any point $(\cos t, \sin t)$ on this circle then takes the general form

$$q(\pi, \hat{n}(t)) = \left[\cos\frac{\pi}{2}, \cos t \sin\frac{\pi}{2}, \sin t \sin\frac{\pi}{2}, 0\right]$$
$$= [0, \cos t, \sin t, 0] \tag{26.8}$$

and the equivalent **SU(2)** matrix from Eq. (26.1) is

$$\left[q(\pi, \hat{n}(t))\right] = \begin{bmatrix} 0 & -iq_1 - q_2 \\ -iq_1 + q_2 & 0) \end{bmatrix} = \begin{bmatrix} 0 & -ie^{-it} \\ -ie^{+it} & 0 \end{bmatrix}.$$

We recall that the equatorial D_k vertices in Eq. (25.3) were chosen so the first edge center aligned with the $\hat{x}$-axis, $e(z = 0; 0) = [1, 0, 0]$, so we can always choose the flip to be a 180-degree rotation about the $\hat{x}$-axis, that is, $t = 0$ in Eq. (26.8). The corresponding flipping action $q(\pi, \hat{x})$ fixing the $\hat{x}$-axis, in the three notations, quaternion, **SU(2)**, and 3D matrix form, is simply

$$b = (0, 1, 0, 0), \quad B = -i\begin{bmatrix} 0 & 1 \\ 1 & 0 \end{bmatrix}, \quad R(b) = \begin{bmatrix} 1 & 0 & 0 \\ 0 & -1 & 0 \\ 0 & 0 & -1 \end{bmatrix}. \tag{26.9}$$

From the 3D rotation $R(b)$ in Eq. (26.9), we see that a North–South pole flip that leaves the x-axis fixed will change the signs of *both* the y- and z-axes. This is in fact required to preserve the parity or handedness of the orientation.

The two fundamental sets of D_k operations are the discrete versions of these continuous transformations:

- *Simple cyclic rotations:* The rotations $q(\theta, \hat{z})$ fixing the $\hat{z}$-axis, as in Eqs. (26.4) and (26.6).
- *Simple cyclic rotations after a flip:* The rotations $q(\theta, \hat{z}) \star q(\pi, \hat{x})$, combining the cyclic rotations with the pole-exchanging flip $q(\pi, \hat{x})$ of Eq. (26.8) produced by a 180-degree rotation about the $\hat{x}$-axis.

The full D_k dihedral group has $2(k + 1)$ elements, constructed from one copy of the same elements as the cyclic A_k group supplemented by a copy of A_k modified by the pole flip b in Eq. (26.9). We will construct the second set of $k + 1$ actions for D_k by acting first with the flip quaternion b, and then with $q(\theta_{k,n}, \hat{z})$, that is,

$$q_{\text{flip}}(\theta_{k,n}, \hat{z}) = q(\theta_{k,n}, \hat{z}) \star b. \tag{26.10}$$

Writing $\bar{v} = v \star b = v \star [0, 1, 0, 0]$, we find that the elements of the D_k quaternion group take the form

$$q(D_k) = \begin{cases} q_{\text{ID}} : [1, 0, 0, 0], \\ v(1), \ldots, v(k) : \left[\cos\left(\frac{n\pi}{(k+1)}\right), 0, 0, \sin\left(\frac{n\pi}{(k+1)}\right)\right], \\ q_{\text{flip:ID}} : [0, 1, 0, 0], \\ \bar{v}(1), \ldots, \bar{v}(k) : \left[0, \cos\left(\frac{n\pi}{(k+1)}\right), \sin\left(\frac{n\pi}{(k+1)}\right), 0\right]. \end{cases} \tag{26.11}$$

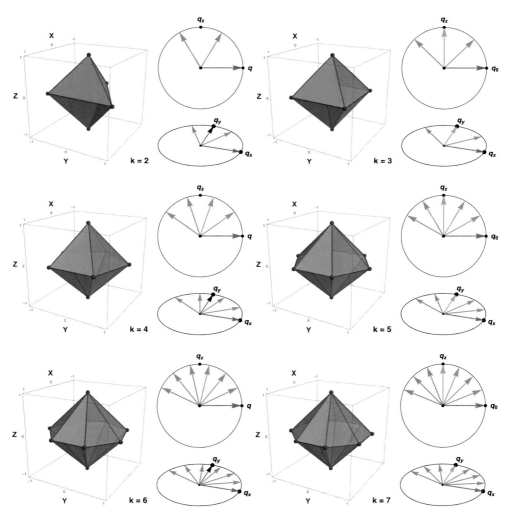

FIGURE 26.7

We show the geometry of D_k for $k = 2, \ldots, 7$, paired with the two sets of quaternions that make up the D_k group. The $q_0 q_z$ set of quaternions is a clone of the A_k quaternions, containing the identity element $q_0 = 1$, and constrained to discrete points on a circle in the $q_0 q_z$-plane. The second set is a matched set, flipped by 180 degrees about the $\hat{\mathbf{x}}$-axis of the first edge center, and thus constrained to a circle in the orthogonal $q_x q_y$-plane. In the flipped set of quaternions, the image of the identity quaternion is no longer an identity, but a bare rotation about the $\hat{\mathbf{x}}$-axis inverting the direction of $\hat{\mathbf{z}}$. The actions of the D_k group are clearly shown here to consist of a pair of $k + 1$ quaternions lying in disjoint 4D planes.

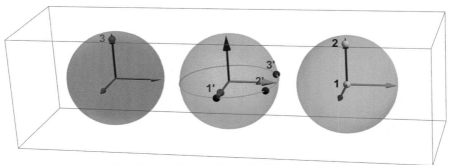

D_k : k = 2: 3 cyclic and 3 x–axis flipped quaternions

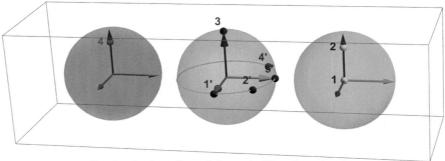

D_k : k = 3: 4 cyclic and 4 x–axis flipped quaternions

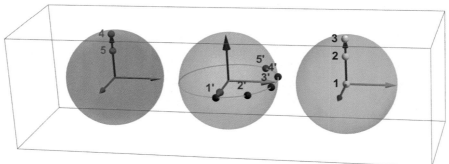

D_k : k = 4: 5 cyclic and 5 x–axis flipped quaternions

FIGURE 26.8

Quaternion coordinates for the actions of the Dihedral groups for $k = \{2, 3, 4\}$, showing only one of the two equivalent signs. The spheres, left to right, present **q** for $q_0 < 0$, the equator $q_0 = 0$, and $q_0 > 0$. All $k + 1$ cyclic group actions for both A_k and D_k appear with plain numbers, including the identity numbered "1" at the origin of the $q_0 > 0$ sphere. All the $k + 1$ *additional* "flipped" actions of the dihedral group are on the $q_x q_y$ equator of the $q_0 = 0$ equatorial sphere. Note that the identity element $(1, 0, 0, 0)$, labeled "1," is mapped to this equatorial sphere as the element $(0, 1, 0, 0)$, labeled "1'," which flips the North pole to the South pole keeping the $\hat{\mathbf{x}}$-axis fixed.

We observe that the elements that coincide with A_k all lie in the $q_0 q_z$-plane of the quaternion $\mathbf{S}^3$, and the other $k + 1$ elements consist of a *paired* action whose quaternion lies in the *disjoint* quaternion $q_x q_y$-plane. Note that the flipped group element corresponding to the identity in A_k is now nontrivial and, obviously, is just the flipping quaternion element b itself, lying along the chosen axis (such as the q_x-axis) in the $q_x q_y$-plane.

In Fig. 26.7, we display the basic polyhedral geometric elements of D_k from Fig. 25.1 and Fig. 26.2, and combine them with a display adjoining the A_k group elements in the $q_0 q_z$-plane with the matched set of additional D_k group elements in the $q_x q_y$-plane. The quaternion visualizations of the D_k group elements give us a remarkably clear intuition for the context of this discrete group.

Finally, in Fig. 26.8 we present the D_k group elements as global points in a full three-part quaternion volume display as suggested in Chapter 4. We will see shortly that this visualization method helps us to tease out the properties of the much more complex actions of E_6, E_7, and E_8. Fig. 26.8 exploits the full quaternion $\mathbf{S}^3$ hyperhemisphere technique, with a red 3D solid ball for the South pole, $q_0 < 0$, a separate blue 3D solid ball for the North pole, $q_0 > 0$, and a distinct equatorial $\mathbf{S}^2$ 2D shell where the two solid balls touch each other for the special case of $q_0 = 0$. The latter happens to be particularly important to reveal the quaternion geometry of D_k because fully *half* of the D_k rotational elements fall exactly on the $q_0 = 0$ equator, and not in the interior of either the North or South hyperhemispheres. We will see shortly that for the $E_{6,7,8}$ groups, the edge-center symmetry transformation is exactly analogous to the D_k flip quaternion b, with the 180-degree flip preserving $q_0 = 0$ and thus all such elements are in the orange equatorial sphere.

26.4.3 The E_6 tetrahedral quaternion group

Our final collection of groups, acting on the tetrahedron, the octahedron, and the icosahedron comprising the complete collection of dual merged Platonic solids, is both simpler than the infinite-dimensional A_k and D_k groups we have just examined, and more complex. While they have a much more diverse set of elements, and correspond to much more complicated mathematical constructions, the structure of their symmetry groups corresponds to a very regular set of rules based on their vertices, edge centers, and face centers.

We first study the simplest of the exceptional groups, the E_6 Kleinian group, which is based on the self-dual tetrahedral geometry. In Fig. 26.3, we depict the standard 3D rotations derived from the E_6 tetrahedral group acting on the vertices, edge centers, and face centers of the tetrahedron. Each possible rotation is represented by its fixed axis and a length given by the quaternion weight $\sin(\theta/2)$ in the pure imaginary direction. The information here is redundant in some places and invisible due to overlaps in others, so this image provides intuition, but gives an incomplete picture of the actual unique set of rotations. To go further, first we write down the 12 fundamental quaternion elements of the tetrahedral group. The first part of the construction is based on the four vertex elements, whose normalized directions $\hat{\mathbf{v}}_i$, $i = 1, 2, 3, 4$, we will exploit as the axes of the axis-angle rotation notation. Each $\hat{\mathbf{v}}_i$ possesses a three-fold symmetry, which corresponds to the A_2 group with an identity plus two rotations of 120 degrees and 240 degrees; we handle the identity separately, as it is the same for all of the subgroups, so we need only the two nonidentity actions of A_2, which we denote by the quaternion

$$a_2(n) = \left(\cos\left(\frac{n\pi}{3}\right), 0, 0, \sin\left(\frac{n\pi}{3}\right) \right),$$

for $n = 1, 2$. To each of our basic formulas for symmetry groups, we will be adding an alternative notation, using the utility quaternion

$$d(\hat{\mathbf{p}}) = d(\hat{\mathbf{p}} \to \hat{\mathbf{z}}) = \left(\cos\left(\frac{\theta}{2}\right), \sin\left(\frac{\theta}{2}\right) \frac{\hat{\mathbf{p}} \times \hat{\mathbf{z}}}{\|\hat{\mathbf{p}} \times \hat{\mathbf{z}}\|} \right), \tag{26.12}$$

where $\cos\theta = \hat{\mathbf{p}} \cdot \hat{\mathbf{z}}$. Conjugating by the quaternion $d(\hat{\mathbf{p}})$ aligns any vector $\hat{\mathbf{p}}$ on S^2 with the $\hat{\mathbf{z}}$-axis, unless $|\hat{\mathbf{p}} \cdot \hat{\mathbf{z}}| = 1$, when obviously we just rotate about $\hat{\mathbf{p}}$. This alignment allows us to apply an element of A_k to produce a symmetry action at the pole before reversing the action of $d(\hat{\mathbf{p}})$. Thus the vertex quaternions for E_6 are

$$v(i : n) = \left(\cos\left(\frac{n\pi}{3}\right), \sin\left(\frac{n\pi}{3}\right) \hat{\mathbf{v}}_i \right)$$
$$= \overline{d}(\hat{\mathbf{v}}_i) \star a_2(n) \star d(\hat{\mathbf{v}}_i). \tag{26.13}$$

There is a face center opposite each vertex, so the vertices are not doubled, and there are no additional group actions arising from the face centers. The edge-center symmetries are given by any three distinct edge centers $\hat{\mathbf{e}}_i$, chosen from the six available edge centers. Edges have only a single 180-degree flip from the action of the single A_1 element $a_1 = (0, 0, 0, 1)$, so the remainder of the E_6 group is simply

$$e(i) = \left(0, \hat{\mathbf{e}}_i \right)$$
$$= \overline{d}(\hat{\mathbf{e}}_i) \star a_1 \star d(\hat{\mathbf{e}}_i). \tag{26.14}$$

Since $q_0 = 0$, all the edge-center action quaternions lie on the equator of the 3-sphere quaternion plot. Choosing the four tetrahedron vertex coordinates Eq. (25.9), aligned with a top diagonal of the cube and its orthogonal bottom cube diagonal, we apply a rotation by the angle $2n\pi/3$, with $n = 1, 2$ the nontrivial actions, corresponding to the $a_2(n)$ polar A_2 action. This action pairs with the same action on the opposite face centers Eq. (25.11) of the tetrahedron, which we ignore as redundant. The remaining actions are the set of six 180-degree edge flips about both ends of the Cartesian axes Eq. (25.10), of which we choose the three positive axes, the negative axes again being redundant. Adjoining the identity to these, we can write the 12 unique quaternion elements of E_6 (up to sign reversal of any elements) as

$$q(E_6) = \left\{ \begin{array}{rl} q_{\mathrm{ID}} : & [1, 0, 0, 0], \\[2mm] v(1:1), v(1:2) : & \left[\frac{1}{2}, \frac{1}{2}, -\frac{1}{2}, \frac{1}{2}\right], \left[-\frac{1}{2}, \frac{1}{2}, -\frac{1}{2}, \frac{1}{2}\right], \\[2mm] v(2:1), v(3:2) : & \left[\frac{1}{2}, \frac{1}{2}, \frac{1}{2}, -\frac{1}{2}\right], \left[-\frac{1}{2}, \frac{1}{2}, \frac{1}{2}, -\frac{1}{2}\right], \\[2mm] v(3:1), v(3:2) : & \left[\frac{1}{2}, -\frac{1}{2}, \frac{1}{2}, \frac{1}{2}\right], \left[-\frac{1}{2}, -\frac{1}{2}, \frac{1}{2}, \frac{1}{2}\right], \\[2mm] v(4:1), v(4:2) : & \left[\frac{1}{2}, -\frac{1}{2}, -\frac{1}{2}, -\frac{1}{2}\right], \left[-\frac{1}{2}, -\frac{1}{2}, -\frac{1}{2}, -\frac{1}{2}\right], \\[2mm] e(1) : & [0, 1, 0, 0], \\[2mm] e(2) : & [0, 0, 1, 0], \\[2mm] e(3) : & [0, 0, 0, 1]. \end{array} \right. \tag{26.15}$$

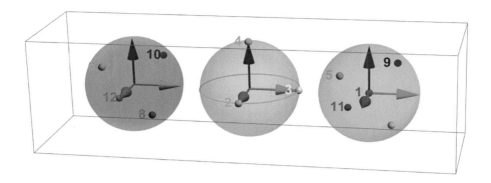

E6: **Tetrahedral Group: 12 unique**

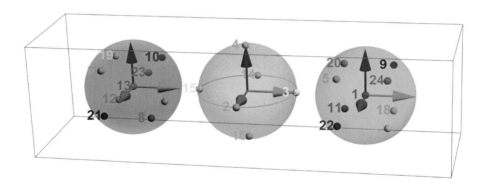

E6: **Tetrahedral Group: 24 all**

FIGURE 26.9

(top) Twelve distinct E_6 tetrahedral group quaternion actions, each having an omitted equivalent quaternion sign-reversed mirror partner. The "1" is the identity as usual, the three edge flips are on the ends of the Cartesian axes in the center, all having $q_0 = 0$ at 180 degrees, and the four two-fold actions spinning around the four vertices are paired left and right with the same **q**, but opposite signs of q_0. (bottom) Including the entire (sign-doubled) set of E_6 binary group quaternions. Elements with a common fixed vertex rotation are drawn with matching colors, though those colors can be distorted by the background sphere's coloring.

In Fig. 26.9, we show the triple sphere **q** representation of the unique 12 quaternions of the E_6 tetrahedral group, accompanied by the full 24 element binary group elements including both signs. Like colors originate with the same fixed vertex, so we see pairs in the top image, and sign-doubled quartets in the bottom. (Note that colors inside the spheres may be distorted by the sphere background color.) The quaternions on the three Cartesian axes in the orange center equatorial $\mathbf{S}^2$ with $q_0 = 0$ are the 180-degree edge flips, and in the 12-element depiction, we see four axis-based rotations in each of the $q_0 < 0$ and $q_0 > 0$ solid balls. Remember that the *center* of each solid sphere on the left and right is the point where $\pm q_0 \equiv 1$.

26.4.4 The E_7 octahedral quaternion group

In Fig. 26.4, we showed the standard 3D versions of the E_7 octahedron–hexahedron pair, with all the possible 3D axes of rotation. This figure provides a reasonable visualization of the types of rotations that give symmetries, but has a number of uncomfortable compromises: although we can technically identify the rotation axes aligned with each vertex, the center of each edge, and the center of each face, and even the amount of the discrete rotation, we have trouble establishing uniqueness – we can detect that there is redundancy in actions on diametrically opposite vertex, face, and edge pairs, but the unique group elements are not easy to identify in one single viewpoint, and distinct group elements aligned on the same invariant axis cannot be easily distinguished. However, we can represent a unique set of the 24 distinct rotations about the vertices, edge centers, and faces in quaternion form (remembering we have a sign ambiguity in choosing any individual action) by inspecting Fig. 26.4. Again, we remind ourselves that every vertex, edge, and face of the octahedron has mirror image pairing, so, for a unique set of group elements, we need to choose half of each set; we can pick any sign we want, but we cannot include diametric opposites. The group elements can be constructed from rotations about 3 of the 6 octahedral vertices $\hat{\mathbf{v}}_i$, 6 of the 12 edge centers $\hat{\mathbf{e}}_i$, and 4 of the 8 face centers $\hat{\mathbf{f}}_i$. The construction we used for E_6, transforming an axis to the $\hat{\mathbf{z}}$-axis and applying the appropriate A_k discrete group elements, works here as follows:

- The *vertices* have a four-fold symmetry, to which we apply the three nonidentity elements of $a_3(n)$ of the A_3 cyclic group.
- The *edge centers* have only the 180-degree a_1 flip operation of A_1 applied to each edge center.
- The *face centers* have a three-fold symmetry implemented by the two actions $a_2(n)$ from the A_2 group.

With the help of our $\hat{\mathbf{z}}$-axis-aligning quaternion conjugation tool $d(\hat{\mathbf{p}})$ from Eq. (26.12), we can obtain our entire set of E_7 quaternion group elements in two equivalent intuitive formulations, one just applying the rotations directly to the corresponding axis using Eq. (2.16), and the other using the conjugation construction:

$$v(i:n) = \left(\cos\left(\frac{n\pi}{4}\right), \sin\left(\frac{n\pi}{4}\right) \hat{\mathbf{v}}_i\right)$$
$$= \overline{d}(\hat{\mathbf{v}}_i) \star a_3(n) \star d(\hat{\mathbf{v}}_i) \qquad \{n = 1, 2, 3\}, \tag{26.16}$$

$$e(i) = \left(0, \hat{\mathbf{e}}_i\right)$$
$$= \overline{d}(\hat{\mathbf{e}}_i) \star a_1 \star d(\hat{\mathbf{e}}_i), \tag{26.17}$$

$$f(i:n) = \left(\cos\left(\frac{n\pi}{3}\right), \sin\left(\frac{n\pi}{3}\right) \hat{\mathbf{f}}_i\right)$$
$$= \overline{d}(\hat{\mathbf{f}}_i) \star a_2(n) \star d(\hat{\mathbf{f}}_i) \qquad \{n = 1, 2\}. \tag{26.18}$$

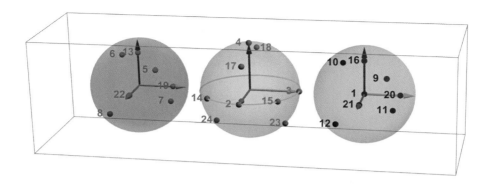

E7: Octahedral Group: 24 unique

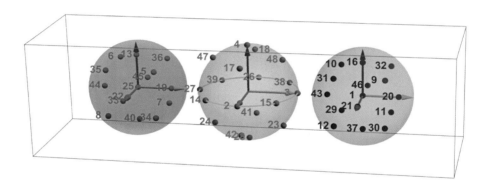

E7: Octahedral Group: 48 all

FIGURE 26.10

(top) A selection of 24 unique E_7 octahedral group quaternion actions displayed in the 3-spheres method with **q**-coordinates, with $q_{ID} = (1, 0, 0, 0)$ at the center of the right-hand sphere. The equatorial orange spherical surface contains three components from the 180-degree vertex-fixing rotations and six more from the edge flips of half of the 12 edges, all producing a quaternion with $q_0 = 0$ and thus lying on the orange equatorial sphere. (bottom) Adding in all of the opposite-signed pairs to show the full set of 48 binary group elements.

Choosing the octahedron vertices aligned with the Cartesian axes, we have the 6 vertex axes of Eq. (25.12), the 12 edge-center axes of Eq. (25.13), and the 8 face-center axes of Eq. (25.14). Omitting duplicate mirror-image vertices, edges, and faces, these form the basis for the explicit 24 unique E_7 octahedral quaternion group elements:

$$q(E_7) = \begin{cases}
\quad q_{\text{ID}} : [\; 1 \quad 0 \quad 0 \quad 0 \;], \\[4pt]
v(1:1), v(1:2), v(1:3) : [\tfrac{1}{\sqrt{2}}, \tfrac{1}{\sqrt{2}}, 0, 0], \quad [0, 1, 0, 0], \quad [-\tfrac{1}{\sqrt{2}}, \tfrac{1}{\sqrt{2}}, 0, 0], \\[4pt]
v(2:1), v(2:2), v(2:3)) : [\tfrac{1}{\sqrt{2}}, 0, \tfrac{1}{\sqrt{2}}, 0], \quad [0, 0, 1, 0], \quad [-\tfrac{1}{\sqrt{2}}, 0, \tfrac{1}{\sqrt{2}}, 0], \\[4pt]
v(3:1), v(3:2), v(3:3)) : [\tfrac{1}{\sqrt{2}}, 0, 0, \tfrac{1}{\sqrt{2}}], \quad [0, 0, 0, 1], \quad [-\tfrac{1}{\sqrt{2}}, 0, 0, \tfrac{1}{\sqrt{2}}], \\[4pt]
\quad e(1) : [0 \quad -\tfrac{1}{\sqrt{2}} \quad \tfrac{1}{\sqrt{2}} \quad 0], \\[4pt]
\quad e(2) : [0 \quad -\tfrac{1}{\sqrt{2}} \quad 0 \quad -\tfrac{1}{\sqrt{2}}], \\[4pt]
\quad e(3) : [0 \quad -\tfrac{1}{\sqrt{2}} \quad 0 \quad \tfrac{1}{\sqrt{2}}], \\[4pt]
\quad e(4) : [0 \quad -\tfrac{1}{\sqrt{2}} \quad -\tfrac{1}{\sqrt{2}} \quad 0], \\[4pt]
\quad e(5) : [0 \quad 0 \quad \tfrac{1}{\sqrt{2}} \quad -\tfrac{1}{\sqrt{2}}], \\[4pt]
\quad e(6) : [0 \quad 0 \quad \tfrac{1}{\sqrt{2}} \quad \tfrac{1}{\sqrt{2}}], \\[4pt]
f(1:1), f(1:2) : [\tfrac{1}{2}, \tfrac{1}{2}, \tfrac{1}{2}, \tfrac{1}{2}], \quad [-\tfrac{1}{2}, \tfrac{1}{2}, \tfrac{1}{2}, \tfrac{1}{2}], \\[4pt]
f(2:1), f(2:2) : [\tfrac{1}{2}, \tfrac{1}{2}, -\tfrac{1}{2}, -\tfrac{1}{2}], \quad [-\tfrac{1}{2}, \tfrac{1}{2}, -\tfrac{1}{2}, -\tfrac{1}{2}], \\[4pt]
f(3:1), f(3:2) : [\tfrac{1}{2}, -\tfrac{1}{2}, \tfrac{1}{2}, -\tfrac{1}{2}], \quad [-\tfrac{1}{2}, -\tfrac{1}{2}, \tfrac{1}{2}, -\tfrac{1}{2}], \\[4pt]
f(4:1), f(4:2) : [\tfrac{1}{2}, -\tfrac{1}{2}, -\tfrac{1}{2}, \tfrac{1}{2}], \quad [-\tfrac{1}{2}, -\tfrac{1}{2}, -\tfrac{1}{2}, \tfrac{1}{2}].
\end{cases} \tag{26.19}$$

These quaternion group actions are shown in the full triple-sphere display of Fig. 26.10, where we can see, for example, the nine elements with $q_0 = 0$ on the surface of the equatorial $\mathbf{S}^2$ in the center, and the matched pairs with different q_0 signs in the left and right solid spheres, with the positive identity quaternion at the origin of the right sphere. Each of the three Cartesian axes appears three times, once in each sphere, representing one unique A_3 set of the vertex-based actions, each of the four face diagonal directions with A_2 symmetry appears twice, and the six single A_1 edge flips appear alone. Thus we have $3 \times 3 + 4 \times 2 + 6 +$ one identity $= 24$ symmetries as expected. In the bottom part of Fig. 26.10, we can observe the entire doubled E_7 group, with all 48 elements shown.

26.4.5 The E_8 icosahedral quaternion group

In Fig. 26.5, we showed the standard 3D versions of the E_8 icosahedron–dodecahedron pair, with all the possible 3D axes of rotational symmetry. Again, even though the number of elements in the tessellation is quite large, we can see clearly how the discrete symmetry operations assemble themselves into actions on the vertices, edge centers, and face centers, and how the exact same operations appear in the dual dodecahedral polyhedron. This is the most complicated set of full symmetries of the sphere, and there is nothing more complicated, so that is somewhat comforting. Nevertheless, again, there is no easy way to make all the diametrically opposite redundancies visible using the 3D space rotation axes representation of Fig. 26.5, and in particular it is very difficult to unambiguously expose the multiple angles of rotation occurring at each rotation axis. So, as before, we go to the quaternion representation for a more satisfactory set of visualization strategies. Now, however, we have 60 unique elements, 120 with the $q \to -q$ redundancy, and that is a lot of information. Based on the icosahedral symmetry rotation axes, the 12 vertices $\hat{\mathbf{v}}_i$ in Eq. (25.15), the 30 edges $\hat{\mathbf{e}}_i$ in Eq. (25.16), and the 20 faces $\hat{\mathbf{f}}_i$ in Eq. (25.17), we can assemble the components of the entire group, the doubled binary E_8 group. Based on this information, the full set of unique quaternion-valued symmetry transformations for the icosahedral group takes the following form:

- **Twelve vertices with four rotations plus the identity.** These have the same quaternion geometry as the $k = 4$ cyclic group A_4, with group elements $a_4(n)$, now starting from 12 different vertex directions:

$$a_4(n) \ [n = 1, 2, 3, 4] = \left(\cos \left(\frac{n\pi}{5} \right), 0, 0, \sin \left(\frac{n\pi}{5} \right) \right) ,$$

$$d(\hat{\mathbf{v}}_i) \ [i = 1, \ldots, 12] = \text{Tilt the icosahedral } i\text{-th vertex direction to align with } \hat{\mathbf{z}}\text{-axis} , \quad (26.20)$$

$$\text{Vertex action:} \quad v(n; i) = \overline{d}(\hat{\mathbf{v}}_i) \star a_4(n) \star d(\hat{\mathbf{v}}_i) .$$

We note that all the transformations incorporating the pentagonal symmetry of the icosahedron involve at some level the *golden ratio* ϕ,

$$\phi = \frac{1 + \sqrt{5}}{2} , \qquad \qquad \frac{1}{\phi} = \frac{-1 + \sqrt{5}}{2} ,$$

appearing in

$$\cos \left(\frac{\pi}{5} \right) = \frac{\phi}{2} = \frac{1}{4} \left(1 + \sqrt{5} \right) , \qquad \cos \left(\frac{2\pi}{5} \right) = \frac{1}{2\phi} = \frac{1}{4} \left(-1 + \sqrt{5} \right) ,$$

$$\cos \left(\frac{3\pi}{5} \right) = -\frac{1}{2\phi} = \frac{1}{4} \left(1 - \sqrt{5} \right) , \qquad \cos \left(\frac{4\pi}{5} \right) = -\frac{\phi}{2} = \frac{1}{4} \left(-1 - \sqrt{5} \right) .$$

This construction tips the direction of the i-th icosahedral vertex to the vertical direction, spins it by applying the n-th A_4 cyclic group action, and returns the spun vertex to its original orientation. Each vertex has a mirror opposite counterpart, and thus including both vertices of any such pair does not result in a new group element, but *does* give a new quaternion, and we will show all the quaternions in some representations.

- **Thirty edges with a single 180-degree rotation plus the identity.** These have the same quaternion geometry as the cyclic group with $k = 1$, the group A_1, basically a flipping rotation $a_1 = [0, 0, 0, 1]$, but starting in 30 different directions from the identity:

$$a_1 = \left(\cos \left(\frac{\pi}{2} \right), 0, 0, \sin \left(\frac{\pi}{2} \right) \right) = [0, 0, 0, 1] ,$$

$$d(\hat{\mathbf{e}}_i) \ [i = 1, \ldots, 30] = \text{Tilt the icosahedral } i\text{-th edge center } \hat{\mathbf{e}}_i \text{ to the } \hat{\mathbf{z}}\text{-axis} , \quad (26.21)$$

$$\text{Edge-center action:} \quad e(n) = \overline{d}(\hat{\mathbf{e}}_i) \star a_1 \star d(\hat{\mathbf{e}}_i) .$$

This construction tips the direction of the i-th icosahedral edge to the vertical direction, applies the a_1 cyclic group 180-degree rotation about the $\hat{\mathbf{z}}$-axis, and returns the spun edge center to its original orientation. Again each edge has a mirror opposite counterpart yielding the same 3D rotation, but that we can display as a distinct quaternion.

- **Twenty triangular faces with two rotations plus the identity.** These have the same quaternion geometry as the cyclic group with $k = 2$, the group A_2, but starting in 20 different directions toward the face centers from the identity:

$$a_2(n) \ [n = 1, 2] = \left(\cos \left(\frac{\pi n}{3} \right), 0, 0, \sin \left(\frac{\pi n}{3} \right) \right)$$

$$= \left\{ \left[\begin{array}{cccc} \frac{1}{2} & 0 & 0 & \frac{\sqrt{3}}{2} \end{array} \right], \left[\begin{array}{cccc} -\frac{1}{2} & 0 & 0 & \frac{\sqrt{3}}{2} \end{array} \right] \right\}, \tag{26.22}$$

$$d(\hat{\mathbf{f}}_i) \ [i = 1, \ldots, 20] = \text{Tilt the icosahedral } i\text{-th face center } \hat{\mathbf{f}}_i \text{ to the } \hat{\mathbf{z}}\text{-axis },$$

$$\text{Face-center action:} \quad f(n : i) = \overline{d}(\hat{\mathbf{f}}_i) \star a_2(n) \star d(\hat{\mathbf{f}}_i) .$$

Opposite faces again are equivalent, and so just one of each pair must be taken to make a given list unique, while we can display the doubled quaternion elements distinctly if we choose.

The 3D space $\mathbf{SO}(3)$ rotation group for the icosahedral discrete symmetries thus consists of $40/2 = 20$ group actions on the icosahedral faces, $30/2 = 15$ on the edges, and $48/2 = 24$ on the vertices, plus the identity, so we have $20 + 15 + 24 + 1 = 60$ unique group actions. These choices are not unique, as any rotation axis could be replaced by its mirrored counterpart without changing the $\mathbf{SO}(3)$ group elements. Nevertheless, the complete doubled 120-element symmetry appears if we include the redundant mirror-opposite actions on paired vertices, edges, and faces, or, more relevantly, the complete 120-element quaternion plot that includes all opposite-sign quaternions. In order to complete the story and create visual representations for the icosahedral group elements, we next need to choose actual coordinates for the vertices of the icosahedron. Although any orientation of the fundamental vertices will do, as the resulting quaternions are changed only by a global rotation of $\mathbf{S}^3$ in Euclidean 4D space, we have reasons for choosing a particular global orientation.

We recall that in Eq. (25.15), we chose to write the 12 icosahedral vertices, using the golden ratio $\phi = (1 + \sqrt{5})/2$, as

$$v_1 = \frac{1}{\sqrt{\phi^2 + 1}} \begin{bmatrix} 1 & \phi & 0 \end{bmatrix}, \qquad v_2 = \frac{1}{\sqrt{\phi^2 + 1}} \begin{bmatrix} -1 & \phi & 0 \end{bmatrix},$$

$$v_3 = \frac{1}{\sqrt{\phi^2 + 1}} \begin{bmatrix} 0 & 1 & \phi \end{bmatrix}, \qquad v_4 = \frac{1}{\sqrt{\phi^2 + 1}} \begin{bmatrix} 0 & -1 & \phi \end{bmatrix},$$

$$v_5 = \frac{1}{\sqrt{\phi^2 + 1}} \begin{bmatrix} \phi & 0 & 1 \end{bmatrix}, \qquad v_6 = \frac{1}{\sqrt{\phi^2 + 1}} \begin{bmatrix} \phi & 0 & -1 \end{bmatrix},$$

$$v_7 = \frac{1}{\sqrt{\phi^2 + 1}} \begin{bmatrix} -1 & -\phi & 0 \end{bmatrix}, \qquad v_8 = \frac{1}{\sqrt{\phi^2 + 1}} \begin{bmatrix} 1 & -\phi & 0 \end{bmatrix}, \tag{26.23}$$

$$v_9 = \frac{1}{\sqrt{\phi^2 + 1}} \begin{bmatrix} 0 & -1 & -\phi \end{bmatrix}, \qquad v_{10} = \frac{1}{\sqrt{\phi^2 + 1}} \begin{bmatrix} 0 & 1 & -\phi \end{bmatrix},$$

$$v_{11} = \frac{1}{\sqrt{\phi^2 + 1}} \begin{bmatrix} -\phi & 0 & -1 \end{bmatrix}, \qquad v_{12} = \frac{1}{\sqrt{\phi^2 + 1}} \begin{bmatrix} -\phi & 0 & 1 \end{bmatrix}.$$

With this orientation, the 12 vertices are aligned with three Cartesian plane golden-ratio rectangles, shown in Fig. 25.5, and have no vertices aligned with the North–South $\hat{\mathbf{z}}$-axis. This is a suitable, but of course not unique, choice for our list of axes as the directions invariant under quaternion rotations implementing the five-fold subset of icosahedral vertex symmetry actions.

We can as usual compute the edge centers and the face centers from appropriate averages of the bounding pairs or triples of these vertices, and use those as the basis for the E_8 group actions. However, the complexity of the algebraic forms for these elements becomes quite unwieldy, and we omit the explicit forms here, referring the reader to the simple algorithms of Eq. (26.20) and Eq. (26.21) exploiting the edges and faces computed from the vertex locations in Eq. (26.23). The basic list of 60 symbolic group actions, choosing appropriate half-subsets that exclude duplicate mirror-image vertex, edge, and face actions, now takes this symbolic form – the pentagonal symmetry at each vertex is handled by the four nonidentity elements of A_4, the 180-degree flip at each edge (giving $q_0 \equiv 0$) is handled by A_1, and the two nonidentity elements of A_2 take care of the triangular symmetry of each face:

$$
q(E_8) = \left\{
\begin{array}{rl}
q_{\mathrm{ID}}: & [\ 1 \quad 0 \quad 0 \quad 0\ ], \\
v(1:1),\ v(1:2),\ v(1:3),\ v(1:4): & \text{Four elements of } A_4 \text{ action}, \\
v(2:1),\ v(2:2),\ v(2:3),\ v(2:4): & \text{`` ''} \\
v(3:1),\ v(3:2),\ v(3:3),\ v(3:4): & \text{`` ''} \\
v(4:1),\ v(4:2),\ v(4:3),\ v(4:4): & \text{`` ''} \\
v(5:1),\ v(5:2),\ v(5:3),\ v(5:4): & \text{`` ''} \\
v(6:1),\ v(6:2),\ v(6:3),\ v(6:4): & \text{`` ''} \\
\hline
e(1): & \text{Single flip action of } A_1, \\
e(2): & \text{`` ''} \\
e(3): & \text{`` ''} \\
\vdots\ : & \vdots \\
e(13): & \text{`` ''} \\
e(14): & \text{`` ''} \\
e(15): & \text{`` ''} \\
\hline
f(1:1),\ f(1:2): & \text{Two elements of } A_2 \text{ action}. \\
f(2:1),\ f(2:2): & \text{`` ''} \\
f(3:1),\ f(3:2): & \text{`` ''} \\
\vdots\ : & \vdots \\
f(8:1),\ f(8:2): & \text{`` ''} \\
f(9:1),\ f(9:2): & \text{`` ''} \\
f(10:1),\ f(10:2): & \text{`` ''}
\end{array}
\right.
\tag{26.24}
$$

These 60 quaternions are shown in the triple-sphere display of Fig. 26.11(top). Since the choice of quaternion signs is not deterministic, we show in Fig. 26.11(bottom) the entire doubled 120-element quaternion space, including both signs of the identity, thus revealing all possible symmetries with no arbitrariness. The central equatorial orange spherical surface shows all 30 $q_0 = 0$ edge-flip elements, for example.

As an additional attempt at showing the complex information of the E_8 icosahedral discrete groups, we present one more set of triple-sphere plots in Fig. 26.12 that *splits* the quaternion group elements into vertex symmetries with 12 sets of four at the top, edge symmetries (all 30 with $q_0 = 0$) in the middle, and the 20 pairs of face symmetries at the bottom. For both vertex and face symmetries, sequences of actions on one vertex or one face cross between the $q_0 < 0$ sector and the $q_0 > 0$ sector, and those

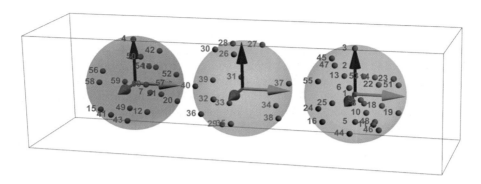

E8: **Icosahedral Group: 60 unique elements**

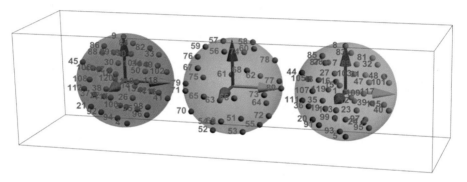

E8: **Icosahedral Group: all 120 elements**

FIGURE 26.11

(top) A choice of 60 unique quaternion actions of the E_8 icosahedral group. As usual, the "Southern" hemisphere solid ball B^3, with $q_0 < 0$, is the solid magenta sphere on the left; the "South pole" $q = (-1, 0, 0, 0)$ (not part of the unique set) is at the center. The "equator" empty sphere S^2, with $q_0 \equiv 0$, is the surface of the orange ball in the middle; this is the "skin" completing the outer surface of both the left and right open solid balls, and contains all A_1 180-degree edge-flip actions, with $q_0 = 0$. The "Northern" hemisphere solid ball B^3, with $q_0 > 0$, is the solid cyan sphere on the right; the "North pole" identity quaternion $q_{ID} = (+1, 0, 0, 0)$ is at the center. (bottom) Adding in the entire set of possible quaternions of the binary group, showing all 120 possible rotations, including all opposite sign partners. This is a more symmetric picture since all 120 quaternions are present, with no arbitrary choices.

connections are drawn as orange lines. Since there are four total actions at each vertex, the two at $q_0 < 0$ are connected with a red line, and the two at $q_0 > 0$ are connected with a blue line, plus the region-crossing orange line. The important feature of the quaternion display in Fig. 26.12 is that each discrete quaternion group action is visible, unique, and distinguishable as originating from a vertex, edge center, or face center.

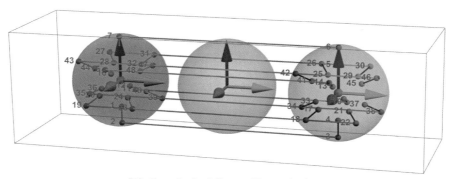

E8: **Icosahedral Group Vertex Action**

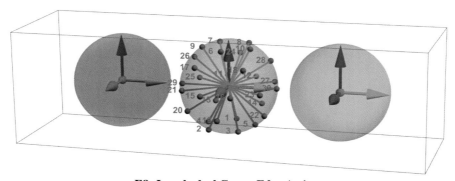

E8: **Icosahedral Group Edge Action**

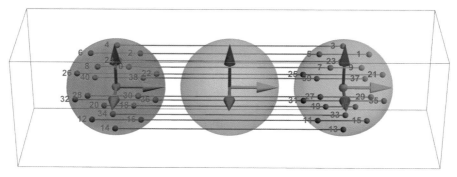

E8: **Icosahedral Group Face Action**

FIGURE 26.12

(top) The doubled *vertex symmetry* quaternions of the E8 icosahedral group. The horizontal lines across spheres connect actions on the same vertex that change signs of q_0 in the rotation sequence. (middle) The doubled *edge symmetry* quaternions of the E8 icosahedral group. Here the diameter-crossing lines connect the sign-reversed $q_0 = 0$ edge-flip quaternion pairs. (bottom) The doubled *face symmetry* quaternions of the E8 icosahedral group. The horizontal lines across spheres connect actions on the same vertex that change signs of q_0 in the rotation sequence.

The quaternion presentation form of the discrete ADE groups

27.1 ADE presentation diagrams: the I, J, K prototype

We have now seen the full structure of the *geometry* of the symmetric tessellations of the 3D spherical surface $\mathbf{S}^2$, the choice of *coordinate systems* of vertices, edges, and faces that elegantly expose the spatial structure of those tessellations, and the *quaternion group actions* based on cyclic rotations around fixed axes determined by those vertices, edges, and faces. That knowledge now prepares us to study an elegant group-theoretical construction that reduces the entire ADE discrete group system to a description requiring no more than three quaternions for each type, A_k, D_k, E_6, E_7, and E_8. The executive summary is that for each case there exists a minimal structure of group elements that form the *Presentation* basis from which *all the group elements* can be derived by repeated action of combinations of these elements (see, e.g., Wikipedia (2023c,h,f,e,d); Coxeter and Moser (1980)).

In order to create a mnemonic for remembering these structures, we will borrow a graphing method from the Dynkin diagrams or Coxeter diagrams of classical group theory, but without any claim that we are actually using any of the deeper group-theoretical properties of, say, Dynkin diagrams. However, the diagrams we introduce will be seen to be very useful as a memory aid and a conceptual guide to a few pieces of the group-theoretical foundations buried in the ADE framework.[1]

The quaternion group diagram and the generators of SU(2). We will now argue that every one of the ADE groups possesses a structure remarkably like a warped quaternion. To make this comparison, we first observe that we can look at the Hamilton imaginaries **I**, **J**, **K** as forming a kind of orthogonal coordinate system that is the basis of the Lie algebra **su(2)** (which exponentiates to the full Lie group **SU(2)**), in the following way:

$$\mathbf{II} = \mathbf{JJ} = \mathbf{KK} = \mathbf{IJK} = -1 \,, \tag{27.1}$$

$$\left.\begin{array}{l} \mathbf{I} = \mathbf{JK} \\ \mathbf{J} = \mathbf{KI} \\ \mathbf{K} = \mathbf{IJ} \end{array}\right\} \,. \tag{27.2}$$

These are the relations among the equations that Hamilton legendarily carved into the stone of Broome Bridge. We notice immediately that the full quaternions represented by **I**, **J**, **K** are simply *quaternion*

---

[1] We thank Nigel Hitchin for bringing our attention to the possibilities of these methods, noting that any mathematical errors and shortcomings are totally our own.

Visualizing More Quaternions. https://doi.org/10.1016/B978-0-32-399202-2.00042-3

rotations by 180 degrees about the $\hat{\mathbf{x}}$-, $\hat{\mathbf{y}}$-, and $\hat{\mathbf{z}}$-axes:

$$\begin{aligned}
\mathbf{I} &= q(\pi, \hat{\mathbf{x}}) = [0, 1, 0, 0] \, , \\
\mathbf{J} &= q(\pi, \hat{\mathbf{y}}) = [0, 0, 1, 0] \, , \\
\mathbf{K} &= q(\pi, \hat{\mathbf{z}}) = [0, 0, 0, 1] \, .
\end{aligned} \tag{27.3}$$

One thing we know very well is that these exact relationships must hold automatically for the three unitary 2×2 Pauli matrices, with the identification

$$\mathbf{I} = -i\sigma_x \, , \qquad \mathbf{J} = -i\sigma_y \, , \qquad \mathbf{K} = -i\sigma_z \, .$$

Thus, in the context of the quaternion algebra, we have three ways to see the fundamental identity for a sequence of the three 180-degree rotations:

$$\left. \begin{aligned}
\mathbf{IJK} &= -1 \\
q(\pi, \hat{\mathbf{x}}) \star q(\pi, \hat{\mathbf{y}}) \star q(\pi, \hat{\mathbf{z}}) &= -q_{\mathrm{ID}} = (-1, 0, 0, 0) \\
(-i\sigma_x) \cdot (-i\sigma_y) \cdot (-i\sigma_z) &= -\begin{bmatrix} 1 & 0 \\ 0 & 1 \end{bmatrix} = -I_2
\end{aligned} \right\} . \tag{27.4}$$

That is, re-expressing $(\mathbf{I}, \mathbf{J}, \mathbf{K})$ as the Pauli matrices $(-i\sigma_x, -i\sigma_y, -i\sigma_z)$, the **SU(2)** matrices obey precisely the same algebra as $(\mathbf{I}, \mathbf{J}, \mathbf{K})$ except that now the identity quaternion becomes the 2×2 identity matrix I_2.

We also note that, although there are three basic objects in Eq. (27.1), only *two* are independent: according to Eq. (27.2) (which is derivable from Eq. (27.1)), any one can be eliminated by expressing it in terms of the other two. Finally, the fundamental common object is the *negative* identity – every expression in Eq. (27.1) needs to be repeated twice to reach the actual identity, e.g., $\mathbf{IIII} = +1$, since elementary pure imaginary quaternions have the same properties as $\exp(i\pi/2) = \sqrt[4]{+1}$.

In general, a family of relations of this type is known as a *Presentation* of a group, a minimal set of operations from which all others in the group can be assembled by repetition. So, we might ask, if Hamilton's relations Eq. (27.1) and Eq. (27.2) are fundamental for **SU(2)**, what happens if we examine the same expressions for 3D **SO**(3) matrix rotations by 180 degrees? Suppose we look at

$$\begin{aligned}
R_x &= R(\pi, \hat{\mathbf{x}}) = \begin{bmatrix} 1 & 0 & 0 \\ 0 & -1 & 0 \\ 0 & 0 & -1 \end{bmatrix} , \\
R_y &= R(\pi, \hat{\mathbf{y}}) = \begin{bmatrix} -1 & 0 & 0 \\ 0 & 1 & 0 \\ 0 & 0 & -1 \end{bmatrix} , \\
R_z &= R(\pi, \hat{\mathbf{z}}) = \begin{bmatrix} -1 & 0 & 0 \\ 0 & -1 & 0 \\ 0 & 0 & 1 \end{bmatrix} .
\end{aligned} \tag{27.5}$$

Then one can easily calculate that the 3D 180-degree rotations (R_x, R_y, R_z) obey exactly the same Presentation relations as $(\mathbf{I}, \mathbf{J}, \mathbf{K})$, with one critical exception in the final matrix,

$$R_x \cdot R_x = R_y \cdot R_y = R_y \cdot R_y = R_x \cdot R_y \cdot R_z = +I_3 = + \begin{bmatrix} 1 & 0 & 0 \\ 0 & 1 & 0 \\ 0 & 0 & 1 \end{bmatrix}. \tag{27.6}$$

Instead of yielding four copies of the *negative* identity, we get four copies of the *positive* identity. This is the essence of the difference between the generators of the **SO**(3) ADE groups and the corresponding *binary* discrete groups.

The quaternion presentation diagram: the central identity. We now introduce an *ad hoc* diagram, in the style of a Coxeter–Dynkin diagram (Wikipedia, 2023g), but with a distinct interpretation that visually summarizes the essential properties of our ADE quaternion actions. We will exploit our diagrams to reveal the basic structure that repeats itself throughout the ADE group generators that we derived in Section 26.4. This type of diagram appears prominently in the theory of group properties, but exploring those is beyond our scope: our constructions here are focused on creating a visualization of the properties of the ADE groups that is sound and useful, and, for that purpose, these *ad hoc* diagrams are very successful.

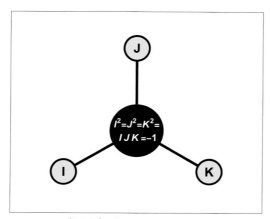

$(\mathbf{I}, \mathbf{J}, \mathbf{K})$: **Quaternion group**

FIGURE 27.1

The quaternion algebra presentation. Any repeated leg collapses to the negative identity quaternion at the origin, as do all three actions together in an even permutation.

The $(\mathbf{I}, \mathbf{J}, \mathbf{K})$ presentation. In Fig. 27.1, we show our diagramatic representation of the Hamilton equations in Eq. (27.1), an object that one might call the graph of the "**SU**(**2**) $\sim$ binary **SO**(3)" group. Each element of the quaternion basis can be interpreted here in (at least) three ways. With each branch being a Hamilton imaginary, the graph stands for $\mathbf{I}^2 = \mathbf{J}^2 = \mathbf{K}^2 = \mathbf{I}\mathbf{J}\mathbf{K} = -1$. The key unusual idea here is the position of the black disk, denoting the *central identity*. Each Hamilton imaginary is a vertex of a two-element multiplication table, with the negative identity at the center the result of squaring the quaternion. That is, if we take **I** as its yellow disk, then its quaternion square **II** must be −1, the common black circle, where "−1" is precisely the negative form of the identity quaternion. Take any Hamilton imaginary element and multiply it by itself, and you simply go from that element to the point −1 traveling along the line between them. The (negative) identity is shared, and all three Hamilton imaginaries share that identity with one another in this two-part cycle. The second, now obvious, idea is that if you multiply together *all three* elements that are one graph line from the identity, *the even-permutation triple product* also terminates on the quaternion identity "−1" (the odd permutations of **I J K** are clearly "+1").

The quaternion imaginaries as a binary symmetry group. The 180-degree action interpretation of the basic Hamilton imaginaries, (**I**, **J**, **K**), can be interpreted as an ADE-like discrete **SO**(3) symmetry group acting on a degenerate lune, something like a $k = 0$ geometric object in the A_k series. In Fig. 27.2 we show a simple example of how the symmetry group would act on that geometry.

An alternative quaternion geometry plot. Next, we introduce an alternative graphical styling of essentially the same information as Fig. 27.1, except that now we relate our quaternion actions to the three Euclidean axes to which they correspond. Fig. 27.3 assigns **I**, **J**, and **K** to the $\hat{\mathbf{x}}$-, $\hat{\mathbf{y}}$-, and $\hat{\mathbf{z}}$-axes represented as a red, green, or blue sphere, respectively. Each sphere stands for a single Hamilton imaginary element multiplying the yellow sphere at the origin, which represents the positive identity quaternion $+q_{\mathrm{ID}}$. Multiplying twice, e.g., $\mathbf{I}^2 = -1$ moves each element along its axis to the black dot representing the negative identity quaternion $-q_{\mathrm{ID}}$ on the surface of the unit sphere. Multiplying once more brings us to the third sphere on each line, (−**I**, −**J**, −**K**), and a fourth multiplication returns us to the positive identity quaternion's yellow spheres, shown at twice the radius of the foundational sphere. We add to this picture the remarkable additional feature that each element multiplied *once* by the others *also* returns to the negative identity, **I J K** = −1. We will see next that each of the ADE groups will possess generators with properties very similar to the quaternion basis of Hamilton imaginaries, and will admit symbolic depictions with striking parallels to Fig. 27.1 and Fig. 27.3.

This same relationship among nodes and links will appear, slightly more complicated, as a way of representing the essence of each ADE group in the context of its Presentation, a minimal set of group generators. Just as we can think of each quaternion entry **I**, **J**, or **K** to be a 180-degree rotation about the $\hat{\mathbf{x}}$- (or $\hat{\mathbf{y}}$- or $\hat{\mathbf{z}}$-)axis, we can make a graph of *any ADE edge flip*, which is a two-element 180-degree rotation, that exactly matches the graph of, say, (**J** ↔ −1). The other members of the ADE minimal generators will vary from case to case, but often the parallel structure to the bare quaternion algebra will be striking. We now work through each of the five sets of quaternion ADE Presentations. Each (binary) quaternion group has been assigned a notation by Coxeter (Wikipedia, 2023b) of the form $\langle n_{\mathrm{vert}} \rangle$ or $\langle n_{\mathrm{edge}}, n_{\mathrm{face}}, n_{\mathrm{vert}} \rangle$ denoting the number of total cycles at each VEF symmetry axis. A parallel notation with round brackets denotes the **SO**(3) subgroups corresponding to the ADE symmetries.

27.2 A_k cyclic group *presentation*: $\langle k + 1 \rangle$

The $k + 1$ unique quaternion elements of the binary A_k cyclic group are exceptionally simple, with only one cycle of elements transforming the digons into one another, with all values of Eq. (26.6) generated by the single smallest polar quaternion repeated from zero (the identity) to k times. The edges and

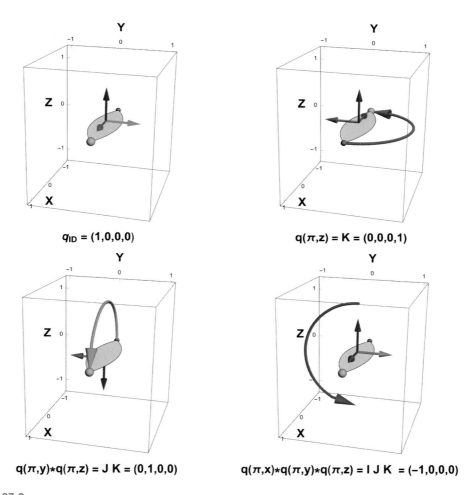

$q_{\text{ID}} = (1,0,0,0)$

$q(\pi,z) = K = (0,0,0,1)$

$q(\pi,y) \ast q(\pi,z) = J\,K = (0,1,0,0)$

$q(\pi,x) \ast q(\pi,y) \ast q(\pi,z) = I\,J\,K = (-1,0,0,0)$

FIGURE 27.2

The quaternion algebra geometry as a symmetry of a degenerate lune. The $k=0$ geometry of the A_k cyclic group would act on a single lune, the most structureless polygon symmetry we can imagine. Here we show how the 180-degree action interpretation of the Hamilton $(\mathbf{I}, \mathbf{J}, \mathbf{K})$ quaternions manifest themselves as discrete symmetries of a single lune. After acting in sequence, we recover the result of the binary group $\mathbf{IJK} = -1$, which is the identity orientation in 3D.

faces of the digons are ignored here, which is why the Coxeter notation exceptionally has only one element (Wikipedia, 2023c). We thus draw our A_k diagram in Fig. 27.4 as a single line of $k+1$ nodes, starting with the negative identity as a black circle on the right, followed by the smallest vertex-acting generator $V = q(A_k) = q(2\pi/(k+1), \hat{\mathbf{z}})$, and repeating for each group element power until the group is exhausted and all elements of A_k have appeared. This is consistent with ignoring the quaternion origin of V and observing that an *ordinary complex root of unity* would behave in exactly this way.

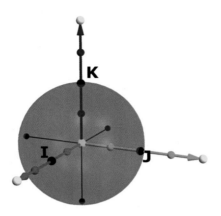

(I, J, K): Quaternion group

FIGURE 27.3

The quaternion algebra presentation in the form of the three Cartesian axes standing for the quaternion multiplication table, e.g., $\{(+\text{ID}), \mathbf{I}, (-\text{ID}), -\mathbf{I}, (+\text{ID})\}$ along each axis, with the implicit extra constraint that $\mathbf{IJK} = -1$. The Hamilton imaginaries form an *orthogonal* basis in this way. The ADE groups, in particular the E_6, E_7, and E_8 Platonic polyhedral groups, will have extremely similar properties, but will instead form nonorthogonal bases.

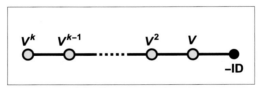

A_k: Cyclic group $< k + 1 >$

FIGURE 27.4

This graph corresponds to the simply laced Dynkin diagram for the A_k cyclic group, showing symbolically how the group generator repeated actions collect to produce the whole group. The labels are symbolic for the purposes of exhibiting the actions of the generators; the actual significance of these diagrams in group theory has many more meanings.

Technically, one might place the *positive* identity q_{ID} on the right end as the zeroth power of V, and the negative $(-q_{\text{ID}})$ on the left and as the value of V^{k+1}. However, we will want to note the appearance of the analog of $\mathbf{IJK} = -1$ for the other groups, which suggests leaving $-q_{\text{ID}}$ on the right. As one last observation, we note that we do in fact have a parallel to this last feature if we delete $\mathbf{J}$ and identify $V \to \mathbf{I}$ to the left of $-q_{\text{ID}}$ and $V^k \to \mathbf{K}$ as implicitly living to the *right* of $-q_{\text{ID}}$: then $V \star V^k = -q_{\text{ID}}$ is precisely the desired special case of $\mathbf{IJK} = -1$ for the cyclic group's Presentation.

27.3 $\mathbf{D_k}$ dihedral group *Presentation*: $\langle 2, 2, k+1 \rangle$

The $2(k+1)$ unique quaternion elements of the binary D_k dihedral group are essentially two copies of the A_k elements, with one equatorial plane direction quaternion b flipping one set acting at the North pole to a mirrored set acting at the South pole. The latter happens only *after* the flip action, so those new $k+1$ actions are distinct maps of the dihedral isosceles triangles into one another (Wikipedia, 2023h), and there are no others, so no other edge-center or face-center symmetries appear in the group. However, for the purposes of the D_k Presentation, the single edge flip *exactly parallels* the role of the $\mathbf{J}$ quaternion element, and so we have an extra dimension we can add to our diagram. The important fact is that if we start with just the A_k North pole action's smallest element $V = q(2\pi/(k+1), \hat{\mathbf{z}})$ as one generator, forming a k-element power series to the left of the $-q_{\text{ID}}$ node, and placing the flip quaternion, say $b = [0, 1, 0, 0]$, like $\mathbf{J}$, above the $-q_{\text{ID}}$ node, we can do more: just as we could place the $\mathbf{K}$ quaternion generator and the A_k element V^k to the right of $-q_{\text{ID}}$ to complete a pattern analogous to $\mathbf{IJK} = -1$, we can add a *third* supplementary generator $c = q(2\pi/(k+1), \hat{\mathbf{z}}) \star b$ to the D_k diagram that is redundant, and causes no extraneous group elements, but completes the triple cycle $\mathbf{IJK} = -1$. We thus assign to the binary D_k group the three-part Presentation system

$$\left. \begin{array}{c} V = q(2\pi/(k+1), \hat{\mathbf{z}}) = \left[\cos\dfrac{\pi}{k+1}, \ 0, \ 0, \ \sin\dfrac{\pi}{k+1} \right] \\[2mm] E = b = [0, \ 1, \ 0, \ 0] \\[2mm] F = c = V \star E = \left[0, \ \cos\dfrac{\pi}{k+1}, \ \sin\dfrac{\pi}{k+1}, \ 0 \right] \end{array} \right\} . \tag{27.7}$$

(We remind the reader of the convention of Eq. (2.16) that the θ in $q(\theta, \hat{\mathbf{n}})$ is the 3D rotation matrix angle, not the arccosine of q_0.) These obey the desired quaternion-like Presentation algebra,

$$V^{k+1} = E^2 = F^2 = V \star E \star F = -q_{\text{ID}} = [-1, 0, 0, 0], \tag{27.8}$$

and we can thus represent the binary D_k group using the generators Eq. (27.7) with the elegant diagram Fig. 27.5. Note that since $c^2 = c \star c = -q_{\text{ID}}$, we conform exactly to the Coxeter symbol description $\langle 2, 2, k+1 \rangle$ noted at the beginning.

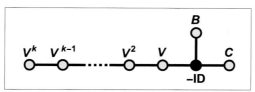

$\mathbf{D_k}$: Dihedral group $< 2, 2, k+1 >$

FIGURE 27.5

This graph corresponds to the simply laced Dynkin diagram for the Dk dihedral group, showing symbolically how the group generator repeated actions collect to produce the whole group. The labels are symbolic for the purposes of exhibiting the actions of the generators; the actual significance of these diagrams in group theory has many more meanings.

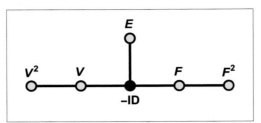

E_6: Tetrahehedral group $< 2, 3, 3 >$

FIGURE 27.6

The pattern of the simply laced Dynkin diagram for E_6 adapted to exhibiting the third-order Vertex symmetry V, the second-order edge generator E, and the third-order face symmetry F, all sharing their final action terminating at the central shared negative quaternion identity in black.

27.4 E_6 tetrahedral group *presentation* $\langle 2, 3, 3 \rangle$

The 24 quaternion elements of the binary E_6 tetrahedral group can be generated by a three-element selection of the elements, one each for the vertex, edge-center, and face-center symmetry actions. These are traditionally denoted by the symbols R, S, and T (Wikipedia, 2023f), while we choose to rearrange them as TRS and then relabel that form as VEF to establish a more direct and intuitive connection to our tetrahedral geometry. We now have the geometry of the tetrahedral vertices, edge centers, and face centers in Eq. (25.9), Eq. (25.10), and Eq. (25.11). We also have the mechanism for collecting the entire set of vertex-fixing, edge-center-fixing, and face-center-fixing discrete binary tetrahedral quaternion group elements from Eq. (26.15). We now choose a particular triple of elements, based on the traditional binary octahedral Presentation. Identifying R as the edge-flip operator $E(e)$, always of order two, S as the order three triangular face operator $F(f)$, and T as the order three vertex-spin operator $V(v)$, we can write the binary tetrahedral Presentation as

$$\begin{aligned}
V &= [\tfrac{1}{2} \quad \tfrac{1}{2} \quad -\tfrac{1}{2} \quad \tfrac{1}{2}], \\
E &= [0 \quad 1 \quad 0 \quad 0], \\
F &= [\tfrac{1}{2} \quad \tfrac{1}{2} \quad \tfrac{1}{2} \quad \tfrac{1}{2}].
\end{aligned} \tag{27.9}$$

We easily confirm that these quaternions have the same identity properties as the Hamilton $(\mathbf{I}, \mathbf{J}, \mathbf{K})$, but limited cyclic properties, which might be interpreted as the fact that (V, E, F) is not an orthonormal basis like the quaternions:

$$V \star V \star V = E \star E = F \star F \star F = V \star E \star F = [-1, 0, 0, 0], \tag{27.10}$$

$$\left.\begin{aligned}
V &\neq E \star F \\
E &= F \star V \\
F &\neq V \star E
\end{aligned}\right\}. \tag{27.11}$$

The two-fold flip operator E corresponds exactly to the 180-degree rotation property of $\mathbf{J}$, and we place the edge-flip operator E there in our E_6 tetrahedral Presentation diagram, Fig. 27.6. The vertex and face generators are essentially cube roots of -1 instead of square roots of -1 like $\mathbf{I}$ and $\mathbf{K}$, and so

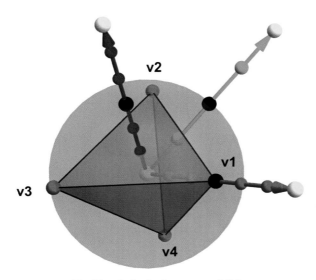

E_6: Tetrahehedral group $< 2, 3, 3 >$

FIGURE 27.7

Here we show an alternative view of the way the three generators of the presentation of E_6 discrete symmetry group can be understood: each generator is shown in the direction of its rotation-fixed vector, the fundamental vertex vector $\hat{\mathbf{v}}$ in red, the edge-center single-flip vector $\hat{\mathbf{e}}$ in green, and the face-center vector $\hat{\mathbf{f}}$ in blue. These three suffice to generate the entire E_6 tetrahedral symmetry group via multiple mutual actions, possessing a specific nonorthonormal generalization of the $(\mathbf{I}, \mathbf{J}, \mathbf{K})$. Instead of centering the initial actions of V, E, and F around $-q_{\mathrm{ID}}$ as we did in Fig. 27.6, we start from the symbolic positive identity denoted by a yellow sphere at the origin, and apply each generator three, two, and three times to reach the $-q_{\mathrm{ID}}$ marked by black spheres after the generators have gone *half way*, that is, up to an angle of 2π, through the quaternion space. Then we repeat the same cycle, arriving at the positive $+q_{\mathrm{ID}}$ shown as a yellow sphere at the completion of a full 4π quaternion cycle. This shows clearly the fundamental properties of the tetrahedron's binary quaternion group in a way that illustrates exactly how it relates to the similar, but orthogonal, quaternion $(\mathbf{I}, \mathbf{J}, \mathbf{K})$ cycle.

their lengths are each extended to a total of three units in Fig. 27.6, corresponding to (E, F, V) in the Coxeter symbol $\langle 2, 3, 3 \rangle$. We can now construct an alternative way of looking at this same information relative to the actual tetrahedral geometry description that is analogous to the quaternion $(\mathbf{I}, \mathbf{J}, \mathbf{K})$ depiction in Fig. 27.3. For the tetrahedron, our alternative image in Fig. 27.7 shows the alignment of the (V, E, F) presentation generators with a particular vertex, edge center, and face center on the tetrahedron. The yellow spheres at the origin and at twice the radius of the surrounding sphere denote the positive identity quaternion $+q_{\mathrm{ID}}$, while the black spheres at unit radius are the negative identity quaternion $-q_{\mathrm{ID}}$. The negative identity is reached precisely after multiplying the origin (the internal yellow sphere) by the V generator three times (red dots), the F generator three times (blue dots), and the E generator twice (green dots, the same as any quaternion generator). Doubling this action obviously returns each generator to the positive quaternion identity at the outer yellow dots. What is special about this choice is that multiplying *just these three* in the order $V \star E \star F$ also ends on the negative identity: any vertex, edge, and face actions will be the same as those shown in Fig. 27.7, but only very special choices will have the cyclic property of the E_6 group generators of Eq. (27.9).

27.5 E_7 octahedral group *presentation* $\langle 2, 3, 4 \rangle$

The 48 quaternion elements of the binary E_7 octahedral group can be generated by a three-element selection of the elements, one each for the vertex, edge-center, and face-center symmetry actions. Note that these are special choices: only a fraction of such triples generate the entire group. These are traditionally denoted by the symbols R, S, and T (Wikipedia, 2023e), which correspond to our edge, face, and vertex variables, that is, $RST \sim EFV$. We rearrange the formulas with our Presentation variables to correspond to the (V, E, F) variables to establish a more direct and intuitive connection to our octahedral geometry. We now have the geometry of the octahedral vertices, edge centers, and face centers in Eq. (25.12), Eq. (25.13), and Eq. (25.14). We also have the mechanism for collecting the entire set of vertex-fixing, edge-center-fixing, and face-center-fixing discrete binary octahedral quaternion group elements from Eq. (26.16). We now choose a particular triple of elements, based on the traditional binary octahedral Presentation (Wikipedia, 2023e). Identifying R as the edge-flip operator $E(e)$, always of order two, S as the order three triangular face operator $F(f)$, and T as the order four vertex-spin operator $V(v)$, we can write the binary octahedral Presentation as

$$\left. \begin{array}{l} V = [\,\frac{1}{\sqrt{2}} \quad \frac{1}{\sqrt{2}} \quad 0 \quad 0\,] \\[2mm] E = [\,0 \quad \frac{1}{\sqrt{2}} \quad \frac{1}{\sqrt{2}} \quad 0\,] \\[2mm] F = [\,\frac{1}{2} \quad \frac{1}{2} \quad \frac{1}{2} \quad \frac{1}{2}\,] \end{array} \right\} . \tag{27.12}$$

These quaternions have the same identity properties as the Hamilton $(\mathbf{I}, \mathbf{J}, \mathbf{K})$ algebra, but limited cyclic properties, which can again be interpreted as (V, E, F) not being an orthonormal basis like the quaternions:

$$V \star V \star V \star V = E \star E = F \star F \star F = V \star E \star F = [-1, 0, 0, 0] \,, \tag{27.13}$$

$$\left. \begin{array}{l} V \neq E \star F \\ E = F \star V \\ F \neq V \star E \end{array} \right\} . \tag{27.14}$$

The two-fold flip operator E as usual corresponds exactly to the 180-degree rotation property of $\mathbf{J}$, and we place the edge-flip operator E there in our E_7 octahedral Presentation diagram (Fig. 27.8). The face generators for all the Platonic polygons are cube roots of -1 and so again have their length extended to a total of three units in Fig. 27.8, but the vertex generators now are quartic roots of -1, and have four elements in their leftmost sector of Fig. 27.8. An alternative representation is shown in Fig. 27.9 that adds a geometric perspective to the abstraction of the graph in Fig. 27.8. In particular, we place a yellow sphere at the origin and at the doubled quaternion points representing the cyclic repeated positive identity. We place a black sphere, denoting the negative quaternion identity, at the power 2 of the edge symmetry (green), at the power 3 of the face symmetry (blue), and at the power 4 of the vertex symmetry (red). Repeating each of those power sequences along the same symmetry axes yields the next cyclic positive identity in yellow.

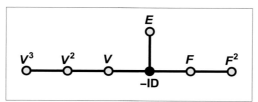

E_7: Octahehedral group $< 2, 3, 4 >$

FIGURE 27.8

This graph corresponds to the simply laced Dynkin diagram for the E_7, octahedral discrete symmetry group (Wikipedia, 2023b). Our heuristic application of this graph interprets the nodes in terms of the number of times one of the basic V, E, or F group operations must be repeated to recover the binary group identity element " -1 " $= -q_{ID} = (-1, 0, 0, 0)$. The mixed triple sequence $V \star E \star F = -1$ follows the pattern of the Hamilton imaginaries.

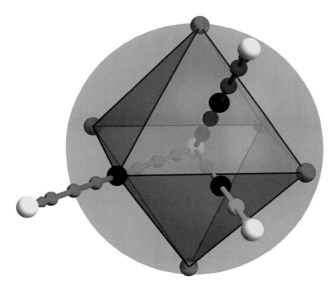

E_7: Octahehedral group $< 2, 3, 4 >$

FIGURE 27.9

Here we show an alternative view of the way the three generators of the presentation of the E_7 discrete symmetry group can be understood. Instead of centering the initial actions of V, E, and F around $-q_{ID}$, we start from the symbolic positive identity denoted by a yellow sphere at the origin, and apply each generator four, two, and three times to reach the $-q_{ID}$ marked by black spheres after the generators have gone *half way*, that is, up to an angle of 2π, through the quaternion space. Then we repeat the same cycle, arriving at the positive $+q_{ID}$ shown as a yellow sphere at the completion of a full 4π quaternion cycle. This shows clearly the fundamental properties of the octahedron's binary quaternion group in a way that illustrates exactly how it relates to the similar, but orthogonal, quaternion cycle.

27.6 E_8 icosahedral group *presentation*: $\langle 2, 3, 5 \rangle$

The fundamental structure of E_8 can be generated from three quaternions selected from the discrete group of symmetry-preserving actions on the vertices, edges, and faces of the icosahedron. We know already that for this final symmetry group, the quaternion vertex action V on the icosahedron has a fifth-order repetition to return to the identity, and E, the edge reflection quaternion, has a second-order action, while the triangular face action F has a third-order action returning to the identity. Among the limited set of specific triples that actually generate the entire icosahedral symmetry group, a conventional choice (see Wikipedia (2023d)) is the following:

$$\left. \begin{array}{l} V = \dfrac{1}{2}\left[\phi, \dfrac{1}{\phi}, 1, 0\right] \\[2mm] E = \dfrac{1}{2}\left[0, \dfrac{1}{\phi}, \phi, 1\right] \\[2mm] F = \left[\dfrac{1}{2}, \dfrac{1}{2}, \dfrac{1}{2}, \dfrac{1}{2}\right] \end{array} \right\} , \tag{27.15}$$

where as usual $\phi = (1 + \sqrt{5})/2$ is the golden ratio. We find it interesting to explore these fundamental generators of the E_8 group in a little more detail: writing each of the (V, E, F) elements in the quaternion axis-angle form with $q_0 = \cos(\theta/2)$ and $\mathbf{q} = \hat{\mathbf{n}}\sin(\theta/2)$, we see that these elements correspond to rotation angles in 3D space with the values required by the fifth-order, second-order, and third-order vertex, edge, and triangular face symmetries:

$$\theta_V = \pi/5 = 72° , \qquad \theta_E = \pi = 180° , \qquad \theta_F = \pi/3 = 120° . \tag{27.16}$$

The unnormalized directions $\mathbf{n}$ of the rotation axes are less predictable and take the forms

$$\mathbf{n}_V = (1, \phi, 0) , \qquad \mathbf{n}_E = (1, \phi^2, \phi) , \qquad \mathbf{n}_F = (1, 1, 1) , \tag{27.17}$$

while the normalized directions, the $\hat{\mathbf{n}}$ in $q(\theta, \hat{\mathbf{n}})$, are

$$\hat{\mathbf{n}}_V = \frac{1}{\sqrt{1+\phi^2}}(1, \phi, 0) , \qquad \hat{\mathbf{n}}_E = \left(\frac{1}{2\phi}, \frac{\phi}{2}, \frac{1}{2}\right) = \frac{1}{2\phi}\left(1, \phi^2, \phi\right) , \qquad \hat{\mathbf{n}}_F = \frac{1}{\sqrt{3}}(1, 1, 1) . \tag{27.18}$$

The group generators in Eq. (27.15) obey the expected Presentation identities for the binary icosahedral group,

$$V \star V \star V \star V \star V = E \star E = F \star F \star F = V \star E \star F = -q_{\text{ID}} = [-1, 0, 0, 0] , \tag{27.19}$$

which we can abbreviate as $V^5 = E^2 = F^3 = VEF = -1$. We note that we have arranged our coordinate system for the icosahedron in Eq. (25.15) so that the first vertex $v(1)$ corresponds to the fixed axis of the generator V in Eq. (27.15). The edge-flip action of the quaternion E was chosen conveniently to flip the sign of the x component of v_1 to produce the next vertex v_2, so that this pair of necessity lies in the $z = 0$ plane. All the rest of the vertices are then seen to group into three sets of four, the first set in the $z = 0$ plane, with four signs of the (x, y) components, the next in the $x = 0$ plane, and the last in the $y = 0$ plane. The 12 icosahedral vertices following from this construction, displayed in Fig. 25.5, include the directions of the fixed rotation axes of the V, E, F quaternions that are generators of the

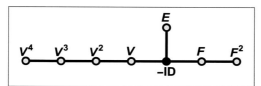

E_8: Icosahehedral group $< 2, 3, 5 >$

FIGURE 27.10

This graph, closely related to the simply laced Dynkin diagram for the E_8 discrete symmetry group (Wikipedia, 2023b). For our purposes, we interpret it as a way of relating the three quaternion generators Eq. (27.15) forming the Presentation of E_8 to their broader context as a generalized quaternion basis. The five-fold icosa-hedral vertex symmetry operators extend to the left from the central negative quaternion identity, with four powers until the identity repeats itself. The single-flip edge symmetry and the three-fold triangular face symme-tries have one action plus the identity, and two actions plus the identity as for the other Platonic polyhedra.

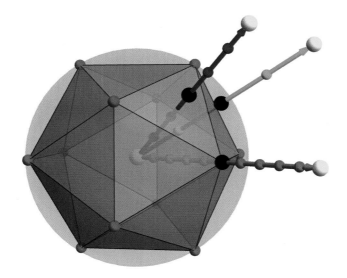

E_8: Icosahehedral group $< 2, 3, 5 >$

FIGURE 27.11

Here we show an alternative view of the way the three generators of the presentation of the E_8 discrete sym-metry group can be understood very geometrically. Instead of centering the initial actions of V, E, and F around $-q_{ID}$, we start from the symbolic positive identity denoted by a yellow sphere at the origin, and apply each generator five, two, and three times to reach the $-q_{ID}$ marked by black spheres after the generators have gone *half way*, that is, up to an angle of 2π, through the quaternion space. Then we repeat the same cycle, arriving at the positive $+q_{ID}$ shown as a yellow sphere at the completion of a full 4π quaternion cycle. This shows clearly the fundamental properties of the icosahedron's binary quaternion group in a way that illustrates exactly how it relates to the similar, but orthogonal, quaternion cycle.

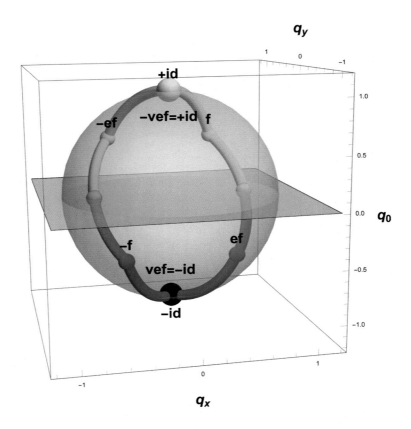

E$_6$: Tetrahedral group V $\star$ E $\star$ F $= -1$

FIGURE 27.12

The three generators for the tetrahedral group that we call V, E, and F (often written as RST in the order EFV in the literature) actually combine in a closed cycle $V \star E \star F = -q_{ID}$ in Fig. 27.6 and Fig. 27.7. This visualization is specially designed to combine the different pieces of the circular path of the three generators in quaternion space into one sphere with axes (q_0, q_x, q_y) that combines the elements of the 3-spheres display to expose more clearly the path when q_0 forms part of the circle. q_z is of course projected out and must be inferred from the $q \cdot q = 1$ constraint.

full icosahedral group obeying the quaternion algebra Eq. (27.19). The elements V, E, F that comprise our chosen Presentation of the icosahedral group (Wikipedia, 2023b) are symbolized in the diagram shown in Fig. 27.10. We present also a more geometrically oriented visualization in Fig. 27.11 that shows the increasing deviation of the icosahedral group generators from the quaternion orthogonality. Similarly to our other Platonic geometric group visualizations, each generator originates at the shared positive identity, proceeds through one 2π quaternion cycle of each generator to reach the negative quaternion identity at the surface of the enclosing sphere, and then finishes out a 4π total quaternion action returning to the positive identity.

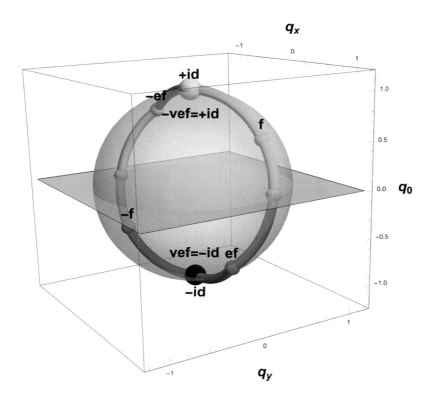

$$E_7: \text{ Octahedral group } V \star E \star F = -1$$

FIGURE 27.13

The three generators for the octahedral group that we call V, E, and F combine in a closed cycle $V \star E \star F = -q_{ID}$ in Fig. 27.8 and Fig. 27.9. This visualization is specially designed to combine the different pieces of the circular path of the three generators in quaternion space into one sphere with axes (q_0, q_x, q_y) that combines the elements of the 3-spheres display to expose more clearly the path when q_0 forms part of the circle. q_z is of course projected out and must be inferred from the $q \cdot q = 1$ constraint.

27.7 The VEF cycle

The $V \star E \star F = -1$ quaternion circle. Finally, utilizing the representation exposing the entire quaternion structure, we show how the three-part presentation cycle can be displayed directly in quaternion space, using the projection (q_0, q_x, q_y) to better expose the role of q_0 in this circular path. In particular we consider the paths among the three acting Presentation quaternions using the smooth `slerp` interpolator connecting each to the next on a geodesic curve in quaternion space. We note that the edge action is the longest path, 180 degrees, and cuts through the equator, represented as an orange plane, with the crossing point marked with an orange ball in the middle of the 180-degree path. We show these triple

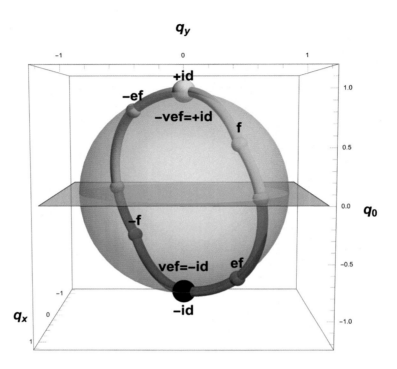

E$_8$: Icosahedral group $V \star E \star F = -1$

FIGURE 27.14

It is difficult to see how the three generators that we call V, E, and F (often written as RST in the order EFV in the literature) actually combine in a closed cycle $V \star E \star F = -q_{\text{ID}}$ in Fig. 27.10 and Fig. 27.11. In this 3-spheres visualization approach, we can see the full circle, doubling every element in the right-hand $q_0 > 0$ ball on the right with a pair in the left-hand $q_0 < 0$, with the unique points passing through the $q_0 = 0$ equator noted as the closures of the two open semicircular quaternion paths. The first element to act on the identity, the right-hand central yellow sphere, is F, followed by the path multiplying by E to $E \star F$, and completing the 2π half-quaternion circuit with the third left-multiplication by V to arrive as promised at $-q_{\text{ID}}$. Multiplying by F again actually produces $-F$ (we multiplied by $-q_{\text{ID}}$) and completing the cycle returns us to the $+q_{\text{ID}}$. So the traditional $V \star E \star F = -1$, viewed in this full quaternion space plot, shows us additional detail about just how special this particular choice of three group elements for the icosahedral group presentation turns out to be: of the many possible combinations, only a select few have this property as one can easily see. This visualization is specially designed to combine the different pieces of the circular path of the three generators in quaternion space into one sphere with axes (q_0, q_x, q_y). This approach combines the elements of the 3-spheres display to expose more clearly the involvement of q_0; q_z is of course projected out and must be inferred from the $q \cdot q = 1$ constraint.

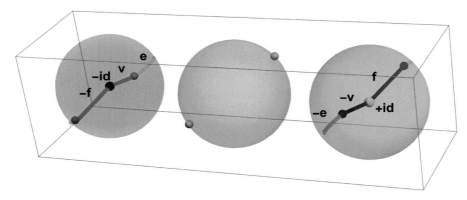

E_8: **Icosahehedral group in 3-sphere display** $V \star E \star F = -1$

FIGURE 27.15

The smooth circular paths in Figs. 27.12, 27.13, and 27.14 are shown here in our other standard 3-spheres display to show how a circle in quaternion space appears in the 3-spheres method in comparison to the (q_0, q_x, q_y) method. In this 3-spheres visualization approach, we can see the full circle, doubling every element in the right-hand $q_0 > 0$ ball on the right with a pair in the left-hand $q_0 < 0$ ball, with the unique points passing through the $q_0 = 0$ equator noted as the closures of the two open semicircular quaternion paths. The quaternion elements of the three-part presentation basis are the vertex, edge, and face actions denoted in the plot as v, e, and f; the path from the identity, the central yellow sphere in the right-hand $q_0 > 0$ ball, to the negative identity, the central black sphere in the left-hand $q_0 < 0$ ball, is the triple product $v \star e \star f = -1$, and the action is depicted *in that order*, with f acting *first*. So the first path from $+q_{\text{ID}}$ to f is a straight line in the (x, y, z)-coordinate projection, interpolating f incrementally to the blue point. This is followed by the path multiplying by e, interpolating through the $q_0 = 0$ equator to the quaternion point $e \star f$, the red point in the left-hand $q_0 < 0$ ball; we show this curve in orange and mark the $q_0 = 0$ equatorial point where it passes through. We complete the 2π half of the circuit with the third left-multiplication by v to arrive as promised at the $v \star e \star f = -q_{\text{ID}}$ negative identity at the black sphere. Multiplying by f again actually produces $-f$ (since we multiplied by $-q_{\text{ID}}$) and completing the cycle returns us to $(v \star e \star f)^2$, which is $+q_{\text{ID}}$. So the traditional $v \star e \star f = -1$, viewed in this full quaternion space plot, shows us additional detail about just how special this particular choice of three group elements for the icosahedral group presentation turns out to be: of the many possible triple combinations, only a select few have this property.

product circles, doubled to return to the positive identity, for all three cases, the E_6 tetrahedral cycle, the E_7 octahedral cycle, and the E_8 icosahedral cycle, in Figs. 27.12, 27.13, and 27.14, respectively. The colors show essentially where a given part of the path would lie in the 3-spheres method, which we show for comparison in Fig. 27.15 using the icosahedral case: in the single-ball (q_0, q_x, q_y) figures, the cyan curves lie in the right-hand blueish solid ball with the positive identity (shown with the usual yellow sphere) at the origin, the magenta curves lie in the left-hand reddish solid ball with the negative identity (black sphere) at the origin, and the orange plane stands for the $q_0 = 0$ equatorial surface forming the boundary between the solid quaternion volumes with $q_0 > 0$ and $q_0 < 0$.

The Klein invariants of the ADE discrete quaternion groups

28.1 Introduction

In this chapter, we will explore one final important feature of the ADE groups, the ability to adapt complex variable theory to explore interesting properties following from the construction of complex equations that incorporate the group invariants. These equations follow naturally from the approach to the ADE groups studied very early by Klein (1884) and subsequently by many others. Equations for the associated singular polynomials and the technology of removing their singularities are the subject of longstanding mathematical investigations. An extensive early exploration of these issues for the ADE groups appears in three comprehensive papers by Du Val (Du Val, 1934a,b,c; Wikipedia, 2023i), while a very detailed contemporary treatment is that of Katz and Morrison (1992). There is an even more sophisticated body of mathematical literature pertaining to the role of the ADE groups, e.g., in gravity theory, that exploits twistor theory, twistor data, hyperkähler structures, and so on. We refer the reader to the literature for further details of the massive amount of information that has been accumulated on these various related topics, which are far beyond the scope of what we can possibly treat here. Nevertheless, there is much to be examined within the relatively elementary context that we have established leading up to this chapter, and we will now study a number of interesting examples and their visualizations arising from the ADE framework.

Our basic framework for the construction of the ADE symmetry groups started with quaternions described by axis-angle parameterizations, that is, rotations by an angle θ that keep a symmetry axis $\hat{\mathbf{n}}$ in 3D space *fixed*, and that will also be our starting point for finding invariants. We recall that

$$q(\theta, \hat{\mathbf{n}}) = \left[\cos\frac{\theta}{2}, \, \hat{\mathbf{n}}\sin\frac{\theta}{2} \right], \tag{28.1}$$

where $\hat{\mathbf{n}} \cdot \hat{\mathbf{n}} = 1$, θ has the range $[0, 4\pi)$, while q and $-q$ (thought of as the ranges $[0, 2\pi)$ and $[2\pi, 4\pi)$, respectively) correspond to the same 3D rotation. This quaternion was then recast in the form of 2×2 matrices, that is, the **SU(2)** form based on the Pauli matrices,

$$[Q(q)] = q_0 I_2 - i\mathbf{q} \cdot \boldsymbol{\sigma} = \begin{bmatrix} q_0 - iq_z & -iq_x - q_y \\ -iq_x + q_y & q_0 + iq_z \end{bmatrix}. \tag{28.2}$$

As we argued in Chapter 3, we can interpret a modified version of the matrix Eq. (28.2), that is

$$L(a, b, c, d) = \begin{bmatrix} a & b \\ c & d \end{bmatrix} = [Q(q)^*] = \begin{bmatrix} q_0 + iq_z & +iq_x - q_y \\ +iq_x + q_y & q_0 - iq_z \end{bmatrix}, \tag{28.3}$$

Visualizing More Quaternions. https://doi.org/10.1016/B978-0-32-399202-2.00043-5

as representing the space of rotation-creating linear transformations. As this matrix acts on the homogeneous variables (z_1, z_2) of the complex projective space $\mathbf{CP}^1$, commonly referred to as the Riemann sphere, it also backprojects to a rotation with a corresponding fixed point, acting on the points of the two-sphere $\mathbf{S}^2$ embedded in 3D Euclidean space. This action on $\mathbf{CP}^1$ takes the form

$$
\begin{bmatrix} z_1' \\ z_2' \end{bmatrix} = L(a, b, c, d) \cdot \begin{bmatrix} z_1 \\ z_2 \end{bmatrix} = \begin{bmatrix} az_1 + bz_2 \\ cz_1 + dz_2 \end{bmatrix}, \tag{28.4}
$$

and this in turn defines the linear fractional transformation corresponding to the inhomogeneous $z_2 \neq 0$ $\mathbf{CP}^1$ parameterization $z = z_1/z_2$, finally leading to the inhomogeneous linear fractional transformation for z:

$$
z' = \frac{az + b}{cz + d} . \tag{28.5}
$$

For further details about complex analysis, the Riemann sphere, and the roles of homogeneous and inhomogeneous variable for the complex projective space $\mathbf{CP}^1$, the reader is referred to standard textbook treatments, (see, e.g., Griffiths and Harris, 1978; Brown and Churchill, 1989).

The net effect is that, with the matrix $[Q(q)^*]$ in Eq. (28.3) defining the behavior under rotation of points in the complex plane $z = x + iy$ via Eq. (28.5), we can study the action of quaternions on fixed points in the complex plane. *That is, all the fixed rotation axes defining the ADE symmetry groups on $\mathbf{S}^2$ can now be studied using complex analysis.*

Summary. Our framework for studying the ADE groups began with the work of Klein (1884) on the discrete $\mathbf{SU(2)}$ subgroups of the projective group $\mathbf{PGL}(2, \mathbb{C})$. We established this connection by reinterpreting the topological sphere $\mathbf{S}^2$ in terms of the fundamental complex projective space $\mathbf{CP}^1$, also known as the *Riemann sphere*. Next, we will identify all of the points on $\mathbf{S}^2$ that are invariant under the action of each ADE group, and project them from that sphere, now the Riemann sphere $\mathbf{CP}^1$, into the infinite complex plane. This process will allow us to study the discrete $\mathbf{S}^2$ group actions as a subset of the linear fractional transformations, and to completely restructure the way we look at the geometric symmetries of $\mathbf{S}^2$: our task has thus been transformed into the study of *complex polynomials in the complex plane* and sets of points in that plane that are invariant under the ADE group actions projected to the complex plane.

> This approach allows us to study quaternion symmetries algebraically using complex analysis in a way we have not previously exploited.

28.2 The projection from the Riemann sphere

The projection from the unit sphere $\mathbf{S}^2$ to the complex plane can be represented as a line drawn from a pole through a unit-length point $\hat{\mathbf{p}}$ on the sphere to a point (x, y) in the 3D plane through the origin corresponding to the complex point $z = x + iy$ (see Fig. 28.1). We take the $\mathbf{S}^2$ corresponding to the Riemann sphere to be centered at the origin $(0, 0, 0)$, and we trace projections starting from the North pole

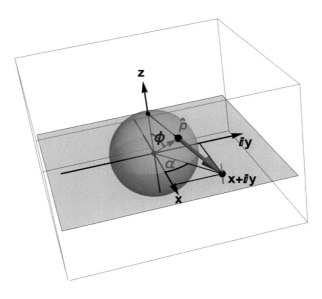

FIGURE 28.1

The origin-centered Riemann sphere $\mathbf{CP}^1$ (that is, $\mathbf{S}^2$), showing the path of a projection through a spherical point $\hat{\mathbf{p}}$, typically a fixed rotation axis, to its corresponding point $x + iy$ in the complex plane. The sphere's North pole $(0, 0, +1)$ projects to infinity, and the South pole $(0, 0, -1)$ projects to the origin.

$(0, 0, +1)$, which corresponds to the point at infinity in the complex plane. Reviewing the mathematics, we begin by parameterizing an arbitrary unit vector $\hat{\mathbf{p}}$ on the sphere as it sits in 3D space as

$$\hat{\mathbf{p}} = \begin{bmatrix} \cos\alpha \sin\phi \\ \sin\alpha \sin\phi \\ \cos\phi \end{bmatrix} .$$

Then the projection along a line running from the North pole at $(0, 0, +1)$ through a generic point $\hat{\mathbf{p}}$ on the unit sphere $\mathbf{S}^2$ intersects the complex plane at the point

$$x = \frac{p_x}{1 - p_z} , \tag{28.6}$$

$$y = \frac{p_y}{1 - p_z} , \tag{28.7}$$

[as z-plane complex variable] $\quad "z" = x + iy = e^{i\alpha} \frac{\sin\phi}{1 - \cos\phi} , \tag{28.8}$

as illustrated in Fig. 28.1. Note that if we are considering the special case of interest when $\hat{\mathbf{p}}$ is a fixed point $\hat{\mathbf{n}}$ of a 3D rotation acting on the sphere, there is actually a pair of antipodal fixed points, at $\pm\hat{\mathbf{p}}$, with quite distinct corresponding z-plane fixed points

$$z_\pm = \frac{\pm(n_x + in_y)}{1 \mp n_z} . \tag{28.9}$$

Note on infinity. The homogeneous pair of projective coordinates $(z_1, z_2) \sim (\lambda z_1, \lambda z_2)$ parameterizes the nontrivial spherical manifold of the $\mathbf{CP}^1$ Riemann sphere using two coordinate charts, $z = z_1/z_2$ for $z_2 \neq 0$ and $z = z_2/z_1$ for $z_1 \neq 0$, providing a valid coordinate system, in local pieces, for the entire sphere. One notes that we have chosen the coordinate $z = z_1/z_2$ for $z_2 \neq 0$ for our default system, and thus there might be a problem studying vertices of the Kleinian symmetric ADE polygons that could lie at $(0, 0, +1)$. Here is where the wisdom of thinking about our problem in terms of homogeneous as well as the inhomogeneous coordinates comes into play: whenever we have a polynomial in z that is missing a needed factor representing a vertex, edge center, or face center at "∞", that is $(0, 0, +1)$, we adjoin a factor of z_2 to represent that point, and typically multiply by enough powers of z_2 to convert the z powers into powers of z_1. The result is typically an elegant symmetric expression in z_1 and z_2 as we shall see.

28.3 Symmetries and invariant polynomials of the Kleinian groups

Much of the modern interest in applications of the Kleinian groups follows from Klein's original enumeration of the invariant polynomials of the ADE symmetries manifested in the complex plane corresponding to the Riemann sphere, that is, the complex projective space $\mathbf{CP}^1$. The fundamental tool of the correspondence is that each 3D symmetry point of an embedded symmetric tessellation of $\mathbf{S}^2$ maps to a point $z = x + iy$ in the complex plane using the methods we have just outlined. Given such a mapping, there are some interesting and sophisticated properties of which we can take advantage.

- $Z(\mathbf{z_1, z_2})$ **The vertex equation.** Suppose we are given a set of 3D, unit-length vertices $\{v[i]\}$ belonging to an ADE tessellation on the unit sphere. By the elementary projective geometry identifications we have now discussed at length, we can project each vertex, except possibly the pole of the projection, to a complex point $w[i]$. If a vertex is at the pole, as it can be for any of the groups (though we have avoided that for our choices of the tetrahedral and icosahedral groups), we can simply use two poles, with the points at $w = z = z_1/z_2$ supplemented with $w' = z_2/z_1$. The key here is to include *both* homogeneous coordinates when we have both poles in the vertex set, so $(z_1 - 0) \times (z_2 - 0) = z_1 z_2$ is the simultaneous algebraic equation for *both* the North and South poles of the complex analytic 2-sphere. The rest of the vertices of the symmetric polyhedra reduce to

$$Z(z_1, z_2) = \textbf{non-North-pole vertex locations}$$

$$= \prod_{k=1}^{K} (z - w[k]) = \prod_{k=1}^{K} \left(\frac{z_1}{z_2} - w[k] \right)$$

$$\overset{\textbf{cancel powers of z}_2}{=} \prod_{k=1}^{K} (z_1 - w[k]\, z_2) \,. \tag{28.10}$$

If one of the roots is $w[k] = 0$, implying an overall factor of z_1, the vertex at the South pole, then in the ADE context that means there is a hidden paired vertex at the North pole, the infinity of our complex plane. Hence in this case, when the geometric structure includes the two poles, this polynomial simply needs to be multiplied by z_2 to add that North pole vertex in homogeneous coordinates, thus the overall polynomial has a symmetric factor $z_1 z_2$. Our resulting homogeneous coordinate expression is thus always homogeneous of degree n_V, the number of vertices in the symmetric tessellation, and is conventionally denoted by $Z(z_1, z_2)$. Its (complex) zeroes

$$Z(z_1, z_2) = 0 \tag{28.11}$$

describe the corresponding rotation-fixed vertices of the spherical polyhedron in our new language of complex variables, where the homogeneous form allows us to deal nicely with possible polar vertex locations. We now have a remarkable new tool, the ability to study quaternion actions using the analytic properties of the complex plane.

- $Y(z_1, z_2)$ **The face-center equation.** There is no obstacle at this point to simply computing the locations of the directions of the unit face-center vectors that are fixed under the polyhedron-face set of symmetries. However, there are more interesting ways to find that polynomial, which is traditionally called $Y(z_1, z_2)$, where solving $Y(z_1, z_2) = 0$ yields the locations of the face-center invariants in the complex plane. As pointed out by Klein, deep properties of the Hessian matrix produce the equations for the face centers without having to do any geometric calculations at all. Up to a scale, which we will generally remove in practice so that the coefficient of the highest power of z_i is unity, the solution for the face centers is given by the Hessian determinant of the vertex polynomial,

$$Y(z_1, z_2) = \textbf{face-center locations}$$

$$= \det \begin{bmatrix} \dfrac{\partial^2 Z(z_1, z_2)}{\partial z_1{}^2} & \dfrac{\partial^2 Z(z_1, z_2)}{\partial z_1 \, \partial z_2} \\[2ex] \dfrac{\partial^2 Z(z_1, z_2)}{\partial z_1 \, \partial z_2} & \dfrac{\partial^2 Z(z_1, z_2)}{\partial z_2{}^2} \end{bmatrix}. \tag{28.12}$$

We remind ourselves that, in order to have meaningful solutions to $Y(z_1, z_2) = 0$, we will find that $Y(z_1, z_2)$ is homogeneous in z_1 and z_2 of degree n_F, the number of faces in the polyhedron.

- $X(z_1, z_2)$ **The edge-center equation.** Again, we could simply compute the locations of the directions of the unit edge-center vectors that are fixed under the corresponding set of symmetries. But as before, there are more interesting ways to find that polynomial, which is traditionally called $X(z_1, z_2)$, where solving $X(z_1, z_2) = 0$ yields the locations of the edge-center invariants in the complex plane, and $X(z_1, z_2)$ is of course homogeneous in z_1 and z_2 of degree n_E, the number of edges in the polyhedron. Klein invokes what he terms as the *functional determinant* of $Y(z_1, z_2)$ and $Z(z_1, z_2)$, a kind of differential cross product of the two gradients. This calculation produces the equations for the edge centers of the appropriate degree directly, and again we typically scale the result to normalize the highest power in the polynomial. The solution for the edge centers is given by the functional determinant

$$X(z_1, z_2) = \textbf{edge-center locations}$$

$$= \det \begin{bmatrix} \dfrac{\partial Y(z_1, z_2)}{\partial z_1} & \dfrac{\partial Z(z_1, z_2)}{\partial z_1} \\[2ex] \dfrac{\partial Y(z_1, z_2)}{\partial z_2} & \dfrac{\partial Z(z_1, z_2)}{\partial z_2} \end{bmatrix}. \tag{28.13}$$

- **Remarks.** With some labor, one can cross-check that X, Y, and Z indeed are the same as taking the product $\prod_k (z - \omega[k])$ over the z-plane locations of all the edge centers, face centers, and vertices, projected from their $\mathbf{S}^2$ geometry. The argument that each of these polynomials is invariant under elements of the group is a consequence of the fact that the directions of the points $v[k]$, $e[k]$, and $f[k]$ that are fixed under the rotation group actions of the ADE symmetries are still fixed when projected to the z-plane: after any group action, the set of the points in the z-plane consists of the same points. Finally, we note a critical fact that we will immediately make use of, namely that we know the numbers, n_E, n_F, and n_V, of edge centers, face centers, and vertices corresponding precisely to the degrees of $X(z_1, z_2)$, $Y(z_1, z_2)$, and $Z(z_1, z_2)$, respectively. Since this can be confusing in the traditional notations, let us summarize the notations for polynomial properties that we have encountered in this simple table:

Geometric object	Multiplicity: degree	Klein polynomial	Presentation generators	Group action power
V: Vertex	n_V	$Z(z_1, z_2)$	T	Number of edges or faces at vertex
E: Edge	n_E	$X(z_1, z_2)$	R	Always 2
F: Face	n_F	$Y(z_1, z_2)$	S	Number of edges or vertices of face

The feature of this information that we will use next is that, knowing the degrees of the polynomials, we can look for the least common powers that match, and those will lead us directly to the legendary Kleinian singular polynomials in three complex dimensions.

Now we follow Klein's argument that, for each of the ADE groups, one can define a vanishing polynomial relating the three invariant polynomials X, Y, and Z, that we can treat as an abstraction in its own right, with remarkable consequences. We propose to find expressions satisfying

$$f(X, Y, Z) = 0 \tag{28.14}$$

relating our polynomials, whose zeroes locate the rotation axes of our vertices, edges, and faces in the complex plane.

The first example: A_k invariants. For A_k and D_k, we can construct the invariants and the constraints more or less by inspection. Remembering that rotations in the complex plane itself are full 3D rotations mapped from their paths in the $\mathbf{S}^2$ Riemann sphere, we conclude that each homogeneous complex variable transforms under a rotation according to the map of Eq. (28.3), or as the ratio of those half-angle maps for the corresponding inhomogeneous variables z. Since the only acting rotation for $\mathbf{A_k}$ is a rotation about the z-axis, we apply the rotations

$$\left. \begin{aligned} z_1' &= \exp\left(+i\frac{n\pi}{k+1}\right) z_1 \\ z_2' &= \exp\left(-i\frac{n\pi}{k+1}\right) z_2 \end{aligned} \right\}, \tag{28.15}$$

where $n = (0, 1, \ldots, 2k - 1)$ to cover the entire quaternion angular domain. The set of physical rotations in the z-plane and on $\mathbf{S}^2$ itself is covered by the inhomogeneous ratio of these,

$$z' = \frac{z_1'}{z_2'} = \exp\left(i\frac{2n\pi}{k+1}\right) z, \tag{28.16}$$

with $n = (0, 1, \ldots, k - 1)$. We choose the product $z_1 z_2$ of the two homogeneous opposite-pole fixed-point positions of the cyclic rotation as the vertex set $Z(z_1, z_2)$ invariant under Eq. (28.15). In the absence of actual fixed points under the transformation for the rest of the A_k geometry, we construct the other two expected invariants by hand. Alongside the fixed-vertex term $Z(z_1, z_2)$, we define two variables X and Y with powers of k that have only a possible sign dependence under the action of the rotations Eq. (28.15),

$$X(z_1, z_2) = z_1^{(k+1)} \,, \tag{28.17}$$

$$Y(z_1, z_2) = z_2^{(k+1)} \,, \tag{28.18}$$

$$Z(z_1, z_2) = z_1 z_2 \,. \tag{28.19}$$

The form $Z(z_1, z_2) = 0$ is clearly the homogeneous complex equation with the rotation-invariant North pole and South pole as its solutions; these are not only the sole vertices but also the sole points of any kind that stay fixed under symmetries of the multi-digon partition of $\mathbf{S}^2$ into equal orange slices. Next, we form a quadratic construction that manifestly removes the sign issue in the individual functions X and Y, leading as required to invariants under Eq. (28.15). Starting with the expressions

$$\widetilde{X}(z_1, z_2) = \frac{1}{2}\left(z_1^{(k+1)} + z_2^{(k+1)}\right) = \frac{1}{2}\left(X(z_1, z_2) + Y(z_1, z_2)\right) \,, \tag{28.20}$$

$$\widetilde{Y}(z_1, z_2) = \frac{1}{2}\left(z_1^{(k+1)} - z_2^{(k+1)}\right) = \frac{1}{2}\left(X(z_1, z_2) - Y(z_1, z_2)\right) \,, \tag{28.21}$$

we see that obviously the *squares* of $\widetilde{X}(z_1, z_2)$ and $\widetilde{Y}(z_1, z_2)$ are manifestly invariant under Eq. (28.15). Thus the following equation is an acceptable polynomial each term of which is invariant under the A_k symmetry group:

$$\widetilde{X}^2 - \widetilde{Y}^2 - Z^{k+1} = 0 \,. \tag{28.22}$$

Now we can justify collapsing Eq. (28.22) to cancel out the squares of X and Y, leaving only the cross-term in our chosen A_k polynomial

$$f_{(A,k)} = X Y - Z^{k+1} = 0 \,. \tag{28.23}$$

One can of course use Eq. (28.22) directly if desired. We will see shortly that even our apparently simple A_k Klein invariant relation can have significant structure.

The D_k dihedral invariants. The dihedral D_k geometry and its group are quite different from the A_k case in that there are clearly planar polyhedra involved in the construction. Thus, though the main set of symmetries is the same as the A_k group, we can see a little more involvement of all the vertices, edges, and faces. Spinning about the $\hat{\mathbf{z}}$-axis is clearly a symmetry inherited from A_k, but now in addition we can flip 180 degrees about either an equatorial vertex or an equatorial edge center. However, unlike the $E_{6,7,8}$ groups, the nonpolar edges for the dihedra are not all going to contribute, in fact only one flip is needed plus the A_k group to exhaust all possible symmetries: more flips would result in redundant group elements. We thus supplement the principal rotation transformations by the $\hat{\mathbf{x}}$-axis flip from Eq. (26.9),

$$\left.\begin{array}{l} z_1{}' = -i z_2 \\ z_2{}' = -i z_1 \end{array}\right\} \,, \tag{28.24}$$

which create maps such as $z_1 z_2 \to z_1' z_2' = -z_1 z_2$. With these insights, again more or less by inspection, we can add factors of $z_1 z_2$ and sums and differences of powers of $k + 1$, leading to the choices

$$X(z_1, z_2) = \frac{1}{2} z_1 z_2 \left(z_1^{2(k-2)} - z_2^{2(k-2)} \right),$$ (28.25)

$$Y(z_1, z_2) = \frac{1}{2} \left(z_1^{2(k-2)} + z_2^{2(k-2)} \right),$$ (28.26)

$$Z(z_1, z_2) = (z_1 z_2)^2.$$ (28.27)

We can see that $Z = 0$ is clearly the homogeneous complex equation for the D_k fixed points at the North and the South poles, with the addition of a power of two to compensate for the flip action of Eq. (28.24), and an additional $z_1 z_2$ in X to preserve the signs under the action of Eq. (28.24). The constraint relating the D_k invariants can then be verified as

$$f_{(D,k)} = X^2 - ZY^2 + Z^{k-1} = 0.$$ (28.28)

The factor of Z in the second term multiplying Y must be included to maintain the homogeneity of the equation in the presence of the $\sqrt{Z}$ sign-maintaining factor in X. One sometimes sees added factors of $i = \sqrt{-1}$ to adjust the overall signs.

The E_6 tetrahedral invariants. Moving on to the Platonic polyhedra, our task is made somewhat more systematic thanks to insights dating back at least to Felix Klein's work. If we simply start with the basic E_6 tetrahedral shapes from Fig. 25.1, with our vertex coordinate system from Eq. (25.9), we can take the invariant directions of vertex-based E_6 group operations as

$$v_1 = \frac{1}{\sqrt{3}} [\; 1 \quad -1 \quad 1\;],$$

$$v_2 = \frac{1}{\sqrt{3}} [\; 1 \quad 1 \quad -1],$$

$$v_3 = \frac{1}{\sqrt{3}} [-1 \quad 1 \quad 1\;],$$

$$v_4 = \frac{1}{\sqrt{3}} [-1 \quad -1 \quad -1],$$

and transform them to points on the Riemann sphere. Using Eq. (28.6) to perform this projection, we find the four corresponding points in the complex plane to be

$$p_1 = \frac{1}{2}(1 + \sqrt{3}) - \frac{i}{2}(1 + \sqrt{3}),$$

$$p_2 = -\frac{1}{2}(1 - \sqrt{3}) - \frac{i}{2}(1 - \sqrt{3}),$$

$$p_3 = -\frac{1}{2}(1 + \sqrt{3}) + \frac{i}{2}(1 + \sqrt{3}),$$

$$p_4 = \frac{1}{2}(1 - \sqrt{3}) + \frac{i}{2}(1 - \sqrt{3}).$$

The algebraic equation whose solutions are the positions of the tetrahedral vertices is constructed by using these values as roots of a fourth-order polynomial in the z-plane, and then homogenizing the

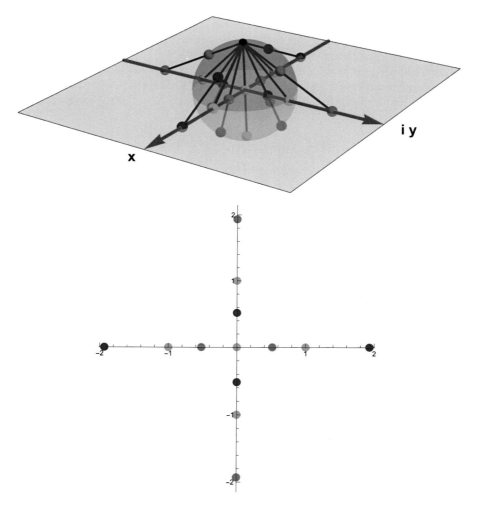

FIGURE 28.2

The E_6 tetrahedron invariant vertex, edge-center, and face-center axes converted via Klein's Riemann surface projection into the corresponding symmetry points in the complex plane. The zeroes of the complex fourth-degree polynomial $Z(z_1, z_2)$ are the z-plane projections of the four tetrahedron vertices, shown in red in the bottom figure. The zeroes of the edge-center and face-center polynomials $X(z_1, z_2)$ and $Y(z_1, z_2)$ are shown in green and blue, respectively. Note that the top edge-center symmetry point is at the North pole, so only five of the expected six green points appear in this projection. The three polynomials are not independent, and produce the Kleinian constraint equation $12i\sqrt{3}X^2 + Y^3 - Z^3 = 0$.

coordinates by replacing $z \to z_1/z_2$ and multiplying by $z_2{}^4$, yielding

$$Z(z_1, z_2) = \prod_{k=1}^{4} (z_1 - p_k z_2) = z_1{}^4 + 2i\sqrt{3}z_1{}^2 z_2{}^2 + z_2{}^4 . \tag{28.29}$$

The four roots of this quartic equation are the locations of the projected invariant rotation directions of the tetrahedron. Note that rotating the coordinate system will generally alter the overall coefficients, leaving the main features of the equation intact.

Eq. (28.29) is all we need in order to complete the calculation of the face-center invariant of degree four, and edge-center invariant of degree six from Eq. (28.12) and Eq. (28.13). Recalling that $X(z_1, z_2)$ corresponds to the tetrahedron edges, $Y(z_1, z_2)$ to the faces, and $Z(z_1, z_2)$ to the vertices, we find these corresponding results:

$$X(z_1, z_2) = (z_1 z_2)\left(z_1{}^4 - z_2{}^4\right), \tag{28.30}$$

$$Y(z_1, z_2) = z_1{}^4 - 2i\sqrt{3}z_1{}^2 z_2{}^2 + z_2{}^4, \tag{28.31}$$

$$Z(z_1, z_2) = z_1{}^4 + 2i\sqrt{3}z_1{}^2 z_2{}^2 + z_2{}^4. \tag{28.32}$$

The zeroes of these equations are the invariant complex plane positions of the tetrahedral edge centers, face centers, and vertices, respectively, and Fig. 28.2 illustrates the projection of the E_6 group invariant points on $\mathbf{S}^2$ onto the complex plane, where, as foreshadowed, the North pole top edge-center singularity needs to be supplied by hand to obtain the factor $(z_1 z_2)$ in $X(z_1, z_2)$. From the degrees of the invariant equations, we can see that the compatible powers of X, Y, and Z are 2, 3, and 3, resulting in each case in a polynomial of degree 12. Summing the three powers and working out the constants, we find the vanishing E_6 Kleinian polynomial constraint

$$\tilde{f}_{E_6} = 12i\sqrt{3}X^2 + Y^3 - Z^3 = 0. \tag{28.33}$$

For our applications of the constraint equation, the constants can be incorporated into the invariants, and we will typically use the simplified E_6 equation

$$f_{E_6} = X^2 + Y^3 - Z^3 = 0. \tag{28.34}$$

The E_7 octahedral invariants. The octahedron of the E_7 discrete group contains six vertices, four of which are the fourth roots of unity around the equator and are easily transformed to the complex plane, one of which is an overall zero of z_1 at the South pole, and the other is at the North pole, and needs to be added by hand as a factor of z_2 as discussed earlier. Thus we will need to remove a projection singularity by using the homogeneous binomial $z_1 z_2$ to accommodate the poles, combined with the four points that are the fourth roots of unity around the equator, which are of the form

$$z^4 - 1 = \left(\frac{z_1}{z_2}\right)^4 - 1.$$

Thus our algebraic equation for the positions of the six octahedral vertices in the complex plane combines these terms, homogenizing the coordinates to have the same powers of z_1 and z_2, to yield the result

$$Z(z_1, z_2) = z_2{}^5 z_1 \prod_{k=0}^{3} \left(\frac{z_1}{z_2} - \exp\left(\frac{2\pi i k}{4}\right)\right) \tag{28.35}$$

$$= (z_1 z_2)(z_1{}^4 - z_2{}^4). \tag{28.36}$$

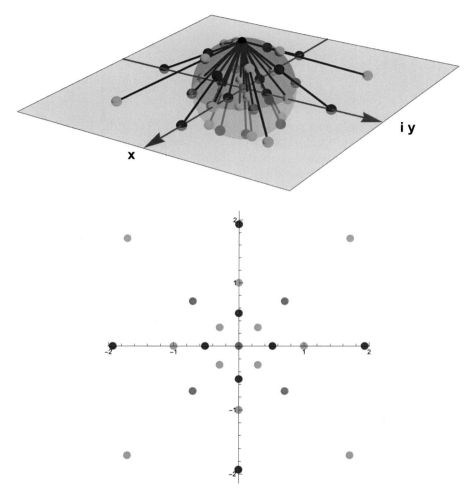

FIGURE 28.3

The E_7 tetrahedron invariant vertex, edge-center, and face-center axes converted via Klein's Riemann surface projection into the corresponding symmetry points in the complex plane. The zeroes of the complex sixth-degree polynomial $Z(z_1, z_2)$ are the z-plane projections of the six octahedron vertices, shown as red points in the bottom figure. Since the North pole projects to infinity, only five points are seen in this image. Similarly, the zeroes of the edge-center and face-center polynomials $X(z_1, z_2)$ and $Y(z_1, z_2)$ are shown in green and blue, respectively. These three polynomials are not independent, obeying the Kleinian constraint equation $X^2 - Y^3 + 108\,Z^4 = 0$.

It is now straightforward to compute Y and X, the invariant complex plane positions of the octahedral face centers and edge centers, from Eq. (28.12) and Eq. (28.13) (or directly if one is so inclined). The projection of the $\mathbf{S}^2$ invariant rotation axes of the octahedron onto the complex plane are shown in Fig. 28.3.

The E_7 octahedral Klein invariants then become

$$X(z_1, z_2) = z_1{}^{12} - 33 z_1{}^8 z_2{}^4 - 33 z_1{}^4 z_2{}^8 + z_2{}^{12} , \tag{28.37}$$

$$Y(z_1, z_2) = z_1{}^8 + 14 z_1{}^4 z_2{}^4 + z_2{}^8 , \tag{28.38}$$

$$Z(z_1, z_2) = z_1 z_2 (z_1{}^4 - z_2{}^4) . \tag{28.39}$$

We may also remark that the vertex term $Z(z_1, z_2)$ in Eq. (28.39) for the E_7 invariants shares the form of the edge term $X(z_1, z_2)$ in Eq. (28.30) for the E_6 invariants, since both geometries have six invariant rotation centers aligned with the Cartesian axes. We note as before that rotating the coordinate system may change various parts of the coefficients. From the degree of the invariants, we can see that the degrees of X, Y, and Z are 12, 8, and 6, the numbers of edges, faces, and vertices of the octahedron. Thus the compatible degrees of these invariant polynomials going into a compatible same-degree constraint will be 2, 3, and 4, with a total overall degree of 24. Working out the constants, we find the vanishing E_7 Kleinian polynomial constraint

$$\tilde{f}_{(E_7)} = X^2 - Y^3 + 108 Z^4 = 0 . \tag{28.40}$$

It is noted in some literature that "108" can be considered a fundamental constant of the octahedral geometry, although the constants depend on the orientation. In the applications for which we will use the constraint equation, the constant can be incorporated into the invariants, and we will typically use the simplified E_7 equation

$$f_{(E_7)} = X^2 - Y^3 + Z^4 = 0 . \tag{28.41}$$

The E8 icosahedral invariants. We turn finally to determining the Klein invariants for the icosahedral group E_8, along with the corresponding constraint equation relating the invariants. The traditional set of equations for E_8 is based on an embedding orientation of the icosahedron vertices that is slightly different from the coordinates Eq. (25.15) or Eq. (26.23) that we chose for compatibility with the RST Presentation literature for the E_8 binary group. Since rotations of the symmetry axes on the sphere $\mathbf{S}^2$ do not significantly change the structure of our equations, we will begin for the sake of connection to the Klein literature with this more traditional set of coordinates.

We first take the 12 vertices of the icosahedron oriented to include a pair of vertices at the North and South poles. The coordinates we will use are

$$
\left.
\begin{array}{ll}
v_1 = [\quad 0 \qquad 0 \qquad 1 \quad], & v_2 = [\quad 0 \qquad 0 \qquad -1 \quad] \\[4pt]
v_3 = [\ -\frac{2}{\sqrt{5}} \qquad 0 \quad -\frac{1}{\sqrt{5}} \ ], & v_4 = [\ \frac{2}{\sqrt{5}} \qquad 0 \qquad \frac{1}{\sqrt{5}} \ ] \\[4pt]
v_5 = [\ \frac{\phi}{\sqrt{5}} \quad -\frac{1}{\sqrt[4]{5}\sqrt{\phi}} \quad -\frac{1}{\sqrt{5}} \ ], & v_6 = [\ \frac{\phi}{\sqrt{5}} \quad \frac{1}{\sqrt[4]{5}\sqrt{\phi}} \quad -\frac{1}{\sqrt{5}} \ ] \\[4pt]
v_7 = [\ -\frac{\phi}{\sqrt{5}} \quad -\frac{1}{\sqrt[4]{5}\sqrt{\phi}} \quad \frac{1}{\sqrt{5}} \ ], & v_8 = [\ -\frac{\phi}{\sqrt{5}} \quad \frac{1}{\sqrt[4]{5}\sqrt{\phi}} \quad \frac{1}{\sqrt{5}} \ ] \\[4pt]
v_9 = [\ -\frac{1}{\sqrt{5}\phi} \quad -\frac{\sqrt{\phi}}{\sqrt[4]{5}} \quad -\frac{1}{\sqrt{5}} \ ], & v_{10} = [\ -\frac{1}{\sqrt{5}\phi} \quad \frac{\sqrt{\phi}}{\sqrt[4]{5}} \quad -\frac{1}{\sqrt{5}} \ ] \\[4pt]
v_{11} = [\ \frac{1}{\sqrt{5}\phi} \quad -\frac{\sqrt{\phi}}{\sqrt[4]{5}} \quad \frac{1}{\sqrt{5}} \ ], & v_{12} = [\ \frac{1}{\sqrt{5}\phi} \quad \frac{\sqrt{\phi}}{\sqrt[4]{5}} \quad \frac{1}{\sqrt{5}} \ ]
\end{array}
\right\} , \tag{28.42}
$$

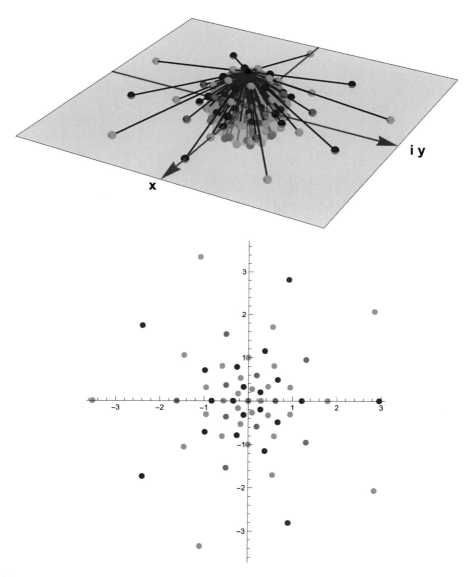

FIGURE 28.4

The E_8 icosahedron invariant vertex, edge-center, and face-center axes based on Eq. (28.42) converted via Klein's Riemann surface projection into the corresponding symmetry points in the complex plane. The zeroes of the complex 12th-degree polynomial $Z(z_1, z_2)$ are the z-plane projections of the 12 icosahedron vertices. The bottom figure shows the complex plane icosahedral vertices in red, the edge-center zeroes of $X(z_1, z_2)$ in green, and the face-center zeroes of $Y(z_1, z_2)$ in blue. The polynomials are not independent, and are related by the Kleinian E_8 equation $X^2 - Y^3 - 1728\, Z^5 = 0$.

where $\phi = (1 + \sqrt{5})/2$ is the golden ratio as usual. Projected from the Riemann sphere and adjusted for the addition of the poles, the complex plane vertex invariants take the form of the zeroes of

$$Z(z_1, z_2) = (z_1 z_2)\left(z_1^{10} + 11 z_1^5 z_2^5 - z_2^{10}\right). \tag{28.43}$$

Again applying Eq. (28.12) and Eq. (28.13), we find our final pair of invariant equations for $Y(z_1, z_2)$ and $X(z_1, z_2)$, whose zeroes correspond to the complex plane positions of the icosahedral face centers and edge centers. The projections of the $\mathbf{S}^2$ invariant rotation axes of the icosahedron onto the complex plane are shown in Fig. 28.4. Our full set of E_8 icosahedral invariants becomes

$$X(z_1, z_2) = z_1^{30} + 522 z_1^{25} z_2^5 - 10005 z_1^{20} z_2^{10} - 10005 z_1^{10} z_2^{20} - 522 z_1^5 z_2^{25} + z_2^{30}, \tag{28.44}$$

$$Y(z_1, z_2) = z_1^{20} - 228 z_1^{15} z_2^5 + 494 z_1^{10} z_2^{10} + 228 z_1^5 z_2^{15} + z_2^{20}, \tag{28.45}$$

$$Z(z_1, z_2) = z_1 z_2 \left(z_1^{10} + 11 z_1^5 z_2^5 - z_2^{10}\right). \tag{28.46}$$

As noted, rotations to new coordinate orientations may change the positions of the projected points, thus modifying the polynomials, as we will see next, but the fundamental structure will be preserved. From the degree of the invariants, we can see that the degrees of X, Y, and Z are 30, 20, and 12, the numbers of edges, faces, and vertices of the icosahedron. Thus the degrees of these invariant polynomials summed into a compatible same-degree constraint will be 2, 3, and 5, with a total degree of 60. Working out the constants, we find the vanishing E_8 Kleinian polynomial constraint

$$\tilde{f}_{(E_8)} = X^2 - Y^3 - 1728 Z^5 = 0. \tag{28.47}$$

Some sources suggest that the number "1728" can be considered as a fundamental constant of the icosahedral geometry, but in fact we shall see that the constants in Eq. (28.47) depend strongly on the 3D orientation of the source icosahedron. We will typically renormalize the constants to create a simpler equation, so we will usually use the simplified E_8 equation

$$f_{(E_8)} = X^2 - Y^3 - Z^5 = 0. \tag{28.48}$$

Alternative embedding for the E_8 icosahedral invariants. Since we have an alternative set of icosahedron coordinates Eq. (25.15) that we have used extensively in these chapters, we can work out an alternative set of equations. Projecting the vertices, edge centers, and face centers of these coordinates,

$$
\begin{aligned}
v_1 &= \frac{1}{\sqrt{\phi^2 + 1}} [\ 1 \quad \phi \quad 0\ ], & v_2 &= \frac{1}{\sqrt{\phi^2 + 1}} [\ -1 \quad \phi \quad 0\ ] \\
v_3 &= \frac{1}{\sqrt{\phi^2 + 1}} [\ 0 \quad 1 \quad \phi\ ], & v_4 &= \frac{1}{\sqrt{\phi^2 + 1}} [\ 0 \quad -1 \quad \phi\ ] \\
v_5 &= \frac{1}{\sqrt{\phi^2 + 1}} [\ \phi \quad 0 \quad 1\ ], & v_6 &= \frac{1}{\sqrt{\phi^2 + 1}} [\ \phi \quad 0 \quad -1\ ] \\
v_7 &= \frac{1}{\sqrt{\phi^2 + 1}} [\ -1 \quad -\phi \quad 0\ ], & v_8 &= \frac{1}{\sqrt{\phi^2 + 1}} [\ 1 \quad -\phi \quad 0\ ] \\
v_9 &= \frac{1}{\sqrt{\phi^2 + 1}} [\ 0 \quad -1 \quad -\phi\ ], & v_{10} &= \frac{1}{\sqrt{\phi^2 + 1}} [\ 0 \quad 1 \quad -\phi\ ] \\
v_{11} &= \frac{1}{\sqrt{\phi^2 + 1}} [\ -\phi \quad 0 \quad -1\ ], & v_{12} &= \frac{1}{\sqrt{\phi^2 + 1}} [\ -\phi \quad 0 \quad 1\ ]
\end{aligned}
\right\}, \tag{28.49}
$$

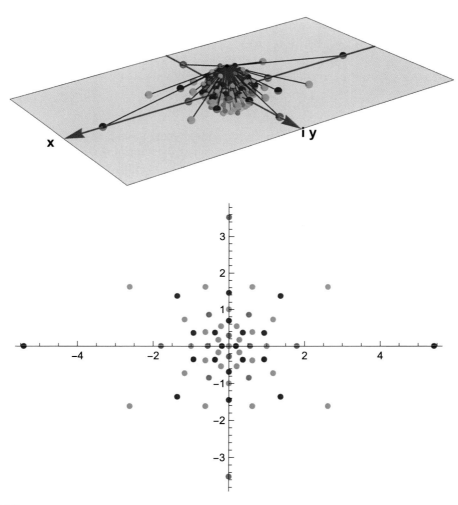

FIGURE 28.5

The E_8 icosahedron invariant vertex, edge-center, and face-center axes based on Eq. (28.49) converted via Klein's Riemann surface projection into the corresponding symmetry points in the complex plane. The zeroes of the complex 12th-degree polynomial $Z(z_1, z_2)$ are the z-plane projections of the 12 icosahedron vertices. The bottom figure shows the complex plane icosahedral vertices in red, the edge-center zeroes of $X(z_1, z_2)$ in green, and the face-center zeroes of $Y(z_1, z_2)$ in blue. The polynomials are related by the alternative Kleinian E_8 equation $4X^2 + 25Y^3 - 135Z^5 = 0$.

with four vertices each in one of three Cartesian plane golden-ratio rectangles, and now we have a pair of edge centers at the poles instead of vertices, as shown earlier in Fig. 26.5.

The vertices themselves, projected to the complex plane, satisfy the equation

$$Z(z_1, z_2) = \sqrt{5}\,z_1{}^{12} + 22\,z_1{}^{10}z_2{}^2 - 33\sqrt{5}\,z_1{}^8z_2{}^4 - 44\,z_1{}^6z_2{}^6 - 33\sqrt{5}\,z_1{}^4z_2{}^8 + 22\,z_1{}^2z_2{}^{10} + \sqrt{5}\,z_2{}^{12}\,. \tag{28.50}$$

This comes from a symbolic expression for which the ratio of the second coefficient to the first one is

$$\frac{4\phi^{10} + 10\phi^8 - 10\phi^4 - 4\phi^2}{\phi^8 + 2\phi^6 + \phi^4} = \frac{22}{\sqrt{5}}\,,$$

and thus there is not a great simplification in carrying through the formula in terms of the golden ratio ϕ. We note how different this is from Eq. (28.43). Computing the remaining equations as usual, we find the set of three invariant equations for the golden-ratio embedding of the icosahedron to be

$$X(z_1, z_2) = z_1 z_2 \left(z_1{}^4 - z_2{}^4\right)\left(225\sqrt{5}\,z_1{}^{24} - 2900\,z_1{}^{22}z_2{}^2 + 16146\sqrt{5}\,z_1{}^{20}z_2{}^4 + 101500\,z_1{}^{18}z_2{}^6\right.$$
$$+ 106191\sqrt{5}\,z_1{}^{16}z_2{}^8 - 98600\,z_1{}^{14}z_2{}^{10} + 676476\sqrt{5}\,z_1{}^{12}z_2{}^{12} - 98600\,z_1{}^{10}z_2{}^{14}$$
$$\left.+ 106191\sqrt{5}\,z_1{}^8z_2{}^{16} + 101500\,z_1{}^6z_2{}^{18} + 16146\sqrt{5}\,z_1{}^4z_2{}^{20} - 2900\,z_1{}^2z_2{}^{22} + 225\sqrt{5}\,z_2{}^{24}\right)\,, \tag{28.51}$$

$$Y(z_1, z_2) = \left(z_1{}^8 + 14\,z_1{}^4z_2{}^4 + z_2{}^8\right)\left(3\sqrt{5}\,z_1{}^{12} - 190\,z_1{}^{10}z_2{}^2 - 99\sqrt{5}\,z_1{}^8z_2{}^4 + 380\,z_1{}^6z_2{}^6\right.$$
$$\left.- 99\sqrt{5}\,z_1{}^4z_2{}^8 - 190\,z_1{}^2z_2{}^{10} + 3\sqrt{5}\,z_2{}^{12}\right)\,, \tag{28.52}$$

$$Z(z_1, z_2) = \sqrt{5}\,z_1{}^{12} + 22\,z_1{}^{10}z_2{}^2 - 33\sqrt{5}\,z_1{}^8z_2{}^4 - 44\,z_1{}^6z_2{}^6 - 33\sqrt{5}\,z_1{}^4z_2{}^8 + 22\,z_1{}^2z_2{}^{10} + \sqrt{5}\,z_2{}^{12}\,. \tag{28.53}$$

Note the appearance of the Cartesian axes from both the E_6 and E_7 equations in the edge formula $X(z_1, z_2)$, and the same form as the E_7 octahedral faces in the face formula $Y(z_1, z_2)$. We show the complex plane image of the solutions of these equations in Fig. 28.5.

Finally, we gather together the expressions for $X(z_1, z_2)$, $Y(z_1, z_2)$, and $Z(z_1, z_2)$ to discover the resulting form of the Kleinian constraint equation to be

$$\tilde{f}^{\text{alt}}_{(E_8)} = 4\,X^2 + 25\,Y^3 - 135\,Z^5 = 0\,. \tag{28.54}$$

While this is superficially much different from our first E_8 Kleinian polynomial Eq. (28.47), we will of course simply rescale the variables and use the simpler universal E_8 polynomial of Eq. (28.48).

28.4 Resolving singularities of the ADE polynomial real subspaces

We now have a comprehensive picture of the characteristics of the ADE symmetry groups from a broad spectrum of viewpoints:

- **Geometry.** We know the geometry of the tessellations of $\mathbf{S}^2$ into identical polygons that support the action of symmetry transformations, selected rotations whose actions leave the appearance of the shape unchanged.
- **Rotational symmetries in SO(3).** We have seen how specific rotations (in the group $\mathbf{SO}(3)$) acting on the tessellated sphere's vertex directions, edge midpoints, and face midpoints leave the shape unchanged.

- **Binary rotational symmetries from quaternion SU(2) action.** We know how the symmetric actions of 3D rotations correspond to quaternion multiplications and to the matrix action of 2×2 discretely acting matrices in the group **SU**(2).
- **Complex plane representation of the vertex-, edge-, and face-based symmetries.** We have followed the path of Klein's original method to turn the invariant properties of the **SO**(3) and quaternion rotations around fixed axes into polynomials whose zeroes in the complex plane provide an alternate realization of the invariances.
- **Invariant polynomials.** We have built homogeneous complex polynomials that are explicitly invariant under the ADE discrete rotations mapped into the complex plane. For the each ADE group, we found exactly three invariants, corresponding roughly with symmetries based on the three geometric features, the vertices, edge centers, and face centers, of the associated symmetric tessellations of S^2.
- **Kleinian constraint equations among the invariant polynomials.** For each ADE group, we found *constraint equations* relating the three invariants. We now move on to viewing each group's constraint as a vanishing polynomial in three complex dimensions; these in fact describe singular 4-manifolds embedded in $\mathbb{C}^3$, and it is those 4-manifolds that are of interest for gravitational instantons. In particular, the asymptotic forms correspond to the geometry of the instanton Einstein spaces as their boundaries approach infinity, and the *resolution of the singularities of the Kleinian complex constraint equations near the origin* must be understood to guarantee the regularity of each Einstein manifold as a whole. We will consider the complex equations to be beyond our scope, but, for visualization purposes, we will now simplify these complex equations by considering their real parts, which can be represented as surfaces embedded in $\mathbb{R}^3$ that are still quite interesting.

In this final section, we explore the beginnings of the deeper mathematics related to our discovery in the preceding section of the family of three-complex-dimensional constraint equations, one for each ADE group. We summarize these constraint equations in Table 28.1. What we will do for our last set of visualizations related to the ADE symmetries is to study each of these equations, which technically denote two constraints (one real, one imaginary) imposed upon an equation of three complex or six real variables, and thus are *4D manifolds*. We will not consider these full 4D structures here, leaving them perhaps for another time. But what we can do quite effectively is to look at the *real subspace* of each equation, which reduces to one real constraint on three real variables, resulting in an easily drawn *surface* embedded in three dimensions.[1]

However, even in this simplified context, there remain enormous complications: in fact, the real versions of the equations in Table 28.1 are highly *singular*, and include a complicated internal structure that is hidden in the initial form; our task now will be to *resolve these singularities* to the best of our ability to expose this structure, that in turn being the first step towards understanding more completely what might be hidden in the more challenging 4D complex domains. These singular polynomials and the technology of removing their singularities have long been the subject of mathematical investigations. Extensive treatments of these equations can be found in Du Val (Du Val, 1934a,b,c; Wikipedia, 2023i) and Katz and Morrison (1992), among many others. Extensions to the so-called "quiver diagrams" play a significant role in theoretical physics, as described, for example, in Lindström et al. (2000). From the mathematical treatments of, e.g., Hitchin (1979) and Kronheimer (1989a,b), we know that these polynomials and the surfaces they represent *after* the singularities are resolved correspond

---

[1] We are indebted to Tamas Hausel for private communications regarding the utility of examining these real surfaces.

Table 28.1 Kleinian ADE Invariants.	
Groups	**Singular Invariant Polynomials**
A_k	$XY - Z^{k+1}$
D_k	$X^2 - ZY^2 + Z^{k-1}$
E_6	$X^2 + Y^3 - Z^3$
E_7	$X^2 - Y^3 + Z^4$
E_8	$X^2 - Y^3 - Z^5$

to smooth Einstein manifolds with boundary that admit nontrivial Ricci-flat Einstein metrics. Only one class of these ALE (asymptotically locally Euclidean) metrics, the A_k series, has actually been found explicitly (Eguchi and Hanson, 1978; Gibbons and Hawking, 1978; Eguchi and Hanson, 1979; Eguchi et al., 1980). Our following, and final, chapter will explore the Eguchi–Hanson metric and its relationship to the A_1 symmetry group along with quaternion-based methods for embedding the corresponding Einstein manifold to explore and visualize its properties.

Images of the Klein ADE invariant polynomials. We now look at each of the Kleinian equations in Table 28.1 in turn to understand their implicit shapes. First, we look at the common character of the singularities that are intrinsic to all of the polynomials; these singularities reflect the effect of each of the three invariant terms in the constraint polynomial collapsing to zero in concert with the other terms. Then we will look for the *cause* of the singularity, and attempt techniques such as splitting up high powers, which are a frequent source of singular behavior, into products of separated roots. In this way, we can start to get an intuitive understanding of the underlying smooth manifolds that we believe correspond to smooth Ricci-flat 4D Einstein spaces. We now look at the simplified real Kleinian equations for each ADE group in turn, restricting our view to the real 2D manifold embedded in $\mathbb{R}^3$. For each case, we draw first the singular manifold corresponding exactly to Table 28.1, and then we propose and execute techniques for blowing up those singularities to exhibit the underlying smooth continuous surface.

The A_k Kleinian spaces. We begin with the A_k real singular surfaces described by

$$XY = Z^{k+1} . \tag{28.55}$$

We plot these for $k = 0, 1, 2, 3, 4, 5$ in Fig. 28.6. These 2D real subspaces of the full 4D manifolds exhibit some interesting properties. For one thing, $k = 0$ is completely smooth, yet clearly related to the other manifolds. All the surfaces with odd powers of k show a point singularity where two branches collide, and the shape of the colliding surfaces gets progressively squared off with higher powers. For k even, we have one continuous surface with pinch points that also square off for higher powers. Presumably the outer boundaries also have interesting properties as they go to infinity, and in fact we will see next, in Chapter 29, that for $k = 1$, the full complex equations at their boundary have the topology of the real 3D projective space, $\mathbf{RP}^3$, the asymptotic form of the cotangent space of the 2-sphere, denoted $\mathbf{T}^*\mathbf{S}^2$. For $k > 1$, more exotic transformations collapse $\mathbf{S}^3$ into a family of *lens spaces* (see, e.g, Adams, 1994; Rolfson, 1976).

There is a relatively simple way to resolve the singularities of Eq. (28.55) that we see in Fig. 28.6, which gives us significant insights into the general nature of singularity resolution. Our approach is to

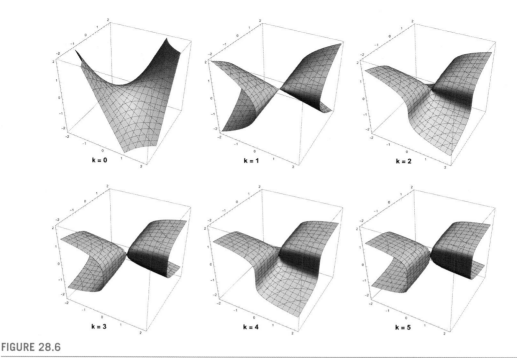

FIGURE 28.6

The equations of the bare A_k Klein invariant polynomials $xy = z^{k+1}$ are singular in interesting ways, shown here for $k = 0, 1, 2, 3, 4, 5$. Observe that $k = 0$ is essentially just a flat hyperbolic space, and that even and odd k behave very differently in terms of how the singularities appear.

transform the isolated power Z^{k+1} into a product of $k + 1$ distinct roots of a characteristic polynomial (e.g., a polynomial that might result from seeking the eigenspectrum of a nondegenerate $(k + 1)$-dimensional matrix). A simple and effective approach is to look at the expansion of Z^{k+1} into a complex unit circle, $Z^{k+1} - 1$, which is actually a product of roots of the form

$$(Z^{k+1} - 1) = \prod_{n=0}^{k} (Z - \omega^n)$$

where we impose the condition $\omega^{k+1} = 1$ (28.56)

so that $\omega = \exp \dfrac{2\pi i}{k + 1}$.

To adapt to our needs for real singularity resolution, we can effectively "unroll" the unit circle onto the real line with the transformation

$$Z^{k+1} \rightarrow \prod_{n=0}^{k} (Z - c_n) : c_n = \{-k, -k + 2, \ldots, k - 2, k\} .$$ (28.57)

In Fig. 28.7, we plot how this solution looks for $k = 1, 2, 3$, and for $k = 4$ marking the locations of the five real roots c_n on the Z-axis with red spheres.

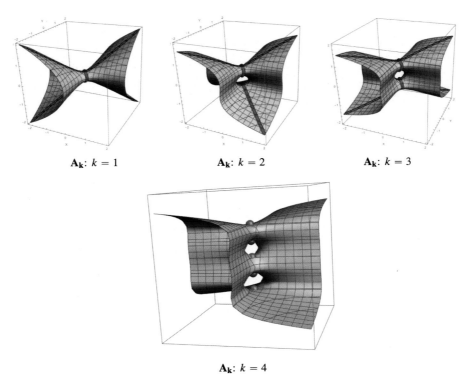

$\mathbf{A_k}$: $k = 1$ $\mathbf{A_k}$: $k = 2$ $\mathbf{A_k}$: $k = 3$

$\mathbf{A_k}$: $k = 4$

FIGURE 28.7

Resolving the A_k invariant real polynomial singularities. $k = 1$: Real implicit surface solving $xy = z^2 \rightarrow z^2 - 1$. $k = 2$: Real implicit surface solving $xy = z^3 \rightarrow z(z^2 - 2)$. $k = 3$: Real implicit surface solving $xy = z^4 \rightarrow (z^2 - 1)(z^2 - 3)$. The paths of the real circles that should be $\mathbf{CP}^1$ complex Riemann spheres in the full complex resolution are marked with ribbons. For $k = 4$, we show the five points on the real z-axis used for the resolution displayed as red spheres.

For $k = 1$, the equation of the resolved singularity is simply

$$x y = (z - 1)(z + 1) = z^2 - 1 \, ,$$

and its image in Fig. 28.7 appears to show one smooth hole introduced surrounding the origin together with a pair of perpendicular asymptotic curves, marked with an embedded red ribbon. The three roots of the $k = 2$ case's cubic polynomial produce a pair of orthogonal holes, for $k = 3$, we see a triple of orthogonal holes, and the pattern repeats. Adjusting the parameters easily confirms that the k local circles collapse to the singular points shown in Fig. 28.6 as all the $c_n \rightarrow 0$.

The D_k Kleinian spaces. Next we examine the series of D_k real singular surfaces described by

$$X^2 - Y^2 Z + Z^{k-1} = 0 \, . \tag{28.58}$$

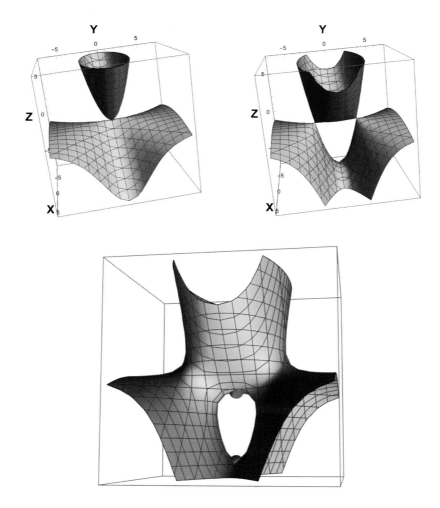

D_3: Singular, partially resolved, and fully resolved

FIGURE 28.8

The real D_3 Klein polynomial shows two different singular forms as one adjusts the parameters. Splitting out the z singularities on the real line at the two points marked by red spheres with $X^2 - Y^2Z + bY + c(Z - a_0) * (Z - a_1)$ resolves the singularities.

We plot the $k = 3$ example in Fig. 28.8, showing two different levels of singularities, one partially resolved, and then using the resolution expansion

$$X^2 - Y^2Z + bY + c(Z - a_0) * (Z - a_1) = 0 \tag{28.59}$$

for the bottom $k = 3$ resolved figure. The b term seems to be inconsequential, so we will generally omit it.

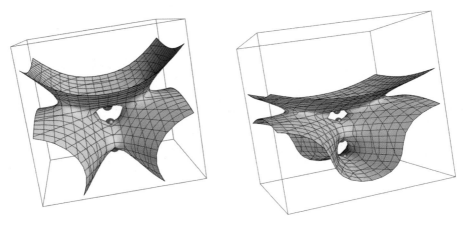

D$_4$ Resolved   **D$_5$ Resolved**

FIGURE 28.9

Resolving the D_k invariant real polynomial singularities starting from $X^2 - Y^2 Z + Z^{k-1} = 0$, then resolving for D_4 with $X^2 - Y^2 z + c(Z - a_0) * (Z - a_1) * (Z - a_2) = 0$ and for D_5 with $X^2 - Y^2 z + c(Z - a_0) * (Z - a_1) *$ $(Z - a_2) * (Z - a_3) = 0$. The locations of the resolution parameters a_i on the real line of the z-axis are shown as red spheres. Note that there are two additional circular "wings" branching out in a "Y" pattern from the top hole in the structure.

Further examples of the resolved D_k real constraint equation for $k = 4$ and $k = 5$ are given in Fig. 28.9, using the resolution forms, following suggestions by Katz and Morrison (1992),

$$X^2 - Y^2 Z + c(Z - a_0) * (Z - a_1) * (Z - a_2) = 0$$

for $D = 4$ and

$$X^2 - Y^2 Z + b_0 Y + b_1 XY + b_2 X + c(Z - a_0) * (Z - a_1) * (Z - a_2) * (Z - a_3) = 0$$

for D_5. Again, the b terms do not seem to have significant effect on singularity resolution. Finally, we present one more appealing example of both the singular and the resolved D_6 real manifold in Fig. 28.10. As k increases, the apparent dominance of a "Y"-shaped structure of rings is clear: this appears to be the topological signature of the Kleinian D_k polynomials.

The E$_6$ Kleinian space. The first Platonic solid group is the E_6 tetrahedral group whose real singular surfaces obey the constraint equation

$$X^2 + Y^3 - Z^3 = 0 \,. \tag{28.60}$$

We first plot the unresolved singular E_6 surface on the left in Fig. 28.11, observing the high degree of singularity at the origin. We attempt a resolved figure, at the right of Fig. 28.11, by expanding both the Y cubic powers and the Z cubic powers symmetrically into cubic products of distinct roots, taking the form

$$X^2 + t(Y^3 - 1) - s(Z^3 - 1) = 0 \,. \tag{28.61}$$

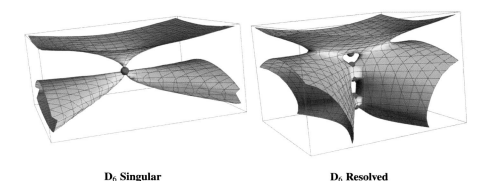

D₆ Singular　　　　　　　　　**D₆ Resolved**

FIGURE 28.10

The D_6 invariant real polynomial singularities starting from the singular form $X^2 - Y^2Z + Z^{k-1} = 0$, then resolving for D_6 with five separated vertices, plotted as red spheres, on the real line. The "Y"-shaped structure at the top of the line of red-sphere vertices is prominent. They are of course collapsed to a single point in the singular form.

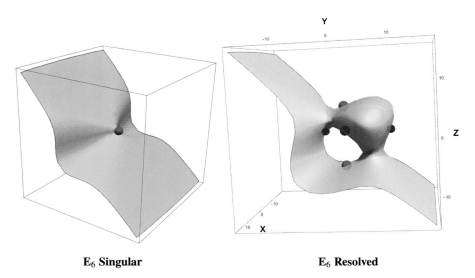

E₆ Singular　　　　　　　　　**E₆ Resolved**

FIGURE 28.11

The E_6 invariant real polynomial singularities starting from the singular form $X^2 + Y^3 - Z^3 = 0$, then resolving, at least partially, with the real form $X^2 + b(Y^3 - 1) - a(Z^3 - 1)$.

Unrolling the cubic complex roots of unity onto the real line as usual in the fashion of Eq. (28.57), we do see the three red dots at the Z roots and the three blue dots at the Y roots. Thus it is plausible that this is a true resolution of the E_6 Kleinian invariant polynomial. These processes are very delicate, however, and other authors have suggested more complex approaches than what we implemented here.

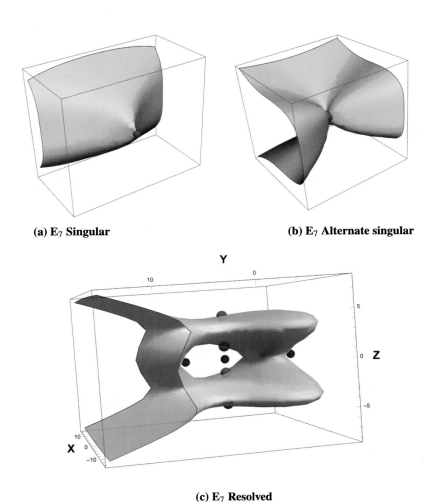

(a) E_7 **Singular**

(b) E_7 **Alternate singular**

(c) E_7 **Resolved**

FIGURE 28.12

The structure of the E_7 invariant real polynomial $X^2 - Y^3 + Z^4 = 0$. (a) Real singular surface from the given polynomial. (b) Since the full manifold is complex, we can change any signs by changing the chosen real subspace in any variable, so, for example this shows an equally valid singular surface of $X^2 + Y^3 - Z^4 = 0$. (c) A resolution of the usual polynomial form, possibly partial, with the real form obtained from flattening the roots of unity of $X^2 - t(Y^3 - 1) + s(Z^4 - 1)$. Red dots show the locations of the Z domain singularities and blue dots show influential locations of the three Y domain singularities. While it is difficult to determine if there is additional structure, we can clearly see the positions of the seven resolving variables as the border what should be spheres in complex space, so this is a reasonable initial candidate for the resolved E_7 Kleinian polynomial.

The E_7 Kleinian space. The next Platonic solid group is the E_7 octahedral group whose real singular surfaces obey the constraint equation

$$X^2 - Y^3 + Z^4 = 0 . \tag{28.62}$$

In Fig. 28.12, we first perform a small experiment on the bare unresolved E_7 surface: in (a) at the upper left, we plot the canonical singular E_7 surface $X^2 - Y^3 + Z^4 = 0$, while in (b) at the upper right, we swap imaginary parts by plotting the equally valid real singular surface $X^2 + Y^3 - Z^4 = 0$. Finally, we resolve both the edge- and vertex-related singularities using real-line polynomial flattening of the roots of unity based on the generic equation

$$X^2 - a(Y^3 - 1) + b(Z^4 - 1) + cX(Z^2 - 1) = 0 \tag{28.63}$$

The resolved E_7 result shown in Fig. 28.12(c) is based on the most informative heuristic parameter set we could find. This is a sensitive process, and may well require other adjustments that we have missed to expose the full structure, but we have achieved our objective of revealing the general structure of the singularity resolution process in the real cross-section.

The E_8 Kleinian space. Our final invariant corresponds to the E_8 icosahedral group whose real singular surfaces obey the constraint equation

$$X^2 - Y^3 - Z^5 = 0 . \tag{28.64}$$

We first plot a pair of unresolved singular E_8 surfaces (a) and (b) at the top of Fig. 28.13, related by making a complex phase transformation on the variables. We perceive familiar-looking yet unique conformations of the singular surface. We show our candidate resolved figure, using the formula

$$X^2 - a(Y^3 - 1) - b(Z^5 - 1) = 0 , \tag{28.65}$$

in (c) at the bottom of Fig. 28.13. As usual, we unroll the roots of unity onto points on the real line shown as red dots for the Z roots and blue for Y. The three $(Y^3 - 1)$ roots and five $(Z^5 - 1)$ roots give us eight candidate parameters, the minimal number that is technically expected to be present in the E_8 resolution.

Other terms have been suggested by theoretical arguments (see, e.g., Du Val, 1934a,b,c; Katz and Morrison, 1992), and in Fig. 28.14 we show an example of the more topologically complex resolution we obtained from making these possible additions to the equation:

$$X^2 - a(Y^3 - 1) - b(Z^5 - 1) + cY(Z^4 - 1) - d(Z^4 - 1) = 0 . \tag{28.66}$$

Since E_8 is very complicated, it is difficult to be sure that all possible topologies are included in either of these resolutions, but we basically expect the five constants on the real Z domain to define four finite holes and at least one going to infinity, with three more due to the real Y domain, plus the possible introduction of deformations. If our eventual objective is to understand the geometry of the smoothed central manifold replacing the singularity at the origin of the hypothesized E_8 gravitational instanton, the simpler shape in Fig. 28.13 with eight visible parameters could be a reasonable candidate for the resolved real Kleinian polynomial for E_8.

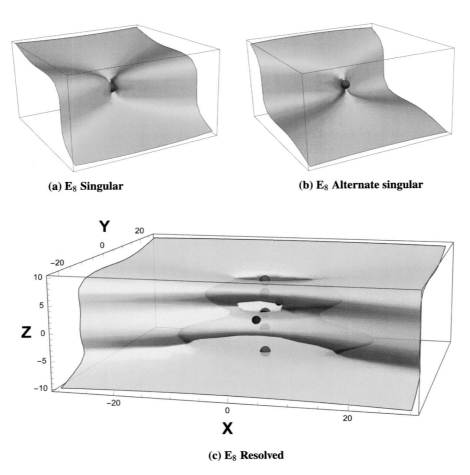

(a) E_8 **Singular**

(b) E_8 **Alternate singular**

(c) E_8 **Resolved**

FIGURE 28.13

The E_8 invariant real polynomial singularities starting from the singular form $X^2 - Y^3 - Z^5 = 0$, then resolving with the real form $X^2 - a(Y^3 - 1) - b(Z^5 - 1) = 0$. In (a) and (b), we show two singular forms related by applying a phase transformation to the real variables. In (c), we obtain a basic singularity-resolved surface by unrolling the $Y^3 - 1$ and $Z^5 - 1$ roots of unity onto their real lines as usual, so the resulting real surface passes exactly through all the roots. There are eight parameters, with the five red dots indicating the basic pulling-apart action that resolves the real Z^5 singularities, and the three blue dots marking the locations of influence of the Y^3 singularities.

28.5 Remarks

In this chapter, we have provided a basic introduction to the relationships between quaternions and the many phenomena that have emerged from Felix Klein's work on the discrete ADE symmetries of the Riemann sphere S^2. We have worked to identify circumstances where quaternions can clarify our

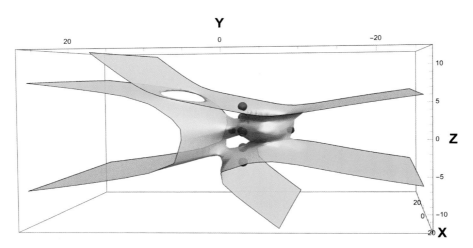

E_8 with alternative resolution terms

FIGURE 28.14

Results of resolving the E_8 invariant real polynomial singularities adding additional lower-order terms with inconsistent internal (z_1, z_2) asymptotic polynomial order. Extra complexity is produced by introducing terms such as the following: $X^2 - a(Y^3 - 1) - b(Z^5 - 1) + cY(Z^4 - 1) - d(Z^4 - 1) = 0$. Accompanying this additional topological structure, the exact alignment of the surface with the singularities on the Z-axis and on the Y-axis is deformed. The resulting branched structure contains the potential for additional symmetries, but whether those are essential, e.g., for the resolved singularity at the origin of the hypothesized E_8 gravitational instanton, is unclear.

understanding, and developed visual insights into the ADE groups themselves and their quaternion forms. While a more complete treatment of all the implications for mathematics and physics buried in the ADE groups and their related polynomials is far beyond the scope of this book, we have been able to experience a taste of the flavor of this subject. In the next and final chapter, we will explore the relationship of the very simplest $k = 1$ case of the A_k cyclic group to the solution of Einstein's Euclidean 4D equations known as "gravitational instantons." Remarkably, quaternion techniques that we have used throughout our studies in this treatise allow us to produce an actual visualization of the manifold whose embedding produces exactly the solution of Einstein's equations for the singularity-resolved space implicit in the A_1 Kleinian group.

Quaternion-based embedding and visualization of the $\mathbf{A}_1$ gravitational instanton ☆

29.1 Context

Our goal here is to explore a quaternion-inspired embedding of a geometric object that automatically encodes the metric of the simplest so-called "gravitational instanton," which is the unique Ricci-flat ALE (asymptotically locally Euclidean) Einstein metric on the topological 4-manifold $\mathbf{T}^*\mathbf{S}^2$. Our treatment is based on the methods introduced in Hanson and Sha (2017). The manifold we will discover corresponds to the $k = 1$ case of the A_k (cyclic group) gravitational instanton series, which admits a metric with a self-dual Riemann tensor and thus automatically satisfies Einstein's equations in 4D Euclidean space. The source of our treatment is the Eguchi–Hanson (1978) metric, which is also known to be the $n = 1$ case of a family of $2n$-complex-dimensional Ricci-flat spaces described later by Calabi (1979). Subsequently Gibbons and Hawking (1978) produced a series of ALE Einstein metrics, parameterized by $k + 1$ 3D points, that turned out to describe each of the possible A_k instantons, with the $k = 1$ case corresponding again precisely to the Eguchi–Hanson metric (see, e.g., Prasad (1979)). Here we detail the application of quaternion variables to the task of obtaining a perfect isometric embedding of the Eguchi–Hanson metric.

The remaining ALE metrics and the discrete quaternion groups. Upon the discovery of the A_k ALE Einstein solutions (Eguchi and Hanson, 1978; Gibbons and Hawking, 1978), it was soon realized that there was a significant pattern relating *all* the discrete groups of $\mathbf{SU}(2)$ acting on $\mathbf{S}^2$ to ALE solutions of Euclidean gravity. The connection between these group actions, studied by Felix Klein in 1884 (Klein, 1884), and the class of ALE self-dual Einstein spaces was recognized by Hitchin, in his landmark "Polygons and Gravitons" paper (Hitchin, 1979), appearing immediately after the work of Eguchi–Hanson and Gibbons–Hawking. Hitchin conjectured but could not prove that there existed Einstein manifolds corresponding to the entire "ADE" set of discrete $\mathbf{SU}(2)$ subgroups, two infinite discrete series, A_k (cyclic) and D_k (dihedral), along with the three isolated discrete elements, E_6 (tetrahedral), E_7 (octahedral), and E_8 (icosahedral); it seemed almost certain that these exhausted the list of possible ALE gravitational instantons. After some years of continuing interest based on the relationship discovered between the ADE metrics and supersymmetry, Lindström and Roček (1983) introduced the basic version of a useful idea known as the hyperkähler quotient method. As further insights developed that revealed associations among supersymmetry, hyperkähler quotients, Dynkin diagrams, quiver

---

☆ This chapter closely follows some of the material in Hanson and Sha (2017).

Visualizing More Quaternions. https://doi.org/10.1016/B978-0-32-399202-2.00044-7

diagrams, twistor methods, and the ADE spaces, Hitchin, Karlhede, Lindström, and Roček (1987) ultimately were able to draw together the various pieces into a coherent mathematical framework incorporating hyperkähler quotient methods as well as twistor methods. Finally, Kronheimer (1989a) was able to wrap things up and show, via twistor methods and the hyperkähler quotient mechanism, that the ADE spaces indeed exhausted the possible ALE self-dual Einstein spaces. Among subsequent investigations, Lindström, Roček, and von Unge (2000) showed how, starting directly with quiver diagrams and the implied moment-map constraints, one could more deeply understand the role of the Kleinian ADE algebraic varieties, along with the singularity-resolved forms associated with Du Val (1934a,b,c) and further elaborated, e.g., by Katz and Morrison (1992).

Within this intricate context, we will limit ourselves to exploring how quaternions can be exploited to obtain an isometric embedding of the A_1 metric. We will see how a parameterization with a close correspondence to the quaternion adjugate variables, introduced in Chapter 12, allows us to find an embedding of the A_1 gravitational instanton manifold in $\mathbb{R}^{11}$ whose induced metric is exactly the Ricci-flat ALE Eguchi–Hanson metric.

29.2 Features of the Eguchi–Hanson $\mathbf{A_1}$ metric

We begin with a review of some of the structures we will refer to in the process of building an isometric embedding for the Ricci-flat A_1 metric. We will be looking for collections of coefficients of the Maurer–Cartan forms, which encode the structure of $\mathbf{S}^3$ as 1-forms. We will write these in terms of the generic quaternion element g parameterized by the point $\{w, x, y, z\}/r$ in $\mathbf{SU}(2)$ (or $\mathbf{S}^3$) as

$$g = \frac{1}{r}(wI_2 - i\,\Sigma \cdot [x, y, z])$$

$$= \frac{1}{r}\begin{bmatrix} w - iz & -ix - y \\ y - ix & w + iz \end{bmatrix},$$

where Σ denotes the conventional Pauli matrices (we reserve the symbol σ for other purposes below), and $r^2 = w^2 + x^2 + y^2 + z^2$. The form of g is chosen to reproduce precisely a standard right-handed quaternion algebra when two distinct matrices are multiplied together. We then define the Maurer–Cartan forms in our Euclidean coordinates as

$$g^{-1}dg = \begin{bmatrix} \sigma_x \\ \sigma_y \\ \sigma_z \end{bmatrix} = \frac{1}{r^2}\begin{bmatrix} -x & w & z & -y \\ -y & -z & w & x \\ -z & y & -x & w \end{bmatrix} \cdot \begin{bmatrix} dw \\ dx \\ dy \\ dz \end{bmatrix}.$$

We note for reference the commonly used polar form (see Eq. (29.10)) for the Maurer–Cartan forms:

$$\begin{bmatrix} \sigma_x \\ \sigma_y \\ \sigma_z \end{bmatrix} = \begin{bmatrix} \frac{1}{2}(\cos\psi\,d\theta + \sin\psi\,\sin\theta\,d\phi) \\ \frac{1}{2}(-\sin\psi\,d\theta + \cos\psi\,\sin\theta\,d\phi) \\ \frac{1}{2}(d\psi + \cos\theta\,d\phi) \end{bmatrix}.$$

These forms obey the fundamental structure equations

$$d\sigma_x + 2\sigma_y \wedge \sigma_z = 0 , \tag{29.1}$$
$$d\sigma_y + 2\sigma_z \wedge \sigma_x = 0 , \tag{29.2}$$
$$d\sigma_z + 2\sigma_x \wedge \sigma_y = 0 . \tag{29.3}$$

(This follows from taking the exterior derivative of $I = g^{-1}g$ and observing that $d(g^{-1}dg) + g^{-1}dg \wedge g^{-1}dg = 0$; the factor of 2, an algebraic identity in the 1-forms, comes from the sum over Pauli matrices in the matrix form of g.) The attentive reader will note that there are some alternate choices of signs and factors of 2 in the literature (Eguchi et al., 1980; Gibbons and Pope, 1979); this is our chosen convention for this presentation.

Maurer–Cartan basis for the metric. If we now take the Maurer–Cartan basis of 1-forms for $\mathbf{SU}(2)$ using the above form for the σ's, then the Eguchi–Hanson solution (Eguchi and Hanson, 1978) for the self-dual metric can be written as $d\tau^2 = (e_0)^2 + (e_1)^2 + (e_2)^2 + (e_3)^2$ with the following vierbeins,

$$e = \left\{ \frac{1}{\sqrt{1 - \left(\frac{s}{r}\right)^4}} dr, \ r\sigma_x, \ r\sigma_y, \ \sqrt{1 - \left(\frac{s}{r}\right)^4} \, r\sigma_z \right\} , \tag{29.4}$$

where s is a constant.

The connection 1-forms $\omega^a{}_b$ are the solutions to the Levi-Civita torsion-free conditions

$$de^a + \omega^a{}_b \wedge e^b = 0$$

and the curvature 2-forms $R^a{}_b$ derived from the connections are (see, e.g., Eguchi et al., 1980)

$$R^a{}_b = d\omega^a{}_b + \omega^a{}_c \wedge \omega^c{}_b .$$

In this gauge, the solutions for the connections themselves take the self-dual form

$$\omega_x = \omega_{23} = \omega_{01} = -\sqrt{1 - \left(\frac{s}{r}\right)^4}\, \sigma_x ,$$
$$\omega_y = \omega_{31} = \omega_{02} = -\sqrt{1 - \left(\frac{s}{r}\right)^4}\, \sigma_y , \tag{29.5}$$
$$\omega_z = \omega_{12} = \omega_{03} = -\left(1 + \left(\frac{s}{r}\right)^4\right) \sigma_z ,$$

while the Riemann curvature 2-forms $R^a{}_b$ become

$$R_x = R_{23} = R_{01} = -\frac{2s^4}{r^6}\left(e^2 \wedge e^3 + e^0 \wedge e^1\right) ,$$
$$R_y = R_{31} = R_{02} = -\frac{2s^4}{r^6}\left(e^3 \wedge e^1 + e^0 \wedge e^2\right) , \tag{29.6}$$
$$R_z = R_{12} = R_{03} = +\frac{4s^4}{r^6}\left(e^1 \wedge e^2 + e^0 \wedge e^3\right) .$$

The explicit self-duality of the connection 1-forms $\omega^a{}_b$ in (a, b) produces an automatically self-dual Riemann curvature 2-form, and that in turn implies the vanishing of the Ricci tensor, so this is a Ricci-flat Einstein space in four Euclidean dimensions. Observe that changing the order of $\{w, x, y, z\}$ or the signs of $\{\sigma_x, \sigma_y, \sigma_z\}$ can change various signs in Eq. (29.5) and Eq. (29.6) and can interchange self-dual and antiself-dual labeling. Restricting the parameters to $\mathbf{RP}^3$ instead of a Euclidean $\mathbf{S}^3$ at infinity removes the cone singularity at the core $\mathbf{S}^2$ as $r \to s$, and one can verify by explicit integration that, with that choice of integration volume, the Euler integral of the volume is 3/2, while the Chern surface term is 1/2, giving the total Euler number $\chi = 3/2 + 1/2 = 2$ as required by the topology of $\mathbf{T}^*\mathbf{S}^2$; the signature, which has no surface correction, can similarly be shown to be $\tau = -1$ (see Eguchi and Hanson, 1979).

Polar form of the metric. To provide additional context for interpreting our embedding in the next section, we now rewrite the Eguchi–Hanson solution Eq. (29.4) as an abstraction (which in fact is a general Bianchi type IX metric) of the form

$$e = \left\{ \sqrt{f(r)}dr, \; r\sqrt{g(r)}\sigma_x, \; r\sqrt{g(r)}\sigma_y, \; r\sqrt{h(r)}\sigma_z \right\} , \tag{29.7}$$

where the solution of Eq. (29.4) of course can be written

$$f(r) = \left(1 - \left(\frac{s}{r}\right)^4\right)^{-1} , \qquad g(r) = 1 , \qquad h(r) = \left(1 - \left(\frac{s}{r}\right)^4\right) . \tag{29.8}$$

From $d\tau^2 = e \cdot e = dx^\mu g_{\mu\nu} dx^\nu$, we can extract a convenient Cartesian form of the metric that we will be able to match with the Cartesian form of our anticipated embedding, with coordinates $x^\mu = \{w, x, y, z\}$ and $r^2 = w^2 + x^2 + y^2 + z^2$:

$$g_{\mu\nu} = \frac{1}{r^2} \begin{bmatrix} w^2 f(r) + \left(x^2 + y^2\right) g(r) + z^2 h(r) & wx f(r) + (-wx + yz) g(r) - yz h(r) \\ wx f(r) + (-wx + yz) g(r) - yz h(r) & x^2 f(r) + \left(w^2 + z^2\right) g(r) + y^2 h(r) \\ wy f(r) - (wy + xz) g(r) + xz h(r) & xy(f(r) - h(r)) \\ wz(f(r) - h(r)) & xz f(r) - (wy + xz) g(r) + wy h(r) \end{bmatrix}$$

$$\begin{bmatrix} wy f(r) - (wy + xz) g(r) + xz h(r) & wz(f(r) - h(r)) \\ xy(f(r) - h(r)) & xz f(r) - (wy + xz) g(r) + wy h(r) \\ y^2 f(r) + \left(w^2 + z^2\right) g(r) + x^2 h(r) & yz f(r) + (wx - yz) g(r) - wx h(r) \\ yz f(r) + (wx - yz) g(r) - wx h(r) & z^2 f(r) + \left(x^2 + y^2\right) g(r) + w^2 h(r) \end{bmatrix} . \tag{29.9}$$

Using the polar coordinates

$$w = r \cos \frac{\theta}{2} \cos \frac{\phi + \psi}{2} ,$$

$$x = r \sin \frac{\theta}{2} \cos \frac{\phi - \psi}{2} ,$$

$$y = r \sin \frac{\theta}{2} \sin \frac{\phi - \psi}{2} , \tag{29.10}$$

$$z = r \cos \frac{\theta}{2} \sin \frac{\phi + \psi}{2} ,$$

with the order $\{r, \theta, \phi, \psi\}$, $0 \leqslant \theta < \pi$, $0 \leqslant \phi < 2\pi$, $0 \leqslant \psi < 2\pi$ (for $\mathbf{S}^3$, we would have $0 \leqslant \psi < 4\pi$), we can write the metric in polar form as

$$
\begin{bmatrix}
f(r) & 0 & 0 & 0 \\
0 & \frac{1}{4}r^2 g(r) & 0 & 0 \\
0 & 0 & \frac{1}{4}r^2 \left(h(r)\cos^2\theta + g(r)\sin^2\theta\right) & \frac{1}{4}r^2 h(r)\cos\theta \\
0 & 0 & \frac{1}{4}r^2 h(r)\cos\theta & \frac{1}{4}r^2 h(r)
\end{bmatrix} .
\tag{29.11}
$$

29.3 Embedding $\mathbf{T}^*\mathbf{S}^2$ (the $\mathbf{A}_1$ manifold) in $\mathbb{R}^{11}$

We now are ready to begin the process of deriving the metric of the A_1 member of the ADE family of Ricci-flat Euclidean Einstein spaces via an embedding in 11D Euclidean space. Our result corresponds exactly to the Eguchi–Hanson (1978) form for the A_1 metric, which has properties quite different from the $k = 1$ form of the Gibbons–Hawking (1978; 1979) $(k + 1)$-point parameterized solution for the whole family that has subsequently become known as the A_k gravitational instanton metric family. (An explicit correspondence for $k = 1$ is given by Prasad (1979).) Starting out with *only* the topological space $\mathbf{T}^*\mathbf{S}^2$, using no other knowledge, we now show how to derive an isometric embedding of $\mathbf{T}^*\mathbf{S}^2$ in the Euclidean space $\mathbb{R}^{11}$ that corresponds to the $k = 1$ Ricci-flat metric (Hanson and Sha, 2017).

29.3.1 Construction of the embedding map

We first need to find an embedding of $\mathbf{RP}^3$, or equivalently $\mathbf{SO}(3)$, which is the asymptotic boundary of $\mathbf{T}^*\mathbf{S}^2$, along with some way of expressing the fact that $\mathbf{T}^*\mathbf{S}^2$ collapses topologically to $\mathbf{S}^2$ at the origin. One example of an embedding of $\mathbf{SO}(3)$ is the standard 3D rotation matrix or the equivalent quaternion quadratic form. The classic form for the rotation matrix, given in Eq. (2.17), is the rotation $\mathbf{R}(\theta, \hat{\mathbf{n}})$ by an angle θ about a fixed unit-norm axis $\hat{\mathbf{n}} = [n_1, n_2, n_3]$. The general element of $\mathbf{SO}(3)$ in Eq. (2.17) is exactly equivalent to a quadratic quaternion form constructed from the unit quaternion

$$
q = (\cos(\theta/2),\, n_1\sin(\theta/2),\, n_2\sin(\theta/2),\, n_3\sin(\theta/2)) .
\tag{29.12}
$$

With a generic Cartesian quaternion parameterization $q = \{w, x, y, z\}$ obeying $q \cdot q = 1$ ($q \in \mathbf{S}^3$), this quadratic form is just

$$
\mathbf{R}(q) =
\begin{bmatrix}
w^2 + x^2 - y^2 - z^2 & 2xy - 2wz & 2xz + 2wy \\
2xy + 2wz & w^2 - x^2 + y^2 - z^2 & 2yz - 2wx \\
2xz - 2wy & 2yz + 2wx & w^2 - x^2 - y^2 + z^2
\end{bmatrix} .
\tag{29.13}
$$

Since $\mathbf{R}(q)$ is an orthogonal matrix, each row and each column has unit length when $q \cdot q = 1$ is imposed, and all pairs of rows as well as pairs of columns are orthogonal. The important fact is that, since topologically $\mathbf{R}(q)$ is just $\mathbf{RP}^3$, the nine quadratic expressions that make up $\mathbf{R}(q)$ are precisely a parameterized embedding mapping $\mathbf{S}^3$ to $\mathbf{RP}^3$ in $\mathbb{R}^9$, or, more generally, in $\mathbb{R}^{10}$ if we relax $q \cdot q = 1 \rightarrow r^2$ to allow scaling. (Note that $\mathbf{S}^3$ double covers $\mathbf{RP}^3$, so we must restrict the parameter domain accordingly to half of $\mathbf{S}^3$.)

Next we untangle the combinations of quadratic forms in $\mathbf{R}(q)$ into a list that is precisely our 10 quaternion adjugate variables from Chapter 12, now reinterpreted as an embedding of $\mathbf{RP}^3$. This 10-vector of quadratic forms also is recognizable as the extension to the next dimension of the Veronese map traditionally used to parameterize $\mathbf{RP}^2$. We now exploit these quaternion adjugate forms to define two related maps from $\mathbf{S}^3$ to $\mathbf{RP}^3$ embedded in $\mathbb{R}^{10}$ as follows:

$$
\mathbf{p}(w, x, y, z) =
\begin{bmatrix}
w^2 \\
x^2 \\
y^2 \\
z^2 \\
wx \\
wy \\
wz \\
yz \\
zx \\
xy
\end{bmatrix}
, \qquad
\hat{\mathbf{p}}(w, x, y, z) =
\begin{bmatrix}
w^2 \\
x^2 \\
y^2 \\
z^2 \\
\sqrt{2}wx \\
\sqrt{2}wy \\
\sqrt{2}wz \\
\sqrt{2}yz \\
\sqrt{2}zx \\
\sqrt{2}xy
\end{bmatrix}
. \tag{29.14}
$$

The scaled version $\hat{\mathbf{p}}$ is constructed to define a radius of constant length, $\hat{\mathbf{p}} \cdot \hat{\mathbf{p}} = \left(w^2 + x^2 + y^2 + z^2\right)^2 = r^4$, so it is explicitly a point on a round $\mathbf{S}^9$. Not only is each element equivalent to an adjugate quaternion variable as introduced for other purposes in Chapter 12, but in addition $\hat{\mathbf{p}}$ is also related to the generalized Hopf fibration noted in Eq. (22.23) in our chapter on quantum computing, a fact that we will exploit in a moment.

Now that we have $\mathbf{RP}^3$ embedded to define the asymptotic boundary of $\mathbf{T}^*\mathbf{S}^2$, as well as the ability to scale it into the interior, we must find a way to terminate that ingoing mapping smoothly on the "origin," which is topologically $\mathbf{S}^2$. That should be easy, considering the fact that we know another very nice map, the Hopf fibration, from $\mathbf{S}^3$ to either $\mathbb{R}^3$ or $\mathbb{R}^4$, whose image is an embedded $\mathbf{S}^2$. In fact we have six explicit such maps embodied in Eq. (29.13), since each column and each row has unit length and hence is a map from the three Euler angles of $\mathbf{S}^3$ to a two-parameter $\mathbf{S}^2$. We choose our map from the elements of the last column of Eq. (29.13),

$$
\text{(last column)} \quad \mathbf{m}(w, x, y, z) =
\begin{bmatrix}
\sqrt{2}(w^2 + z^2) \\
\sqrt{2}(x^2 + y^2) \\
2(wx - yz) \\
2(wy + xz)
\end{bmatrix}
, \tag{29.15}
$$

where $(1/2)\mathbf{m} \cdot \mathbf{m} = \left(w^2 + x^2 + y^2 + z^2\right)^2 = r^4$, defining an $\mathbf{S}^2$ of radius $R = q \cdot q = r^2$. For reasons that will become clear, we have used the 4D version of the Hopf fibration that corresponds to the last column of Eq. (29.13) after a rotation by 45 degrees to the desired axis component $w^2 + z^2 - (x^2 + y^2)$ of the 3D column's subspace (the remaining, orthogonal, direction is just $w^2 + x^2 + y^2 + z^2$). We check that this is indeed the fibration corresponding to fibering out the ψ angular variable in Eq. (29.10) by explicit substitution:

$$\mathbf{R}_4\left(\frac{\pi}{4}, \text{1-2 plane}\right) \cdot \mathbf{m}(r, \theta, \phi, \psi) = r^2 \begin{bmatrix} \cos\theta \\ 1 \\ \cos\phi\,\sin\theta \\ \sin\phi\,\sin\theta \end{bmatrix}. \tag{29.16}$$

These are the standard $\mathbf{S}^2$ spherical coordinates $\{\cos\theta, \cos\phi\,\sin\theta, \sin\phi\,\sin\theta\}$, which we need at the "origin" of $\mathbf{T}^*\mathbf{S}^2$ in the embedding.

We now proceed to generate a parameterized interpolation from Eq. (29.14) to Eq. (29.15) that will become the sought-for isometric embedding of $\mathbf{T}^*\mathbf{S}^2$. However, it turns out that 10 dimensions is actually just slightly too rigid to get an isometric embedding of the metric in this context, and we will have to add an 11th dimension that we take to be parameterized by the scaling radius r. We thus consider this map, interpolating from $\mathbf{RP}^3$ to $\mathbf{S}^2$ at the "origin" $r = s$, which we will use to pull back a metric on the 4D manifold parameterized by $\{w, x, y, z\}$:

$$\mathbf{p}(w, x, y, z) = \frac{1}{r^2} \begin{bmatrix} \frac{1}{\sqrt{2}}(a(r)w^2 + b(r)z^2) \\ \frac{1}{\sqrt{2}}(a(r)x^2 + b(r)y^2) \\ \frac{1}{\sqrt{2}}(a(r)y^2 + b(r)x^2) \\ \frac{1}{\sqrt{2}}(a(r)z^2 + b(r)w^2) \\ a(r)wx - b(r)yz \\ a(r)yz - b(r)wx \\ a(r)wy + b(r)xz \\ a(r)xz + b(r)wy \\ (a(r) - b(r))wz \\ (a(r) - b(r))xy \\ \frac{1}{\sqrt{2}}c(r) \end{bmatrix}. \tag{29.17}$$

If $a(r)$ and $b(r)$ are two monotonic positive smooth functions defined for $s \leqslant r < \infty$ with the properties that $a(s) = b(s)$, $a(r \to \infty) \to r$, and $b(r \to \infty) \to 0$, then we should have a smooth embedding of $\mathbf{T}^*\mathbf{S}^2$ incorporating interpolation functions in r that could permit deformation of the path in a way that produces a Ricci-flat induced metric.

As $r \to \infty$ (with $a(r) \to r$ and $b(r) \to 0$), the entire column corresponds to the $\mathbf{RP}^3$ boundary. In the $\mathbf{S}^2$ limit, $a(s) = b(s)$, we can check using our polar coordinates from Eq. (29.10) that we have a Hopf-fibered $\mathbf{S}^2$ embedded in the 11 vector components,

$$\left(\frac{1}{\sqrt{2}}\cos^2\frac{\theta}{2}, \frac{1}{\sqrt{2}}\sin^2\frac{\theta}{2}, \frac{1}{\sqrt{2}}\sin^2\frac{\theta}{2}, \frac{1}{\sqrt{2}}\cos^2\frac{\theta}{2}, \right.$$

$$\left. \frac{1}{2}\cos\phi\,\sin\theta, -\frac{1}{2}\cos\phi\,\sin\theta, \frac{1}{2}\sin\phi\,\sin\theta, \frac{1}{2}\sin\phi\,\sin\theta, 0, 0, \frac{1}{\sqrt{2}}c(s)\right).$$

29.3.2 Identifying and solving the interpolation functions

Employing the Cartesian parameterization $v^\mu = \{w, x, y, z\}$ and Eq. (29.17) for the vector $\mathbf{p}_i(w, x, y, z)$, $i = 1 \dots 11$, in $\mathbb{R}^{11}$, we can now compute the induced metric from the usual formula

$$g_{\mu\nu} = \sum_{i=1}^{11} \frac{\partial \mathbf{p}_i(w,x,y,z)}{\partial v^{\mu}} \frac{\partial \mathbf{p}_i(w,x,y,z)}{\partial v^{\nu}} . \tag{29.18}$$

Using $r^2 = w^2 + x^2 + y^2 + z^2$ to simplify the notation, we find for the first column,

$$\frac{1}{2r^4} \left[\begin{array}{c} w^2 r^2 \left(a'(r)^2 + b'(r)^2 + c'(r)^2 \right) + 2 \left(r^2 - w^2 \right) \left(a(r)^2 + b(r)^2 \right) - 4z^2 a(r) b(r) \\[2mm] -2wx \left(a(r)^2 + b(r)^2 \right) + 4yz\, a(r) b(r) + wx\, r^2 \left(a'(r)^2 + b'(r)^2 + c'(r)^2 \right) \\[2mm] -2wy(a(r)^2 + b(r)^2) - 4xz\, a(r) b(r) + wy\, r^2 \left(a'(r)^2 + b'(r)^2 + c'(r)^2 \right) \\[2mm] wz\, r^2 \left(a'(r)^2 + b'(r)^2 + c'(r)^2 \right) - 2wz(a(r) - b(r))^2 \end{array} \right] ,$$

and the other three columns follow this pattern. We can already see the groupings of the unknown interpolation terms into factors identifiable with f, g, h in Eq. (29.9). The algebraic forms become clearer if we go to polar coordinates and compare with Eq. (29.11):

$$\left[\begin{array}{cccc} \frac{1}{2} \left(a'^2 + b'^2 + c'^2 \right) & 0 & 0 & 0 \\[2mm] 0 & \frac{1}{4} \left(a^2 + b^2 \right) & 0 & 0 \\[2mm] 0 & 0 & \frac{1}{4} \left(a^2 + b^2 - 2ab \cos^2 \theta \right) & \frac{1}{4} (a-b)^2 \cos\theta \\[2mm] 0 & 0 & \frac{1}{4} (a-b)^2 \cos\theta & \frac{1}{4} (a-b)^2 \end{array} \right] . \tag{29.19}$$

Collecting corresponding terms, we discover exactly three groups of the embedding interpolation functions $a(r)$, $b(r)$, and $c(r)$ and their first derivatives that correspond to $f(r)$, $g(r)$, and $h(r)$ in Eqs. (29.9) and (29.11),

$$a'(r)^2 + b'(r)^2 + c'(r)^2 \rightarrow 2f(r) , \tag{29.20}$$

$$a(r)^2 + b(r)^2 \rightarrow r^2 g(r) , \tag{29.21}$$

$$(a(r) - b(r))^2 \rightarrow r^2 h(r) . \tag{29.22}$$

If we solve Eqs. (29.21) and (29.22) for $a(r)$ and $b(r)$ using Eq. (29.8), we find

$$a(r) = \frac{\sqrt{r^4 + \sqrt{r^8 - s^8}}}{\sqrt{2}r} , \tag{29.23}$$

$$b(r) = \frac{s^4}{\sqrt{2}r \sqrt{r^4 + \sqrt{r^8 - s^8}}} , \tag{29.24}$$

with $a(s) = s/\sqrt{2}$, $a(r \rightarrow \infty) \rightarrow r$ and $b(s) = s/\sqrt{2}$, $b(r \rightarrow \infty) \rightarrow 0$. From Eqs. (29.20), (29.23), and (29.24), we can define $c(r)$ by its differential equation:

$$c'(r) = \sqrt{\frac{3s^4 + r^4}{s^4 + r^4}} . \tag{29.25}$$

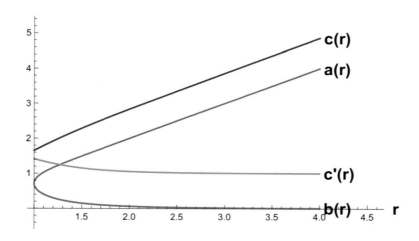

FIGURE 29.1

The interpolation functions for the Ricci-flat $\mathbb{R}^{11}$ isometric embedding of $\mathbf{T}^*\mathbf{S}^2$ for $\mathbf{S}^2$ radius $s = 1$. (red) $a(r)$; (green) $b(r)$; (blue) $c(r)$; (cyan) $c'(r)$.

We can then solve Eq. (29.25) for $c(r)$ with an interpolating function or use the explicit form

$$c(r) = \sqrt{3}r\, F_1\left(\frac{1}{4}; \frac{1}{2}, -\frac{1}{2}; \frac{5}{4}; -\frac{r^4}{s^4}, -\frac{r^4}{3s^4}\right) , \tag{29.26}$$

where F_1 is the first Appell function. $c(r)$ has the following properties:

$$\begin{aligned}
c(s) &= 1.65069s , \\
c(r \to \infty) &= r + \text{const}(s) , \\
c'(s) &= \sqrt{2} , \\
c'(\infty) &= 1 .
\end{aligned} \tag{29.27}$$

These same expressions can be obtained without reference to Eqs. (29.20), (29.21), and (29.22), with some effort, by computing the Ricci tensor directly and solving the resulting differential equations for $a(r)$, $b(r)$, and $c(r)$.

With these results, Eq. (29.17) now defines a smooth map from $\mathbf{T}^*\mathbf{S}^2$ into $\mathbb{R}^{11}$ whose induced metric is Ricci-flat.

29.3.3 Visual representations of the $\mathbb{R}^{11}$ embedding

Now that we have explicit forms for the interpolation functions that create an isometric embedding of the Ricci-flat geometry with the topology of $\mathbf{T}^*\mathbf{S}^2$, we can examine the shapes of these functions. In Fig. 29.1, we plot the forms of $a(r)$, $b(r)$, and $c(r)$ as well as $c'(r)$. We see that for the purposes of the embedding, the 11th dimension described by $c(r)$ is almost a straight line with unit slope, although it plays a *critical* role in the behavior of $f(r)$ near the origin to enforce the Ricci-flat condition in that

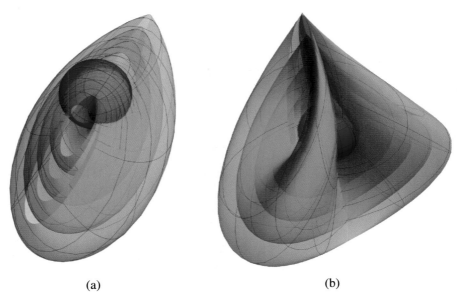

(a)   (b)

FIGURE 29.2

The $\mathbf{S}^2$ "core" at $r = s$ of the isometric embedding of $\mathbf{T}^*\mathbf{S}^2$ and its behavior for sampled values $s \leqslant r < 2s$. (a) View with several cutaway layers at sampled values with small r. (b) View of the entire layers deforming away from $\mathbf{S}^2$ at sampled r.

neighborhood. Past a radius of about twice the radius s of the $\mathbf{S}^2$ at the origin, the shape of the constant-r cross-section is already essentially in the asymptotic form of a canonical $\mathbf{RP}^3$ corresponding to the quadratic Veronese map of the underlying $\mathbf{S}^3$. This $\mathbf{RP}^3$ embodies the ALE property of our manifold.

Next, we present some 2D cross-sections of our isometrically embedded 4-manifold projected somewhat arbitrarily from $\mathbb{R}^{11}$ to 3D to give an impression of the shape. In Fig. 29.2(a), we show a cutaway of surfaces sampled in r moving out from the $\mathbf{S}^2$ "core" at $r = s$. The full shape that is swept out becomes quite complex even with r sampled near to s, as shown in Fig. 29.2(b). Fig. 29.3(a) picks a selection of latitude–longitude samples *on the surface* of $\mathbf{S}^2$, and shows the disks formed by sweeping out a segment in r in the collapsed Hopf variable ψ (see Eqs. (29.10) and (29.16)). As the maximum disk radius r moves outward from the $\mathbf{S}^2$ surface, we see in Fig. 29.3(b) that the boundary circles of the disks begin to delineate a sampling of the ALE boundary 3-manifold $\mathbf{RP}^3$ parameterized by ψ and the $\mathbf{S}^2$ variables θ and ϕ.

Finally, in Fig. 29.4(a), we present just the sampled $\mathbf{RP}^3$ described by the boundary circles for large radius r with origin at sampled latitude–longitude pairs on the $\mathbf{S}^2$. These are essentially $\mathbb{Z}_2$ identifications of the more familiar Hopf fiber rings embedded in $\mathbf{S}^3$, which we show in Fig. 29.4(b) for comparison. All this follows from the fact that our Veronese map of Eq. (29.14) is quadratic in the coordinates of $\mathbf{S}^3$, and hence double covered in $\mathbf{RP}^3$, so each boundary circle in Fig. 29.4(a) corresponds in parameter space to one half of the corresponding circle in Fig. 29.4(b).

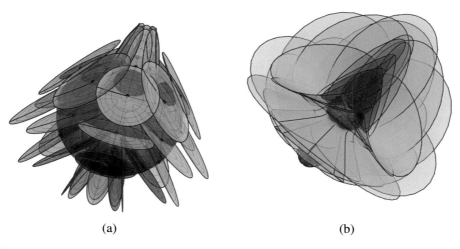

(a)   (b)

FIGURE 29.3

The A_1 4-manifold represented by disks in the radial and Hopf-fibration variables, sampled at values of the $\mathbf{S}^2$ latitude and longitude. (a) The nascent disks close to the $\mathbf{S}^2$ samples. (b) Expanding the disks away from the $\mathbf{S}^2$ "core" for larger values of sampled latitude/longitude, beginning to show the shape of $\mathbf{RP}^3$ in the circles bounding the disks.

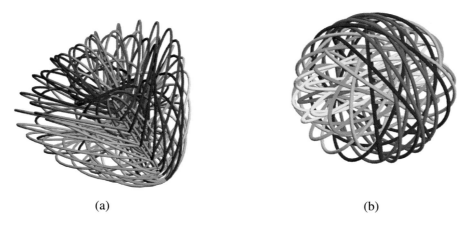

(a)   (b)

FIGURE 29.4

(a) Far away from the sampled latitude–longitude points on $\mathbf{S}^2$, we tessellate the asymptotic $\mathbf{RP}^3$ ALE boundary 3-manifold at a fixed large value of r, using rings in the fiber variable ψ that collapses and disappears on $\mathbf{S}^2$. (b) For comparison, these are the corresponding Hopf fiber rings tessellating $\mathbf{S}^3$ using similar $\mathbf{S}^2$ samples. These are essentially the double covers of the rings in (a).

29.4 Summary

The properties of metrics on the ADE manifolds have been extensively studied since Hitchin (1979) originally introduced the concept that all the possible 4D ALE self-dual solutions to Einstein's equations could be identified with the Kleinian groups, or, equivalently, to the corresponding ADE lens spaces $\mathbf{S}^3/G$ of $\mathbf{S}^3$. At this moment, only the A_k metrics have explicit solutions, and embedding-based treatments are very difficult for $k > 1$. Here we have shown how the $k = 1$ Eguchi–Hanson case can be explicitly embedded in $\mathbb{R}^{11}$ by employing a quaternion-coordinate-based embedding of the topological space $\mathbf{T}^*\mathbf{S}^2$ of the A_1 gravitational instanton. Not only do we see explicitly how this isometric embedding embodies an induced metric with vanishing Ricci curvature, thus satisfying Einstein's equations nontrivially in empty Euclidean 4-space, but we are also able to exploit the embedding to create explicit visualizations of the exact shape of the topological object that has these remarkable properties.

Conclusion

We now bring this episode of our mathematical odyssey through the land of quaternions to its close. Almost everything we have seen traces back to the wonderfully fundamental themes of Chapter 2, whose brevity and elegance we now see as deceptively simple. From the fundamental relations

$$p \star q = (p_0, p_1, p_2, p_3) \star (q_0, q_1, q_2, q_3)$$

$$= \begin{bmatrix} p_0 q_0 - p_1 q_1 - p_2 q_2 - p_3 q_3 \\ p_1 q_0 + p_0 q_1 + p_2 q_3 - p_3 q_2 \\ p_2 q_0 + p_0 q_2 + p_3 q_1 - p_1 q_3 \\ p_3 q_0 + p_0 q_3 + p_1 q_2 - p_2 q_1 \end{bmatrix} = \begin{bmatrix} p_0 & -p_1 & -p_2 & -p_3 \\ p_1 & p_0 & -p_3 & p_2 \\ p_2 & p_3 & p_0 & -p_1 \\ p_3 & -p_2 & p_1 & p_0 \end{bmatrix} \begin{bmatrix} q_0 \\ q_1 \\ q_2 \\ q_3 \end{bmatrix}$$

$$= (p_0 q_0 - \mathbf{p} \cdot \mathbf{q}, \ p_0 \mathbf{q} + q_0 \mathbf{p} + \mathbf{p} \times \mathbf{q}),$$

$$p \cdot q = [p_0, p_1, p_2, p_3] \cdot [q_0, q_1, q_2, q_3]$$

$$= p_0 q_0 + p_1 q_1 + p_2 q_2 + p_3 q_3$$

$$= p_0 q_0 + \mathbf{p} \cdot \mathbf{q},$$

$$q \cdot q = (q_0)^2 + (q_1)^2 + (q_2)^2 + (q_3)^2 \ = \ (q_0)^2 + \mathbf{q} \cdot \mathbf{q} = 1,$$

$$\bar{q} = (q_0, -q_1, -q_2, -q_3) \ = \ (q_0, -\mathbf{q}),$$

$$q \star \bar{q} = (q \cdot q, \mathbf{0}) \ = \ (1, \mathbf{0}),$$

$$R(q) = \begin{bmatrix} q_0^2 + q_1^2 - q_2^2 - q_3^2 & 2q_1 q_2 - 2q_0 q_3 & 2q_1 q_3 + 2q_0 q_2 \\ 2q_1 q_2 + 2q_0 q_3 & q_0^2 - q_1^2 + q_2^2 - q_3^2 & 2q_2 q_3 - 2q_0 q_1 \\ 2q_1 q_3 - 2q_0 q_2 & 2q_2 q_3 + 2q_0 q_1 & q_0^2 - q_1^2 - q_2^2 + q_3^2 \end{bmatrix},$$

we have found an enormous number of questions that can be asked and properties that can be computed by exploiting the power of the quaternion to represent spatial orientation, and even spacetime. We have found novel ways to discover and represent spatial ideas and to extract information from data, and we have extensively explored how to represent what we have found by exploiting visualization tools.

There is undoubtedly much more that quaternions can tell us and help us to analyze, compute, and understand. We hope that this book, along with its predecessor, will serve as a guide along the path to even more ways to visualize quaternions.

Visualizing More Quaternions. https://doi.org/10.1016/B978-0-32-399202-2.00045-9

Notation

If the reader is conversant with the conventions of 3D vector notation and has used complex variables, this appendix should be elementary and can be skipped. Nevertheless, for anyone who might benefit from a quick summary, or who might be accustomed to substantially different notational conventions from the author, the summary presented here may be essential in order to follow some of the notation in the main body of the book.

A.1 Vectors

A vector $\mathbf{x}$ is a set of real numbers that we typically write in the form

$$\mathbf{x} = (x, y) \tag{A.1}$$

for 2D vectors, and as

$$\mathbf{x} = (x, y, z) \tag{A.2}$$

for 3D vectors. Technically, we should treat this notation as a shorthand for a column vector because we are thinking in the back of our minds of multiplying these vectors by rotation matrices to transform them to a new orientation. That is, in proper matrix notation,

$$\mathbf{x} = x\,\hat{\mathbf{x}} + y\,\hat{\mathbf{y}}$$
$$= x \begin{bmatrix} 1 \\ 0 \end{bmatrix} + y \begin{bmatrix} 0 \\ 1 \end{bmatrix} = \begin{bmatrix} x \\ y \end{bmatrix}$$

for 2D vectors and

$$\mathbf{x} = x\,\hat{\mathbf{x}} + y\,\hat{\mathbf{y}} + z\,\hat{\mathbf{z}}$$
$$= x \begin{bmatrix} 1 \\ 0 \\ 0 \end{bmatrix} + y \begin{bmatrix} 0 \\ 1 \\ 0 \end{bmatrix} + z \begin{bmatrix} 0 \\ 0 \\ 1 \end{bmatrix} = \begin{bmatrix} x \\ y \\ z \end{bmatrix}$$

for 3D vectors. Since writing column vectors in a line of narrative text is awkward, we frequently use a "column as a transposed row vector" notation such as $[x, y, z]^t$, while in certain contexts a less busy "coordinate set" notation $[x, y, z]$ or (x, y, z) may alternatively be employed for the components of a vector. These notations obviously extend to any dimension.

489

A.2 Length of a vector

The *length* of a Euclidean vector is computed from the Pythagorean theorem, generalized to higher dimensions. Several equivalent notations are common for the squared length of a Euclidean vector, e.g.,

$$\|\mathbf{u}\|^2 = \mathbf{u}^2 = \mathbf{u} \cdot \mathbf{u} = x^2 + y^2$$

in 2D, and

$$\|\mathbf{u}\|^2 = \mathbf{u}^2 = \mathbf{u} \cdot \mathbf{u} = x^2 + y^2 + z^2$$

in 3D; the inner product or "dot product" notation is generalized to arbitrary pairs of vectors below.

The *length* of the vector $\mathbf{u}$ is then the square root of its squared length, and is typically written using double vertical bars as

$$\|\mathbf{u}\| = \sqrt{x^2 + y^2}$$

in 2D and

$$\|\mathbf{u}\| = \sqrt{x^2 + y^2 + z^2}$$

in 3D. Note that we distinguish the norm or length of a *vector*, $\|\mathbf{u}\|$, from the *modulus* of a real number,

$$|x| = |-x| = \sqrt{x^2}$$

or of a complex number (see below)

$$|z| = |x + iy| = \sqrt{x^2 + y^2} = r \ .$$

A.3 Unit vectors

A unit vector $\hat{\mathbf{u}}$ is the vector that results when we divide a (nonzero) vector $\mathbf{u} = (x, y)$ or $\mathbf{u} = (x, y, z)$ by its Euclidean length, $\|\mathbf{u}\|$, that is,

$$\hat{\mathbf{u}} = \frac{\mathbf{u}}{\|\mathbf{u}\|} \ ,$$

and we see that obviously the Euclidean length of $\hat{\mathbf{u}}$ is one, or "unity," and hence the terminology "unit vector."

A.4 Polar coordinates

As a basis for some of the concepts that we will use to understand both complex variables and quaternions, we note that the Cartesian coordinates defined in Eqs. (A.1) and (A.2) have alternative forms based on their *magnitude* $\|\mathbf{u}\|$ which we write as a "radius" $r = \|\mathbf{u}\|$. Then the 2D polar form of a Euclidean vector uses elementary trigonometry to express the components in terms of r and the angle θ

between the vector $\|\mathbf{u}\|$ and the $\hat{\mathbf{x}}$-axis, namely

$$\mathbf{u} = (x,\, y) = (r\cos\theta,\, r\sin\theta)\,.$$

This is a point on a circle of radius r. A 3D polar coordinate then becomes a point on a sphere of radius r, expressible in terms of trigonometric functions as

$$\mathbf{u} = (x,\, y,\, z) = (r\cos\theta\sin\phi,\, r\sin\theta\sin\phi,\, r\cos\phi)\,.$$

Beyond 2D, there are various alternatives for polar coordinates, and quaternions will be expressed in several equivalent 4D polar coordinate systems.

A.5 Spheres

Spheres are labeled by the dimension of the space that results if you cut out a bit of the sphere in every direction around the North pole and flatten it out. A circle then has dimension one, and a balloon dimension two (an exploded balloon can be flattened like a sheet of paper). Thus a circle is the "1-sphere" $\mathbf{S}^1$, a balloon is a "2-sphere" $\mathbf{S}^2$, and so on. Now that we know some sample equations for polar coordinates, we can start from there and define a sphere mathematically as a set of points that all lie at a fixed radius from the center: the circle, or 1-sphere $\mathbf{S}^1$ embedded in two dimensions, then obeys the equation

$$\|\mathbf{u}\| = x^2 + y^2 = r^2 \simeq \text{constant}\,, \tag{A.3}$$

and the "ordinary sphere," or 2-sphere $\mathbf{S}^2$ embedded in three dimensions, obeys the equation

$$\|\mathbf{u}\| = x^2 + y^2 + z^2 = r^2 \simeq \text{constant}\,. \tag{A.4}$$

In fact, quaternions obey the equation of the unit hypersphere,

$$\|q\|^2 = q_0{}^2 + q_1{}^2 + q_2{}^2 + q_3{}^2 = 1 \simeq \text{constant} \tag{A.5}$$

describing the 3-sphere, written formally as $\mathbf{S}^3$. This is the basic origin of all of quaternion geometry.

A.6 Matrix transformations

In our context, matrices are rectangular arrays of real numbers. We will typically deal with square matrices. A square matrix $\mathbf{R}$ acting by right-multiplication on a vector $\mathbf{x}$ produces a new vector $\mathbf{x}'$ as follows:

$$
\begin{aligned}
\mathbf{x}' &= \mathbf{R} \cdot \mathbf{x} \\
&= \begin{bmatrix} r_{11} & r_{12} \\ r_{21} & r_{22} \end{bmatrix} \begin{bmatrix} x \\ y \end{bmatrix} = \begin{bmatrix} x\,r_{11} + y\,r_{12} \\ x\,r_{21} + y\,r_{22} \end{bmatrix} \\
&= \begin{bmatrix} x' \\ y' \end{bmatrix}
\end{aligned}
$$

in 2D and

$$\mathbf{x}' = \mathbf{R} \cdot \mathbf{x}$$

$$= \begin{bmatrix} r_{11} & r_{12} & r_{13} \\ r_{21} & r_{22} & r_{23} \\ r_{31} & r_{32} & r_{33} \end{bmatrix} \begin{bmatrix} x \\ y \\ z \end{bmatrix} = \begin{bmatrix} x\,r_{11} + y\,r_{12} + z\,r_{13} \\ x\,r_{21} + y\,r_{22} + z\,r_{23} \\ x\,r_{31} + y\,r_{32} + z\,r_{33} \end{bmatrix} = \begin{bmatrix} x' \\ y' \\ z' \end{bmatrix}$$

for the 3D case.

A.7 Features of square matrices

Matrices in general are rectangular arrays of numbers; square matrices are used to transform vectors into similar vectors, and thus have unique features. The first of two features we will need to use on occasion is the *trace*, which is the sum of the diagonal elements,

$$\text{Trace } \mathbf{R} \equiv \text{tr}\,\mathbf{R} = \sum_{i=1}^{n} r_{ii} \,,$$

so that in 2D

$$\text{Trace } \mathbf{R} \equiv \text{tr}\,\mathbf{R} = r_{11} + r_{22}$$

and in 3D

$$\text{Trace } \mathbf{R} \equiv \text{tr}\,\mathbf{R} = r_{11} + r_{22} + r_{33} \,.$$

The other frequently encountered property is the *determinant*. Determinants in general have elegant expressions in terms of totally antisymmetric products, for which we refer the interested reader to Appendix L. For the special cases of square 2D and 3D matrices, the determinant can easily be given explicitly; in 2D, we write

$$\text{Determinant } \mathbf{R} \equiv \det \mathbf{R} = r_{11}\,r_{22} - r_{12}\,r_{21} \,,$$

and in 3D

$$\begin{aligned} \text{Determinant } \mathbf{R} \equiv \det \mathbf{R} = \; & r_{11}\,r_{22}\,r_{33} - r_{11}\,r_{23}\,r_{32} \\ & - r_{12}\,r_{21}\,r_{33} + r_{12}\,r_{23}\,r_{31} \\ & + r_{13}\,r_{21}\,r_{32} - r_{13}\,r_{22}\,r_{31} \,. \end{aligned}$$

A.8 Orthogonal matrices

An orthogonal matrix is a square matrix whose transpose is its own inverse. What this means is that if we let I_2 and I_3 be the 2×2 and 3×3 identity matrices, respectively, where

$$I_2 = \begin{bmatrix} 1 & 0 \\ 0 & 1 \end{bmatrix}$$

and

$$I_3 = \begin{bmatrix} 1 & 0 & 0 \\ 0 & 1 & 0 \\ 0 & 0 & 1 \end{bmatrix},$$

then, if the superscript "t" denotes the transposed matrix, and $\mathbf{R}_N$ is an orthogonal $N \times N$ matrix,

$$\mathbf{R}_2(\mathbf{R}_2)^t = I_2$$

and

$$\mathbf{R}_3(\mathbf{R}_3)^t = I_3 .$$

A.9 Vector products

There are two products of vectors that concern us because they have particularly useful properties when we transform the component vectors by applying an orthogonal matrix $\mathbf{R}$. As above, we use the notation $\mathbf{x}_i = (x_i, y_i)$ for 2D vectors and $\mathbf{x}_i = (x_i, y_i, z_i)$ for 3D vectors.

2D dot product. In 2D, the first of these is the inner product or *dot product*,

$$\mathbf{x}_1 \cdot \mathbf{x}_2 = x_1 x_2 + y_1 y_2 ,$$

which is *invariant* under the action of orthogonal matrix multiplication:

$$\begin{aligned} \mathbf{x}_1' \cdot \mathbf{x}_2' &= \mathbf{R}\mathbf{x}_1 \cdot \mathbf{R}\mathbf{x}_2 \\ &= (\mathbf{x}_1 \mathbf{R})^t \cdot (\mathbf{R}\mathbf{x}_2) \\ &= \mathbf{x}_1 \cdot I_2 \cdot \mathbf{x}_2 \\ &= \mathbf{x}_1 \cdot \mathbf{x}_2 . \end{aligned}$$

The dot product may be thought of as the generalization of the squared Euclidean length to a pair of vectors. It can be shown that, in any dimension, the dot product is invariant under the action of orthogonal matrices and is proportional to the *cosine* of the angle between the two vectors,

$$\mathbf{x}_1 \cdot \mathbf{x}_2 = \|\mathbf{x}_1\| \|\mathbf{x}_2\| \cos\theta_{12} .$$

2D cross product. The concept of the 2D cross product is unconventional, but, in retrospect, extremely natural from the point of view of N-dimensional geometry (see, e.g., Hanson (1994)). Here, we can simply define the *2D cross product* as the procedure that generates a new vector $\mathbf{c}$ that is *perpendicular* to a given 2D vector $\mathbf{x}$:

$$\begin{aligned} \mathbf{c} &= \times \mathbf{x} \\ &= \det \begin{bmatrix} x & \hat{\mathbf{x}} \\ y & \hat{\mathbf{y}} \end{bmatrix} \end{aligned}$$

$$= \begin{bmatrix} -y \\ x \end{bmatrix} .$$

One can easily verify that $\mathbf{c} \cdot \mathbf{x} = 0$.

3D dot product. In 3D, a construction analogous to the 2D case yields the invariant dot product, again proportional to the cosine,

$$\mathbf{x}_1 \cdot \mathbf{x}_2 = x_1 x_2 + y_1 y_2 + z_1 z_2 = \|\mathbf{x}_1\| \|\mathbf{x}_2\| \cos\theta_{12} .$$

3D cross product. The cross product in 3D is defined as

$$\mathbf{c} = \mathbf{x}_1 \times \mathbf{x}_2$$

$$= \det \begin{bmatrix} x_1 & x_2 & \hat{\mathbf{x}} \\ y_1 & y_2 & \hat{\mathbf{y}} \\ z_1 & z_2 & \hat{\mathbf{z}} \end{bmatrix}$$

$$= (y_1 z_2 - y_2 z_1,\ z_1 x_2 - z_2 x_1,\ x_1 y_2 - x_2 y_1) ,$$

where now we find that $\mathbf{c}$ is perpendicular to *each* of the component vectors:

$$\mathbf{c} \cdot \mathbf{x}_1 = \mathbf{c} \cdot \mathbf{x}_2 = 0 .$$

The *magnitude* of the 3D cross product is proportional to the *sine* of the angle between the vectors and is thus equal to the *area* of the parallelogram defined by the two vectors in the plane they span:

$$\text{Area of parallelogram} = \|\mathbf{c}\| = \|\mathbf{x}_1 \times \mathbf{x}_2\| = \|\mathbf{x}_1\| \|\mathbf{x}_2\| \sin\theta_{12} .$$

A.10 Complex variables

Complex variables were once thought to be so unnatural that they became known as *imaginary numbers*, because, presumably, there was no way they could be connected with reality. In fact nothing could be further from the truth. Quantum mechanics, the best theoretical description yet developed to predict wide ranges of important, very real, physical processes, depends in essential and inescapable ways on complex numbers. Perhaps a more traditional concrete example is the fact that quadratic algebraic equations cannot be solved for all ranges of their parameters unless complex numbers are allowed. The essence of complex numbers thus comes directly from examining the two equations

$$z^2 = +1 ,$$
$$z^2 = -1 ,$$

and attempting to solve them: the only way we can solve both of these innocent-looking equations is to invent a new object, the pure "imaginary" number "i," which is endowed with the remarkable property that

$$i^2 = -1 \qquad \Rightarrow \qquad i = \sqrt{-1} .$$

The solutions of the two equations introduced above are thus

$$z = \pm 1 \,,$$
$$z = \pm i \,,$$

respectively. With some effort, one can show that, by including "i" in our world, we can miraculously express the solution of any algebraic equation of one variable. If we allow only real variables, this is no longer possible.

We can represent a general complex number z, which can be an arbitrary combination of a real part x and an imaginary part proportional to $i = \sqrt{-1}$, in the following alternative ways (observe that one of these choices is actually the columns of a 2×2 *rotation matrix*):

$$z = x + iy \,,$$
$$z = r \cos\theta + i r \sin\theta \,, \tag{A.6}$$
$$z = \begin{bmatrix} x & -y \\ y & x \end{bmatrix} = r \begin{bmatrix} \cos\theta & -\sin\theta \\ \sin\theta & \cos\theta \end{bmatrix} \,.$$

We note that the complex identity

$$e^{i\theta} = \cos\theta + i\sin\theta$$

is one of the most fundamental equations in all of mathematics, and is often referred to as "Euler's identity." With this identity, we take the polar coordinate expression just given and re-express it as

$$z = r\cos\theta + i r \sin\theta = r e^{i\theta} \,.$$

We also may use the *modulus* or length of a complex number

$$|z| = |x + iy| = \sqrt{x^2 + y^2} = r \,.$$

The algebra of complex multiplication follows from any of the formulas shown in Eq. (A.6), either by explicitly using $i^2 = -1$, or from matrix multiplication. Thus, for example,

$$\begin{aligned} z_1 z_2 &= (x_1 + iy_1)(x_2 + iy_2) \\ &= (x_1 x_2 - y_1 y_2) + i(x_1 y_2 + x_2 y_1) \,, \\ z_1 z_2 &= r_1 e^{i\theta_1} r_2 e^{i\theta_2} \\ &= r_1 r_2 e^{i(\theta_1 + \theta_2)} \\ &= r_1 r_2 \cos(\theta_1 + \theta_2) + i r_1 r_2 \sin(\theta_1 + \theta_2) \,, \\ z_1 z_2 &= \begin{bmatrix} x_1 & -y_1 \\ y_1 & x_1 \end{bmatrix} \cdot \begin{bmatrix} x_2 & -y_2 \\ y_2 & x_2 \end{bmatrix} \\ &= \begin{bmatrix} x_1 x_2 - y_1 y_2 & -(x_1 y_2 + x_2 y_1) \\ x_1 y_2 + x_2 y_1 & x_1 x_2 - y_1 y_2 \end{bmatrix} \,, \end{aligned}$$

$$z_1 z_2 = \begin{bmatrix} x_1 & -y_1 \\ y_1 & x_1 \end{bmatrix} \begin{bmatrix} x_2 \\ y_2 \end{bmatrix}$$

$$= \begin{bmatrix} x_1 x_2 - y_1 y_2 \\ x_1 y_2 + x_2 y_1 \end{bmatrix}.$$

Thus we see that all of these forms are exactly equivalent to the more abstract statement that the *complex multiplication algebra* is defined by associating the complex product of two pairs of real numbers with a *third pair* constructed from their elements as follows:

$$z_1 \star z_2 = (x_1, y_1) \star (x_2, y_2) = \begin{bmatrix} x_1 \\ y_1 \end{bmatrix} \star \begin{bmatrix} x_2 \\ y_2 \end{bmatrix}$$

$$= (x_1 x_2 - y_1 y_2, \ x_1 y_2 + x_2 y_1)$$

$$= \begin{bmatrix} x_1 x_2 - y_1 y_2 \\ x_1 y_2 + x_2 y_1 \end{bmatrix}.$$

Quaternion methods

In this appendix, we work out a family of miscellaneous algorithms and methods that we have found useful from time to time.

B.1 Quaternion logarithms and exponentials

If we parameterize a quaternion in a simplified way, with $0 \leqslant \theta < 4\pi$, to cover the entire $\mathbf{S}^3$ as $q(\theta, \hat{\mathbf{n}}) = (\cos(\theta/2), \sin(\theta/2)\hat{\mathbf{n}})$, then we can show using the quaternion algebra that this parameterization of $\mathbf{S}^3$ follows from the exponential series

$$\exp\left(0, \frac{\theta}{2}\hat{\mathbf{n}}\right) = \left(\cos\frac{\theta}{2}, \sin\frac{\theta}{2}\hat{\mathbf{n}}\right) .$$

The logarithm of a structure is the object that, when exponentiated, produces the generic structure; hence we must have

$$\log q = \log\left(\cos\frac{\theta}{2}, \sin\frac{\theta}{2}\hat{\mathbf{n}}\right) = \left(0, \frac{\theta}{2}\hat{\mathbf{n}}\right) .$$

Raising a quaternion to a power is defined via the exponential power series as well:

$$q^t = e^{t\log q} = e^{\left(0, t\frac{\theta}{2}\hat{\mathbf{n}}\right)}$$
$$= \left(\cos\frac{t\theta}{2}, \sin\frac{t\theta}{2}\hat{\mathbf{n}}\right) .$$

B.2 The quaternion square root trick

Quaternion square roots are often needed in various calculations, and there are several approaches of varying elegance.

The most obvious is simply to express the quaternion in eigenvector coordinates and take one half of the total rotation:

$$q(\theta, \hat{\mathbf{n}}) = \left(\cos\frac{\theta}{2}, \hat{\mathbf{n}}\sin\frac{\theta}{2}\right) ,$$
$$p = \sqrt{q} \text{ iff } p \star p = q ,$$

$$p = \left(\cos \frac{\theta}{4}, \hat{\mathbf{n}} \sin \frac{\theta}{4} \right).$$

However, there is also a more elegant algebraic approach, motivated by the observation that if x is the cosine of an angle, then the cosine of half the angle is

$$y = \frac{\sqrt{1+x}}{\sqrt{2}}$$
$$= \frac{1+x}{\sqrt{2(1+x)}}.$$

The half-angle formula in fact has an exact quaternion analog solving the quaternion formula $p \star p = q$ that allows us to write the square root of q in the form

$$p = \frac{1+q}{\sqrt{2(1+q_0)}}, \tag{B.1}$$

where $1 + q = (1 + q_0, q_x, q_y, q_z) = (1 + q_0, \mathbf{q})$. We can further simplify this by dividing through by the denominator,

$$p = \left(\sqrt{\frac{1+q_0}{2}}, \hat{\mathbf{n}} \sqrt{\frac{1-q_0}{2}} \right). \tag{B.2}$$

The case $q_0 = -1$ is special, since a quaternion rotation $(0, \hat{\mathbf{n}})$ by π about any direction $\hat{\mathbf{n}}$ will square to $q = (-1, 0, 0, 0)$, so no unique square root exists. Letting $q = (q_0, \mathbf{q})$, and noting the usual relation

$$\mathbf{q} \cdot \mathbf{q} = 1 - (q_0)^2,$$

we can confirm the validity of Eq. (B.1) by computing

$$(1+q) \star (1+q) = ((1+q_0)^2 - \mathbf{q} \cdot \mathbf{q}, \; 2(1+q_0)\mathbf{q})$$
$$= (2q_0(1+q_0), \; 2(1+q_0)\mathbf{q})$$
$$= 2(1+q_0)\, q.$$

The identity $p \star p = q$ thus follows at once if $q \cdot q = 1$. Note that if the quaternion is not normalized, e.g., with $q \cdot q = r^2$, the formula still works replacing "1" by "r":

$$p(r) = \frac{r+q}{\sqrt{2(r+q_0)}}.$$

B.3 Gram–Schmidt spherical interpolation

To find an interpolated unit vector that is guaranteed to remain on the sphere, and thus preserve its length of unity, we first assume that one vector, say $\mathbf{q}_0$, is the starting point of the interpolation. In order to apply a standard rotation formula using a rigid, length-preserving orthogonal transformation in the

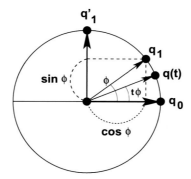

FIGURE B.1

The length-preserving spherical interpolation framework, showing the construction of a new orthonormal basis $(\mathbf{q}_0, \mathbf{q}'_1)$ that allows us simply to rotate the unit vector rigidly using sines and cosines.

plane of $\mathbf{q}_0$ and $\mathbf{q}_1$, we need a unit vector *orthogonal* to $\mathbf{q}_0$ that contains some portion of the direction $\mathbf{q}_1$. Defining

$$\mathbf{q}'_1 = \frac{\mathbf{q}_1 - \mathbf{q}_0(\mathbf{q}_0 \cdot \mathbf{q}_1)}{\|\mathbf{q}_1 - \mathbf{q}_0(\mathbf{q}_0 \cdot \mathbf{q}_1)\|} \; ,$$

we see that by construction

$$\mathbf{q}'_1 \cdot \mathbf{q}_0 = 0 \; ,$$
$$\mathbf{q}'_1 \cdot \mathbf{q}'_1 = 1 \; ,$$

where we of course continue to require $\mathbf{q}_0 \cdot \mathbf{q}_0 = 1$. The denominator of this expression has the curious property that

$$\|\mathbf{q}_1 - \mathbf{q}_0(\mathbf{q}_0 \cdot \mathbf{q}_1)\|^2 = 1 - 2\cos^2 \phi + \cos^2 \phi$$
$$= \sin^2 \phi \; ,$$

where we recall that $\cos \phi = \mathbf{q}_0 \cdot \mathbf{q}_1$. Note that since we have imposed $0 \leqslant \phi < \pi$, the sine is always nonnegative, and we can replace $|\sin \phi|$ by $\sin \phi$, which we will find convenient below. When $\phi = 0$, there is no interpolation to be done in any event.

Referring to the graphical construction in Fig. B.1, we next rephrase the unit-length-preserving rotation using the angle $t\phi$, where $0 \leqslant t \leqslant 1$ takes us from a unit vector aligned with $\mathbf{q}_0$ at $t = 0$ to one aligned with $\mathbf{q}_1$ at $t = 1$, using our new orthonormal basis:

$$\mathbf{q}(t) = \mathbf{q}_0 \cos t\phi + \mathbf{q}'_1 \sin t\phi$$
$$= \mathbf{q}_0 \cos t\phi + (\mathbf{q}_1 - \mathbf{q}_0 \cos \phi) \frac{\sin t\phi}{\sin \phi}$$

$$= \mathbf{q}_0 \frac{\cos t\phi \sin\phi - \sin t\phi \cos\phi}{\sin\phi} + \mathbf{q}_1 \frac{\sin t\phi}{\sin\phi}$$

$$= \mathbf{q}_0 \frac{\sin(1-t)\phi}{\sin\phi} + \mathbf{q}_1 \frac{\sin t\phi}{\sin\phi} \;.$$

This is the `slerp` formula, which guarantees that

$$\mathbf{q}(t) \cdot \mathbf{q}(t) \equiv 1$$

by construction. The `slerp` interpolator rotates one unit vector into another, keeping the intermediate vector in the mutual plane of the two limiting vectors while also guaranteeing that the interpolated vector preserves its unit length throughout, and therefore always remains on the sphere. The formula is true *in any dimension whatsoever* because it depends only on the local 2D plane determined by the two limiting vectors.

B.4 Direct solution for spherical interpolation

Let q be a unit vector on a sphere. Assume q is located partway between two other unit vectors q_0 and q_1, with the location defined by some constants c_0 and c_1:

$$q = c_0 q_0 + c_1 q_1 \;. \tag{B.3}$$

As shown in Fig. B.2, q must partition the angle ϕ between q_0 and q_1, where $\cos\phi = q_0 \cdot q_1$, into two subangles, ϕ_0 and ϕ_1, where $\cos\phi_0 = q \cdot q_1$ and $\cos\phi_1 = q \cdot q_0$ and $\phi = \phi_0 + \phi_1$; the labeling is chosen so that $\phi_0 = \phi$ makes $q = q_0$, and $\phi_1 = \phi$ makes $q = q_1$. No matter what the dimension of the unit-length q's, taking two dot products reduces this to a solvable linear system:

$$q \cdot q_0 = \cos\phi_1 = c_0 + c_1 \cos\phi \;,$$
$$q \cdot q_1 = \cos\phi_0 = c_0 \cos\phi + c_1 \;.$$

Using Cramer's rule, we immediately find c_0 and c_1:

$$c_0 = \frac{\det \begin{bmatrix} \cos\phi_1 & \cos\phi \\ \cos\phi_0 & 1 \end{bmatrix}}{\det \begin{bmatrix} 1 & \cos\phi \\ \cos\phi & 1 \end{bmatrix}} = \frac{\cos\phi_1 - \cos\phi_0 \cos\phi}{1 - \cos^2\phi} \;,$$

$$c_1 = \frac{\det \begin{bmatrix} 1 & \cos\phi_1 \\ \cos\phi & \cos\phi_0 \end{bmatrix}}{\det \begin{bmatrix} 1 & \cos\phi \\ \cos\phi & 1 \end{bmatrix}} = \frac{\cos\phi_0 - \cos\phi_1 \cos\phi}{1 - \cos^2\phi} \;.$$

Making the substitution $\phi = \phi_0 + \phi_1$ and carrying out some trigonometry,

$$\cos\phi_1 - \cos\phi_0 \cos\phi = \cos\phi_1 - \cos\phi_0 (\cos\phi_0 \cos\phi_1 - \sin\phi_0 \sin\phi_1)$$

$$= \cos\phi_1 \left(1 - \cos^2\phi_0\right) + \sin\phi_1 \cos\phi_0 \sin\phi_0$$

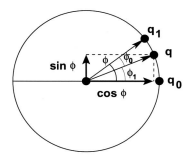

FIGURE B.2

Alternate length-preserving spherical interpolation framework, applying linear algebra to a partition of the angles with $\phi = \phi_0 + \phi_1$.

$$= \sin\phi_0 \, (\sin\phi_0 \cos\phi_1 + \sin\phi_1 \cos\phi_0)$$
$$= \sin\phi_0 \, \sin\phi \,,$$

we find

$$c_0 = \frac{\sin\phi_0 \, (\cos\phi_1 \sin\phi_0 + \cos\phi_0 \sin\phi_1)}{\sin^2\phi} = \frac{\sin\phi_0}{\sin\phi}$$
$$= \frac{\sin t_0\phi}{\sin\phi} \,,$$
$$c_1 = \frac{\sin\phi_1 \, (\cos\phi_1 \sin\phi_0 + \cos\phi_0 \sin\phi_1)}{\sin^2\phi} = \frac{\sin\phi_1}{\sin\phi}$$
$$= \frac{\sin t_1\phi}{\sin\phi} \,.$$

Here we have defined $t_0 = \phi_0/\phi$ and $t_1 = \phi_1/\phi$ to obtain a partition of unity, $t_0 + t_1 = 1$. Choosing, for example, $t_0 = 1 - t$ and $t_1 = t$, we recover the standard `slerp` formula,

$$\mathbf{q}(t) = \mathbf{q}_0 \frac{\sin(1-t)\phi}{\sin\phi} + \mathbf{q}_1 \frac{\sin t\phi}{\sin\phi} \,. \tag{B.4}$$

Thus the `slerp` formula can be verified in a number of different ways.

B.5 Converting linear algebra to quaternion algebra

The equation describing a linear Euclidean interpolation

$$x(t) = a + t(b - a) \tag{B.5}$$

transforms into the quaternion relation

$$q(t) = a \left(a^{-1} b \right)^t \tag{B.6}$$

when we apply these simple rules:

- convert sums to products;
- convert minus signs to products with the inverse;
- convert multiplicative scale factors to exponents.

Since quaternions are not necessarily commutative, however, one must be careful about the order of multiplication; the order given in Eq. (B.6) will generally give consistent results.

Eigensystem of the rotation matrix

One of our themes is constructing and understanding eigensystems of interesting matrices, so here we expand our analysis of the basic 3D rotation matrices to include some additional details. First, note that we have two ways of writing the 3D rotation, as $R_3(\theta, \hat{n})$ and as $R_3(q)$. Thus there are two ways to write the eigenvalues, which we can compute to be

$$\left\{ \begin{array}{c} 1 \\ e^{i\theta} \\ e^{-i\theta} \end{array} \right\}, \qquad \left\{ \begin{array}{c} 1 \\ (q_0^2 - q_1^2 - q_2^2 - q_3^2 + 2iq_0\sqrt{q_1^2 + q_2^2 + q_3^2}) \\ (q_0^2 - q_1^2 - q_2^2 - q_3^2 - 2iq_0\sqrt{q_1^2 + q_2^2 + q_3^2}) \end{array} \right\}, \qquad (C.1)$$

respectively, where the two columns are of course identical, but we have chosen expressions in q (along with an implicit choice of square root sign determining $\sin(\theta/2)$) that match exactly with the $R_3(q)$ eigenvectors. The unnormalized eigenvectors corresponding to each eigenvalue in the rows of Eq. (C.1) can be written as the columns

$$\left\{ \begin{bmatrix} n_1 \\ n_2 \\ n_3 \end{bmatrix} \begin{bmatrix} -in_2 - n_1 n_3 \\ in_1 - n_2 n_3 \\ n_1^2 + n_2^2 \end{bmatrix} \begin{bmatrix} +in_2 - n_1 n_3 \\ -in_1 - n_2 n_3 \\ n_1^2 + n_2^2 \end{bmatrix} \right\},$$

$$\left\{ \begin{bmatrix} q_1 \\ q_2 \\ q_3 \end{bmatrix} \begin{bmatrix} -q_1 q_3 - iq_2\sqrt{q_1^2 + q_2^2 + q_3^2} \\ -q_2 q_3 + iq_1\sqrt{q_1^2 + q_2^2 + q_3^2} \\ q_1^2 + q_2^2 \end{bmatrix} \begin{bmatrix} -q_1 q_3 + iq_2\sqrt{q_1^2 + q_2^2 + q_3^2} \\ -q_2 q_3 - iq_1\sqrt{q_1^2 + q_2^2 + q_3^2} \\ q_1^2 + q_2^2 \end{bmatrix} \right\},$$

$$(C.2)$$

where we emphasize that in general R_3 has only one real eigenvalue (which is unity), whose eigenvector is the direction of the 3D axis $\hat{n}$ invariant under that particular rotation. Since any quaternion can be written in the form Eq. (2.16), the trace of any rotation can be written as

$$\text{tr } R_3 = 3q_0^2 - q_1^2 - q_2^2 - q_3^2 = 4q_0^2 - 1 = 1 + 2\cos\theta, \qquad (C.3)$$

which follows from the half-angle formula. This means that in the task of Chapter 8, that is, maximizing expressions of the form $\text{tr}(R \cdot E)$, if E is an identity matrix, the rotation giving the maximal trace corresponds to R_3 being the identity matrix, $\theta = 0$, while if E is a rotation matrix, the maximal trace occurs when the product of the two matrices has vanishing angle θ for the *composite* matrix produced by the product of one quaternion with the inverse of the other, so the optimal rotation matrix R_3 is

503

the inverse of E. This property is exploited in Chapter 9, where we investigate the Bar-Itzhack algorithm (Bar-Itzhack, 2000) for extracting the nearest perfect rotation matrix from a noisy approximate rotation.

Lie algebra origins of the rotation matrix

The traditional derivation of the expression Eq. (2.14) for the 3×3 3D rotation matrix $R(q)$ in terms of quaternions,

$$[0, R(q) \cdot \mathbf{x}] = q \star [0, \mathbf{x}] \star \bar{q} ,$$

is technically an example of *conjugation* by elements of a group, which produces the *adjoint* representation that, significantly, has the same dimension as the group itself. In the context of how quaternions correspond to rotations, the relevant group is $\mathbf{SU(2)}$, which has an algebra corresponding to the quaternion algebra, and has dimension three, corresponding to the degrees of freedom of the unit quaternion that we use throughout. The corresponding mathematical language replaces the quaternion 4-vector $[0, \mathbf{x}]$ in Eq. (2.13) by an element of *the corresponding Lie algebra*, and the unit quaternion 4-vector generating the transformation by its $\mathbf{SU(2)}$ matrix.

To accomplish this, we write down the following complex, 2×2 special unitary matrices (unit determinant, inverse being the complex conjugate transpose) defining the group $\mathbf{SU(2)}$:

$$S_0 = \sigma_0 = \begin{bmatrix} 1 & 0 \\ 0 & 1 \end{bmatrix} ,$$

$$S_1 = -i\sigma_1 = \begin{bmatrix} 0 & -i \\ -i & 0 \end{bmatrix} , \quad S_2 = -i\sigma_2 = \begin{bmatrix} 0 & -1 \\ 1 & 0 \end{bmatrix} , \quad S_3 = -i\sigma_3 = \begin{bmatrix} -i & 0 \\ 0 & i \end{bmatrix} , \tag{D.1}$$

where the σ_i are the *Pauli matrices*. From the observation that the last three matrices $\mathbf{S}$ obey the relations

$$S_i \cdot S_j = -\delta_{ij} S_0 + \sum_{k=1}^{3} \epsilon_{ijk} S_k , \tag{D.2}$$

where ϵ_{ijk} is the totally antisymmetric Levi-Civita symbol (see Appendix L), we can easily verify that these matrices represent the quaternion algebra,

$$(q \cdot S) \cdot (p \cdot S) = (q \star p) \cdot S .$$

(Here we use the notation $q \cdot S \equiv \sum_{i=0}^{3} q_i S_i = q_0 S_0 + \mathbf{q} \cdot \mathbf{S}$.)

The second step is to note that, while $q \cdot S$ represents an element $G(q)$ of the *Lie group* $\mathbf{SU(2)}$, what is actually needed for the construction of the adjoint representation is an element of the *Lie algebra*.

The Lie algebra is formed from the *three* matrices $\mathbf{S} = [S_1, S_2, S_3]$ by exponentiation,

$$G(\theta, \hat{\mathbf{n}}) = \exp\left(-\tfrac{\mathrm{i}}{2}\theta\,\boldsymbol{\sigma}\cdot\hat{\mathbf{n}}\right) = \exp\left(\tfrac{1}{2}\theta\,\mathbf{S}\cdot\hat{\mathbf{n}}\right)$$

$$= \left[\cos(\tfrac{\theta}{2})S_0 + \sin(\tfrac{\theta}{2})\hat{\mathbf{n}}\cdot\mathbf{S}\right], \tag{D.3}$$

$$G(q) = q \cdot S.$$

Now we can write down the action of $\mathbf{SU(2)}$ group conjugation on an element $\mathbf{X}$ of the 3D Lie algebra (which we write henceforth as $\mathbf{S}\cdot\mathbf{X}$), as required by Lie group constructs unknown in Hamilton's time:

$$\begin{bmatrix}\text{Derivation of adjoint}\\ \text{representation } R(q)\end{bmatrix} = G(q)\cdot(\mathbf{S}\cdot\mathbf{X})\cdot G^\dagger(q)$$

$$= (R(q)\cdot\mathbf{X})\cdot\mathbf{S}. \tag{D.4}$$

Here the important, and somewhat nonintuitive, facts are:

- While the $\mathbf{SU(2)}$ group element $G(q)$ is a sum of four matrices, the requirement that it have unit determinant forces $q \cdot q = 1$, and so the group has only three parameters.
- Therefore, since the group has three parameters, the four matrices S_i are not the adjoint representation, which must have the same dimension as the number of free parameters of the group by definition.
- The adjoint representation construction Eq. (D.4) results in a (3×3)-dimensional matrix $R(q)$ that acts on the 3D Lie algebra, and that construction produces precisely our fundamental equation Eq. (2.14) as a function of the unit quaternion parameters q.
- That matrix function happens in this case to be the 3×3 adjoint representation of $\mathbf{SU(2)}$, which is in turn the element of the extremely important group $\mathbf{SO}(3)$ that rotates physical objects in 3D Euclidean space.

Derivation of rotation group representations

There is yet another way to understand the deeper origins of Eq. (2.14) that is completely independent of the traditional Eq. (2.13), and which in fact allows us to derive an infinite set of $(2j + 1)$-dimensional representations labeled by $j \in \{1/2, 1, 3/2, 2, \ldots\}$, with half-integers for even-dimensional representations and integers for odd-dimensional representations. We start with the concept of a *parameterized 2D basis* upon which an **SU(2)** element acts. We write this process as

$$\begin{bmatrix} x' \\ y' \end{bmatrix} = (q \cdot S) \cdot \begin{bmatrix} x \\ y \end{bmatrix} , \tag{E.1}$$

where the matrix S is defined in Appendix D, and the pair (x, y) parameterizes the basis on which the rotation is acting. Now we consider the polynomial

$$(x + y)^{2j} = x^{2j} + 2jx^{2j-1}y + \cdots + 2jxy^{2j-1} + y^{2j} = \sum_{k=0}^{2j} \binom{2j}{k} x^k y^{2j-k} , \tag{E.2}$$

where the coefficients in the expansion are the elements of Pascal's triangle. Now we simply assign an abstract algebraic variable $a_k : k \in \{0, \ldots, 2j\}$ to each of the $2j + 1$ terms in Eq. (E.2) (recall that $2j + 1$ is the expected dimension of representations labeled by j). The indices of the a_k basically correspond to the powers of y, and we get the polynomial expression

$$P(x, y; a_k) = a_0 x^{2j} + 2ja_1 x^{2j-1}y + \cdots + 2ja_{2j-1}xy^{2j-1} + a_{2j}y^{2j} = \sum_{k=0}^{2j} \binom{2j}{k} a_{2j-k} x^k y^{2j-k} . \tag{E.3}$$

This is an algebraic way of studying the action of a group as it generates a family of representations. The process is to apply Eq. (E.1) to each pair (x, y) to generate a modified pair (x', y'), which we substitute into the expression to get $P(x', y'; a_k)$, and then we expand the results to obtain a polynomial *reassembled into the original terms in* (x, y). If we call that expression $P(x, y; a'_k)$, we see that each a'_k is a linear combination of the original coefficients a_k, where each has been modified by a $(2j)$-th power of q. This remarkable set of linear combinations can now be expressed as a $(2j + 1) \times (2j + 1)$-dimensional *matrix* relating the transformed coefficients a'_k to the original coefficients as follows:

507

$$
\begin{bmatrix} a'_0 \\ a'_1 \\ \vdots \\ a'_{2j-1} \\ a'_{2j} \end{bmatrix} = D^j(q) \cdot \begin{bmatrix} a_0 \\ a_1 \\ \vdots \\ a_{2j-1} \\ a_{2j} \end{bmatrix} .
\tag{E.4}
$$

The matrix $D^j(q)$ is the $(2j+1)$-dimensional representation of $\mathbf{SU}(2)$, *expressed explicitly* in terms of $(2j)$-th powers of quaternions. The corresponding formula is well known from classic work on quantum mechanics and the theory of the rotation group, and takes the form (Biedenharn and Louck, 1984)

$$
\begin{aligned}
D^{(j)}_{m'm}(q) = {} & [(j+m')!(j-m')!(j+m)!(j-m)!]^{1/2} \\
& \times \sum_s \frac{(q_0 - iq_3)^{j+m-s}(-iq_1 - q_2)^{m'-m+s}(-iq_1 + q_2)^s(q_0 + iq_3)^{j-m'-s}}{(j+m-s)!(m'-m+s)!s!(j-m'-s)!} ,
\end{aligned}
\tag{E.5}
$$

where s ranges over the set of legal values determined by j, and j can in principle range over either half-integer representation labels or integer representation labels (note that one typically uses l instead of j when restricting to integers only). This matrix is an *explicit* realization of the quaternion algebra, and has the required property of a group representation, now exposed more clearly in terms of quaternions: the multiplication rule for $D^{(j)}_{m'm}(q)$ considered as a matrix $\mathbf{D}^j(q)$ is simply

$$
\mathbf{D}^j(p) \cdot \mathbf{D}^j(q) = \mathbf{D}^j(p \star q) .
\tag{E.6}
$$

This can be verified by repeating the process leading to Eq. (E.4) twice and collecting and rearranging the terms into collections appearing in $p \star q$.

The 3D adjoint representation. Finally, we work out the details for the case of specific interest, Eq. (2.14), showing a significant alternative, group-theoretically rigorous, derivation that arises without reference to the conjugation process. The expression Eq. (E.3) for the 3D case ($j = 1$, giving dimension $2j+1 = 3$) becomes

$$
P_3(x, y; a_k) = a_0 x^2 + 2a_1 xy + a_2 y^2 .
\tag{E.7}
$$

Transforming (x, y) to (x', y') using the matrix $q \cdot S$ and collecting terms in P_3, we find

$$
\begin{aligned}
a'_0 &= a_0 \left(q_0^2 - 2iq_0 q_3 - q_3^2\right) + a_1(-2iq_0 q_1 + 2q_0 q_2 - 2q_1 q_3 - 2iq_2 q_3) + a_2 \left(-q_1^2 - 2iq_1 q_2 + q_2^2\right) , \\
a'_1 &= a_0(-2iq_0 q_1 - 2q_0 q_2 - 2q_1 q_3 + 2iq_2 q_3) + a_1 \left(2q_0^2 - 2q_1^2 - 2q_2^2 + 2q_3^2\right) \\
&\quad + a_2(-2iq_0 q_1 + 2q_0 q_2 + 2q_1 q_3 + 2iq_2 q_3) , \\
a'_2 &= a_0 \left(-q_1^2 + 2iq_1 q_2 + q_2^2\right) + a_1(-2iq_0 q_1 - 2q_0 q_2 + 2q_1 q_3 - 2iq_2 q_3) + a_2 \left(q_0^2 + 2iq_0 q_3 - q_3^2\right) ,
\end{aligned}
\tag{E.8}
$$

or, in matrix form,

$$
[a'] = [D] \cdot [a] = \begin{bmatrix} D_{+1,+1} & D_{+1,\,0} & D_{+1,-1} \\ D_{\,0,+1} & D_{\,0,\,0} & D_{\,0,-1} \\ D_{-1,+1} & D_{-1,\,0} & D_{-1,-1} \end{bmatrix} \cdot [a] .
\tag{E.9}
$$

This representation has the property that it obeys the required constraint Eq. (E.6) to be a representation, but it is not in the form we need to act on ordinary 3-vectors because the rows and columns are in fact labeled by the possible projections of the value of j onto the z-axis of wave functions expanded, e.g., in terms of spherical harmonics. Specifically, the three indices of each row are $m = +1$, $m = 0$, $m = -1$ instead of $(1, 2, 3)$ or (x, y, z). The $m = 0$ index corresponds essentially to the z direction, and the $m = \pm 1$ correspond to $x \pm iy$. With some hindsight, the matrix components can be regrouped to a new matrix with *all real* components that have a basis in orthogonal 3D (x, y, z) space instead of complex $m = (+1, 0, -1)$ components, and that transformation, familiar in, e.g., the settings of quantum mechanics and electromagnetism, can be written as

$$
\begin{bmatrix}
D_{+1,+1} & D_{+1,\,0} & D_{+1,-1} \\
D_{0,+1} & D_{0,\,0} & D_{0,-1} \\
D_{-1,+1} & D_{-1,\,0} & D_{-1,-1}
\end{bmatrix} \longrightarrow
$$

$$
\begin{bmatrix}
\frac{1}{2}(D_{+1,+1} - D_{+1,-1} + D_{-1,-1} - D_{-1,+1}) & \frac{1}{2i}(D_{+1,+1} - D_{+1,-1} - D_{-1,-1} + D_{-1,+1}) & \frac{1}{2}(D_{0,-1} - D_{0,+1}) \\
\frac{i}{2}(D_{+1,+1} + D_{+1,-1} - D_{-1,-1} - D_{-1,+1}) & \frac{1}{2}(D_{+1,+1} + D_{+1,-1} + D_{-1,-1} + D_{-1,+1}) & \frac{1}{2i}(D_{0,+1} + D_{0,-1}) \\
\frac{1}{2}(D_{-1,\,0} - D_{+1,\,0}) & \frac{i}{2}(D_{+1,\,0} + D_{-1,\,0}) & \frac{1}{2}D_{0,\,0}
\end{bmatrix} . \quad \text{(E.10)}
$$

Substituting the coefficients of Eq. (E.8) for the elements of $D_{m,m'}(q)$ into Eq. (E.10), we recover exactly Eq. (2.14). This completes our verification of the larger context of the form of Eq. (2.14) in terms of the adjoint representation of the group **SU(2)** that embodies the group-theoretical context of our subject, namely unit quaternions.

Quaternion from double 3D Clifford reflection

Here we describe the double-reflection formula that permits a nice construction for a quaternion in the context of Clifford algebras, extending the intuitive framework behind the treatment in Hanson (2006, Chapter 31).

F.1 Abstract description

We can see with a trivial calculation that a faithful representation of the parameters of an arbitrary quaternion can be written using two unit-length 3-vectors $\mathbf{A}$ and $\mathbf{B}$, or equivalently, directions on the celestial sphere with $\mathbf{A} \cdot \mathbf{A} = 1$, $\mathbf{B} \cdot \mathbf{B} = 1$, by examining

$$q = (\mathbf{A} \cdot \mathbf{B}, \ \mathbf{A} \times \mathbf{B}) \,. \tag{F.1}$$

Checking the length of q, we see immediately that it lies on $\mathbf{S}^3$ and is thus a valid general quaternion,

$$q \cdot q = (\mathbf{A} \cdot \mathbf{B})^2 + \Big((\mathbf{A} \cdot \mathbf{A})(\mathbf{B} \cdot \mathbf{B}) - (\mathbf{A} \cdot \mathbf{B})^2\Big)$$

$$= (\mathbf{A} \cdot \mathbf{A})(\mathbf{B} \cdot \mathbf{B}) = 1 \,, \tag{F.2}$$

where we have taken advantage of Eq. (L.4) from Appendix L. Clearly

$$\left. \begin{aligned} \mathbf{A} \cdot \mathbf{B} &= \cos\left(\frac{\theta}{2}\right) \\ \mathbf{A} \times \mathbf{B} &= \frac{\mathbf{A} \times \mathbf{B}}{\sqrt{(\mathbf{A} \times \mathbf{B}) \cdot (\mathbf{A} \times \mathbf{B})}} \sqrt{1 - (\mathbf{A} \cdot \mathbf{B})^2} = \widehat{\mathbf{A} \times \mathbf{B}} \sqrt{1 - \cos^2(\theta/2)} = \hat{\mathbf{n}} \sin\left(\frac{\theta}{2}\right) \end{aligned} \right\} \tag{F.3}$$

makes Eq. (F.1) compatible with our basic quaternion representation Eq. (2.16).

Thus Eq. (F.1) parameterizes an arbitrary unit quaternion in terms of a pair of unit 3-vectors $\{\mathbf{A}, \mathbf{B}\}$. However, we know that if $\mathbf{A}$ and $\mathbf{B}$ are arbitrary 3D unit vectors, they have a total of four degrees of freedom, but our quaternions, although they are 4D, have only three independent degrees of freedom. What about this one unwanted degree of freedom? In fact this is evidence of a *fibration* that is essentially a circle, with one degree of freedom. This redundant degree of freedom can be elegantly removed by observing that our expression Eq. (F.1) for q is invariant precisely under the family of 3D rotations that leave the axis $(\mathbf{A} \times \mathbf{B})$ fixed; meanwhile, the q_0 term $\mathbf{A} \cdot \mathbf{B}$ is obviously invariant under any 3D rotation. We can thus eliminate the fourth degree of freedom by choosing any convenient value for the initial rotation angle, say, ϕ, of the coordinate frame containing the rotation-fixed axis $\hat{\mathbf{n}} \propto \mathbf{A} \times \mathbf{B}$.

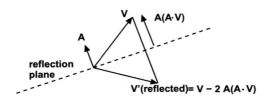

FIGURE F.1

Clifford algebra conjugation results in a reflection of the generic 3D vector $\mathbf{V}$ around the plane through the origin perpendicular to the unit vector $\mathbf{A}$, so if $\mathbf{x}$ is a point in the reflection plane, $\mathbf{A} \cdot \mathbf{x} = 0$.

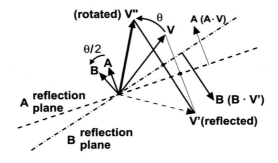

FIGURE F.2

Double reflection in Clifford algebra rotates an arbitrary vector by *twice the angle* $\theta/2 = \arccos \mathbf{A} \cdot \mathbf{B}$ between the plane normals $\mathbf{A}$ and $\mathbf{B}$.

F.2 Relation to Clifford algebra

The quaternion representation $q = (\mathbf{A} \cdot \mathbf{B}, \mathbf{A} \times \mathbf{B})$ does not in fact have to be guessed, but can be derived deterministically from Clifford algebra. We know from Hanson (2006, Chapter 31) that $\mathbf{A}$ can be used as a Clifford reflection parameter to reflect a vector $\mathbf{V}$ in the plane through the origin defined by

$$\mathbf{A} \cdot \mathbf{x} = 0 . \tag{F.4}$$

The reflection of a vector $\mathbf{V}$ in that plane is

$$\mathbf{V}' = \mathbf{V} - 2\mathbf{A} (\mathbf{V} \cdot \mathbf{A}) ,$$

as shown in Fig. F.1 Furthermore the *pair* $\{\mathbf{A}, \mathbf{B}\}$, when used to produce a *double reflection* in the corresponding pair of planes, results in an $\mathbf{SO}(3)$ rotation matrix realizing an ordinary 3D rotation on any vector, that is, $\mathbf{V}'' = \mathbf{R}(\mathbf{A}, \mathbf{B}) \cdot \mathbf{V}$. The relation of Eq. (F.1) to the Clifford algebra double reflection becomes obvious when we draw the graphical representation of the rotation shown in Fig. F.2. One can see purely from plane geometry identities that $\mathbf{A} \cdot \mathbf{B} = \cos(\theta/2)$, where θ is the angle of rotation experienced by the vector $\mathbf{V}'' = $ (double reflection). Then, from ordinary geometry, we know also that $\|\mathbf{A} \times \mathbf{B}\| = |\sin(\theta/2)|$, and we also can see that the rotation happens in the plane spanned by $\{\mathbf{A}, \mathbf{B}\}$, and therefore the vector part of the corresponding quaternion must be the normalized direction $\hat{\mathbf{n}} = $

(normalize) $(\mathbf{A} \times \mathbf{B}) = \widehat{\mathbf{A} \times \mathbf{B}}$. *Strictly from geometry* we are therefore able to construct the fact that Eq. (F.1) represents the quaternion

$$q = (\mathbf{A} \cdot \mathbf{B}, \mathbf{A} \times \mathbf{B}) = \left(\cos(\theta/2), \hat{\mathbf{n}}\sin(\theta/2)\right)$$

for the Clifford double rotation $\mathbf{V} \to \mathbf{V}''$ via a reflection in the plane perpendicular to $\mathbf{A} = [a_1, a_2, a_3]$ followed by a reflection in the plane perpendicular to $\mathbf{B} = [b_1, b_2, b_3]$. The explicit algebraic form of this double-reflection rotation becomes

$$\mathbf{R}(\mathbf{A}, \mathbf{B}) = \begin{bmatrix} 1 + 4a_1{}^2 b_1{}^2 - 2a_1{}^2 + 4a_1 a_2 b_1 b_2 + 4a_1 a_3 b_1 b_3 - 2b_1{}^2 \\ 4a_1{}^2 b_1 b_2 + 4a_1 a_2 b_2{}^2 - 2a_1 a_2 + 4a_1 a_3 b_2 b_3 - 2b_1 b_2 \\ 4a_1{}^2 b_1 b_3 + 4a_1 a_2 b_2 b_3 + 4a_1 a_3 b_3{}^2 - 2a_1 a_3 - 2b_1 b_3 \end{bmatrix}$$

$$\begin{matrix} 4a_1 a_2 b_1{}^2 - 2a_1 a_2 + 4a_2{}^2 b_1 b_2 + 4a_2 a_3 b_1 b_3 - 2b_1 b_2 \\ 1 + 4a_1 a_2 b_1 b_2 + 4a_2{}^2 b_2{}^2 - 2a_2{}^2 + 4a_2 a_3 b_2 b_3 - 2b_2{}^2 \\ 4a_1 a_2 b_1 b_3 + 4a_2{}^2 b_2 b_3 + 4a_2 a_3 b_3{}^2 - 2a_2 a_3 - 2b_2 b_3 \end{matrix}$$

$$\begin{matrix} 4a_1 a_3 b_1{}^2 - 2a_1 a_3 + 4a_2 a_3 b_1 b_2 + 4a_3{}^2 b_1 b_3 - 2b_1 b_3 \\ 4a_1 a_3 b_1 b_2 + 4a_2 a_3 b_2{}^2 - 2a_2 a_3 + 4a_3{}^2 b_2 b_3 - 2b_2 b_3 \\ 1 + 4a_1 a_3 b_1 b_3 + 4a_2 a_3 b_2 b_3 + 4a_3{}^2 b_3{}^2 - 2a_3{}^2 - 2b_3{}^2 \end{matrix} \Biggr].\qquad \text{(F.5)}$$

Finally, we can confirm that if we insert Eq. (F.1) into the standard equation Eq. (2.14) constructing an **SO**(3) rotation matrix from a quaternion, the matrix reduces exactly to the Clifford formula Eq. (F.5) for $\mathbf{R}(\mathbf{A}, \mathbf{B})$ after some manipulations to enforce $\mathbf{A} \cdot \mathbf{A} = \mathbf{B} \cdot \mathbf{B} = 1$. We thus see the remarkable fact that, while we considered the quaternion formula Eq. (2.14) defining the 3D rotation $R(q)$ as a fundamental quadratic expression implying the quaternion was a *square root* of a rotation, it would be equally valid to consider Eq. (F.5) as even more fundamental, displaying the property of the Clifford representation as the *fourth root* of a rotation.

Example code for rotation to quaternion algorithm

This appendix reviews the long-standard algorithm for determining a usable quaternion from an input consisting of exactly orthonormal 3D rotation matrix data (see, e.g., Shepperd (1978); Shuster and Natanson (1993); Sarabandi and Thomas (2019), or Section 16.1 of Hanson (2006)). This basic procedure accounts for all possible anomalies, however rare, checking for assorted zeroes and small numbers, but it depends on *exact data*, and may not perform well on rotation data with significant measurement errors. For errorful data, one should use instead the Bar-Itzhack algorithm in Chapter 9, which always returns the quaternion (or adjugate quaternion) producing the orthonormal $\mathbf{SO}(3)$ rotation matrix closest to the approximate rotation matrix data. The standard algorithm is documented below in four forms: abstract pseudocode, a C language algorithm, a Python program, and a Mathematica language implementation.

G.1 Pseudocode

Compute the trace of R: $\mathrm{tr} = \mathrm{Trace}(R)$, noting that $\mathrm{tr} = 1 + 2\cos\theta$.

(if)   *if* $\mathrm{tr} > 0$, then set $q_0 = \dfrac{\sqrt{\mathrm{tr}+1}}{2} = \cos(\theta/2)$.

Compute $s = (2\sqrt{\mathrm{tr}+1})^{-1} = (4\cos(\theta/2))^{-1}$, then
set $q_1 = s * (m_{32} - m_{23})$, $q_2 = s * (m_{13} - m_{31})$, $q_3 = s * (m_{21} - m_{12})$

(else)   *else* check m_{ii}:

if m_{11} is largest, set $s = \sqrt{m_{11} - m_{22} - m_{33} + 1}$
   set $q_1 = s/2$, set $s = 1/(2s)$
   set $q_0 = s * (m_{3,2} - m_{2,3})$, $q_2 = s * (m_{2,1} - m_{1,2})$, $q_3 = s * (m_{1,3} - m_{3,1})$
if m_{22} is largest, set $s = \sqrt{m_{22} - m_{33} - m_{11} + 1}$
   set $q_2 = s/2$, set $s = 1/(2s)$
   set $q_0 = s * (m_{1,3} - m_{3,1})$, $q_3 = s * (m_{3,2} - m_{2,3})$, $q_1 = s * (m_{2,1} - m_{1,2})$
if m_{33} is largest, set $s = \sqrt{m_{33} - m_{11} - m_{22} + 1}$
   set $q_3 = s/2$, set $s = 1/(2s)$
   set $q_0 = s * (m_{2,1} - m_{1,2})$, $q_1 = s * (m_{1,3} - m_{3,1})$, $q_2 = s * (m_{3,2} - m_{2,3})$
Normalize to unity.

G.2 C code

```c
typedef struct tag_Quat {double w, x, y, z;} Quat;
/* quat->w is the scalar component, translated here
   internally as q[3] to facilitate the manipulation of
   the vector, or imaginary component, using indices 0,1,2 */

MatToQuat(double m[4][4], Quat * quat)
{ double  tr, s, q[4];    int  i, j, k;    int nxt[3] = {1, 2, 0};

  tr = m[0][0] + m[1][1] + m[2][2];

  if (tr > 0.0) {  /* the diagonal  is positive */
    s = sqrt (tr + 1.0);
    quat->w = s / 2.0;
    s = 0.5 / s;
    quat->x = (m[2][1] - m[1][2]) * s;
    quat->y = (m[0][2] - m[2][0]) * s;
    quat->z = (m[1][0] - m[0][1]) * s;
  } else {     /* the diagonal is nonnegative */
    i = 0;
    if (m[1][1] > m[0][0]) i = 1;
    if (m[2][2] > m[i][i]) i = 2;
    j = nxt[i];
    k = nxt[j];

    s = sqrt ((m[i][i] - (m[j][j] + m[k][k])) + 1.0);

    q[i] = s * 0.5;

    if (s != 0.0) s = 0.5 / s;

    q[3] = (m[k][j] - m[j][k]) * s;
    q[j] = (m[i][j] + m[j][i]) * s;
    q[k] = (m[i][k] + m[k][i]) * s;

    quat->x = q[0];
    quat->y = q[1];
    quat->z = q[2];
    quat->w = q[3];
  }
}
```

G.3 Python code

(Adapted from https://github.com/cvlab-epfl/single-stage-pose/blob/master/dataset.py.)

```python
import numpy as np
from torch.utils.data import Dataset
import torch
import math
import random
import os

def rotation2quaternion(M):
    tr = np.trace(M)
    m = M.reshape(-1)
    if tr > 0:
        s = np.sqrt(tr + 1.0) * 2
        w = 0.25 * s
        x = (m[7] - m[5]) / s
        y = (m[2] - m[6]) / s
        z = (m[3] - m[1]) / s
    elif m[0] > m[4] and m[0] > m[8]:
        s = np.sqrt(1.0 + m[0] - m[4] - m[8]) * 2
        w = (m[7] - m[5]) / s
        x = 0.25 * s
        y = (m[1] + m[3]) / s
        z = (m[2] + m[6]) / s
    elif m[4] > m[8]:
        s = np.sqrt(1.0 + m[4] - m[0] - m[8]) * 2
        w = (m[2] - m[6]) / s
        x = (m[1] + m[3]) / s
        y = 0.25 * s
        z = (m[5] + m[7]) / s
    else:
        s = np.sqrt(1.0 + m[8] - m[0] - m[4]) * 2
        w = (m[3] - m[1]) / s
        x = (m[2] + m[6]) / s
        y = (m[5] + m[7]) / s
        z = 0.25 * s
    Q = np.array([w, x, y, z]).reshape(-1)
    return Q
```

G.4 Mathematica code

```
(* This code randomizes the sign of q0 in the final step. This is optional. *)

RotToQuat[mat_] := Module[{qinit, q0, q1, q2, q3, trace, s, t1, t2, t3},
    trace = Sum [mat[[i, i]], {i,1,3}];
    If[trace > 0,
        s = Sqrt[trace + 1];
        q0 = s/2; s = 1/(2 s);
        q1 = (mat[[3, 2]] - mat[[2, 3]]) s;
        q2 = (mat[[1, 3]] - mat[[3, 1]]) s;
        q3 = (mat[[2, 1]] - mat[[1, 2]]) s,
        If[mat[[1, 1]] >= mat[[2, 2]] && mat[[1, 1]] >= mat[[3, 3]],
            s = Sqrt[mat[[1, 1]] - mat[[2, 2]] - mat[[3, 3]] + 1];
            q1 = s/2; s = 1/(2 s);
            q0 = (mat[[3, 2]] - mat[[2, 3]]) s;
            q2 = (mat[[2, 1]] + mat[[1, 2]]) s;
            q3 = (mat[[1, 3]] + mat[[3, 1]]) s,
            If[mat[[1, 1]] < mat[[2, 2]] && mat[[1, 1]] >= mat[[3, 3]],
                s = Sqrt[mat[[2, 2]] - mat[[3, 3]] - mat[[1, 1]] + 1 ];
                q2 = s/2; s = 1/(2 s);
                q0 = (mat[[1, 3]] - mat[[3, 1]]) s;
                q3 = (mat[[3, 2]] + mat[[2, 3]]) s;
                q1 = (mat[[2, 1]] + mat[[1, 2]]) s,
                s = Sqrt[mat[[3, 3]] - mat[[1, 1]] - mat[[2, 2]] + 1 ];
                q3 = s/2; s = 1/(2 s);
                q0 = (mat[[2, 1]] - mat[[1, 2]]) s;
                q1 = (mat[[1, 3]] + mat[[3, 1]]) s;
                q2 = (mat[[3, 2]] + mat[[2, 3]]) s]]];
    qinit = N[{q0, q1, q2, q3}];
    qinit = If[Abs[q0]  < 1/(10^10) , qinit, Sign[q0] qinit];
    qinit = (1 - 2 RandomInteger[{0, 1}]) qinit;
    Normalize[qinit]]
```

Selected details of the 3D perspective solution

In Chapter 15, we presented two alternative loss expressions for perspective projection pose estimation that avoided the difficulties of the usual perspective division. We review these here for convenience, using $\hat{\mathbf{v}} = [u, v, 1]^t/\sqrt{1 + u^2 + v^2}$ to define the LHM loss Eq. (15.60)

$$S_{\text{3D LHM perspective}}(q) = \sum_{k=1}^{K} \left\| (R(q) \cdot \mathbf{x}_k + \mathbf{C}) - \hat{\mathbf{v}}_k \left(\hat{\mathbf{v}}_k \cdot (R(q) \cdot \mathbf{x}_k + \mathbf{C}) \right) \right\|^2 \tag{H.1}$$

which employs the Gram-Schmidt nearest-point distance in object space in place of the image-space distance in the $z = 1$ plane. The Z-prime loss Eq. (15.65) is simpler, needing only $\mathbf{v} = [u, v, 1]^t$ to define a horizontal error displacement measure, also in object space,

$$S_{\text{3D Z-prime perspective}}(q) = \sum_{k=1}^{K} \left\| (R(q) \cdot \mathbf{x}_k + \mathbf{C}) - \mathbf{v}(D(q) \cdot \mathbf{x}_k + c) \right\|^2 . \tag{H.2}$$

The abstract symbolic problem to be minimized over the quaternion adjugate rotation variables q_{ij} takes the exact same form for both loss expressions,

$$S_{\text{3D perspective}}(q_{ij}) =$$

$$a_{01} + a_{02}q_{12} + a_{03}q_{12}^2 + a_{04}q_{13} + a_{05}q_{12}q_{13} + a_{06}q_{13}^2 + a_{07}q_{23} + a_{08}q_{12}q_{23} +$$
$$a_{09}q_{13}q_{23} + a_{10}q_{23}^2 + a_{11}q_{03} + a_{12}q_{03}q_{12} + a_{13}q_{03}q_{13} + a_{14}q_{03}q_{23} + a_{15}q_{03}^2 + a_{16}q_{02} +$$
$$a_{17}q_{02}q_{12} + a_{18}q_{02}q_{13} + a_{19}q_{02}q_{23} + a_{20}q_{02}q_{03} + a_{21}q_{02}^2 + a_{22}q_{01} + a_{23}q_{01}q_{12} + a_{24}q_{01}q_{13} +$$
$$a_{25}q_{01}q_{23} + a_{26}q_{01}q_{03} + a_{27}q_{01}q_{02} + a_{28}q_{01}^2 + a_{29}q_{33} + a_{30}q_{12}q_{33} + a_{31}q_{13}q_{33} + a_{32}q_{23}q_{33} +$$
$$a_{33}q_{03}q_{33} + a_{34}q_{02}q_{33} + a_{35}q_{01}q_{33} + a_{36}q_{33}^2 + a_{37}q_{22} + a_{38}q_{12}q_{22} + a_{39}q_{13}q_{22} + a_{40}q_{22}q_{23} +$$
$$a_{41}q_{03}q_{22} + a_{42}q_{02}q_{22} + a_{43}q_{01}q_{22} + a_{44}q_{22}q_{33} + a_{45}q_{22}^2 + a_{46}q_{11} + a_{47}q_{11}q_{12} + a_{48}q_{11}q_{13} +$$
$$a_{49}q_{11}q_{23} + a_{50}q_{03}q_{11} + a_{51}q_{02}q_{11} + a_{52}q_{01}q_{11} + a_{53}q_{11}q_{33} + a_{54}q_{11}q_{22} + a_{55}q_{11}^2 + a_{56}q_{00} +$$
$$a_{57}q_{00}q_{12} + a_{58}q_{00}q_{13} + a_{59}q_{00}q_{23} + a_{60}q_{00}q_{03} + a_{61}q_{00}q_{02} + a_{62}q_{00}q_{01} +$$
$$a_{63}q_{00}q_{33} + a_{64}q_{00}q_{22} + a_{65}q_{00}q_{11} + a_{66}q_{00}^2 , \tag{H.3}$$

with the caveat that the 66 expressions $a_{ij}(xyz{:}uv)$ for the two cases are entirely different functions of the generalized cross-covariances built from the input cloud and image-plane data. Solving the minimization problem of Eq. (H.3) for the q_{ij} as functions of the abstract a_{ij} gives the same result for both cases, and yet when the distinct numbers $a_{ij}(xyz{:}uv)$ are inserted, the numerical results for the q_{ij} with error-free input data are identical.

519

Details of the LHM coefficients. Here we expand in full the coefficients in the 3D LHM object space perspective projection loss function of Eq. (15.61), expressing the algebraic sums in Eq. (15.63), for example, as

$$\mathrm{uy} = \sum_{k=1}^{K} \frac{u_k y_k}{u_k^2 + v_k^2 + 1} \,, \qquad \mathrm{uxy} = \sum_{k=1}^{K} \frac{u_k x_k y_k}{u_k^2 + v_k^2 + 1} \,, \qquad \mathrm{uvxy} = \sum_{k=1}^{K} \frac{u_k v_k x_k y_k}{u_k^2 + v_k^2 + 1} \,,$$

in the code-like format of Eq. (15.64). Note that we assume that the center of mass $\mathbf{C} = [0, 0, c]^t$ of the rotated cloud is a supplied parameter like $\mathbf{X} = \{x, y, z\}$ and $\mathbf{U} = \{u, v\}$. We have

$$a_{01} = c**2 * (\mathrm{uu} + \mathrm{vv}) \,,$$
$$a_{02} = -4 * c * (\mathrm{uy} + \mathrm{vx}) \,,$$
$$a_{03} = 4 * (\mathrm{uuxx} - 2 * \mathrm{uvxy} + \mathrm{vvyy} + \mathrm{xx} + \mathrm{yy}) \,,$$
$$a_{04} = 4 * c * (\mathrm{uux} - \mathrm{uz} + \mathrm{vvx}) \,,$$
$$a_{05} = -8 * (\mathrm{uvxz} + \mathrm{uxy} - \mathrm{vvyz} + \mathrm{vxx} - \mathrm{yz}) \,,$$
$$a_{06} = 4 * (\mathrm{uuxx} - 2 * \mathrm{uxz} + \mathrm{vvxx} + \mathrm{vvzz} + \mathrm{zz}) \,,$$
$$a_{07} = 4 * c * (\mathrm{uuy} + \mathrm{vvy} - \mathrm{vz}) \,,$$
$$a_{08} = 8 * (\mathrm{uuxz} - \mathrm{uvyz} - \mathrm{uyy} - \mathrm{vxy} + \mathrm{xz}) \,,$$
$$a_{09} = 8 * (\mathrm{uuxy} - \mathrm{uvzz} - \mathrm{uyz} + \mathrm{vvxy} - \mathrm{vxz}) \,,$$
$$a_{10} = 4 * (\mathrm{uuyy} + \mathrm{uuzz} + \mathrm{vvyy} - 2 * \mathrm{vyz} + \mathrm{zz}) \,,$$
$$a_{11} = 4 * c * (\mathrm{uy} - \mathrm{vx}) \,,$$
$$a_{12} = 8 * (\mathrm{uuxx} - \mathrm{vvyy} + \mathrm{xx} - \mathrm{yy}) \,,$$
$$a_{13} = -8 * (\mathrm{uvxz} - \mathrm{uxy} + \mathrm{vvyz} + \mathrm{vxx} + \mathrm{yz}) \,,$$
$$a_{14} = 8 * (\mathrm{uuxz} + \mathrm{uvyz} + \mathrm{uyy} - \mathrm{vxy} + \mathrm{xz}) \,,$$
$$a_{15} = 4 * (\mathrm{uuxx} + 2 * \mathrm{uvxy} + \mathrm{vvyy} + \mathrm{xx} + \mathrm{yy}) \,,$$
$$a_{16} = -4 * c * (\mathrm{uux} + \mathrm{uz} + \mathrm{vvx}) \,,$$
$$a_{17} = 8 * (-\mathrm{uvxz} + \mathrm{uxy} + \mathrm{vvyz} + \mathrm{vxx} + \mathrm{yz}) \,,$$
$$a_{18} = -8 * (\mathrm{uuxx} + \mathrm{vvxx} - \mathrm{vvzz} - \mathrm{zz}) \,,$$
$$a_{19} = -8 * (\mathrm{uuxy} + \mathrm{uvzz} + \mathrm{uyz} + \mathrm{vvxy} - \mathrm{vxz}) \,,$$
$$a_{20} = -8 * (\mathrm{uvxz} + \mathrm{uxy} + \mathrm{vvyz} - \mathrm{vxx} + \mathrm{yz}) \,,$$
$$a_{21} = 4 * (\mathrm{uuxx} + 2 * \mathrm{uxz} + \mathrm{vvxx} + \mathrm{vvzz} + \mathrm{zz}) \,,$$
$$a_{22} = 4 * c * (\mathrm{uuy} + \mathrm{vvy} + \mathrm{vz}) \,,$$
$$a_{23} = -8 * (\mathrm{uuxz} - \mathrm{uvyz} + \mathrm{uyy} + \mathrm{vxy} + \mathrm{xz}) \,,$$
$$a_{24} = 8 * (\mathrm{uuxy} + \mathrm{uvzz} - \mathrm{uyz} + \mathrm{vvxy} + \mathrm{vxz}) \,,$$
$$a_{25} = 8 * (\mathrm{uuyy} - \mathrm{uuzz} + \mathrm{vvyy} - \mathrm{zz}) \,,$$
$$a_{26} = -8 * (\mathrm{uuxz} + \mathrm{uvyz} - \mathrm{uyy} + \mathrm{vxy} + \mathrm{xz}) \,,$$
$$a_{27} = -8 * (\mathrm{uuxy} - \mathrm{uvzz} + \mathrm{uyz} + \mathrm{vvxy} + \mathrm{vxz}) \,,$$
$$a_{28} = 4 * (\mathrm{uuyy} + \mathrm{uuzz} + \mathrm{vvyy} + 2 * \mathrm{vyz} + \mathrm{zz}) \,,$$
$$a_{29} = 2 * c * (\mathrm{uuz} + \mathrm{ux} + \mathrm{vvz} + \mathrm{vy}) \,,$$
$$a_{30} = -4 * (\mathrm{uuxy} - \mathrm{uvxx} - \mathrm{uvyy} + \mathrm{uyz} + \mathrm{vvxy} + \mathrm{vxz} + 2 * \mathrm{xy}) \,,$$

$$a_{31} = 4 * (\text{uuxz} + \text{uvyz} + \text{uxx} - \text{uzz} + \text{vxy} - \text{xz}) \,,$$

$$a_{32} = 4 * (\text{uvxz} + \text{uxy} + \text{vvyz} + \text{vyy} - \text{vzz} - \text{yz}) \,,$$

$$a_{33} = -4 * (\text{uuxy} - \text{uvxx} + \text{uvyy} - \text{uyz} - \text{vvxy} + \text{vxz}) \,,$$

$$a_{34} = -4 * (\text{uuxz} - \text{uvyz} + \text{uxx} + \text{uzz} + 2 * \text{vvxz} + \text{vxy} + \text{xz}) \,,$$

$$a_{35} = 4 * (2 * \text{uuyz} - \text{uvxz} + \text{uxy} + \text{vvyz} + \text{vyy} + \text{vzz} + \text{yz}) \,,$$

$$a_{36} = \text{uuyy} + \text{uuzz} - 2 * \text{uvxy} + 2 * \text{uxz} + \text{vvxx} + \text{vvzz} + 2 * \text{vyz} + \text{xx} + \text{yy} \,,$$

$$a_{37} = -2 * c * (\text{uuz} - \text{ux} + \text{vvz} + \text{vy}) \,,$$

$$a_{38} = 4 * (\text{uuxy} + \text{uvxx} - \text{uvyy} + \text{uyz} - \text{vvxy} + \text{vxz}) \,,$$

$$a_{39} = -4 * (\text{uuxz} + \text{uvyz} - \text{uxx} - \text{uzz} + 2 * \text{vvxz} + \text{vxy} + \text{xz}) \,,$$

$$a_{40} = 4 * (\text{uvxz} + \text{uxy} - \text{vvyz} - \text{vyy} + \text{vzz} + \text{yz}) \,,$$

$$a_{41} = 4 * (\text{uuxy} + \text{uvxx} + \text{uvyy} - \text{uyz} + \text{vvxy} + \text{vxz} + 2 * \text{xy}) \,,$$

$$a_{42} = 4 * (\text{uuxz} - \text{uvyz} - \text{uxx} + \text{uzz} + \text{vxy} - \text{xz}) \,,$$

$$a_{43} = -4 * (2 * \text{uuyz} + \text{uvxz} - \text{uxy} + \text{vvyz} + \text{vyy} + \text{vzz} + \text{yz}) \,,$$

$$a_{44} = -2 * (\text{uuyy} + \text{uuzz} - \text{vvxx} + \text{vvzz} + 2 * \text{vyz} - \text{xx} + \text{yy}) \,,$$

$$a_{45} = \text{uuyy} + \text{uuzz} + 2 * \text{uvxy} - 2 * \text{uxz} + \text{vvxx} + \text{vvzz} + 2 * \text{vyz} + \text{xx} + \text{yy} \,,$$

$$a_{46} = -2 * c * (\text{uuz} + \text{ux} + \text{vvz} - \text{vy}) \,,$$

$$a_{47} = 4 * (-\text{uuxy} - \text{uvxx} + \text{uvyy} + \text{uyz} + \text{vvxy} + \text{vxz}) \,,$$

$$a_{48} = 4 * (-\text{uuxz} + \text{uvyz} - \text{uxx} + \text{uzz} + \text{vxy} + \text{xz}) \,,$$

$$a_{49} = -4 * (2 * \text{uuyz} + \text{uvxz} + \text{uxy} + \text{vvyz} - \text{vyy} - \text{vzz} + \text{yz}) \,,$$

$$a_{50} = -4 * (\text{uuxy} + \text{uvxx} + \text{uvyy} + \text{uyz} + \text{vvxy} - \text{vxz} + 2 * \text{xy}) \,,$$

$$a_{51} = 4 * (\text{uuxz} + \text{uvyz} + \text{uxx} + \text{uzz} + 2 * \text{vvxz} - \text{vxy} + \text{xz}) \,,$$

$$a_{52} = 4 * (\text{uvxz} - \text{uxy} - \text{vvyz} + \text{vyy} - \text{vzz} + \text{yz}) \,,$$

$$a_{53} = 2 * (\text{uuyy} - \text{uuzz} - 2 * \text{uxz} - \text{vvxx} - \text{vvzz} - \text{xx} + \text{yy}) \,,$$

$$a_{54} = -2 * (\text{uuyy} - \text{uuzz} + 2 * \text{uvxy} + \text{vvxx} - \text{vvzz} + \text{xx} + \text{yy}) \,,$$

$$a_{55} = \text{uuyy} + \text{uuzz} + 2 * \text{uvxy} + 2 * \text{uxz} + \text{vvxx} + \text{vvzz} - 2 * \text{vyz} + \text{xx} + \text{yy} \,,$$

$$a_{56} = 2 * c * (\text{uuz} - \text{ux} + \text{vvz} - \text{vy}) \,,$$

$$a_{57} = 4 * (\text{uuxy} - \text{uvxx} - \text{uvyy} - \text{uyz} + \text{vvxy} - \text{vxz} + 2 * \text{xy}) \,,$$

$$a_{58} = 4 * (\text{uuxz} - \text{uvyz} - \text{uxx} - \text{uzz} + 2 * \text{vvxz} - \text{vxy} + \text{xz}) \,,$$

$$a_{59} = 4 * (2 * \text{uuyz} - \text{uvxz} - \text{uxy} + \text{vvyz} - \text{vyy} - \text{vzz} + \text{yz}) \,,$$

$$a_{60} = 4 * (\text{uuxy} - \text{uvxx} + \text{uvyy} + \text{uyz} - \text{vvxy} - \text{vxz}) \,,$$

$$a_{61} = -4 * (\text{uuxz} + \text{uvyz} - \text{uxx} + \text{uzz} - \text{vxy} - \text{xz}) \,,$$

$$a_{62} = 4 * (\text{uvxz} - \text{uxy} + \text{vvyz} - \text{vyy} + \text{vzz} - \text{yz}) \,,$$

$$a_{63} = -2 * (\text{uuyy} - \text{uuzz} - 2 * \text{uvxy} + \text{vvxx} - \text{vvzz} + \text{xx} + \text{yy}) \,,$$

$$a_{64} = 2 * (\text{uuyy} - \text{uuzz} + 2 * \text{uxz} - \text{vvxx} - \text{vvzz} - \text{xx} + \text{yy}) \,,$$

$$a_{65} = -2 * (\text{uuyy} + \text{uuzz} - \text{vvxx} + \text{vvzz} - 2 * \text{vyz} - \text{xx} + \text{yy}) \,,$$

$$a_{66} = \text{uuyy} + \text{uuzz} - 2 * \text{uvxy} - 2 * \text{uxz} + \text{vvxx} + \text{vvzz} - 2 * \text{vyz} + \text{xx} + \text{yy} \,.$$

Details of the Z-prime coefficients. The alternative 3D perspective loss function Eq. (H.2) produces exactly the same abstract algebraic loss function, Eq. (H.3), as does Eq. (H.1), and thus *exactly* the same closed form algebraic solution for the q_{ij} rotation parameters in terms of the a_{ij}. However, many of the coefficients differ. The individual data-content terms lack the divisor and, for example, take the form

$$\text{uy} = \sum_{k=1}^{K} u_k y_k \,, \qquad \text{uxy} = \sum_{k=1}^{K} u_k x_k y_k \,, \qquad \text{uvxy} = \sum_{k=1}^{K} u_k v_k x_k y_k \,.$$

Again in the code-like format of Eq. (15.64), the corresponding full list of the algebraic sums in Eq. (15.67) for the Z-prime system is the following:

$$a_{01} = c **2 * (\text{uu} + \text{vv}) \,,$$
$$a_{02} = -4 * c * (\text{uy} + \text{vx}) \,,$$
$$a_{03} = 4 * (\text{xx} + \text{yy}) \,,$$
$$a_{04} = 4 * c * (\text{uux} - \text{uz} + \text{vvx}) \,,$$
$$a_{05} = -8 * (\text{uxy} + \text{vxx} - \text{yz}) \,,$$
$$a_{06} = 4 * (\text{uuxx} - 2 * \text{uxz} + \text{vvxx} + \text{zz}) \,,$$
$$a_{07} = 4 * c(\text{uuy} + \text{vvy} - \text{vz}) \,,$$
$$a_{08} = -8 * (\text{uyy} + \text{vxy} - \text{xz}) \,,$$
$$a_{09} = 8 * (\text{uuxy} - \text{uyz} + \text{vvxy} - \text{vxz}) \,,$$
$$a_{10} = 4 * (\text{uuyy} + \text{vvyy} - 2 * \text{vyz} + \text{zz}) \,,$$
$$a_{11} = -4 * c * (\text{vx} - \text{uy}) \,,$$
$$a_{12} = 8 * (\text{xx} - \text{yy}) \,,$$
$$a_{13} = -8 * (-\text{uxy} + \text{vxx} + \text{yz}) \,,$$
$$a_{14} = -8 * (-\text{uyy} + \text{vxy} - \text{xz}) \,,$$
$$a_{15} = 4 * (\text{xx} + \text{yy}) \,,$$
$$a_{16} = -4 * c * (\text{uux} + \text{uz} + \text{vvx}) \,,$$
$$a_{17} = 8 * (\text{uxy} + \text{vxx} + \text{yz}) \,,$$
$$a_{18} = -8 * (\text{uuxx} + \text{vvxx} - \text{zz}) \,,$$
$$a_{19} = -8 * (\text{uuxy} + \text{uyz} + \text{vvxy} - \text{vxz}) \,,$$
$$a_{20} = 8 * (-\text{uxy} + \text{vxx} - \text{yz}) \,,$$
$$a_{21} = 4 * (\text{uuxx} + 2 * \text{uxz} + \text{vvxx} + \text{zz}) \,,$$
$$a_{22} = 4 * c * (\text{uuy} + \text{vvy} + \text{vz}) \,,$$
$$a_{23} = -8 * (\text{uyy} + \text{vxy} + \text{xz}) \,,$$
$$a_{24} = 8 * (\text{uuxy} - \text{uyz} + \text{vvxy} + \text{vxz}) \,,$$
$$a_{25} = 8 * (\text{uuyy} + \text{vvyy} - \text{zz}) \,,$$
$$a_{26} = -8 * (-\text{uyy} + \text{vxy} + \text{xz}) \,,$$
$$a_{27} = -8 * (\text{uuxy} + \text{uyz} + \text{vvxy} + \text{vxz}) \,,$$
$$a_{28} = 4 * (\text{uuyy} + \text{vvyy} + 2 * \text{vyz} + \text{zz}) \,,$$
$$a_{29} = 2 * c * (\text{uuz} + \text{ux} + \text{vvz} + \text{vy}) \,,$$

$$a_{30} = -4 * (uyz + vxz + 2 * xy) \,,$$
$$a_{31} = 4 * (uuxz + uxx - uzz + vvxz + vxy - xz) \,,$$
$$a_{32} = 4 * (uuyz + uxy + vvyz + vyy - vzz - yz) \,,$$
$$a_{33} = -4 * (vxz - uyz) \,,$$
$$a_{34} = -4 * (uuxz + uxx + uzz + vvxz + vxy + xz) \,,$$
$$a_{35} = 4 * (uuyz + uxy + vvyz + vyy + vzz + yz) \,,$$
$$a_{36} = uuzz + 2 * uxz + vvzz + 2 * vyz + xx + yy \,,$$
$$a_{37} = -2 * c * (uuz - ux + vvz + vy) \,,$$
$$a_{38} = 4 * (uyz + vxz) \,,$$
$$a_{39} = -4 * (uuxz - uxx - uzz + vvxz + vxy + xz) \,,$$
$$a_{40} = -4 * (uuyz - uxy + vvyz + vyy - vzz - yz) \,,$$
$$a_{41} = 4 * (-uyz + vxz + 2 * xy) \,,$$
$$a_{42} = 4 * (uuxz - uxx + uzz + vvxz + vxy - xz) \,,$$
$$a_{43} = -4 * (uuyz - uxy + vvyz + vyy + vzz + yz) \,,$$
$$a_{44} = 2 * (-uuzz - vvzz - 2 * vyz + xx - yy) \,,$$
$$a_{45} = uuzz - 2 * uxz + vvzz + 2 * vyz + xx + yy \,,$$
$$a_{46} = -2 * c * (uuz + ux + vvz - vy) \,,$$
$$a_{47} = 4 * (uyz + vxz) \,,$$
$$a_{48} = -4 * (uuxz + uxx - uzz + vvxz - vxy - xz) \,,$$
$$a_{49} = -4 * (uuyz + uxy + vvyz - vyy - vzz + yz) \,,$$
$$a_{50} = -4 * (uyz - vxz + 2 * xy) \,,$$
$$a_{51} = 4 * (uuxz + uxx + uzz + vvxz - vxy + xz) \,,$$
$$a_{52} = -4 * (uuyz + uxy + vvyz - vyy + vzz - yz) \,,$$
$$a_{53} = -2 * (uuzz + 2 * uxz + vvzz + xx - yy) \,,$$
$$a_{54} = -2 * (-uuzz - vvzz + xx + yy) \,,$$
$$a_{55} = uuzz + 2 * uxz + vvzz - 2 * vyz + xx + yy \,,$$
$$a_{56} = 2 * c * (uuz - ux + vvz - vy) \,,$$
$$a_{57} = 4 * (-uyz - vxz + 2 * xy) \,,$$
$$a_{58} = 4 * (uuxz - uxx - uzz + vvxz - vxy + xz) \,,$$
$$a_{59} = 4 * (uuyz - uxy + vvyz - vyy - vzz + yz) \,,$$
$$a_{60} = -4 * (vxz - uyz) \,,$$
$$a_{61} = -4 * (uuxz - uxx + uzz + vvxz - vxy - xz) \,,$$
$$a_{62} = 4 * (uuyz - uxy + vvyz - vyy + vzz - yz) \,,$$
$$a_{63} = -2 * (-uuzz - vvzz + xx + yy) \,,$$
$$a_{64} = -2 * (uuzz - 2 * uxz + vvzz + xx - yy) \,,$$
$$a_{65} = 2 * (-uuzz - vvzz + 2 * vyz + xx - yy) \,,$$
$$a_{66} = uuzz - 2 * uxz + vvzz - 2 * vyz + xx + yy \,.$$

Exact cloud matching rotations in any dimension

I

In the main text, we have shown that for the error-free 3D-to-3D rotation matching problem, and the closely related 3D-to-orthographically-projected-2D pose extraction problem, we can solve the least-squares loss problem explicitly with ratios of cross-covariance-related determinants. In this appendix, we prove that for the cloud matching and projected pose matching problems in *any dimension* N, we can also immediately write down the optimal $\mathbf{SO}(N)$ rotations that align error-free data in terms of such determinants. While exact solutions are available for noisy data in 2D, 3D, and 4D, in general only the SVD and matrix-square-root approaches are available to obtain provably optimal rotations for noisy data in all dimensions. However, if we are able to restrict ourselves to error-free data and orthographic projections, our 3D exact methods extend to any dimension.

The N-dimensional rotation matching problem. The argument is as follows: again, assume a reference data set $\{\mathbf{X}\}$, which we now take to be a cloud of K N-dimensional points in Euclidean space, and then assume a rotation R in $\mathbf{SO}(N)$ that creates a measured set $\{\mathbf{U}\}$ of K N-dimensional test points by rotations as $u^k{}_a = \sum_b R_{ab} x^k{}_b$. Our problem is to take the measured points $\{\mathbf{U}\}$ as input and discover the optimal rotation R without knowing it in advance. Our first task is to treat the test set $\{\mathbf{U}\}$ as N-dimensional, so we have a cloud-to-cloud alignment problem between a matched pair of N-dimensional clouds. By induction from the determinants in Eq. (15.23) and their assembly into the rotation matrix Eq. (15.22), we hypothesize that the desired rotation matrix can be computed from the $(2N \times 2N)$-dimensional symmetric cross-covariance matrix; since we will not need the self-covariances of the $\{\mathbf{U}\}$ variables, we actually only use the top half. Defining the inner products over the K individual vector components as

$$x_{ij} = \sum_{k=1}^{K} [x^k]_i\,[x^k]_j\,, \qquad u_{ij} = \sum_{k=1}^{K} [x^k]_i\,[u^k]_j\,, \tag{I.1}$$

the needed $N \times 2N$ cross-covariance matrix extending Eq. (15.18) to clouds of size K embedded in N spatial dimensions can be written

$$C_N = \begin{bmatrix} x_{11} & x_{12} & \cdots & x_{1N} & u_{11} & u_{12} & \cdots & u_{1N} \\ x_{21} & x_{22} & \cdots & x_{2N} & u_{21} & u_{22} & \cdots & u_{2N} \\ \vdots & \vdots & \ddots & \vdots & \vdots & \vdots & \ddots & \vdots \\ x_{N1} & x_{N2} & \cdots & x_{NN} & u_{N1} & u_{N2} & \cdots & u_{NN} \end{bmatrix}. \tag{I.2}$$

For reference, we define the determinant of the N-dimensional self-covariance matrix,

$$d_0 = \det \begin{bmatrix} x_{11} & x_{12} & \cdots & x_{1N} \\ x_{21} & x_{22} & \cdots & x_{2N} \\ \vdots & \vdots & \ddots & \vdots \\ x_{N1} & x_{N2} & \cdots & x_{NN} \end{bmatrix}. \tag{I.3}$$

It is now clear that the form of the proof in Eq. (15.27) can be generalized to any dimension as follows: consider this $N \times N$ matrix and its determinant

$$d_{ab} = \det \begin{bmatrix} x_{11} & x_{12} & \cdots & x_{1,b-1} & u_{1,a} & x_{1,b+1} & \cdots & x_{1N} \\ x_{21} & x_{22} & \cdots & x_{2,b-1} & u_{2,a} & x_{2,b+1} & \cdots & x_{2N} \\ \vdots & \vdots & \ddots & \vdots & \vdots & \vdots & \ddots & \vdots \\ x_{N1} & x_{N2} & \cdots & x_{N,b-1} & u_{N,a} & x_{N,b+1} & \cdots & x_{NN} \end{bmatrix}, \tag{I.4}$$

where d_{ab} is a number, not a matrix, and in the determinant, the a-th set of cross-covariances $u_{c,a}$ replaces the b-th column, that is, $u_{c,a} = \sum_k x_c u_a$ replaces $x_{c,b} = \sum_k x_c x_b$.

We then substitute the explicit construction of $u_a = \sum_{c=1}^{N} r_{ac} x_c$, where $[R]_{ab} = r_{ab}$ are the elements of an orthonormal $N \times N$ **SO**(N) rotation matrix, for each occurrence of u. Clearly multiples of the $N-1$ columns excepting column b can be subtracted from column b, removing all traces of r_{ij} except for r_{ab}, which multiplies the entire column $[x_{1b}, x_{2b}, \ldots, x_{Nb}]^t$. Factoring the coefficient r_{ab} out of the determinant leaves simply

$$d_{ab} = r_{ab} \, d_0 . \tag{I.5}$$

The optimal 3D cloud aligning rotation matrix for error-free data is thus simply

$$R_{\text{opt}}(x_1, \cdots, x_N; u_1, \cdots, u_N)_{ab} = \frac{d_{ab}}{d_0} . \tag{I.6}$$

The N-dimensional orthographic pose problem. All the arguments we made in Section 15.4 go through basically unchanged for N-dimensional reference cloud data projected orthographically to an $(N-1)$-dimensional image after applying a rotation and dropping the last coordinate. Even though we are missing the last set of cloud coordinates $u^k{}_N$, the assumption of orthonormality of the first $(N-1)$ rows for the rotation matrix R requires that the final row can be constructed from the N-dimensional cross product of the first $N-1$ rows from the symbolic vector-valued determinant

$$\begin{bmatrix} r_{N1} & r_{N2} & \cdots & r_{N,N-1} & r_{NN} \end{bmatrix} = \det \begin{bmatrix} r_{11} & r_{12} & \cdots & r_{1,N-1} & r_{1N} \\ r_{21} & r_{22} & \cdots & r_{2,N-1} & r_{2N} \\ \vdots & \vdots & \ddots & \vdots & \vdots \\ r_{N-1,1} & r_{N-1,2} & \cdots & r_{N-1,N-1} & r_{N-1,N} \\ \hat{e}_1 & \hat{e}_2 & \cdots & \hat{e}_{N-1} & \hat{e}_N \end{bmatrix}, \tag{I.7}$$

where the $\hat{\mathbf{e}}_i$ are the Euclidean basis vectors. Alternatively, we can use the construction of the final row of the rotation matrix without using $u^k{}_N$ via the determinants of the form

$$
d_{Na} = \det \begin{bmatrix} u_{11} & u_{12} & \cdots & u_{1,a-1} & x_{1,a} & u_{1,a+1} & \cdots & u_{1N} \\ u_{21} & u_{22} & \cdots & u_{2,a-1} & x_{2,a} & u_{2,a+1} & \cdots & u_{2N} \\ \vdots & \vdots & \ddots & \vdots & \vdots & \vdots & \ddots & \vdots \\ u_{N1} & u_{N2} & \cdots & x_{N,a-1} & x_{N,a} & u_{N,a+1} & \cdots & u_{NN} \end{bmatrix} , \tag{I.8}
$$

for the elements of row N and columns that have the same x_a reference variable repeated in the column a. Replacing each u_a in $u_{b,a} = \sum x_b u_a$ by its original definition $u_a = \sum_{c=1}^{N} r_{ac} x_c$ results in a determinant that factors out the cross product in Eq. (I.7), so

$$
d_{Na} = r_{Na}\, d_0 . \tag{I.9}
$$

Therefore we also have a complete closed form solution in terms of simple ratios of determinants for the N-dimensional cloud rotated and orthographically projected to an $(N-1)$-dimensional image for error-free measurements of the image data.

Details of the Cardano fourth-order eigenvalue solution[☆]

J

Quaternions are 4D unit vectors, and therefore several of the applications that we have described exploit fourth-order eigensystems and their fourth-degree characteristic equations to determine the value of a quaternion. A remarkable fact of mathematics is that one can solve algebraic equations up to the fourth order in terms of elementary functions, but no higher-order equation possesses such a solution. Thus the solution of the quartic published by Cardano in 1545 and investigated further by Euler (1733); Bell (2008 (1733)); Nickalls (2009) (see also Abramowitz and Stegun (1970); Weisstein (2019b); Nickalls (1993); Wikipedia (2023a)) was the highest-order complex algebraic polynomial solution that could ever be found. We will now examine the various properties of the Cardano solution and write down the complete algebraic forms in ways that are useful for solving a variety of mathematical problems involving quaternions. This treatment is based on the "Supporting Information" accompanying Hanson (2020).

Eigenvalue problems. The fundamental context for applications of the Cardano quartic equations is in finding the *algebraic* eigenvalues of various 4×4 matrices that come up in quaternion applications. Being able to examine these eigensystems and solve their quartic polynomial characteristic equations for their eigenvalues *algebraically* can provide deeper insights into the structure of the problem in many ways, for example, by providing differentiable functions for the solutions. Our objective now is to study some features of these results in more detail, and in particular we emphasize *real symmetric matrices*, with and without a trace, since almost every relevant problem that we have encountered reduces to finding the maximal eigenvalues of a matrix in that category.

The eigenvalue expansions. We begin by writing down the characteristic equation $\chi(M)$ of an arbitrary real 4D matrix M as

$$\chi(M) = [M - eI_4] . \tag{J.1}$$

From standard linear algebra (Golub and van Loan, 1983), we know that there are four, not necessarily distinct, eigenvalues of this equation that are determined by solving the polynomial emerging from setting its determinant to zero,

$$\det \chi(M) = \det[M - eI_4] = 0 . \tag{J.2}$$

Here e denotes a generic eigenvalue and I_4 is the 4D identity matrix. Our task is to express the eigenvalues, that is, the solutions to Eq. (J.2), in terms of the elements of the matrix M, and also to find the corresponding eigenvectors.

[☆] This appendix derives much of its content from Hanson (2020).

It is important to note that if M is a real symmetric matrix, as most of our important examples are, the eigenvalues are real. In numerical computations on symmetric matrices, even with a complete algebraic solution, a computer's numerical precision limits can introduce small imaginary components that have to be dealt with in an acceptable way, e.g., by picking a numerical precision and removing imaginary components less than that limit in order to guarantee real eigenvalues that have real eigenvectors.

By expanding Eq. (J.2) in powers of e, we see how the four eigenvalues $e = \epsilon_{k=1,\dots,4}$ depend on the known components of the matrix M and correspond to the solutions of the quartic equations that we can express in two useful forms,

$$e^4 + e^3 p_1 + e^2 p_2 + e p_3 + p_4 = 0 , \tag{J.3}$$

$$(e - \epsilon_1)(e - \epsilon_2)(e - \epsilon_3)(e - \epsilon_4) = 0 . \tag{J.4}$$

Here the p_k are homogeneous polynomials of order k that can be expressed alternatively employing elements of M or elements of E for the 3D and 4D spatial data, or with the corresponding orientation frame data. At this point we want to be as general as possible, and so we note the form valid for all 4×4 matrices M in the expansion of Eq. (J.2) and Eq. (J.3):

$$p_1(M) = -\operatorname{tr}[M] , \tag{J.5}$$

$$p_2(M) = -\frac{1}{2} \operatorname{tr}[M \cdot M] + \frac{1}{2} (\operatorname{tr}[M])^2 , \tag{J.6}$$

$$p_3(M) = -\frac{1}{3} \operatorname{tr}[M \cdot M \cdot M] + \frac{1}{2} \operatorname{tr}[M \cdot M] \operatorname{tr}[M] - \frac{1}{6} (\operatorname{tr}[M])^3 , \tag{J.7}$$

$$p_4(M) = -\frac{1}{4} \operatorname{tr}[M \cdot M \cdot M \cdot M] + \frac{1}{3} \operatorname{tr}[M \cdot M \cdot M] \operatorname{tr}[M] + \frac{1}{8} \operatorname{tr}([M \cdot M])^2$$

$$- \frac{1}{4} \operatorname{tr}[M \cdot M] (\operatorname{tr}[M])^2 + \frac{1}{24} (\operatorname{tr}[M])^4$$

$$= \det[M] . \tag{J.8}$$

Remember that, for our problem, M is just a real symmetric numerical matrix, and the four expressions $p_k(M)$ are just real numbers.

Matching the coefficients of powers of e in Eqs. (J.3) and (J.4), we can also eliminate e to express the matrix data expressions p_k in terms of the symmetric polynomials of the eigenvalues ϵ_k as (Abramowitz and Stegun, 1970)

$$\left. \begin{aligned} p_1 &= -\epsilon_1 - \epsilon_2 - \epsilon_3 - \epsilon_4 \\ p_2 &= \epsilon_1\epsilon_2 + \epsilon_1\epsilon_3 + \epsilon_2\epsilon_3 + \epsilon_1\epsilon_4 + \epsilon_2\epsilon_4 + \epsilon_3\epsilon_4 \\ p_3 &= -\epsilon_1\epsilon_2\epsilon_3 - \epsilon_1\epsilon_2\epsilon_4 - \epsilon_1\epsilon_3\epsilon_4 - \epsilon_2\epsilon_3\epsilon_4 \\ p_4 &= \epsilon_1\epsilon_2\epsilon_3\epsilon_4 \end{aligned} \right\} . \tag{J.9}$$

Both Eq. (J.3) and Eq. (J.9) can in principle be solved directly for the eigenvalues in terms of the matrix data using the solution of the quartic published by Cardano in 1545, given in Abramowitz and Stegun (1970), and investigated further in the references mentioned in the introduction. Applying, e.g., the Mathematica function

```
Solve[ myQuarticEqn[e] == 0, e, Quartics -> True ]
```
$$\tag{J.10}$$

to Eq. (J.3) immediately returns a usable algebraic formula. However, applying `Solve[ ]` to Eq. (J.9) is in fact unsuccessful, although invoking

$$\texttt{Reduce[pkofepsEqns,\{}\epsilon\texttt{[1],}\epsilon\texttt{[2],}\epsilon\texttt{[3],}\epsilon\texttt{[4]\}, Quartics -> True, Cubics -> True]} \qquad \text{(J.11)}$$

can solve Eq. (J.9) iteratively and produces the same final answer that we obtain from Eq. (J.3), as does using a Gröbner basis based on Eq. (J.9).

In Chapter 8, we presented a robust algebraic solution that could be evaluated numerically for the quaternion eigenvalues in the special case of a symmetric traceless 4×4 profile matrix $M(E)$ based on the 3D cross-covariance matrix E; we will complete the steps deriving that solution below. But first we will study general 4×4 real matrices, and then specialize to symmetric matrices with and without a trace, as all of our cases of interest are of this latter type. We note (Golub and van Loan, 1983) that any nonsingular real matrix that can be written in the form $[S^t \cdot S]$ is itself symmetric and has only positive real eigenvalues; in general, the symmetric matrices $[S^t \cdot S]$ and $[S \cdot S^t]$ share one set of eigenvalues, but have distinct eigenvectors. Thus, even if we study only symmetric matrices, we can get significant information about *any* matrix S as long as we can recast our investigation to exploit the associated symmetric matrices $[S^t \cdot S]$ and $[S \cdot S^t]$.

The basic structure: standard algebraic solutions for 4D eigenvalues. When we solve Eq. (J.3) directly using the textbook quartic solution without explicitly imposing restrictions, we find that the general structure for the eigenvalues $e = \epsilon_k(p_1, p_2, p_3, p_4)$ takes the form

$$
\begin{aligned}
\epsilon_1(p) &= -\frac{p_1}{4} + F(p) + G_+(p), & \epsilon_2(p) &= -\frac{p_1}{4} + F(p) - G_+(p) \\[2mm]
\epsilon_3(p) &= -\frac{p_1}{4} - F(p) + G_-(p), & \epsilon_4(p) &= -\frac{p_1}{4} - F(p) - G_-(p)
\end{aligned}
\right\} . \qquad \text{(J.12)}
$$

Here $-p_1 = (\epsilon_1 + \epsilon_2 + \epsilon_3 + \epsilon_4)$ is the trace, and we can see that the "canonical form" for the quartic Eq. (J.3), with a missing cubic term in e, results from simply changing variables from $e \to e + (\epsilon_1 + \epsilon_2 + \epsilon_3 + \epsilon_4)/4$ to effectively add $1/4$ of the trace to each eigenvalue. The other two types of terms have the following explicit expressions in terms of the four independent coefficients p_k:

$$
\left.
\begin{aligned}
F(p_1, p_2, p_3, p_4) &= \sqrt{\frac{p_1^2}{16} - \frac{p_2}{6} + \frac{1}{12}\, t(a, b)} \\[4mm]
G_\pm(p_1, p_2, p_3, p_4) &= \sqrt{\frac{3p_1^2}{16} - \frac{p_2}{2} - F^2(p) \pm \frac{s(p)}{32\, F(p)}} \\[4mm]
&= \sqrt{\frac{p_1^2}{8} - \frac{p_2}{3} - \frac{1}{12}\, t(a, b) \pm \dfrac{s(p)}{32\sqrt{\dfrac{p_1^2}{16} - \dfrac{p_2}{6} + \dfrac{1}{12}\, t(a, b)}}}
\end{aligned}
\right\} \qquad \text{(J.13)}
$$

with

$$
\left.
\begin{aligned}
t(a, b) &= \left(\sqrt[3]{a + \sqrt{-b^2}} + \frac{r^2}{\sqrt[3]{a + \sqrt{-b^2}}} \right) \\
r^2(p_1, p_2, p_3, p_4) &= p_2{}^2 - 3 p_1 p_3 + 12 p_4 = \sqrt[3]{a^2 + b^2} \\
a(p_1, p_2, p_3, p_4) &= p_2{}^3 + \frac{9}{2} \left(3 p_3{}^2 + 3 p_1{}^2 p_4 - p_1 p_2 p_3 - 8 p_2 p_4 \right) \\
b^2(p_1, p_2, p_3, p_4) &= r^6(p) - a^2(p) \\
s(p_1, p_2, p_3, p_4) &= 4 p_1 p_2 - p_1{}^3 - 8 p_3
\end{aligned}
\right\}.
\tag{J.14}
$$

For general real matrices, which may have complex conjugate pairs of eigenvalues, the sign of r^2 can play a critical role, so giving in to the temptation to write

$$
\frac{r^2}{\sqrt[3]{a + \sqrt{-b^2}}} \;\rightarrow\; \sqrt[3]{a - \sqrt{-b^2}}
$$

leads to anomalies; in addition, b^2 can take on any value, so evaluating this algebraic expression numerically while getting the phases of all the roots right can be problematic. So far as we can confirm, setting aside matrices with individual peculiarities, Eq. (J.12) yields correct complex eigenvalues for all real matrices, though the numerical order of the eigenvalues can be irregular. When we restrict our attention to real symmetric matrices, a number of special constraints come into play that significantly improve the numerical behavior of the algebraic solutions, as well as allowing us to simplify the algebraic expression itself. The real symmetric matrices are all that concern us for any of the alignment problems.

Symmetric matrices. We restrict our attention from here on to general symmetric 4×4 real matrices, for which the eigenvalues must be real, and so the roots of the matrix's quartic characteristic polynomial must be real. A critical piece of information comes from the fact that the quartic roots are based on an underlying cube root solution (a careful examination of how this works can be found, for example, in Coutsias et al. (2004); Coutsias and Wester (2019); Nickalls (2009)). As noted, e.g., in Abramowitz and Stegun (1970), the roots of this cubic are *real* provided that a particular discriminant is *negative*. This expression takes the form

$$
q_{\mathrm{AS}}{}^3 + r_{\mathrm{AS}}{}^2 \leqslant 0 ,
$$

where {AS} disambiguates the Abramowitz–Stegun variable names, and the relationship to our parameterization in terms of the eigenequation coefficients p_k is simply

$$
q_{\mathrm{AS}} = -\frac{1}{9} r^2(p_1, p_2, p_3, p_4) , \qquad r_{\mathrm{AS}} = \frac{1}{27} a(p_1, p_2, p_3, p_4) .
\tag{J.15}
$$

Thus we can see from Eq. (J.14) that

$$
b^2(p_1, p_2, p_3, p_4) = r^6(p) - a^2(p) = -9^3 \left(q_{\mathrm{AS}}{}^3 + r_{\mathrm{AS}}{}^2 \right) ,
\tag{J.16}
$$

and hence for symmetric real matrices we must have $b^2(p) \geqslant 0$. Therefore for this case we can always write

$$\left(a(p) + \sqrt{-b(p)^2} \right) \longrightarrow (a + ib) ,$$

(J.17)

and then we can rephrase our general solution from Eqs. (J.12), (J.13), and (J.14) as

$$
\left.
\begin{aligned}
F(p_1, p_2, p_3, p_4) &= \sqrt{\frac{p_1{}^2}{16} - \frac{p_2}{6} + \frac{1}{6} r(p)\, c(a, b)} \\[2mm]
G_\pm(p_1, p_2, p_3, p_4) &= \sqrt{\frac{p_1{}^2}{8} - \frac{p_2}{3} - \frac{1}{6} r(p)\, c(a, b) \pm \frac{s(p)}{32 \sqrt{\frac{p_1{}^2}{16} - \frac{p_2}{6} + \frac{1}{6} r(p)\, c(a, b)}}} \\[2mm]
&= \sqrt{\frac{3p_1{}^2}{16} - \frac{p_2}{2} - F^2(p) \pm \frac{s(p)}{32\, F(p)}}
\end{aligned}
\right\},
$$

(J.18)

where the cube root terms can now be reduced to real-valued trigonometry:

$$
\left.
\begin{aligned}
r(p)\, c(a, b) &= r(p) \cos\left(\frac{\arg(a + ib)}{3} \right) = \frac{1}{2}\left((a + ib)^{1/3} + (a - ib)^{1/3} \right) \\
r^2(p) &= p_2{}^2 - 3p_1 p_3 + 12 p_4 = \sqrt[3]{a^2 + b^2} = (a + ib)^{1/3}(a - ib)^{1/3} \\
r^6(p) &= a^2(p) + b^2(p) \\
s(p) &= 4 p_1 p_2 - p_1{}^3 - 8 p_3
\end{aligned}
\right\}.
$$

(J.19)

Alternative method: the cube root triples method and its properties. Our first general method above corresponds directly to Abramowitz and Stegun (1970), and consists of combinations of signs in two blocks of expressions. The second method that we are about to explore uses sums of three expressions in all four eigenvalues, with each term having a square-root ambiguity; this is fundamentally Euler's solution, discussed, for example, in Coutsias et al. (2004); Coutsias and Wester (2019) and Nickalls (2009). The correspondence between this triplet and the four expressions in Eq. (J.18) is delicate, but deterministic, and we will show the argument leading to the equations we introduced in the main text.

The "cube root triple" method follows from the observation that if we break up the general form of the four quartic eigenvalues into a trace part and a sum of three identical parts whose signs are arranged to be traceless, we find an equation that can be easily solved, and which (under some conditions that we will remove) evaluates numerically to the same eigenvalues as Eq. (J.18), but can be expressed in terms of a one-line formula for the eigenvalue system. The Ansatz that we start with is closely related to the SVD method as noted in Appendix N, and takes the following form:

$$
\left.
\begin{aligned}
\epsilon_1 &\overset{?}{=} -\frac{p_1}{4} + \sqrt{X} + \sqrt{Y} + \sqrt{Z} \\[4pt]
\epsilon_2 &\overset{?}{=} -\frac{p_1}{4} + \sqrt{X} - \sqrt{Y} - \sqrt{Z} \\[4pt]
\epsilon_3 &\overset{?}{=} -\frac{p_1}{4} - \sqrt{X} + \sqrt{Y} - \sqrt{Z} \\[4pt]
\epsilon_4 &\overset{?}{=} -\frac{p_1}{4} - \sqrt{X} - \sqrt{Y} + \sqrt{Z}
\end{aligned}
\right\} .
\tag{J.20}
$$

If we now insert our expressions for $\epsilon_k(p_1, X, Y, Z)$ from Eq. (J.20) into Eq. (J.9), we see that the p_k equations are transformed into a quartic system of equations that can in principle be solved for the components of the eigenvalues:

$$
\left.
\begin{aligned}
p_1 &= p_1 \\[6pt]
p_2 &= \frac{3p_1^{\,2}}{8} - 2(X + Y + Z) \\[6pt]
p_3 &= \frac{p_1^{\,3}}{16} - 8\sqrt{XYZ} - p_1(X + Y + Z) \\[6pt]
p_4 &= \frac{p_1^{\,4}}{256} + X^2 + Y^2 + Z^2 - 2(YZ + ZX + XY) - p_1\sqrt{XYZ} - \frac{p_1^{\,2}}{8}(X + Y + Z)
\end{aligned}
\right\} .
\tag{J.21}
$$

While our original equation Eq. (J.9) does not respond to `Solve[... , {`$\epsilon_1, \epsilon_2, \epsilon_3, \epsilon_4$`}, ... ]`, and Eq. (J.21) with $X \to u^2$, $Y \to v^2$, $Z \to w^2$ does not respond to `Solve[... , {u,v,w}, ... ]`, for some reason Eq. (J.21) with X, Y, Z as the free variables responds immediately to

$$\texttt{Solve[pkEqnList , \{X,Y,Z\}, Quartics -> True],}$$

and produces a solution for $X(p)$, $Y(p)$, and $Z(p)$ that we can manipulate into the following form:

$$
F_f(p) = \frac{p_1^{\,2}}{16} - \frac{p_2}{6} - \frac{1}{12}\left(\phi(f)\left(a(p) + \sqrt{-b^2(p)}\right)^{1/3} + \frac{r^2(p)}{\phi(f)\left(a(p) + \sqrt{-b^2(p)}\right)^{1/3}} \right) .
\tag{J.22}
$$

Here $F_f(p)$ with $f = (x, y, z)$ represents $X(p)$, $Y(p)$, or $Z(p)$ corresponding to one of the three values of the cube roots $\phi(f)$ of (-1) given by

$$
\phi(x) = -1, \qquad \phi(y) = \tfrac{1}{2}\left(1 + i\sqrt{3}\right), \qquad \phi(z) = \tfrac{1}{2}\left(1 - i\sqrt{3}\right),
\tag{J.23}
$$

and the utility functions are defined as above in Eq. (J.14). Once again, because we have symmetric real matrices with real eigenvalues, we know that the discriminant condition for real solutions requires $b^2(p) \geqslant 0$, so we can again apply Eq. (J.17) to transform each $\left(a(p) \pm \sqrt{-b^2(p)}\right)$ term into the form $(a(p) \pm i b(p))$. This time we get a slightly different formula because there is a different $\sqrt[3]{-1}$ phase incorporated into each of the X, Y, Z terms, and we obtain the following intermediate result:

$$
\begin{aligned}
F_f(p) &= \frac{p_1{}^2}{16} - \frac{p_2}{6} - \frac{1}{12}\left(\phi(f)\,(a+ib)^{1/3} + r^2(p)\frac{1}{\phi(f)\,(a+ib)^{1/3}}\right) \\
&= \frac{p_1{}^2}{16} - \frac{p_2}{6} - \frac{1}{6}\left(\phi(f)(a+ib)^{1/3} + \overline{\phi(f)}\,(a-ib)^{1/3}\right) \\
&= \frac{p_1{}^2}{16} - \frac{p_2}{6} - \frac{1}{6}\left(\phi(f)(a+ib)^{1/3} + \overline{\phi(f)(a+ib)^{1/3}}\right),
\end{aligned}
\tag{J.24}
$$

where $\overline{\phi(f)}$, etc., denotes the complex conjugate, and we took advantage of the relation $\sqrt[3]{a^2+b^2} = r^2(p)$. The cube root terms again reduce to real trigonometry, giving our final result (remember that $\phi(x) = -1$, changing the sign)

$$
F_f(p_1, p_2, p_3, p_4) = \frac{p_1{}^2}{16} - \frac{p_2}{6} + \frac{1}{6}\left(r(p)\cos_f(p)\right),
\tag{J.25}
$$

where we remember that $f = (x, y, z)$. Now with the direct incorporation of the three phases of $\sqrt[3]{-1}$ from Eq. (J.23) (see, e.g., Nickalls (1993)), we get nothing but phase-shifted real cosines,

$$
\cos_{x, y, z}(p) = \cos\left(\frac{\arg(a+ib)}{3} + \psi_{x, y, z}\right) \quad \text{with} \quad \psi_{x, y, z} = \left\{0, -\frac{2\pi}{3}, +\frac{2\pi}{3}\right\}.
\tag{J.26}
$$

The needed subset of the utility functions now reduces to

$$
\left.
\begin{aligned}
r^2(p_1, p_2, p_3, p_4) &= p_2{}^2 - 3p_1 p_3 + 12 p_4 = \sqrt[3]{a^2+b^2} = (a+ib)^{1/3}(a-ib)^{1/3} \\
a(p_1, p_2, p_3, p_4) &= p_2{}^3 + \frac{9}{2}\left(3p_3{}^2 + 3p_1{}^2 p_4 - p_1 p_2 p_3 - 8 p_2 p_4\right) \\
b^2(p_1, p_2, p_3, p_4) &= r^6(p) - a^2(p)
\end{aligned}
\right\}.
\tag{J.27}
$$

Repairing anomalies in the cube root triple form. We are not quite finished, as our X, Y, Z triplets acquire an ambiguity due to possible alternate sign choices when we take the square roots of X, Y, Z to construct the eigenvalues themselves using the Ansatz of Eq. (J.20). As long as all the terms of one part change sign together, the tracelessness of the X, Y, Z segment of the eigenvalue system is maintained, so there are a number of things that could happen with the signs without invalidating the general properties of Eq. (J.20). We can check that, with random symmetric matrix data, Eq. (J.20) with Eq. (J.25) will yield the correct eigenvalues about half the time, while Eq. (J.12) with Eq. (J.18) always works. Inspecting Eq. (J.18) and Eq. (J.25) with Eq. (J.26), we observe that $F(p_1, p_2, p_3, p_4) = \sqrt{F_X(p_1, p_2, p_3, p_4)} = \sqrt{X}$; we can also see that Eq. (J.18) suggests that a relation of the following form should hold,

$$
G_\pm(p_1, p_2, p_3, p_4) \sim \sqrt{Y} \pm \sqrt{Z},
$$

so we can immediately conjecture that something is going wrong with the sign choice of the root $\sqrt{Z}$. It turns out that $G_+(p)$ changes its algebraic structure to essentially that of $G_-(p)$ when the numerator $s(p) = (4p_1 p_2 - p_1{}^3 - 8p_3)$ inside the square root in Eq. (J.18) changes sign. That tells us exactly where there is a discrepancy with the choice $\sqrt{Y} + \sqrt{Z}$. If we define the following sign test,

$$\sigma(p_1, p_2, p_3, p_4) = \text{sign}\left(4p_1 p_2 - p_1^3 - 8p_3\right), \tag{J.28}$$

we discover that we can make Eq. (J.20) agree exactly with the robust $G_\pm(p)$ from Eq. (J.18) for all the random symmetric numerical matrices we were able to test, provided we make the following simple change to the final form of the X, Y, Z formula for the eigenvalue solutions:

$$\left.\begin{aligned}
\epsilon_1 &= -\frac{p_1}{4} + \sqrt{X} + \sqrt{Y} + \sigma(p)\sqrt{Z} \\[4pt]
\epsilon_2 &= -\frac{p_1}{4} + \sqrt{X} - \sqrt{Y} - \sigma(p)\sqrt{Z} \\[4pt]
\epsilon_3 &= -\frac{p_1}{4} - \sqrt{X} + \sqrt{Y} - \sigma(p)\sqrt{Z} \\[4pt]
\epsilon_4 &= -\frac{p_1}{4} - \sqrt{X} - \sqrt{Y} + \sigma(p)\sqrt{Z}
\end{aligned}\right\} . \tag{J.29}$$

Algebraic equivalence of standard and cube root triple form. With the benefit of hindsight, we now complete the picture by working out the algebraic properties of Eq. (J.12) and Eq. (J.13) that confirm our heuristic derivation of Eq. (J.29). First, we look back at Eq. (J.21) and discover that, using the relations for p_2 and p_3, we can incorporate $X + Y + Z = 3p_1^3/16 - p_2/2$ into p_3 to get a very suggestive form for our expression $s(p)$ from Eq. (J.14) in terms of the only square-root ambiguity in our original equations that we used to solve for $(X(p), Y(p), Z(p))$, which is

$$s(p_1, p_2, p_3, p_4) = 4p_1 p_2 - p_1^3 - 8p_3 = 64\sqrt{X(p)Y(p)Z(p)} . \tag{J.30}$$

Already we see that this is potentially nontrivial because $s(p)$ does not have a deterministic sign, but $\sqrt{X(p)Y(p)Z(p)}$ will always be positive unless we have a deterministic reason to choose the negative root.

Next, using Eq. (J.22), we recast Eq. (J.13) in a form that uses $F(p) \equiv \sqrt{X(p)} \equiv \sqrt{F_x(p)}$, as well as Eq. (J.30), to give

$$\left.\begin{aligned}
F(p_1, p_2, p_3, p_4) &= \sqrt{X(p_1, p_2, p_3, p_4)} \\[6pt]
&= \sqrt{\frac{p_1^2}{16} - \frac{p_2}{6} + \frac{1}{12}\left(\sqrt[3]{a - \sqrt{-b^2}} + \sqrt[3]{a + \sqrt{-b^2}}\right)} \\[10pt]
G_\pm(p_1, p_2, p_3, p_4) &= \sqrt{\frac{3p_1^2}{16} - \frac{p_2}{2} - F^2(p) \pm \frac{s(p)}{32\,F(p)}} \\[6pt]
&= \sqrt{A(p_1, p_2, p_3, p_4) \pm B(p_1, p_2, p_3, p_4)}
\end{aligned}\right\}, \tag{J.31}$$

where in fact we know a bit about how $B(p)$ should look:

$$B(p) = \frac{s(p)}{32\sqrt{X(p)}} . \tag{J.32}$$

Now we solve the equations

$$\sqrt{A(p) \pm B(p)} = \sqrt{Y} \pm \sigma(p)\sqrt{Z} \tag{J.33}$$

for $A(p)$ and $B(p)$, to discover

$$A(p) = Y(p) + \sigma^2 Z(p)$$
$$= Y(p) + Z(p), \tag{J.34}$$
$$B(p) = 2\sigma\sqrt{Y(p)Z(p)}, \tag{J.35}$$

where we note that these useful relations are nontrivial to discover *directly* from our original expressions for $F(p)$ and $G_\pm(p)$. Finally, using Eq. (J.32), we conclude that

$$s(p) = 64\,\sigma(p)\sqrt{X(p)Y(p)Z(p)}, \tag{J.36}$$

which confirms that the appearance of

$$\sigma(p) = \text{sign}(s(p)) = \text{sign}(4p_1 p_2 - p_1^3 - 8p_3) \tag{J.37}$$

in the (X, Y, Z) expression of Eq. (J.29) is rigorous and inevitable, as it can be deduced directly from its appearance in $B(p)$.

Alternative reduction of the quartic solution. Perhaps a more explicit way to connect the $(F, G_\pm)$ and (X, Y, Z) forms, and one we might have used from the beginning with further insight, is to observe that $G_\pm$ is actually the square root of a perfect square,

$$\left. \begin{aligned} G_\pm &= \sqrt{\left(\sqrt{Y} \pm \sigma\sqrt{Z}\right)^2} \\[2mm] &= \sqrt{Y + Z \pm 2\sigma\sqrt{YZ}} \\[2mm] &= \sqrt{Y + Z \pm 2\sigma\frac{\sqrt{XYZ}}{\sqrt{X}}} \\[2mm] &= \sqrt{Y + Z \pm 2\sigma\frac{64\sqrt{XYZ}}{64\sqrt{X}}} \\[2mm] &= \sqrt{Y + Z \pm 2\sigma\frac{|s(p)|}{64\sqrt{X}}} \\[2mm] &= \sqrt{Y + Z \pm \frac{s(p)}{32\sqrt{X(p)}}} \end{aligned} \right\} , \tag{J.38}$$

where we used the fact that $\sigma(p)|s(p)| = s(p)$. As long as the sign with which $G_\pm$ enters into the solution is consistent, the alternative overall signs of the radicals in Eq. (J.38) will be included correctly.

The traceless triple form. The explicitly traceless X, Y, Z triplet form that corresponds to a set of eigenvalues in descending magnitude order that we introduced for the 3D RMSD problem in the main text is obtained by imposing the traceless condition, $p_1 = 0$, obeyed by the 3D profile matrix $M_3(E_3)$:

$$\left. \begin{aligned} \epsilon_1 &= +\sqrt{X} + \sqrt{Y} + \sigma(p)\sqrt{Z} \\ \epsilon_2 &= +\sqrt{X} - \sqrt{Y} - \sigma(p)\sqrt{Z} \\ \epsilon_3 &= -\sqrt{X} + \sqrt{Y} - \sigma(p)\sqrt{Z} \\ \epsilon_4 &= -\sqrt{X} - \sqrt{Y} + \sigma(p)\sqrt{Z} \end{aligned} \right\} . \tag{J.39}$$

Then Eq. (J.21) simplifies to

$$p_1 = 0 , \tag{J.40}$$
$$p_2 = -2(X + Y + Z) , \tag{J.41}$$
$$p_3 = -8\sigma(p)\sqrt{XYZ} , \tag{J.42}$$
$$p_4 = X^2 + Y^2 + Z^2 - 2(YZ + ZX + XY) , \tag{J.43}$$

and the solutions for $X(p)$, $Y(p)$, and $Z(p)$ (and thus for $\epsilon_k(p)$) reduce to

$$F_f(p_2, p_3, p_4) = +\frac{1}{6}\left(r(p)\cos_f(p) - p_2\right) , \tag{J.44}$$

where the phased cosine terms retain their form

$$\cos_{x,y,z}(p) = \cos\left(\frac{\arg(a + ib)}{3} + \psi_{x,y,z}\right) \quad \text{with} \quad \psi_{x,y,z} = \left\{0, -\frac{2\pi}{3}, +\frac{2\pi}{3}\right\} . \tag{J.45}$$

Here $F_f(p)$ with $f = (x, y, z)$ as always represents $X(p)$, $Y(p)$, or $Z(p)$ and the utility functions simplify to

$$\left. \begin{aligned} \sigma(p_3) &= \text{sign}(-p_3) \\ r^2(p_2, p_3, p_4) &= p_2^2 + 12p_4 = \sqrt[3]{a^2 + b^2} = (a + ib)^{1/3}(a - ib)^{1/3} \\ a(p_2, p_3, p_4) &= p_2^3 + \frac{9}{2}\left(3p_3^2 - 8p_2p_4\right) \\ b^2(p_2, p_3, p_4) &= r^6(p) - a^2(p) \\ &= \frac{27}{4}\left(16p_4p_2^4 - 4p_3^2p_2^3 - 128p_4^2p_2^2 + 144p_3^2p_4p_2 - 27p_3^4 + 256p_4^3\right) \end{aligned} \right\} . \tag{J.46}$$

Summary: *We therefore have two alternate robust expressions, Eq. (J.12) with Eq. (J.18) and Eq. (J.29) with Eq. (J.25), for the entire eigenvalue spectrum of any real, symmetric 4×4 matrix M characterized by its four intrinsic eigenequation coefficients (p_1, p_2, p_3, p_4). For the simpler traceless case, we can take advantage of Eq. (J.39) with Eq. (J.44).*

A quaternion context for traditional Ramachandran protein plots☆

In this appendix, we complete the overall picture relating quaternion protein maps to the Ramachandran plots that may be more familiar to some readers, describing in detail various relationships between the traditional 2D Ramachandran plot and our 4D quaternion maps.

Review of Ramachandran definitions. The standard triple of Ramachandran angles is determined by a sliding set of six atom positions as defined in Fig. K.1. A convenient labeling, including the neighboring residues, is the following:

$$\overset{\phi \quad \psi \quad \omega}{}$$

Atom:	N	C_α	C	N	C_α	C	N	C_α	C
ID number:	−1	0	1	2	3	4	5	6	7

standard frame

The Ramachandran starting position is the carbonyl carbon obtained by dropping the first two atomic positions (N and C_α) of the residue to the left of the residue that is our central focus, adjoining the $NC_\alpha C$ atoms of that residue, and appending the first two atoms (NC_α) of the residue to the right. We number this $CNC_\alpha CNC_\alpha$ sequence as 123456, with 234 being the atoms of the central residue, the one we have already used to define our standard quaternion frame parameters. The angle ϕ is associated with the 23 axis, ψ with the 34 axis, and ω with the 45 axis; however, we need to be careful about the signs, as described below.

In this group of six atoms, each set of three atomic positions from a PDB file defines a plane, and each pair of these triangles forms something that may be thought of as a bent hinge with the middle two atoms being the axis of the hinge (e.g., the vector $(\mathbf{3} \to \mathbf{2})$ is the hinge of 1234). We may then label the normals to each of the triangles by the ordered triple of vertex indices (see Fig. K.1), where we define the corresponding normal to be the result of the cross product formed by the ordered vertex differences labeled as follows:

$$\hat{\mathbf{n}}(123) = \frac{(\mathbf{2} - \mathbf{1}) \times (\mathbf{3} - \mathbf{2})}{\|(\mathbf{2} - \mathbf{1}) \times (\mathbf{3} - \mathbf{2})\|} \, ,$$

$$\hat{\mathbf{n}}(234) = \frac{(\mathbf{3} - \mathbf{2}) \times (\mathbf{4} - \mathbf{3})}{\|(\mathbf{3} - \mathbf{2}) \times (\mathbf{4} - \mathbf{3})\|} \, ,$$

$$\hat{\mathbf{n}}(345) = \frac{(\mathbf{4} - \mathbf{3}) \times (\mathbf{5} - \mathbf{4})}{\|(\mathbf{4} - \mathbf{3}) \times (\mathbf{5} - \mathbf{4})\|} \, ,$$

$$\hat{\mathbf{n}}(456) = \frac{(\mathbf{5} - \mathbf{4}) \times (\mathbf{6} - \mathbf{5})}{\|(\mathbf{5} - \mathbf{4}) \times (\mathbf{6} - \mathbf{5})\|} \, .$$

☆ This appendix follows closely the treatment in Hanson and Thakur (2012).

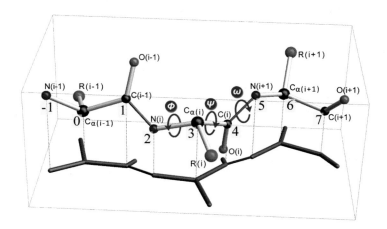

FIGURE K.1

Amino acid neighboring structure. Triples of atoms $(i-1)$, (i), and $(i+1)$ correspond to a single amino acid residue. The group of six atoms, $C:N:C_\alpha:C:N:C_\alpha$, starting at label "1" in the figure for the first C, defines the Ramachandran angles as "hinge" angles of the three groups of four atoms in the sequence of six. The planes of the peptide bonds connecting to adjacent amino acids define the ψ and ϕ dihedral angles. The angle ω describes the normally negligible torsion of the peptide bond, which is relatively rigid. The central tetrahedron has the alpha-carbon at its top center, and we note that the orientation is the dominant L-form: the implicit hydrogen points upwards and the [CO]-R-N triangle goes clockwise.

Remember that $\hat{n}(234)$ is the $\hat{z}$-axis in Fig. 16.2. The cosine of each Ramachandran angle is given by the *inner* product of the pair of adjacent normals $\hat{n}$, and the sign of the sine is given by the inner product of the hinge axis $\hat{A} = A/\|A\|$ with the *cross* product of the two normals. Alternatively, writing $a = 2 - 1$, $b = 3 - 2$, and $c = 4 - 3$, and

$$
\begin{aligned}
x &= (a \times b) \cdot (b \times c) \\
&= (a \cdot b)(b \cdot c) - (a \cdot c)(b \cdot b) , \\
y &= (a \times b) \times (b \times c) \cdot b/\|b\| \\
&= \|b\| (a \cdot (b \times c)) ,
\end{aligned}
$$

(see Appendix L) we can determine the correctly signed cosine and sine from

$$
\cos \phi = \frac{x}{\sqrt{x^2 + y^2}} ,
$$

$$
\sin \phi = \frac{y}{\sqrt{x^2 + y^2}} ,
$$

where we cycle from 1234 through 2345 and 3456 to get ψ and ω, respectively.

While this basic geometry is well known for the computation of the Ramachandran angles, we need the notation in order to proceed with the quaternion definitions that will allow us to gain some additional insights. First, we define the base coordinate system $(\hat{\mathbf{x}}, \hat{\mathbf{y}}, \hat{\mathbf{z}})$ as usual (from Fig. 16.2) for the vertices 234, with atoms $NC_{\alpha}C$. Then there exist three special rotations relative to that frame that we can write in axis-angle form as the 3D rotations

$$\mathbf{R}_1 = \mathbf{R}\left(\phi, \hat{\mathbf{A}}_1 = -\hat{\mathbf{A}}_{23}\right) ,$$

$$\mathbf{R}_2 = \mathbf{R}\left(\psi, \hat{\mathbf{A}}_2 = +\hat{\mathbf{A}}_{34}\right) ,$$

$$\mathbf{R}_3 = \mathbf{R}\left(\omega, \hat{\mathbf{A}}_3 = +\hat{\mathbf{A}}_{45}\right) .$$

Here $\hat{\mathbf{A}}_{ij}$ is the normalized unit vector constructed from the atomic coordinates $\mathbf{j} - \mathbf{i}$, and the angles are the Ramachandran angles, the "hinge" angles of right-handed rotations leaving fixed the $\hat{\mathbf{A}}_{ij}$-axes. We need these because next we are going to define the corresponding quaternions whose positions in $\mathbf{S}^3$ represent the rotations that have to be applied to $\hat{\mathbf{n}}(234) = \hat{\mathbf{z}}$, the z-axis of our standard $NC_{\alpha}C$ frame, to change its direction to match the other three normals generated by the triangles in the 1234567 sequence.

$Q_1 = \left(\cos(\phi/2), \hat{\mathbf{A}}_1 \sin(\phi/2)\right)$	Rotates $\hat{\mathbf{z}}$ to align with the direction of $\hat{\mathbf{n}}(123)$, the inverse of the actual Ramachandran ϕ rotation.
$Q_2 = \left(\cos(\psi/2), \hat{\mathbf{A}}_2 \sin(\psi/2)\right)$	Rotates $\hat{\mathbf{z}}$ to align with the direction of $\hat{\mathbf{n}}(345)$, the Ramachandran ψ rotation.
$Q_3 = \left(\cos(\omega/2), \hat{\mathbf{A}}_3 \sin(\omega/2)\right)$	Rotates $\mathbf{R}_2 \cdot \hat{\mathbf{z}}$ to align with the direction of $\hat{\mathbf{n}}(456)$, the normally ignored Ramachandran rotation that *precedes* the final rotation taking $\hat{\mathbf{z}}$ of the current $NC_{\alpha}C$ frame to $\hat{\mathbf{z}}'$, the z-axis of the *next* $NC_{\alpha}C$ frame in the protein.
$Q_4 = \left(\cos(\phi'/2), \hat{\mathbf{A}}_{56}\left[= -\hat{\mathbf{A}}_{1'}\right] \sin(\phi'/2)\right)$	Rotates $\mathbf{R}_3 \cdot \mathbf{R}_2 \cdot \hat{\mathbf{z}}$ to align with the direction of $\hat{\mathbf{n}}(456) = \hat{\mathbf{z}}'$; this is the positive actual Ramachandran ϕ' rotation.

We can now choose an example protein representation, such as the PDB file for the mostly helical 1AIE with 31 residues, or the more complex 1AO5 with 357 residues, and plot a variety of quantities for comparison.

- **Ramachandran angles.** We have ϕ and ψ, and so we can show the standard Ramachandran plots in Fig. K.2, with clusters of points near $\phi \approx -60$ and $\psi \approx -40$ as is typical of the alpha-helices contained in 1AIE and 1AO5.
- **I: xy quaternion Cartesian sum map.** First we take the 3-vector parts of the quaternions Q_1 and Q_2 defined above and refer them to our standard C_{α} residue frame, so that the ϕ-rotation axis and the ψ-rotation axis lie in the same local reference frame, that is, the local xy-plane (by definition, the ψ-rotation axis is the x-axis). The plot of these quantities in the 3D quaternion space, as shown in Fig. K.3, follows from simply adding the quaternion vectors, and this gives a quaternion-scaled 2D plot that is for all practical purposes indistinguishable from the Ramachandran plot. The most natural way to think of these 2D coordinates is as quaternion lengths arising from a single-axis rotation; they are also closely related (by replacing $\sin(\phi/2)$ with $\sin(\phi)$) to the axis-angle coordinates

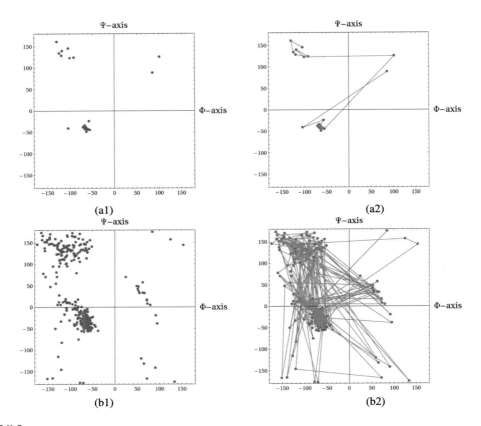

FIGURE K.2

(a1,a2) Standard Ramachandran plots of 31-residue 1AIE, with disconnected points and with adjacency-ordered line segments. (b1,b2) Same plots for the more extensive 357-residue 1AO5 protein.

sometimes used in orientation analysis. The quaternions embed a rigorous definition of distance between rotations, while axis-angle coordinates are *ad hoc*.

- **II. xy quaternion product map.** The quaternion maps in Fig. K.3 correspond essentially to the Ramachandran plot, and are constructed as a Cartesian sum of vectors that can be added in any order. This is not the way rotations actually act, and the rules of quaternion rotation representation are violated: while each single 3-vector Q_1 and Q_2 in Fig. K.3 is a part of a unit-length quaternion (remember that we can calculate the missing scalar part from the visible 3-vector), the Euclidean sum is not. However, we can correct that by performing a quaternion product, $Q_2 \star Q_1^{-1}$, and the result will be a quaternion that rotates the normal of the 123 triangle by the angle ϕ to the normal of the frame 234, and then rotates that normal by ψ to align with the normal of the 345 triangle; that is, the resulting quaternion represents the total rotation carried out when rotating by both Ramachandran angles to get approximately to the first "leaf" of the next ϕ frame.

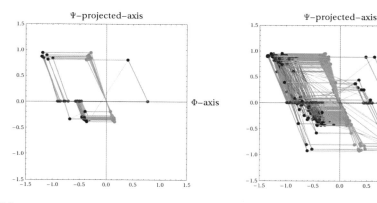

FIGURE K.3

Quaternion geometry corresponding to the standard Ramachandran plots for 1AIE (left) and 1AO5 (right). Red dots on the horizontal axis are the ϕ-related quaternion points. Green dots are the ψ-related quaternion points relative to the quaternion coordinate frame. Blue dots are the Cartesian (Euclidean) sum of this pair of coordinates for each residue, with the red and green lines showing the components of each sum. Blue lines connecting the residues as in Fig. K.2 appear where they do not interfere with the red-green summation lines.

Fig. K.4 shows the results of this action on 1AIE and a 200-residue portion of 1AO5, rotating the 123 normal until it aligns with the 345 normal. Reversing the order (distinct from using the inverse) results in a quaternion that differs by a sign in the z component of the resulting quaternion.

- **III. Quaternion action of the three Ramachandran angles, and the missing twist.** The Ramachandran angles provide sufficient information to define the transition from the plane of a given $NC_\alpha C$ frame to the next, provided we split them up so that the ψ and ω quaternions of the given frame are composed with the ϕ' frame of its successor. There are two steps needed to finally compare the Ramachandran data to the quaternion data in a fully quantitative fashion:

  - We can find the value of the *new normal* for the next $NC_\alpha C$ frame, which we call $\hat{z}'$, by applying the neighboring (split up) Ramachandran rotations in ordinary 3D space, and we can also express that complete rotation in quaternion form as follows:

  $$\hat{z}' = \mathbf{R}_{1'} \cdot \mathbf{R}_3 \cdot \mathbf{R}_2 \cdot \hat{z},$$
  $$Q_{z \to z'} = Q_{1'} \star Q_3 \star Q_2.$$

  However, all this tells us is the *orientation of the perpendicular* to the plane of the next $NC_\alpha C$ frame: it is powerless to tell us the *entire frame*. This is a deficiency of the Ramachandran approach.

  - The final step necessary for complete understanding of the protein geometry, and one of our fundamental points in this treatment of protein orientation frames, is the addition of one final *spin* about the $\hat{z}'$-axis! This is then the final relation between quaternion frames and Ramachandran angles: in Fig. K.5(a), we show the location of a typical given $NC_\alpha C$ frame and use it as the identity reference frame, i.e., as a point at the origin of the xyz quaternion projection; then we plot the three quaternion arcs $Q_{1'} \star Q_3 \star Q_2$ in sequence taking that frame's normal $\hat{z}$ to the next

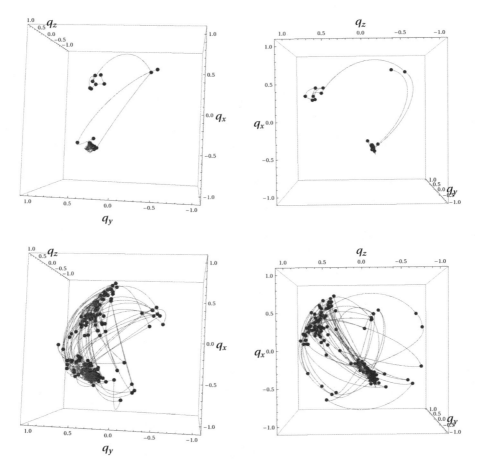

FIGURE K.4

Quaternion geometry with the action of the Ramachandran rotations represented as full quaternion products, with correct unit-length quaternion results, again for 1AIE (top, two viewpoints) and 1AO5 (bottom, two viewpoints).

$\hat{\mathbf{z}}'$. But now we also plot the quaternion value of the *next* $NC_\alpha C$ frame, and see that it differs from the result of the Ramachandran transformation. The difference is simply a rotation by an angle σ about the $\hat{\mathbf{z}}'$-axis that can be computed in a number of ways, e.g.,

$$\mathbf{F}_{1'} = q(\sigma, \hat{\mathbf{z}}') \star Q_{z \to z'},$$

where the quaternion frame $\mathbf{F}_{1'}$, or the $NC_\alpha C$ atoms having $\hat{\mathbf{z}}'$ as the normal to their plane, is computed in the coordinate system that has the original $NC_\alpha C$ frame $\mathbf{F}$ as the identity frame.

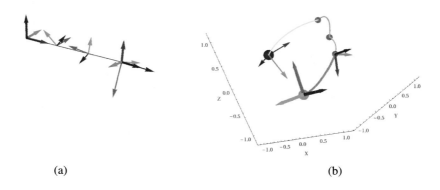

(a)   (b)

FIGURE K.5

Relationship between the rotations defined by the three Ramachandran angles taking one representative $NC_\alpha C$ frame to the next, and the quaternion value representing the transformation between the same two frames. (a, b) The axial rotation performing the final frame alignment shown at the end of the path in the left image (a), and as a quaternion path in the right image (b); this orientation gap cannot be represented using the Ramachandran angles, which therefore lack crucial information.

To conclude, in Fig. K.5, we can lay out a plot of the global locations of all the quaternion frames for the entire protein in two equivalent forms: as the single quaternion arcs from $\mathbf{F}_i$ to $\mathbf{F}_{i+1}$, *or* as the *pair* of quaternion arcs consisting of the Ramachandran composite arc (from the total value of $Q_{z \to z'}$, composed with the $\hat{\mathbf{z}}'$-axis spin $q(\sigma, \hat{\mathbf{z}}')$). This last small "spin" arc is plotted in a thick line to emphasize the distinction between the global frame orientation and the information available from the Ramachandran angles.

Determinants and the Levi-Civita symbol

One of the unifying features of the geometry we use to study quaternion-related problems, and which is extensible to higher dimensions, is the ubiquitous appearance of *inner products* that generate scalars, and *determinants*, which we will see are key to many operations that also generate tensors. In this appendix, we review a useful mathematical tool for exploiting determinants, the Levi-Civita symbol. Original references for this are hard to locate; the author learned these techniques by apprenticeship while studying general relativity, but even classic texts like Møller (1972) contain only passing mention of the methods; somewhat more detail is given in hard-to-find sources such as Efimov and Rozendorn (1975), though with the advent of thorough encyclopedic web-based coverage of mathematics, this material is now more generally accessible.

L.1 The basics of Euclidean tensor algebra

In this first section, we define the basic structures in terms of which we can write any inner product and any important tensor related to determinants.

First we define the Kronecker delta, δ_{ij},

$$\delta_{ij} = 1 , \qquad i = j , $$
$$\delta_{ij} = 0 , \qquad i \neq j , $$

where $(i, j) = \{1, \ldots, N\}$, and which is thus essentially an algebraic version of the N-dimensional Euclidean identity matrix I_N. The particular property that makes this matrix important is that it is invariant when conjugated by $\mathbf{SO}(N)$ rotation matrices,

$$\sum_{j=1, k=1}^{N,N} R_{ij}\delta_{jk}R_{kl} = \delta_{il} , $$

$$R \cdot I_N \cdot R^{\text{t}} = I_N . $$

Next, we define the universally useful Levi-Civita symbol, $\epsilon_{ijk\ldots}$, which is the totally antisymmetric pseudotensor with the properties

$$\epsilon_{ijk\ldots} = 1 , \qquad i, j, k, \ldots \text{ in an even permutation of cyclic order} , $$
$$= -1 , \qquad i, j, k, \ldots \text{ in an odd permutation of cyclic order} , $$
$$= 0 , \qquad \text{when any two indices are equal} . $$

547

All indices are assumed to range from 1 to N, i.e., $i = \{1, 2, \ldots, (N-1), N\}$, so that, for example, 1234, 1342, 4132, and 4321 are even permutations and 1324, 2134, 1243, and 4312 are odd permutations. We remark that $\epsilon_{ijk\ldots}$ is referred to as a *pseudotensor* because, unlike a *tensor*, which transforms under rotations as $\mathbf{A}' = R \cdot \mathbf{A}$, if we apply N rotation matrices to the Levi-Civita symbol, with one matrix contracted on the right with each index of $\epsilon_{ijk\ldots}$, we generate a new copy of $\epsilon_{ijk\ldots}$ *multiplied by the determinant of the matrix R*, as we show in a moment. Thus for rotation matrices with a negative signature, the resulting rotated Levi-Civita tensor is multiplied by a minus sign.

Since we are going to be summing over nearly every index in sight, we now adopt the "Einstein summation convention," namely, that any time we *repeat* an index, we mean to *sum* that index from 1 to N. With this convention, we can use the Kronecker delta to write the dot product between two N-dimensional vectors as

$$\mathbf{A} \cdot \mathbf{B} = A_i \delta_{ij} B_j = A_i B_i \ ,$$

and the contraction of a rotation matrix and its transpose with the identity matrix between them as

$$R \cdot R^{\mathrm{t}} = R_{ij} \delta_{jk} R_{lk} = R_{ij} R_{lj} = \delta_{il} \ .$$

Similarly, the Levi-Civita symbol allows us to write the determinant of a matrix $[M]$ as

$$\det[M] = \epsilon_{i_1 i_2 \ldots i_N} M_{1, i_1} M_{2, i_2} \cdots M_{N, i_N} \ .$$

The fundamental formula for the product of two Levi-Civita symbols is

$$\epsilon_{i_1 i_2 \ldots i_N} \epsilon_{j_1 j_2 \ldots j_N} = \det \begin{bmatrix} \delta_{i_1 j_1} & \delta_{i_1 j_2} & \cdots & \delta_{i_1 j_N} \\ \delta_{i_2 j_1} & \delta_{i_2 j_2} & \cdots & \delta_{i_2 j_N} \\ \vdots & \vdots & \ddots & \vdots \\ \delta_{i_N j_1} & \delta_{i_N j_2} & \cdots & \delta_{i_N j_N} \end{bmatrix} . \tag{L.1}$$

Note that if we set $\{j_1 j_2 \ldots j_N\} = \{1, 2, \ldots, N\}$, the second Levi-Civita symbol reduces to $+1$, and the resulting determinant is an explicit realization of the Levi-Civita symbol itself and its antisymmetry as a determinant of Kronecker deltas!

L.2 Cross product formulas and normals

The generalized *cross product* $\mathbf{n}$ of a nondegenerate sequence of $(N-1)$ N-dimensional vectors $\{\mathbf{x}^i\}$, $i = \{1, 2, \ldots, (N-1)\}$, in terms of the Euclidean basis $\{\mathbf{e}^i\}$ is conventionally defined in terms of the determinant

$$\mathbf{n} = n_1 \mathbf{e}^1 + n_2 \mathbf{e}^2 + n_3 \mathbf{e}^3 + \cdots + n_N \mathbf{e}^N$$

$$= \det \begin{bmatrix} x_1 & x_2 & \cdots & x_{N-1} & \mathbf{e}^1 \\ y_1 & y_2 & \cdots & y_{N-1} & \mathbf{e}^2 \\ z_1 & z_2 & \cdots & z_{N-1} & \mathbf{e}^3 \\ \vdots & \vdots & \ddots & \vdots & \vdots \\ w_1 & w_2 & \cdots & w_{N-1} & \mathbf{e}^N \end{bmatrix} , \tag{L.2}$$

where we write the N-vector columns as $\mathbf{x}_1 = [x_1, y_1, z_1, \ldots, w_1]^t$, $\mathbf{x}_2 = [x_2, y_2, z_3, \ldots, w_2]^t$, etc., to simplify the notation. Using the Levi-Civita symbol, this can be written explicitly in tensor notation as

$$\mathbf{n} = \sum_{\text{all indices}} \epsilon_{i_1 i_2 \ldots i_{N-1} i_N} x_1^{(i_1)} x_2^{(i_2)} \cdots x_{N-1}^{(i_{N-1})} \mathbf{e}^{(i_N)} ,$$

where the $\mathbf{e}^{(i)}$ are the unit vectors $(\mathbf{e}^1, \mathbf{e}^2, \ldots, \mathbf{e}^N)$ of the Cartesian coordinate system.

The dot product between the normal and another vector simply becomes

$$\mathbf{n} \cdot \mathbf{v} = \sum_{\text{all indices}} \epsilon_{i_1 i_2 i_3 \ldots i_{N-1} i_N} x_1^{(i_1)} x_2^{(i_2)} x_3^{(i_3)} \cdots x_{N-1}^{(i_{N-1})} v^{(i_N)} ,$$

which is just the determinant of Eq. (L.2) with the components of $\mathbf{v}$ replacing the column of $\mathbf{e}^{(i)}$ basis vectors.

For more general cases, we can include an origin $\mathbf{x}_0$ with respect to which the vector lengths are calculated,

$$\mathbf{n} = n_1 \mathbf{e}^1 + n_2 \mathbf{e}^2 + n_3 \mathbf{e}^3 + \cdots + n_N \mathbf{e}^N$$

$$= \det \begin{bmatrix} (x_1 - x_0) & (x_2 - x_0) & \cdots & (x_{N-1} - x_0) & \mathbf{e}^1 \\ (y_1 - y_0) & (y_2 - y_0) & \cdots & (y_{N-1} - y_0) & \mathbf{e}^2 \\ (z_1 - z_0) & (z_2 - z_0) & \cdots & (z_{N-1} - z_0) & \mathbf{e}^3 \\ \vdots & \vdots & \ddots & \vdots & \vdots \\ (w_1 - w_0) & (w_2 - w_0) & \cdots & (w_{N-1} - w_0) & \mathbf{e}^N \end{bmatrix} . \tag{L.3}$$

Examples of common normals. As a first example, the components of a 2D cross product can be written

$$N_k = x^{(i)} \epsilon_{ik}$$
$$= \left(x^{(1)} \epsilon_{11} + x^{(2)} \epsilon_{21}, \; x^{(1)} \epsilon_{12} + x^{(2)} \epsilon_{22} \right)$$
$$= \left(-x^{(2)}, \; +x^{(1)} \right)$$
$$= (-y, \; +x) ,$$

which is sometimes overlooked as a perfect example of generating a 2D vector orthogonal to any given vector. Similarly, a 3D cross product can be written

$$N_k = (\mathbf{A} \times \mathbf{B})_k$$
$$= A^{(i)} B^{(j)} \epsilon_{ijk}$$
$$= \left(A^{(2)} B^{(3)} \epsilon_{231} + A^{(3)} B^{(2)} \epsilon_{321}, \; A^{(1)} B^{(3)} \epsilon_{132} + A^{(3)} B^{(1)} \epsilon_{312}, \; A^{(1)} B^{(2)} \epsilon_{123} + A^{(2)} B^{(1)} \epsilon_{213} \right)$$
$$= \left(A^{(2)} B^{(3)} - A^{(3)} B^{(2)}, \; A^{(3)} B^{(1)} - A^{(1)} B^{(3)}, \; A^{(1)} B^{(2)} - A^{(2)} B^{(1)} \right)$$
$$= \left(A_y B_z - A_z B_y, \; A_z B_x - A_x B_z, \; A_x B_y - A_y B_x \right) ,$$

and so on. The important thing is to get the *free index* of the cross product vector in the right-most position to get the signs correct in *any* dimension: the physics literature on electromagnetism, for example, frequently puts the column of free-index unit vectors on the *left*, which works only for odd dimensions. This exercise also shows that one could leave more than one free index to generate "normal planes" and further similar generalizations.

Rotations of normals: Is the normal a vector? In fact, the normal **n** is *almost* a vector. To check this, we rotate each column vector in the cross product formula using $x'^{(i)} = \sum_{j=1}^{N} R_{ij} x^{(j)}$ and compute the behavior of **n**. Using the identity (Efimov and Rozendorn (1975), p. 203),

$$\epsilon_{i_1 i_2 \ldots i_{N-1} i_N} \det [R] = \sum_{\text{all } j_k \text{ indices}} \epsilon_{j_1 j_2 \ldots j_{N-1} j_N} R_{j_1 i_1} R_{j_2 i_2} \cdots R_{j_{N-1} i_{N-1}} R_{j_N i_N} ,$$

we find

$$n'^{(i)} = \sum_{\substack{\text{all indices} \\ \text{except } i}} \epsilon_{i_1 i_2 \ldots i_{N-1} i} R_{i_1 j_1} x_1^{(j_1)} R_{i_2 j_2} x_2^{(j_2)} \cdots R_{i_{N-1} j_{N-1}} x_{N-1}^{(j_{N-1})}$$

$$= \sum_{j=1}^{N} R_{ij} n^{(j)} \det [R] .$$

Therefore **n** is a *pseudotensor*, and behaves as a vector for ordinary rotations (which have $\det [R] = +1$), but changes sign if $[R]$ contains an odd number of reflections.

Contraction formulas. What happens if we have a dot product between vectors that are *both* constructed of partial determinants, like the normal? In 3D, we recall the familiar identity

$$(\mathbf{A} \times \mathbf{B}) \cdot (\mathbf{A} \times \mathbf{B}) = (\mathbf{A} \cdot \mathbf{A})(\mathbf{B} \cdot \mathbf{B}) - (\mathbf{A} \cdot \mathbf{B})^2 . \tag{L.4}$$

The generalization of this expression to N dimensions starts from this fundamental formula for the product of two Levi-Civita symbols:

$$\epsilon_{i_1 i_2 \ldots i_N} \epsilon_{j_1 j_2 \ldots j_N} = \det \begin{bmatrix} \delta_{i_1 j_1} & \delta_{i_1 j_2} & \cdots & \delta_{i_1 j_N} \\ \delta_{i_2 j_1} & \delta_{i_2 j_2} & \cdots & \delta_{i_2 j_N} \\ \vdots & \vdots & \ddots & \vdots \\ \delta_{i_N j_1} & \delta_{i_N j_2} & \cdots & \delta_{i_N j_N} \end{bmatrix} .$$

We now define **X** and **Y** to be the normal vectors constructed from the cross products of $N - 1$ vectors of dimension N, $\{x_1, \ldots, x_N\}$ and $\{y_1, \ldots, y_N\}$, respectively. Contracting on the two free indices in **X** and **Y**, the N-dimensional analog of Eq. (L.4), we then get

$$\mathbf{X} \cdot \mathbf{Y} = x_1^{(i_1)} x_2^{(i_2)} \ldots x_{N-1}^{(i_{N-1})} y_1^{(j_1)} y_2^{(j_2)} \ldots y_{N-1}^{(j_{N-1})} \det \begin{bmatrix} \delta_{i_1 j_1} & \delta_{i_1 j_2} & \cdots & \delta_{i_1 j_{N-1}} \\ \delta_{i_2 j_1} & \delta_{i_2 j_2} & \cdots & \delta_{i_2 j_{N-1}} \\ \vdots & \vdots & \ddots & \vdots \\ \delta_{i_{N-1} j_1} & \delta_{i_{N-1} j_2} & \cdots & \delta_{i_{N-1} j_{N-1}} \end{bmatrix}$$

as the generalization of the dot product of two cross products.

A partial determinant of $(N - K)$ N-dimensional vectors can be expressed as a Levi-Civita symbol expression with K free indices: the general form for the dot products of such tensors can be expanded as scalar products (i.e., as Kronecker deltas) in the following way:

$$\epsilon_{i_1 i_2 \ldots i_{N-K} i_{N-K+1} \ldots i_N} \epsilon_{j_1 j_2 \ldots j_{N-K} i_{N-K+1} \ldots i_N}$$

$$= K! \det \begin{bmatrix} \delta_{i_1 j_1} & \delta_{i_1 j_2} & \cdots & \delta_{i_1 j_{N-K}} \\ \delta_{i_2 j_1} & \delta_{i_2 j_2} & \cdots & \delta_{i_2 j_{N-K}} \\ \vdots & \vdots & \ddots & \vdots \\ \delta_{i_{N-K} j_1} & \delta_{i_{N-K} j_2} & \cdots & \delta_{i_{N-K} j_{N-K}} \end{bmatrix},$$

where we emphasize that the last K repeated indices are summed over all values from 1 to N, while the initial indices $\{i_1 i_2 \ldots i_{N-K}\}$ and $\{j_1 j_2 \ldots j_{N-K}\}$ take some fixed but unspecified value, as they are normally summed with a matching set of column vectors. The expression Eq. (L.4) for the dot product of two cross products (or two normals) is a special case of this general formula.

Examples. There are many frequently occurring 3D formulas such as Eq. (L.4) that arise from the Levi-Civita product formula of Eq. (L.1). Here we list a few of those for reference:

$$\mathbf{A} \cdot (\mathbf{B} \times \mathbf{C}) = \mathbf{C} \cdot (\mathbf{A} \times \mathbf{B}) = \mathbf{B} \cdot (\mathbf{C} \times \mathbf{A})$$

$$\mathbf{A} \times (\mathbf{B} \times \mathbf{C}) = \mathbf{B}(\mathbf{A} \cdot \mathbf{C}) - \mathbf{C}(\mathbf{A} \cdot \mathbf{B})$$

$$(\mathbf{A} \times \mathbf{B}) \cdot (\mathbf{C} \times \mathbf{D}) = \mathbf{A} \cdot (\mathbf{B} \times (\mathbf{C} \times \mathbf{D}))$$

$$= (\mathbf{A} \cdot \mathbf{C})(\mathbf{B} \cdot \mathbf{D}) - (\mathbf{A} \cdot \mathbf{D})(\mathbf{B} \cdot \mathbf{C})$$

$$(\mathbf{A} \times \mathbf{B}) \times (\mathbf{C} \times \mathbf{D}) = \mathbf{C}(\mathbf{D} \cdot (\mathbf{A} \times \mathbf{B})) - \mathbf{D}(\mathbf{C} \cdot (\mathbf{A} \times \mathbf{B}))$$

$$= \mathbf{B}(\mathbf{A} \cdot (\mathbf{C} \times \mathbf{D})) - \mathbf{A}(\mathbf{B} \cdot (\mathbf{C} \times \mathbf{D}))$$

$$\mathbf{A} \times (\mathbf{B} \times (\mathbf{C} \times \mathbf{D})) = (\mathbf{A} \times \mathbf{C})(\mathbf{B} \cdot \mathbf{D}) - (\mathbf{A} \times \mathbf{D})(\mathbf{B} \cdot \mathbf{C}).$$

The dual basis

In this appendix, we present and prove the properties of the *dual basis* of a nonorthonormal set of coordinates. We will assume that we are given an N-dimensional Euclidean space $\mathbb{R}^N$ and a nondegenerate set of N unit vectors $\{\mathbf{v}_n\}$ spanning that space. In the special case where the set of vectors is orthonormal, we will write the basis as $\{\mathbf{e}_n\}$, with $\mathbf{e}_i \cdot \mathbf{e}_j = \delta_{ij}$. In the orthonormal basis, any vector $\mathbf{P}$ can be decomposed as

$$
\begin{aligned}
\mathbf{P} &= \mathbf{e}_1 \left(\mathbf{P} \cdot \mathbf{e}_1\right) + \mathbf{e}_2 \left(\mathbf{P} \cdot \mathbf{e}_2\right) + \cdots + \mathbf{e}_{N-1} \left(\mathbf{P} \cdot \mathbf{e}_{N-1}\right) + \mathbf{e}_N \left(\mathbf{P} \cdot \mathbf{e}_N\right) \\
&= \mathbf{e}_1 P_1 + \mathbf{e}_2 P_2 + \cdots + \mathbf{e}_{N-1} P_{N-1} + \mathbf{e}_N P_N \,,
\end{aligned}
\tag{M.1}
$$

where the orthonormality of the $\{\mathbf{e}_n\}$ basis immediately allows us to compute the Euclidean length of $\mathbf{P}$ as

$$
\|\mathbf{P}\|^2 = \mathbf{P} \cdot \mathbf{P} = \sum_{n=1}^{N} P_n P_n \,.
\tag{M.2}
$$

Introducing a nonorthonormal basis. This gets more interesting if, instead of an orthonormal basis $\{\mathbf{e}_n\}$, we have a nondegenerate set of unit vectors $\{\mathbf{v}_n\}$ that is a nonorthonormal basis. Clearly, since the $\{\mathbf{v}_n\}$ span $\mathbb{R}^N$, it must be possible to express $\mathbf{P}$ in the alternative basis. Note that, while we ignored the origin in the $\{\mathbf{e}_n\}$ basis, the origin served as an $(N+1)$-st point relative to which the unit vectors were positioned. We will leave the origin implicit for the moment, but note that, e.g., for N-simplex-based barycentric coordinates, one needs to incorporate an arbitrary origin as an additional coordinate in the context of barycentric coordinates for $\mathbf{P}$. In our current context, we consider our N vectors $\mathbf{v}_i$ to be rays from the origin, and define the $N \times N$ matrix $\mathbf{V}$, whose columns are the vectors $\mathbf{v}_i$; next, we look at the adjugate $\mathrm{Adj}(\mathbf{V})$, and how the components of the adjugate combine to form the determinant of $\mathbf{V}$ as well as the inverse of $\mathbf{V}$ for nonsingular matrices:

$$
\left.
\begin{aligned}
\mathbf{V} \cdot \mathrm{Adj}(\mathbf{V}) &= \det \mathbf{V} \, I_N \\
\mathbf{V}^{-1} &= \frac{\mathrm{Adj}(\mathbf{V})}{\det \mathbf{V}}
\end{aligned}
\right\} \,.
\tag{M.3}
$$

We note the important fact that $\mathrm{Adj}(\mathbf{V})$ is still perfectly well defined even if $\det \mathbf{V} = 0$; in fact we exploit this feature repeatedly in the main text.

The generalized cross product. Now we need a construction that is based on the generalization of the cross product, written in the Cartesian basis as

$$\mathbf{A} \times \mathbf{B} = (A_y B_z - A_z B_y)\,\mathbf{e}_x + (A_z B_x - A_x B_z)\,\mathbf{e}_y + (A_x B_y - A_y B_x)\,\mathbf{e}_z = \det \begin{bmatrix} A_x & B_x & \mathbf{e}_x \\ A_y & B_y & \mathbf{e}_y \\ A_z & B_z & \mathbf{e}_z \end{bmatrix}.$$

Our generalization of the cross product corresponding to $\mathbf{V}$ is to replace each column of the determinant in turn with the column of *Cartesian* basis vectors

$$\textbf{Cartesian basis:} \qquad \mathbf{E} = \begin{bmatrix} \mathbf{e}_1 \\ \mathbf{e}_2 \\ \vdots \\ \mathbf{e}_{N-1} \\ \mathbf{e}_N \end{bmatrix}. \tag{M.4}$$

Leaving out questions of normalization for the moment, we then construct our *dual basis* of generalized cross products as follows (see, e.g., Möbius, 1846):

$$\textbf{Dual basis:} \qquad \mathbf{N} = \{\mathbf{n}_i\} = \left\{ \det \begin{bmatrix} v_{1,x} & v_{2,x} & \cdots & \mathbf{e}_1 & \cdots & v_{N,x} \\ v_{1,y} & v_{2,y} & \cdots & \mathbf{e}_2 & \cdots & v_{N,y} \\ v_{1,z} & v_{2,z} & \cdots & \mathbf{e}_3 & \cdots & v_{N,z} \\ \vdots & \vdots & \ddots & \vdots & \ddots & \vdots \\ v_{1,v} & v_{2,v} & \cdots & \mathbf{e}_{N-1} & \cdots & v_{N,v} \\ v_{1,w} & v_{2,w} & \cdots & \mathbf{e}_N & \cdots & v_{N,w} \end{bmatrix} \right\}. \tag{M.5}$$

The elements of Eq. (M.5) have several notable properties: first, the i-th element $\mathbf{n}_i$ is *orthogonal* to every vector $\mathbf{v}_j$ for $j \neq i$, and if contracted with $\mathbf{v}_i$, the result is the determinant $\det \mathbf{V}$. In other words, the dual basis is effectively an expansion of the *adjugate elements* that compose the inverse of the matrix $\mathbf{V}$ whose columns are the unit vectors $\{\mathbf{v}_i\}$. Using (i, j) to label the individual vectors, $\mathbf{v}_i$ or $\mathbf{n}_j$, in our collection, and (a, b) to label the N Cartesian components of one vector, $\mathbf{P} = \sum_a (\mathbf{P})^a\, \mathbf{e}_a$, we know

$$\left. \begin{aligned} \sum_{a=1}^{N} (\mathbf{v}_i)^a\, (\mathbf{n}_j)^a &= \delta_{ij} \det \mathbf{V} \\ \mathbf{v}_i \cdot \mathbf{n}_j &= \delta_{ij} \det \mathbf{V} \\ \mathbf{V}^{t} \cdot \frac{\mathbf{N}}{\det \mathbf{V}} &= I_N \end{aligned} \right\}. \tag{M.6}$$

These coefficients also obey a *dual relation*: if we sum over a product of the elements $\mathbf{v}_i$ and $\mathbf{n}_i$ labeled by $i = 1, 2, 3, \ldots$ for distinct choices of the $(x, y, z, \ldots, v, w)$ components labeled by (a, b), we find that

$$\sum_{i=1}^{N} \left(v_i{}^a n_i{}^b \right) = \delta_{ab} \det \mathbf{V}. \tag{M.7}$$

Nonorthogonal unit vector basis expansion. What we are looking for is a way to expand any unit vector $\mathbf{P}$ in a nonorthogonal basis of unit vectors, such as the columns of $\mathbf{V}$ can be, in a way that guarantees that the unit length of $\mathbf{P}$ is preserved. That is, we want to express any point on a unit sphere in terms of a minimal set of other unit vectors on a sphere. We write our proposal tentatively as

$$\mathbf{P} = a_1(\mathbf{P})\,\mathbf{v}_1 + a_2(\mathbf{P})\,\mathbf{v}_2 + \cdots + a_{N-1}(\mathbf{P})\,\mathbf{v}_{N-1} + a_N(\mathbf{P})\,\mathbf{v}_N . \tag{M.8}$$

If we contract one of our dual basis elements $\mathbf{n}_i$ with this expression for $\mathbf{P}$, all the terms except the i-th term *vanish*, leaving

$$\mathbf{P} \cdot \mathbf{n}_i = a_i(\mathbf{P})\,\mathbf{n}_i \cdot \mathbf{v}_i = a_i(\mathbf{P})\,\det \mathbf{V} ,$$

or

$$a_i(\mathbf{P}) = \frac{\mathbf{P} \cdot \mathbf{n}_i}{\det \mathbf{V}} . \tag{M.9}$$

This is basically the same algebraic form as the Euclidean partition of unity in Eq. (17.17) and Eq. (17.18), except that, since it refers to a nonorthogonal coordinate system constrained to a sphere, it is *not a partition of unity*. We are now working with spherical bases, so for the remainder of our treatment, we will assume that $\mathbf{v}_i \cdot \mathbf{v}_i = 1$ for each i; thus we have our main result, that $\mathbf{P}$ can be written alternatively in both the $\mathbf{e}_i$ basis and in the $\mathbf{v}_i$ basis with $\mathbf{n}_i$-related weights as

$$\mathbf{P} = \sum_{i=1}^{N} \mathbf{e}_i\,(\mathbf{e}_i \cdot \mathbf{P}) = \sum_{i=1}^{N} \mathbf{e}_i\,P_i = \sum_{i=1}^{N} \mathbf{v}_i\left(\frac{\mathbf{n}_i \cdot \mathbf{P}}{\det \mathbf{V}}\right) . \tag{M.10}$$

We can verify that

$$\mathbf{P} \cdot \mathbf{P} = \sum_{i}(\mathbf{P} \cdot \mathbf{v}_i)\left(\frac{\mathbf{n}_i \cdot \mathbf{P}}{\det \mathbf{V}}\right) = \frac{1}{\det \mathbf{V}}\sum_{a,b} P_a P_b \sum_{i}\left(v_i{}^a n_i{}^b\right) \equiv \mathbf{P} \cdot \mathbf{P} \tag{M.11}$$

due to the fact that the componentwise sum of $\mathbf{n}$ and $\mathbf{v}$ collapses to a Kronecker delta identity matrix in the $(x, y, z, \ldots)$ elements due to Eq. (M.7).

Interpretation as a generalized Shoemake "slerp" basis. Consider the case where the vector $\mathbf{P}$ slides from a general position into the 2D plane of just two basis vectors, and let us call those $\mathbf{v}_1$ and $\mathbf{v}_2$ without loss of generality. Then these are essentially the Shoemake `slerp` decomposition written as a kind of Gram–Schmidt basis. We recall that the usual `slerp` formula has the form

$$\mathbf{P}(t) = \mathbf{v}_1\frac{\sin((1-t)\theta)}{\sin\theta} + \mathbf{v}_2\frac{\sin(t\theta)}{\sin\theta} . \tag{M.12}$$

But if we look at the geometry shown in Fig. M.1, we see

$$\cos\theta = \mathbf{v}_1 \cdot \mathbf{v}_2 , \tag{M.13}$$

$$(\sin\theta)^2 = 1 - (\mathbf{v}_1 \cdot \mathbf{v}_2)^2 = \det\begin{bmatrix} \mathbf{v}_1 \cdot \mathbf{v}_1 & \mathbf{v}_1 \cdot \mathbf{v}_2 \\ \mathbf{v}_1 \cdot \mathbf{v}_2 & \mathbf{v}_2 \cdot \mathbf{v}_2 \end{bmatrix} = (\det \mathbf{V})^2 , \tag{M.14}$$

since we assumed the $\mathbf{v}_i$ were unit vectors on the unit circle. Therefore $\sin\theta = \det \mathbf{V}$. Now we consider θ to be subdivided into $\theta = \theta_1 + \theta_2$, as shown in Fig. M.1, where $\theta_1 = (1-t)\theta$ is, counterintuitively, the angle between the point $\mathbf{P}(t)$ and v_2, and $\theta_2 = t\theta$ is the angle between the point v_1 and $\mathbf{P}(t)$. Then

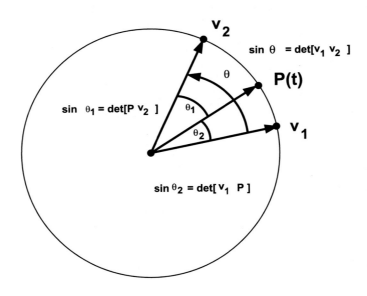

FIGURE M.1

The `slerp` geometry is shown here in terms of a dual basis viewpoint, with sines shown in the text to be equivalent to determinants of the unit vectors of the problem.

we see that the `slerp` formula Eq. (M.12) actually takes the form of ratios of determinants composing a 2D dual basis – that explains precisely why $\sin(\theta_1)$ involves $\mathbf{v}_2$ and $\sin(\theta_2)$ involves $\mathbf{v}_1$. Working out the explicit formulas for the trigonometric components solely in terms of linear algebra, we find the correspondences

$$\cos\theta_1 = \mathbf{P}\cdot\mathbf{v}_2 , \qquad\qquad (\sin\theta_1)^2 = \det\left[\mathbf{P}^a\mathbf{v}_2{}^b\right]^2 = 1 - (\mathbf{P}\cdot\mathbf{v}_2)^2 , \qquad\qquad \text{(M.15)}$$

$$\rightarrow\ \sin\theta_1 = \det[\mathbf{P}\ \mathbf{v}_2] , \qquad\qquad\qquad\qquad\qquad\qquad\qquad\qquad \text{(M.16)}$$

$$\cos\theta_2 = \mathbf{P}\cdot\mathbf{v}_1 , \qquad\qquad (\sin\theta_2)^2 = \det\left[\mathbf{v}_1{}^a\mathbf{P}^b\right]^2 = 1 - (\mathbf{v}_1\cdot\mathbf{P})^2 , \qquad\qquad \text{(M.17)}$$

$$\rightarrow\ \sin\theta_2 = \det[\mathbf{v}_1\ \mathbf{P}] , \qquad\qquad\qquad\qquad\qquad\qquad\qquad\qquad \text{(M.18)}$$

and thus

$$\rightarrow\rightarrow\ \sin\theta = \sin(\theta_1 + \theta_2) = \det[\mathbf{v}_1\ \mathbf{v}_2] = \det\mathbf{V} . \qquad\qquad \text{(M.19)}$$

The `slerp` is thus seen to have exactly the form Eq. (M.10) of our 2D dual basis expansion

$$\mathbf{P}(t) = \mathbf{v}_1\frac{\det[\mathbf{P}\ \mathbf{v}_2]}{\det\mathbf{V}} + \mathbf{v}_2\frac{\det[\mathbf{v}_1\ \mathbf{P}]}{\det\mathbf{V}} . \qquad\qquad \text{(M.20)}$$

We may thus argue that the dual basis has all the properties we can reasonably expect to generalize the `slerp` to describe an N-dimensional point $\mathbf{P}$ on an $(N-1)$-dimensional sphere $\mathbf{S}^{N-1}$ positioned in the midst of an arbitrary, usually nonorthogonal, basis of N unit vectors $\mathbf{v}_i$ on the sphere. Note that if the basis is orthonormal, the denominator is just unity, and, since $\theta_1 + \theta_2 = \pi/2$, the numerators are just the projections of $\mathbf{P}$ onto the matching Cartesian axes:

$$\sin\theta_2 = \sin\left(\frac{\pi}{2} - \theta_1\right) = \cos\theta_1 = \mathbf{P} \cdot \mathbf{v}_1 \, ,$$
$$\sin\theta_1 = \sin\left(\frac{\pi}{2} - \theta_2\right) = \cos\theta_2 = \mathbf{P} \cdot \mathbf{v}_2 \, .$$

Equivalence of SVD and quaternion eigensystem methods

We have often mentioned in the main text that we can find the 3D rotation best aligning two matched sets of points (the RMSD solution) using either quaternions (e.g., Horn, 1987, etc.) or alternatively using singular-value-decomposition (SVD) (e.g., Kabsch, 1976; Schönemann, 1966; Golub and van Loan, 1983, etc.). In this appendix, we sketch the argument that shows exactly how the two approaches are equivalent, as worked out, e.g., by Coutsias et al. (2004, 2005); Coutsias and Wester (2019). In this appendix, we will restrict our attention to the 3D case, noting, however, that closely related material used to analyze the 4D matching problem is given in Chapter 20.

Quaternion form. The quaternion form starts with an arbitrary 3×3 real matrix E and converts the maximizing squared-error function into a quaternion 4×4 matrix problem, $\Delta(E) = \operatorname{tr} R(q) \cdot E \Rightarrow q \cdot M(E) \cdot q$. The fundamental object of study then becomes the profile matrix,

$$M(E) = \begin{bmatrix} E_{xx} + E_{yy} + E_{zz} & E_{yz} - E_{zy} & E_{zx} - E_{xz} & E_{xy} - E_{yx} \\ E_{yz} - E_{zy} & E_{xx} - E_{yy} - E_{zz} & E_{xy} + E_{yx} & E_{zx} + E_{xz} \\ E_{zx} - E_{xz} & E_{xy} + E_{yx} & -E_{xx} + E_{yy} - E_{zz} & E_{yz} + E_{zy} \\ E_{xy} - E_{yx} & E_{zx} + E_{xz} & E_{yz} + E_{zy} & -E_{xx} - E_{yy} + E_{zz} \end{bmatrix}, \quad \text{(N.1)}$$

where $M(E)$ is a traceless, symmetric 4×4 matrix with necessarily real eigenvalues. By expanding $M(E)$'s characteristic polynomial $\chi_4(M, e) = \det[M - eI_4] = 0$ in powers of e, with solutions being the four eigenvalues $e = \lambda_{k=1,\ldots,4}$, we find

$$e^4 + e^3 p_1 + e^2 p_2 + e p_3 + p_4 = 0 \quad \text{(N.2)}$$

$$(e - \lambda_1)(e - \lambda_2)(e - \lambda_3)(e - \lambda_4) = 0, \quad \text{(N.3)}$$

so we can express the polynomials $p_k(E)$ directly in terms of totally symmetric polynomials of the eigenvalues in the form

$$\left. \begin{aligned} p_1 &= -\lambda_1 - \lambda_2 - \lambda_3 - \lambda_4 \\ p_2 &= \lambda_1\lambda_2 + \lambda_1\lambda_3 + \lambda_2\lambda_3 + \lambda_1\lambda_4 + \lambda_2\lambda_4 + \lambda_3\lambda_4 \\ p_3 &= -\lambda_1\lambda_2\lambda_3 - \lambda_1\lambda_2\lambda_4 - \lambda_1\lambda_3\lambda_4 - \lambda_2\lambda_3\lambda_4 \\ p_4 &= \lambda_1\lambda_2\lambda_3\lambda_4 \end{aligned} \right\}. \quad \text{(N.4)}$$

Here we will confine our attention to the case where $M(E)$ is traceless and symmetric, so $p_1 = 0$, and the remaining p_k coefficients simplify and can be expressed as the following functions of either M

or E:

$$
\left.\begin{aligned}
p_1 &= -\operatorname{tr}[M] = 0 \\
p_2 &= -\frac{1}{2}\operatorname{tr}[M \cdot M] = -2\operatorname{tr}[E \cdot E^{\mathrm{t}}] \\
&= -2\left(E_{xx}^2 + E_{xy}^2 + E_{xz}^2 + E_{yx}^2 + E_{yy}^2 + E_{yz}^2 + E_{zx}^2 + E_{zy}^2 + E_{zz}^2\right) \\
p_3 &= -\frac{1}{3}\operatorname{tr}[M \cdot M \cdot M] = -8\det[E] \\
&= 8\left(E_{xx}\,E_{yz}\,E_{zy} + E_{yy}\,E_{xz}\,E_{zx} + E_{zz}\,E_{xy}\,E_{yx}\right) \\
&\quad - 8\left(E_{xx}\,E_{yy}\,E_{zz} + E_{xy}\,E_{yz}\,E_{zx} + E_{xz}\,E_{zy}\,E_{yx}\right) \\
p_4 &= \det[M] = 2\operatorname{tr}[E \cdot E^{\mathrm{t}} \cdot E \cdot E^{\mathrm{t}}] - \left(\operatorname{tr}[E \cdot E^{\mathrm{t}}]\right)^2
\end{aligned}\right\}.
\tag{N.5}
$$

Since $M(E)$ itself does not appear in the SVD form of the optimization problem that we will address next, we are mainly interested in the particular forms $p_i(E)$. The characteristic polynomial of the profile matrix then becomes

$$
\begin{aligned}
\chi_4(E) &= e^4 + p_2(E)\,e^2 + p_3(E)\,e + p_4(E) \\
&= e^4 - 2\operatorname{tr}[E \cdot E^{\mathrm{t}}]\,e^2 - 8\det[E]\,e + \det[M(E)] .
\end{aligned}
\tag{N.6}
$$

Interestingly, the polynomial $M(E)$ is arranged so that $-p_2(E)/2$ is the (squared) Frobenius norm of E, and $-p_3(E)/8$ is its determinant. Our task now is to express the four eigenvalues $e = \lambda_k(p_1, p_2, p_3, p_4)$, $k = 1, \ldots, 4$, usefully in terms of the matrix elements, and also to find their eigenvectors; we are of course particularly interested in the maximal eigenvalue λ_{opt}.

SVD form. The SVD form of the problem starts with E itself, and exploits the fact that this can be converted into a real eigensystem by constructing the real symmetric 3×3 matrix pair

$$
F(E) = E \cdot E^{\mathrm{t}} \qquad F^{\mathrm{t}}(E) = E^{\mathrm{t}} \cdot E ,
\tag{N.7}
$$

which have identical eigenvalues but distinct eigenvector systems, producing the two distinct SVD orthogonal matrices. The SVD system is defined in our case by finding the set of eigenvalues $\{\mu_1, \mu_2, \mu_3\}$ of $F(E)$, and the corresponding distinct pair of left/right eigenvectors. For non-degenerate eigenvalues, the orthogonal sets of left and right eigenvectors (or, equivalently, the eigenvectors of $F(E)$ and $F^{\mathrm{t}}(E)$) can be made orthonormal, and thus comprise the two rotation matrices U and V that make up the SVD of E, taking the form

$$
E = U \cdot S(\mu) \cdot V^{\mathrm{t}} ,
\tag{N.8}
$$

where $S(\mu) = \operatorname{diag}(\sqrt{\mu_1}, \sqrt{\mu_2}, \sqrt{\mu_3})$ contains the eigenvalues in question. Note: if there are any degenerate eigenvalues, special care must be taken in the SVD algorithm to create an orthonormal basis of eigenvectors for U and V, as that feature may not be automatic for numerical eigensystem utilities. In particular, if E itself is a noise-free rotation matrix, so $E \cdot E^{\mathrm{t}} = R \cdot R^{\mathrm{t}} = I_3$ is the identity matrix with three identical unit eigenvalues, special treatment is required, but this case is handled correctly by SVD algorithms in our experience.

We define the characteristic polynomial for $F(E)$ with a minus sign to keep the largest power positive,

$$\det \chi_3(E) = -\det(F(E) - z\,I_3) = 0 \tag{N.9}$$

$$= z^3 + s_1(E)\,z^2 + s_2(E)\,z + s_3(E) = 0 \tag{N.10}$$

$$= (z - \mu_1)(z - \mu_2)(z - \mu_3) = 0 . \tag{N.11}$$

The three eigenvalue solutions $\{\mu_1, \mu_2, \mu_3\}$ are functions of E, and are determined by the coefficients

$$\left. \begin{aligned} s_1(E) &= -\operatorname{tr} F(E) \\ s_2(E) &= \frac{1}{2}(\operatorname{tr} F(E))^2 - \frac{1}{2}(\operatorname{tr} F(E) \cdot F(E)) \\ s_3(E) &= -\det F(E) \end{aligned} \right\} . \tag{N.12}$$

The parallel to the $M(E)$ relation Eq. (N.4) for the $F(E)$ eigensystem is

$$\left. \begin{aligned} s_1 &= -\mu_1 - \mu_2 - \mu_3 \\ s_2 &= \mu_1\mu_2 + \mu_1\mu_3 + \mu_2\mu_3 \\ s_3 &= -\mu_1\mu_2\mu_3 \end{aligned} \right\} . \tag{N.13}$$

Merging the quaternion and SVD forms. *(Following Coutsias et al. (2004).)* We now have two eigensystems for the same optimal-rotation problem. This sets the stage for us to exploit a classical result from, e.g., Weisner (1938), pages 140–143, that spells out the relationship between eigenvalue solutions to the characteristic polynomial equation from a traceless quartic matrix and the eigenvalue solutions to a certain *resolvent* cubic polynomial. Showing that the latter in fact corresponds precisely to our SVD-based characteristic polynomial is all we need to complete our argument that the quartic quaternion eigensystem and the cubic SVD eigensystem for the optimal cloud-matching rotation matrix are identical.

- **Quaternion context.** From the quaternion side, we have the 4×4 quaternion profile matrix $M(E)$, with characteristic polynomial of Eq. (N.6) with coefficients $\{p_i(E)\}$ of degree i in E. From that information, we can find the quaternion solution q_{opt} (generating the rotation $R_{\text{opt}} = R(q_{\text{opt}})$) as the eigenvector of the largest eigenvalue λ_{opt} of $M(E)$'s four eigenvalues $\{\lambda_i\}$, $i = (1, 2, 3, 4)$.
- **SVD context.** From SVD side, we have an immediate solution for the rotation matrix R_{opt} based on the method of Schönemann (1966, etc.) and the work of Kabsch (1976, 1978, etc.), since, given the SVD form $E = U \cdot S(\mu) \cdot V^{\text{t}}$ in Eq. (N.8), we can express the optimal rotation solution as $R_{\text{opt}} = V \cdot D(\mu) \cdot U^{\text{t}}$. The relevant observation is that the derivation of the SVD form follows from constructing the eigensystems of the symmetric matrices $F(E)$ and $F^{\text{t}}(E)$ in Eq. (N.7); thus we have a second characteristic polynomial Eq. (N.10) in the problem, with three more eigenvalues $\{\mu_1, \mu_2, \mu_3\}$.

The eigenvalue relations. As noted in Coutsias et al. (2004), the two sets of eigenvalues are related. First, any 4D eigensystem with eigenvalues $\{\lambda_i\}$ that is traceless, obeying the constraint $\sum_{i=1}^{4} \lambda_i = 0$, can be parameterized with three parameters (which we presciently label as μ_i, noting that Eqs. (8.24) and (J.29), for example, follow the same Ansatz) as follows:

$$\lambda_i = \sum_{j=1}^{3} \sigma^i{}_j \sqrt{\mu_j} \,. \tag{N.14}$$

Here $\sigma^i{}_j = \pm 1$, and for each λ_i, $i = (1, 2, 3, 4)$, $\sigma_1 \sigma_2 \sigma_3 = \text{sign} \det E$. Thus we can generally choose to write, in agreement with Eqs. (8.24) and (J.29), eigenvalues sorted from largest to smallest, and $\sigma = \text{sign} \det E$,

$$\left.\begin{aligned}
\lambda_1 &= \sqrt{\mu_1} + \sqrt{\mu_2} + \sigma\sqrt{\mu_3} \\
\lambda_2 &= \sqrt{\mu_1} - \sqrt{\mu_2} - \sigma\sqrt{\mu_3} \\
\lambda_3 &= -\sqrt{\mu_1} + \sqrt{\mu_2} - \sigma\sqrt{\mu_3} \\
\lambda_4 &= -\sqrt{\mu_1} - \sqrt{\mu_2} + \sigma\sqrt{\mu_3}
\end{aligned}\right\} \,. \tag{N.15}$$

Degenerate eigenvalues may require special attention, as discussed in Coutsias and Wester (2019). Here the square root is clearly is related to the fact that the eigenvalues of $F(E)$ follow from the square of the 3×3 E matrix that appears only linearly in $M(E)$, and the signs and parity of the square roots can vary depending on the sign of $\det E$. The relationships between these eigenvalues can be shown (Weisner, 1938) by the correspondence between the coefficients of the traceless quartic characteristic polynomial,

$$\phi_4 = e^4 + 6p\,e^2 + 4q\,e + r = 0 \,, \tag{N.16}$$

and its resolvent cubic

$$\phi_3 = z^3 + 3p\,z^2 + \frac{1}{4}(9p^2 - r)\,z - \frac{1}{4}q^2 = 0 \,. \tag{N.17}$$

The final step of our investigation is to expose the relation between the $M(E)$ and $F(E)$ eigensystems to close the gap on the claimed relationships of Eq. (N.15). Again following Coutsias et al. (2004) and the references invoked therein, we set up the following context:

	characteristic equation	characteristic polynomial
Quaternion	$\det(M(E) - e\,I_4) = 0,$	$e^4 + p_2(E)e^2 + p_3(E)e + p_4(E) = 0$
4th order form		$\phi_4 := \quad e^4 + 6p\,e^2 + 4q\,e + r = 0$
SVD	$\det(F(E) - z\,I_4) = 0,$	$z^3 + s_1(E)\,z^2 + s_2(E)\,z + s_3(E) = 0$
3rd order resolvent		$\phi_3 := \quad z^3 + 3p\,z^2 + \frac{1}{4}(9p^2 - r)\,z - \frac{1}{4}q^2 = 0$

(N.18)

Here the critical new piece of information is that, with the ϕ_3 expression being the resolvent of the ϕ_4 rephrasing of the traceless $M(E)$ characteristic equation, the solution of $\phi_4 = 0$ implies that the new variables p, q, and r, appearing in both ϕ_3 and ϕ_4 must be identical to the functions of E implied by both the $M(E)$ and $F(E)$ characteristic polynomials given by the indicated vanishing determinants. We observe that actually computing the algebraic forms of the $\{\lambda_i\}$ and $\{\mu_i\}$ to verify Eq. (N.15) is possible, but extremely difficult, though it is quite easy to verify numerically. This theorem with deep algebraic origins concerning the cubic resolvent constraints on quartic equations provides us with a "shortcut" that allows us to prove the effective equivalence of the quaternion $M(E)$ eigensystem and the SVD's $F(E)$ eigensystem without actually checking Eq. (N.15) algebraically.

The equivalence equations. Now all that remains is to take the explicit forms of $p_i(E)$ and $s_i(E)$ from the characteristic polynomials, computed directly from the 4×4 matrix $M(E)$, whose leading eigenvector generates R_{opt}, and the 3×3 symmetrized quadratic matrix $F(E)$ that enables the derivation of the SVD version of R_{opt}. Comparing first the $p_i(E)$ terms in the $M(E)$ characteristic polynomial Eq. (N.6) with ϕ_4, we find these explicit constraints

$$\left. \begin{aligned} p &= \frac{1}{6}\, p_2(E) = -\frac{1}{3}\, \text{tr}[E \cdot E^{\text{t}}] \\ q &= \frac{1}{4}\, p_3(E) = -2 \det[E] \\ r &= p_4(E) = \det[M(E)] = 2\,\text{tr}[E \cdot E^{\text{t}} \cdot E \cdot E^{\text{t}}] - \left(\text{tr}[E \cdot E^{\text{t}}]\right)^2 \end{aligned} \right\} \qquad (\text{N}.19)$$

All that remains is to compare the terms $s_i(E)$ in the characteristic polynomial, derived directly from $F(E)$ to those in ϕ_3 with the functions of E that result when Eq. (N.19) is substituted into ϕ_3. We find that these constraints

$$s_1(E) = 3p = -\text{tr}[E \cdot E^{\text{t}}] \quad \text{:from } M(E), \text{ Eq. (N.5) and Eq. (N.19)}$$
$$= -\text{tr}[F(E)] \qquad \text{:from } F(E) \text{ Eq. (N.12)}$$
$$s_2(E) = \frac{1}{4}(9p^2 - r) = \frac{1}{2}\left(\left(\text{tr}[E \cdot E^{\text{t}}]\right)^2 - \text{tr}[E \cdot E^{\text{t}} \cdot E \cdot E^{\text{t}}]\right) \quad \text{:from } M(E), \text{ Eq. (N.5) and Eq. (N.19)}$$
$$= \frac{1}{2}\left((\text{tr}[F(E)])^2 - \text{tr}[F(E) \cdot F(E)]\right) \qquad \text{:from } F(E) \text{ Eq. (N.12)}$$
$$s_3(E) = -\frac{1}{4}q^2 = -\det[E]^2 \quad \text{:from } M(E), \text{ Eq. (N.5) and Eq. (N.19)}$$
$$= -\det[F(E)] \qquad \text{:from } F(E) \text{ Eq. (N.12)}$$

are satisfied, confirming the effective identity of the quaternion eigensystem and SVD eigensystem algebraic solutions to the 3D cloud matching problem.

References

Abramowitz, M., Stegun, I., 1970. Handbook of Mathematical Functions. Dover Publications Inc., New York, pp. 17–18.

Adams, C.C., 1994. The three-sphere and lens spaces, sec. (9.2). In: The Knot Book: An Elementary Introduction to the Mathematical Theory of Knots. W. H. Freeman, New York, pp. 246–256.

Adler, Stephen L., 1995. Quaternionic Quantum Mechanics and Quantum Fields. Oxford University Press.

Albrecht, K., Hart, J., Alex, S., Dunker, A.K., 1996. Quaternion contact ribbons: a new tool for visualizing intra- and intermolecular interactions in proteins. Pacific Symposium on Biocomputing 94, 41–52.

Alfeld, P., Neamtu, M., Schumaker, L.L., 1996. Bernstein-Bézier polynomials on spherical and sphere-like surfaces. CAGD Journal 13, 333–349. URL http://www.math.vanderbilt.edu/~schumake/sbc.ps.

Altmann, S.L., 1986. Rotations, Quaternions, and Double Groups. Oxford University Press.

Altmann, S.L., 1989. Hamilton, Rodrigues, and the quaternion scandal. Mathematics Magazine 62 (5), 291–308. https://doi.org/10.2307/2689481.

Antonuccio, Francesco, 2013. 4-spinors and 5D spacetime. arXiv:1307.2551 [hep-th].

Antonuccio, Francesco, 2015. Split quaternions and the Dirac equation. Advances in Applied Clifford Algebras 25, 13–29.

Arun, K.S., Huang, T.S., Blostein, S.D., 1987. Least-squares fitting of two 3D point sets. IEEE Transactions on Pattern Analysis and Machine Intelligence PAMI-9 (5), 698–700. https://doi.org/10.1109/TPAMI.1987.4767965.

Ball, R.S., 1900. A Treatise on the Theory of Screws. Cambridge University Press. Reprinted in 2008 by Kessinger Publishing.

Bar-Itzhack, Itzhack Y., 2000. New method for extracting the quaternion from a rotation matrix. Journal of Guidance, Control, and Dynamics 23 (6), 1085–1087. https://doi.org/10.2514/2.4654. URL https://arc.aiaa.org/doi/abs/10.2514/2.4654.

Belavin, A.A., Polyakov, A.M., Schwartz, A.S., Tyupkin, Yu.S., 1975. Pseudoparticle solutions of the Yang-Mills equations. Physics Letters B (ISSN 0370-2693) 59 (1), 85–87. https://doi.org/10.1016/0370-2693(75)90163-X. URL https://www.sciencedirect.com/science/article/pii/037026937590163X.

Bell, J., 2008 (1733). A conjecture on the forms of the roots of equations. URL arXiv:0806.1927v1 [math.HO]. http://arxiv.org/abs/0806.1927. An English translation of L. Euler, De formis radicum aequationum cujusque ordinis conjectatio, Commentarii academiae scientiarum imperialis Petropolitianae 6 (1733) (pub. 1738) 216–231.

Berger, M., 1987. Geometry I, II. Springer Verlag, Berlin.

Berry, T., Visser, M., 2021. Lorentz boosts and Wigner rotations: self-adjoint complexified quaternions. Physics 3, 352–366. https://doi.org/10.3390/physics3020024.

Biedenharn, L.C., Louck, J.D., 1984. Angular momentum in quantum physics. In: Rota, G.-C. (Ed.), Encyclopedia of Mathematics and Its Applications, vol. 8. Cambridge University Press, New York.

Blaschke, W., 1960. Kinematik und Quaternionen (Kinematics and Quaternions). VEB Deutscher Verlag der Wissenschaften, Berlin. URL http://www.neo-classical-physics.info/uploads/3/4/3/6/34363841/blaschke_-_kinematics_and_quaternions.pdf. Translated PDF version by D.H. Delphenich.

Bojovic, Viktor, Sovic, Ivan, Bacic, Andrea, Lucic, Bono, Skala, Karolj, 2011. A novel tool/method for visualization of orientations of side chains relative to the protein's main chain. In: MIPRO'11, pp. 242–245.

Boyer, C.B., Merzbach, U.C., 1991. A History of Mathematics, 2nd edition. Wiley, New York.

Branden, C., Tooze, J., 1999. Introduction to Protein Structure, 2nd edition. Garland Publishing, New York and London.

Brown, J.L., Worsey, A.J., 1992. Problems with defining barycentric coordinates for the sphere. Mathematical Modelling and Numerical Analysis 26, 37–49.

Brown, James, Churchill, Ruel, 1989. Complex Variables and Applications. McGraw-Hill, New York. ISBN 0-07-010905-2.

Buss, S.R., Fillmore, J.P., 2001. Spherical averages and applications to spherical splines and interpolation. ACM Transactions on Graphics (ISSN 0730-0301) 20 (2), 95–126. URL http://doi.acm.org/10.1145/502122.502124.

Calabi, E., 1979. Métriques Kählériennes et fibrés holomorphes. Annales Scientifiques de l'Ecole Normale Supérieure 12, 269–294.

Cayley, A., 1845. On certain results relating to quaternions. Philosophical Magazine 26, 141–145.

Chambers, J.M., Cleveland, W.S., Kleiner, B., Tukey, P.A., 1983. Graphical Methods for Data Analysis. Wadsworth, Belmont, CA.

Chen, Michael, Mountford, S. Joy, Sellen, Abigail, 1988. A study in interactive 3-d rotation using 2-d control devices. In: Proceedings of SIGGRAPH 1988. Computer Graphics 22, 121–130.

Cho, Y., Kim, H., 1999. On the volume formula for hyperbolic tetrahedra. Discrete & Computational Geometry 22, 347–366. URL https://link.springer.com/content/pdf/10.1007/PL00009465.pdf.

Cliff, N., 1966. Orthogonal rotation to congruence. Psychometrika 31, 33–42.

Clifford, W.K., 1873. Preliminary sketch of bi-quaternions. Proceedings of the London Mathematical Society 4, 381–395.

Clifford, W.K., 1882. Preliminary sketch of bi-quaternions. In: Tucker, R. (Ed.), Mathematical Papers. Macmillan, London.

Coutsias, E.A., Wester, M.J., 2019. RMSD and symmetry. Journal of Computational Chemistry 40 (15), 1496–1508. https://doi.org/10.1002/jcc.25802.

Coutsias, E.A., Seok, C., Dill, K.A., 2004. Using quaternions to calculate RMSD. Journal of Computational Chemistry 25 (15), 1849–1857. URL http://www.ncbi.nlm.nih.gov/pubmed/15376254.

Coutsias, E.A., Seok, C., Dill, K.A., 2005. Rotational superposition and least squares: the SVD and quaternions approaches yield identical results. Reply to the preceding comment by G. Kneller. Journal of Computational Chemistry 26 (15), 1663–1665.

Coxeter, H.S.M., 1935. The functions of Schlafli and Lobatchefsky. Quarterly Journal of Mathematics 6, 13–29. https://doi.org/10.1093/qmath/os-6.1.13.

Coxeter, H.S.M., Moser, W.O.J., 1980. Generators and Relations for Discrete Groups. Springer-Verlag, New York. ISBN 0-387-09212-9.

Cross, R.A., Hanson, A.J., 1994. Virtual reality performance for virtual geometry. In: Proceedings of Visualization '94. Los Alamitos, CA. IEEE Computer Society Press, pp. 156–163.

Crowe, M.J., 1994. A History of Vector Analysis: the Evolution of the Idea of a Vectorial System. Dover, Mineola, NY.

Daniilidis, K., 1999. Hand-eye calibration using dual quaternions. The International Journal of Robotics Research 18, 286–298.

Davenport, P.B., 1968. A vector approach to the algebra of rotations with applications. Technical Report TN D-4696. NASA: Goddard Space Flight Center, Greenbelt, Maryland.

Denton, Peter B., Park, Stephen J., Tao, Terence, Zhang, Xining, 2019. Eigenvectors from eigenvalues: a survey of a basic identity in linear algebra. URL https://arxiv.org/abs/1908.03795v2.

Descartes, René, 1637 (1954). Book III: On the Construction of Solid and Supersolid Problems, the Geometry of René Descartes. Dover, with facsimile of the first edition. ISBN 0-486-60068-8. Translated by David Eugene Smith and Marcia L. Latham.

Diaconis, P., Shahshahani, M., 1987. The subgroup algorithm for generating uniform random variables. Probability in the Engineering and Informational Sciences 1, 15–32.

Diamond, R., 1988. A note on the rotational superposition problem. Acta Crystallographica. Section A 44, 211–216.

Dirac, P.A.M., 1930. The Principles of Quantum Mechanics. Oxford University Press.

Doran, Chris, Lasenby, Anthony, 2003. Geometric Algebra for Physicists. Cambridge University Press.

Dorst, L., Doran, C., Lasenby, J., 2002. Applications of Geometric Algebra in Computer Science and Engineering. Birkhäuser, Boston, MA.

Dorst, L., Fontijne, D., Mann, S., 2007. Geometric Algebra for Computer Science. Elsevier/Morgan-Kaufmann.

Doyle, P., Leibon, G., 2003 & 2018. 23040 symmetries of hyperbolic tetrahedra. URL https://arxiv.org/pdf/math/0309187.pdf.

Du Val, Patrick, 1934a. On isolated singularities of surfaces which do not affect the conditions of adjunction. I. Proceedings of the Cambridge Philosophical Society (ISSN 0008-1981) 30, 453–459.

Du Val, Patrick, 1934b. On isolated singularities of surfaces which do not affect the conditions of adjunction. II. Proceedings of the Cambridge Philosophical Society (ISSN 0008-1981) 30, 460–465.

Du Val, Patrick, 1934c. On isolated singularities of surfaces which do not affect the conditions of adjunction. III. Proceedings of the Cambridge Philosophical Society (ISSN 0008-1981) 30, 483–491.

Dunker, A.K., Garner, E., Guilliot, S., Romero, P., Albrecht, K., Hart, J., Obradovic, Z., 1998. Protein disorder and the evolution of molecular recognition: theory, predictions, and observations. Pacific Symposium on Biocomputing 3, 473–484.

Efimov, N.V., Rozendorn, E.R., 1975. Linear Algebra and Multi-Dimensional Geometry. Mir Publishers, Moscow.

Eguchi, T., Hanson, A.J., 1978. Asymptotically flat self-dual solutions to Euclidean gravity. Physics Letters B 74, 249–251.

Eguchi, T., Hanson, A.J., 1979. Self-dual solutions to Euclidean gravity. Annals of Physics 120, 82–106.

Eguchi, T., Gilkey, P., Hanson, A.J., 1980. Gravitation, gauge theories and differential geometry. Physics Reports 66 (6), 213–393.

Eriksson, Folke, 1978. The law of sines for tetrahedra and n-simplices. Geometriae Dedicata 7 (1), 71–80. URL https://link.springer.com/article/10.1007/BF00181352.

Eriksson, Folke, 1990. On the measure of solid angles. Mathematics Magazine 63 (3), 184–187. URL http://www.jstor.org/stable/2691141.

Euler, L., 1733. De formis radicum aequationum cujusque ordinis conjectatio. Commentarii academiae scientiarum imperialis Petropolitianae 6, 216–231. URL http://www.eulerarchive.org/pages/E030.html.

Fanea, E., Carpendale, S., Isenberg, T., 2005. An interactive 3D integration of parallel coordinates and star glyphs. In: Stasko, John, Ward, Matt (Eds.), Proceedings of the IEEE Symposium on Information Visualization. InfoVis 2005, October 23–25, 2005, Minneapolis, Minnesota, USA. IEEE Computer Society, Los Alamitos, CA, pp. 149–156. URL http://doi.ieeecomputersociety.org/10.1109/INFOVIS.2005.5.

Faugeras, Olivier, Hebert, Martial, 1983. A 3D recognition and positioning algorithm using geometrical constraints between primitive surfaces. In: Proc. 8th Joint Conf. on Artificial Intell. IJCAI'83. Morgan Kaufmann, pp. 996–1002. URL http://dl.acm.org/citation.cfm?id=1623516.1623603.

Faugeras, Olivier, Hebert, Martial, 1986. The representation, recognition, and locating of 3D objects. The International Journal of Robotics Research 5, 27. https://doi.org/10.1177/027836498600500302.

Fillmore, David W., Fillmore, Jay P., 2018. Whitney forms for spherical triangles I: the Euler, Cagnoli, and Tuynman area formulas, barycentric coordinates, and construction with the exterior calculus. URL https://arxiv.org/pdf/1404.6592.pdf.

Fisher, G.P., 1972. The Thomas precession. American Journal of Physics 40, 1772–1780.

Floater, M.S., Hormann, K., Kós, G., 2006. A general construction of barycentric coordinates over convex polygons. Advances in Computational Mathematics 24, 311–331.

Floater, Michael S., 2003. Mean value coordinates. Computer Aided Geometric Design 20 (1), 19–27.

Flower, D.R., 1999. Rotational superposition: a review of methods. Journal of Molecular Graphics & Modelling 17, 238–244.

Fogolari, F., Foumthuim, C.J.D., Fortuna, S., Soler, M.A., Corazza, A., Esposito, G., 2016. Accurate estimation of the entropy of rotation-translation probability distributions. Journal of Chemical Theory and Computation 12 (1), 1–8. https://doi.org/10.1021/acs.jctc.5b00731. PMID: 26605696.

Förstner, Wolfgang, 1999. On estimating rotations. In: Heipke, C., Mayer, H. (Eds.), Festschrift für Prof. Dr.-Ing. Heinrich Ebner zum 60. Geburtstag. Lehrstuhl für Photogrammetrie und Fernerkundung, Technische Universität München, p. 12. http://pajarito.materials.cmu.edu/documents/Kisa.Papers/on-estimating-rotations.pdf.

Frenkel, I., Libine, M., 2008. Quaternionic analysis, representation theory and physics. Advances in Mathematics 218, 1806–1877. URL https://arxiv.org/0711.2699.

Frenkel, I., Libine, M., 2011. Split quaternionic analysis and the separation of the series for sl(2,r). Advances in Mathematics 228, 678–763. URL https://arxiv.org/1009.2532.

Frenkel, I., Libine, M., 2021. Quaternionic analysis, representation theory and physics II. Advances in Theoretical and Mathematical Physics 25 (2). URL https://arxiv.org/1907.01594.

Gibbons, G.W., Hawking, S.W., 1978. Gravitational multi-instantons. Physics Letters B 78, 430–432.

Gibbons, G.W., Pope, C.N., 1979. The positive action conjecture and asymptotically Euclidean metrics in quantum gravity. Communications in Mathematical Physics 66 (3), 267–290.

Gibson, W.A., 1960. Orthogonal from oblique transformations. Educational and Psychological Measurement 20 (4), 713–721. https://doi.org/10.1177/001316446002000407.

Golub, G.H., van Loan, C.F., 1983. Matrix Computations, 1st edition. Johns Hopkins University Press, Baltimore, MD. Sec 12.4.

Gray, J.J., 1980. Olinde Rodrigues' paper of 1840 on transformation groups. Archive for the History of the Exact Sciences 21, 376–385.

Green, Bert F., 1952. The orthogonal approximation of an oblique structure in factor analysis. Psychometrika 17, 429–440. https://doi.org/10.1007/BF02288918.

Greiter, M., Schuricht, D., 2003. Imaginary in all directions: an elegant formulation of special relativity and classical electrodynamics. European Journal of Physics 24, 397–401.

Griffiths, Phillip, Harris, Joseph, 1978. Principles of Algebraic Geometry. John Wiley & Sons, New Jersey. ISBN 0-471-32792-1.

Grove, K., Karcher, H., Ruh, E.A., 1974. Jacobi fields and Finsler metrics on compact Lie groups with an application to differentiable pinching problem. Mathematische Annalen 211, 7–21.

Hamilton, W.R., 1853. Lectures on Quaternions. Cambridge University Press, Cambridge, UK.

Hamilton, W.R., 1866. Elements of Quaternions. Cambridge University Press, Cambridge, UK. Republished by Chelsea, 1969.

Hanson, A.J., 1992. The rolling ball. In: Kirk, David (Ed.), Graphics Gems III. Academic Press, Cambridge, MA, pp. 51–60.

Hanson, A.J., 1994. Geometry for N-dimensional Graphics. In: Heckbert, Paul (Ed.), Graphics Gems IV. Academic Press, Cambridge, MA, pp. 149–170.

Hanson, A.J., 1995. Rotations for N-dimensional Graphics. In: Paeth, Alan (Ed.), Graphics Gems V. Academic Press, Cambridge, MA, pp. 55–64.

Hanson, A.J., 1998. Constrained optimal framings of curves and surfaces using quaternion Gauss maps. In: Proceedings of Visualization. IEEE Computer Society Press, pp. 375–382.

Hanson, A.J., 2020. The quaternion-based spatial-coordinate and orientation-frame alignment problems. Acta Crystallographica. Section A 76 (4), 432–457. https://doi.org/10.1107/S2053273320002648.

Hanson, A.J., Hanson, S.M., 2022. Exploring the adjugate matrix approach to quaternion pose extraction. URL https://arxiv.org/pdf/2205.09116.pdf; URL https://arxiv.org/abs/2205.09116. arXiv:2205.0911v1 [eess.IV], 17 May 2022.

Hanson, A.J., Ma, H., 1994. Visualizing flow with quaternion frames. In: VIS '94: Proceedings of the Conference on Visualization '94. Los Alamitos, CA, USA. IEEE Computer Society Press. ISBN 0-7803-2521-4, pp. 108–115 (SOFTBOUND).

Hanson, A.J., Ma, H., 1995. Quaternion frame approach to streamline visualization. IEEE Transactions on Visualization and Computer Graphics (ISSN 1077-2626) 1 (2), 164–174. https://doi.org/10.1109/2945.468403.

Hanson, A.J., Sha, J.P., 2017. Isometric embedding of the A_1 gravitational instanton. In: Phua, K.K., Low, H.B., Xiong, C. (Eds.), Memorial Volume for Kerson Huang. World Scientific Pub. Co., Singapore, pp. 95–111. ISBN-13: 978-9813207424, ISBN-10: 9813207426.

Hanson, A.J., Thakur, S., 2012. Quaternion maps of global protein structure. Journal of Molecular Graphics & Modelling 38, 256–278.

Hanson, A.J., Ortiz, G., Sabry, A., Tai, Y.-T., 2013. Geometry of discrete quantum computing. Journal of Physics A: Mathematical and Theoretical 46 (18), 185301. https://doi.org/10.1088/1751-8113/46/18/185301.

Hanson, Andrew J., 2006. Visualizing Quaternions. Morgan-Kaufmann/Elsevier.

Hanson, R., Kohler, D., Braun, S., 2011. Quaternion-based definition of protein secondary structure straightness and its relationship to Ramachandran angles. Proteins: Structure, Function, and Bioinformatics (ISSN 1097-0134) 79 (7), 2172–2180. https://doi.org/10.1002/prot.23037.

Haralick, Robert M., Joo, Hyonam, Lee, Chung-Nan, Zhuang, Xinhua, Vaidya, Vinay G., Kim, Man Bae, 1989. Pose estimation from corresponding point data. IEEE Transactions on Systems, Man and Cybernetics 19 (6), 1426–1446. https://doi.org/10.1109/21.44063.

Hartley, R., Aftab, K., Trumpf, J., 2011. L1 rotation averaging using the Weiszfeld algorithm. In: Proceedings of the IEEE Computer Society Conference on Computer Vision and Pattern Recognition, pp. 3041–3048.

Hartley, R., Trumpf, J., Dai, Y., Li, H., 2013. Rotation averaging. International Journal of Computer Vision 103 (3), 267–305. https://doi.org/10.1007/s11263-012-0601-0.

Hebert, Martial, 1983. Reconnaissance de formes tridimensionnelles. Ph.D. thesis. University of Paris South. ISBN 2-7261-0379-0. Available as INRIA Tech. Rep.

Hestenes, D., 2000. New Foundations for Classical Mechanics, 2nd edition. Reidel.

Hestenes, D., Sobczyk, G., 1984. Clifford Algebra to Geometric Calculus, a Unified Language for Mathematics and Physics. Kluwer, Dordrecht/Boston.

Hitchin, Nigel, 1979. Polygons and gravitons. Mathematical Proceedings of the Cambridge Philosophical Society 85, 465–476.

Hitchin, Nigel J., Karlhede, A., Lindström, U., Roček, M., 1987. Hyperkähler metrics and supersymmetry. Communications in Mathematical Physics 108, 535.

Hormann, K., Floater, M.S., 2006. Mean value coordinates for arbitrary planar polygons. ACM Transactions on Graphics 25 (4), 1424–1441. https://doi.org/10.1145/1183287.1183295.

Horn, B.K.P., 1987. Closed-form solution of absolute orientation using unit quaternions. Journal of the Optical Society of America. A 4, 629–642. URL https://www.osapublishing.org/josaa/viewmedia.cfm?uri=josaa-4-4-629&seq=0.

Horn, Berthold K.P., Hilden, Hugh M., Negahdaripour, Shahriar, 1988. Closed-form solution of absolute orientation using orthonormal matrices. Journal of the Optical Society of America. A 5 (7), 1127–1135. https://doi.org/10.1364/JOSAA.5.001127. URL http://josaa.osa.org/abstract.cfm?URI=josaa-5-7-1127.

Hornik, Kurt, Stinchcombe, Maxwell, White, Halbert, 1989. Multilayer feedforward networks are universal approximators. Neural Networks (ISSN 0893-6080) 2 (5), 359–366. https://doi.org/10.1016/0893-6080(89)90020-8. URL https://www.sciencedirect.com/science/article/pii/0893608089900208.

Huang, T.S., Blostein, S.D., Margerum, E.A., 1986. Least-squares estimation of motion parameters from 3D point correspondences. In: Proc. IEEE Conf. Computer Vision and Pattern Recognition. IEEE Computer Society, pp. 24–26.

Huggins, David J., 2014. Comparing distance metrics for rotation using the k-nearest neighbors algorithm for entropy estimation. Journal of Computational Chemistry 35, 377–385. https://doi.org/10.1002/jcc.23504.

Hurwitz, Adolf, 1923. Über die komposition der quadratischen formen. Mathematische Annalen 88 (1–2), 1–25. https://doi.org/10.1007/bf01448439.

Huynh, Du Q., 2009. Metrics for 3d rotations: comparison and analysis. Journal of Mathematical Imaging and Vision (ISSN 0924-9907) 35 (2), 155–164. https://doi.org/10.1007/s10851-009-0161-2.

Immel, S., Köck, M., Reggelin, M., 2018. Configurational analysis by residual dipolar coupling driven floating chirality distance geometry calculations. Chemistry 24 (52), 13918–13930.

Immel, S., Köck, M., Reggelin, M., 2019. Configurational analysis by residual dipolar couplings: a critical assessment of diastereomeric differentiabilities. Chirality 31 (5), 384–400.

Inselberg, A., 2009. Parallel Coordinates: Visual Multidimensional Geometry and Its Applications. Springer.

Ju, T., Schaefer, W., Warren, J., 2005a. Mean value coordinates for closed triangular meshes. ACM Transactions on Graphics 24 (3), 561–566.

Ju, T., Schaefer, W., Warren, J., Desbrun, M., 2005b. A geometric construction of coordinates for convex polyhedra using polar duals. In: Proceedings of the Symposium on Geometry Processing, pp. 181–186.

Kabsch, W., 1976. A solution for the best rotation to relate two sets of vectors. Acta Crystallographica. Section A 32, 922–923. https://doi.org/10.1107/S0567739476001873.

Kabsch, W., 1978. A discussion of the solution for the best rotation to relate two sets of vectors. Acta Crystallographica. Section A 34, 827–828. https://doi.org/10.1107/S0567739478001680.

Karcher, H., 1977. Riemannian center of mass and mollifier smoothing. Communications on Pure and Applied Mathematics 30 (5), 509–541.

Karney, Charles F.F., 2007. Quaternions in molecular modeling. Journal of Molecular Graphics & Modelling (ISSN 1093-3263) 25 (5), 595–604. https://doi.org/10.1016/j.jmgm.2006.04.002.

Katz, Sheldon, Morrison, David R., 1992. Gorenstein threefold singularities with small resolutions via invariant theory for Weyl groups. arXiv:alg-geom/9202002v1.

Kavan, Ladislav, Collins, Steven, Žára, Jiří, O'Sullivan, Carol, 2008. Geometric skinning with approximate dual quaternion blending. ACM Transactions on Graphics (ISSN 0730-0301) 27 (4). https://doi.org/10.1145/1409625.1409627.

Kearsley, S.K., 1989. On the orthogonal transformation used for structural comparisons. Acta Crystallographica. Section A 45 (2), 208–210. https://doi.org/10.1107/S0108767388010128.

Kearsley, S.K., 1990. An algorithm for the simultaneous superposition of a structural series. Journal of Computational Chemistry 11, 1187–1192.

Klein, F., 1884. Vorlesungen über das Ikosaeder und die Auflösung der Gleichungen vom fünften Grade. Teubner, Leipzig. Reprinted Birkhäuser, Basel, 1993 (edited by P. Slodowy); Translated as Lectures on the icosahedron and the solution of equations of the fifth degree, Kegan Paul, London, 1913 (2nd edition); Reprinted by Dover, 1953.

Kneller, G.R., Calligari, P., 2006. Efficient characterization of protein secondary structure in terms of screw motions. Acta Crystallographica. Section D, Biological Crystallography 62, 302–311.

Kneller, Gerald R., 1991. Superposition of molecular structures using quaternions. Molecular Simulation 7 (1–2), 113–119. URL http://www.tandfonline.com/doi/abs/10.1080/08927029108022453.

Kronheimer, P.B., 1989a. The construction of ALE spaces as hyperkähler quotients. Journal of Differential Geometry 29, 665–683.

Kronheimer, P.B., 1989b. A Torelli-type theorem for gravitational instantons. Journal of Differential Geometry 29, 685–697.

Langer, T., Belyaev, A., Seidel, H.-P., 2006. Spherical barycentric coordinates. In: Siggraph/Eurographics Symposium Geom. Processing, p. 81088. URL http://domino.mpi-inf.mpg.de/intranet/ag4/ag4publ.nsf/0/9144C5FF262D3F9CC12571BE00348DDF/$file/paper.pdf.

Lesk, A.M., 1986. A toolkit for computational molecular biology 2. On the optimal superposition of 2 sets of coordinates. Acta Crystallographica. Section A 42, 110–113.

Lin, Chen, Hanson, Andrew J., Hanson, Sonya M., 2023. Algebraically rigorous quaternion framework for the neural network pose estimation problem. In: Proceedings of the IEEE/CVF International Conference on Computer Vision (ICCV), pp. 14097–14106.

Lindström, U., Roček, M., 1983. Scalar tensor duality and N=1, N=2 nonlinear sigma models. Nuclear Physics. B 222, 285–308.

Lindström, Ulf, Roček, Martin, von Unge, Rikard, 2000. Hyperkähler quotients and algebraic curves. The Journal of High Energy Physics 2000 (01), 022. URL http://stacks.iop.org/1126-6708/2000/i=01/a=022.

Liu, P., Agrafiotis, D.K., Theobald, D.L., 2010. Fast determination of the optimal rotational matrix for macromolecular superpositions. Journal of Computational Chemistry 31, 1561–1563. https://doi.org/10.1002/jcc.21439.

Lu, Chien-Ping, Hager, Gregory D., Mjolsnes, Eric, 2000. Fast and globally convergent pose estimation from video images. IEEE Transactions on Pattern Analysis and Machine Intelligence 22 (6).

Mackay, A.L., 1984. Quaternion transformation of molecular orientation. Acta Crystallographica. Section A 40, 165–166.

MacLachlan, A.D., 1982. Rapid comparison of protein structures. Acta Crystallographica. Section A 38, 871–873. https://doi.org/10.1107/S0567739482001806.

Magarshak, Y., 1993. Quaternion representation of RNA sequences and tertiary structures. Biosystems (ISSN 0303-2647) 30 (1–3), 21–29. https://doi.org/10.1016/0303-2647(93)90059-L. URL http://www.sciencedirect.com/science/article/pii/030326479390059L.

Manton, J.H., 2004. A globally convergent numerical algorithm for computing the centre of mass on compact Lie groups. In: Proc. 8th Intern. Conf. on Control, Automation, Robotics, and Vision. Kunming, China, pp. 2211–2216.

Markley, F.L., 1988. Attitude determination using vector observations and the singular value decomposition. The Journal of the Astronautical Sciences 38 (2), 245–258.

Markley, F.L., Mortari, D., 2000. Quaternion attitude estimation using vector observations. The Journal of the Astronautical Sciences 48 (2), 359–380.

Markley, F.L., Cheng, Y., Crassidis, J.L., Oshman, Y., 2007. Averaging quaternions. Journal of Guidance, Control, and Dynamics 30 (4), 1193–1197.

McCarthy, J.M., 1990. Introduction to Theoretical Kinematics. MIT Press.

Milnor, John, 1994. The Schlafli differential equality. In: John Milnor Collected Papers: Volume 1: Geometry. American Mathematical Society, pp. 281–295.

Moakher, M., 2002. Means and averaging in the group of rotations. SIAM Journal on Matrix Analysis and Applications 24 (1), 1–16. https://doi.org/10.1137/S0895479801383877.

Möbius, A.F., 1846. Uber eine neue behandlungsweise der analyltischen sphärik. Abhandlungen bei Begründung der Königl. Sächs Gesellschaft der Wissenschaften, pp. 45–86. Republished in A.F. Möbius, Gesammelte Werke, F. Klein (Ed.), vol. 2, Leipzig, 1886, pp. 1–54.

Möbius, August Ferdinand, 1827. Der barycentrische Calcul. Verlag von Johann Ambrosius Barth, Leipzig.

Mohanty, Yana Zilberberg, 2002. Hyperbolic polyhedra: volume and scissors congruence. PhD thesis. UC San Diego. URL http://math.ucsd.edu/~thesis/thesis/ymohanty/ymohanty.pdf.

Møller, C., 1972. The Theory of Relativity. Clarendon Press, Oxford.

Morris, A.L., MacArthur, M.W., Hutchinson, E.G., Thornton, J.M., 1992. Stereochemical quality of protein structure coordinates. Proteins 12, 345–364.

Müller, Hans Robert, 1962. Sphärische Kinematik. VEB Deutscher Verlag der Wissenschaften, Berlin.

Muller-Kirsten, Harald J.W., Wiedemann, A., 1987. Supersymmetry: An Introduction with Conceptual and Calculational Details. World Scientific.

Murakami, J., Yano, M., 2005. On the volume of a hyperbolic and spherical tetrahedron. Communications in Analysis and Geometry 13, 379–400. URL https://www.intlpress.com/site/pub/files/_fulltext/journals/cag/2005/0013/0002/CAG-2005-0013-0002-a005.pdf.

Murakami, Jun, 2012. Volume formulas for a spherical tetrahedron. Proceedings of the American Mathematical Society 140 (9), 3289–3295. URL https://www.ams.org/journals/proc/2012-140-09/S0002-9939-2012-11182-7/S0002-9939-2012-11182-7.pdf.

Nickalls, R.W.D., 1993. A new approach to solving the cubic: Cardan's solution revealed. Mathematical Gazette 77, 354–359. URL http://www.jstor.org/stable/3619777.

Nickalls, R.W.D., 2009. The quartic equation: invariants and Euler's solution revealed. Mathematical Gazette 93, 66–75. URL http://www.nickalls.org/dick/papers/maths/quartic2009.pdf.

Park, F.C., Ravani, Bahram, 1997. Smooth invariant interpolation of rotations. ACM Transactions on Graphics (ISSN 0730-0301) 16 (3), 277–295. https://doi.org/10.1145/256157.256160. URL http://doi.acm.org/10.1145/256157.256160.

PDB, 2011. The RCSB Protein Data Bank: site functionality and bioinformatics use cases.

Penrose, R., 1959. The apparent shape of a relativistically moving sphere. Mathematical Proceedings of the Cambridge Philosophical Society 55 (1), 137–139. https://doi.org/10.1017/S0305004100033776.

Peretroukhin, Valentin, Giamou, Matthew, Greene, W. Nicholas, Rosen, David, Kelly, Jonathan, Roy, Nicholas, 2020. A smooth representation of belief over SO(3) for deep rotation learning with uncertainty. In: Robotics: Science and Systems XVI.

Prasad, M.K., 1979. Equivalence of Eguchi-Hanson metric to two center Gibbons-Hawking metric. Physics Letters B 83, 310.

Quine, J.R., 1999. Helix parameters and protein structure using quaternions. Journal of Molecular Structure 460, 53–66.

Ramachandran, G.N., Ramakrishnan, C., Sasisekharan, V., 1963. Stereochemistry of polypeptide chain configurations. Journal of Molecular Biology 7, 95–99.

Rodrigues, O., 1840. Des lois géométriques qui régissent les déplacements d'un système solide dans l'espace, et la variation des coordonnées provenant de ses déplacments consideérés indépendamment des causes qui peuvent les produire. Journal de Mathématiques Pures et Appliquées 5, 380–440.

Rolfson, D., 1976. Knots and Links. Publish or Perish Press, Wilmington, DE.

Roth, B., Bottema, O., 2012. Theoretical Kinematics. Dover.

Sansò, Fernando, 1973. An exact solution of the roto-translation problem. Photogrammetria (ISSN 0031-8663) 29 (6), 203–216. https://doi.org/10.1016/0031-8663(73)90002-1. https://www.sciencedirect.com/science/article/pii/0031866373900021.

Sarabandi, Soheil, Thomas, Federico, 2019. Accurate computation of quaternions from rotation matrices. In: Lenarcic, J., Parenti-Castelli, V. (Eds.), Advances in Robot Kinematics 2018. ARK 2018. In: Springer Proceedings in Advanced Robotics, vol. 8. Springer, Cham, pp. 1–8.

Sarabandi, Soheil, Thomas, Federico, 2022. On closed-form solutions to the 4D nearest rotation matrix problem. Mathematical Methods in the Applied Sciences (Special Issue), 1–9. https://doi.org/10.1002/mma.8524. URL https://onlinelibrary.wiley.com/doi/abs/10.1002/mma.8524.

Sarabandi, Soheil, Perez-Gracia, Alba, Thomas, Federico, 2018. Singularity-free computation of quaternions from rotation matrices in $\mathbb{E}^4$ and $\mathbb{E}^3$. In: Conference on Applied Geometric Algebras in Computer Science and Engineering, pp. 23–27.

Sarabandi, Soheil, Shabani, Arya, Porta, Josep M., Thomas, Federico, 2020. On closed-form formulas for the 3D nearest rotation matrix problem. IEEE Transactions on Robotics 36 (4), 1333–1339. URL http://www.iri.upc.edu/people/thomas/papers/IEEE TRO 2020.pdf.

Saxena, A., Driemeyer, J., Ng, A.Y., 2009. Learning 3D object orientation from images. In: IEEE Inter. Conf. Robotics and Automation (ICRA '09), pp. 794–800.

Schläfli, L., 1860. On the multiple integral $\int \cdots \int dx\,dy\cdots dz$, whose limits are $p_1 = a_1 x + b_1 y + \cdots + h_1 z > 0$, $p_2 > 0, \cdots, p_n > 0$, and $x^2 + y^2 + \cdots + z^2 < 1$. Quarterly Journal of Mathematics 140 (3), 54–68, 97–108.

Schönemann, P.H., 1966. A generalized solution of the orthogonal procrustes problem. Psychometrika 31, 1–10. https://doi.org/10.1007/BF02289451.

Schut, G.H., 1960. On exact linear equations for the computation of the rotational elements of absolute orientation. Photogrammetria (ISSN 0031-8663) 17 (1), 34–37. https://doi.org/10.1016/S0031-8663(60)80029-4. https://www.sciencedirect.com/science/article/pii/S0031866360800294.

Shepperd, S.W., 1978. Quaternion from rotation matrix. Journal of Guidance and Control 1 (3), 223–224.

Shoemake, K., 1985. Animating rotation with quaternion curves. In: Proceedings of SIGGRAPH 1985. Computer Graphics 19 (3), 245–254. https://doi.org/10.1145/325165.325242.

Shoemake, K., 1992. Uniform random rotations. In: Kirk, David (Ed.), Graphics Gems III. Academic Press, Cambridge, MA, pp. 124–132.

Shoemake, K., 1994. Arcball rotation control. In: Heckbert, Paul (Ed.), Graphics Gems IV. Academic Press, Cambridge, MA, pp. 175–192.

Shuster, M.D., Natanson, G.A., 1993. Quaternion computation from a geometric point of view. The Journal of the Astronautical Sciences 41 (4), 545–556.

Siminovitch, D.J., 1997a. Rotations in NMR: part I. Euler-Rodrigues parameters and quaternions. Concepts in Magnetic Resonance 9, 149–171.

Siminovitch, D.J., 1997b. Rotations in NMR: part II. Applications of Euler-Rodrigues parameters. Concepts in Magnetic Resonance 9, 211–225.

Snyder, John P., 1993. Flattening the Earth: Two Thousand Years of Map Projections. Univ. of Chicago Press. ISBN 0-226-76747-7.

Srinivasan, R., Geetha, V., Seetharaman, J., Mohan, S., 1993. A unique or essentially unique single parametric characterisation of biopolymetric structures. Journal of Biomolecular Structure & Dynamics 11, 583–596.

Study, E., 1891. Von den bewegungen und umlegungen. Mathematische Annalen 39, 441–566.

Tait, P.G., 1867. An Elementary Treatise on Quaternions. Cambridge University Press, Cambridge, UK.

Tait, P.G., 1873. Introduction to Quaternions. Cambridge University Press, Cambridge, UK.

Tait, P.G., 1890. An Elementary Treatise on Quaternions, revised edition. Cambridge University Press, Cambridge, UK.

Terrell, James, 1959. Invisibility of the Lorentz contraction. Physical Review 116, 1041–1045. https://doi.org/10.1103/PhysRev.116.1041. URL https://link.aps.org/doi/10.1103/PhysRev.116.1041.

Theobald, Douglas, 2005. Rapid calculation of RMSDs using a quaternion-based characteristic polynomial. Acta Crystallographica. Section A 61, 478–480. URL http://scripts.iucr.org/cgi-bin/paper?S0108767305015266.

Thomas, L.H., 1926. The motion of the spinning electron. Nature 117, 514.

Thompson, E.H., 1958. An exact linear solution of the problem of absolute orientation. Photogrammetria (ISSN 0031-8663) 15 (4), 163–179. https://doi.org/10.1016/S0031-8663(58)80023-X. https://www.sciencedirect.com/science/article/pii/S003186635880023X.

Tuynman, Gijs M., 2013. Areas of spherical and hyperbolic triangles in terms of their midpoints.

Umeyama, Shinji, 1991. Least-squares estimation of transformation parameters between two point patterns. IEEE Transactions on Pattern Analysis and Machine Intelligence 13 (4), 376–380. https://doi.org/10.1109/34.88573.

van der Waerden, B.L., 1976. Hamilton's discovery of quaternions. Mathematics Magazine 49, 227–234.

Wachspress, E.L., 1975. A Rational Finite Element Basis, vol. 114. Academic Press.

Wahba, G., 1965. Problem 65-1, a least squares estimate of spacecraft attitude. SIAM Review 7 (3), 409.

Warren, J., 1996. Barycentric coordinates for convex polytopes. Advances in Computational Mathematics 6 (2), 97–108.

Weiskopf, Daniel, Kobras, Daniel, Ruder, Hanns, 2000. Real-world relativity: image-based special relativistic visualization. In: Proceedings of the Conference on Visualization '00. VIS '00, Washington, DC, USA. IEEE Computer Society Press. ISBN 158113309X, pp. 303–310.

Weisner, L., 1938. Introduction to the Theory of Equations. University of Michigan. ISBN 9780598438645. Reprinted in 1947 by Macmillan. URL https://archive.org/details/in.ernet.dli.2015.212184/page/n151/mode/2up.

Weisstein, Eric W., 2019a. Cubic formula. [Online]. http://mathworld.wolfram.com/CubicFormula.html. (Accessed 12 May 2019).

Weisstein, Eric W., 2019b. Quartic equation. [Online]. http://mathworld.wolfram.com/QuarticEquation.html. (Accessed 12 May 2019).

Wigner, E., 1939. On unitary representations of the inhomogeneous Lorentz group. Annals of Mathematics 40, 149–204. https://doi.org/10.2307/1968551.

Wikipedia, 2023a. Ars Magna (Cardano book) — Wikipedia, the free encyclopedia. [Online]. http://en.wikipedia.org/w/index.php?title=Ars Magna (Cardano book)&oldid=1146935672. (Accessed 31 July 2023).

Wikipedia, 2023b. ADE classification — Wikipedia, the free encyclopedia. [Online]. http://en.wikipedia.org/w/index.php?title=ADE classification&oldid=1104095599. (Accessed 31 July 2023).

Printed in the United States
by Baker & Taylor Publisher Services